Library of Congress Control Number: 2015936110

Print ISBN: 978-1-93951-205-5

Electronic ISBN: 978-1-61444-617-0

Printed in the United States of America

Current Printing (last digit):
10 9 8 7 6 5 4 3 2 1

An Invitation
to
Real Analysis

An Invitation to Real Analysis

Luis F. Moreno
SUNY Broome Community College

Published and distributed by
The Mathematical Association of America

MAA TEXTBOOKS

Bridge to Abstract Mathematics, Ralph W. Oberste-Vorth, Aristides Mouzakitis, and Bonita A. Lawrence

Calculus Deconstructed: A Second Course in First-Year Calculus, Zbigniew H. Nitecki

College Calculus: A One-Term Course for Students with Previous Calculus Experience, Michael E. Boardman and Roger B. Nelsen

Combinatorics: A Guided Tour, David R. Mazur

Combinatorics: A Problem Oriented Approach, Daniel A. Marcus

Complex Numbers and Geometry, Liang-shin Hahn

A Course in Mathematical Modeling, Douglas Mooney and Randall Swift

Cryptological Mathematics, Robert Edward Lewand

Differential Geometry and its Applications, John Oprea

Distilling Ideas: An Introduction to Mathematical Thinking, Brian P. Katz and Michael Starbird

Elementary Cryptanalysis, Abraham Sinkov

Elementary Mathematical Models, Dan Kalman

An Episodic History of Mathematics: Mathematical Culture Through Problem Solving, Steven G. Krantz

Essentials of Mathematics, Margie Hale

Field Theory and its Classical Problems, Charles Hadlock

Fourier Series, Rajendra Bhatia

Game Theory and Strategy, Philip D. Straffin

Geometry Revisited, H. S. M. Coxeter and S. L. Greitzer

Graph Theory: A Problem Oriented Approach, Daniel Marcus

An Invitation to Real Analysis, Luis F. Moreno

Knot Theory, Charles Livingston

Learning Modern Algebra: From Early Attempts to Prove Fermat's Last Theorem, Al Cuoco and Joseph J. Rotman

Lie Groups: A Problem-Oriented Introduction via Matrix Groups, Harriet Pollatsek

Mathematical Connections: A Companion for Teachers and Others, Al Cuoco

Mathematical Interest Theory, Second Edition, Leslie Jane Federer Vaaler and James W. Daniel

Mathematical Modeling in the Environment, Charles Hadlock

Mathematics for Business Decisions Part 1: Probability and Simulation (electronic textbook), Richard B. Thompson and Christopher G. Lamoureux

Mathematics for Business Decisions Part 2: Calculus and Optimization (electronic textbook), Richard B. Thompson and Christopher G. Lamoureux

Mathematics for Secondary School Teachers, Elizabeth G. Bremigan, Ralph J. Bremigan, and John D. Lorch

The Mathematics of Choice, Ivan Niven

The Mathematics of Games and Gambling, Edward Packel

Math Through the Ages, William Berlinghoff and Fernando Gouvea

Noncommutative Rings, I. N. Herstein

Non-Euclidean Geometry, H. S. M. Coxeter

Number Theory Through Inquiry, David C. Marshall, Edward Odell, and Michael Starbird

Ordinary Differential Equations: from Calculus to Dynamical Systems, V. W. Noonburg

A Primer of Real Functions, Ralph P. Boas
A Radical Approach to Lebesgue's Theory of Integration, David M. Bressoud
A Radical Approach to Real Analysis, 2nd edition, David M. Bressoud
Real Infinite Series, Daniel D. Bonar and Michael Khoury, Jr.
Topology Now!, Robert Messer and Philip Straffin
Understanding our Quantitative World, Janet Andersen and Todd Swanson

MAA Service Center
P.O. Box 91112
Washington, DC 20090-1112
1-800-331-1MAA FAX: 1-301-206-9789

To my wife, patient donor of many hours of quiet time, to my parents,
and ultimately, to Jesus of Nazareth, Son of the most high God.

Call unto Me, and I will answer thee, and shew thee great
and mighty things, which thou knowest not.

Jeremiah 33:3

Contents

To the Student

We are all students, or pupils (*mathetés* in Greek), and should never cease to be so. This book is an invitation and an introduction to real analysis, a branch of mathematics that most students of calculus find to be quite unexpected. Broadly speaking, real analysis studies the foundational principles of the real number system, and the properties of real functions and series, often with regard to limits. As such, one of its accomplishments was to create the theoretical foundations for calculus. Thus, some of the theorems that were—do we dare say—old friends from Calculus I and II, are presented in a more general setting, and always with proofs. But there will be many new and brilliant theorems about sets, functions, series, and other topics. Also, real analysis is where mathematics finally came to terms with the infinite. This aspect has always fascinated me; I am sure that you will feel the same way. No other human endeavor studies infinity quantitatively and in so many different aspects. I invite you to enter this fascinating, meticulous subject, and meet the people who matured it.

Some comments from those who have accepted the invitation and stepped through the door are, (1) "Do we really need to prove that, isn't it pretty obvious?" Also, (2) "It's almost all about #*x@ inequalities!" (We may call "#*x@" a "variable adjective," since its meaning changes by different word substitutions.) And, (3) "Where did all the nice pictures of things go?" And sometimes, (4) "That was tough to understand, but it's awesome!" And from my memory of my first real analysis course, (5) "Who were those guys who came up with these amazing ideas?" Each of the students' comments above carries worthwhile insight.

(1) Regarding the first one, you already know that some things are obvious only because they haven't been considered very carefully. Analysis is in the business of establishing the truth of statements by deduction, not by "hand-waving." And sometimes, an abstract mathematical claim becomes clear to you only after conquering its proof. Writing your own proof is almost an art, full of hunches, competing viewpoints, dead ends, and the like. You will learn to devise proofs that are logically sound (avoiding circular reasoning, for instance) and even elegant, but this comes only with patient practice, so that homework cannot be shortchanged.

In this book, to show (or demonstrate) that something is true means the same as to prove that it is true. Alternatively, the words "convince yourself" and "verify that" allow less rigorous, intuitive, thinking.

(2) Regarding the second comment, while most of calculus concentrates on equations, many concepts in real analysis require inequalities. Your thinking about how to preserve the size

order of things will have to be strengthened. By doing the exercises, you will learn to dissect, construct, and apply inequalities effectively.

(3) The comment about illustrations is quite interesting. Pictures may be misleading in mathematics, as you will soon see (for this reason, some advanced texts have absolutely no figures at all). Thus, although there are plenty of them in this book, the illustrations are intended to be supportive, suggestive, and schematic. How to draw figures that clarify an issue but don't assume too much is something to be learned. Thus, draw many pictures as part of your diligently done homework.

(4) With regard to difficulty of understanding, we know that some things are difficult just because they are communicated badly. But in analysis, some things are difficult to understand because they are genuinely profound, in fact so profound that mere prose cannot be used to communicate the concepts. This is why the marvelous symbolic language of mathematics is necessary. Entering real analysis, you are as a young artist who has just been handed a palette of colors never seen before. The symbols are the colors with which mathematics is painted. In order to become handy with them, you must use them often in homework exercises.

It seems that homework is the string that ties the whole bundle securely. I think this is close to the truth. Consider the following table with four combinations of ingredients. The upper headings of the table are the two answers to a question we always have about a topic, call it Topic X: "Do I understand topic X?" To keep things manageable, we won't allow tentative answers to this question. This question may be answered with certainty *after* the fact of learning, for example, "Do I know how to multiply three-digit whole numbers?" But for topics that you are in the *process* of learning ("Do I understand how to detect a cluster point?"), the answer won't be known for certain. A graded exam, for instance, may determine the answer quite clearly, but by then, it may be too late. Down the left side of the table, you must ask yourself whether you *think* that you understand Topic X. This is your subjective appraisal of your understanding of (or skill with) Topic X, at the present time. Again, limit your answer to a simple "yes" or "no." This is an honest appraisal; you are not dealing with unknowns here.

	I actually do understand Topic X	I actually don't understand Topic X
I think that I understand Topic X	I	II
I don't think that I understand Topic X	III	IV

We now consider the situations marked I, II, III, and IV. What is the effect of situation I? Here, you are confident that you understand Topic X, and, in fact, you actually do know the topic. This is the case where effective study of homework exercises, discussion with the instructor, and so forth, have made you competent in the topic. It is very likely that you will do well on a test of Topic X, and you will remember it well later on. Situation I lowers anxiety, and it is generally a good position to be in. Situation IV isn't bad either, assuming that there is time to rectify the problem. As you see, you have identified the lack of understanding, and you may get together with a couple of friends and a pot of strong coffee to "figure it out." At some imperceptible moment, you will cross over from IV to III, and soon after, you will arrive at situation I. Situation III is problematic only in that you might spend too much time grinding away at Topic X and not budget time for other necessary things.

Finally, consider situation II, where you are falsely assuming that you know Topic X. This case results in a false sense of security, which is usually catastrophic in mathematics. Perhaps Topic X was clear when you left the lecture. A few of the (as it turned out) easier homework exercises were casually done. But when the topic had to be applied in a later context, you brushed aside the details in order to get the main job done. The truth about Topic X is rudely discovered only at test time. Fortunately, effectively done homework can suddenly push you from II to IV, and then slowly but steadily to I.

A Homework Strategy

There are over 600 exercises in this book. The table above supports the idea that studying mathematics and doing homework effectively are learned behaviors. You surely know some techniques that help you grasp ideas more efficiently in any subject. This book will suggest several methods to gain understanding (to move you towards situation I), and you should try them all. For example, after creating a proof that a function is continuous at a point, an exercise may ask you to change the function slightly to see how this affects the proof. Another strategy is that of simplifying the assumptions given in an exercise, and then seeing what you can deduce. A third strategy is to switch to a different form of proof (remember proof by contradiction?). In time, strategies such as these will become part of your "solver's toolkit" for attacking problems in general.

Solutions to odd-numbered exercises appear at the end of each chapter. But how do you know that a proof written by you is "airtight?" Or, when the solution isn't listed, could your solution have an undetected fatal flaw? My suggestions to prevent such things from happening are as follows:

(1) Do as much of the exercise as you can by applying the concepts that you have just studied, along with past concepts as needed, without looking at a hint or solution. You may need to apply one of the tools in your toolkit. If you succeed in answering the exercise, then congratulations, the wrestling match was worth it! What's more, your solution or proof may be more elegant than the text's, and that is the best of all situations.
(2) Often, hints are attached to the exercises. If you can't see a first step, or can't even grasp the problem, look at the hint, or a part of the hint. Pursue any line of attack that the hint suggests. That may be all it takes to jump-start your solution. Then, compare your work with that of others in the class. Together, you may find an error of some kind. Be not discouraged, for this is a superb way to learn. Finally, compare your work with the given solution (if available).
(3) If the hint takes you only part way, or nowhere at all, you have nevertheless begun to unravel the knot. Now, look at the *first sentence* of the solution, if it is there. Notice how the hint is applied, and try to proceed to the solution.
(4) If the solution is supplied, and if step (3) didn't help completely, then look at more of the solution, but not all of it. At this point, you will see more of the path to the conclusion, but it is important for you to follow the signs in a logical manner. You may come to realize that the solution is staring right at you! If the solution is not supplied, it is time to ask about it in class. Your instructor may have the full solutions manual.

Of course, the worst thing to do is to make a dispirited attempt at the exercise, and then read the full solution. That's comparable to reading the spoiler to a suspenseful film, and, just as bad, it draws you into situation II.

The People of Mathematics

Comment number five is addressed here. Mathematics shouldn't be presented as a disembodied collection of facts. Common parlance, especially what comes out of Hollywood, has cheapened the meaning of genius. I hope the short passages included here will reinstate the proper sense of the word. Most of these men—and more recently, women—of mathematics displayed an insight into the unknown that will seem to you almost unearthly. Even now, with over a century of hindsight on their work, the vision of mathematicians such as Bernhard Riemann and Georg Cantor is still stunning, especially because the mathematical analysis they wrote about exists outside of physical, sensory existence. Nevertheless, many of these geniuses were also superb at applied mathematics, in fields such as astronomy, thermodynamics, and engineering.

The visual arts handle poorly both the work of mathematicians and the individuals themselves. The cinematic industry regularly displays its incompetence in portraying mathematicians (and scientists in general). Only recently has an occasional documentary realistically discussed the struggles and triumphs of a few mathematicians (I'm thinking of Andrew Wiles (1953–)). And so, enjoy the selections herein and look into the references under "History of Mathematics" in the bibliography for the real story.

To the Instructor

With this book I hope to ease a student's transition from what may be called a "consumer of mathematics" up through calculus II, into one beginning participate in its creative process. Specifically I mean that with your guidance, this book will help a student to

(1) be able to transform a faulty proof into a good, maybe elegant, one,
(2) correctly use symbols, and correctly apply definitions and theorems,
(3) become aware of the underlying laws of logic,
(4) understand the axiomatic development of real numbers,
(5) have a deeper understanding of the core topics of real analysis, and
(6) appreciate the brilliant minds who, in creating this subject, brought clarity and rigor to the rapidly growing bundle of mathematical techniques of the late 18th century.

A somewhat abbreviated coverage of this book would encompass Chapters 4 through 50, skipping Chapters 11 and 32. This covers (some of) the foundations of real numbers, countable and uncountable sets, sequences, series, the properties of functions, differentiation, and Riemann integration. Those chapters, most of which are short, may be covered in a 14-week semester, with six weeks covering four sections per week and eight weeks with three per week, leaving time for testing. In a 15-week semester, all the chapters can be covered, with a cursory look at Chapter 1. This would then include a more complete treatment of the axioms, an investigation of transfinite cardinals, and an introduction to metric spaces, the topology of the real line, and the Cantor Ternary set.

A diverse, annotated bibliography is broken up into four sections to encourage students to use it.

Over 600 exercises are included. A complete solutions manual is available to the instructor.

Several techniques address objective (1), the ability of students to recognize when their written proofs are valid and understandable. Some theorems have proofs with a parallel case that is left as an exercise, since the given case then acts as a scaffold for the student. Certain corollaries' proofs are also placed as exercises for the same reason. Some proofs to be done as exercises have a step-by-step form that is (I think) pedagogically effective.

Regarding objective (2), there is occasional redundancy and repetition, especially if the concept in question hasn't been used in a while. New or unfamiliar symbols are related to past concepts, and collected in an appendix. Many figures are employed. Some definitions (e.g., that of a function unbounded on an interval) are included for clarity and reference, and to reinforce goal (3). Many exercises have hints, to be used judiciously per the preface to the student.

As for objective (3), understanding valid inference is the purpose of an introduction to logic in Chapter 1. I don't just mention it and then leave it behind. The difference between necessity and sufficiency is constantly stressed. The negation of quantifiers appears in step-by-step fashion where appropriate.

Objective (4) is accomplished in a unified way by forming a pedigree of axioms, so that the student can see at a glance which axioms are inherited from the previous number set, and therefore, which one (or ones) form the discriminating factor for the next, richer, set. In this way, a student can see both the richness and the limitation of $\mathbb{Q}$, for instance.

Objective (5) is, of course, the main purpose of this book. A few slightly less traditional topics have been included, some as appendices. As examples, a subsection on quadratic extensions adds a little flesh to what is often just mentioned as "algebraic numbers." A theorem about the division of power series seems natural after the work on products of series. The full proof of Newton's binomial series theorem is left to a more advanced text; instead, the convergence behavior at the endpoints is analyzed since it requires good use of previous concepts (and it's not often found in other texts). The proof that $\lim_{\theta \to 0} \frac{\sin\theta}{\theta} = 1$ is different, using the sequential criterion for convergence. Continued fractions are introduced in an appendix. Another appendix contains the full proof of l'Hospital's rule (again, something not often found). And surely, students will be intrigued by the curious Farey sequences found in another appendix.

Regarding objective (6), short notes about many mathematicians are interspersed throughout the text. I hope this helps to break the theorem-proof-example cadence which can't be sustained for long periods.

Numerous references to articles in *The College Mathematics Journal*, *Mathematics Magazine*, *The American Mathematical Monthly*, and other sources in mathematics may be fruitfully assigned to students in order to deepen their understanding of analysis and the mathematicians themselves.

Acknowledgments

I would like to thank my long-time friend and colleague Paul O'Heron for his helpful advice at so many times, blended with his constant support throughout the creation of this book. Also instrumental was S.T.E.M. Dean Dr. Kelli Ligeikis, who exhibited great faith in her support.

0

Paradoxes?

Surely a chapter before Chapter One must be a prelude or introduction of some kind. This is the case here, except that if this book were about mountain climbing, then this chapter would tell us of nothing but impending disaster. A more discouraging start couldn't be imagined, and yet, mathematics is well able to withstand such an introduction. This is so because this subject, perhaps more than any other, flourishes under the most withering scrutiny from its explorers.

Having taken some calculus courses, you are becoming familiar with a large body of theorems and applications. All of those topics in calculus are worthy of concentrated study. To that end, they are presented in polished, finished form. But surprisingly, there seems to be paradoxes lurking about. How can this be? Is not mathematics the subject that deals in certainties? The very word is derived from the Greek *máthesis*, meaning "the means of getting knowledge"—with the connotation of *sure* knowledge. Thus, a paradox, an enigma, in mathematics presents a tougher challenge to mathematicians than Mount Everest did to mountaineers prior to 1953. But in this landscape, we must conquer the peak every time: the paradox must always be resolved.

So, let us see if mathematics is as capable as it seems. A mountain range of six formidable paradoxes looms before you, the intrepid mathematical pilgrim.

Paradox 0.1 The simple alternating series $1 - 1 + 1 - 1 + \cdots$ can be rewritten as $(1 - 1) + (1 - 1) + \cdots$, from which we conclude that the ultimate sum must be zero. Just as reasonably, we can rewrite $1 - 1 + 1 - 1 + \cdots$ as $1 + (-1 + 1) + (-1 + 1) + \cdots$, giving us this time a sum of one. Therefore, $0 = 1$. ■

Maybe this isn't as much a paradox as it is a chintzy parlor trick, but where is the error? We had better be careful in our explanation, for the co-founder of calculus, Gottfried Wilhelm Leibniz, was certain that the series added to $\frac{1}{2}$, and *his* thinking was wrong.

Paradox 0.2 Consider two typical intervals of real numbers: $[1, 5]$ and $(5, 6)$. Recall that the first interval is closed, that is, it contains its two endpoints. The second interval is open. We clearly see that they have no point in common, and in particular, $5 \in [1, 5]$, and $5 \notin (5, 6)$. Question: how close are these two intervals? They are not separated by any positive distance; that is, they are at zero distance from each other, so they are touching. Now, it is true that if two points on the number line are at zero distance from each other, then they identify the same number. But if 5, clearly belonging to $[1, 5]$, is at zero distance from the interval $(5, 6)$, then there must be a point in $(5, 6)$ that equals 5. This is not possible. Hence the intervals are some positive distance apart, so they are not touching. But they are at zero distance, but then . . . ? ■

That is a tricky paradox. No amount of rhetoric will unravel it. What is needed is careful mathematical analysis. And the next paradox is terrible indeed!

Paradox 0.3 Here, Galileo Galilei (1564–1642) presents some triangles from plane geometry. In Figure 0.1, observe that every single point of segment AB is in one-to-one correspondence with a point of segment CD, as demonstrated by the cevian that matches points P and Q. Aha, but then how could the length of CD be nearly twice that of AB? Yes, there are infinitely many points in each segment, but it would seem that CD must have more points than AB. How can we conceive of more than infinity? The paradox is strengthened when one compares AB to congruent segment CE in the next figure, where $ABEC$ is a parallelogram. Clearly, CE is a proper subset of CD, and yet, their constituent points match up perfectly in pairs! Is this a flaw in geometry? ■

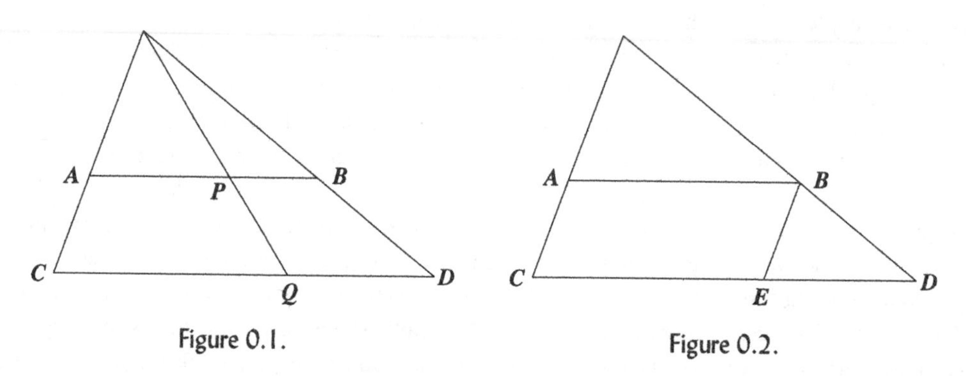

Figure 0.1.

Figure 0.2.

Galileo, and the geometers before him, sensed the borders of insanity in such deliberations, and just avoided the issue. But we need an explanation. It is found in mathematical analysis.

Paradox 0.4 Around 450 B.C., a Greek philosopher named Zeno of Elea rocked the current views of motion and mathematics by proposing some incisive paradoxes. The most famous one is couched in the story of a race between Greek superhero Achilles and a lowly tortoise. Let us move the scene to the modern era, and imagine that Achilles can run 10 meters per second, while the tortoise can muster one centimeter per second—both realistic speeds. To be fair, the tortoise gets an enormous head start of 100 m. All is ready, and the gun goes off. Ten seconds into the race, Achilles has effortlessly covered the 100 m. But in the same time, the tortoise has moved 10 cm forward, and is clearly ahead of the warrior. Achilles now covers those 10 cm in a hundredth of a second, but in that time span, our mighty tortoise has moved 0.01 cm, and high-speed photography shows that it is still in the lead. Zeno (who has wagered heavily on the tortoise), invites us to repeat this procedure without end, leading us to the conclusion that Achilles will never catch up to the reptile. Since this is absurd, Zeno leaps—as philosophers sometimes do—to an amazing conclusion: that all motion is just an illusion. ■

Mathematical analysis rectifies this paradox, and what is interesting is that to do so involves the infinite in one way or another. We should have been suspicious of this as soon as the bearded sage asked us to repeat his procedure *ad infinitum*.

Paradox 0.5 The following case is presented before a Court of Arbitration. Lawyer A asks the court to consider the isosceles right triangle in Figure 0.3. Divide its diagonal into two equal parts, and construct the vertical and horizontal step segments shown in Figure 0.4, in heavy line. Clearly, the total length of the segments is 2. Divide each half-diagonal again in two, constructing four smaller steps as in Figure 0.5, with a total length 2. Continue the constructions (Figure 0.6), so that iteration n yields 2^n steps, always with total length 2. As $n \to \infty$, the step function converges towards the diagonal. Therefore, the length of the limiting curve, which is the length of the hypotenuse, is also 2, and the attorney rests his case. But lawyer B, representing Pythagoras, simply states the unimpeachable truth of his client's theorem: the hypotenuse has length $\sqrt{2}$, and nothing else. And so, $\sqrt{2} = 2$, but the judge is not amused! ■

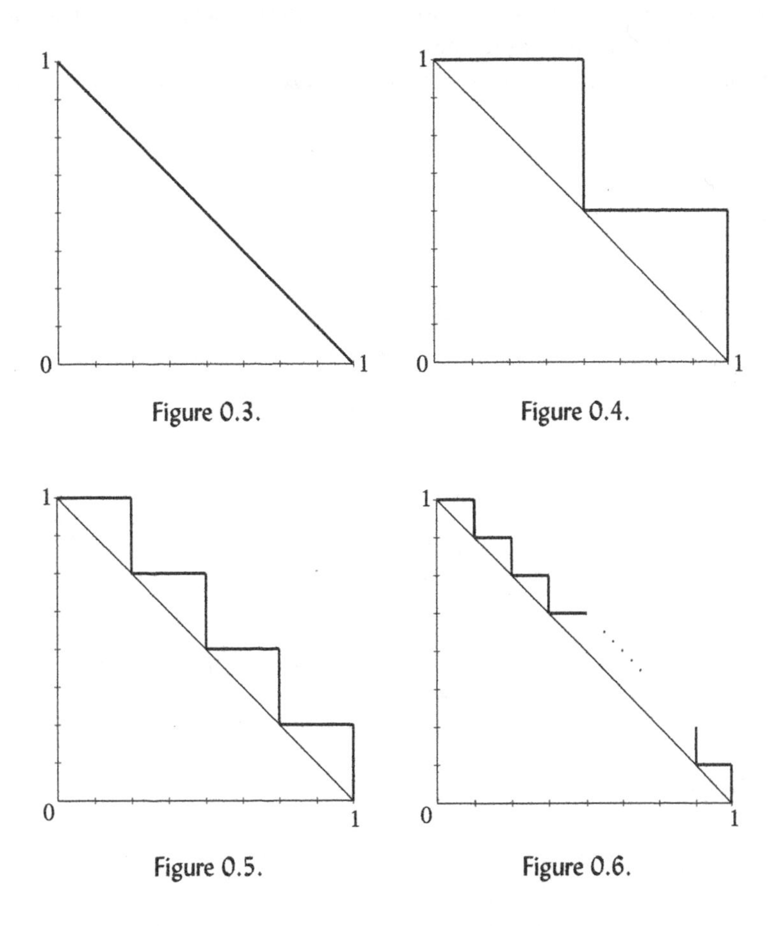

Figure 0.3.

Figure 0.4.

Figure 0.5.

Figure 0.6.

Enough of infinity. Next, we shall "prove" that all the elements of any finite set S of numbers are equal.

Paradox 0.6 Let $S_n = \{x_1, x_2, \ldots, x_{n-1}, x_n\}$ be a finite set of n real numbers. We proceed by induction. Our assertion is clearly true for S_1, a set with one element. Assume as true that all the elements of any set of n elements are equal. Now, consider the set $T_n = \{x_2, x_3, \ldots, x_n, z\}$ with n elements. Using the induction hypothesis, all the elements of T_n must be equal, $x_2 = x_3 = \cdots = z$. But since $x_1 = x_2$, we now create a set of $n + 1$ equal numbers: $S_{n+1} = \{x_1, x_2, \ldots, x_n, z\}$. Since S_n was completely arbitrary, it follows that all sets S_{n+1} have the sameness property. By induction, all finite sets of numbers must have nothing but equal numbers! ■

By now, mathematics seems to be battered, shaken by serious paradoxes. What can be done? A lot is at stake here. As Albert Einstein said, "it is mathematics that offers the exact natural sciences a certain measure of security which, without mathematics, they could not attain." Physics, chemistry, engineering, and the technologies all depend on that security. So, if mathematics admits doubt, all these other endeavors are left without theoretical support.

Problems like these began to be tackled around the turn of the 19th century. Many unstated assumptions were slowly clarified, and mathematical analysis was born. Roughly speaking, real analysis studies the foundational principles of the real number system, and the properties of functions and series. As such, it is the foundation beneath calculus, which sustains all the modern sciences upon its powerful shoulders.

As you study the chapters, one and all of the paradoxes will be addressed, unravelled, and neatly laid to rest.

1

Logical Foundations

Logic is the hygiene the mathematician practices to keep his ideas healthy and strong.

Hermann Weyl (1885–1955)

It shouldn't seem odd to begin a book about mathematics with a chapter on logic. Even though much of our early learning about mathematics involved experience, some of us instinctively knew that this subject was held together by something far more persuasive than experience, authority, culture, or law.

Our job in this chapter will be to investigate certain **rules of inference** by which a conclusion is found to depend without any doubt from one or more accepted **premises**. Mathematics enjoys its certainty because all of its conclusions follow *necessarily* from its premises, by rules of inference. This way of arriving at conclusions is called **deductive reasoning**. We shall see how logic uncovers some rules of inference.

1.1 Categorical Propositions

The sentences "Prime numbers after three are expressible as $6n \pm 1$, where n is a natural number," and "Plural English words end with an 's'," are starkly different in content, but how do their structures compare? Aristotle (384–322 B.C.) established logic to analyze such a question.

In logic, the two sentences are called statements. A **statement** is any sentence that expresses a claim that we can decide to be either true or false. Without any angst, we shall henceforth exclude the myriad other sentences that express such things as commands, questions, and wishes, since they do not have assignable truth values. Next, we shall concentrate on statements expressing how one category (or set or collection) of things relates to another category. The first category, is generically symbolized "S," and the second category, generically "P."

Aristotle identified exactly four statements relating category S to category P, and called them **categorical propositions.** They appear below, and will form the basis of all our important statements. Modern set notation, unknown to Aristotle, is included since we will use it very often. Note that $\varnothing$ represents the empty set or category, and $\overline{P}$ is the complement of set P. Logically speaking, $\overline{P}$ is everything in the universe of discourse that is not in category P. (Set theory itself will be studied in Chapter 5.)

CATEGORICAL PROPOSITION	*EQUIVALENT SET NOTATIONS*		*EQUIVALENT TRANSLATIONS*
All S are P.	If $x \in S$, then $x \in P$.	$S \subseteq P$.	If S, then P; each S is a P.
No S are P.	If $x \in S$, then $x \notin P$.	$S \cap P = \varnothing$.	If S, then not P; none of S is a P.
Some S are P.	$x \in S$ exists, and $x \in P$.	$S \cap P \neq \varnothing$.	At least one S is a P; there exists one S that is a P.
Some S aren't P.	$x \in S$ exists, and $x \notin P$.	$S \cap \overline{P} \neq \varnothing$.	At least one S isn't a P; there exists one S that isn't a P.

The power of Aristotle's logic lay in his observation that an enormous body of statements in language could be reduced to these four categorical propositions. (It would take a bit of practice to transform a statement such as, "Today's traffic spoiled our plans to visit Dr. Hawking" into a form appropriate for logical analysis. Fortunately, we won't run across this type of statement in our studies.) Let's apply Aristotelian logic to the grammatical rule mentioned above.

Example 1.1 "Plural English words end with an 's' " is not a categorical proposition, since Aristotelian logic recognizes only the verbs "are" and "aren't." We therefore change the sentence's structure without changing its meaning, as follows: "Plural English words are words that end with an 's'." Now we have a legitimate categorical proposition, with $S =$ {*plural English words*}, and $P =$ {*words that end with an 's'*}. Notice that both S and P are indeed sets of things. But we need more information. If the author meant the statement to be true in every case, then we obtain the proposition

$$\text{"All } \underbrace{\text{plural English words}}_{S} \text{ are } \underbrace{\text{words that end with an 's'}}_{P}\text{."}$$

Equivalently, we have "If $x \in$ {*plural English words*}, then $x \in$ {*words that end with an 's'*}."

English grammar being what it is, these propositions are false. What's more, we need only one exception to prove their falsehood. Thus, we try a translation that affirms only particular cases, and arrive at what was surely meant:

$$\text{"Some } \underbrace{\text{plural English words}}_{S} \text{ are } \underbrace{\text{words that end with an 's'}}_{P}\text{."}$$

And we will sacrifice style in pursuit of logical precision: "There exists at least one $x \in$ {*plural English words*}, such that $x \in$ {*words that end with an 's'*}." ■

Example 1.2 Now consider the algebraic statement, "Prime numbers after 3 are expressible as $6n \pm 1$, where n is a natural number." In this case, $S =$ {*prime numbers after 3*}, and $P =$ {*numbers expressible as* $6n \pm 1$, *where n is a natural number*}. We have an important choice in translation again. "Some S are P" is the proposition that affirms only particular cases. By logic, this means that there exists at least one member of S that belongs to P. To verify this proposition's truth we need to find only one instance of such a prime. That's easy (11 for instance), so this algebraic statement is trivially true.

But the writer clearly had a better idea in mind: "All S are P." This is harder to prove than the particular proposition, but the fruits are worth the effort. If you can prove this universal proposition, then the particular one becomes a simple consequence. Try to work out a proof for yourself. ■

Two categorical propositions are called **contradictories** of each other when they have opposite truth values. Consider the next two examples.

Example 1.3 Let's look at the statement "No place on Venus is habitable by people." We use the categories $V = \{places\ on\ Venus\}$ and $H = \{places\ habitable\ by\ people\}$ to translate it into the negative proposition "No V are H." Whether "No V are H" is true or not, Aristotle saw that "Some V are H" must have the *opposite* truth value. We can see this by comparing "Some places on Venus are habitable by people," with the original statement. Likewise, the affirmative "All places on Venus are habitable by people" (terribly false!), must have the opposite truth value to the fourth proposition in the table above, "Some places on Venus aren't habitable by people" (quite true). These two are contradictories, too. ■

Example 1.4 What is the contradictory statement to "Many quadratic polynomials have real roots"? First, we must convert the given statement into a categorical proposition in proper form without changing its meaning. The adjective "many" must become "some," and to obtain categories connected by one of the two allowed verbs, we arrive at "Some quadratic polynomials (Q) are polynomials with real roots (P)." Next, in order to contradict this, we would have to assert that every single quadratic polynomial does not have real roots (absurd!). Thus, we might write, "All quadratic polynomials (Q) aren't polynomials with real roots (P)." But Aristotelian logic forbids this form. Hence, the contradictory statement must be "No quadratic polynomials (Q) are polynomials with real roots (P)," which is a proposition found in the table. Comparing this statement and the original one, exactly one is true and one is false, since they are contradictories. ■

1.2 Some Rules of Inference

The contradictory relation is a rule of inference. With it, for example, we can start by affirming (claiming as true) the statement "No widgets are gizmos" and conclude without a doubt that "Some widgets are gizmos" is false. If someone objects that we don't know what "widgets" and "gizmos" are, we happily exclaim, "The point is that we don't need to know!" The power of a rule of inference is that the conclusion is *necessarily* true, once the premise is affirmed to be true. Do Exercise 1.1 at this time.

Note 1: Logicians since the second century A.D. have drawn a diagram like the one here, called the Square of Opposition, to gather certain rules of inference among the four categorical propositions. For the complete Square of Opposition and a wider presentation of Aristotelian logic, see the very readable text [41].

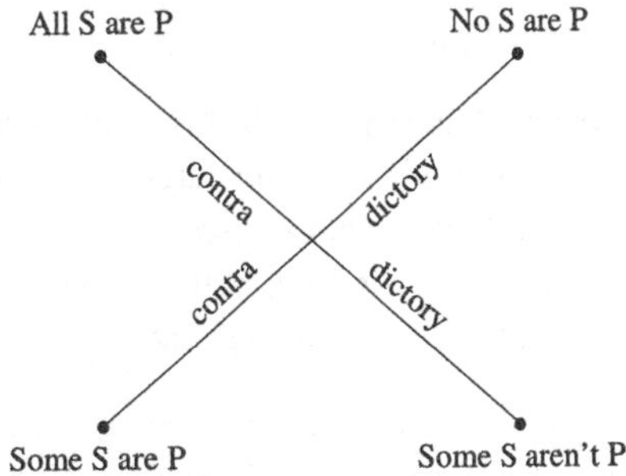

Classical logic deals not only with the contradictory relation, but also with three operations: the **converse**, the **obverse**, and the **contrapositive** (the nouns conversion, obversion, and contraposition are also used). They are defined by how they alter the four categorical propositions, as seen in the table below.

	CATEGORICAL PROPOSITION			
	All S are P.	No S are P.	Some S are P.	Some S aren't P.
CONVERSE	All P are S.	No P are S.	Some P are S.	Some P aren't S.
OBVERSE	No S are *non-P*.	All S are *non-P*.	Some S aren't *non-P*.	Some S are *non-P*.
CONTRAPOSITIVE	All *non-P* are *non-S*.	No *non-P* are *non-S*.	Some *non-P* are *non-S*.	Some *non-P* are S.

A boxed statement always carries the *same* truth value as the original categorical proposition in the heading above it: either both true, or both false. Hence, only those operations establish new rules of inference. For instance, the converse of "Some S are P" always yields a valid conclusion.

Note 2: In the table, contraposing "Some S aren't P" yields "Some non-P aren't non-S," from which the displayed result follows.

Example 1.5 It is true that "Some quadrilaterals can't be inscribed in a circle." This is rewritten as

$$\text{"Some } \underbrace{\text{quadrilaterals}}_{S} \text{ aren't } \underbrace{\text{figures inscribable in a circle}}_{P}\text{."}$$

The valid contrapositive would then be

$$\text{"Some } \underbrace{\text{figures not inscribable in a circle}}_{non\text{-}P} \text{ are } \underbrace{\text{quadrilaterals}}_{S}\text{."}$$

Notice how this form brings out an alternative point of view. ■

Continuing with the table, statements not in boxes *must not be used* as inferences, unless they have been independently verified. Those statements have truth values undeterminable from the original statement's truth value. The most notorious of these fallacious missteps is to assume that the converse of "All S are P" is true. Haven't we all believed that "All continuous functions are differentiable functions," at least for a little while? We quickly learned that this was the false converse of the theorem "All differentiable functions are continuous functions."

Did you notice that the obversion operation preserves truth value in all cases? Consider the next example.

Example 1.6 Begin with the statement "Some intervals of real numbers aren't closed sets" (Some S aren't P). This happens to be a true statement. Obversion gives "Some intervals of real numbers are non-closed sets" (Some S are *non-P*). Obversion has given us a different but equivalent way to look at the statements. This is valuable when we need to better understand highly abstract statements. ■

We can chain successive valid inferences to form what is called a **proof sequence** (this is the core of a proof, as defined in the next chapter). For example, following the obversion just above, we could very well apply conversion to get "Some non-closed sets are intervals of real numbers" (Some *non-P* are S). This is true precisely because the rules of inference preserved the truth value of the original statement. (You are right if you imagine that logic could be used to start with a false premise and derive a false conclusion. This forms the basis of a technique of proof discussed in the next chapter.)

1.3 Conditional Statements; Necessity and Sufficiency

We have seen that a categorical proposition such as "All squirrels (S) are mammals (M)," is equivalent to asserting "If $x \in S$, then $x \in M$." The "if... then" structure identifies a **conditional statement**. In the first table of this chapter, we saw that "If $x \in S$, then $x \in M$" also highlighted the subset relationship between S and M, namely, $S \subseteq M$.

We amplify the conditional statement "If $x \in S$, then $x \in P$" by saying, "Belonging to S is a sufficient condition for belonging to P." This is the sense that is conveyed when we say, "In a deck of cards, having a club is a sufficient condition for having a black card."

Now, the contrapositive of "All S are P" is "All *non-P* are *non-S*," or, "If $x \notin P$, then $x \notin S$." This brings to us another relationship between categories P and S. Rather than stating that $x \notin P$ is a sufficient condition for $x \notin S$, we say "Belonging to P is a necessary condition for belonging to S." The sense is that $x \in P$ is indispensable, but not sufficient, for claiming that $x \in S$. Returning to our two examples, we say that being a mammal is a necessary condition for being a squirrel, just as having a black card in your hand is a necessary condition for the card to be a club. (Of course, being a mammal is not a sufficient condition for claiming that the beast is a squirrel, nor is having a black card sufficient for a sure bet that it is a club.)

In the following summary, note how categories S and P exchange positions:

These are all equivalent:

- All S are P.
- If $x \in S$, then $x \in P$.
- $x \in S$ is a **sufficient condition** for $x \in P$.

- $x \in P$ is a **necessary condition** for $x \in S$.
- If $x \notin P$, then $x \notin S$.
- All *non-P* are *non-S* (contraposition of "All S are P").

1.4 Extending Aristotelian Logic

Let's pursue these ideas even further. In the conditional statement "If $x \in S$, then $x \in P$," observe that $x \in S$ is not a category, but a statement in itself, as is $x \in P$. These statements are of course each either true or false. In modern times, Aristotelian logic was extended by allowing arbitrary statements Q and R to appear in the conditional (not just statements about categories), as in

$$\text{"If } \underbrace{\text{a triangle lies on a plane,}}_{Q} \text{ then } \underbrace{\text{its angles add to } 180^\circ}_{R}\text{."}$$

The conditional "If Q, then R" becomes a rule of inference the moment it is proved that the occurrence of Q is a sufficient condition for the occurrence of R. Put another way, conditional statements in mathematics are used as rules of inference when they express theorems. In the conditional "If a triangle lies on a plane, then its angles add to 180°," any time we assume the truth of the statement, "The triangle I am working with lies on a plane," we may then confidently conclude that "its angles add to 180°." The inference is valid because the conditional statement is, in fact, a theorem.

The discussion now opens the possibility that Q may be false. What can be said about R? Logic offers four possible arrangements of truth and falsehood within a conditional statement, as follows.

	CONDITIONAL STATEMENT	*TRUTH VALUE*
I	If Q (is true), then R (is true).	True
II	If Q (is true), then R (is false).	False
III	If Q (is false), then R (is true).	True
IV	If Q (is false), then R (is false).	True

The last column indicates the truth value of the conditional statement as a whole. Mathematical theorems invariably deal with case I implications. Case II is the situation with invalid deductions in mathematics—we may perhaps call them "anti-theorems"! Cases III and IV are quite unsettling, because together they imply that beginning with the falsehood of Q, we can affirm or deny R as a conclusion. In the instance just above, if we begin with a triangle that is not drawn on a plane (Q is false), then nothing certain can be said about the angle sum of the triangle (indeterminate truth value of R). Only rarely do we encounter cases III and IV.

Note 3: The table is precisely the truth table for what logicians call "material implication." We will use the terms "conditional statement" and "**implication**" synonymously.

The conditional statement, its equivalent contrapositive, and its negation are listed together below for future reference. The negation is interesting because it isn't a conditional statement at all. To derive it, we use the following proof sequence: (1) "If Q, then R" is equivalent to "All instances that Q is true are instances that R is true," (2) the negation of this is "Some instances that Q is true are instances that R is false" (see the Square of Opposition in Note 1), (3) hence, at least one instance exists where statement Q is true and simultaneously statement $not\text{-}R$ is also true.

CONDITIONAL STATEMENT	*CONTRAPOSITIVE*	*NEGATION*
If Q, then R.	If $not\text{-}R$, then $not\text{-}Q$.	Q and $not\text{-}R$.

For example, the conditional "If Bill Clinton was President, then he was impeached" is a true statement. Its negation, "Bill Clinton was President and he was not impeached," must therefore be false.

1.5 Equal Categories; Equivalent Statements

The highly important case when *both* the statement "All S are P" and its converse "All P are S" have been proved true means that categories S and P are **equal**: $S = P$. Next, remember that for statements Q and R, "If Q, then R," and "If R, then Q" are converses. Whenever both of these conditionals have been proved true, we say that Q and R are **equivalent** statements, written $Q \equiv R$. The commonly used form "Q **if and only if** R" has exactly the same meaning as "Q is equivalent to R."

Reintroducing sufficiency and necessity, we have the following translations:

CATEGORIES	S and P are equal categories.	$x \in S$ is a necessary and sufficient condition for $x \in P$.	$x \in S$ if and only if $x \in P$.	$S = P$
STATEMENTS	Q and R are equivalent statements.	Q is a necessary and sufficient condition for R.	Q if and only if R.	$Q \equiv R$

It stands to reason that proving two categories to be equal requires two distinct sections of proof. To prove $S = P$ requires us to prove not only that (1) $S \subseteq P$, but also that (2) $P \subseteq S$. The logic in the case of a proof of equivalence $Q \equiv R$ is similar. A proof of "Q if and only if R" splits into the proofs of (1) the sufficiency of R section, "If R, then Q," and (2) the necessity of R section, "If Q, then R." We summarize this as follows.

> Proving equivalence always requires proving a conditional statement and its converse.

Example 1.7 Suppose we must prove the equivalence

$$\text{“}\underbrace{\text{The side lengths of a triangle in a plane satisfy } a^2+b^2=c^2}_{Q}\text{ if and only if }\underbrace{\text{it is a right triangle.}}_{R}\text{”}$$

This is how we would arrange the proof:

(1) Proof of necessity [If Q, then R]: Suppose planar triangle $\mathbb{E}$ has side lengths that satisfy $a^2 + b^2 = c^2$. [the proof continues].... We conclude that $\mathbb{E}$ is a right triangle.
(2) Proof of sufficiency [If R, then Q]: Conversely, suppose planar triangle $\mathbb{E}$ is a right triangle. [the proof continues].... We conclude that $\mathbb{E}$ has side lengths that satisfy $a^2 + b^2 = c^2$. Our theorem has now been proved [$Q \equiv R$]. ■

The same format (or elaborations of it) is found in all proofs of equivalence. The rose of proof always finds secure hold on the strong trellis of logic.

Note 4: A definition of a term or expression is an equivalence statement. Thus, the sentence "A means B" is a rendition of the symbol $\equiv$. When we see statements such as "$\Delta x \equiv x_1 - x_0$," or "$f^{(0)}(x) \equiv f(x)$," they define one expression or symbol to mean the same as the other one. You may always rewrite "A means B" as "B means A." It is therefore valid to substitute B for A, and vice versa. Such is the logical nature of definitions.

1.6 Compound Statements

Logic also applies to statements that have a substructure within themselves. These are generally called **compound statements**. A common example is the conditional "If Q, then R" in which Q is itself a statement of the form "H and K," called a **conjunction**.

In Example 1.7, the theorem had the form "(H and K) if and only if R," using statements $H =$ *a triangle is in a plane*, $K =$ *the triangle's sides satisfy* $a^2 + b^2 = c^2$, and $R =$ *the triangle is a right triangle*.

Note 5: We understand by context that the parentheses just encountered in "(H and K) if and only if R" are grouping symbols for the conjunction, and not simple punctuation marks. A different conjunction would occur if we wrote "H and (K if and only if R)."

Example 1.8 "For rational $x \neq 0$, no value of e^x is rational, in other words, all such values of e^x are irrational" (true statements, actually). Translation into logic:

$$\text{“If } \underbrace{x \text{ is rational}}_{R} \text{ and } \underbrace{x \neq 0}_{N}, \text{ then } \underbrace{e^x \text{ is irrational}}_{I}.\text{”}$$

This is the compound statement "If (R and N), then I." See Exercise 1.4 for a bit of further insight. ■

1.7 Major and Minor Premises

In some theorems, certain premises provide background information for the theorem, such as a universal category from which we must not stray. We shall call these the **minor premises** of the theorem, for while they are necessary ingredients for the proof, we shouldn't get confused and treat them as things to be proved. On the other hand, any statements that are central to the point of the theorem shall be called **major premises**. In passing, we note that all the premises of a theorem should be appealed to in the proof; hence, irrelevant statements must be kept to a minimum, functioning only as clarifying information.

Example 1.9 Return to the outline in Example 1.7. The theorem was found to have the form "(H and K) if and only if R," using statements $H =$ *a triangle is in a plane*, $K =$ *the triangle's sides satisfy* $a^2 + b^2 = c^2$, and $R =$ *the triangle is a right triangle*. It becomes clear that in the sufficiency part of the proof we are proving that triangle $\mathbb{E}$ follows the Pythagorean theorem, not that $\mathbb{E}$ lies on a plane. Statement H is necessary, but it is a background requirement. We therefore identify H as the minor premise. This produces a streamlined proof that focuses on the key idea:

Let $\mathbb{E}$ be a triangle in a plane [minor premise H].

(1) Proof of necessity [If K, then R]: Suppose $\mathbb{E}$ has side lengths that satisfy $a^2 + b^2 = c^2$ [major premise K]. [the proof continues].... We conclude that $\mathbb{E}$ is a right triangle.
(2) Proof of sufficiency [If R, then K]: Conversely, suppose $\mathbb{E}$ is a right triangle [major premise R]. [the proof continues].... We conclude that $\mathbb{E}$ has side lengths that satisfy $a^2 + b^2 = c^2$.
Our theorem has now been proved [$K \equiv R$]. ■

1.8 Quantification

Aristotelian logic was further enriched near the end of the 1800s by logician Gottlob Frege (1848–1925), who focused on the quantifiers "all," "some," and "no." The proposition "If $x \in S$, then $x \in P$" can be expanded to say "For any x, if $x \in S$, then $x \in P$." Frege introduced the **universal quantifier** $\forall x$, meaning literally "for any x" or "for all x," and with it we can translate "All S are P" into "$\forall x$, if $x \in S$, then $x \in P$." Frege also introduced the **existential quantifier** $\exists x$, meaning "for some x" or "there exists an x," to use for statements indicating particular cases. In this system of logic, "Some S are P" becomes "$\exists x$ such that $x \in S \cap P$." Of course, this also means that "Some P are S."

To interpret "No S are P," Frege negated its contradictory, this way: "There does not exist an x such that $x \in S \cap P$." In symbols, this would be "*not*-($\exists x$ such that $x \in S \cap P$)." To see how to simplify this, suppose a person asserts, "Some students remembered their homework." In Aristotelian form, we say

$$\text{"Some } \underbrace{\text{students}}_{S} \text{ are } \underbrace{\text{people who remembered their homework.}}_{R}\text{"}$$

The correct negation would be "No students are people who remembered their homework," by the contradictory relation. By way of obversion, the negation becomes "All students are people who forgot their homework." The quantified logical statement for this sounds strange: "For all x, if x is a student, then x forgot its homework." In symbols, this is "$\forall x$, if $x \in S$, then $x \in \overline{R}$." In general, then, the translation of "No S are P" is "$\forall x$, if $x \in S$, then $x \in \overline{P}$."

Common usage commonly negates universal statements in a fallacious way. To deny "No students remembered their homework," people tend to say, "That's wrong, they all remembered their homework." But this is invalid! Common language is often fast and loose with logic.

The four categorical propositions and their negations appear in the following table. The negations still adhere to the contradictory relations in the Square of Opposition. For statements Q and R to be relevant, they must pertain to the "thing" x, often a number or a function. Now, there is occasionally an issue regarding categories that are empty. Propositions of types I and II do not in themselves imply that S is nonempty. Thus, if the existence of elements in S is demanded, then we must state it, often as a minor premise. On the other hand, propositions of types III and IV certainly imply the existence of at least one member of S.

TYPE	*CATEGORICAL PROPOSITION*	*QUANTIFIED FORMS*		*NEGATION*
I	All S are P.	$\forall x$, if $x \in S$, then $x \in P$.	$\forall x$, if Q then R.	To negate I, use IV.
II	No S are P.	$\forall x$, if $x \in S$, then $x \in \overline{P}$.	$\forall x$, if Q then *not*-R.	To negate II, use III.
III	Some S are P.	$\exists x$ such that $x \in S \cap P$.	$\exists x$ such that Q and R.	To negate III, use II.
IV	Some S aren't P.	$\exists x$ such that $x \in S \cap \overline{P}$.	$\exists x$ such that Q and *not*-R.	To negate IV, use I.

Note 6: Logicians are in their element when they render a statement entirely in symbolic form. To that end, they use a small set of symbols such as the following.

$Q \to R$	if Q, then R	$Q \wedge R$	Q and R	$\neg Q$	not Q
$Q \equiv R$	Q if and only if R	$Q \vee R$	Q or R		

With these, "If Q then R" can be rendered entirely in symbols as "$\forall x(Q \to R)$," and "If Q then *not*-R" becomes "$\forall x(Q \to \neg R)$." Also, "x exists such that Q and R" becomes "$\exists x(Q \wedge R)$," and "x exists such that Q and *not*-R" becomes "$\exists x(Q \wedge \neg R)$." You are welcome to use these extra symbols in any translations, if they seem helpful.

Quantifiers are useful for precisely stating many theorems and definitions. The following example explores a classic definition. We study it here only from the logical point of view.

Example 1.10 A function f is continuous at a point a in its domain if and only if given any $\varepsilon > 0$, there exists a corresponding $\delta > 0$ such that for all x in the domain with $|x - a| < \delta$, we have $|f(x) - f(a)| < \varepsilon$. What on earth is this saying?

To start, let's symbolize the domain of f by $\mathcal{D}_f$, and let statements $D \equiv x \in \mathcal{D}_f$, $Q \equiv |x - a| < \delta$, and $R \equiv |f(x) - f(a)| < \varepsilon$. The statement that $a \in \mathcal{D}_f$ is a minor premise (a is a constant). Now, "Given any $\varepsilon > 0$, there exists a corresponding $\delta > 0$," translates as "$\forall\varepsilon$, if $\varepsilon > 0$, then $\exists\delta > 0$." Next, "such that for all $x \in \mathcal{D}_f$ with $|x - a| < \delta$," translates as "such that $\forall x$, (if D and Q)." Thus, the definition finally becomes "function f is continuous at $a \in \mathcal{D}_f$ if and only if $\forall\varepsilon$, (if $\varepsilon > 0$, then ($\exists\delta > 0$ such that $\forall x$, (if $D$and Q, then R)))." At the very least, this brings out the structure of the definition. In the chapter about continuity, we shall study the *implications* of all this! ■

Remember, all definitions are—by definition—logical equivalences. Accordingly, in order to describe non-continuity (discontinuity) at some point, we will have to negate the definition that we so carefully constructed.

Example 1.11 We determine what discontinuity implies. First, negate the definition of continuity, and then use the table above in a step-by-step fashion to get the following proof sequence.

Negate the definition's statement:	*not*-($\forall\varepsilon$, (if $\varepsilon > 0$, then ($\exists\delta > 0$ such that $\forall x$, (if $D$and Q, then R)))).
Negate type I by using type IV:	$\exists\varepsilon > 0$ such that *not*-($\exists\delta > 0$ such that $\forall x$, (if D and Q, then R)).
Negate type III by using type II:	$\exists\varepsilon > 0$ such that $\forall\delta$, (if $\delta > 0$, then *not*-($\forall x$, (if D and Q, then R))).
Negate type I by using type IV:	$\exists\varepsilon > 0$ such that $\forall\delta$, (if $\delta > 0$, then ($\exists x$ such that D and Q, and *not*-R)).

Replacing D, Q, and R by the statements they represent, we state, "Let $a \in \mathcal{D}_f$ (minor premises can't be forgotten). A function f is not continuous at a if and only if $\exists\varepsilon > 0$ such that $\forall\delta > 0$, ($\exists x \in \mathcal{D}_f$ and $|x - a| < \delta$ and $|f(x) - f(a)| \geq \varepsilon$)."

Putting more words into it, we arrive at the logical negation of the continuity criterion: f is not continuous at a in its domain if and only if for some $\varepsilon > 0$, and for any $\delta > 0$, there exists some x in the domain of f such that $|x - a| < \delta$, and yielding $|f(x) - f(a)| \geq \varepsilon$. (This negation will be extended in scope in Chapter 24.) ■

1.9 Exercises

1. Translate the statement "Some actors in the movie *Forbidden Planet* have died" into one of the four categorical propositions. Use the contradictory relation as a rule of inference to deduce a valid conclusion from it. You have to do absolutely no research about this fine classic to arrive at a valid conclusion!

2. Find the contradictory proposition and its truth value for each of these statements:
   a. All rational numbers are real.
   b. No real numbers are rational.
   c. Some integrable functions aren't differentiable.
   d. All visible things are things with mass.

3. Find the obverse of the four propositions above, and compare the truth values with those of the originals.

4. In Example 1.8, the original statement

$$\text{"No } \underbrace{\text{values of } e^x \text{ with rational } x \neq 0}_{S} \text{ are } \underbrace{\text{rational numbers}}_{P}\text{"}$$

was followed by "in other words, all such values of e^x are irrational." What rule of inference did I apply to get those other words?

5. Find the converse of the four propositions above, and compare the truth values with those of the originals.

6. The conditional statement "If Q, then R" can also be translated as "*Not-*Q, or R." Use this form to translate the theorem that all differentiable functions are continuous functions. Note the interesting sense it gives.

7. Why are the statements "T is true if and only if U is true" and "T is false if and only if U is false", equivalent?

8. The following theorem was discovered by the great amateur mathematician Pierre de Fermat (1601–1665): if $f(x)$ has a local extremum at $x = c$, and $f'(c)$ exists, then $f'(c) = 0$.
   a. Symbolize the theorem, using $Q = \{f(x) \textit{ has a local extremum at } x = c\}$, $R = \{f'(c)$ *exists*$\}$, *and* $S = \{f'(c) = 0\}$.
   b. State the contrapositive of Fermat's theorem (is it true or false?).
   c. Write the negation of Fermat's theorem, using the existential quantifier.

9. Assume the truth of "All numbers larger than any given number are infinite quantities." Then "Some numbers larger than any given number are infinite quantities" must also be true. This means that there does exist at least one number that is larger than any given number. But that is false. Where did we go wrong?

10. Let's analyze the mean value theorem from the point of view of logic. It says, "If a function f is continuous over an interval $[a, b]$ and differentiable over (a, b), then there exists a number $c \in (a, b)$ such that $f'(c) = \frac{f(b)-f(a)}{b-a}$." This is in essence a proposition of the "All S are P" type, with $S = \{$*functions* f *that are continuous over* $[a, b]$ *and differentiable over* $(a, b)\}$ and $P = \{$*there exists a number* $c \in (a, b)$ *such that* $f'(c) = \frac{f(b)-f(a)}{b-a}\}$.

    But S is itself the conjunction of two categories, $S_1 = \{$*functions* f *that are continuous over* $[a, b]\}$ and $S_2 = \{$*functions* f *that are differentiable over interval* $(a, b)\}$. Furthermore, P has the structure "Some Q are R," with $Q = \{$*numbers* $c \in (a, b)\}$ and $R = \{$*numbers* c *such that* $f'(c) = \frac{f(b)-f(a)}{b-a}\}$. So, we can symbolize the mean value theorem as "If (S_1 and S_2) then (some Q are R)," or "All (cases of S_1 and S_2) are (cases that some Q are R)."
    a. Use the form "If (S_1 and S_2) then (some Q are R)" to write the contrapositive of the mean value theorem.
    b. We now perform a sequence of logical inferences on the results of part a. (1) Use the contradictory relation to simplify the statement "*not*-(some Q are R)." (2) The statement "*not*-(S_1 and S_2)" is equivalent to "*not*-S_1 or *not*-S_2," by a theorem called De Morgan's laws (to be proved in a later chapter). Thus, recast the mean value theorem in this newer logical form, and write it in good English form.
    c. The mean value theorem in quantified form is: $\forall f \left[\text{if } f \in (S_1 \text{ and } S_2) \text{ then } \exists c \text{ such that } (c \in (a, b) \text{ and } f'(c) = \frac{f(b)-f(a)}{b-a})\right]$. What is its converse, and is it true?

2

Proof, and the Natural Numbers

It is a far finer gambit than any chess gambit: a chess player may offer the sacrifice of a pawn or even a piece, but a mathematician offers the game.

G. H. Hardy (1877–1947), *A Mathematician's Apology*

2.1 Axiomatic Systems and Proof

It was mentioned at the beginning of the previous chapter that the two statements, "Prime numbers after three are expressible as $6n \pm 1$, where n is a natural number," and "Plural English words end with an 's'," are starkly different in content. There is a more profound difference. A novice in English, if asked to test the truth of the grammatical rule about forming plurals, would have to obtain the largest dictionary possible, and check the plural forms of every noun, looking for an exception. Granting that it wouldn't take long to find an exception, the point is that there is no better way to test this—or any—grammatical rule. In contrast, the algebraic statement can be proved for an *infinitude* of primes, without resorting to the impossible task of checking each prime individually.

Why is this possible, and what is proof? This brings us to issues that are at the core of mathematics. The classical algebra and geometry that you have learned are examples of **deductive axiomatic systems**. Such systems are composed of (1) a small list of undefined terms, (2) a list of definitions, (3) a small list of **axioms**, or accepted relations among the definitions and terms, and (4) many statements that are all ultimately deduced from the lists of definitions and axioms. A **proof** of a mathematical proposition is a finite sequence of statements invoking axioms, definitions, or already proved statements, deriving all its other statements by rules of inference, and ending with the proposition in question. Upon discovering such a proof sequence, you triumphantly label the original statement, the proposition that you wanted to prove, as a **theorem**. In this way, the entire structure of algebra or geometry is made to rest on axioms and definitions. We will showcase five major techniques of proof at the same time as we develop the axiomatic system of natural numbers. (We have already seen several proof sequences in the previous chapter.)

2.2 The Natural Numbers

One of the invitations of this book is to comprehend in some depth the set $\mathbb{R}$ of real numbers. But before we can get a good grip on $\mathbb{R}$, it is necessary to understand rational numbers clearly. And to understand rationals well, we must know precisely what a "natural number" is. But here we get into deep water, because, for example, one must define what "three" means without resorting to "threeness" within the definition (in which case the definition becomes mathematically useless). It's enjoyable to try to write a rigorous definition of "three," and then look at the circular definitions that dictionaries settle with.

Let us begin by answering the question, "What are natural numbers?" Many cultures have gotten by with little more than those numbers. For instance, the Romans had their numerals, which were efficient for counting things, and for not much else. But simplicity doesn't imply shallowness. To describe clearly the capabilities of the numbers that we have known since childhood, we need a set of axioms and some deductive reasoning. In fact, there is no way to define natural numbers except through a deductive axiomatic system.

The following is a typical axiomatic structure for the **natural numbers**. This structure holds everything necessary and sufficient about natural numbers. (A similar development is found in Section 7.3 of Eves' and Newsom's fascinating and detailed book about the foundations of mathematics [12].) As undefined terms, we have equality, substitution, a set $\mathbb{N} = \{a, b, c, \ldots\}$ of unspecified things that we shall call "natural numbers," and two operations $+$ and $\times$ on pairs of elements of $\mathbb{N}$. These operations will respectively and suggestively be called "addition" and "multiplication." It's convenient to denote $a \times b$ by $a{\cdot}b$ or ab. The list of axioms is:

N1: For all $a, b \in \mathbb{N}$, $a + b$ and ab are unique elements of $\mathbb{N}$. (Closure and uniqueness)

N2: For all $a, b \in \mathbb{N}$, $a + b = b + a$ and $ab = ba$. (Commutativity)

N3: For all $a, b, c \in \mathbb{N}$, $a + (b + c) = (a + b) + c$ and $a\,(bc) = (ab)\,c$. (Associativity)

N4: For all $a, b, c \in \mathbb{N}$, $a\,(b + c) = ab + ac$. (Distributivity of $\times$ across $+$)

N5: There exists a unique number $1 \in \mathbb{N}$ such that for any $a \in \mathbb{N}$, $1a = a$. (Existence of a multiplicative identity, named "one")

N6: Let M be a set of natural numbers such that $1 \in M$, and whenever $k \in M$, then $k + 1 \in M$. Then $M = \mathbb{N}$, that is, M contains all the natural numbers. (Mathematical induction)

N7: For all $a, b \in \mathbb{N}$, exactly one of the following holds: $a = b$, or $a < b$, or $b < a$. (Trichotomy)

Notice that zero does not exist here! We can't move on until we define clearly what **inequality** means, since the symbol was used in N7. Let $a, b, c \in \mathbb{N}$. Then

$a < b$ and $b > a$ both mean that $a + n = b$ for some $n \in \mathbb{N}$.
$a \leq b$ and $b \geq a$ both mean that either $a < b$ or $a = b$.
$a < b < c$ (a triple inequality) means that $a < b$, $b < c$, and $a < c$ simultaneously.
$a \neq b$ means that either $a < b$ or $b < a$.

To illustrate the workings of this axiomatic system, we prove two lemmas and two theorems. A **lemma** is a theorem with a (usually) short proof that is (usually) used to help prove a more important theorem. A **corollary** is a direct but noteworthy consequence of a theorem, often with a short proof due to the close relationship with the theorem. However, there is useful latitude in the application of these two nouns.

Our first lemma uses a **direct proof**, that is, a proof that proceeds from the premises directly to the conclusion by rules of inference.

Lemma 2.1 For all $a, b, c \in \mathbb{N}$, if $a < b$, then $c + a < c + b$.

Proof sequence. A direct proof.

Statement	Reason
(1) There exists a natural number x such that $a + x = b$.	Definition of $<$.
(2) $(c + a) + x = c + (a + x)$.	N3.
(3) $(c + a) + x = c + b$.	Substitution.
(4) $c + a < c + b$.	Definition of $<$.

QED

Notice that the definition of $<$ is being used in both its direct and converse forms, since all definitions are logical equivalences.

Note 1: QED = *Quod Erat Demonstrandum* = *which was to be proved*, indicated that a proof was complete in the days when mathematics was written longhand, in Latin, the universal scholarly language. "QED" is a salute to all those mathematicians from long ago.

It is sometimes the case that the contrapositive of a proposition is actually easier to prove than the original. Thus, instead of directly proving "If Q, then R," we would prove "If *not*-R, then *not*-Q." Devising such a proof is an effective strategy called **proof by contraposition**. The following lemma uses this technique to prove a well-known property of equations.

Lemma 2.2 If $c + a = c + b$, then $a = b$, that is, cancellation is valid for addition.

Proof sequence. A proof by contraposition. We prove that if $a \neq b$, then $c + a \neq c + b$.

Statement	Reason
(1) Assume that $a \neq b$.	Premise for proof by contraposition.
(2) Case 1: $a < b$.	Possible consequence of $\neq$.
(3) $c + a < c + b$.	Lemma 2.1.
(4) $c + a \neq c + b$.	Definition of $\neq$.
(5) Case 2: $b < a$.	Possible consequence of $\neq$.
(6) $c + b < c + a$.	Lemma 2.1.
(7) $c + b \neq c + a$.	Definition of $\neq$.
(8) The lemma as stated is true.	(1), (4), and (7), and the contrapositive operation.

QED

Nothing has been implied about subtraction in the lemma and proof, since subtraction has not been defined at this point.

The next theorem states what seems childishly trivial: 1 is the smallest natural number! Why is a proof necessary? Recall that $\mathbb{N}$ is populated by as yet *unknown* elements, so that we cannot assume any common numerical properties about them beyond what can be proved. From this point of view, Theorem 2.1 will yield an interesting conclusion, namely, that the natural number identified by N5 as the "identity" is precisely the smallest natural number. Do you see that what seems trivial is sometimes quite profound and tricky to prove? This is a theme that runs throughout the history of mathematics.

Theorem 2.1 will use N6, the axiom justifying the highly important process of **mathematical induction**. To use it, note that M must first be shown to be nonempty (step 2), then the conditional statement "if $a \in M$, then $1 + a \in M$" is proved (steps 3–11), and only then is the conclusion of N6 drawn. We will investigate induction more deeply in Chapter 4.

Theorem 2.1 For all $a \in \mathbb{N}$, $1 \leq a$.

Proof sequence. A proof using mathematical induction.

Statement	Reason
(1) Let M be the set of natural numbers a such that $1 \leq a$.	Defining M.
(2) Let $1 = a$. Then $a \in M$.	$1 \leq 1$, and definition of M.
(3) $1 + 1 = 1 + a$.	Substitution.
(4) $1 < 1 + a$.	Definition of $<$.
(5) $1 + a \in M$.	Definition of M.
(6) Let $1 < a$. Then $a \in M$.	Definition of M.
(7) $1 + 1 < 1 + a$.	Lemma 2.1.
(8) $(1 + 1) + u = 1 + a$ for some $u \in \mathbb{N}$.	Definition of $<$.
(9) $1 + (1 + u) = 1 + a$.	N3.
(10) $1 < 1 + a$.	N1 and definition of $<$.
(11) $1 + a \in M$.	Definition of M.
(12) If $1 \leq a$, then $1 + a \in M$.	(2), (5), (6) and (11).
(13) $M = \mathbb{N}$.	(2), (12) and N6.
(14) For all $a \in \mathbb{N}$, $1 \leq a$.	(1) and (13).

QED

Recall from Chapter 1 that the contradictory relation is a rule of inference. It is most effectively used in the technique called **proof by contradiction**. In this strategy, we assume the truth of the *contradictory* to the statement that we want to prove. Since it is impossible for a statement and its contradictory to be both true in a deductive system, the contradictory statement is employed to derive a violation of any other mathematically true statement. This violation proves the contradictory statement to be false, and consequently, the original statement must be true. We shall see many examples of this method of inference, beginning here. Consider another "obvious" property: natural numbers are spaced one apart. The precise formulation appears in the following theorem.

Theorem 2.2 For all $a \in \mathbb{N}$, there is no natural number z such that $a < z < a + 1$.

Proof sequence. Proof by contradiction. The form of the theorem is that of a minor premise, $a \in \mathbb{N}$, followed by the categorical proposition "No S are P," where $S = \{z \in \mathbb{N}\}$ and $P = \{z \text{ such that } a < z < a + 1\}$. We prove that the contradictory proposition, "Some S are

P," is false. That is, "Some natural number z exists such that $a < z < a + 1$," will be proved false.

Statement	Reason
(1) Assume $z \in \mathbb{N}$ exists such that $a < z < a + 1$.	Premise for proof by contradiction.
(2) There exists natural numbers n and m such that $a + n = z$ and $z + m = a + 1$.	Definition of $<$.
(3) $a + (n + m) = (a + n) + m$.	N3.
(4) $a + (n + m) = z + m$.	Substitution.
(5) $a + (n + m) = a + 1$.	Substitution.
(6) $n + m = 1$.	Lemma 2.2.
(7) $n < 1$.	Definition of $<$.
(8) Theorem 2.1 is violated!	(7) and (2) together deny Theorem 2.1.
(9) Assumption (1) must be false.	From (1) we deduced the denial of Theorem 2.1.
(10) The theorem as stated is true.	(9) and the contradictory relation.

QED

Note 2: We see that proof by contradiction is easily explained in Aristotelian logic. Some authors use quantification logic, which for comparison proceeds as follows.

Statement to be proved (Theorem 2.2): $\forall a$, if $a \in \mathbb{N}$, then not-($\exists z \in \mathbb{N}$ such that $a < z < a + 1$).

Assume that its negation is true: *not*-[$\forall a$, if $a \in \mathbb{N}$, then not-($\exists z \in \mathbb{N}$ such that $a < z < a + 1$)].

Proceeding to negate the bracketed conditional statement, $\exists a \in \mathbb{N}$ such that not-not-($\exists z \in \mathbb{N}$ such that $a < z < a + 1$).

This immediately becomes $\exists a \in \mathbb{N}$ such that ($\exists z \in \mathbb{N}$ such that $a < z < a + 1$), which is equivalent to $\exists a, z \in \mathbb{N}$ such that $a < z < a + 1$. It's much easier to jump from a statement to its contradictory, and go for the proof.

Despite how precise the several proof sequences above are, their format is increasingly cumbersome. We therefore strive to write proofs as understandable, concise, clear arguments, using English as the connective tissue. As John Quincy Adams said in a broader sense, we want "reason, clothed with speech." Mathematical reasoning is the most exacting reasoning of all, and the skill of writing clear proofs is refined by much homework and perseverance!

Let's recast the proof by contradiction of Theorem 2.2 in a reader-friendly format.

Proof. On the contrary, suppose that some natural number z exists such that $a < z < a + 1$. We will derive a contradiction. By the definition of a triple inequality, there exist $n, m \in \mathbb{N}$ such that $a + n = z$ and $z + m = a + 1$. Consider that $a + (n + m) = (a + n) + m$ by the associative Axiom N3. Then, by substitution, $(a + n) + m = z + m = a + 1$. Hence, we have $a + (n + m) = a + 1$. By cancellation Lemma 2.2, we deduce that $n + m = 1$. This means that $n < 1$. But n is a natural number, which contradicts Theorem 2.1 requiring that $1 \leq n$ for all $n \in \mathbb{N}$. This contradiction was created by assuming the existence of the natural number z. We conclude that there does not exist a natural number z with the claimed property, and Theorem 2.2 is proved. **QED**

Defining "2" as the unique number $1 + 1$, Theorem 2.2 assures us that we haven't overlooked a natural number between 1 and 2. Next, $3 \equiv 1 + 2$ is the very next member of $\mathbb{N}$, and so on. It is amazing that all the properties of the natural numbers $\{1, 2, 3, \ldots\}$ can be derived by proof sequences such as those above, thus indicating that algebra in $\mathbb{N}$ is a deductive axiomatic system. The exercises for this chapter will ask you to prove a few more, basic, properties of $\mathbb{N}$. Look into [12] for still more proofs and exercises.

If the proof sequences appear similar to those you learned in plane geometry, it is for the reason that geometry is also a deductive axiomatic system. Even more astonishing is that all of mathematical analysis (including calculus) over the set of real numbers $\mathbb{R}$ was likewise shown to be a deductive axiomatic system in the 19th century. The axioms needed to do this will be studied in succeeding chapters. Keep in mind that Axioms N1–N7 will not be forgotten, but rather will be absorbed into the larger collection of axioms for $\mathbb{R}$.

Note 3: Theorem 2.2 indicated (but did not require) the proposition that for all $a \in \mathbb{N}$, $a < a + 1$. Reconsidering the identity $a + 1 = (a + 1)$ will immediately prove the proposition. This now affirms that $1 < 2 < 3 < \cdots$. Seems trivial? Finance, for instance, depends upon those inequalities. If whole dollar amounts were, say, vector quantities, then we wouldn't even be sure that \$1 < \$2!

2.3 A Fifth Technique of Proof

So far, we have worked with four different techniques of proof: direct, contrapositive, induction, and contradiction. A fifth technique, which seems invalid because it initially assumes the truth of the conclusion, is just a strategy to create a direct proof when the steps are obscure at first. The following example illustrates this new method, proves a nice lemma, and provides a good workout with inequalities.

Note 4: Example 2.1 involves rational numbers, which will be properly introduced in the next chapter. For now, we will work with them as if we knew their properties well.

Example 2.1 Let p be any fixed, natural number except 1. Prove that the **factorial** of a natural number, $n! = (1)(2)\cdots(n-1)(n)$ by definition, will exceed n^p for a large enough n. That is, prove that $n^p < n!$ for some large enough n. (When $p = 1$ we get the easy-to-understand inequality $n \leq n!$.) ■

Proof. None of our four previous techniques seem to help. To begin the new technique, take $n^p < n!$ *as if it were true!* Suppose that $n = p$ (a reasonable first trial). Then our assumption produces

$$\underbrace{p \cdot p \cdot p \cdot \cdots \cdot p}_{p \text{ factors}} < \underbrace{1 \cdot 2 \cdot 3 \cdot \cdots \cdot (p-1) \cdot p}_{p \text{ factorial}},$$

and dividing,

$$\underbrace{\frac{p}{1} \cdot \frac{p}{2} \cdot \frac{p}{3} \cdot \cdots \cdot \frac{p}{p-1} \cdot \frac{p}{p}}_{p \text{ fractions}} < 1.$$

This is false, even when $n = p = 2$. We may get discouraged at this point, since there is only sketchy calculator evidence to support our belief that $n^p < n!$ for large n.

But the first trial suggests another one: $n = 2p$. Then $(2p)^p < (2p)!$ implies that

$$\underbrace{\frac{2p}{1} \cdot \frac{2p}{2} \cdot \frac{2p}{3} \cdot \dots \cdot \frac{2p}{p-1} \cdot \frac{2p}{p}}_{p \text{ fractions}} < \underbrace{(p+1)(p+2)\cdots(2p-1)(2p)}_{p \text{ factors}},$$

and we shall compare factors left and right, after we divide $2p$ from both sides. We find that

$$\frac{2p}{p} < \frac{2p}{p-1} < \cdots < \frac{2p}{3} < \frac{2p}{2} < p+1 < p+2 < \cdots < 2p-1$$

are all true inequalities when $p \geq 2$. Nevertheless, what if $(2p)^p < (2p)!$ is false in the first place? We can't base a proof on a false assumption!

But now the new technique takes over. The steps above have blazed a logical trail which we can traverse backwards. Thus, begin with the above string of inequalities, which are *true* for all $p \geq 2$. Multiply the corresponding sides of the inequalities, starting with $\frac{2p}{2} < p + 1$ and $\frac{2p}{3} < p + 2$, to get $\frac{2p}{2} \cdot \frac{2p}{3} < (p + 1)(p + 2)$, and so on, successively. Finally, we arrive at the true inequality

$$\underbrace{\frac{2p}{2} \cdot \dots \cdot \frac{2p}{p}}_{p-1 \text{ fractions}} < \underbrace{(p+1)\cdots(2p-1)}_{p-1 \text{ factors}}.$$

The next step backwards would be to multiply both sides by $2p$, giving us

$$\frac{2p}{1} \cdot \frac{2p}{2} \cdot \dots \cdot \frac{2p}{p} < (p+1)\cdots(2p-1)(2p).$$

From this we conclude that $(2p)^p < (2p)!$. We are done, for given any $p \geq 2$, we now know that $n = 2p$ is a sufficiently large n to guarantee that $n^p < n!$. (Is $n^p < n!$ true for every n larger than $2p$? See Exercise 2.9.) **QED**

This technique of temporarily assuming the truth of a statement will be used heavily later on in proving theorems about limits. Try your hand at the various techniques as you do the exercises. Observe that every exercise in this book that asks you to "prove" or "show" something actually yields a little (or sometimes big) theorem in analysis.

2.4 Exercises

1. You have just read the reader-friendly version of Theorem 2.2. In the same fashion,
   a. write a reader-friendly version of Lemma 2.2, and
   b. write a reader-friendly version of Theorem 2.1.
2. Prove the following, using proofs by contradiction. You do not have to write them as formal proof sequences.
   a. If $a = b$, then $c + a = c + b$ (the reverse operation to Lemma 2.2), and
   b. if $a = b$, then $ca = cb$.

3. Show that in the definition of the triple inequality $a < b < c$, the third inequality $a < c$ can be derived from the other two. Speaking theoretically, you will have proved that the inequality relation is **transitive**, that is, if $a < b$ and $b < c$, then $a < c$. Now, recast this into a formal proof sequence, in table form.

   Note 5: In the most abstract setting, an arbitrary set X enjoys a transitive relation R when it can be asserted of any three elements α, β, γ in X that $\alpha R \beta$ and $\beta R \gamma$ imply $\alpha R \gamma$.

4. Prove the multiplicative analogue to Lemma 2.1: if $a < b$, then $ca < cb$ (remember, zero and negative quantities do not exist yet). No formal proof sequence is needed. Hint: Use Exercise 2.2b.

5. Prove the multiplicative analogue to Lemma 2.2: if $ca = cb$, then $a = b$, so that "cancellation" is valid under multiplication. Note that no warning about division by zero is needed yet. Hint: Use Exercise 2.4.

6. Prove that if $a < b$ and $c < d$, then $a + c < b + d$, and similarly, prove that if $a < b$ and $c < d$, then $ac < bd$ (the second inference will become false when negative numbers enter the discussion).

7. **Powers** of natural numbers are defined by what is termed a "recursive definition" : define $a^1 \equiv a$, and then $a^{n+1} \equiv a^n a$, where a and n are any numbers in $\mathbb{N}$.
   a. Let a be a natural number. Prove that for all $n \in \mathbb{N}$, $a^n \geq a$. Hint: You will need N6.
   b. A basic "law of exponents" is that for any $a, n, m \in \mathbb{N}$, $a^n a^m = a^{n+m}$. Prove it, again using N6.

   Note: For Exercises 8 and 9, use properties of positive rational numbers as in Example 2.1.

8. Prove that for all $n \in \mathbb{N}$, $n^n \leq (n!)^2$. Hint: Prove that $\frac{n^n}{n!} \leq n!$.

9. This exercise extends Example 2.1 to a stronger lemma that $n^p \leq n!$ for all $n \geq 2p$. The example showed that $n^p < n!$ for any even number $n = 2p$ starting with $n = 4$ (when $n = 2$, $2^1 = 2!$ occurs). We only need to prove that $(2p+1)^p < (2p+1)!$ in order to encompass all odd numbers n as well (here, $p \in \mathbb{N}$). Follow these steps.
   a. To use the fifth technique, begin by assuming that $(2p+1)^p < (2p+1)!$ is true. Derive that

$$\frac{2p+1}{(1)(2p+1)} \cdot \frac{2p+1}{(2)(2p)} \cdot \frac{2p+1}{(3)(2p-1)} \cdot \cdots \cdot \frac{2p+1}{(p)(p+2)} < p+1.$$

   Note that the left side displays exactly p fractions.
   b. Now prove that

$$\frac{2p+1}{(2)(2p)} \cdot \frac{2p+1}{(3)(2p-1)} \cdot \cdots \cdot \frac{2p+1}{(p)(p+2)} < p+1$$

   is true. Hint: $\frac{2p+1}{(2)(2p)} < 1$, where this fraction exists only when $p \geq 2$. Prove that each of the other fractions there are also less than 1.
   c. Reverse your steps to prove that $n^p < n!$ for all odd $n \geq 3$. Conclude with the lemma that $n^p \leq n!$ for all $n \geq 2p$. (Equality holds when $n = p = 1$ or when $n = 2$ and $p = 1$.)

10. *The Random House Dictionary*, 2nd ed., defines "three" as "a cardinal number, 2 plus 1." Why is this definition terribly unsatisfactory for mathematical analysis?

11. It happens that axiom set N1–N7 is not the most primitive characterization of the natural numbers. Giuseppe Peano (1858–1932) discovered in 1889 a famous set of five axioms from which all the properties of the set of natural numbers $\mathbb{N}$ could be derived. Find Peano's axioms (or postulates) in [12], for example, and notice how they build up the set $\mathbb{N}$.

3

The Integers, and the Ordered Field of Rational Numbers

The axiomatic method is, without doubt, the single most important contribution of ancient Greece to mathematics.

"Rigor and Proof in Mathematics: A Historical Perspective" by Israel Kleiner, in *Mathematics Magazine*, vol. 64, no. 5, pg. 293.

3.1 The Integers

So far, we have a firm foundation for the set of natural numbers $\mathbb{N}$. But advanced thinking in any culture requires the use of fractions (see Chapters I and II of [8], or the longer exposition in the first three chapters of [3]). In 1202, Leonardo of Pisa (ca. 1180–1250, also known as Fibonacci) imported into Europe the ten Hindu-Arabic numerals, which included the novel number " 0." This would revolutionize arithmetic.

A strong motivation to enlarge the collection of things we call numbers is that with only $\mathbb{N}$, we aren't able to solve a simple equation such as $x + 5 = 2$. The new class of numbers $\mathbb{Z}$ shall be called the "integers," and we intend to have $\mathbb{N} \subset \mathbb{Z}$. How can we guarantee this? By taking advantage of the deductive axiomatic system that we have been working with. We decide upon three new axioms to include with N1–N6, which will allow us to define "negative numbers" without losing any properties belonging to $\mathbb{N}$. Axioms N1–N6 now will be designated "ZN1–ZN6" to remind us of their origin in the axiomatic system of $\mathbb{N}$.

Create a set of elements $\mathbb{Z} = \{a, b, c, \ldots\}$ called **integers** along with two operations $+$ and $\times$, that follows this list of axioms (as before, $a \times b$ is usually written ab):

ZN1: For all $a, b \in \mathbb{Z}$, $a + b$ and ab are unique elements of $\mathbb{Z}$. (Closure and uniqueness)

ZN2: For all $a, b \in \mathbb{Z}$, $a + b = b + a$ and $ab = ba$. (Commutativity)

ZN3: For all $a, b, c \in \mathbb{Z}$, $a + (b + c) = (a + b) + c$ and $a\,(bc) = (ab)\,c$. (Associativity)

ZN4: For all $a, b, c \in \mathbb{Z}$, $a\,(b + c) = ab + ac$. (Distributivity of $\times$ across $+$)

ZN5: There exists a unique number $1 \in \mathbb{Z}$ such that for any $a \in \mathbb{Z}$, $1a = a$. (Existence of a multiplicative identity, named "one")

ZN6: Let M be a set of natural numbers such that $1 \in M$, and whenever $k \in M$, then $k + 1 \in M$. Then $M = \mathbb{N}$, that is, M contains all the natural numbers. (Mathematical induction)

Z7: There exists a unique number $0 \in \mathbb{Z}$, not equal to 1, such that for any $a \in \mathbb{Z}$, $0 + a = a$. (Existence of an additive identity, named "zero")

Z8: For any $a \in \mathbb{Z}$, there exists a unique, corresponding number $-a \in \mathbb{Z}$ such that $-a + a = 0$. (Existence of opposites)

Z9: There exists a subset $\mathbb{Z}^+$ of $\mathbb{Z}$ such that (1) for all nonzero $a \in \mathbb{Z}$, either $a \in \mathbb{Z}^+$ or $-a \in \mathbb{Z}^+$ but not both, and (2) for any $a, b \in \mathbb{Z}^+$, both $a + b \in \mathbb{Z}^+$ and $ab \in \mathbb{Z}^+$. (Existence of a positive class)

Let us now see where the three new axioms lead. All the statements of inequality from the last chapter are now based on the following one, which is only slightly modified to fit Axiom Z9.

$a < b$ and $b > a$ both mean that $a + n = b$ for some $n \in \mathbb{Z}^+$.

The positive class $\mathbb{Z}^+$ that exists by Axiom Z9 is called the set of **positive integers**. We define negative integers by using Z8: for any $a \in \mathbb{Z}^+$, the corresponding opposite integer $-a$ shall be called a **negative integer**. The set of negative integers is denoted $\mathbb{Z}^-$. The existence of opposites allows us to define the **subtraction** operation: $a - b \equiv a + (-b)$. Hence, subtraction is closed in $\mathbb{Z}$.

Note 1: A set is *partitioned* into a collection of subsets when the subsets are nonempty, pairwise disjoint (mutually exclusive), and their union is the full set, much like a continent is partitioned into countries. This idea is officially introduced in Chapter 5, but it is useful in the following corollary.

Lemma 3.1 $\mathbb{Z} = \mathbb{Z}^- \cup \{0\} \cup \mathbb{Z}^+$, and these three sets partition $\mathbb{Z}$.

Proof. $\mathbb{Z}^+$ and $\mathbb{Z}^-$ account for every integer except zero, by Z9. Also, Axiom Z9 prevents the possibility that $-a = a$ for any nonzero a. Hence, excluding zero, we are sure that $\mathbb{Z}^+$ is disjoint from $\mathbb{Z}^-$. When $a = 0$, Z8 affirms the existence of the unique integer -0 such that $-0 + 0 = 0$, and Z7 yields $0 + (-0) = -0$. So we find that $-0 = 0$. By Z9, it follows that $0 \notin \mathbb{Z}^+$ and $0 \notin \mathbb{Z}^-$. Consequently, $\mathbb{Z} = \mathbb{Z}^- \cup \{0\} \cup \mathbb{Z}^+$ forms a partition of $\mathbb{Z}$. **QED**

A proof of the statement "zero times any number is zero" is part of our first lemma. The necessity statement must be proved, because it is false for instance in matrix algebra, where two nonzero matrices may have a zero product.

Corollary 3.1 For any $a, b \in \mathbb{Z}$, $ab = 0$ if and only if either $a = 0$ or $b = 0$.

Proof of sufficiency. Without loss of generality (see Note 2), we only prove that $0b = 0$. Begin with $1b = 0 + b$ by ZN5 and Z7. This implies that $(0 + 1)b = 0 + b$ by Z7 and substitution. By ZN4, $0b + 1b = 0 + b$, giving $0b + b = 0 + b$, by ZN5 again. Lemma 2.2, which extends to integers, then implies that $0b = 0$. Hence, if $a = 0$ (or just as well, if $b = 0$), then $ab = 0$.
Proof of necessity. To prove that if $ab = 0$ then either $a = 0$ or $b = 0$, assume on the contrary that $a \neq 0$ and $b \neq 0$. Suppose first that $a, b \in \mathbb{Z}^+$. Then $ab \in \mathbb{Z}^+$ by Z9, but $ab = 0 \notin \mathbb{Z}^+$, as we proved in Lemma 3.1. This is a contradiction, so suppose that at least one of a or b is outside of $\mathbb{Z}^+$. Without loss of generality, if $a \notin \mathbb{Z}^+$ and $b \in \mathbb{Z}^+$, then $-a \in \mathbb{Z}^+$, so that $(-a)b \in \mathbb{Z}^+$.

But now, $(-a)\,b = (-1)\,(ab) = (-1)\,0 = 0$ by Exercises 3.4 and the sufficiency proof just done. This produces the same contradiction as before. As the last recourse, assume that neither a nor b are in $\mathbb{Z}^+$. This leads to $(-a)\,(-b) = (-1)\,(-1)\,ab = (-1)\,(-1)\,0 = 0$, so the contradiction remains. We conclude that if $ab = 0$, then either $a = 0$ or $b = 0$. **QED**

Note 2: Whenever a statement is made *without loss of generality*, it means that proving any alternative, pertinent statements will lead only to trivial alternative proof sequences that need not be written out. In the sufficiency part of the proof, proving the alternative case that $a \times 0 = 0$ leads to a proof sequence that trivially exchanges a with b. No new ground will be covered.

Axiom N7 has been removed because trichotomy can now be derived from the new idea of a positive class.

Theorem 3.1 Trichotomy For all $a, b \in \mathbb{Z}$, exactly one of the following is true: $a = b$, $a < b$, or $b < a$.

Proof. We use the existence of a positive class by Axiom Z9. To begin, assume that $a = b$ and $a < b$ are simultaneously true. Then there exists an $n \in \mathbb{Z}^+$ such that $a + n = b$. Given that $a = b$, we have $a + n = a$, which implies that $n = 0$. But we have seen that $0 \notin \mathbb{Z}^+$. Hence, $a = b$ and $a < b$ can't be true simultaneously. A similar argument proves that $a = b$ and $b < a$ cannot be simultaneously true.

As another possibility, assume that $a < b$ and simultaneously $b < a$. Then there exist numbers $n, m \in \mathbb{Z}^+$ such that both $a + n = b$ and $b + m = a$. Adding corresponding sides, $a + b + n + m = a + b$, which implies that $n + m = 0$. This is impossible, since $n + m \in \mathbb{Z}^+$ by Z9, but $0 \notin \mathbb{Z}^+$. Hence, $a < b$ and $b < a$ can't be true together.

Finally, all three relations can't be simultaneously false (a short reason?). Therefore, trichotomy follows. **QED**

Theorem 3.2 $\mathbb{N} \subset \mathbb{Z}$.

Proof. Focus on the positive class $\mathbb{Z}^+$ established by Axiom Z9. All its members satisfy axioms ZN1–ZN6, and Theorem 3.1 demonstrates that $\mathbb{Z}^+$ obeys trichotomy. This implies that $\mathbb{Z}^+$ is mathematically indistinguishable from the set of natural numbers. We may therefore declare that $\mathbb{Z}^+ = \mathbb{N}$, and hence, $\mathbb{N} \subseteq \mathbb{Z}$. Next, for each $a \in \mathbb{Z}^+$, $-a \notin \mathbb{Z}^+$, by Z9. It follows that $\mathbb{N} \subset \mathbb{Z}$. **QED**

At this point, we characterize the sets of positive and negative integers by convenient inequalities.

Lemma 3.2 $\mathbb{Z}^+ = \{a \in \mathbb{Z} : a > 0\}$ and $\mathbb{Z}^- = \{a \in \mathbb{Z} : a < 0\}$.

Proof. To demonstrate the first equality, recall (from Chapter 1) that sets A and B are equal if and only if when $x \in A$, then $x \in B$, and conversely. Accordingly, let $b \in \mathbb{Z}^+$. To prove that $b > 0$, we must show that some $n \in \mathbb{Z}^+$ exists such that $0 + n = b$. Clearly, n is found to be b. Hence,

$b \in \{a \in \mathbb{Z} : a > 0\}$. Conversely, let $b \in \{a \in \mathbb{Z} : a > 0\}$. This means that there does exist an $n \in \mathbb{Z}^+$ such that $0 + n = b$. Then $b = n \in \mathbb{Z}^+$. It follows that $\mathbb{Z}^+$ is equal to $\{a \in \mathbb{Z} : a > 0\}$.

A parallel argument shows that $\mathbb{Z}^- = \{a \in \mathbb{Z} : a < 0\}$. **QED**

Theorem 3.3 For all $a \in \mathbb{Z}$, there is no integer z such that $a < z < a + 1$.

Proof. Assume that some integer z exists such that $a < z < a + 1$. Mimic the proof of Theorem 2.2 to find a contradiction to Theorem 2.1, and this theorem will be proved. **QED**

Nonnegative powers of nonzero integers are a straightforward generalization of the case in $\mathbb{N}$ (Exercise 2.7). The new case here is that for all $a \in \mathbb{Z}^+$ and all $m \in \mathbb{N} \cup \{0\}$, define $(-a)^m$ as $(-1)^m a^m$, by virtue of Exercise 3.4.

Prime numbers, those natural numbers that have exactly two distinct positive factors, arise from your knowledge of factoring integers. (We will take for granted that every natural number beginning with 2 has, in fact, *some* factorization into primes.) A fundamental property of natural numbers appears in the next theorem, which sometimes carries the regal name "the fundamental theorem of arithmetic." The theorem easily extends to all integers $n \leq -2$.

Theorem 3.4 The Unique Factorization Theorem Every natural number $n \geq 2$ can be factored as a product of unique primes p_i raised to unique natural number powers, as in $n = p_1^{a_1} p_2^{a_2} \cdots p_k^{a_k}$.

Proof. Let us assume that there are two possibly different prime factorizations of n, namely $P = p_1^{a_1} p_2^{a_2} \cdots p_k^{a_k}$ and $Q = q_1^{b_1} q_2^{b_2} \cdots q_m^{b_m}$, where $n = Q = P$ and all exponents are in $\mathbb{N}$. We may assume without loss of generality that $k \leq m$. Organizing things a bit, let the primes in P appear in size order, that is, $p_1 < p_2 < \cdots < p_k$, and the same for the primes in Q. Now, if p_1 does not appear in Q, then n is both divisible and not divisible by p_1. This is absurd, so p_1, and $p_1^{a_1}$ as well, must be factors of Q, meaning that $q_1^{a_1} = p_1^{a_1}$. The same reasoning can be applied to every other factor $p_2^{a_2}, p_3^{a_3}, \ldots, p_k^{a_k}$. At this point, we have deduced that

$$Q = p_1^{a_1} p_2^{a_2} \cdots p_k^{a_k} \left(p_1^{b_1 - a_1}\right)\left(p_2^{b_2 - a_2}\right) \cdots \left(p_k^{b_k - a_k}\right) q_{k+1}^{b_{k+1}} \cdots q_m^{b_m}.$$

Thus, we now have

$$n = Q = P = P\left[\left(p_1^{b_1 - a_1}\right)\left(p_2^{b_2 - a_2}\right) \cdots \left(p_k^{b_k - a_k}\right) q_{k+1}^{b_{k+1}} \cdots q_m^{b_m}\right].$$

Since $P \neq 0$, the factor in brackets must equal one. But all the p_i and q_j are greater than one. These facts tell us first that $k = m$ (otherwise, the prime q_{k+1} would exist), and second, that all $b_i = a_i$. The conclusion is that P and Q are identical factorizations, and the factorization of n is unique. **QED**

This wasn't a proof by contradiction, for we did not assume that the factorizations were actually different and thereby derive a contradiction. To press on, we will not prove that further properties of equations and inequalities proved in Chapter 2 for system $\mathbb{N}$ can be extended to system $\mathbb{Z}$. (If you are interested in more details about this, see [12], pp. 194–209.)

3.2 The Rational Numbers

We are slowly but surely building the set of real numbers. A solid mathematical theory of rational numbers was first developed by the Greek mathematician, astronomer, physician, legislator, and geographer, Eudoxus of Cnidus (ca. 408 B.C.–ca. 355 B.C.). The theory was soon put to good use in Euclid's monumental exposition of algebra and geometry, simply named the *Elements* (ca. 300 B.C.). One motivation for having these numbers is that integers cannot provide a solution to as simple an equation as $3x = 7$. This time, we introduce one new axiom and a set $\mathbb{Q}$ of things called "rational numbers" that follow familiar properties, and we intend that $\mathbb{Z} \subset \mathbb{Q}$. To emphasize the lineage from past axioms inherited in the following algebraic structure, some are designated "QZN" or " QZ."

Create a set of ordered pairs $\mathbb{Q} = \{\langle a, b\rangle : a \in \mathbb{Z} \; and \; b \in \mathbb{N}\}$called **rational numbers** along with two operations $+$ and $\times$, that follows the list of axioms below. For ease of notation, the ordered pairs are denoted by r, s, or t. Very often $r \times s$ is written as rs, and occasionally as $r \cdot s$.

QZN1: For all $r, s \in \mathbb{Q}$, $r + s$ and rs are unique elements of $\mathbb{Q}$. (Closure and uniqueness)

QZN2: For all $r, s \in \mathbb{Q}$, $r + s = s + r$ and $rs = sr$. (Commutativity)

QZN3: For all $r, s, t \in \mathbb{Q}$, $r + (s + t) = (r + s) + t$ and $r(st) = (rs)t$. (Associativity)

QZN4: For all $r, s, t \in \mathbb{Q}$, $r(s + t) = rs + rt$. (Distributivity of $\times$ across $+$)

QZN5: There exists a unique number $1 \in \mathbb{Q}$ such that for any $r \in \mathbb{Q}$, $1r = r$. (Existence of a multiplicative identity, named " one")

QZN6: Let M be a set of natural numbers such that $1 \in M$, and whenever $k \in M$, then $k + 1 \in M$. Then $M = \mathbb{N}$, that is, M contains all the natural numbers. (Mathematical induction)

QZ7: There exists a unique number $0 \in \mathbb{Q}$, not equal to 1, such that for any $r \in \mathbb{Q}$, $0 + r = r$. (Existence of an additive identity, named " zero")

QZ8: For any $r \in \mathbb{Q}$, there exists a unique, corresponding number $-r \in \mathbb{Q}$ such that $-r + r = 0$. (Existence of opposites)

QZ9: There exists a subset $\mathbb{Q}^+$ of $\mathbb{Q}$ such that (1) for all nonzero $r \in \mathbb{Q}$, either $r \in \mathbb{Q}^+$ or $-r \in \mathbb{Q}^+$, but not both, and (2) for any $r, s \in \mathbb{Q}^+$, both $r + s \in \mathbb{Q}^+$ and $rs \in \mathbb{Q}^+$. (Existence of a positive class)

Q10: For any $r \in \mathbb{Q}$ except 0, there exists a unique, corresponding number $r^{-1} \in \mathbb{Q}$ such that $r^{-1}r = 1$. (Existence of reciprocals)

All the statements of inequality from the last chapter are now based on the following one, which is only slightly upgraded to fit Axiom QZ9.

$$s < t \text{ and } t > s \text{ both mean that } s + r = t \text{ for some } r \in \mathbb{Q}^+.$$

Looking at the lineage of the axioms, we observe that the only new one is Q10. This axiom therefore contains the distinguishing characteristic of the rational numbers! Let us see where it leads. First of all, we are now able to define the **division** operation $\frac{r}{s} \equiv rs^{-1}$ provided that $s \neq 0$ (in particular, $\frac{1}{s} = s^{-1}$). The familiar properties $\frac{r}{t} \cdot \frac{s}{q} = \frac{rs}{tq}$ and $\frac{r}{t} + \frac{s}{q} = \frac{rq+st}{tq}$ can be derived from this definition.

The positive class $\mathbb{Q}^+$ that exists by Axiom QZ9 is called the set of **positive rationals**. We then define negative numbers by using QZ8: for each positive rational r, the corresponding opposite number $-r$ is called a **negative rational.** The set of negative rationals is denoted $\mathbb{Q}^-$. As in the system $\mathbb{Z}$, $0 \notin \mathbb{Q}^+$, and $0 \notin \mathbb{Q}^-$. Also, $\mathbb{Q} = \mathbb{Q}^- \cup \{0\} \cup \mathbb{Q}^+$ is a partition of $\mathbb{Q}$ (see Exercise 3.3a regarding trichotomy).

We extend the **laws of exponents** by defining $r^m r^n \equiv r^{m+n}$ for any $m, n \in \mathbb{Z}$. Putting $m = 0$ then gives $r^0 r^n = r^n$, which indicates that $r^0 = 1$ as long as $r \neq 0$, by the uniqueness of the multiplicative identity in QZN5. Putting $m = -n$ gives $r^{-n} r^n = r^0 = 1$, implying that $r^{-n} = \frac{1}{r^n} = (r^n)^{-1}$ by the uniqueness of reciprocals in Q10. Furthermore, since $\frac{1}{r^n} = \left(\frac{1}{r}\right)^n = (r^{-1})^n$, then $(r^n)^{-1} = (r^{-1})^n$. Last, replacing n by $-n$ gives $r^m r^{-n} = r^{m-n}$, but also $r^m r^{-n} = \frac{r^m}{r^n}$, from which we derive that $r^{m-n} = \frac{r^m}{r^n}$. Yet another law of exponents is left for you to prove in Exercise 3.5f.

Note 3: We have just mentioned that $r^0 = 1$ provided $r \neq 0$. What if $r = 0$? In that case, take $n \neq 0$ in the equation $r^0 r^n = r^n$, and we have $0^0\,(0) = 0$. This equation doesn't yield a unique value for 0^0, per Corollary 3.1. 0^0 is therefore an "indeterminate form." No contradiction arises if we assign a value to 0^0, and we often find $0^0 \equiv 1$ to be very convenient.

The following example (a corollary) shows how to use the definition of inequality in system $\mathbb{Q}$. It extends Exercise 2.4 of the previous chapter.

Example 3.1 Choose any $r, s, t \in \mathbb{Q}$ with $r < s$. Show that (1) if $t > 0$, then $tr < ts$, but (2) if $t < 0$, then $tr > ts$.

We are given that $r + q = s$ for some $q \in \mathbb{Q}^+$. To prove the first statement, begin with $tr + tq = ts$. Since $t > 0$, by Exercise 3.2 and QZ9, $tq \in \mathbb{Q}^+$. It follows that $tr < ts$.

For the second statement, let $t < 0$. By Exercise 3.5b, $tq < 0$. Now, $tr + tq = ts$ yields $tr = ts + -(tq)$. By Exercise 3.2 and trichotomy (Exercise 3.3a), $-(tq) \in \mathbb{Q}^+$. Therefore, $tr > ts$. ■

Up to this point, there seems to be little connection between $\mathbb{Q}$ and $\mathbb{Z}$. But the connection is inherent in the definition of $\mathbb{Q}$ as a set of ordered pairs $\langle a, b\rangle$ where a is any integer and b is any natural number. It turns out that the element denoted by $\langle a, b\rangle$ is nothing other than the ratio $\frac{a}{b}$. As examples, $\langle -5, 8\rangle = \frac{-5}{8}$, $\langle 1, 9\rangle^{-1} = \langle 9, 1\rangle$, and $\langle -1, 9\rangle^{-1} = \langle -9, 1\rangle$. (Developing the correspondence between $\langle a, b\rangle$ and $\frac{a}{b}$ would distract us a bit. If you wish to read the intriguing details, see [12], pp. 208–209.)

Theorem 3.5 $\mathbb{Z} \subset \mathbb{Q}$.

Proof. According to the correspondence discussed above, the subset of $\mathbb{Q}$ given by $\{\langle a, 1\rangle : a \in \mathbb{Z}\}$ is the set of divisions $\frac{a}{1} = a \times 1^{-1} = a \times 1 = a$. Thus, $\mathbb{Z} \subseteq \mathbb{Q}$, but are there any elements of $\mathbb{Q}$ other than those in $\mathbb{Z}$, as we should hope?

Given $n \in \mathbb{Z}$ with $1 < n$, Exercise 3.5e indicates that $0 < \frac{1}{n} < \frac{n}{n} = 1$. But Theorem 2.1 states that 1 is the smallest natural number, which is to say that 1 is the smallest positive integer. Therefore, whenever $1 < n$, $\frac{1}{n} \notin \mathbb{Z}$. In view of the fact that $\frac{1}{n} = \langle 1, n\rangle \in \mathbb{Q}$, we have $\mathbb{Z} \subset \mathbb{Q}$. **QED**

According to QZ9, when $x \neq 0$, exactly one member of $\{x, -x\}$ must belong to $\mathbb{Q}^+$. The member that does is defined as the **absolute value** of x. Using Exercise 3.3a, the following well-known definition can be formed:

$$|x| = \begin{cases} x & \text{if } x > 0 \\ 0 & \text{if } x = 0 \\ -x & \text{if } x < 0 \end{cases}.$$

It follows that $|x| \geq 0$ without exception. This definition will extend without change to all real numbers x later on.

Once again, we will not prove more algebraic properties; rather, we will use them with confidence in their truth. The exercises below select a few of them for you to demonstrate. (Many more properties of integers, rational numbers, equations, and inequalities appear in [12], sections 7.4 and 7.5, and [18], sections 103–108.)

3.3 The Rational Numbers as an Ordered Field

Recall that our axioms for $\mathbb{Q}$ were applied to sets of unspecified objects $\{r, s, t, \ldots\}$. As we developed theorems and symbols for using these objects, we realized that the fractions of our youth behaved suspiciously like these objects. Hence, we adopted this deductive axiomatic system as the system defining the rational numbers. It's as if two people were given a set of mysterious playing pieces and told to play against each other according to certain rules. After a while, one of them correctly observes, "Say, isn't this just a game of chess with weird pieces?" In the same way, Fibonacci's numerals $0, 1, 2, \ldots$ have no intrinsic meaning, and we could have opted to use binary numerals or Mayan numerals to develop $\mathbb{N}$, $\mathbb{Z}$, and $\mathbb{Q}$. It is the rules that count!

Note 4: We have kept a distinction between the sets

$$\mathbb{N} = \{1, 2, 3, \ldots\}$$
$$\mathbb{Z} = \{\ldots, -2, -1, 0, 1, 2, \ldots\}$$
$$\mathbb{Q} = \left\{\frac{a}{b} : a \in \mathbb{Z}, b \in \mathbb{N}\right\}$$

and the axiomatic systems using them. But no ambiguity should occur if we use the letters to denote both the sets and the corresponding axiomatic systems.

Mathematicians have capitalized a great deal on the versatility of axiomatic systems. Let's consider what happens when certain axioms are deleted. First of all, we see that cutting off Axiom Q10 (the existence of reciprocals) from the system of rational numbers $\mathbb{Q}$ immediately drops us down to the system of integers $\mathbb{Z}$. Doing some more surgery, reinsert Q10, and remove induction Axiom QZN6 and positive class Axiom QZ9. The remaining axioms—specifically QZN1–QZN5, QZ7, QZ8, and Q10—when applied to any set of objects, form an axiomatic system called a **field**. Mathematicians saw that fields worked very much like the number system $\mathbb{Q}$. But a strange difference arose. Having cut off QZN6, it suddenly became possible to do algebra quite comfortably with just a *finite* set of numbers (16th century mathematicians would have been shocked). See Exercise 3.9 for a classic example of such a field.

The effect of removing the positive class Axiom QZ9 is to leave behind a system of numbers that has no size order! An example of such a field is the set of complex numbers (see Exercise

3.10). $\mathbb{N}$ exhibits size order in the sense that it obeys the trichotomy Theorem 3.2, but it is not what is called an " ordered set." The distinction is that **ordered number sets** have a zero element, a positive class, and a negative class, that partition the set (see Note 1). Clearly, $\mathbb{Z}$ and $\mathbb{Q}$ are both ordered.

Accepting only Axioms QZN1–QZN6 and trichotomy brings us back to the natural number system $\mathbb{N}$. Both sets $\mathbb{N}$ and $\mathbb{Z}$ are too meager to be fields (why?), even though they have infinitely many elements. As we have just noted, $\mathbb{Z}$ goes one step beyond $\mathbb{N}$ in that $\mathbb{Z}$ is ordered. (Beware of terminology: a *well-ordered* set is something quite different, to be studied in the next chapter.)

An axiomatic system with QZN1–Q10 is named an **ordered field**, and the first example of one is the system of rational numbers $\mathbb{Q}$. The name is logically correct, for ordered fields are fields that are also ordered. Moreover, $\mathbb{Q}$ is an infinite ordered field, and in fact, no ordered field can be finite (convince yourself of this proposition). Another example of an infinite ordered field is $\mathbb{R}$.

3.4 Exercises

1. Show that the definition of inequality for $\mathbb{Z}$ is a special case of the definition of inequality for $\mathbb{Q}$.

2. Show that $\mathbb{Q}^+ = \{r \in \mathbb{Q} : r > 0\}$, and that $\mathbb{Q}^- = \{r \in \mathbb{Q} : r < 0\}$.

3. a. Prove that $\mathbb{Q}$ satisfies trichotomy. That is, all rationals r fall into exactly one of these categories: (1) $r < 0$, (2) $r = 0$, or (3) $r > 0$. We conclude that $\mathbb{Q}$ is partitioned as $\mathbb{Q}^- \cup \{0\} \cup \mathbb{Q}^+$.
   b. If we define $\mathbb{Q}^+ \equiv \left\{\frac{m}{n} : m, n \in \mathbb{N}\right\}$, show that $\mathbb{Q}^+$ is closed under addition and multiplication, so that part of QZ9 need not be stated as an axiom. Hint: $r > 0$ means that some $\frac{m}{n} \in Q^+$ exists such that $0 + \frac{m}{n} = r$.

4. Prove the important lemma that $-a = (-1)(a)$ in $\mathbb{Z}$ and $\mathbb{Q}$, keeping an eye on details. Hint: Show that $(-1)(a) + a = 0$.

5. Prove as many of the following as time allows (r, s, t, and q are rational numbers):
   a. $(-1)(-1) = 1$.
   b. If $t < 0$ and $q > 0$, then $tq < 0$.
   c. $-(-s) = s$.
   d. $(-r)(-s) = rs$. In particular, $(-r)^2 = r^2$, and since $r^2 = r \times r$, we get $r^2 \geq 0$.
   e. If $0 < r < s$, then $0 < \frac{1}{s} < \frac{1}{r}$.
   f. $(r^m)^n = r^{mn}$ for $r \neq 0$ and any $m, n \in \mathbb{Z}$, given that this is true for any $m, n \in \mathbb{N}$.

6. Prove that if $r, s, t, u \in \mathbb{Q}^+$, and $r < s$ and $t \leq u$, then $0 < rt < su$. Also prove that $0 < \frac{r}{u} < \frac{s}{t}$.

7. Prove that for $r, s \in \mathbb{Q}$, $r^2 + s^2 \geq 0$. Also, if $r^2 + s^2 = 0$, then $r = s = 0$.

8. Given $r, s \in \mathbb{Q}$, show that we may identify the larger and smaller of them by the following operations: $\max\{r, s\} = \frac{1}{2}(r + s + |r - s|)$, and $\min\{r, s\} = \frac{1}{2}(r + s - |r - s|)$. What happens if $r = s$? These operations naturally extend to all real numbers, and will be used in Chapters 6 and 32.

9. a. A classic example of a finite field is generated by "clock arithmetic." We will define new operations $\oplus$ and $\otimes$ on a clock face that displays the set of only five "hours" $F = \{0, 1, 2, 3, 4\}$. Let $a \oplus b = a + b \pmod 5$ and $a \otimes b = ab \pmod 5$. The "mod 5" command means to calculate $a + b$ and ab within $\mathbb{Z}$, and output the remainder after division by five. For instance, $4 \otimes (3 \oplus 2) = 4 \otimes (5 \pmod 5) = 4 \otimes 0 = 0 \pmod 5 = 0$, while the distributive law would yield $4 \otimes (3 \oplus 2) = (4 \otimes 3) \oplus (4 \otimes 2) = (12 \pmod 5) \oplus (8 \pmod 5) = 2 \oplus 3 = 5 \pmod 5 = 0$. So QZN4 seems to be working. More: $3^{-3} = (3^3)^{-1} = (27 \pmod 5)^{-1} = 2^{-1}$. To simplify 2^{-1}, solve $2 \otimes x = 1$. We find that $x = 3$. Thus, $3^{-3} = 3$ in this field! Pick any one of the field axioms and prove that it is true in this finite field. Hint: Any integer n can be expressed uniquely as $n = 5q + s$, where $q \in \mathbb{Z}$, and $0 \leq s \leq 4$ is the remainder after division by 5.
   b. Show that no positive class exists and that trichotomy fails in F. Therefore, F is not an ordered set.

10. The statements in Exercise 3.7 hold true for any ordered field, not just $\mathbb{Q}$. Prove that the field of complex numbers is not an ordered field.

4

Induction and Well-Ordering

Having established that mathematicians are concerned with things that do not exist in reality, but only in their thoughts, let us examine the statement... that mathematics gives us more reliable and more trustworthy knowledge than does any other branch of science.

Alfréd Rényi (1921–1970), *Dialogues on Mathematics*, pp. 9–10

Axiom QZN6 of Chapter 3 gives us mathematical induction, a basis for proving statements to be true about all natural numbers. The form in which it appears there is supplemented by the common—and equivalent—form called the principle of mathematical induction. It isn't difficult to derive this form as a theorem from QZN6 (see Exercise 4.10), and it is often used as the axiom of induction itself.

4.1 The Principle of Mathematical Induction

Theorem 4.1 The Principle of Mathematical Induction Let $P(n)$ be a statement that involves natural number n (or is indexed by n). Suppose that

(1) $P(1)$ is true, and
(2) assuming $P(n)$ to be true, we use it to prove that the succeeding statement $P(n+1)$ is also true.

Then statement $P(n)$ is true for every $n \in \mathbb{N}$.

Proof by induction is a proof sequence that uses QZN6 or Theorem 4.1. The three examples that follow will use the new form. Proving $P(1)$ to be true is usually quite easy; it is called the **anchor** of the induction. Assuming $P(n)$ to be true is called the **induction hypothesis**, and this hypothesis *must* be used in the proof of statement $P(n+1)$. Sometimes, a different initial statement $P(k)$ may serve as the anchor, where $k > 1$. In such a case, the induction concludes that $P(n)$ is true for all $n \geq k$.

Example 4.1 A pyramid of same-sized oranges is stacked in a grocery store window. There are n tiers of oranges, with the topmost tier containing one orange, the second tier four oranges, the next one nine, and so on. Prove by induction that the total number of oranges in n tiers is $T(n) = \frac{1}{3}n^3 + \frac{1}{2}n^2 + \frac{1}{6}n$.

The statement that we want to prove is $P(n) = \{$*the total number of oranges in n tiers is* $T(n) = \frac{1}{3}n^3 + \frac{1}{2}n^2 + \frac{1}{6}n\}$. The anchor is obviously true: for the $n = 1$ tier, $P(1)$ says that $T(1) = 1$. As our induction hypothesis, *assume* that the formula is true for n tiers (that is, assume $P(n)$ is true). Then we must use the induction hypothesis to prove statement $P(n+1)$: that the total number of oranges is $T(n+1) = \frac{1}{3}(n+1)^3 + \frac{1}{2}(n+1)^2 + \frac{1}{6}(n+1)$.

It is observed that to preserve the pyramidal form, each tier adds a perfect square number of oranges. Thus, $T(n+1) = T(n) + (n+1)^2 = \frac{1}{3}n^3 + \frac{1}{2}n^2 + \frac{1}{6}n + (n+1)^2$, where the last equality can only be justified by *using the induction hypothesis*. All that remains now is to use a bit of algebra to show that $\frac{1}{3}n^3 + \frac{1}{2}n^2 + \frac{1}{6}n + (n+1)^2 = \frac{1}{3}(n+1)^3 + \frac{1}{2}(n+1)^2 + \frac{1}{6}(n+1)$. Do so in Exercise 4.3.

By mathematical induction, $T(n) = \frac{1}{3}n^3 + \frac{1}{2}n^2 + \frac{1}{6}n$ is correct for all $n \in \mathbb{N}$. ■

Note 1: Simply substituting $n+1$ for n in the formula $T(n) = \frac{1}{3}n^3 + \frac{1}{2}n^2 + \frac{1}{6}n$ and then claiming that $T(n+1)$ is the new total number of oranges, misses the whole point. We would then be assuming that the formula $T(n)$ is already true, without justification! The thrust of the induction is to prove that $T(n+1)$ does in fact give the total number of oranges in $n+1$ tiers whenever $T(n)$ does.

Notice how curious the polynomial $T(n)$ is. Even though its coefficients are fractions, it always evaluates to a whole number, regardless of the value of n. Convince yourself that this is so.

Example 4.2 Prove the statement $P(n) = \{n^2 < 2^n$ for all $n \geq 5\}$. In this example, the anchor must be at $n = 5$, and $5^2 < 2^5$ is true. The induction hypothesis is to assume that $P(n)$ is true. We must then prove that $(n+1)^2 < 2^{n+1}$. Using the hypothesis (as we must), we can say that $2n^2 < 2^{n+1}$. Is it possible that $(n+1)^2 < 2n^2$ for $n \geq 5$? We prove this by doing some algebra with $(n+1)^2 < 2n^2$, to arrive at $2 + \frac{1}{n} < n$, which is true for $n \geq 5$. Now we can affirm that for such n, $(n+1)^2 < 2n^2 < 2^{n+1}$, from which we see that $(n+1)^2 < 2^{n+1}$. By induction, therefore, $n^2 < 2^n$ for all $n \geq 5$. ■

Our next application is in the proof of a theorem first brought to light by the great mathematician Jacob Bernoulli (1654–1705) in 1689. It's amazing that between this year and 1700, much of our elementary differential and integral calculus was developed through the prodigious (and often contentious) correspondence among Jacob, Gottfried Wilhelm Leibniz (1646–1716), and Johann Bernoulli (1667–1748, Jacob's equally talented brother). We'll call upon this important inequality several times in this book.

Theorem 4.2 Bernoulli's Inequality Let $p \geq -1$ be a constant. Then for all $n \in \mathbb{N}$, $(1+p)^n \geq 1 + np$.

Proof. The anchor for induction is case $n = 1$, at which the statement is trivially true. Next, the induction hypothesis is to assume the case n to be true, that is, we assume the statement to be true exactly as given. Next, write the statement for the case $n+1$: for $p \geq -1$, $(1+p)^{n+1} \geq 1 + (n+1)p$, where $n \in \mathbb{N}$. This is the statement that we must prove (officially, we should have written "where $n + 1 \in \mathbb{N}$," but this immediately follows from $n \in \mathbb{N}$). Begin rather obviously: $(1 + p)^{n+1} = (1 + p)(1 + p)^n$. There are several ways to proceed from here, but look at what must be proved! We need to simplify things, and now is the time to use the induction hypothesis. Thus, $(1 + p)(1 + p)^n \geq (1 + p)(1 + np)$, for $p \geq -1$. But now we have $(1 + p)(1 + np) = 1 + np + p + np^2 \geq 1 + np + p = 1 + (n + 1)p$.

Hence, $(1 + p)^{n+1} \geq 1 + (n + 1)p$, and the case $n + 1$ has been proved. By induction, Bernoulli's inequality is proved for all $n \in \mathbb{N}$. **QED**

When inequalities (such as Bernoulli's) are not strict, a proof of when equality holds is often included. See Exercises 4.6 and 4.8.

Note 2: Despite its name, mathematical induction is actually a deductive form of proof. Remember, only deductive reasoning is allowed in proofs, because conclusions necessarily follow from premises only through this type of reasoning.

Inductive reasoning is the process of deriving conclusions that are probably true from premises that are assumed to be true. This process is commonly used when people dispute in a "logical" way. Often, such arguments need only rudimentary arithmetic, or no quantitative content at all. For example, a mechanic finds a nearly dry radiator in a car with a seized-up engine. The conclusion is that disregarding the antifreeze level probably caused the engine to overheat and destroy itself. This is an induction based on cause and effect.

Here is another example of inductive reasoning. A software designer discovers that an application she is working on has a consistent bug. She justifiably concludes that users will probably experience problems with the application. This is induction by generalization, and it is used extensively within and outside of science.

One reason that a human endeavor may be called a science—in contrast, say, to an art—is that it uses statistical methods as tools for advancement. Statistics is deeply involved with inferring probable conclusions about a population by studying only a sample of the population. Although the conclusions are almost never certain, statistics uses probability theory to tell the researcher the level of confidence that they will carry. As such, statistics provides us with a sophisticated method of induction, generalizing from particular observations.

Mathematical induction extends the concept of generalization to the ultimate. When using mathematical induction, we affirm a statement to be true for only a finite set of cases (usually only one), but the conclusion affirms that the statement is true for an infinite set of cases, enumerated by the set of natural numbers.

Example 4.3 You are familiar with summation notation. A similar notation called **product notation** is used when we need to indicate products of factors: $(x_1)(x_2)(x_3)\cdots(x_n) \equiv \prod_{i=1}^{n} x_i$, where the new symbol is the capital Greek pi. A very famous inequality in mathematics is the **arithmetic mean–geometric mean inequality**. It states that

$$\sqrt[n]{x_1 x_2 \cdots x_n} \leq \frac{x_1 + x_2 + \cdots + x_n}{n}$$

for any set of nonnegative real numbers $\{x_1, x_2, \ldots, x_n\}$. In words, the geometric mean $\sqrt[n]{x_1 x_2 \cdots x_n}$ of n nonnegative real numbers is never more than their arithmetic mean. The arithmetic mean–geometric mean inequality may be written as

$$\sqrt[n]{\prod_{i=1}^{n} x_i} \leq \frac{1}{n}\sum_{i=1}^{n} x_i.$$

One proof of it uses an unusual form of induction, as follows.

The anchor for induction is case $n = 2$. It is easy to prove that $\sqrt{x_1 x_2} \leq \frac{1}{2}(x_1 + x_2)$ by squaring both sides and arriving at the identity $0 \leq (x_1 - x_2)^2$. Now jump to the case $n = 4$ by letting $x_1 = \sqrt{u_1 u_2}$ and $x_2 = \sqrt{u_3 u_4}$ (the four us are arbitrary nonnegative values, just as the two xs are). We have

$$\begin{aligned}\sqrt[4]{u_1 u_2 u_3 u_4} &= \sqrt{\sqrt{u_1 u_2}\sqrt{u_3 u_4}} \overset{\checkmark}{\leq} \frac{1}{2}(\sqrt{u_1 u_2} + \sqrt{u_3 u_4}) \overset{\checkmark}{\leq} \frac{1}{2}\left(\frac{1}{2}(u_1 + u_2)\right) + \frac{1}{2}(u_3 + u_4)) \\ &= \frac{1}{4}(u_1 + u_2 + u_3 + u_4),\end{aligned}$$

where the anchor was used to justify the checked inequalities. Hence, the case $n = 4$ is true.

But what about the case $n = 3$? Since u_4 is nonnegative arbitrary, we wish to give it a value that will make

$$\frac{1}{4}(u_1 + u_2 + u_3 + u_4) = \frac{1}{3}(u_1 + u_2 + u_3).$$

Thus,

$$u_4 = \frac{4}{3}(u_1 + u_2 + u_3) - (u_1 + u_2 + u_3) = \frac{1}{3}(u_1 + u_2 + u_3),$$

a nonnegative value to be sure. Substituting into the established case $n = 4$,

$$\sqrt[4]{\frac{1}{3}u_1 u_2 u_3(u_1 + u_2 + u_3)} \leq \frac{1}{4}(u_1 + u_2 + u_3 + u_4) = \frac{1}{3}(u_1 + u_2 + u_3).$$

This yields

$$\frac{1}{3}u_1 u_2 u_3(u_1 + u_2 + u_3) \leq \left(\frac{1}{3}(u_1 + u_2 + u_3)\right)^4, \quad \text{or} \quad \sqrt[3]{u_1 u_2 u_3} \leq \frac{1}{3}(u_1 + u_2 + u_3),$$

by impressive algebra! Hence, the case $n = 3$ is true.

We now prove the case $n = 8$ by a similar strategy as was done for $n = 4$. Let

$$u_1 = \sqrt{v_1 v_2}, \quad u_2 = \sqrt{v_3 v_4}, \quad u_3 = \sqrt{v_5 v_6}, \quad \text{and} \quad u_4 = \sqrt{v_7 v_8}.$$

Proceeding as before, and using the case $n = 4$,

$$\begin{aligned}\sqrt[8]{\prod_{i=1}^{8} v_i} &= \sqrt[4]{u_1 u_2 u_3 u_4} \leq \frac{1}{4}(u_1 + u_2 + u_3 + u_4) \\ &= \frac{1}{4}(\sqrt{v_1 v_2} + \sqrt{v_3 v_4} + \sqrt{v_5 v_6} + \sqrt{v_7 v_8}) \overset{\checkmark}{\leq} \frac{1}{8}\sum_{i=1}^{8} v_i.\end{aligned}$$

Again, the checked inequality uses the anchor case.

We are now faced with the proof of cases 5 through 7. The path isn't hard to find, only tedious. The case $n = 7$ is proved by setting $v_8 = \frac{1}{7}\sum_{i=1}^{7} v_i$. Then

$$\frac{1}{8}\sum_{i=1}^{8} v_i = \frac{1}{8}\left(\sum_{i=1}^{7} v_i + \frac{1}{7}\sum_{i=1}^{7} v_i\right) = \frac{1}{7}\sum_{i=1}^{7} v_i.$$

Once it is proved, it is used to prove the case 6, which itself is then used to prove the case 5.

The next jump is to $n = 16$, proved by appealing to the case $n = 8$. We then fill in the proofs of the cases 15 down to 9, at least in our mathematical imagination.

This unusual application of mathematical induction finally convinces us that

$$\sqrt[n]{\prod_{i=1}^{n} x_i} \le \frac{1}{n}\sum_{i=1}^{n} x_i$$

is true for every natural number n. ■

Note 3: The proof that you have just read is by Augustin-Louis Cauchy, a mathematician from whom you will learn a great deal as you study any book on mathematical analysis, for he is one of the subject's founders. It is on page 457 of his lengthy textbook, the *Cours d'analyse de l'École royale polytechnique*, "Course of Analysis of the Royal Polytechnic School" (1821). See more of Cauchy's work in Note 1 of Chapter 35 and in Example 36.2.

The shortest proof of the arithmetic mean-geometric mean inequality may be the little gem by Norman Schaumberger appearing on pg. 68 of *The College Mathematics Journal*, vol. 31, no. 1 (January 2000). His proof rests on the inequality $\ln x \le x - 1$. This inequality is proved in Chapter 35, but it ultimately depends upon the arithmetic mean-geometric mean inequality. Alas, using Schaumberger's proof here would be a devastating example of circular reasoning.

In contrast to this, a stand-alone proof worthy of Cauchy himself appears in "A New Proof of the Arithmetic-Geometric Mean Inequality," by Mihály Bencze and Norman Schaumberger, in *Mathematics Magazine*, vol. 66, no. 4 (Oct. 1993), pg. 245.

4.2 The Well-Ordering Principle

A property of $\mathbb{N}$ that isn't often recognized is that every nonempty subset of $\mathbb{N}$ must have a least element. The situation is completely different for $\mathbb{Q}$. The only reasonable answer to the question, "Is there a least member in $\{r \in \mathbb{Q} : 1 < r < 2\}$?" is, "No." In the proof that follows, it will be more convenient to use QZN6 rather than the principle of mathematical induction.

Theorem 4.3 The Well-Ordering Principle Every nonempty subset of $\mathbb{N}$ has a least element. That is, every nonempty $T \subseteq \mathbb{N}$ has a number $t \in T$ with the property that $t \le n$ for every $n \in T$.

Proof. Assume for the purpose of contradiction that there exists a nonempty subset T of natural numbers that does not have a smallest element. Define the set $S = \{s \in \mathbb{N} : t > s \text{ for all } t \in T\}$. By Theorem 2.1, $t > 1$ for all members of T, and it follows that $1 \in S$. This begins a proof sequence for induction. Assume that $n \in S$. Then $t > n$ for all members of T. Were it possible for some $t = n + 1$, then $n + 1$ would be the smallest element of T, which we have assumed does not exist. Thus, $n + 1 \in S$. By induction Axiom QZN6, $S = \mathbb{N}$. But the definition of S demands that all $t \in T$ be larger than any member of S. Since S now contains all the natural numbers, it forces T to be empty.

The original assumptions are denied; therefore, either $T = \varnothing$, or T always has a smallest element. **QED**

Sets that are **well-ordered**, meaning that each of their nonempty subsets has a least element, are not to be confused with ordered sets. $\mathbb{Q}$ is an ordered set (meaning that it has a positive class, a negative class, and zero), but it is *not* well-ordered, as we have just seen with $\{r \in \mathbb{Q} : 1 < r < 2\}$. In contrast, $\mathbb{N}$ is well-ordered, but it certainly *isn't* an ordered set. The moral here might be that the inventors of names have no pity for harried students!

4.3 Exercises

1. Prove that $3^n < n!$ for $n \geq 7$.

2. The following formulas give useful closed forms for certain summations. Prove them by induction:
   a. the "sum of whole numbers formula" $\sum_{j=1}^{n} j = \frac{1}{2}n(n+1)$.
   b. the "sum of squares formula" $\sum_{j=1}^{n} j^2 = \frac{1}{6}n(n+1)(2n+1)$. How does this relate to Example 4.1?
   c. the "sum of cubes formula" $\sum_{j=1}^{n} j^3 = \left(\sum_{j=1}^{n} j\right)^2 = \left(\frac{1}{2}n(n+1)\right)^2$.

3. Fill in the needed algebra to establish that

$$\frac{1}{3}n^3 + \frac{1}{2}n^2 + \frac{1}{6}n + (n+1)^2 = \frac{1}{3}(n+1)^3 + \frac{1}{2}(n+1)^2 + \frac{1}{6}(n+1)$$

   in Example 4.1.

4. The grocery store across the street sells same-sized oranges arranged in a tetrahedral pile, with one orange in tier 1, three oranges in tier 2, six in tier 3, and so on. Prove that n tiers will contain a total of $T(n) = \frac{1}{6}n^3 + \frac{1}{2}n^2 + \frac{1}{3}n$ oranges.

5. In Bernoulli's inequality, take $p = 1$. Use it to prove that $\sqrt[n]{n} < 2$ for all $n \in \mathbb{N}$. (Then take a look at the graph of $y = \sqrt[x]{x}$.)

6. Prove that equality holds in Bernoulli's inequality if and only if $p = 0$. Hint: For the "only if" implication, begin with $(1+p)^n = 1 + np$ for all $n \in N$.

7. Prove Bernoulli's inequality for $p \geq 0$ without induction, by means of the **binomial theorem**: for any $n \in \mathbb{N}$,

$$(a+b)^n = a^n + na^{n-1}b + \frac{n(n-1)}{2!}a^{n-2}b^2 + \frac{n(n-1)(n-2)}{3!}a^{n-3}b^3 + \cdots + nab^{n-1} + b^n.$$

   In truth, we have not escaped induction, for how would one prove the binomial theorem?

8. Prove Bernoulli's inequality as a corollary of the arithmetic mean–geometric mean inequality. Hint: $\sqrt[n]{x} = \sqrt[n]{1\cdot 1\cdot 1\cdots x}$, where there are $n-1$ 1 factors.

9. Let a be any real number except 1. Prove that the finite geometric series of n terms $\sum_{k=1}^{n} a^k$ has the sum $a\frac{1-a^n}{1-a}$.

10. Show that the principle of mathematical induction can be derived from Axiom QZN6. Hint: Let M be the set of natural numbers for which $P(n)$ is true. We are given that $P(1)$ is true.

11. Demolish the outrageous Paradox 0.6: for any finite set of real numbers, all its members are equal.

12. Consider this subset of $\mathbb{N}$: $\{\ldots, 18, 16, 14, \ldots, 11, 9, 7\}$. Are there two least elements, despite the well-ordering principle? Is there anything wrong with this set?

13. The well-ordering principle can be written using quantification symbols from Chapter 1 as follows: $(\forall T \subseteq \mathbb{N})(\exists t \in T)(if\ n \in T \text{ then } t \leq n)$. The negation (per Chapter 1), is $(\exists T \subseteq \mathbb{N})(\forall t \in T)(\exists n \in T \text{ such that } t > n)$, and it must be false, by the contradictory relation. Take $T = \{10, 12, 14, \ldots\}$ and show that this is not the T that the negation claims to exist. Can you find such a T?

14. a. (See the product notation in Example 4.3.) Prove the **Weierstrass-Bernoulli inequality**: for $a_k \geq 0$, $\prod_{k=1}^{n}(1+a_k) \geq 1 + \sum_{k=1}^{n} a_k$. How is this a generalization of Bernoulli's inequality? Hint: $(1+a_1)(1+a_2) = 1 + a_1 + a_2 + a_1a_2 \geq 1 + a_1 + a_2$. Next, $(1+a_1)(1+a_2)(1+a_3) \geq (1+a_1+a_2)(1+a_3)$.

    b. Prove that the Weierstrass-Bernoulli inequality becomes an equality if and only if at most one a_k is positive. Hint: Proving that equality holds if at most one a_k is positive is easy. For the converse, look at a specific example to find patterns that generalize $\prod_{k=1}^{3}(1+a_k) = (1+a_1)(1+a_2)(1+a_3) = 1 + a_1 + a_2 + a_3 + a_1a_2 + a_1a_3 + a_2a_3 + a_1a_2a_3 = 1 + \sum_{k=1}^{3} a_k + (a_1a_2 + a_1a_3 + a_2a_3) + (a_1a_2a_3)$.

5
Sets

> *We cannot define the concept of a set in terms of more fundamental concepts, because there are no more fundamental concepts.*
>
> George Pólya, *Mathematical Discovery*, vol. 1, pg. 20.

Sets, or synonymously, categories, collections, or families, have their own mathematical development, which we shall look at now. We have been defining the sets $\mathbb{N}$, $\mathbb{Z}$, $\mathbb{Q}$, and others, without considering what a set actually is. Alas, it is far more difficult to define a set than to define the natural numbers (which we did in Chapter 2). Indeed, "set" may not be a logically definable term, as the eminent mathematician George Pólya (1887–1985) observed in the quote above. Taking his wise advice, we shall move on.

5.1 Properties of Sets; Types of Sets

Previous chapters have already used the relationships of element, equality, and subset. For the record, we state the definitions. The items that belong to a set are called **elements** or **members** of the set. If an item x is an element of a set S, we write $x \in S$. The only alternative is non-inclusion, written $x \notin S$.

All sets are imbued with three properties. These are:

(a) A set requires a rule that unambiguously describes which elements (if any) are included in it. A mathematically precise type of rule often is a statement $P(x)$ which is true if and only if x belongs to the set. For instance, the interval $[4, 25]$ is actually $\{x : x \text{ is real and } 4 \leq x \leq 25\}$, where the statement $P(x)$ is found after the colon (the colon is a shorthand for "such that"). The rational numbers set $\{\frac{m}{n} : n \in \mathbb{N},\ m \in \mathbb{Z}\}$ is a short form of $\{x : x = \frac{m}{n} \text{ and } n \in \mathbb{N} \text{ and } m \in \mathbb{Z}\}$, and again, $P(x)$ is seen after the colon. In {*20th century monarchs of England*}, the colon and "x" have been removed without harm. The rule may be replaced by a display of all the items, as in $\{a^2, b^3, c^4\}$ and $\{\clubsuit, \spadesuit, \diamondsuit, \heartsuit\}$, or enough of them that the rule of inclusion is understood, as in {*Alabama*, *Alaska*, ..., *Wyoming*} and $\{1, \frac{1}{2}, \frac{1}{4}, \frac{1}{8}, \ldots\}$.

(b) A set is not altered by rearrangements of its elements (if any).

(c) A set must have distinguishable elements (if any). For instance, $\{a, a, b\}$ will not be a set (duplicate elements are the business of "multisets"). This does not mean that elements of a set can't be equal; the set $S = \{\frac{m}{n} : \frac{m}{n} = \frac{2}{3}\}$ is composed of equal, but distinguishable, elements.

Two sets S and P are **equal** if and only if they contain the same elements. More precisely, $S = P$ means that whenever $x \in S$, then $x \in P$, and conversely, whenever $x \in P$, then $x \in S$. Hence, to prove equality, *both* conditional statements must be proved. S is a **subset** of P if and only if we assert one of those conditionals: whenever $x \in S$, then $x \in P$. We then write $S \subseteq P$, and this is often verbalized as "S is contained in P." Occasionally, this is also written $P \supseteq S$, meaning "P contains S."

The subset relationship does not rule out the possible equality of the two sets. The expression $S \subset P$ does rule out equality, and means that not only is S a subset of P, but also that there exists some $x \in P$ that does not belong to S. In this case, we say that S is a **proper subset** of P.

Lemma 5.1 If $S \subseteq P$ and $P \subseteq S$, then $S = P$.

Proof. This direct application of the definition of subset is left for you to prove in Exercise 5.3. **QED**

The set with no elements at all is the **empty set**, denoted by $\varnothing$. It is true that $\varnothing$ is a subset of every set, for assuming that a set S exists with the property that $\varnothing \not\subseteq S$ would imply that there must be an $x \in \varnothing$ such that $x \notin S$. But $x \in \varnothing$ is impossible.

One step up from $\varnothing$, so to speak, is a **singleton set**, which is any set containing exactly one element, as in $\{x\}$.

A set in mathematics is often understood to be a subset of some **universal set** U, which is the background set relevant to the discussion at hand. For example, the Cartesian plane is a common universal set. But if a mathematician is studying functions of complex variables, the universal set is usually the complex (or Argand, or Gaussian) plane. The most important universal set in this book is $\mathbb{R}$, the set of real numbers.

The **complement** of set S, written $\overline{S}$, is the set of elements in the universal set that are not contained in S: $\overline{S} = \{x \in U : x \notin S\}$. For example, $\overline{\mathbb{Q}} = \{x \in \mathbb{R} : x \notin \mathbb{Q}\}$, the set formed from the universal set $\mathbb{R}$by removing all rational numbers. When the universal set is taken for granted, the notation is just $\overline{S} = \{x : x \notin S\}$.

Note 1: In classical, Aristotelian, logic, a set was thought of as a repository of things for the purpose of classification. Thus, for Scholastics (and even to this day), a *genus* was a set of things all of which had a distinguishing property. Subsets of a genus were called *species*, and each species was definable by a specific difference within the genus (consider the etymology of the word "specific"). For example, the species "blue whale" would be clearly defined by a specific difference within the genus "whale." Evidently, a set such as {the Pyramid of Khufu, a statement by Socrates, an equilateral triangle} would have been pointless to Aristotle—and the empty set, totally nonsensical. Our modern concept of a set is a comparatively recent development, brought to you by Gottlob Frege and Georg Cantor (1845–1918) in the 1800s.

A universal set should not be confused with the idea of a set of all things. The "set of everything" may be a cognitive illusion, and certainly is not a set according to our requirements, since it would have to contain infinitely many duplicates—of everything! Nor should this be confused with the physical universe, which we are confident does exist.

Intervals of real numbers are handy sets in real analysis. A bracket always indicates the inclusion of an endpoint; a parenthesis indicates exclusion of it. Closed intervals always include their endpoints, while open intervals never include them.

INTERVALS			
BOUNDED		*UNBOUNDED*	
$[a, b] \equiv \{x : a \leq x \leq b\}$	Closed	$[a, \infty) \equiv \{x : x \geq a\}$	Closed
$(a, b) \equiv \{x : a < x < b\}$	Open	$(a, \infty) \equiv \{x : x > a\}$	Open
$[a, b) \equiv \{x : a \leq x < b\}$	Neither open nor closed. (Half-open and half-closed.)	$(-\infty, b] \equiv \{x : x \leq b\}$	Closed
$(a, b] \equiv \{x : a < x \leq b\}$	Neither open nor closed. (Half-open and half-closed.)	$(-\infty, b) \equiv \{x : x < b\}$	Open

Illustrating the use of these symbols, the complement of the interval (a, ∞) is written $\overline{(a, \infty)} = (-\infty, a]$. (It must be remembered that ∞ is not a real number, but just the symbol for "unbounded.") Logically speaking, $(-\infty, \infty) = \mathbb{R}$, whereas $[-\infty, \infty]$ is an abuse of the bracket symbol, and is meaningless. At the opposite extreme, $(a, a) = \varnothing$, and $[a, a] = \{a\}$, a singleton set.

> To distinguish between the open interval (a, b) and an ordered pair of numbers (a **point** on the Cartesian plane), we shall use angle brackets for the point: $\langle a, b \rangle$.

Sets can also be categorized by the number of elements that they contain, their "size" roughly speaking. There are three types. A **finite set** can always have its elements numbered, as follows: $S = \{s_1, s_2, \ldots, s_m\} = \{s_n\}_{n=1}^{m}$. We are using what is called a **finite subscripting set**, $\{1, 2, 3, \ldots, m\}$, to keep track of the elements s_n, with subscript $n \in \{1, 2, 3, \ldots, m\}$.

Extending the subscripting set to all of $\mathbb{N}$, we can define a **denumerable set** as one whose elements can be numbered like this: $S = \{s_1, s_2, \ldots, s_n, \ldots\} = \{s_n\}_{n=1}^{\infty}$, and we see that subscript is now $n \in \mathbb{N}$. A typical example of such a set is $\{\frac{n-1}{n}\}_{n=1}^{\infty} = \{0, \frac{1}{2}, \frac{2}{3}, \ldots\}$, with $s_n = \frac{n-1}{n}$. The key difference between this set and the sequence that looks like it is that the set can be rearranged in any order whatsoever, but the sequence must maintain the defined order.

Finally, a set can be described as uncountably infinite. As it indicates, this type of set is unable to be counted. This amazing possibility will be informally discussed under its own subheading below, and will be properly defined in Chapter 11.

5.2 Three Operations for Sets

Given sets S and T, we define three operations for them. Their **union** (featuring the "cup" symbol), is the set $S \cup T \equiv \{x : x \in S \text{ or } x \in T\}$, where the inclusive "or" is intended. Their **intersection** (with the "cap" symbol), is the set $S \cap T \equiv \{x : x \in S \text{ and } x \in T\}$. Their **relative complement** is the set $S \setminus T \equiv \{x : x \in S \text{ and } x \notin T\}$, that is, all elements not in T, but still in S. Hence, this is a generalization of the complement idea. The diagrams in Figure 5.1 are the invention of the logician John Venn (1834–1923). Viewed together, they could become a celebrated work of modern art.

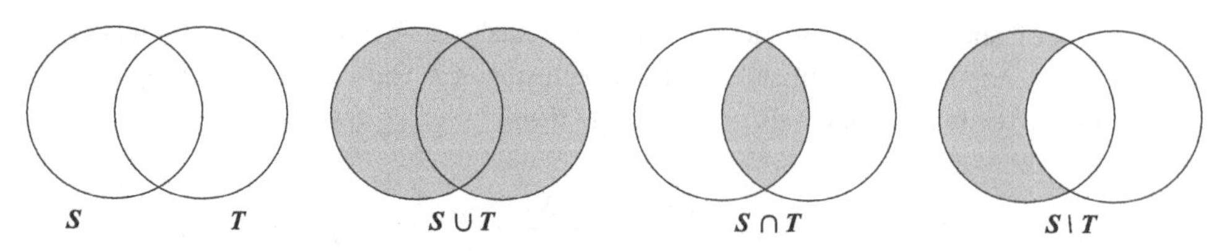

Figure 5.1.

While $S \cup T = T \cup S$ and $S \cap T = T \cap S$, the relative complement is not commutative, that is, $S \setminus T \neq T \setminus S$.

If two sets V and W have no common elements, then their intersection is the empty set $\varnothing$. Thus, $V \cap W = \varnothing$. We say that V and W are **disjoint** or **mutually exclusive** sets.

Our first theorem for unions and intersections reminds us of the distributive law. As has been said, to prove equality, two conditional statements will be proved.

Theorem 5.1 For any three sets, $R \cap (S \cup T) = (R \cap S) \cup (R \cap T)$ and $R \cup (S \cap T) = (R \cup S) \cap (R \cup T)$. That is, intersection is distributive over union, and union is distributive over intersection.

Proof. To prove the first equality, begin with $x \in R \cap (S \cup T)$. This means that $x \in R$, and either $x \in S$ or $x \in T$. It follows that either $x \in R \cap S$ or $x \in R \cap T$(or both). In any case, $x \in (R \cap S) \cup (R \cap T)$. We now know that if $x \in R \cap (S \cup T)$, then $x \in (R \cap S) \cup (R \cap T)$.

Now, let $x \in (R \cap S) \cup (R \cap T)$. This means that x belongs to either $R \cap S$ or $R \cap T$ (or both). It follows immediately that x must belong to R (why?). It is also true that x belongs to either S or T, so that $x \in S \cup T$. Hence, $x \in R \cap (S \cup T)$. We have proved the converse: if $x \in (R \cap S) \cup (R \cap T)$, then $x \in R \cap (S \cup T)$. Therefore, $R \cap (S \cup T) = (R \cap S) \cup (R \cap T)$.

The proof of the second equality follows a similar train of thought, and is left as Exercise 5.9 for you. **QED**

We now come to an important theorem that allows us to exchange unions for intersections, and vice versa. It is named for the algebraist Augustus De Morgan (1806–1871).

Theorem 5.2 De Morgan's Laws For any sets S and T, $\overline{S \cup T} = \overline{S} \cap \overline{T}$, and $\overline{S \cap T} = \overline{S} \cup \overline{T}$.

Proof. This time, we will use Lemma 5.1. For the first equality, first prove the containment $\overline{S \cup T} \subseteq \overline{S} \cap \overline{T}$. Begin with $x \in \overline{S \cup T}$. This implies that x belongs to neither S nor T. Hence, $x \in \overline{S}$ and $x \in \overline{T}$, and therefore, $x \in \overline{S} \cap \overline{T}$. We now have that $\overline{S \cup T} \subseteq \overline{S} \cap \overline{T}$.

Next, prove the reverse containment, $\overline{S} \cap \overline{T} \subseteq \overline{S \cup T}$. Let $x \in \overline{S} \cap \overline{T}$. It follows that x is an element of both $\overline{S}$ and $\overline{T}$. This means that x cannot belong to S, nor to T, so that it cannot belong to $S \cup T$. Therefore, $x \in \overline{S \cup T}$. We now have that $\overline{S} \cap \overline{T} \subseteq \overline{S \cup T}$. By Lemma 5.1, we conclude that $\overline{S \cup T} = \overline{S} \cap \overline{T}$.

The proof of the second equality is left as Exercise 5.10. **QED**

5.3 Collections of Sets

A set may very well have sets as its elements (so, a set of sets!). To minimize confusion, we may call these collections of sets. Just as for "regular" sets, there are three types of collections of sets. A finite collection of sets $\{S_1, S_2, S_3, \ldots, S_m\}$ has m sets as members. Here, the subscript n belongs to a finite subscripting set $\{1, 2, 3, \ldots, m\}$, allowing us to write the collection as $\{S_n\}_{n=1}^{m}$. If $n \in \mathbb{N}$, then $S_n \in \{S_1, S_2, S_3, \ldots\} = \{S_n\}_{n=1}^{\infty}$, which is a denumerable collection. Suppose 13 is a member of S_6. Then we can write the interesting chain $13 \in S_6 \in \{S_n\}_{n=1}^{\infty}$.

Example 5.1 Let the subscripting set be $\mathbb{N}$, and let $S_n = \{\text{points of the circle } x^2 + y^2 = (\frac{n}{n+1})^2\}$. Then $\{S_n\}_{n=1}^{\infty}$ is the entire (denumerable) collection of concentric circles with radii $\frac{n}{n+1}$, starting with $x^2 + y^2 = (\frac{1}{2})^2$. We notice that since $\frac{n}{n+1} \to 1$, the circles all crowd toward the limiting circle $x^2 + y^2 = 1$, as seen in Figure 5.2. Now, the point $\langle 0, \frac{2}{3}\rangle$ lies on the particular circle $x^2 + y^2 = (\frac{2}{3})^2$. This circle is one of the sets in our collection, namely $S_2 = \{(\langle x, y\rangle : x^2 + y^2 = (\frac{2}{3})^2\}$. We can therefore state that $\langle 0, \frac{2}{3}\rangle \in S_2 \in \{S_n\}_{n=1}^{\infty}$. ■

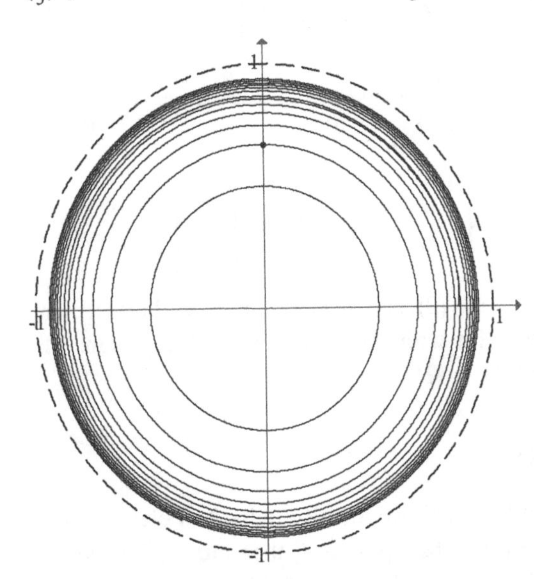

Figure 5.2. The point $\langle 0, \frac{2}{3}\rangle$ is an element of $x^2 + y^2 = \frac{4}{9}$, which is itself an element of the collection of circles beginning with $x^2 + y^2 = \frac{1}{4}$ (the innermost circle), and crowding forever toward $x^2 + y^2 = 1$(the dashed circle). This last circle is not in the collection.

The last type of collection is an uncountably infinite collection of sets. This is expressed as $\{S_\alpha\}$, where the subscripting set for α is no longer $\mathbb{N}$ (nor any subscripting set that is itself countable). More about this is said below, and in Chapter 11.

5.4 Generalized Union and Intersection

Union and intersection need not be restricted to two sets. We shall use the symbol $\bigcup$ to write the **generalized union** of m sets as $\bigcup_{n=1}^{m} S_n = S_1 \cup S_2 \cup \cdots \cup S_m$, and if we have avidly collected an infinite, but denumerable, number of sets, their union would be $\bigcup_{n=1}^{\infty} S_n = S_1 \cup S_2 \cup S_3 \cup \cdots$. For example, with $n \in \mathbb{N}$, if we define sets of two elements $S_n = \{-n, n\}$, then $\bigcup_{n=1}^{\infty} S_n = \{-1, 1\} \cup \{-2, 2\} \cup \cdots = \mathbb{Z} \setminus \{0\}$, since every integer except 0 belongs to at least one set S_n.

The symbol $\bigcap$ indicates **generalized intersection**. Let $S_n = [0, \frac{1}{n}]$, where $n \in \mathbb{N}$. This gives us a denumerable collection of closed intervals (go ahead and list the first few). The notation is employed in these examples:

$$\bigcap_{n=1}^{3} S_n = [0,1] \cap \left[0, \frac{1}{2}\right] \cap \left[0, \frac{1}{3}\right] = \left[0, \frac{1}{3}\right] = S_3,$$

and

$$\bigcap_{n=1}^{m} S_n = \left[0, \frac{1}{m}\right] = S_m,$$

while $\bigcap_{n=1}^{\infty} S_n = S_1 \cap S_2 \cap S_3 \cap \cdots = \{0\}$. Why? The reason is that 0 is the only number common to all the sets S_n. Compare this with the following slight modification. Let $S_n = (0, \frac{1}{n})$ be an open interval for each $n \in \mathbb{N}$. The result now is that $\bigcap_{n=1}^{\infty} S_n = \varnothing$.

Let's summarize. By the definition of union, $x \in \bigcup_{n=1}^{\infty} S_n$ if and only if $x \in S_n$ for at least one S_n. And by the definition of intersection, $x \in \bigcap_{n=1}^{\infty} S_n$ if and only if $x \in S_n$ for *every* S_n.

Example 5.2 Let

$$S_n = \left\{ \text{reduced fractions } \frac{m}{n} : m \in \mathbb{Z}\backslash\{0\} \text{ and } n \in \mathbb{N} \right\}.$$

To begin,

$$S_1 = \{\ldots, -2, -1, 1, 2, \ldots\} = \mathbb{Z}\backslash\{0\}.$$

Next,

$$S_2 = \left\{\ldots, \frac{-1}{2}, \frac{1}{2}, \frac{3}{2}, \frac{5}{2}, \ldots\right\}, S_3 = \left\{\ldots, \frac{-2}{3}, \frac{-1}{3}, \frac{1}{3}, \frac{2}{3}, \frac{4}{3}, \ldots\right\},$$

and so on. Reflecting a bit, we see that $\bigcup_{n=1}^{\infty} S_n = \mathbb{Q}\backslash\{0\}$, and that $\bigcap_{n=1}^{\infty} S_n = \varnothing$. Furthermore, we see that each of these sets is **pairwise disjoint** with every other set, which means that $S_n \cap S_m = \varnothing$ whenever $n \neq m$. (See Exercise 5.11.) ■

5.5 Countable and Uncountable Sets

The subscripting set A (always a subset of $\mathbb{R}$) by which we identify individual elements of a set S becomes the key player in this subsection. If subscript $\alpha \in A$, then our set S is written $\{s_\alpha : \alpha \in A\}$. In other words, an item $s_\alpha \in S$ precisely when $\alpha \in A$. This is actually a very versatile notation. To reiterate, the job of a subscript is to identify each element of a set uniquely.

When S is finite, we wrote $S = \{s_n\}_{n=1}^{m}$, which can now be expressed as $\{s_\alpha : \alpha \in \{1, 2, \ldots, m\}\}$. When S is denumerable, the count is unending, and our notation becomes $S = \{s_n\}_{n=1}^{\infty} = \{s_\alpha : \alpha \in \mathbb{N}\}$. So long as A was $\mathbb{N}$ or a subset of $\mathbb{N}$, we saw in examples above that S could be counted, element by element. Effectively, then, the subscripts not only identified elements, but also allowed us to count them—even if the count was endless. Because of this capability, both finite and denumerable sets are named **countable** sets.

Note 2: Subscripts do not necessarily imply an order for the elements of S. That is, it is possible that $s_1 = 8.2$ and $s_5 = 3.1$, so that $s_5 < s_1$ within a set S.

Now, the subscripting set A need not contain just integers. Choosing $A = [0, 1]$, for instance, lets us form a set $T = \{t_\alpha : \alpha \in [0, 1]\}$. This implies that for example t_1, $t_{0.75}$, and $t_{\pi/6}$ all belong to T (and see Note 2 again). At this point, a completely counter-intuitive property arises. Though the individual elements of T all have unique subscripts, T can never be described in the form $\{t_1, t_2, t_3, \ldots\} = \{t_n\}_{n=1}^{\infty}$, because invariably, some element will be *left out!* We cannot continue to assume that subscripts can count the elements of T. Any set with this amazing property is named an **uncountably infinite**, or **uncountable**, set.

Common sense sternly rejects the claim that sets exist with so many elements that they can't be fully counted. However, the infinite is far beyond common sense. Fortunately, we know that mathematical definitions and theorems allow us to understand things that no amount of speculation can clarify. It was Georg Cantor who unravelled these mysteries, and with marvellous originality established the mathematical theory of infinite sets. We will study some of his discoveries in Chapter 11. For now, Examples 5.3 to 5.7 display some uncountable sets that aren't at all alien.

Example 5.3 Let our subscripting set be the interval $A = [1, 4]$, and let $\alpha \in A$. We shall define $t_\alpha = \alpha^3$, resulting in the set $T = \{\alpha^3 : \alpha \in [1, 4]\} = [1, 64]$. One would probably say, "That's the most round-about way of describing the interval $[1, 64]$ that has ever been dreamed of." However, we can now assign unique subscripts to every element of T. For instance, because $\sqrt[3]{2} \in [1, 4]$, then $t_{\sqrt[3]{2}} = 2$ belongs to T. Again, $\pi \in [1, 4]$, so $t_\pi = \pi^3 \in T$. For a similar reason, $t_{2\pi} \notin T$. The interval $[1, 64]$ is an uncountably infinite set, as is A. ■

Example 5.4 Let $T_\alpha = \{\text{points of the circle } x^2 + y^2 = \alpha^2\}$ and let the subscripting set be $A = [2, 6]$. Then $\{T_\alpha : \alpha \in [2, 6]\}$ is the entire, uncountably infinite, collection of concentric circles starting with $x^2 + y^2 = 4$ and ending with $x^2 + y^2 = 36$. It is clear that the point $\langle 3, 4\rangle$ lies on the specific circle $x^2 + y^2 = 25$. This circle is one of the sets in our collection, namely $T_5 = \{\langle x, y\rangle : x^2 + y^2 = 25\}$. We can therefore state that $\langle 3, 4\rangle \in T_5 \in \{T_\alpha : \alpha \in [2, 6]\}$, as you see in Figure 5.3. ■

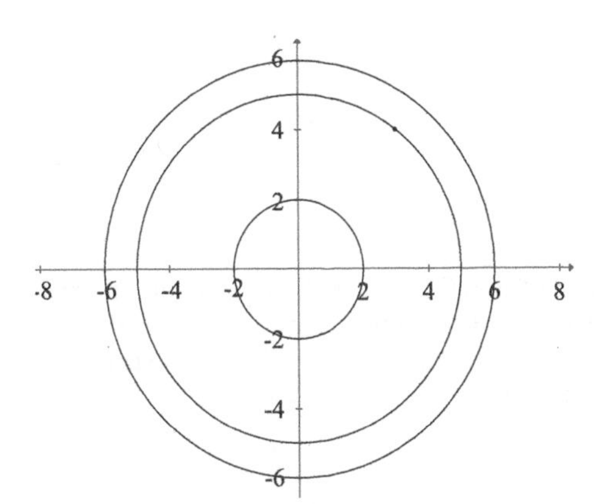

Figure 5.3. The point $\langle 3, 4\rangle$ is an element of $x^2 + y^2 = 25$, which is itself an element of the uncountable collection of circles beginning with $x^2 + y^2 = 4$ (the inner thick circle), and ending with $x^2 + y^2 = 36$, (the outer thick circle).

The notations for generalized intersection and union handle uncountably infinite collections of sets with no problem at all. For an arbitrary subscripting set A (that is, uncountable or countable), we write $\bigcap_{\alpha \in A} S_\alpha$, and $\bigcup_{\alpha \in A} S_\alpha$, where the sets S_α can themselves be either countable or uncountable. As always, by the definition of union, $x \in \bigcup_{\alpha \in A} S_\alpha$ if and only if $x \in S_\alpha$ for at least one S_α. And by the definition of intersection, $x \in \bigcap_{\alpha \in A} S_\alpha$ if and only if $x \in S_\alpha$ for every S_α.

Example 5.5 Let $\alpha \in A = [1, \infty)$, and define closed intervals $T_\alpha = [\frac{1}{\alpha}, 2]$. Then $\{T_\alpha\}$ is an uncountable collection. Because $\frac{1}{\alpha} \to 0$ as α increases, the intervals T_α stretch forever toward the interval $(0, 2]$ (draw a picture). Therefore, their union is $\bigcup_{\alpha \geq 1} T_\alpha = (0, 2]$. Their intersection, as always, contains only those numbers common to all the sets T_α. This leads us to $\bigcap_{\alpha \geq 1} T_\alpha = [1, 2]$. This example demonstrates an uncountable union and intersection of sets that were individually uncountable. ■

Example 5.6 Define open intervals $T_\alpha = (7 - \frac{1}{\alpha}, 8 - \frac{1}{\alpha})$, and choose subscripts $\alpha \geq 1$. Once again, $\{T_\alpha\}$ is an uncountable collection. Then $\bigcup_{\alpha \geq 1} T_\alpha = (6, 8)$, since every real number in $(6, 8)$ belongs to at least one set T_α. The first T_α that doesn't contain $7 - \frac{1}{6} = 6.833\ldots$ is $T_{6.833\ldots}$. Think about why $T_{6.833\ldots}$ may be called the first such set. Finally, $\bigcap_{\alpha \geq 1} T_\alpha = \varnothing$, since no number is common to every T_α (try to find one!). ■

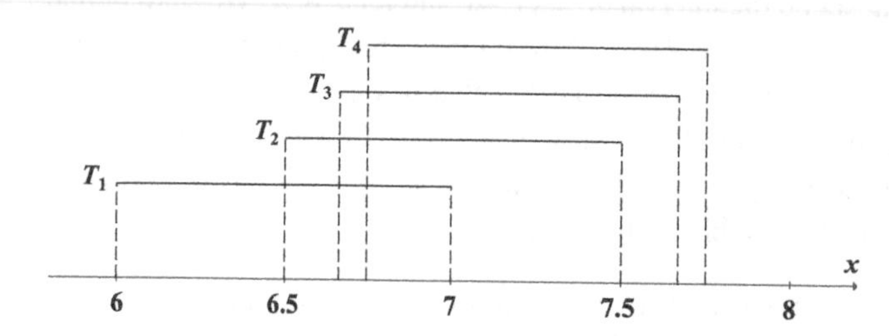

Figure 5.4. T_1, T_2, T_3, and T_4, drawn at different heights for visibility. All members of $\{T_\alpha\}$ have length equal to 1.

Example 5.7 Using the collection from Example 5.4, form the generalized union $\bigcup_{\alpha \in [2,6]} T_\alpha$. This instructs us to unite all the circles with equations $x^2 + y^2 = \alpha^2$, with radii $\alpha \in [2, 6]$. The union is the annular region in Figure 5.5 (compare it with Figure 5.3). And here, for instance, $T_{5.4} \subset \bigcup_{\alpha \in [2,6]} T_\alpha$. ■

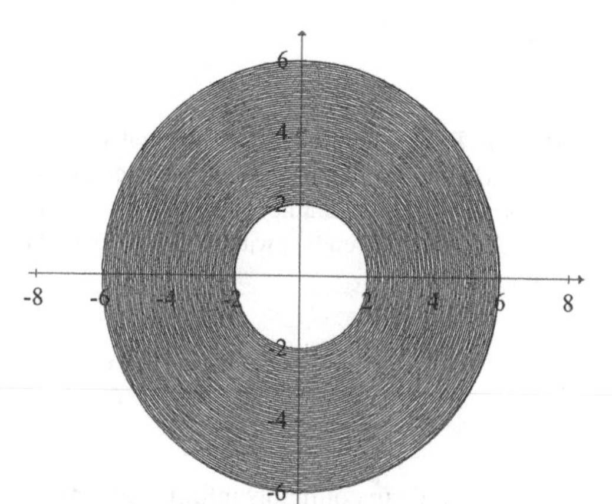

Figure 5.5. The union $\bigcup_{\alpha \in [2,6]} T_\alpha$ forms an annulus with inner radius 2 and outer radius 6. The particular circle $x^2 + y^2 = 5.4^2$ is an element of the union.

A set S is **partitioned** into subsets S_α if and only if the subsets are nonempty, pairwise disjoint, and $\bigcup_{\alpha \in A} S_\alpha = S$. A jigsaw puzzle is a good mental image of a partitioned set. A mathematical example is the partitioning of the annulus in Example 5.7, where each circle is one of the uncountable number of partitions.

Theorems 5.1 and 5.2 can be extended to cover generalized unions and intersections. Prove the extended version of Theorem 5.1 as Exercise 5.14. The extended De Morgan's laws appear next.

Theorem 5.3 Extended De Morgan's Laws Given any subscripting set $A \subseteq \mathbb{R}$ (countable or uncountable), and a collection of sets $\{S_\alpha : \alpha \in A\}$,

A the complement of their union equals the intersection of their complements: $\overline{\bigcup_{\alpha\in A} S_\alpha} = \bigcap_{\alpha\in A} \overline{S_\alpha}$. Also,

B the complement of their intersection equals the union of their complements: $\overline{\bigcap_{\alpha\in A} S_\alpha} = \bigcup_{\alpha\in A} \overline{S_\alpha}$.

Proof. **A** To prove this, first let $x \in \overline{\bigcup_{\alpha\in A} S_\alpha}$. This means that $x \notin S_\alpha$ for all $\alpha \in A$. This implies that $x \in \overline{S_\alpha}$ for every $\alpha \in A$. And this in turn implies that $x \in \bigcap_{\alpha\in A} \overline{S_\alpha}$. Hence, $\overline{\bigcup_{\alpha\in A} S_\alpha} \subseteq \bigcap_{\alpha\in A} \overline{S_\alpha}$.

To show the reverse containment, let $x \in \bigcap_{\alpha\in A} \overline{S_\alpha}$. Then for every $\alpha \in A$, $x \in \overline{S_\alpha}$. This implies that $x \notin S_\alpha$ for every $\alpha \in A$. And this implies that $x \notin \bigcup_{\alpha\in A} S_\alpha$, so that $x \in \overline{\bigcup_{\alpha\in A} S_\alpha}$. Therefore, $\bigcap_{\alpha\in A} \overline{S_\alpha} \subseteq \overline{\bigcup_{\alpha\in A} S_\alpha}$. We conclude (by Lemma 5.1) that $\overline{\bigcup_{\alpha\in A} S_\alpha} = \bigcap_{\alpha\in A} \overline{S_\alpha}$.

B This is proved in a similar fashion, and is left to you in Exercise 5.15. **QED**

When the subscripting set A has two elements, Theorem 5.3 becomes Theorem 5.2. And Theorem 5.3 can be applied to Example 5.7, as follows.

Example 5.8 Starting with the annulus $\bigcup_{\alpha\in[2,6]} T_\alpha$ from Example 5.7, Figure 5.5, we see $\overline{\bigcup_{\alpha\in[2,6]} T_\alpha} = D \cup E$ as the disk $D = \{\langle x, y\rangle : x^2 + y^2 < 2\}$, in union with the region beyond the annulus, going outward without bounds, written $E = \{\langle x, y\rangle : x^2 + y^2 > 6\}$. On the other hand, $\overline{T_\alpha}$ represents all points not on the circle $x^2 + y^2 = \alpha^2$. Intersecting this region with all other such regions identified by $\alpha \in [2, 6]$, we arrive at $\bigcap_{\alpha\in[2,6]} \overline{T_\alpha}$. Convince yourself that this will give exactly the same region $D \cup E$ as before, as guaranteed by Theorem 5.3A.

To use the second part of Theorem 5.3, note that $\bigcap_{\alpha\in[2,6]} T_\alpha = \varnothing$, because the circles are pairwise disjoint. Thus, $\overline{\bigcap_{\alpha\in[2,6]} T_\alpha}$ must be the entire Cartesian plane, our universal set. Convince yourself that $\bigcup_{\alpha\in[2,6]} \overline{T_\alpha}$ yields the same thing. ■

Note 3: This brief introduction is by no means a full basis for set theory. But it is sufficient for what we cover in this book. Edna Kramer's excellent study of mathematics [7] begins discussing sets on page 4. Struggles with the foundations of set theory are ably treated in [39], pp. 147–150.

5.6 Exercises

1. These questions show how tricky the general concept of a set can be:
   a. Why is {large pebbles} not a set?
   b. All electrons are identical items. Is {electrons orbiting a particular oxygen atom} a set?
   c. Under what context can {the tallest building on Manhattan Island (New York), the Empire State Building} be called a set?

d. Suppose $T = \{a, b, c\}$. What is $\{T\}$, and does $T = \{T\}$? And, can a set be a member of itself, symbolically, can $T \in T$?

e. Further, difficult, questions to ponder: Is a set identical to the aggregate of its elements, or is it identical to the rule for deciding membership, or to both? In the absence of intellection, do sets cease to exist? See Note 3 above.

2. Are the sets $\{1, 2, 3, \ldots\}$ and $\{2, 4, 6, \ldots, 1, 3, 5, \ldots\}$ equal?

3. Prove Lemma 5.1. Is its converse true? Incidentally, why is it impossible for $S \subset P$ and $P \subset S$?

4. True or False: Is intersection an associative operation: $A \cap (B \cap C) = (A \cap B) \cap C$? Is union an associative operation: $A \cup (B \cup C) = (A \cup B) \cup C$?

5. Prove that $S \setminus T = S \cap \overline{T}$, where a universal set exists in the background. Hint: Think of the Venn diagrams.

6. Prove that $S \cup T = T \cup (S \setminus T)$. This is important because it allows us to break up a union into the union of two disjoint sets: T and$(S \setminus T)$. Also, convince yourself that $S \cup T = S \cup (T \setminus S)$. Hint: Venn diagrams show it all, but use the previous exercise in a proof.

7. a. Prove that $S \subseteq T$ if and only if $S \cap T = S$.
   b. Prove that $A \cap B \subseteq A \cup B$. Hint: One way to do it is to take $S = A \cap B$ and $T = A \cup B$ in a, and show that $S \cap T = S$.

8. Use Exercise 5.7 to prove again that $\varnothing \subseteq T$ for any set T.

9. Finish proving Theorem 5.1 by showing that $R \cup (S \cap T) = (R \cup S) \cap (R \cup T)$.

10. Finish proving De Morgan's laws by showing that $\overline{S \cap T} = \overline{S} \cup \overline{T}$.

11. It is true that if sets S_n are pairwise disjoint, then $\bigcap_{n=1}^{\infty} S_n = \varnothing$? Give an example of three sets of numbers showing that the converse of the statement is false.

12. Another operation between sets is the **symmetric difference,** $S \ominus T \equiv \{x : x \in S \cup T$ and $x \notin S \cap T\}$. In other words, it is the set that contains elements in exactly one of S or T. The symmetric difference uses the exclusive sense of "or," as opposed to the union operation, which uses the inclusive "or." Draw a Venn diagram representing $S \ominus T$. Prove that $S \ominus T = (S \setminus T) \cup (T \setminus S)$. Hint: Start by writing $\{x : x \in S \cup T$ and $x \notin S \cap T\}$ using only union, intersection, and complement.

13. For the subscripts $\alpha \in [0, 1]$, define closed intervals $S_\alpha = [\alpha^2, \alpha]$. What is $\cup_{0 \le \alpha \le 1} S_\alpha$? What about $\cap_{0 \le \alpha \le 1} S_\alpha$?

14. Consider extending Theorem 5.1 in a step-by-step fashion, as follows.
    a. The first equality can be extended sensibly to say the following:

$$R \cap \bigcup_{n=1}^{\infty} S_n = \bigcup_{n=1}^{\infty} (R \cap S_n).$$

See how this equality works with the example $R = \mathbb{N}$ and $S_n = [n + \frac{1}{2},\ n + 1]$ (where $n \in \mathbb{N}$). Now, prove the equality. Hint: Use induction.

b. **a** can be further extended to $R \cap \bigcup_{\alpha \in A} S_\alpha = \bigcup_{\alpha \in A}(R \cap S_\alpha)$. Prove this equality. Hint: You can't use induction, but Lemma 5.1 still applies.
c. State and prove the extended version of the second equality of Theorem 5.1.

15. Prove part **B** of Theorem 5.3.

16. We have been considering a circle as a set of points (the ancient geometers used the term "locus" of points in a similar sense). Given that interval $[0, 2\pi)$ is an uncountably infinite set, argue that the unit circle $x^2 + y^2 = 1$ contains an uncountable number of points.

17. Let

$$T_n = \left\{ \langle x, y \rangle : x^2 + \left(y - \frac{1}{2^n} \right)^2 \leq \left(\frac{1}{2^n} \right)^2 \right\},$$

where the subscripting set is $\{0, 1, 2, \ldots, n, \ldots\}$. Draw disks T_0, T_1, and T_2, and determine $\bigcup_{n=0}^{\infty} T_n$ and $\bigcap_{n=0}^{\infty} T_n$. This shows a countable union and intersection of sets that are individually uncountable (by Exercise 5.16).

6

Functions

The function $f(x)$ denotes a function completely arbitrary, that is to say a succession of given values, subject or not to a common law, and answering to all the values of x included between 0 and any magnitude X.

Joseph Fourier, *Analytical Theory of Heat*, Article 418.

The modern concept of a function begins perhaps with medieval ecclesiastic Nicole Oresme (ca. 1323–1382), who drew, at successive points on a horizontal line representing moments of time, vertical segments equal to the velocity of an object at the corresponding time. Thus, Oresme had drawn a rudimentary graph of velocity against time, a tremendously advanced idea for his day. To Gottfried Wilhelm Leibniz and Isaac Newton in the 17th century, a function was a curve such as the spiral and the cycloid—in general, the path of a point moved by some physical process and expressed by a formula. James Gregory (1638–1675) established the distinction between algebraic curves such as $y = x^n$, and transcendental curves, for example, $y = \sin x$. Leonhard Euler often thought of a function as any combination of algebraic, trigonometric, and exponential expressions that produced an unbroken curve, including functions defined by processes such as integrals and infinite series, and he introduced functions of several variables. In 1822, Joseph Fourier (1768–1830) published the groundbreaking *Théorie analytique de la chaleur* (Analytical Theory of Heat), in which the infinite series named after him demanded a more general view of functions, including piecewise-defined curves with discontinuities. The concept of a function as a mapping of each x to a unique y was made clear by Peter Gustav Lejeune-Dirichlet (1805–1859). This naturally led to the concepts of domain and range, and these not just of real numbers, but also of more general sets of objects. Today, the concept of function has expanded in a bewildering fashion to answer the demands of fractal geometry, quantum physics, and probability, to mention a few fields.

6.1 Cartesian Products

We begin with a new operation between two sets, D and R. The **Cartesian product** $D \times R$ of these sets is defined as the set of all ordered pairs $\langle x, y \rangle$ where $x \in D$ and $y \in R$. Unless the two sets are equal, $D \times R \neq R \times D$.

Example 6.1 Let $D = \{2, 3, 5, 9\}$ and $R = \{\spadesuit, \clubsuit, \heartsuit, \diamondsuit\}$. Then the Cartesian product contains sixteen elements: $D \times R = \{\langle 2, \spadesuit \rangle, \langle 2, \clubsuit \rangle, \langle 2, \diamondsuit \rangle, \langle 2, \heartsuit \rangle, \langle 3, \spadesuit \rangle, \langle 3, \clubsuit \rangle, \langle 3, \diamondsuit \rangle, \langle 3, \heartsuit \rangle,$

⟨5, ♠⟩, ⟨5, ♣⟩, ⟨5, ♢⟩, ⟨5, ♡⟩, ⟨9, ♠⟩, ⟨9, ♣⟩, ⟨9, ♢⟩, ⟨9, ♡⟩}. These ordered pairs can be placed in an array, as shown in Figure 6.1. ■

♡	2♡	3♡	5♡	9♡
♢	2♢	3♢	5♢	9♢
♣	2♣	3♣	5♣	9♣
♠	2♠	3♠	5♠	9♠
	2	3	5	9

Figure 6.1. The Cartesian product D×R in Example 6.1, conveniently showing all 16 playing cards.

Example 6.2 Let $D = [2, 6]$ and $R = [1, 4)$. Then $D \times R = [2, 6] \times [1, 4) = \{\langle x, y\rangle : 2 \le x \le 6 \text{ and } 1 \le y < 4\}$, which forms a three-sided rectangle with its interior, seen in Figure 6.2. The ordered pairs are now just points of a graph. ■

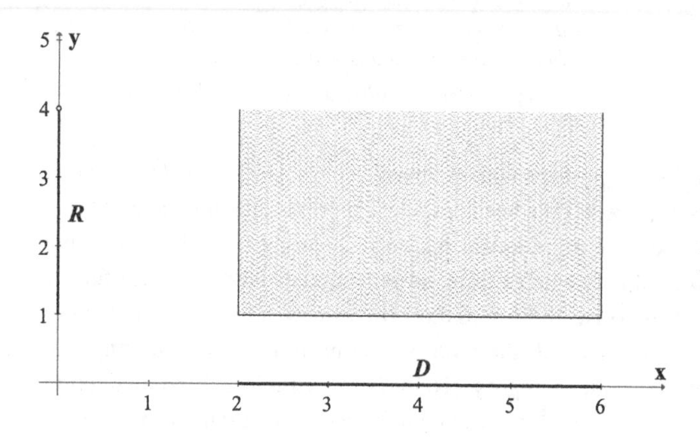

Figure 6.2. The Cartesian product $[2, 6] \times [1, 4)$ in Example 6.2. The intervals D and R are shown with thick lines.

Example 6.3 Let $D = \mathbb{N}$ and $R = \{\sqrt{m} : m \in \mathbb{N}\}$. Then $D \times R = \{\langle n, \sqrt{m}\rangle : n \in \mathbb{N},\ m \in \mathbb{N}\}$, part of which is shown in Figure 6.3. ■

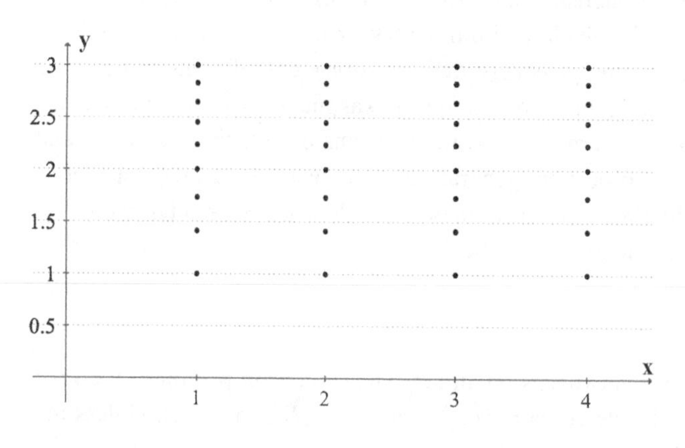

Figure 6.3. The Cartesian product D×R in Example 6.3.

Example 6.4 Now, let $D = \{\sqrt{n} : n \in \mathbb{N}\}$ and $R = \{2\}$. Then $D \times R = \{\langle \sqrt{n}, 2\rangle : n \in \mathbb{N}\}$. The initial portion of this Cartesian product is seen in Figure 6.4. Although not very revealing, it is a graph of the sequence $\langle \sqrt{n}\rangle_{n=1}^{\infty}$. ■

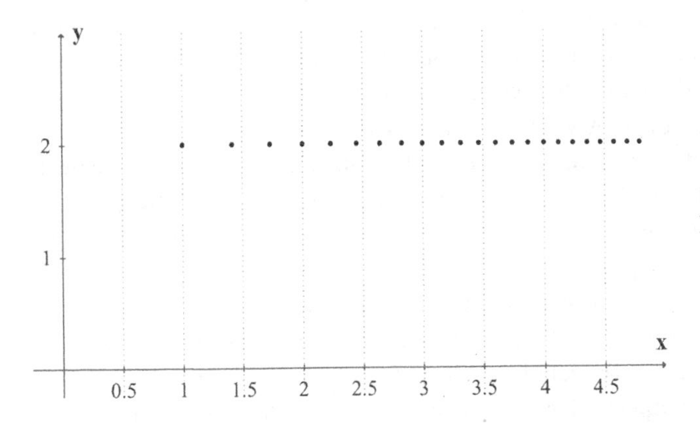

Figure 6.4. The Cartesian product D×R in Example 6.4.

Example 6.5 $\mathbb{R} \times \mathbb{R}$, denoted $\mathbb{R}^2$, is the set of all ordered pairs of real numbers, which is to say, the entire Cartesian plane. Similarly, $\mathbb{R} \times \mathbb{R} \times \mathbb{R} \equiv \mathbb{R}^3$, is the set of all ordered triples $\langle x, y, z\rangle$, which is all of three-dimensional space. We can continue on to spaces of higher dimension in this way. Is $\mathbb{R}^3 = (\mathbb{R} \times \mathbb{R}) \times \mathbb{R}$? See Exercise 6.4b. ■

Note 1: What is a graph? It certainly should be a diagram of all the ordered pairs belonging to a function, since we constantly speak of the "graph of a function." But we don't usually intend a graph to be an array such as the one in Example 6.1. We arrive at a suitable definition: a *graph* is any subset of the Cartesian product $\mathbb{R}^n$, when the elements of the subset are rendered as points ($\mathbb{R}^n$ is n-dimensional space, to be studied in Chapter 52). Thus, Figure 6.2 qualifies as a graph, since it is a subset of $\mathbb{R}^2$. (It certainly isn't the graph of a function, however, as we shall soon see.)

The fascinating possibility of graphs in fractional dimensions, say between one and two, is realized by fractal curves. A beautiful collection of these can be seen in [42], Chapters 5 and 6.

Are there graphs in dimensions $n \geq 4$? How can a four-dimensional object be seen, for instance? Presenting two- or three-dimensional slices of such an object rather sidesteps the issue, but a vector field is at least a partial answer, graphing a two-dimensional arrow at every point on the (two-dimensional) plane. Also, a five-dimensional vector $\langle x, y, z, T, t\rangle$ can represent the position $\langle x, y, z\rangle$ and temperature T of any point in the atmosphere, at any given time t. One can imagine a hologram in which the color of a point is keyed to its temperature, which changes as time goes by. Thus, the colorful influx of a cold front is the graph of a five-dimensional process.

Going beyond the definition here, the position of the beads on an abacus are a representation of a number with graphic qualities. The statistician Herman Chernoff and others have employed humanoid faces to graph multidimensional data sets, in a way that allows the eye to see similarities in complicated data.

6.2 Functions

Given any two sets D and R, a **function** from D **into** R is a subset of $D \times R$ such that if $\langle x, y\rangle$ and $\langle x, z\rangle$ belong to the function, then we can conclude that $y = z$. In other words, whenever the ordered pair $\langle x, y\rangle$ belongs to a function, then x is paired *uniquely* to y.

We often attach a label or a name to a function, such as "f," or "$\sqrt[3]{}$," or "log." Let's choose f as a generic label for a function. The subset of D containing all the first elements x of the ordered pairs of f is called the **domain** of the function and is denoted $\mathcal{D}_f$. The domain $\mathcal{D}_f$ is an essential part of the definition of f. A very suggestive symbolism is $x \xrightarrow{f} y$, which means $\langle x, y\rangle \in f$ (remember, f is a set!). The arrow emphasizes the carrying or mapping of elements $x \in \mathcal{D}_f$ into unique elements $y \in R$. The element y is the **image** of x under f. Often, $R \subseteq \mathbb{R}$, and then we simply write $f(x) = y$ (that is, we evaluate f at x to yield y), but the mapping action is still inherent. In order to emphasize the domain of f, we may write $f: \mathcal{D}_f \to R$. As for terminology, R is sometimes called the **codomain** of function f.

The collection of all elements $y \in R$ that are images of domain elements under f is called the **range** of function f, denoted $\mathcal{R}_f$. By definition, then, $f: \mathcal{D}_f \to \mathcal{R}_f$. As will be seen in the examples below, it frequently happens that $\mathcal{R}_f$ is a proper subset of the codomain R. When both the domain and range of f are subsets of $\mathbb{R}$, and no detail is needed, we write $f: \mathbb{R} \to \mathbb{R}$.

Occasionally (as in the first example below), the set of ordered pairs forming a function is simply listed and labeled. Much more often, the image of each element x is derived by some unambiguous rule, formula, or algorithm (see Exercise 6.10). Unfortunately, the formula for a function is sometimes taken as the function itself. For instance, the formula $\cosh x = \frac{1}{2}(e^x + e^{-x})$ defines the hyperbolic cosine only when $\mathcal{D}_{\cosh} = \mathbb{R}$. A related, but distinct, function is formed when applications require that $\cosh x$ have the domain $[0, \infty)$. We reemphasize the importance of the domain by stating that a function is specified by its domain *and* its rule. Change *either* one, and the function is changed.

Note 2: The hyperbolic functions were carefully investigated as "functions" for the first time by Johann Lambert (1728–1777) in a long paper of 1761, *Mémoire sur quelques propriétés remarquables des quantités transcendantes circulaires et logarithmiques* (Memoir on some remarkable properties of circular and logarithmic transcendental quantities), written to the Berlin Academy of Sciences. But this fine analysis was to become a sideshow to the celebrated main event in the first section, where Lambert proved that π was an irrational number. Lambert wrote diversely, in subjects including astronomy, cartography, and logic, as well as mathematics. Much information about Lambert and the rise of hyperbolic functions may be found in Janet Heine Barnett's article "Enter, Stage Center: the Early Drama of the Hyperbolic Functions," in *Mathematics Magazine*, vol. 77, no. 1 (Feb. 2004), pp. 15–30.

What we have gained in precision, we may have lost in intuitive appeal, for the moment. We need to create several functions to see that old notions have simply been sharpened (or perhaps uprooted!).

Example 6.6 Using the Cartesian product of Example 6.1, we define the function $\hbar: \{2, 3, 5, 9\} \to \{\spadesuit, \clubsuit, \heartsuit, \diamondsuit\}$ as the set $\hbar = \{\langle 2, \diamondsuit\rangle, \langle 3, \diamondsuit\rangle, \langle 5, \spadesuit\rangle, \langle 9, \clubsuit\rangle\}$. It is clear that $\hbar \subset D \times R$ in Figure 6.5. In this case, $\mathcal{D}_\hbar = D$, and heartlessly, $\mathcal{R}_\hbar = \{\spadesuit, \clubsuit, \diamondsuit\}$. Furthermore, $\langle 3, \diamondsuit\rangle \in \hbar$ is also written as $3 \xrightarrow{\hbar} \diamondsuit$. All functions over $\mathcal{D}_\hbar$ have exactly four ordered pairs. ■

Example 6.7 Consider the function $f(x) = -\frac{3}{2}\cos(\pi x) + \frac{5}{2}$ over the domain $\mathcal{D}_f = [2, 6]$. A bit of thinking will yield the range $\mathcal{R}_f = [1, 4]$. Hence, f will not fit in the Cartesian product

♡	2♡	3♡	5♡	9♡
◇	2◇	3◇	5◇	9◇
♣	2♣	3♣	5♣	9♣
♠	2♠	3♠	5♠	9♠
	2	3	5	9

Figure 6.5. The function $\hbar$ is a subset of the Cartesian product $\{2,3,5,9\} \times \{♠,♣,◇,♡\}$.

of $D = [2, 6]$ and $R = [1, 4)$ from Example 6.2. We could modify f by forming the piecewise formula

$$f_2(x) = \begin{cases} 3 & \text{if } x = 3 \text{ or } x = 5 \\ -\frac{3}{2}\cos(\pi x) + \frac{5}{2} & \text{otherwise} \end{cases}$$

seen in Figure 6.6. Now, $\mathcal{R}_{f_2} = [1, 4) = R$ since the former maximum points $\langle 3, 4\rangle$ and $\langle 5, 4\rangle$ are shifted downward.

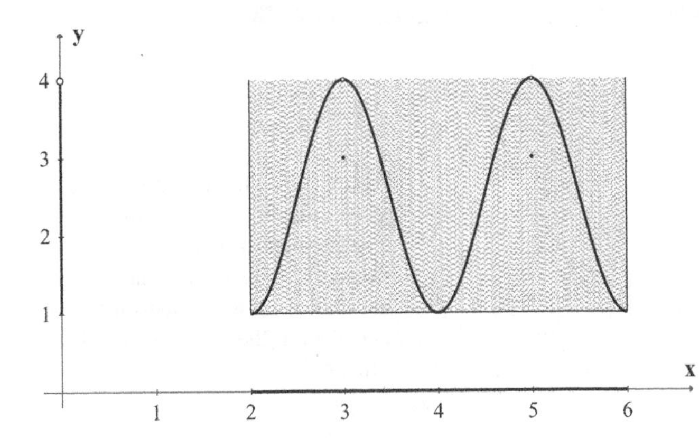

Figure 6.6. The function f_2 is a subset of the Cartesian product $[2, 6] \times [1, 4)$

Once again, any function is changed when its domain is changed. Thus, we can modify f in another way, creating an f_3 that fits in $[2, 6] \times [1, 4)$, by stipulating that $\mathcal{D}_{f_3} = [2, 3) \cup (3, 5) \cup (5, 6] \subset D$. Although the new formula, $f_3(x) = -\frac{3}{2}\cos(\pi x) + \frac{5}{2}$, is identical to that of f, the functions are different. Specifically, this function eliminates the two isolated points in Figure 6.6. Do Exercise 6.8 now. ■

Example 6.8 Define the function $u = \{\langle x, y\rangle : \mathcal{D}_u = [-\frac{3}{2}, -\frac{1}{2}] \cup [2, \sqrt{7})$ and $y = x^2 - 3\}$. This set can be expressed concisely as $\{\langle x,\ x^2 - 3\rangle\}$, and is often written as $u(x) = x^2 - 3$. However, these convenient notations disregard the underlying domain. We may be tempted to include the point $\langle 3, 6\rangle$, until we notice that $3 \neq \mathcal{D}_u$. We also realize that the function cannot exceed $y = (\sqrt{7})^2 - 3 = 4$. Hence, for instance, there is no solution to the equation $5 = x^2 - 3$ within $\mathcal{D}_u$. Also, note that the set R is not given; we are therefore free to create it, with the sole proviso that it not affect the domain (see Exercise 6.9). We could randomly select $R = [-10, 10]$, as in a graphing calculator's default window. Exercising a little more curiosity, we recognize a parabola with a minimum of $y = -3$, and not larger than $y = 4$. So, a more interesting codomain is $R = [-3, 4]$. The graph of u is seen in Figure 6.7, embedded within the Cartesian product $([-\frac{3}{2}, -\frac{1}{2}] \cup [2, \sqrt{7})) \times [-3, 4]$, along with the guiding parabola.

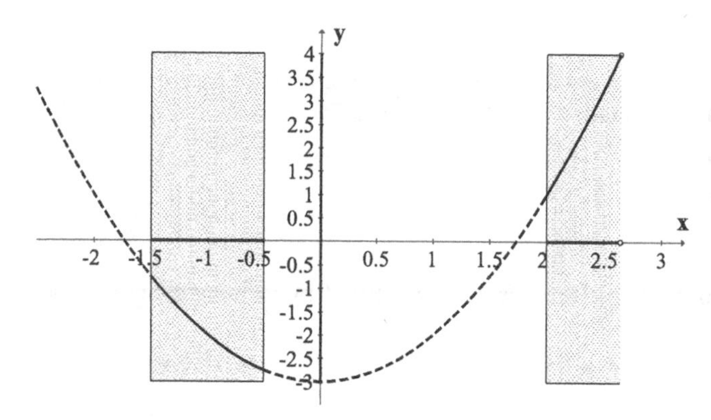

Figure 6.7. The function $u(x) = x^2 - 3$ of Example 6.8 as a subset of the Cartesian product $\mathcal{D}_u \times R$. The domain and a codomain R are shown in thick line on the X and Y axes. Note that $\langle\sqrt{7}, 4\rangle \notin u$ (upper right corner).

Looking more closely, observe that the range of u is not $R = [-3, 4]$. Relying on the graph in Figure 6.8, we see that $\mathcal{R}_u = [u(-\frac{1}{2}), u(-\frac{3}{2})] \cup [u(2), u(\sqrt{7})) = [-\frac{11}{4}, -\frac{3}{4}] \cup [1, 4)$. This final graph shows function u within the Cartesian product of its domain $\mathcal{D}_u$ and range $\mathcal{R}_u$. Certain sides of three of the rectangles must be excluded (explain why). ■

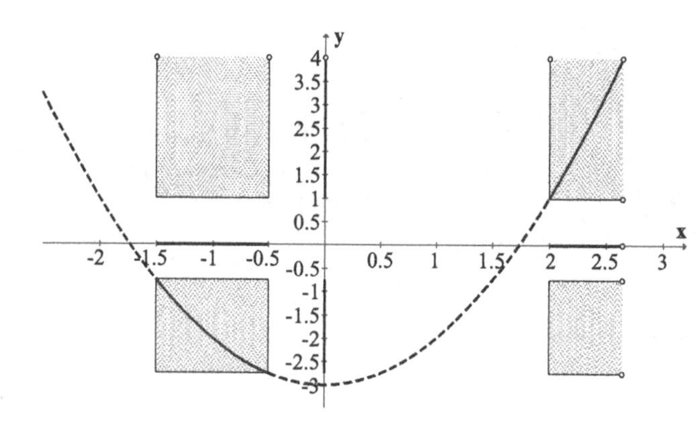

Figure 6.8. The function $u(x) = x^2 - 3$ again, but within the Cartesian product of its domain $\mathcal{D}_u$ and range $\mathcal{R}_u$ The domain and range are shown in thick line on the X and Y axes (they are not part of the function).

Example 6.9 The definition of a function is quite comprehensive. Let $t \in (-\infty, \infty)$, and define the parametric equations $x = x(t) = \int_0^t \cos u^2 du$ and $y = y(t) = \int_0^t \sin u^2 du$. These define a graceful function $C : \mathbb{R} \to \mathbb{R}^2$ variously called Euler's spiral, the cornu, and the **clothoid**. Thus, $C(t) = \langle \int_0^t \cos u^2 du,\ \int_0^t \sin u^2 du\rangle$. ■

6.3 Images and Preimages

We have said that y is the image of x under f when $x \xrightarrow{f} y$. This idea can be extended to the mapping of sets. Thus, the **image** of a set A is the set $f(A) \equiv \{f(x) \in \mathcal{R}_f : x \in A \cap \mathcal{D}_f\}$. We may write this as $A \xrightarrow{f} B$, where B is the image of A. The most encompassing case of this is when $\mathcal{D}_f \subseteq A$, giving us $f(A) = \mathcal{R}_f$, and the most restrictive case is when $A \cap \mathcal{D}_f = \varnothing$, for which we will define $f(A) \equiv \varnothing$. In the special case that A contains just one element $x \in \mathcal{D}_f$, then $f(\{x\}) = \{y\}$ may be written simply as $x \xrightarrow{f} y$. In the even more special case that $f : \mathcal{D}_f \to \mathbb{R}$, we may write $f(x) = y$, but remember that in this general context x may not be a real number. A Venn diagram of these ideas is in Figure 6.10.

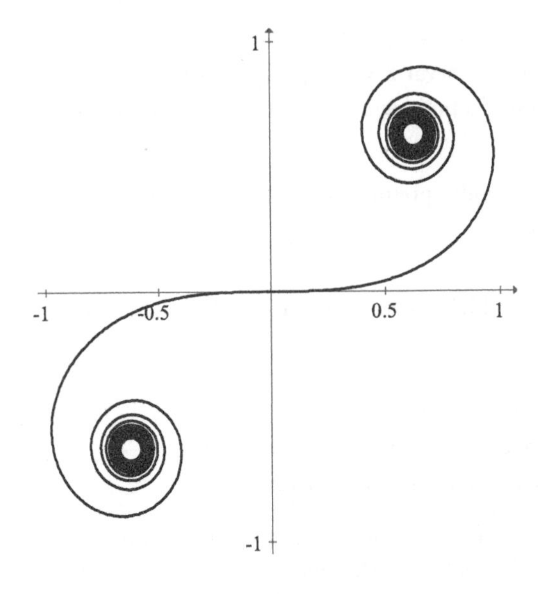

Figure 6.9. The beautiful clothoid of Euler. $\mathcal{D}_C = (-\infty, \infty)$, and what you see are points in the range of C.

In a reverse fashion, the **preimage** of a set E is the set $f^{-1}(E) \equiv \{x \in \mathcal{D}_f : f(x) \in E\}$. A synonym for preimage is **inverse image**. Paralleling what was said above about the domain, we have $f^{-1}(\mathcal{R}_f) = \mathcal{D}_f$. In the special case that E is a set with just one element $y \in \mathcal{R}_f$, its preimage $f^{-1}(\{y\})$ may not have one element (see Example 6.10).

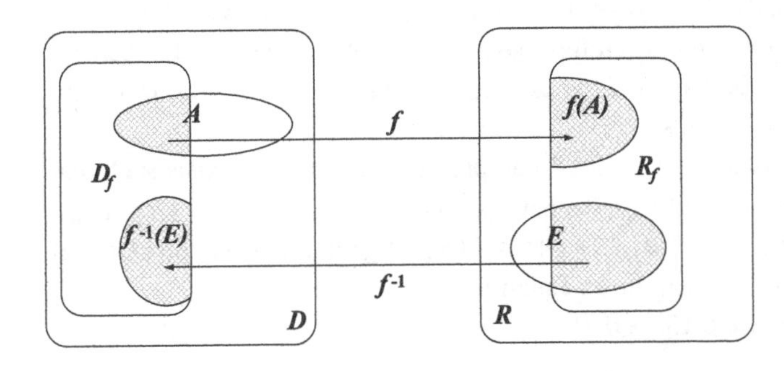

Figure 6.10. Venn diagram of a function f, its domain $\mathcal{D}_f$ and its range $\mathcal{R}_f$. The image of a set A and the preimage of a set E are also shown.

We aren't using, and have not yet defined, an inverse function in this chapter. Any confusion is caused by a historical misuse of the "-1" superscript to represent two similar ideas. To distinguish between the two uses of the superscript in this book, $f^{-1}(\{y\})$, *with braces*, will indicate the preimage of the singleton set $\{y\}$ (whose preimage will always be a set). Beginning in the next chapter, $f^{-1}(y)$, *without braces*, will indicate the use of an inverse function.

Example 6.10 Using the function $\hbar = \{\langle 2, \diamondsuit\rangle, \langle 3, \diamondsuit\rangle, \langle 5, \spadesuit\rangle, \langle 9, \clubsuit\rangle\}$ of Example 6.6, the image of $\{2, 5\}$ is $\hbar(\{2, 5\}) = \{\diamondsuit, \spadesuit\}$. Notice that the preimage of $\{\diamondsuit, \spadesuit\}$ does not return us to $\{2, 5\}$. Instead, we have $\hbar^{-1}(\{\diamondsuit, \spadesuit\}) = \{2, 3, 5\}$. This seems rather asymmetric, and yes, the cause is that the preimage of $\{\diamondsuit\}$ has two elements. ■

Example 6.11 Figure 6.8 from Example 6.8 shows $u([2, \sqrt{7})) = [1, 4)$, and in this case, $u^{-1}([1, 4)) = [2, \sqrt{7})$. Now, consider $X = [-1, \frac{5}{2}]$. This is not a subset of $\mathcal{D}_u$, so the function u can only "look at" numbers in $X \cap \mathcal{D}_u = [-1, -\frac{1}{2}] \cup [2, \frac{5}{2}]$. With this in mind, verify for yourself that $u(X) = u([-1, -\frac{1}{2}] \cup [2, \frac{5}{2}]) = [-\frac{11}{4}, -2] \cup [1, \frac{13}{4}]$. Hence, the image of X jumps over the gap $-2 < y < 1$. Consider the full interval $Y = [-\frac{11}{4}, \frac{13}{4}]$. It is interesting that $u^{-1}(Y) = [-1, -\frac{1}{2}] \cup [2, \frac{5}{2}]$, since the definition of a preimage prevents us from considering y values in the gap $-2 < y < 1$. ■

It may be a little surprising that functions handle preimages better than images of sets. Compare parts B of the following two theorems.

Theorem 6.1 Let f be a function in $D \times R$, and let A and B be subsets of D. Then these relationships hold:

A $f(A \cup B) = f(A) \cup f(B)$ (unions are preserved by functions).
B $f(A \cap B) \subseteq f(A) \cap f(B)$ (intersections may not be preserved by functions).
C If $A \subseteq B$, then $f(A) \subseteq f(B)$ (the subset relation is preserved by functions).

Proof. **A** We are going to prove that the image of a union is the union of the images. As usual, we prove two containments. First, suppose $y \in f(A \cup B)$. This means that there is at least one preimage $x \in A \cup B$ such that $x \xrightarrow{f} y$ (and $x \in \mathcal{D}_f$ to be sure). Now, if $x \in A$, then $f(x) \in f(A)$, and similarly, if $x \in B$, then $f(x) \in f(B)$. Hence, $f(x) \in f(A) \cup f(B)$. We have proved that $f(A \cup B) \subseteq f(A) \cup f(B)$.

Next, let $y \in f(A) \cup f(B)$, so that either $y \in f(A)$, yielding at least one preimage $x_1 \in A$, or else $y \in f(B)$, yielding at least one preimage $x_2 \in B$ (and it is possible for $x_1 = x_2$). Then $x_1, x_2 \in A \cup B$. Consequently, $\{x_1, x_2\} \xrightarrow{f} y \in f(A \cup B)$. We have proved the reverse containment $f(A) \cup f(B) \subseteq f(A \cup B)$. Therefore, statement 6.1A is true. (In this general setting, it may happen that $A \cup B$ is disjoint from $\mathcal{D}_f$. If so, then $f(A \cup B) = \varnothing$. But then, $f(A) = \varnothing = f(B)$, so equality still holds in 6.1A.)

B Suppose $y \in f(A \cap B)$. As in 6.1A, at least one preimage $x \in A \cap B$ exists such that $x \xrightarrow{f} y$. Since $x \in A$, $f(x) \in f(A)$, and simultaneously, since $x \in B$, $f(x) \in f(B)$. Thus, $f(x) \in f(A) \cap f(B)$. Therefore, $f(A \cap B) \subseteq f(A) \cap f(B)$. (Again, it may be the case that $A \cap B$ and $\mathcal{D}_f$ are disjoint. Then, 6.1B will be an equality.)

C This is left for you as Exercise 6.13. **QED**

Exercise 6.14 asks you to create an example for which equality in Theorem 6.1B is false.

Theorem 6.2 Let f be a function in $D \times R$. Also, let E and F be subsets of R. Then these relationships hold:

A $f^{-1}(E \cup F) = f^{-1}(E) \cup f^{-1}(F)$ (unions are preserved by preimages).
B $f^{-1}(E \cap F) = f^{-1}(E) \cap f^{-1}(F)$ (intersections are preserved by preimages).
C If $E \subseteq F$, then $f^{-1}(E) \subseteq f^{-1}(F)$ (the subset relation is preserved by preimages).

Proof. **A** First, suppose that $x \in f^{-1}(E \cup F)$. Then $f(x) \in E \cup F$, implying that either $f(x) \in E$ or $f(x) \in F$. These in turn imply that $x \in f^{-1}(E)$ or $x \in f^{-1}(F)$, which means that $x \in f^{-1}(E) \cup f^{-1}(F)$.

Now for the reverse containment. Let $x \in f^{-1}(E) \cup f^{-1}(F)$, so that either $x \in f^{-1}(E)$ or $x \in f^{-1}(F)$. This implies that either $f(x) \in E$ or $f(x) \in F$. Then $f(x) \in E \cup F$, which means that $x \in f^{-1}(E \cup F)$. The two containments are now proved, and 6.2A is true. (In this general setting, it may happen that $E \cup F$ is disjoint from $\mathcal{R}_f$. If so, then $f^{-1}(E \cup F) = \varnothing$. But then, $f^{-1}(E) = \varnothing = f^{-1}(F)$, so 6.2A is still an equality.)

B As usual, first suppose that $x \in f^{-1}(E \cap F)$, so that $f(x) \in E \cap F$. Finish the proof as Exercise 6.15.

C If $f(x) \in E$, then the preimage $x \in f^{-1}(E)$. But it also true that $f(x) \in F$. Hence $x \in f^{-1}(F)$, so $f^{-1}(E) \subseteq f^{-1}(F)$. (If $E \cap \mathcal{R}_f = \varnothing$, then $f^{-1}(E) = \varnothing \subseteq f^{-1}(F)$, and 6.2C is still true.) **QED**

Example 6.12 Consider the function

$$q(x) = \begin{cases} x+1 & \text{if } x \geq 0 \\ -2 & \text{if } x < 0 \end{cases},$$

graphed in Figure 6.11. We will use the graph and follow the images and preimages carefully to see how the two previous theorems apply.

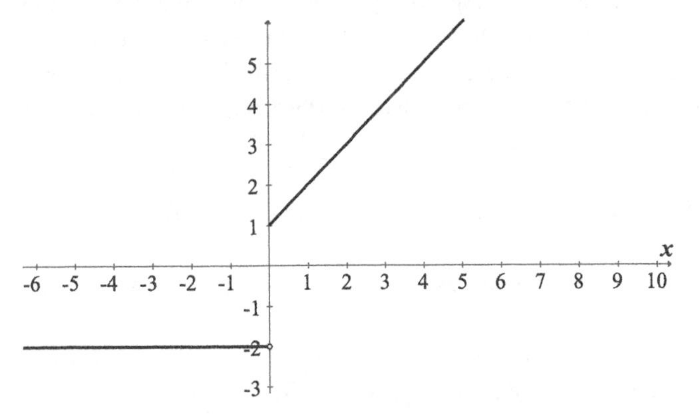

Figure 6.11. The piecewise-defined function q.

Take $A = [-1, 0]$ and $B = [0, 5]$ in the domain. Then $q(A \cup B) = q([-1, 5]) = \{-2\} \cup [1, 6]$, and $q(A \cap B) = q(\{0\}) = \{1\}$, which we usually write as $q(0) = 1$. Upon determining that $q(A) = \{-2, 1\}$ and $q(B) = \{-2\} \cup [1, 6]$, we find that $q(A) \cup q(B) = \{-2\} \cup [1, 6] = q(A \cup B)$, as 6.1A states. Also, $q(A) \cap q(B) = \{-2, 1\}$, so that in agreement with 6.1B, $q(A \cap B) \subset q(A) \cap q(B)$. As for 6.1C, let $C = [0, 1] \subset B$. Then $q(C) = [1, 2] \subset q(B)$, as asserted.

If we now consider $E = \{-2, 1, 2\}$ and $F = [1, 3)$ in the range, we find that $q^{-1}(E \cup F) = (-\infty, 2)$, and $q^{-1}(E \cap F) = \{0, 1\}$. Next, $q^{-1}(E) = (-\infty, 0) \cup \{0, 1\}$, and $q^{-1}(F) = [0, 2)$. Therefore, $q^{-1}(E) \cup q^{-1}(F) = (-\infty, 2) = q^{-1}(E \cup F)$ as 6.2A states. Also, $q^{-1}(E) \cap q^{-1}(F) = \{0, 1\} = q^{-1}(E \cap F)$, as 6.2B asserts. As for 6.2C, let $G = \{-2\} \subset E$. Then $q^{-1}(G) = (-\infty, 0) \subset q^{-1}(E)$, as asserted. ■

6.4 Monotonic Functions

In this subsection, only functions from $\mathbb{R}$ (or a subset of it) into $\mathbb{R}$ will be considered.

If, given any $v, x \in \mathcal{D}_f$ such that $v > x$, it is true that $f(v) \geq f(x)$, then we say that f is an **increasing** function. This definition lets us call the function in Figure 6.11 increasing, which seems a bit strange at first. But if, given any $v, x \in \mathcal{D}_f$ such that $v > x$, it is true that

$f(v) > f(x)$, then f is a **strictly increasing** function, and the function in Figure 6.11 is now excluded. Likewise, if, given any $v, x \in \mathcal{D}_f$ such that $v > x$, it is true that $f(v) \leq f(x)$, then we say that f is a **decreasing** function (constant functions included). But if, given any $v, x \in \mathcal{D}_f$ such that $v > x$, it is true that $f(v) < f(x)$, then f is a **strictly decreasing** function.

Combining the two cases, if f is either increasing or decreasing, it is named a **monotone** or **monotonic** function. And finally, if f is either strictly increasing or strictly decreasing, then we say that f is a **strictly monotone** function.

Clearly, all strictly monotone functions are monotone, but the converse is false. Also, we all know some functions that are increasing on some subsets of their domains, and decreasing on other subsets. And strange indeed, the ruler function introduced in Chapter 25 is an example of a function not monotonic over any interval in its domain!

6.5 Operations with Functions

Let functions f and g have domains $\mathcal{D}_f$ and $\mathcal{D}_g$, respectively. Unless stated otherwise, in this subsection we will assume that the domains are subsets of $\mathbb{R}$, as are the respective ranges. We define four new functions over a domain $\mathcal{D}_f \cap \mathcal{D}_g$ common to f and g, which we should assume is not empty, or else nothing happens (the function in (4) has a further, necessary, restriction). These are

(1) the **sum** $f+g : \mathcal{D}_{f+g} \to R$, found as $(f+g)(x) = f(x) + g(x)$, where $\mathcal{D}_{f+g} = \mathcal{D}_f \cap \mathcal{D}_g$,
(2) the **difference** $f-g : \mathcal{D}_{f+g} \to R$, found as $(f-g)(x) = f(x) - g(x)$, where $\mathcal{D}_{f-g} = \mathcal{D}_f \cap \mathcal{D}_g$,
(3) the **product** $fg : \mathcal{D}_{fg} \to R$, found as $(fg)(x) = f(x)g(x)$, where $\mathcal{D}_{fg} = \mathcal{D}_f \cap \mathcal{D}_g$, and
(4) the **quotient** $\frac{f}{g} : \mathcal{D}_{f/g} \to R$, found as $(\frac{f}{g})(x) = \frac{f(x)}{g(x)}$, where $\mathcal{D}_{f/g} = \mathcal{D}_f \cap \mathcal{D}_g \cap \{x : g(x) \neq 0\}$.

A fifth operation is **composition**, symbolized by $f \circ g$ and defined by $(f \circ g)(x) \equiv f(g(x))$. Hence, f is receiving values from the range of g, so that these range values must fall into the domain of f. To be precise, $\mathcal{D}_{f\circ g} = \{x \in \mathcal{D}_g : g(x) \in \mathcal{D}_f\} = g^{-1}(\mathcal{D}_f)$, the preimage under g of $\mathcal{D}_f$. Composition is often discussed by writing $x \xrightarrow{g} u \xrightarrow{f} y$, where $g(x) = u$. Here, the arrows nicely indicate the two requirements stated for $\mathcal{D}_{f\circ g}$, namely, $x \xrightarrow{g} u$ requires that $x \in \mathcal{D}_g$, and $u \xrightarrow{f} y$ requires that $g(x) = u \in \mathcal{D}_f$. Put together, the requirements mean that $g^{-1}(\mathcal{D}_f)$ can't be empty if $x \xrightarrow{f\circ g} y$ is going to exist. Figure 6.12 shows the relationships involved.

The reverse composition, $g \circ f$, is found as $(g \circ f)(x) = g(f(x))$, where now g is receiving values from the range of f. Symbolically, there is only an exchange of function names, so that $\mathcal{D}_{g\circ f} = \{x \in \mathcal{D}_f : f(x) \in \mathcal{D}_g\} = f^{-1}(\mathcal{D}_g)$, the preimage of $\mathcal{D}_g$ under f.

We consider a few examples to see how effective composition is at creating new functions.

Example 6.13 Let $g(x) = \sqrt{x} - 2$ where $\mathcal{D}_g = [0, \infty)$, and let $f(x) = \ln x$ with $\mathcal{D}_f = (0, \infty)$. Then $(f \circ g)(x) = \ln(\sqrt{x} - 2)$, for which we need a domain. That is, we need $x \in g^{-1}((0, \infty))$, which is the solution set to $0 < \sqrt{x} - 2 < \infty$. Therefore, $\mathcal{D}_{f\circ g} = (4, \infty)$. Of course, this can be easily checked by thinking about the restrictions on x for evaluating $\ln(\sqrt{x} - 2)$.

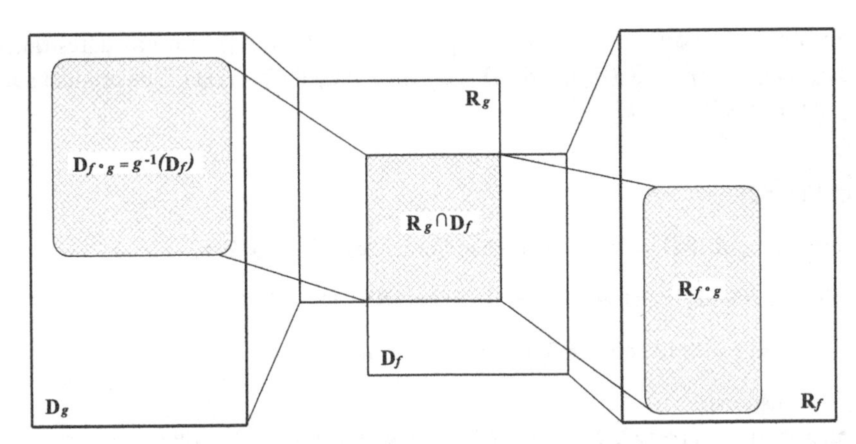

Figure 6.12. A Venn diagram showing the domain and range of $(f \circ g)(x)$, and showing that some range values of g must fall into the domain of f in order for $(f \circ g)(x)$ to exist.

The reverse composition, $(g \circ f)(x) = \sqrt{\ln x} - 2$ has its own domain, namely $\mathcal{D}_{g \circ f} = f^{-1}(\mathcal{D}_g)$. Since $\mathcal{D}_g = [0, \infty)$, we are looking for solutions to $0 \le \ln x < \infty$, which are $x \in [1, \infty)$. Thus, $\mathcal{D}_{g \circ f} = [1, \infty)$. We may check the restrictions on x directly by graphing $y = \sqrt{\ln x} - 2$. ■

Example 6.14 Let $g(n) = \frac{n\pi}{4}$, with $\mathcal{D}_g = \mathbb{N}$, and let $f(x) = \tan x$ with its standard domain. Then $(f \circ g)(n) = \tan \frac{n\pi}{4}$, where $n \in \mathcal{D}_{f \circ g} = g^{-1}(\mathcal{D}_f)$. Since $\mathcal{D}_f = \mathcal{D}_{\tan} = \{u : u \neq \pm\frac{\pi}{2}, \pm\frac{3\pi}{2}, \ldots\}$, it will be easier to first find the preimage under g of $\{\pm\frac{\pi}{2}, \pm\frac{3\pi}{2}, \ldots\}$. Hence, solve $\frac{n\pi}{4} = \frac{(2k-1)\pi}{2}$, $k \in \mathbb{N}$, which gives $n = 4k - 2$, $k \in \mathbb{N}$. Therefore, $\mathcal{D}_{f \circ g} = \{n \in \mathbb{N} : n \neq 4k - 2,\ k \in \mathbb{N}\} = \mathbb{N} \backslash \{2, 6, 10, \ldots\}$.

The function $f(g(n)) = \tan \frac{n\pi}{4}$ produces the interesting sequence $\langle 1, -1, 0, 1, 0, -1, 0, 1, -1, 0, \ldots \rangle$ as its range. ■

Example 6.15 Let $g(x) = \frac{1}{2}(x + |x|)$, and $f(x) = x^2 - 2x$, each with the natural domain $\mathbb{R}$. What is $\mathcal{D}_{g \circ f}$, and what is interesting about $g \circ f$? $\mathcal{D}_{g \circ f} = f^{-1}(\mathcal{D}_g) = f^{-1}(\mathbb{R}) = \mathbb{R}$, as expected since $(g \circ f)(x) = \frac{1}{2}((x^2 - 2x) + |x^2 - 2x|)$ has no restrictions for evaluation.

The graph of $g \circ f$ in Figure 6.13 shows that this composition truncates the negative section of the parabola. Thus, $g \circ f$ has the same effect, or is the same function, as $y = \max\{x^2 - 2x, 0\}$. In general, we call $y = \frac{1}{2}(f(x) + |f(x)|)$ the **nonnegative part** of a function f (see Exercise 6.18 for the nonpositive part).

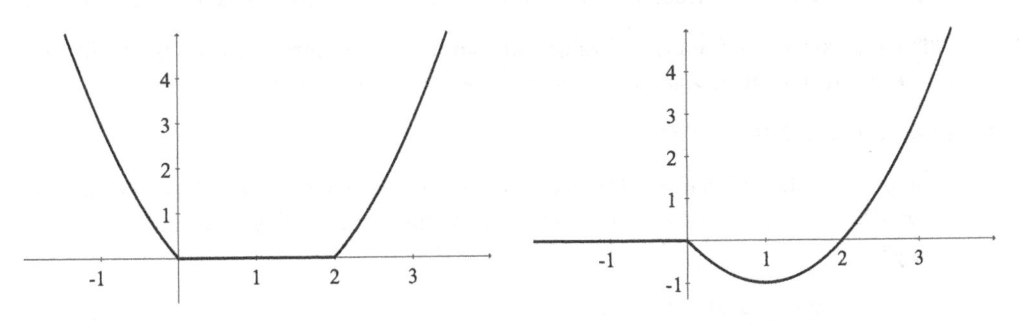

Figure 6.13. Figure 6.14.

The reverse composition $(f \circ g)(x) = \frac{1}{4}(x + |x|)^2 - (x + |x|)$ also has unrestricted domain, $\mathcal{D}_{f \circ g} = g^{-1}(\mathcal{D}_f) = \mathbb{R}$ (Figure 6.14). This time, only the section of the original parabola over $x \geq 0$ remains intact. ■

6.6 Exercises

1. Describe $D \times R$ if $D = R = \mathbb{Z}$ (in other words, describe $\mathbb{Z} \times \mathbb{Z}$).
2. Graph $A \times B$ and $B \times A$ when $A = [2, 5]$ and $B = \{1, 2, 3, 4\}$.
3. In comparison with Example 6.3, find and graph $R \times D$.
4. Prove or disprove:
   a. "distributive property", $A \times (B \cup C) = (A \times B) \cup (A \times C)$, (notice in Exercise 6.2 that $A \times B = (A \times \{1\}) \cup (A \times \{2\}) \cup (A \times \{3\}) \cup (A \times \{4\})$).
   b. $(A \times B) \times C = A \times (B \times C)$, that is, is the Cartesian product associative?
5. How many distinct functions exist in the Cartesian product of Example 6.1 (one of them is in Example 6.6)?
6. True or false: a. $\mathcal{D}_f = D$ b. $\mathcal{D}_f \times \mathcal{R}_f = D \times R$ c. $f = \mathcal{D}_f \times \mathcal{R}_f$
7. For the function $f(\theta) = \cos\theta$ with $\mathcal{D}_{\cos} = \mathbb{R}$, find $f^{-1}(\{-1\})$.
8. In Example 6.7, modify f into a function f_4 with $\mathcal{D}_{f_4} = [2, 6]$ and $\mathcal{R}_{f_4} = [1, 3]$. Hint: At what abscissas does $f(x) = 3$?
9. In Example 6.8, why is it incorrect to select $R = [-3, 2]$?
10. Instead of a formula, the following function is defined by an algorithm. Step 1: take any $x \geq 0$ in decimal form and multiply it by 100. Step 2: Add 0.5 to the result of step 1. Step 3: Truncate all decimal digits from the result of step 2. Step 4: Divide the result of step 3 by 100, and call this output the value of the function, y. Try inputting various values for x. What very useful job does this function do? What is its domain? What happens when we attempt to use it outside of its domain?
11. With regard to the clothoid in Figure 6.9:
    a. Explain how it can be a function, but fail the vertical line test.
    b. Determine the smallest rectangle with sides parallel to the X and Y axes that encloses it, to thousandths place accuracy. Hint: How do you find tangents to parametric curves?
12. Allow that you insert n numerical values on your income tax forms. The output of all your efforts is a real number, your tax. Does this constitute a function, $t : \mathbb{R}^n \to \mathbb{R}$?
13. Prove Theorem 6.1C.
14. Example 6.12 showed that equality may not hold in Theorem 6.1B. Find another function for which $f(A \cap B) \subset f(A) \cap f(B)$. Hint: First, think simply. What happens if $A \cap B$ is empty?
15. Finish proving the equality in 6.2B.

16. Theorems 6.1 and 6.2 continue to hold for generalized unions and intersections. Prove that for an arbitrary subscripting set S,
    a. $f(\bigcup_{\alpha \in S} A_\alpha) = \bigcup_{\alpha \in S} f(A_\alpha)$, and
    b. $f^{-1}(\bigcap_{\alpha \in S} E_\alpha) = \bigcap_{\alpha \in S} f^{-1}(E_\alpha)$, but
    c. $f(\bigcap_{\alpha \in S} A_\alpha) \subseteq \bigcap_{\alpha \in S} f(A_\alpha)$.

17. Let v be a function, let $A \subset \mathcal{D}_v$, and let $B = v^{-1}(v(A))$.
    a. Prove that $A \subseteq B$, so that passing to the range and then returning from it may increase one's assets!
    b. In contrast, let $G \subset \mathcal{R}_v$, and let $H = v(v^{-1}(G))$. Prove that $H = G$.
    c. Demonstrate the statements above with the function $v(\theta) = \cos\theta$, using $A = \{\pi\}$ and $G = \{-1, 1\}$.

18. We have seen the interesting effect of composing with $g(x) = \frac{1}{2}(x + |x|)$ in Example 6.15. Find a similar function h that when composed as $h \circ f$ will truncate the positive sections of f (that is, if $f(x) \geq 0$, then $(h \circ f)(x) = 0$). We call $y = (h \circ f)(x)$ the **nonpositive part** of f. The functions g and h are useful in analysis to isolate the nonnegative and nonpositive parts, respectively, of f. (These considerations are based on Exercise 3.8.)
    What well-known function related to f is given by $(g \circ f)(x) - (h \circ f)(x)$?

19. Is there a function L on $\mathbb{R} \times \mathbb{R}$ such that $L(x + y) = L(x) + L(y)$?

7

Inverse Functions

A necessary and sufficient condition that f be one-to-one is that the inverse image under f of each singleton in the range of f be a singleton in X.
Paul Halmos, *Naive Set Theory*, pg. 38.

We have seen in the previous chapter that functions pair elements $x \in \mathcal{D}_f$ with unique elements $y \in \mathcal{R}_f$. The reverse procedure is often not as neat in that a function can produce a preimage $f^{-1}(\{y\})$ that contains many elements. For instance, the simple function over $\mathbb{R}$ given by $f(x) = 7$ has all $\mathbb{R}$ as $f^{-1}(\{7\})$. It makes no sense at all to evaluate $f^{-1}(7)$ in this case.

7.1 Bijections

It becomes apparent that defining $f^{-1}(y)$ is only possible if the preimage $f^{-1}(\{y\})$ is a set with *exactly one* element x in the domain of f. In this general setting, when we consider all the ordered pairs $\langle x, y\rangle$ belonging to a function f, and find that $f^{-1}(\{y\}) = \{x\}$, a singleton preimage in every case, then we shall call f a **bijection**, or a **bijective function**. We may now simplify the notation and write $y \overset{f^{-1}}{\to} x$, which corresponds to $x \overset{f}{\to} y$. Furthermore, when f^{-1} has its range in $\mathbb{R}$, we won't use the arrow, and write $f^{-1}(y) = x$. A suggestive notation for bijections is $f : \mathcal{D}_f \leftrightarrow \mathcal{R}_f$.

Example 7.1 We consider once more the function from Example 6.6, $\hbar = \{\langle 2, \diamondsuit\rangle, \langle 3, \diamondsuit\rangle, \langle 5, \spadesuit\rangle, \langle 9, \clubsuit\rangle\}$, seen here in Figure 7.1. It is clearly not a bijection, since $\hbar^{-1}(\{\diamondsuit\}) = \{2, 3\}$. The same Cartesian product is displayed in Figure 7.2, but the newly selected ordered pairs now form a bijective function, which we call H. ■

$\heartsuit$	2$\heartsuit$	3$\heartsuit$	5$\heartsuit$	9$\heartsuit$
$\diamondsuit$	[2$\diamondsuit$]	[3$\diamondsuit$]	5$\diamondsuit$	9$\diamondsuit$
$\clubsuit$	2$\clubsuit$	3$\clubsuit$	5$\clubsuit$	[9$\clubsuit$]
$\spadesuit$	2$\spadesuit$	3$\spadesuit$	[5$\spadesuit$]	9$\spadesuit$
	2	3	5	9

Figure 7.1. The function $\hbar$ is a subset of the Cartesian product $\{2,3,5,9\} \times \{\spadesuit, \clubsuit, \diamondsuit, \heartsuit\}$. Here, $\hbar^{-1}(\{\diamondsuit\}) = \{2, 3\}$.

♡	2♡	3♡	5♡	**[9♡]**
◇	2◇	**[3◇]**	5◇	9◇
♣	2♣	3♣	**[5♣]**	9♣
♠	**[2♠]**	3♠	5♠	9♠
	2	3	5	9

Figure 7.2. The bijection H is a subset of the same Cartesian product. Now, $H^{-1}(\diamondsuit) = 3$.

Here is another way to highlight the difference between functions $\hbar$ and H in Example 7.1:

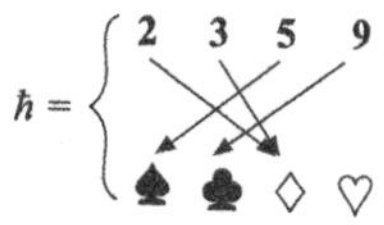

$$\hbar = \left\{ \begin{array}{cccc} 2 & 3 & 5 & 9 \\ \spadesuit & \clubsuit & \diamondsuit & \heartsuit \end{array} \right.$$

Function $\hbar$ from Example 6.6, which is not bijective.

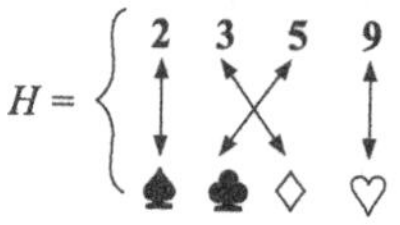

$$H = \left\{ \begin{array}{cccc} 2 & 3 & 5 & 9 \\ \updownarrow & \times & \times & \updownarrow \\ \spadesuit & \clubsuit & \diamondsuit & \heartsuit \end{array} \right.$$

Bijection H.

Figure 7.3.

Example 7.2 Look at the keypad of a cell phone. It defines a function $f : \{\textit{letters of the alphabet}\} \to \mathbb{N}$ that isn't bijective. For instance, $f(m) = 6$, but $f^{-1}(\{6\}) = \{m, n, o\}$ is not a singleton preimage. We can't use the notation $f^{-1}(6)$ since it is reserved for an inverse *function*. The effect of not being bijective is that several mnemonic "phone names" map to the same telephone number. However, it is possible to impose a bijection by defining a similar function F as follows:

$$F = \left\{ \begin{array}{ccccccccl} \{a,b,c\} & \{d,e,f\} & \{g,h,i\} & \{j,k,l\} & \{m,n,o\} & \{p,q,r,s\} & \{t,u,v\} & \{w,x,y,z\} & \} = \mathcal{D}_F \\ \updownarrow & \updownarrow & \updownarrow & \updownarrow & \updownarrow & \updownarrow & \updownarrow & \updownarrow & \\ 2 & 3 & 4 & 5 & 6 & 7 & 8 & 9 & \} = \mathcal{R}_F \end{array} \right.$$

Of course, $\mathcal{D}_F$ now consists of eight *sets* of letters, not the letters themselves. Hence, $F(\{m, n, o\}) = 6$ (an equality of two real numbers) corresponds to $F^{-1}(\{6\}) = \{m, n, o\}$ (an equality of two sets, also written as $\{6\} \overset{F^{-1}}{\to} \{m, n, o\}$).

All of what has been said of course must apply to all the common functions of calculus, but its generality may make things look strange. For instance, we can show that $u : \mathbb{R} \to \mathbb{R}$ given by $u(x) = \frac{x}{x+1}$ is a bijection by solving $y = \frac{x}{x+1}$ for x, thereby discovering the preimage relation $u^{-1}(\{y\}) = \{\frac{y}{1-y}\} = \{x\}$. This means that the preimage of every $\{y\}$ is a singleton $\{x\}$. Therefore, u must be a bijective function $u : \mathbb{R} \leftrightarrow \mathbb{R}$, with $\mathcal{D}_u = \{x \in \mathbb{R} : x \neq -1\}$ and $\mathcal{R}_u = \{y \in \mathbb{R} : y \neq 1\}$. In this case, we simply write $u^{-1}(y) = \frac{y}{1-y}$. (Indeed, u^{-1} will soon be called the inverse function of u.)

As can be seen from the preceding examples, a bijection is a two-directional mapping of every member of its domain and range. We emphasize this special type of mapping by also calling a bijection a **one-to-one function** or a **one-to-one correspondence** between its domain and range.

It is possible (and common) for a function to be one-to-one on only a proper subset A of its domain. In this case, a bijection automatically occurs throughout A. Such is the case in Example 7.5.

Note 1: Having defined a function as a subset of a Cartesian product $D \times R$, we said that f maps elements into R, its codomain, implying that R may have members that aren't images under f. A function maps *onto* a set S whenever every element of S is the image of some domain element. Thus, f may or may not map onto its codomain R, but, by definition, it maps onto its range $\mathcal{R}_f$. It is in this sense that a bijection is said to be simultaneously a one-to-one and onto function.

As for terminology, a synonym for an onto function is a *surjection*, and a one-to-one function is also named an *injection*. Hence, a surjective injection is a bijection, and a bit excessive!

7.2 Inverse Functions

Given a bijection $f : \mathcal{D}_f \leftrightarrow \mathcal{R}_f$, let us replace every $\langle x, y\rangle \in f$ with the reversed ordered pair $\langle y, x\rangle$. Because a bijection is a one-to-one mapping, the collection of ordered pairs $\langle y, x\rangle$ defines a new function named the **inverse function** of f, and denoted f^{-1}. Symbolically, given any bijection f, $\langle y, x\rangle \in f^{-1}$ if and only if $\langle x, y\rangle \in f$. In Exercise 7.2a, you will show that the converse is true: if a function has an inverse, then the function is bijective.

Let's consider these ideas a bit more. Given that f is one-to-one, f^{-1} is also one-to-one, but with a fundamental relationship to f. It is that whenever $x \overset{f}{\to} y$, then immediately, $y \overset{f^{-1}}{\to} x$, and we call x the inverse of y (under f). Hence, $x \overset{f}{\to} y \overset{f^{-1}}{\to} x$, or as a composition, $x \overset{f^{-1}\circ f}{\to} x$. When dealing with functions in real numbers, this looks like $f^{-1}(f(x)) = x$ (as is picturesquely said, "f^{-1} returns x back"). In reverse, we also have that $y \overset{f\circ f^{-1}}{\to} y$ (that is, $f(f^{-1}(y)) = y$). Furthermore, since $\langle y, x\rangle \in f^{-1}$ if and only if $\langle x, y\rangle \in f$, the domain of f is precisely the range of f^{-1}, that is, $\mathcal{D}_f = \mathcal{R}_{f^{-1}}$, and vice versa, $\mathcal{D}_{f^{-1}} = \mathcal{R}_f$. Finally, imagine starting with the function that we are now calling f^{-1}, and then label *its* inverse as $(f^{-1})^{-1}$. This "inverse-inverse" can be none other than the original function f.

To summarize the relationships just discussed,

> A function is bijective if and only if its inverse function exists, and then
> $$\mathcal{D}_f = \mathcal{R}_{f^{-1}},\ \mathcal{D}_{f^{-1}} = \mathcal{R}_f,\ x \overset{f^{-1}\circ f}{\to} x, \text{ and } y \overset{f\circ f^{-1}}{\to} y.$$

Note 2: To make a definition parallel to that of a function in Chapter 6, we would say that a function f on $D\times R$ has a (unique) inverse function f^{-1} if whenever $\langle x, y\rangle$ and $\langle w, y\rangle$ belong to f, then $x = w$.

These considerations extend to the inverse image of a set, now called the set's inverse under f. As first encountered in the last chapter, if $E \subseteq \mathcal{R}_f$, then $f^{-1}(E)$ is exactly the set $A \subseteq \mathcal{D}_f$ such that $f(A) = E$. What's new here is that for every element $e \in E \subseteq \mathcal{R}_f$, there exists a *unique* element $f^{-1}(e) \in A$ in the domain of f, as suggested by the Venn diagram of Figure 7.4.

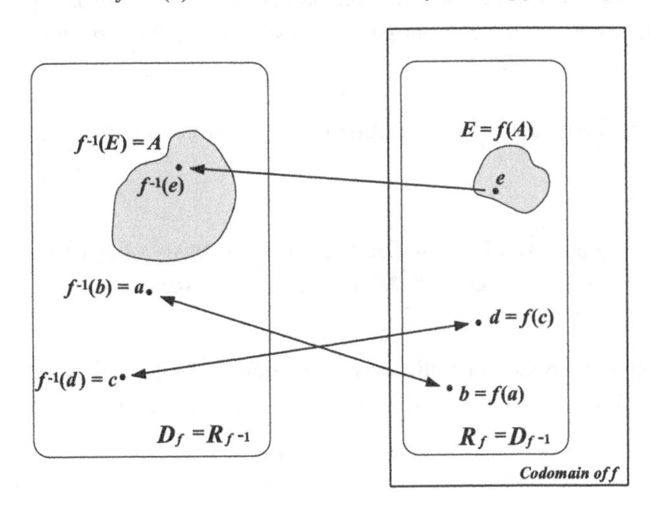

Figure 7.4. A bijection f showing the one-to-one mapping of elements and sets, and the inverse function's behavior.

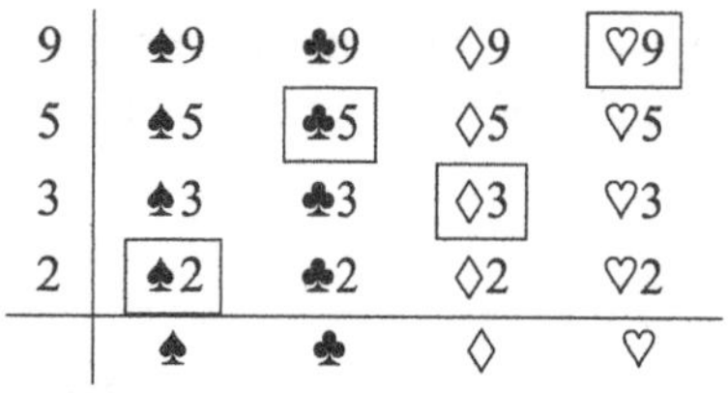

9	♠9	♣9	◇9	[♡9]
5	♠5	[♣5]	◇5	♡5
3	♠3	♣3	[◇3]	♡3
2	[♠2]	♣2	◇2	♡2
	♠	♣	◇	♡

Figure 7.5. The inverse function H^{-1} corresponding to bijection H of Example 7.1, with domain $\mathcal{D}_{H^{-1}} = \mathcal{R}_H$ and $\mathcal{R}_{H^{-1}} = \mathcal{D}_H$. Once again, $H^{-1}(\Diamond) = 3$, for instance.

Example 7.3 Since H in Example 7.1 is a bijection, simply reversing the ordered pairs yields the inverse function $H^{-1} = \{\langle \spadesuit, 2\rangle, \langle \clubsuit, 5\rangle, \langle \Diamond, 3\rangle, \langle \heartsuit, 9\rangle\}$, seen below. Compare this with Figure 7.2. ■

Example 7.4 Let q be the bijection $q(x) = \frac{3x+1}{x-7}$ over its natural domain $\mathcal{D}_q = \mathbb{R}\backslash\{7\}$, with $\mathcal{R}_q = \mathbb{R}\backslash\{3\}$. Writing $y = \frac{3x+1}{x-7}$, we find the inverse function to be $x = q^{-1}(y) = \frac{7y+1}{y-3}$. As expected, $\mathcal{D}_{q^{-1}} = \mathbb{R}\backslash\{3\}$ and $\mathcal{R}_{q^{-1}} = \mathbb{R}\backslash\{7\}$. If we now solve $x = \frac{7y+1}{y-3}$ for y, we will return to the expression $q\,(x)$. In symbolic form, $(q^{-1}(y))^{-1}(x) = \frac{3x+1}{x-7}$, with the same domain and range as the original function q.

Demonstrating Figure 7.4, let $E = [4, 6] \subset \mathcal{R}_q$, and select $e = 5$. Then there is a unique $q^{-1}(5) = 18 \in q^{-1}(E)$. Test yourself by using either the graph of q or of q^{-1} to determine $q^{-1}(E)$.

It should be understood that when graphing an inverse function such as q^{-1}, the equation $x = \frac{7y+1}{y-3}$ gives the full set of ordered pairs $\langle y, x\rangle \in q^{-1}$. Therefore, the y in $\langle y, x\rangle$ is the abscissa for q^{-1}, along the X axis, and x is the corresponding ordinate, along the Y axis. The common—but unnecessary—step of exchanging x and y so as to get $y = \frac{7x+1}{x-3}$ does let us choose x values along the X axis, as we are so fond of doing. The disadvantage of this step is that x and y no longer consistently denote domain and range values of q, respectively. They suddenly switch roles amidst scenes, causing confusion. ■

Example 7.5 Let $y = \sin\theta$. Since this is not a bijection over its natural domain $\mathbb{R}$, no inverse function can exist. But the "principal-valued sine" is the related function $y = \operatorname{Sin}\theta$ equaling $\sin x$ over $\mathcal{D}_{\text{Sin}} = \left[-\frac{\pi}{2}, \frac{\pi}{2}\right]$, and with $\mathcal{R}_{\text{Sin}} = [-1, 1]$, and this *is* a bijection. We therefore define the inverse sine function as $\theta = \operatorname{Sin}^{-1} y$ over $\mathcal{D}_{\text{Sin}^{-1}} = [-1, 1]$, with range $\mathcal{R}_{\text{Sin}^{-1}} = \left[-\frac{\pi}{2}, \frac{\pi}{2}\right]$. The bijection "$\text{Sin}^{-1}$" is usually written without the capital letter, as in $\theta = \sin^{-1} y$, provoking confusion since $y = \sin x$ isn't one-to-one. The common name "arcsin", as in $\theta = \arcsin y$, *does* denote the same function as $\theta = \operatorname{Sin}^{-1} y$. ■

If some function g is proposed as the inverse of a given function f, the following theorem allows us to test the claim.

Theorem 7.1 The function $f : \mathcal{D}_f \to \mathcal{R}_f$ is a bijection if and only if there exists a function $g : \mathcal{R}_f \to \mathcal{D}_f$ such that $g(\{f(x_0)\}) = \{x_0\}$ for each $x_0 \in \mathcal{D}_f$ (g is then recognized as f^{-1}).

Proof of necessity. First, suppose that f is a bijection, with inverse function g. By the definition of the inverse, for each $x_0 \in \mathcal{D}_f$, *the* function $g : \mathcal{R}_f \to \mathcal{D}_f$ yields $x_0 \overset{g\circ f}{\to} x_0$, or in set notation, $g(\{f(x_0)\}) = \{x_0\}$.

Proof of sufficiency. We are given that $g : \mathcal{R}_f \to \mathcal{D}_f$ exists, with $g(\{f(x_0)\}) = \{x_0\}$ as the singleton preimage of $\{f(x_0)\}$ for every $x_0 \in \mathcal{D}_f$. Let $x_0 \xrightarrow{f} y_0$. Then $g(\{f(x_0)\}) = \{x_0\}$ becomes $y_0 \xrightarrow{g} x_0$. We have now established that whenever $\langle x_0, y_0 \rangle \in f$, then $\langle y_0, x_0 \rangle \in g$. This implies that g is, in fact, the inverse of f. Hence, f is a bijection. **QED**

Note 3: Suppose we want to solve $f(x) = y$ for x. We often "take the inverse of both sides," $f^{-1}(f(x)) = f^{-1}(y)$, and arrive at $x = f^{-1}(y)$, because "f^{-1} cancels f." Alas, this convenience is justified only for bijections. For example, to solve $e^x = y$ for x, we can apply the inverse operation to both sides, giving $x = \ln y$, the correct solution regardless of the value of y in the range of the exponential function. In contrast, if we are given $\sin x = y$, applying the inverse function blindly gives us $x = \arcsin y$, which may be false—to the frustration of novices to trigonometry! In terms of Theorem 7.1, there is no function g such that $g(\{\sin x_0\}) = \{x_0\}$, a singleton preimage. See Example 7.5 for the remedy.

Theorem 7.2 If a function $f : \mathbb{R} \to \mathbb{R}$ is strictly monotone over its domain, then f^{-1} exists. Furthermore, both f and f^{-1} have the same type of monotonicity.

Proof. First, to prove that f^{-1} exists, we show that f is a bijection. Assume, on the contrary, that f is not a bijection. Then there exist $x_1, x_2 \in \mathcal{D}_f$ with $x_1 \neq x_2$ such that $f(x_1) = f(x_2)$. This deduction contradicts the strict monotonicity of f, so we must conclude that f is a bijection after all.

Now, suppose that f is strictly increasing. Let $x_1, x_2 \in \mathcal{D}_f$ with $x_1 < x_2$, and therefore $f(x_1) < f(x_2)$. Letting $y_1 = f(x_1)$ and $y_2 = f(x_2)$, we observe that $f^{-1}(y_1) = x_1 < x_2 = f^{-1}(y_2)$. Hence, whenever $y_1 < y_2$ in $\mathcal{D}_{f^{-1}}$, it follows that $f^{-1}(y_1) < f^{-1}(y_2)$. This means that f^{-1} is strictly increasing over its domain. As for the other case—that f is strictly decreasing–parallel reasoning tells us that f^{-1} will also be strictly decreasing. **QED**

Notice that Theorem 7.2 even applies to bijective functions that have gaps in their domains. For instance, defining $y = f(x) = \sqrt[3]{x}$ over $\mathbb{Q}$ leaves infinitely many gaps in the function, one at every irrational abscissa. Nevertheless, $f^{-1}(y) = y^3$ exists, and it is strictly increasing, just as f is (see also Exercise 7.4). The function

$$y = \begin{cases} x & \text{if } x \in [0, 1) \\ 3 - x & \text{if } x \in [1, 2] \end{cases}$$

proves that the converse of this theorem is false. But we may still argue that the two pieces of this function are still strictly monotone. In Exercise 25.4 we'll see a bijection that has absolutely no monotone intervals.

Example 7.6 Some bijective functions have inverses that cannot be explicitly described using the collection of algebraic, trigonometric, and exponential functions, and their inverses. Let $y = f(w) = we^w$ be defined over $\mathcal{D}_f = [-1, \infty)$. You can easily show that f is strictly increasing over this restricted domain, so that by Theorem 7.2, the inverse function $w = W(y) = f^{-1}(y)$ exists. But there is no elementary expression for it (just try solving $y = we^w$ for w). Still, W has many applications, and is called the **Lambert W function**. Seen in Figure 7.6, Lambert W has domain $[-e^{-1}, \infty)$, which you should verify. ■

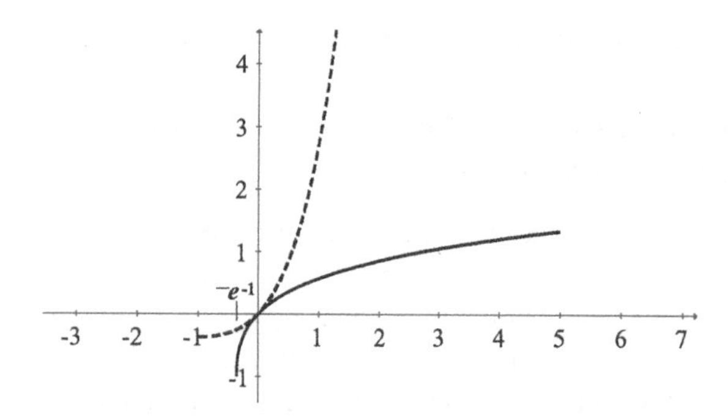

Figure 7.6. The Lambert W function $y = W(x)$ (the solid line), with domain $[-e^{-1}, \infty)$. Its inverse is the explicit function $y = xe^x$ (the broken line), with domain $[-1, \infty)$.

Note 4: The problem of finding the inverse of $y = we^w$ above is hardly different than that of finding the inverse of $y = \sin\theta$ in Example 7.5. In both cases, strictly monotone functions are at hand, but we can't use algebraic methods to explicitly find the inverses, so we simply invent names for the inverses that we know exist. In the latter case, we are comfortable with $f^{-1}(y) = \arcsin y$ only because we know many of its properties and applications.

Here is an application of the Lambert W function. The function $y = x^x$ has not been solvable for x, until now! First of all, this function is not a bijection (graph it over positive x), but verify that it becomes one if we restrict it to $x \in [e^{-1}, \infty)$. Using the fact that $W^{-1}(y) = ye^y$, you can show that the inverse function of the restricted $y = f(x) = x^x$ is $x = f^{-1}(y) = \frac{\ln y}{W(\ln y)}$, where the point $\langle 1, 1\rangle$ must be included. Its domain is $y \geq e^{-1/e}$ (as expected).

A detailed analytical application of the W function appears in "Projectile Motion with Resistance and the Lambert W Function," by Edward W. Packel and David S. Yuen, in *The College Mathematics Journal*, vol. 35, no. 5 (Nov. 2004), pp. 337–350.

Our final theorem tells us how to handle the inverse of a composition of two functions, that is, $(f \circ g)^{-1}$.

Theorem 7.3 Let $u = g(x)$ and $y = f(u)$ be bijections such that $y = (f \circ g)(x) = f(g(x))$ exists. Then the inverse function $(f \circ g)^{-1}$ exists, with $\mathcal{D}_{(f\circ g)^{-1}} = \mathcal{R}_{f\circ g}$, and $\mathcal{R}_{(f\circ g)^{-1}} = \mathcal{D}_{f\circ g}$. Furthermore, $x = (f \circ g)^{-1}(y) = (g^{-1} \circ f^{-1})(y)$, that is, $x = g^{-1}(f^{-1}(y))$. In words, the inverse of a composition is the composition of the inverses, in reverse order.

Proof. Referring to Figure 7.12 in Chapter 6, if both g and f are one-to-one functions, then the preimage of a singleton set $\{y\} \subseteq \mathcal{R}_{f\circ g}$ is guaranteed to be a singleton set $\{x\} \subseteq \mathcal{D}_{f\circ g}$ (and consider what may happen if either function is not a bijection). Thus, $(f \circ g)^{-1}$ exists, and in common with all inverse functions, $\mathcal{D}_{(f\circ g)^{-1}} = \mathcal{R}_{f\circ g}$ and $\mathcal{R}_{(f\circ g)^{-1}} = \mathcal{D}_{f\circ g}$.

Since both g and f are one-to-one functions, $f(g(\{x_0\})) = \{y_0\}$ for each $x_0 \in \mathcal{D}_g$. Since f has an inverse (and $\mathcal{R}_{f\circ g} \subseteq \mathcal{R}_f$), we have $g(\{x_0\}) = f^{-1}(\{y_0\})$. From this, since g has an inverse (and $\mathcal{R}_g \cap \mathcal{D}_f \subseteq \mathcal{R}_g$), we have $x_0 = g^{-1}(f^{-1}(\{y_0\}))$. **QED**

Example 7.7 Let $u = g(x) = x^2$ over $\mathcal{D}_g = [0, e]$ and $y = f(u) = 3 + \ln u$. Since both functions are bijections, $y = (f \circ g)(x) = 3 + \ln(x^2)$ exists over $\mathcal{D}_{f\circ g} = (0, e]$. By Theorem 7.3, therefore, $(f \circ g)^{-1}(y) = g^{-1}(f^{-1}(y)) = \sqrt{e^{y-3}}$, with domain $\mathcal{D}_{(f\circ g)^{-1}} = \mathcal{R}_{f\circ g} = (-\infty, 5]$. ■

7.3 Exercises

1. Let $P(x) = x^2 - 3x$ over all real numbers. Graph it and see that it is not a bijection. Many suitable restrictions to the domain will yield a function that has an inverse. Find one for yourself.

2. a. We have already seen that if a function is bijective, then it has an inverse function. Prove the converse statement, thus making the set of bijective functions equal to the set of functions with inverses.
   b. The sufficient condition for f to be a function is that $f(\{x\}) = \{y\}$ for every $x \in \mathcal{D}_f$. Is this a sufficient condition for the inverse function to exist? If so, prove it. If not, find a counterexample.

3. a. The function $u^{-1}(x) = \frac{x}{1-x}$ was proposed earlier as the inverse function for $u(x) = \frac{x}{x+1}$ (we aren't believing the suggestive notation yet). Check whether $u^{-1}(u(x)) = x$ and $u(u^{-1}(x)) = x$. Is that sufficient to claim that they are inverses of each other?
   b. The function $g(x) = \ln\left(\frac{1+x}{1-x}\right)$ is proposed as the inverse for $f(x) = \frac{e^x-1}{e^x+1}$. Check whether $g(f(x)) = x$ and $f(g(x)) = x$. Is that sufficient to claim that they are inverses of each other? (Later in this book, we will define the natural logarithm function as the inverse of the strictly increasing function $y = e^x$.)

2. Find h^{-1} for the bijection

$$h(x) = \begin{cases} 2x+1 & \text{if } x > 3 \\ \ln x & \text{if } 0 < x \leq 3. \end{cases}$$

   Graph h and h^{-1}, noting their domains and ranges. Note that the two functions have the same monotonicity, per Theorem 7.2.

3. Graph $V(x) = \frac{1}{x} + \frac{1}{x-1}$ and see that it isn't a bijection over its natural domain. However, restrict its domain to $(0, 1)$, and call the new function v. Find a function f such that $f(v(x)) = x$ over $x \in (0, 1)$. Then f is, in fact, the inverse of v, per Theorem 7.1, and we relabel it v^{-1}. In particular, check that your inverse gives $v^{-1}(v(\frac{1}{2})) = \frac{1}{2}$.

4. In Exercise 7.5, you worked with the bijection $v : (0, 1) \leftrightarrow \mathbb{R}$. Find a trigonometric function that is a bijection between $(0, 1)$ and $\mathbb{R}$. Hint: Which trigonometric function has infinite range? Graphing always helps.

5. In the spirit of Lambert's W function, look for a bijection of your own with no elementary expression for its inverse. Hint: Think about an increasing function (or a decreasing one) defined by an integral or a series in calculus.

6. Consider the following expression: $D(f(x)) \equiv \frac{df(x)}{dx}$. D is called the "differential operator," much like Δ is taken to be the "difference operator" in $\Delta y = y_2 - y_1$. Can D be considered a function, and if so, what is its domain and range? If D is a function, does it have an inverse? When we are given $y = e^{3x}$, and take the logarithm of both sides to get $\ln y = 3x$, we may think of the logarithm function as an operator. Are all functions examples of operators, and is the converse true?

7. In reliability theory, one finds functions $R_j : [0, 1] \to [0, 1]$ such as $R_1(x) = 1 - (1 - x)^6$ and $R_2(x) = x^3$. R_1 gives the reliability of a system with six components in parallel, while R_2 gives the reliability of a system with three components in series. Under certain conditions, the reliability of a single component is given by $R(t) = e^{-\lambda t}$, where the domain variable $t \in [0, \infty)$ measures time and λ is a positive constant called the failure rate.
   a. As time progresses, the reliability of the parallel system is given by $y = (R_1 \circ R)(t)$. Establish that R_1 and R are bijections, and find $(R_1 \circ R)^{-1}(y)$ by using Theorem 7.3 (remember, to fully specify a function, its domain must be given, so you must find $\mathcal{D}_{(R_1 \circ R)^{-1}}$). Hint: Consider the derivatives of R_1 and R.
   b. If each component of the parallel system is actually a subsystem of three series components, then the reliability of the full system (18 components in all) is $y = (R_1 \circ R_2 \circ R)(t)$. Since R_2 is clearly a bijection, find $(R_1 \circ R_2 \circ R)^{-1}(y)$ by using Theorem 7.3.

8

Some Subsets of the Real Numbers

We have almost finished proving that $\mathbb{N} \subset \mathbb{Z} \subset \mathbb{Q} \subset \mathbb{R}$, and we know that the set $\mathbb{R} \backslash \mathbb{Q}$ is populated by the irrational numbers. However, there are other very interesting subsets of $\mathbb{R}$ that help us appreciate how rich the set of real numbers actually is.

8.1 Terminating Decimals

Consider the set $\mathbb{T}$ of rational numbers that can be expressed as terminating decimals. For example, $\frac{333}{80} = 4.1625 \in \mathbb{T}$, and we realize that $\mathbb{Z} \subset \mathbb{T} \subset \mathbb{Q}$. It is a good exercise (see Exercise 8.1) to discover exactly what type of integer b will guarantee that the rational number $\frac{a}{b} \in \mathbb{T}$.

The set $\mathbb{T}$ is more than just theoretically interesting. Because computers have only a finite amount of memory in which to store numbers, it stands to reason that they do arithmetic only with elements of $\mathbb{T}$. Furthermore, computers can only work with a rather small, *finite* subset of $\mathbb{T}$, and we rely on the skill of programmers to write algorithms that give reliable output under such tight constraints. (Numerical analysis is a branch of mathematics that studies algorithms from this point of view.)

Thus, it makes sense to ask whether $\mathbb{T}$ is closed under the four arithmetic operations. Addition, subtraction, and multiplication always return terminating decimals when their operands are themselves terminating. So, those operations are closed in $\mathbb{T}$. Ideally, we hope that $\mathbb{T}$ is a field, just as $\mathbb{Q}$ is. Division, however, presents a problem. Suppose that $\frac{a}{b}, \frac{c}{d} \in \mathbb{T}$. It isn't difficult to find cases where $\frac{a}{b} \div \frac{c}{d} \notin \mathbb{T}$. See Exercise 8.2.

8.2 Quadratic Extensions

Going in another direction, define a **quadratic extension** of $\mathbb{Q}$ to be the set of numbers of the form $r + s\sqrt{p}$, where r and s are in $\mathbb{Q}$, and p is a non-square natural number. Denote the set of such numbers by $\mathbb{Q}(\sqrt{p})$. In particular, we shall investigate $\mathbb{Q}(\sqrt{2})$. Then, for example, $-9 - \frac{2}{3}\sqrt{2} \in \mathbb{Q}(\sqrt{2})$ and $\frac{\sqrt{2}}{2} = 0 + \frac{1}{2}\sqrt{2} \in \mathbb{Q}(\sqrt{2})$. But for instance, $\sqrt{5} \notin \mathbb{Q}(\sqrt{2})$, because if $\sqrt{5} = r + s\sqrt{2}$, then $5 = (r + s\sqrt{2})^2$, which soon leads to the embarrassing conclusion that $\sqrt{2}$ is a rational number! The ancient Greeks proved this to be impossible, as we will see in Chapter 12. Thus, it must be that $\mathbb{Q}(\sqrt{2}) \subset \mathbb{R}$.

Suppose that $r + s\sqrt{2}$ and $t + u\sqrt{2}$ are in $\mathbb{Q}(\sqrt{2})$. It is easy to prove that $\mathbb{Q}(\sqrt{2})$ is closed under addition: $(r + s\sqrt{2}) + (t + u\sqrt{2}) = (r + t) + (s + u)\sqrt{2}$, and because both $r + s$

and $t+u$ are rational, $(r+t)+(s+u)\sqrt{2} \in \mathbb{Q}(\sqrt{2})$. In Exercise 8.5, you will show that multiplication is closed in $\mathbb{Q}(\sqrt{2})$.

We next investigate reciprocals in $\mathbb{Q}(\sqrt{2})$. Using a concrete example, consider $\frac{1}{6+5\sqrt{2}}$. The only useful procedure for us here is to rationalize the denominator. This results in $\frac{6-5\sqrt{2}}{-14}$, which we quickly recognize as an element of $\mathbb{Q}(\sqrt{2})$.

With this example behind us, we attempt to find the reciprocal of the general, nonzero, element $r+s\sqrt{2}$. Rationalizing the denominator of $\frac{1}{r+s\sqrt{2}}$ yields

$$\frac{r-s\sqrt{2}}{r^2-2s^2} = \frac{r}{r^2-2s^2} - \frac{s}{r^2-2s^2}\sqrt{2}.$$

This is an element of $\mathbb{Q}(\sqrt{2})$, provided the denominator is never zero. But the denominator's subtraction of positive terms might yield zero. What does $r^2-2s^2=0$ imply? One of two things: either that $r=s=0$, which we disallow because $r+s\sqrt{2}$ was nonzero as a premise, or that $\sqrt{2}=\frac{r}{s}$, a rational number once again (and impossible, once again). Therefore, $r^2-2s^2 \neq 0$. The final conclusion is that *every* nonzero element of quadratic extension $\mathbb{Q}(\sqrt{2})$ has a reciprocal in $\mathbb{Q}(\sqrt{2})$. Moreover, we have an explicit expression for it:

$$\frac{1}{r+s\sqrt{2}} = \left(\frac{r}{r^2-2s^2}\right) + \left(\frac{-s}{r^2-2s^2}\right)\sqrt{2}.$$

The existence of reciprocals immediately proves that $\mathbb{Q}(\sqrt{2})$ is closed under division (verify this). Even better, $\mathbb{Q}(\sqrt{2})$ is, in fact, a field (hence, a **quadratic extension field**). To prove this, we would need to prove that the elements of $\mathbb{Q}(\sqrt{2})$ follow Axioms QZN1–QZN5, QZ7, QZ8, and Q10 from Chapter 3. They are all simple to establish, except Q10. But Q10 was just established in the course of discovering reciprocals in $\mathbb{Q}(\sqrt{2})$.

In showing that $\mathbb{Q}(\sqrt{2})$ is a field, one notices that $p=2$ under the radical is not special; in fact, all the field axioms would hold if p were any non-square natural number. Thus, we have outlined the proof of the following theorem.

Theorem 8.1 Given any non-square natural number p, the quadratic extension $\mathbb{Q}(\sqrt{p})$ is a field.

8.3 Algebraic Numbers

Another observation opens a new door. We will demonstrate that $6+5\sqrt{2}$ is a root of some quadratic polynomial. Let $6+5\sqrt{2}=x$. Then $(5\sqrt{2})^2=(x-6)^2$, so $50=x^2-12x+36$, from which we see that $6+5\sqrt{2}$ is one of the roots of $P(x)=x^2-12x-14$ (what is the other?).

We can do the same procedure with any $r+s\sqrt{2} \in \mathbb{Q}(\sqrt{2})$. Let $r+s\sqrt{2}=x$, and then $(s\sqrt{2})^2=(x-r)^2$, leading us to the quadratic expression $x^2-2rx+(r^2-2s^2)=0$. Hence, $r+s\sqrt{2}$ is a root of $P(x)=x^2-2rx+(r^2-2s^2)$. Thus, all the numbers in $\mathbb{Q}(\sqrt{2})$ are roots of quadratic polynomials. This line of thinking can be greatly generalized, leading to interesting new subsets of $\mathbb{R}$.

An **algebraic number of degree** n is any number (possibly complex, but this won't be our main focus) that is a root of the polynomial $P(x)=a_nx^n+a_{n-1}x^{n-1}+\cdots+a_1x+a_0$, where all the coefficients are integers, $a_n \neq 0$, and the number is not the root of any polynomial of lesser degree with integer coefficients. Such a polynomial is called an **irreducible polynomial** in $\mathbb{Z}$, because $P(x)$ cannot be factorized without introducing irrational coefficients.

Example 8.1 To illustrate the concept, reconsider $6 + 5\sqrt{2}$. It may be algebraic of degree two due to its status as a root of $P(x) = x^2 - 12x - 14$, but is it the root of a polynomial of degree one, with integer coefficients? That is, can this number solve the equation $a_1 x + a_0 = 0$, with integer coefficients, and with $a_1 \neq 0$? Substituting for x, $a_1(6 + 5\sqrt{2}) + a_0 = 0$ gives $\sqrt{2} = -\frac{6a_1 + a_0}{5a_1}$, which is not possible when we remember that a_1 and a_0 are integers. Hence, $P(x)$ is irreducible, and $6 + 5\sqrt{2}$ is algebraic of degree two.

In contrast to this, any rational number $\frac{a}{b}$ is algebraic of degree one, since such a number is the root of $P(x) = bx - a$, with integer coefficients and with $b \neq 0$. ■

In light of the definition of an algebraic number, we can prove the following theorem.

Theorem 8.2 Let p be a non-square natural number. All elements of the quadratic extension field $\mathbb{Q}(\sqrt{p})$ are algebraic of degree two or one.

Proof. Let $r + s\sqrt{p} \in \mathbb{Q}(\sqrt{p})$. If $s = 0$, then immediately, the rational number r is algebraic of degree one. Now take $s \neq 0$. Assume for a moment that $r + s\sqrt{p}$ is algebraic of degree one. Then substitution into the required polynomial $P(x)$ gives

$$P\left(r + s\sqrt{p}\right) = a_1\left(r + s\sqrt{p}\right) + a_0 = 0.$$

Then $p = \left(\frac{a_1 r + a_0}{a_1 s}\right)^2$, which says that p is a perfect square (and perhaps not even in $\mathbb{N}$)! This contradiction proves that $r + s\sqrt{p}$ is not algebraic of degree one when $s \neq 0$. Finally, a derivation like the one above shows that $r + s\sqrt{p}$ is a root of the irreducible $P(x) = x^2 - 2rx + (r^2 - s^2 p)$. **QED**

What do algebraic numbers of higher degrees look like? The next examples illustrate two.

Example 8.2 The number

$$a = \sqrt{1 + \left(\frac{1}{2} - \sqrt{3}\right)^2}$$

is algebraic of degree four. We proceed by clearing radicals:

$$a^2 = 1 + \left(\frac{1}{2} - \sqrt{3}\right)^2 = \frac{17}{4} - \sqrt{3},$$

so that $4a^2 - 17 = -4\sqrt{3}$. Squaring both sides and collecting terms to the left, $16a^4 - 136a^2 + 241 = 0$, implying that $P(x) = 16x^4 - 136x^2 + 241$ is a polynomial of fourth degree that has a as one of its four roots (you may check this on your calculator). Is $P(x)$ an irreducible polynomial? In this case, it is easy to answer affirmatively, since all the factors of $P(x)$ can be explicitly found (how?), and no products of them yield polynomials with strictly integer coefficients. ■

Example 8.3 Find the algebraic degree of $a = \sqrt{10} + \sqrt[3]{7 - \sqrt{5}}$. We begin innocently enough: $(a - \sqrt{10})^3 = 7 - \sqrt{5}$. Expand the left side, collect a few terms, and we arrive at $a^3 + 30a - 7 = (10 + 3a^2)\sqrt{10} + \sqrt{5}$. Squaring both sides and the steps after that are best left to a computer algebra system or to a night of insomnia. The end result is $P(x) = x^{12} - 60x^{10} - 28x^9 + 1500x^8 - 19716x^6 + 16800x^5 + 157320x^4 - 225232x^3 - 409200x^2 + 803040x + 893936$, a polynomial that necessarily has a as one of its 12 roots. Is it irreducible in $\mathbb{Z}$? Since no intermediate steps in the derivation of $P(x)$ yielded a polynomial of lesser degree with strictly integer coefficients, we suspect that the answer is "Yes." Thus, $\sqrt{10} + \sqrt[3]{7 - \sqrt{5}}$ seems to be algebraic of degree 12. ■

Note 1: The argument is not completely convincing, since one would have to show that no sequence of steps whatsoever can yield a polynomial of lesser degree than $P(x)$ with integer coefficients. Here is a helpful theorem: if u is an algebraic number, then a unique irreducible polynomial with lead coefficient 1, with all coefficients rational, and with u as its root, must exist. (The theorem's proof may be read in Chapter 10 of [14].) If our $P(x)$ has any polynomial factors with rational coefficients, then a is of lesser degree than 12. If this is not so, then the theorem tells us that $P(x)$ is the unique polynomial. It turns out that $P(x)$ is the one-and-only, so a is algebraic of degree 12.

So, we are now seeing successively more elaborate algebraic numbers, which were formed by using the four elementary operations of arithmetic, plus integral powers and radicals. It is important to note that although all numbers formed in this way are algebraic, the converse statement is false. An amazing theorem was proved in 1824 by Niels Henrik Abel (1802–1829), stating that polynomials with integer coefficients, starting at the fifth degree, may have algebraic roots that *cannot* be expressed using those six operations a finite number of times. For instance, the polynomial $P(x) = x^5 - 6x - 3$ has three real roots, each of which must therefore be algebraic, but they can't be expressed by any finite combination of the stated operations per a theorem discovered by Leopold Kronecker (1823–1891).

But alas, pursuing these intriguing ideas has taken us out of real analysis and into the theory of equations, so I stop here.

Let us define $\mathbb{A}_m$ as the set of all algebraic numbers of degree less than or equal to m. Thus, we know that

$$\mathbb{Q} = \mathbb{A}_1,\ 6 + 5\sqrt{2} \in \mathbb{A}_2,\ \left\{6 + 5\sqrt{2},\ \sqrt{1 + \left(\tfrac{1}{2} - \sqrt{3}\right)^2}\right\} \subset \mathbb{A}_4,$$

and

$$\left\{\frac{-13}{9},\ 6 + 5\sqrt{2},\ \sqrt{1 + \left(\frac{1}{2} - \sqrt{3}\right)^2},\ \sqrt{10} + \sqrt[3]{7 - \sqrt{5}}\right\} \subset \mathbb{A}_{12}.$$

Because of the definition, we have $\mathbb{Q} \subset \mathbb{A}_2 \subset \mathbb{A}_3 \subset \cdots \subset \mathbb{A}_m \subset \cdots$. From this vantage point, we may ask, "Do the containments ultimately absorb all the real numbers?" In other words, are all real numbers algebraic of some degree or another? It was thought that the famous numbers e and π were not algebraic numbers, that is, that neither of them could be a root of any polynomial whatsoever with integer coefficients. Such numbers, if they existed, would be called **transcendental numbers**, but for many years none could be proved to exist. Finally, in

1844, Joseph Liouville (1809–1882) found that $\mathbb{A}_m \subset \mathbb{R}$ for any m, by proving that a peculiar type of number was not algebraic. Here is one of his transcendental numbers:

$$\sum_{n=1}^{\infty} \frac{1}{10^{n!}} = 0.110001000000000000000001\ldots.$$

Almost 30 years later, in 1873, Charles Hermite (1822–1901) proved e to be transcendental. And in 1882, Ferdinand Lindemann (1852–1939) proved that π was transcendental, too. More amazing things lay ahead in the theory of real numbers. Georg Cantor demonstrated that just about all numbers are transcendental, because the entire set of algebraic numbers (we shall denote it by $\mathbb{A}$) was denumerable! We will see his ingenious proof in Chapter 9.

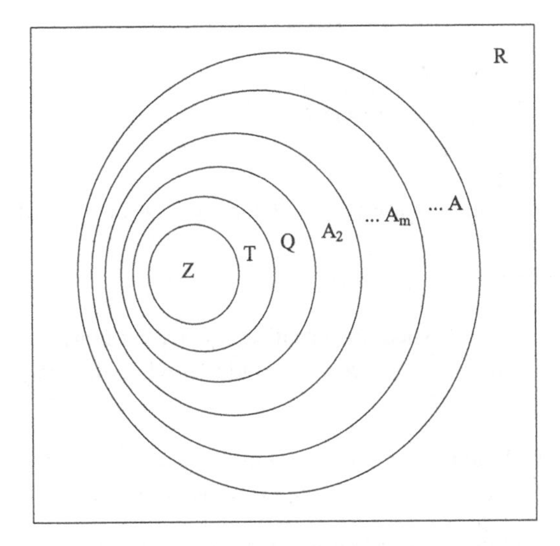

Successively larger subsets of algebraic numbers. Transcendental numbers fall outside the set of all algebraic numbers, $\mathbb{A}$.

Note 2: The proof that Liouville's number above is transcendental is found in chapter seven of [17] by Ivan Niven (1915–1999). Niven also writes a proof of the transcendence of e that is accessible to undergraduates in [36], pp. 25–27. The much more difficult proof of the transcendence of π takes up chapter nine of [36]. A similar, but equally elaborate, proof is in [32], pp. 68–80, by the influential mathematician Felix Klein (1849–1925).

8.4 Exercises

1. Determine the form of the denominators for every number $\frac{a}{b} \in \mathbb{T}$. Specifically, what is the prime factorization of b that distinguishes all, and no other, elements of $\mathbb{T}$? Hint: List a dozen members of T and find the commonality among the denominators. Then, prove your conjecture.

2. The reason why division is not closed in $\mathbb{T}$ is that not all members of $\mathbb{T}$ have reciprocals. Is $\mathbb{T}$ a field? The effect of this is often noticeable. Find a cheap, four-function calculator, and divide 5 by 3. The last displayed digit is either a 6 or a 7 (roughly dependent on how close to \$0 was the calculator's price!) In any case, the display is not exactly $\frac{5}{3}$, and we pay little attention to the round-off error. What number is the closest computational approximation

to $\frac{1}{3}$ that the calculator is able to display? Explain in terms of the properties of $\mathbb{T}$ why, on the calculator, $5 \div 3 \times 3 \div 5$, done left to right, doesn't equal 1.

3. Software designers consider very carefully the set $\mathbb{M}$ of numbers that can be stored in a digital computer's memory. Are any of these true?
a. $\mathbb{M}$ is an infinite set b. $\mathbb{M} \subset \mathbb{Z}$ c. $\mathbb{M} \subset \mathbb{T}$ d. $\mathbb{M} \subset \mathbb{Q}$ e. $\mathbb{N} \subset \mathbb{M}$

4. What, if any, difference does it make in the considerations of the previous exercise if we know that computers (at least early 21st century ones) do not store decimals, but rather the binary representations of decimal numbers?

5. Prove that multiplication is closed in $\mathbb{Q}(\sqrt{2})$.

6. For x, y not both zero, multiply

$$\left(x + y\sqrt{2}\right)\left[\left(\frac{x}{x^2 - 2y^2}\right) + \left(\frac{-y}{x^2 - 2y^2}\right)\sqrt{2}\right].$$

What does this mean?

7. Find polynomials for which the following numbers are roots. Then, give the degrees of these algebraic numbers.
a. $\cos 60^o$ b. the **golden ratio** $\varphi \equiv \frac{1+\sqrt{5}}{2}$ c. $\sqrt{1+\sqrt{2}}$

8. Imagine the set of numbers of the form $r + s\sqrt[3]{n}$, where (as usual), $r, s \in \mathbb{Q}$, but now, n is not a perfect cube. Is this set closed under multiplication? Is it a field? Of what algebraic degree are these numbers?

9. Theorem 8.2 can be proved in another way. What is the discriminant of $P(x) = x^2 - 2rx + (r^2 - s^2 p)$? Make conclusions when $s = 0$ and when $s \neq 0$, using this discriminant.

10. We have seen in Chapter 3 that fields need not have infinitely many elements. Actually, they don't even need to be composed of numbers. We need only display a set $\mathbb{F}$, and an "addition table" and a "multiplication table," to define how the two operations behave with the elements of $\mathbb{F}$. Then we check the axioms of a field to see if our concoction fulfills them. The following example shows a set of four elements, $\mathbb{F} = \{@, \#, \&, \%\}$, with two operations denoted as $+$ and $\cdot$.

+	@	#	&	%
@	@	#	&	%
#	#	&	%	@
&	&	%	@	#
%	%	@	#	&

·	@	#	&	%
@	@	@	@	@
#	@	#	&	%
&	@	&	@	&
%	@	%	&	#

We observe, for example, that $\# + \& = \%$ and $\%^2 = \% \cdot \% = \#$. At first, this seems like complete gibberish, but it serves to show that arithmetic can be defined for objects that have nothing to do with numbers. Let us see whether $\{@, \#, \&, \%\}$, together with these two operations, is a field.

a. Certainly @ is behaving as the additive identity, and # behaves as the multiplicative identity (why, and why?).

b. What quality of the tables shows at a glance that $+$ and $\cdot$ are commutative?
c. Try a case of the distributive law: does $\& \cdot (\% + \#) = (\& \cdot \%) + (\& \cdot \#)$? How many different distributive law cases would have to be tested longhand, given the information from **a** and **b**?
d. Unfortunately, one of the field axioms is not fulfilled. Which one? Whatever this system represents, it is not a field.

9

The Rational Numbers Are Denumerable

Everyone agrees that the set of rational numbers is infinitely numerous. Clearly, there are infinitely many rationals even within the interval $[1, 2]$, and infinitely many more in $[2, 3]$. But in the two intervals there are only three natural numbers. So, to anyone who had ever cared to think about it in the centuries past, there seemed to be enormously more rational numbers than natural numbers, albeit they were both infinitely numerous sets.

In 1895, Georg Cantor wrote an extended article in the journal *Mathematische Annalen* (for a translation from the German, see [27]). There, in a brief paragraph, he indicated a curious listing of the positive rational numbers, and displayed a function from $\mathbb{Q}$ onto $\mathbb{N}$ that showed that $\mathbb{Q}$ was a denumerable set. In other words, Cantor proved that there are just as many rationals as there are natural numbers! Such ideas were so counter-intuitive that even some great mathematicians of the day had dismissed Cantor's work as "fog upon fog." But the argument was, and is, indisputable. We will look at it in close detail.

The proof rests upon discovering a bijection $F : \mathbb{Q} \leftrightarrow \mathbb{N}$. Our strategy is to set up an array that contains all positive rational numbers $\mathbb{Q}^+ = \{\frac{p}{q} : p, q \in \mathbb{N}\}$, reduced or not, written as ordered pairs $\langle p, q\rangle$ (seen at left in Figure 9.1). This array is also $\mathbb{N} \times \mathbb{N}$, and we will prove that it is denumerable. Following this, a bijection between all rationals $\mathbb{Q}$ and $\mathbb{N}$ will prove the theorem.

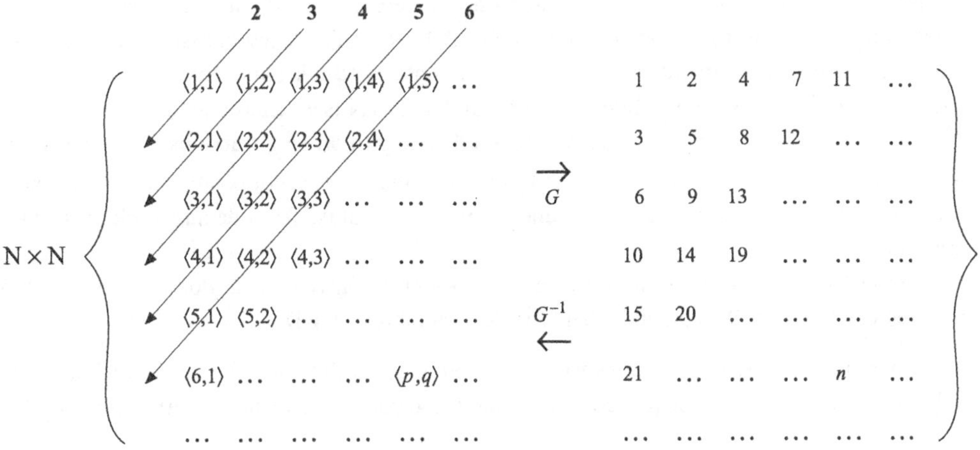

$$
\mathrm{N}\times\mathrm{N}\left\{\begin{pmatrix}
\langle 1,1\rangle & \langle 1,2\rangle & \langle 1,3\rangle & \langle 1,4\rangle & \langle 1,5\rangle & \dots \\
\langle 2,1\rangle & \langle 2,2\rangle & \langle 2,3\rangle & \langle 2,4\rangle & \dots & \dots \\
\langle 3,1\rangle & \langle 3,2\rangle & \langle 3,3\rangle & \dots & \dots & \dots \\
\langle 4,1\rangle & \langle 4,2\rangle & \langle 4,3\rangle & \dots & \dots & \dots \\
\langle 5,1\rangle & \langle 5,2\rangle & \dots & \dots & \dots & \dots \\
\langle 6,1\rangle & \dots & \dots & \dots & \langle p,q\rangle & \dots \\
\dots & \dots & \dots & \dots & \dots & \dots
\end{pmatrix}\right.
\begin{matrix} \xrightarrow{G} \\ \\ \xleftarrow{G^{-1}} \end{matrix}
\begin{pmatrix}
1 & 2 & 4 & 7 & 11 & \dots \\
3 & 5 & 8 & 12 & \dots & \dots \\
6 & 9 & 13 & \dots & \dots & \dots \\
10 & 14 & 19 & \dots & \dots & \dots \\
15 & 20 & \dots & \dots & \dots & \dots \\
21 & \dots & \dots & \dots & n & \dots \\
\dots & \dots & \dots & \dots & \dots & \dots
\end{pmatrix}
$$

(Diagonal lines labelled 2, 3, 4, 5, 6 run through the left array.)

Figure 9.1.

Theorem 9.1 The set of rational numbers $\mathbb{Q}$ is denumerable.

Proof. We begin by constructing a bijection $G : \mathbb{N}\times\mathbb{N} \leftrightarrow \mathbb{N}$; its domain and range are portrayed in Figure 9.1.

Drawing diagonal arrows as shown, we observe that triangles are being formed, all of which have one vertex at $\langle 1, 1\rangle$, and the other two at $\langle k-1, 1\rangle$ and $\langle 1, k-1\rangle$, for $k = 2, 3, 4, \ldots$. Let us call these the kth "corner triangles." Given any $\langle p, q\rangle$ along the kth diagonal arrow, notice that $p + q = k$ (which is why we started at $k = 2$). For example, $\langle 2, 4\rangle$ and $\langle 5, 1\rangle$ lie along the diagonal arrow $k = 6$.

We will need the number of ordered pairs $\langle p, q\rangle$ in the $(k-1)$st corner triangle, that is, in the corner triangle above (or before) diagonal k. Such counts are given by the sequence of **triangular numbers** $0, 1, 3, 6, 10, 15, \ldots$, as illustrated in Figure 9.2. As a function of k, this count is given by $T_k = \frac{1}{2}(k-1)(k-2)$.

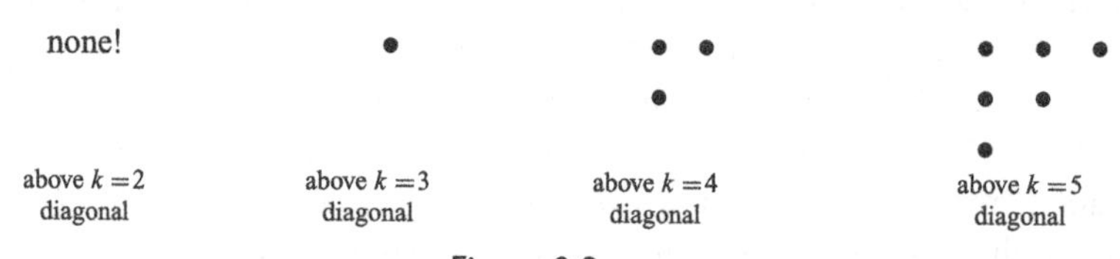

Figure 9.2.

We are finally in a position to construct G. The ordered pair $\langle p, q\rangle$ is located at position p down along the diagonal k. Therefore, $\langle p, q\rangle$ will be mapped *uniquely* to a natural number n if we add its position p to the count of ordered pairs above the diagonal upon which it sits. This count is precisely T_k, where $k = p + q$. Consequently, $G(\langle p, q\rangle)$ is given by

$$G(\langle p, q\rangle) = n = p + T_{p+q} = p + \frac{1}{2}(p+q-1)(p+q-2)\,.$$

This is the function Cantor displayed in his article, on page 494 of the journal.

Now, G is a one-to-one correspondence since different ordered pairs are always assigned different numbers. Moreover, beginning with $G(\langle 1, 1\rangle) = 1$, the function assigns natural numbers successively and without bound, so that it is onto $\mathbb{N}$ (think about this in Exercise 9.4). Therefore, $G : \mathbb{N} \times \mathbb{N} \leftrightarrow \mathbb{N}$ is a bijection, proving that $\mathbb{N}\times\mathbb{N}$ is a denumerable set.

Since the ordered pairs $\langle p, q\rangle$ above are another way of writing fractions $\frac{p}{q}$, we can write $G(\frac{p}{q}) = p + \frac{1}{2}(p+q-1)(p+q-2)$. It follows that the entire set of positive, distinguishable fractions, reduced and not reduced, is a denumerable set. That is, $\mathbb{Q}^+$ is denumerable, as Cantor concluded.

The last step, to construct a bijection $F : \mathbb{Q} \leftrightarrow \mathbb{N}$, is fairly easy to do, and Exercise 9.5 prepares you to do so. We conclude that $\mathbb{Q}$ is denumerable. **QED**

A completely different and unexpected method to count the nonnegative rational numbers is the construction of Farey sequences. Look into these curious sequences in Appendix A.

Example 9.1 The natural numbers corresponding uniquely to $\frac{35}{5}$ and $\frac{7}{1}$ are, respectively $G(\frac{35}{5}) = 35 + \frac{1}{2}(35+7-1)(35+7-2) = 855$, and $G(\frac{7}{1}) = 28$. Thus, G distinguishes between equal fractions. In the first case, $k = 40$ is the number of the diagonal that contains $\frac{35}{5}$ (in the form $\langle 35, 5\rangle$). ■

Upon finishing Exercise 9.5, you will be able to write the full set of rational numbers in this way: $\mathbb{Q} = \{\frac{0}{1}, \frac{1}{1}, \frac{-1}{1}, \frac{1}{2}, \frac{-1}{2}, \ldots\}$. This ordering of the elements is given by the bijection $F\left(\frac{p}{q}\right)$ in that exercise. Because this bijection has counted all fractions, including unreduced ones, the "much smaller" subset of reduced fractions must also be denumerable.

It is interesting to construct the inverse function $G^{-1} : \mathbb{N} \to \mathbb{N} \times \mathbb{N}$ as an algorithm. Looking at our array, we see that every $\langle p, q\rangle$ lies on the diagonal arrow of a unique corner triangle k. Given $n \in \mathbb{N}$, G^{-1} will yield a unique $\langle p, q\rangle$ by first selecting the greatest $T_k < n$. This determines k as well as T_k. Next, p is determined: $p = n - T_k$, representing the position of $\langle p, q\rangle$ down the kth diagonal. Last, the identity $p + q = k$ allows us to find q: $q = k - p = k - (n - T_k)$. Thus, $G^{-1}(n) = \langle n - T_k\,,\ k - (n - T_k)\rangle$, where T_k is uniquely determined by $T_k < n \leq T_{k+1}$, and $T_k = \frac{1}{2}(k-1)(k-2)$ is a triangular number with $k = 2, 3, 4, \ldots$.

Example 9.2 To evaluate $G^{-1}(38)$, first identify the triangular number T_{10} because $T_{10} = 36 < 38 \leq T_{11}$. It follows that $k = 10$. Next, $p = 38 - T_{10} = 2$, from which $q = 10 - 2 = 8$. Thus, $G^{-1}(38) = \langle 2, 8\rangle$, or equivalently, $G^{-1}(38) = \frac{2}{8}$. Notice that $G^{-1}(38) \neq \frac{1}{4}$, that is, G^{-1} distinguishes all fractions. ■

Note 1: In what follows, we will employ a handy operation called the "ceiling of x," denoted $\lceil x\rceil$. For $x \in \mathbb{R}$, $\lceil x\rceil$ equals the next integer larger than or equal to x. Examples: $\lceil 6.2\rceil = 7$, and $\lceil 7\rceil = 7$. (Chapter 13 will introduce $\lceil x\rceil$ as a function.)

With a bit more patience, G^{-1} can be expressed as an explicit function. Any natural number n falls between successive triangular numbers, specifically $T_k < n \leq T_{k+1}$. Thus, for fixed n, $(k-1)(k-2) < 2n \leq (k)(k-1)$. Solving these inequalities for k gives $\frac{1}{2}(1 + \sqrt{1+8n}) \leq k < \frac{1}{2}(3 + \sqrt{1+8n})$. Since the expressions bounding k differ by exactly one, this is equivalent to $k = \lceil \frac{1}{2}(1 + \sqrt{1+8n})\rceil$. Thus,

$$G^{-1}(n) = \langle n - T_k, k - (n - T_k)\rangle = \left\langle n - \frac{1}{2}(k-1)(k-2), k - n + \frac{1}{2}(k-1)(k-2)\right\rangle,$$

where

$$k = k(n) = \left\lceil \frac{1}{2}(1 + \sqrt{1+8n})\right\rceil.$$

Example 9.3 To evaluate $G^{-1}(38)$ explicitly with a calculator, first calculate $k(38) = \lceil \frac{1}{2}(1 + \sqrt{1 + 8(38)})\rceil = 10 = k$. Substituting this value, $G^{-1}(38) = \langle 38 - \frac{1}{2}(9)(8)\,,\ 10 - 38 + \frac{1}{2}(9)(8)\rangle = \langle 2, 8\rangle$. This corresponds to $\frac{2}{8}$ as in the previous example. ■

Note 2: Many other bijections between $\mathbb{Q}^+$ and $\mathbb{N}$ have been devised besides Cantor's original and the one in Appendix A. An interesting example is "Counting the Rationals," by Yoram Sagher in *The American Mathematical Monthly*, vol. 96, no. 9 (Nov. 1989), pg. 823. His method employs the unique prime factorization of natural numbers (Theorem 3.4), so we begin with $m = p_1^{e_1} \cdots p_k^{e_k}$ and $n = q_1^{d_1} \cdots q_\ell^{d_\ell}$. The required bijection $f : \mathbb{Q}^+ \leftrightarrow \mathbb{N}$

is defined as $f\left(\frac{m}{n}\right) = (p_1^{2e_1} \cdots p_k^{2e_k})(q_1^{2d_1-1} \cdots q_\ell^{2d_\ell-1})$, where $\frac{m}{n}$ is in lowest terms, and $f(1) \equiv 1$. For example, $f\left(\frac{2}{3}\right) = 12$, showing that we located $\frac{2}{3}$ as the 12th entry in Sagher's list of positive rationals.

Georg Cantor may have assumed the denumerability of $\mathbb{Q}$ to be indisputable in 1895, thus the bare outline of the proof of Theorem 9.1. Cantor actually had discovered a vastly larger landscape by 1874. As we saw in Chapter 8, $\mathbb{Q}$ forms a minute subset of the set of algebraic numbers $\mathbb{A}$ (recall Figure 8.1 there). Indeed, Theorem 9.1 becomes a corollary of the startling theorem below.

Theorem 9.2 The set of all algebraic numbers $\mathbb{A}$ (of any degree) is denumerable (Cantor, 1874).

Proof. Given an algebraic number β of degree $n \geq 1$, we know from Note 1 that there exists a unique irreducible polynomial $P(x)$ as defined above, with root β (we disregard $-P(x)$ as not essentially different). Hence, for each β, we have only one polynomial of degree n such that $P(\beta) = 0$. Because of the uniqueness of $P(x)$, we (following Cantor) are able to define the **height** of β to be the unique natural number $H = n - 1 + |a_n| + |a_{n-1}| + \cdots + |a_0|$.

Let's find algebraic numbers that have height $H = 1$. Considering $1 = n - 1 + |a_n| + |a_{n-1}| + \cdots + |a_0|$, we find no solutions if $n \geq 3$ (why?). When $n = 2$, $1 = 1 + |a_2| + |a_1| + |a_0|$, which still has no solutions since $a_2 \neq 0$. But if $n = 1$, we have $|a_1| + |a_0| = 1$, which is satisfied by $a_1 = \pm 1$, $a_0 = 0$. Hence, the unique polynomial is $P(x) = \pm 1x + 0$. Setting $P(x) = 0$, the only algebraic number of height one is $\beta_1 = 0$.

Next, algebraic numbers of height $H = 2$. If $n \geq 4$, no solutions exist for $2 = n - 1 + |a_n| + |a_{n-1}| + \cdots + |a_0|$. If $n = 3$, we have $|a_3| + |a_2| + |a_1| + |a_0| = 0$, which once again has no solutions. If $n = 2$, we solve $|a_2| + |a_1| + |a_0| = 1$, which leads to $P(x) = \pm 1x^2 + 0x + 0$, with root $\beta_1 = 0$, which is not of height two. Trying $n = 1$, we seek solutions to $|a_1| + |a_0| = 2$. This gives either $|a_1| = 2$ and $|a_0| = 0$, or $|a_1| = |a_0| = 1$. The former coefficients produce $P(x) = \pm 2x + 0$, yielding $\beta_1 = 0$ again, but the latter coefficients give us polynomials $P(x) = \pm 1x \pm 1$. These four polynomials produce roots $\beta_2 = -1$ and $\beta_3 = 1$, which are therefore the only algebraic numbers of height 2.

It is interesting to find algebraic numbers of greater heights (do this in Exercise 9.7), but it's a diversion. What is crucial is that each successive height contributes only a finite set of algebraic numbers. To prove this, first note that a solution to an equation of the form $\sum_{k=1}^{m} b_k = m$, where $m \in \mathbb{N}$ is a constant and the b_k are nonnegative integers, corresponds to a partitioning of m marbles into m subsets of sizes $b_1, b_2, \ldots, b_m$. (In this context, a partition may be empty.) For example, one of the five solutions to $\sum_{k=1}^{4} b_k = 4$ is visualized as the partition $b_1 = 0$ marbles, $b_2 = b_3 = 1$, and $b_4 = 2$ marbles. Since there is only a finite number of partitions for every m, there are only finitely many polynomials $P(x)$ of the type we want, for each height, each with only finitely many roots. Therefore, each height contributes only a finite collection of algebraic numbers.

We are now able to form a list of all algebraic numbers. The list starts with 0, the single algebraic number with $H = 1$, followed by the two algebraic numbers with $H = 2$ (in either order), followed by those with $H = 3$ (in any order), and so on. The list may begin this way, $\mathbb{A} = \{0, -1, 1, -i, i, -\frac{1}{2}, \ldots\}$. As we can see, Cantor's scheme dutifully picks up all the complex algebraic numbers as well. This listing of $\mathbb{A}$ forms a one-to-one correspondence with $\mathbb{N}$, from which we conclude (hard as it is to believe) that the set of all algebraic numbers is denumerable. **QED**

9.1 Exercises

1. Show that $G(\langle 6, 2\rangle) = 27$, so that $\frac{6}{2}$ is paired uniquely to the natural number 27. Verify it using the array above. Also, use the algorithm above to evaluate $G^{-1}(27)$.

2. One of the close approximations to π is the rational number $\frac{355}{113}$. Determine the unique position that the function G assigns to this fraction in Cantor's sequence of rational numbers.

3. Let y equal the year of your birth (four digit format). Evaluate $G^{-1}(y)$ using the function G^{-1} to find out what fraction is uniquely attached to your year.

4. Show by induction that G is onto $\mathbb{N}$. Hint: If $\langle p, q\rangle$ is not at the bottom of diagonal number k, then use $G(\langle p, q\rangle) = p + T_{p+q} = n$ for the induction. If $\langle p, q\rangle$ is at the bottom of diagonal k, then $\langle p, q\rangle$ is expressed as $\langle k-1, 1\rangle$, and $G(\langle k-1, 1\rangle) = k - 1 + T_k = n$.

5. a. Fill in the missing entries in the table below. The set of all rationals $\mathbb{Q}$ is very conveniently ordered: it follows the paths of the diagonal arrows above, but with corresponding negative fractions inserted. Row ✠ turns these fractions into ordered pairs $\{\langle p, q\rangle : p \in \mathbb{Z},\ q \in \mathbb{N}\} = \mathbb{Z} \times \mathbb{N}$.

$\mathbb{Q}$	0/1	1/1	−1/1	1/2	−1/2	2/1	−2/1	1/3	−1/3	2/2			···
✠	⟨0, 1⟩	⟨1, 1⟩	⟨−1, 1⟩	⟨1, 2⟩									···
$\mathbb{N}$	1	2	3	4	5	6	7	8	9	10	11	12	13 ···

   b. The pattern above should suggest a function F from the full set $\mathbb{Q}$ onto $\mathbb{N}$. Find the function $n = F(\frac{p}{q})$. Its domain is the first row of the table, and its range is the third row. Hint: If $\frac{p}{q} > 0$, then $F(\frac{p}{q}) = 2G(\langle p, q\rangle)$, an even number. What to do with the case $\frac{p}{q} < 0$, and with $\frac{p}{q} = 0$?

   c. Use the result of **b** to find the inverse function $F^{-1}(n)$. This inverse function is onto $\mathbb{Q}$. Conclude that we have discovered a bijection $F : \mathbb{Q} \leftrightarrow \mathbb{N}$, which proves the countability of all rational numbers $\mathbb{Q}$. Hint: For ease of use, let $F(\frac{p}{q}) = n$. Then $\frac{p}{q} = F^{-1}(n)$. To find F^{-1} explicitly, solveeach case of function F in **b** for ordered pair $\langle p, q\rangle$, and then turn the ordered pair into the fraction $\frac{p}{q}$.

6. Show that $G(G^{-1}(n)) = n$, and that $G^{-1}(G(\langle p, q\rangle)) = \langle p, q\rangle$.

7. Find all the algebraic numbers of height $H = 3$. Hint: $3 = n - 1 + |a_n| + |a_{n-1}| + \cdots + |a_0|$ begins to have solutions when $n = 3$. In this case, $|a_3| + |a_2| + |a_1| + |a_0| = 1$. This implies that $|a_3| = 1$ and the rest are zero, which is the only way to partition a set of one marble into four subsets. This yields $P(x) = \pm 1x^3 + 0x^2 + 0x + 0$. Continue!

10

The Uncountability of the Real Numbers

"Je le vois, mais je ne le crois pas!" (I see it, but I don't believe it!)

Georg Cantor in a letter to his friend,
Richard Dedekind, 1877

In 1890–91, Georg Cantor published an article (see Note 1) containing a new proof of one of his astonishing discoveries, a theorem originally from 1874. It was that the set of real numbers $\mathbb{R}$ was not denumerable, in contrast to the set of rational numbers, $\mathbb{Q}$. Since both $\mathbb{R}$ and $\mathbb{Q}$ were unquestionably infinite sets, what Cantor was now demonstrating was that different sizes of infinity existed, and that mere mortals could distinguish among them! His brilliant diagonalization technique has been adapted to other contexts, such as logic and computability, always leading to profound conclusions.

Theorem 10.1 The set of real numbers $\mathbb{R}$ is uncountable.

Proof. Every real number x has a decimal representation, $x = n.d_1d_2d_3\ldots$, where n is the unique integer such that $n \leq x < n + 1$, and d_i represents the ith digit in the representation (see Theorems 3.3 and 28.6). The decimal representation for a nonzero x is unique to it, unless the representation happens to terminate, in which case x has exactly two representations. For example, if $x = \frac{13}{4}$, then $x = 3.25000\ldots = 3.24999\ldots$. To remove the ambiguity in such instances, we will select the representation with endless zeros.

This will be a proof by contradiction, and in particular we will assume that $[0, 1] \subset \mathbb{R}$ is a denumerable set. This means that $[0, 1]$ can be written as a subscripted set this way: $[0, 1] = \{x_1, x_2, x_3, \ldots, x_k, \ldots\}$. Combine this notation with the decimal representation by employing double subscripts, to form a denumerable list:

$$\begin{aligned}
x_1 &= 0.d_{11}d_{12}d_{13}d_{14}\ldots \\
x_2 &= 0.d_{21}d_{22}d_{23}d_{24}\ldots \\
x_3 &= 0.d_{31}d_{32}d_{33}d_{34}\ldots \\
&\ldots \\
x_k &= 0.d_{k1}d_{k2}d_{k3}d_{k4}\ldots \\
&\ldots
\end{aligned} \tag{1}$$

Thus, d_{32} is the second decimal digit of the third real number in $[0, 1]$, and so on.

We emphasize that under our assumption, this list enumerates every single real number in $[0, 1]$. Now, construct a real number $y = 0.y_1y_2y_3\ldots$ by the following rule: for all natural numbers j, if $d_{jj} = 1$, then $y_j = 0$, and if $d_{jj} \neq 1$, then $y_j = 1$. Clearly, $y \in [0, 1]$, so it must appear somewhere in our list. Thus, $y = x_k$ for some $k \in \mathbb{N}$. But this is not possible, since $y_k \neq d_{kk}$. In other words, for all $k \in \mathbb{N}$, $y \neq x_k$ because they differ in the kth decimal place. Hence, y does *not* appear in the list. This contradiction forces us deny the assumption that $[0, 1]$ is denumerable. The interval is therefore uncountable.

Returning to the real number $x = n.d_1d_2d_3\ldots$, we recognize the decimal part $0.d_1d_2d_3\ldots$ as a member of $[0, 1]$. Thus, every interval $[n, n+1]$ is in one-to-one correspondence with $[0, 1]$, and is therefore itself uncountable. And finally, since $\mathbb{R} = \bigcup_{n=0}^{\infty}([n, n+1] \cup [-n-1, -n])$, it follows that $\mathbb{R}$ is uncountable. **QED**

The procedure in Cantor's proof can be used to construct an actual transcendental number. Using Theorem 9.2, list all the real algebraic numbers and choose only those in $[0, 1]$. Then, follow the diagonalization rule for selecting digits y_j. The output of the algorithm is a real number $y = 0.y_1y_2y_3\ldots$ that is not algebraic, because it is not in the list. It must therefore be transcendental.

Note 1: "Über eine elementare Frage der Mannigfaltigkeitslehre"(On a Fundamental Question of the Theory of Manifolds) by Georg Cantor, in the journal *Jahresbericht der Deutschen Mathematiker-Vereinigung*, 1890–91, pp. 75–78.

10.1 Exercises

1. Suppose that $\mathbb{R}$ is denumerable, so that a list such as (1) is available.
   a. Suppose that $\frac{25}{32}$ is the third element in the list. What is d_{33}? Why is the number y constructed in the proof not equal to $\frac{25}{32}$?
   b. Suppose that $\frac{\pi}{4}$ is the 825th element in the list. Why is the number y constructed in the proof not equal to $\frac{\pi}{4}$?

2. Why is the proof of Theorem 10.1 often called Cantor's "diagonalization"?

3. Consider this objection to the proof of Theorem 10.1. Suppose that $\mathbb{R}$ is denumerable, and that the subscripted list (1) happens to have $d_{jj} = 1$ for all j. Then $y_j = 0$ for all j, which implies that $y = 0$. There is no apparent contradiction, for the construction has simply yielded the number $x = 0$. How do you refute this, and obviate the loophole?

4. Suppose that $\mathbb{R}$ is denumerable, and that the subscripted list (1) happens to have $d_{jj} = 0$ for all j. In this scenario, what would be the exact value of y? Why would y not exist in the list?

5. It is interesting to see Cantor's proof in something other than the standard decimal representation of reals. We may express every real number in **binary (base two)** form. In this base, only the digits 0 and 1 exist. Binary integers aren't particularly interesting here, but the binary representation of reals in $[0, 1]$ is. For such representations, only negative powers of two may be employed: $\{2^{-1}, 2^{-2}, 2^{-3}, \cdots\}$. For example, $\frac{5}{8} = 1 \cdot 2^{-1} + 0 \cdot 2^{-2} + 1 \cdot 2^{-3} = 0.101_2$, which terminates.

Now $\frac{3}{10}$, although a terminating decimal in base 10, requires the following fairly self-explanatory algorithm to express it in binary. Divide $\frac{3}{10}$ by the lowest power of two that will give at least 1: $\frac{3}{10} \div 2^{-2} = \frac{6}{5} = 1 + \frac{1}{5}$ (and $\frac{3}{10} \div 2^{-1} < 1$). The expansion begins as $0.01\ldots_2$, where the binary digit 1 is in the 2^{-2} place value. Continue with the remainder $\frac{1}{5}$. Since $\frac{1}{5} \div 2^{-3} = \frac{8}{5} = 1 + \frac{3}{5}$ (and $\frac{1}{5} \div 2^{-2} < 1$), the next binary digit 1 is in the $2^{-2-3} = 2^{-5}$ place value. Thus, we now have $0.01001\ldots_2$. Continue with the remainder $\frac{3}{5}$. We have $\frac{3}{5} \div 2^{-1} = \frac{6}{5} = 1 + \frac{1}{5}$, so a 1 will appear in the $2^{-2-3-1} = 2^{-6}$ place value. The expansion is now $0.010011\ldots_2$. The algorithm starts to cycle without end now, because the remainder $\frac{1}{5}$ has appeared before. Hence, $\frac{3}{10} = 0.010011001100\ldots_2$, a non-terminating binary representation. As might be expected, any rational number in [0, 1] without a power of 2 denominator will yield a non-terminating, repeating, binary representation. An irrational number in [0, 1] will produce a non-terminating, non-repeating, binary representation. This algorithm applies only to reals in [0, 1], but this is not a real (aha!) drawback.

a. Express $\frac{39}{16}$, $\frac{8}{9}$, and $\frac{\sqrt{2}}{2}$ as binary representations. How would they appear in the diagonalization list (1)? In reverse, what decimal number equals $0.1111\ldots_2$? Hint: $\frac{39}{16} = 2 + \frac{7}{16}$, so concentrate on $\frac{7}{16}$. For $\frac{\sqrt{2}}{2}$, try using its decimal form 0.70710678118....
b. Formulate Cantor's diagonalization proof using binary representations for all real numbers.

11

The Infinite

When we've been there ten thousand years,
Bright shining as the sun,
We've no less days to sing God's praise,
Than when we'd first begun.

"Amazing Grace," by John Newton *et al.*

Infinity as an analytical entity was tackled by way of developing set theory, in the work of Georg Cantor during the last quarter of the 19th century. We may begin with a question about counting. When a set is finite, the counting process eventually terminates with a total number of elements. But what happens when a set is infinite? Arriving at a total under this condition becomes physically impossible, but that is inconsequential to mathematical analysis. The fundamentally disturbing problem is that a total number does not exist. For instance, if we are asked, "How many points are inside a square?" or "How many fractions are there?" we just say, "Infinitely many." If pressed to clarify, we drum our fingers against a table and reply, "More than any number you can give." This is actually a good answer from a certain point of view, but it's more likely a ploy that forces the inquirer to answer his own question. As usual, rhetorical devices lead us nowhere. How then can we handle an actual infinity of objects?

11.1 Infinite Sets

The essence of the counting process is in discovering a bijection—a one-to-one correspondence—between the elements of a set and the natural numbers, $\mathbb{N}$ (no wonder that they are often called the counting numbers). A **countable** set is a set that can be used as the domain of a bijection into $\mathbb{N}$. It follows at once that all finite sets are countable, for such sets can be placed into one-to-one correspondence with $\{1, 2, 3, \ldots, n\} \subset \mathbb{N}$, where n is the size of the finite set. If the bijection is *onto* $\mathbb{N}$, then the set is **denumerable**, or **countably infinite**. Hence, finite sets and denumerable sets form a partition of all countable sets. A countable set can always be listed in a sequence using natural number subscripts, since the subscripts are generated by the bijection (examples appear below).

But as we saw in Chapter 10, Cantor proved that it was impossible to form a bijection between $\mathbb{N}$ and the set of real numbers, $\mathbb{R}$. All sets with this behavior were called **uncountable** or **nondenumerable**. Since an uncountable set is infinite but cannot be matched with $\mathbb{N}$, a general definition of "infinite set" is needed, one that doesn't rely on natural numbers. Cantor adopted the definition by an earlier developer of analysis, Bernhard Bolzano (1781–1848): a set is **infinite** if and only if there is a one-to-one correspondence between it and one of its proper subsets. (In Exercise 11.6, you show that a finite set does not fulfill the definition. Indeed, that's what makes it finite.) The following Venn diagram describes the interrelationships among these terms of size.

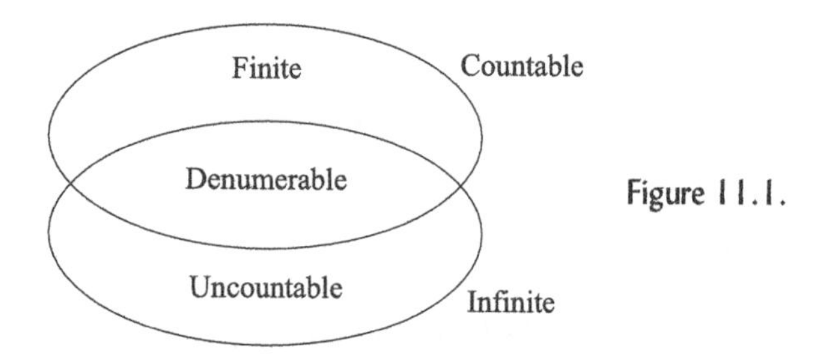

Figure 11.1.

Example 11.1 The set of natural numbers $\mathbb{N}$ is infinite—we know it, but must prove it! The proof is that the bijection $f : \mathbb{N} \leftrightarrow \mathbb{N}\backslash\{1\}$ given by $f(n) = n + 1$ maps $\mathbb{N}$ onto a proper subset of itself, as follows:

$$f = \begin{cases} 1 \quad 2 \quad 3 \quad 4 \quad \cdots \quad n \quad \cdots \longleftarrow \mathcal{D}_f = \mathbb{N} \\ \updownarrow \quad \updownarrow \quad \updownarrow \quad \updownarrow \qquad\qquad \updownarrow \\ 2 \quad 3 \quad 4 \quad 5 \quad \cdots \quad n+1 \quad \cdots \longleftarrow \mathcal{R}_f = \mathbb{N}\backslash\{1\} \end{cases} \blacksquare$$

The bijection $f(n) = n + 1$ in the preceding example is perhaps the simplest one of all. It furnishes the background of a fanciful story called "The Infinite Hotel." Search for it on the Web.

Example 11.2 The set of integers $\mathbb{Z}$ is countable (and denumerable), since $m = f(n) = \begin{cases} 2n & \text{if } n>0 \\ -2n+1 & \text{if } n\le 0 \end{cases}$ is a bijection *onto* $\mathbb{N}$:

$$f = \begin{cases} \cdots \quad -2 \quad -1 \quad 0 \quad 1 \quad 2 \quad 3 \quad \cdots \quad n \quad \cdots \longleftarrow \mathcal{D}_f = \mathbb{Z} \\ \qquad \updownarrow \quad \updownarrow \quad \updownarrow \quad \updownarrow \quad \updownarrow \quad \updownarrow \qquad\qquad \updownarrow \\ \cdots \quad 5 \quad 3 \quad 1 \quad 2 \quad 4 \quad 6 \quad \cdots \quad m \quad \cdots \longleftarrow \mathcal{R}_f = \mathbb{N}. \end{cases}$$

Furthermore, $\mathbb{Z}$ is infinite, since the same bijection carries $\mathbb{Z}$ onto a subset of itself. Notice that the bijection also provides us with subscripts for the set of integers, and hence, a listing all integers is $\mathbb{Z} = \{0_1, 1_2, -1_3, 2_4, -2_5, \ldots\}$. The sequence is fairly obvious; we simply interlace the negative integers with the positive ones. ■

Example 11.3 The set of rational numbers $\mathbb{Q}$ is denumerable, as was shown in Chapter 9. We saw there that Cantor's bijection $G(\langle p, q\rangle) = p + \frac{1}{2}(p+q-1)(p+q-2) = n$ gave us a unique subscript n corresponding to any positive rational number $\frac{p}{q}$. Then, interlacing positive and negative rationals gave us a list of all rationals:

$$\mathbb{Q} = \left\{\frac{0}{1}, \frac{1}{1}, \frac{-1}{1}, \frac{1}{2}, \frac{-1}{2}, \ldots\right\}.$$

This result demolished the notion that there should be many more fractions than whole numbers! And once more, a bijection maps $\mathbb{Q}$ onto a subset of itself (which subset?), making $\mathbb{Q}$ an infinite set (no surprise there). ■

Example 11.4 The enormous set of algebraic numbers was shown to be denumerable in Theorem 9.2. ■

Example 11.5 The set $\mathbb{R}$ is infinite, since we found bijections between $\mathbb{R}$ and $(0, 1) \subset \mathbb{R}$ in Exercises 7.5 and 7.6. But $\mathbb{R}$ is not denumerable, as was proved in Theorem 10.1. Hence, it is uncountable. ■

Example 11.6 The set of points on the unit circle $x^2 + y^2 = 1$ is infinite. To prove this, each point on the circumference has coordinates $(\cos\theta, \sin\theta), \theta \in [0, 2\pi)$. We can form a one-to-one correspondence with the upper semicircle $(\cos\theta, \sin\theta) \leftrightarrow (\cos\frac{\theta}{2}, \sin\frac{\theta}{2}), \theta \in [0, 2\pi)$. This is a bijection of the unit circle with a subset of it.

A second proof was conjectured in Exercise 5.16, and now we can firm it up. The parametrization $(\cos\theta, \sin\theta)$, $\theta \in [0, 2\pi)$, sets up a bijection between the points of the circle and $[0, 2\pi)$. Moreover, a bijection exists between $[0, 2\pi)$ and its subset $[0, 1)$, implying that $[0, 2\pi)$ is an infinite set. Hence, $\{(\cos\theta, \sin\theta) : \theta \in [0, 2\pi)\}$ is an infinite set also; and it must be uncountable. This proof relies on the chain of bijections $(\cos\theta, \sin\theta) \leftrightarrow [0, 2\pi) \leftrightarrow [0, 1)$. The last small step for you to do is to construct a bijection $q : [0, 2\pi) \leftrightarrow [0, 1)$ as Exercise 11.4a. ■

Example 11.7 Paradox 0.3 at the beginning of this book is now resolved. There, Galileo had seen that a line segment *CD* could be put into one-to-one correspondence with a shorter segment *CE* contained within *CD*. This deeply disturbed his intuition, but now we see that he was handling *CD* as an infinite set, and it was therefore behaving true to its nature. ■

Example 11.8 In Example 5.1, we studied the denumerable collection of concentric circles $\{S_n\}_{n=1}^{\infty}$ where $S_n = \{\text{points of the circle } x^2 + y^2 = (\frac{n}{n+1})^2\}$. To prove that $\{S_n\}_{n=1}^{\infty}$ is an infinite collection, use the bijection from Example 11.1, $f(n) = n + 1$ with $\mathcal{D}_f = \mathbb{N}$. This maps S_1 to S_2, S_2 to S_3, and so on, *ad infinitum*. Since $\{S_n\}_{n=2}^{\infty} \subset \{S_n\}_{n=1}^{\infty}$, we know that $\{S_n\}_{n=1}^{\infty}$ is an infinite collection of sets. ■

We now take the next step by discovering different orders of infinity.

11.2 The Cardinal Numbers $\aleph_0$ and c

Sets that are in one-to-one correspondence are called **equivalent** sets. For example, $\{$*Mt. McKinley*, *Mt. Everest*, *Saturn*$\}$, $\{♠, 2e, \cos x\}$, and $\{3\pi, \$, \%\}$, are equivalent. Also, all sets that can be put into one-to-one correspondence with $\mathbb{N}$ are equivalent; they form the collection of denumerable sets. Hence, $\mathbb{N}$, $\mathbb{Z}$, $\mathbb{Q}$, and $\mathbb{A}$ are equivalent sets. In contrast to this, $\mathbb{R}$ is not equivalent to any of them. But $\mathbb{R}$ is equivalent to the set of points forming interval (0, 1) (review Example 11.5).

The **cardinality** of a set S is the common size of all sets equivalent to S. Since all sets with n elements are equivalent, the cardinality of a finite set is therefore just the number n of its elements. The next case is that of denumerable sets. Cantor assigned the first letter of the Hebrew alphabet to the cardinality of a denumerable set: $\aleph_0$, "aleph null." For example, $\mathbb{N}$ and $\mathbb{Q}$ have cardinality $\aleph_0$. Sets with $\aleph_0$ cardinality express the smallest order of infinity, for any subset of a denumerable set is either denumerable again, or finite. Thus, $\aleph_0$ is the **first transfinite cardinal number** (there is no lesser transfinite).

Note 1: Cardinality is an equivalence relation between sets. An equivalence relation always partitions a universal set into subsets called equivalence classes. Here, the equivalence classes are composed of sets with equal cardinality. Thus, the equivalence class determined by "41" is the collection of all sets with 41 elements. $\aleph_0$ represents the equivalence class of all denumerable sets.

The great advantage of equivalence classes is that if a set is equivalent to one member of an equivalence class, then it is equivalent to all members of the class. This will allow us to generalize about cardinalities each time we establish just one case.

Transfinite numbers have their own peculiar form of arithmetic. As the classic example, we write $1 + \aleph_0 = \aleph_0$. No real number behaves that way, since this equation will violate Axiom RQZ7 in the next chapter. The equation actually means that when one element is inserted into a set of cardinality $\aleph_0$, the resulting set still has cardinality $\aleph_0$. In other words, given denumerable set $S = \{s_1, s_2, s_3, \ldots\}$, inserting an element $q \notin S$ produces $S' = \{q, s_1, s_2, s_3, \ldots\}$, which is still denumerable, and hence of the same cardinality as S. Likewise, writing $n + \aleph_0 = \aleph_0$ cannot mean that we are adding n to $\aleph_0$, for $\aleph_0$ doesn't belong to the reals! The equation is just a shorthand for stating that the cardinality of $\{t_1, t_2, \ldots, t_n\} \cup \{s_1, s_2, s_3, \ldots\}$ is still $\aleph_0$.

What about $\aleph_0 + \aleph_0$? The expression $\aleph_0 + \aleph_0$ is just a string of symbols (admittedly suggestive ones), until we breathe meaning into it. So, $\aleph_0 + \aleph_0$ shall indicate the cardinality of the union of two denumerable sets, and we shall use the shorthand $2\aleph_0 \equiv \aleph_0 + \aleph_0$. Extending this notation, $3\aleph_0 \equiv 2\aleph_0 + \aleph_0 = \aleph_0 + \aleph_0 + \aleph_0$, and so forth. The next lemma evaluates any multiple of aleph null.

Lemma 11.1 For any natural number n, $n\aleph_0 = \aleph_0$. That is, the cardinality of the union of n denumerable sets is still $\aleph_0$.

Proof. Let $\{S_j\}_1^n$ be a collection of denumerable sets. As such, we will employ double subscripts and list all the sets in an orderly array of n rows (does it remind you of Theorem 9.1?):

$$\begin{array}{llllllll}
S_1 = \{s_{11} & s_{12} & s_{13} & s_{14} & s_{15} & \cdots\} \\
S_2 = \{s_{21} & s_{22} & s_{23} & s_{24} & \cdots & \cdots\} \\
S_3 = \{s_{31} & s_{32} & s_{33} & \cdots & \cdots & \cdots\} \\
\cdots \\
S_n = \{s_{n1} & s_{n2} & s_{n3} & \cdots & \cdots & \cdots\}.
\end{array}$$

To form a comprehensive list of elements, start at the top left of the array, and select elements along successive diagonals, as follows: $s_{11}, s_{12}, s_{21}, s_{13}, s_{22}, s_{31}, s_{14}, \ldots$. This list is denumerable, with cardinality $\aleph_0$, but it may contain duplicate elements (why?); hence, it may not be the union $\bigcup_{j=1}^n S_j$. Fortunately, in order to obtain the union, we just remove any duplicates we find as we form the list. Thus, the cardinality of $\bigcup_{j=1}^n S_j$ will be at most $\aleph_0$. On the other hand, each S_j has cardinality $\aleph_0$, so that $\bigcup_{j=1}^n S_j$ must have cardinality at least $\aleph_0$. Therefore, the cardinality of $\bigcup_{j=1}^n S_j$ is equal to $\aleph_0$. Symbolically, $n\aleph_0 = \aleph_0$. **QED**

Note 2: Interestingly enough, to think along the lines of a limit problem, $\lim_{n\to\infty}(n+\aleph_0) = \aleph_0 + \aleph_0$, is erroneous, and the alternative $\infty + \aleph_0$ is meaningless. There is a fundamental distinction between the process of increasing without bound—Gauss called it a "potential infinity"—and what Cantor called an "actual infinity." The limit process entails the concept of unboundedness, which we symbolize by ∞, whereas $\aleph_0$ is infinity as an entity in and of itself, a transfinite number.

We can now tackle $\aleph_0 \cdot \aleph_0$, otherwise written as $(\aleph_0)^2$. This symbolism will extend Lemma 11.1, so that $(\aleph_0)^2$ shall indicate the cardinality of the union of a countably infinite collection of denumerable sets, $\bigcup_1^\infty S_j$. It turns out that $(\aleph_0)^2 = \aleph_0$, and in Exercise 11.8 you shall prove this, using a two-dimensional array.

Since $\aleph_0 \cdot \aleph_0$ is associated with a limitless rectangular array of elements (such as the Cartesian product $\mathbb{Z}^2$), we may associate $\aleph_0 \cdot \aleph_0 \cdot \aleph_0 \equiv (\aleph_0)^3$ with a limitless three-dimensional array of elements (such as $\mathbb{Z}^3$). The cardinality of such an array may be determined by noting that $\aleph_0 \cdot \aleph_0 \cdot \aleph_0 = \aleph_0 \cdot (\aleph_0)^2 = \aleph_0 \cdot \aleph_0 = \aleph_0$. Hence, $(\aleph_0)^3 = \aleph_0$. In other words, $(\aleph_0)^3 = \aleph_0$ is the cardinality of a denumerable union of denumerable unions of denumerable sets. Imagine an infinitely large box stacked full of children's toy blocks; if a rich uncle paid \$1 for each block, the cost of such a gift would be the *same* as the cost of an infinitely long single line of blocks, merely $\$\aleph_0$. (Now, do Exercise 11.13.) Following this thread of reasoning, we can continue onward through successively higher dimensions, and use mathematical induction to state that $(\aleph_0)^n = \aleph_0$, for any natural number n.

Note 3: Did you notice that by writing $\aleph_0 \cdot \aleph_0 \cdot \aleph_0 = \aleph_0 \cdot (\aleph_0)^2$, we tacitly assumed that the associative law works? Not being real numbers, that assertion may be groundless. Going back to definitions, the left side of this equation is the cardinality of $\bigcup_{i,j\in\mathbb{N}} S_{i,j}$, where each $S_{i,j}$ is a denumerable set in itself, and the sets aren't chosen in any particular order to enter into the union. The right side of the equation is $\aleph_0 \cdot (\aleph_0 \cdot \aleph_0)$, which expresses the cardinality of $\bigcup_{j=1}^\infty(\bigcup_{i=1}^\infty S_{i,j})$. This time, the sets enter into the union in the particular order $S_{1,1} \cup S_{2,1} \cup S_{3,1} \cup \cdots \cup S_{1,2} \cup S_{2,2} \cup S_{3,2} \cup \cdots \cup S_{1,3} \cup \cdots$. But regardless of their order, $\bigcup_{i,j\in\mathbb{N}} S_{i,j} = \bigcup_{j=1}^\infty(\bigcup_{i=1}^\infty S_{i,j})$, and therefore, $\aleph_0 \cdot \aleph_0 \cdot \aleph_0 = \aleph_0 \cdot (\aleph_0 \cdot \aleph_0) = \aleph_0 \cdot (\aleph_0)^2$. Recall that for Cartesian products, $\mathbb{Z}\times\mathbb{Z}\times\mathbb{Z} \neq \mathbb{Z}\times\mathbb{Z}^2$, but fortunately the corresponding cardinalities are equal.

The following theorem makes an especially interesting connection between a denumerable set and an uncountable set. Recall that a partition of a set S is a collection of *nonempty* and disjoint subsets of S whose union is S itself.

Theorem 11.1 Every infinite set S contains a denumerable subset.

Proof. In the case that S is denumerable, $S = \{s_1, s_2, s_3, \ldots\}$ answers the claim. (A typical denumerable proper subset of S is $D = \{s_2, s_3, \ldots\}$.) In the case that S is uncountable, let $\mathbb{H} = \{S_1, S_2, \ldots, S_n, \ldots\}$ be a partition of S into a denumerable infinity of nonempty subsets. Select S_1 requiring only that $S\backslash S_1$ be infinite, then S_2 such that $S\backslash(S_1 \cup S_2)$ is still infinite, and so forth. From each subset S_i, select a single element $s_i \in S_i$. Then the set $D = \{s_1, s_2, \ldots, s_n, \ldots\}$ is the required denumerable subset of S. **QED**

Note 4: It happens that in the proof just finished, a seemingly innocent step actually requires a completely new axiom of higher analysis! It is the unsettling *Axiom of Choice*: let $\mathbb{H}$ be any collection of nonempty sets $\mathbb{H} = \{S_a, S_b, \ldots\}$; then there always exists some set $M \subseteq \bigcup S_\alpha$ that is composed of exactly one element from each member subset of $\mathbb{H}$. Thus, M acts as a congress of one representative from each state $S_a, S_b, \ldots$ in the $\mathbb{H}$ nation. No one thought this selection process would be problematic, and it isn't for finite collections $\mathbb{H}$. But the axiom may be necessary when $\mathbb{H}$ is infinite, which occurs in the proof of Theorem 11.1.

Take $\mathbb{R}$ as our universal set, and let $\mathbb{H}$ be a partition of $\mathbb{R}$. Three examples of partitions $\mathbb{H}$ are (1) the finite partition $\mathbb{H} = \{\mathbb{Q}, \mathbb{I}\}$, (2) the denumerable partition $\mathbb{H} = \{[n, n+1) : n \in \mathbb{Z}\}$, and (3) the uncountable partition $\mathbb{H} = \{S_u\}$ with subscripts $u \in [0, 1)$, where the subset $S_u = \{u + n : n \in \mathbb{Z}\} = \{\ldots, u-2, u-1, u, u+1, \ldots\}$. For case (1), a typical $M = \{1, \pi\}$; for (2), a typical $M = \mathbb{Z}$; and for (3), a typical $M = [0, 1)$. However, a partition of $\mathbb{R}$ into more complex subsets may not allow us to formulate the set M, so we must establish its existence through the axiom of choice.

Read [12] Appendix III, [7] pp. 595–597, and [22] Section 5.5, to see what the axiom of choice entails.

Theorem 10.1 proved that no one-to-one correspondence existed between $\mathbb{N}$ and $\mathbb{R}$. Thus, the cardinality of $\mathbb{R}$ is not $\aleph_0$. Cantor labeled the cardinality of the real numbers "c," because the real number line was called the "continuum" at the time. Any set that is equivalent to $\mathbb{R}$ will have cardinality c. We now consider how to compare different cardinalities.

Let A have cardinality α and B have cardinality β. Suppose that A can be put into one-to-one correspondence with a *proper* subset of B, but no one-to-one correspondence exists between A and all of B. We shall then state that $\alpha < \beta$.

This is true whenever sets A and B are finite. One transfinite case is known already: $\aleph_0 < c$, since $\mathbb{N}$ can be put into one-to-one correspondence with a proper subset of $\mathbb{R}$ (name such a subset), but no such correspondence exists between $\mathbb{N}$ and all of $\mathbb{R}$. Thus, c is a *larger* transfinite number than $\aleph_0$. In fact, c is so vast that $\aleph_0$ behaves relative to c as does a finite set relative to $\aleph_0$! For instance, compare $n + \aleph_0 = \aleph_0$ to what the following lemma says.

Lemma 11.2 Let S be an infinite set, with cardinality α. Then $\aleph_0 + \alpha = \alpha$. In particular, $\aleph_0 + c = c$.

Proof. We must show that the union of a denumerable set and a set of cardinality α results in a set with cardinality α once again. Let S have cardinality α, and let $D = \{d_1, d_2, d_3, \ldots\}$ have cardinality $\aleph_0$, and let S and D be disjoint. Extract from S a denumerable subset $E = \{e_1, e_2, e_3, \ldots\}$, by virtue of Theorem 11.1. Then S can be partitioned as $S = \{e_1, e_3, \ldots\} \cup \{e_2, e_4, \ldots\} \cup S\backslash E$. Also, we have $D \cup S = D \cup E \cup S\backslash E$. There now exists the following bijection between S and $D \cup S$:

$$\begin{array}{cccccccccccccccc}
\{e_1 & e_3 & e_5 & \cdots\} & \cup & \{e_2 & e_4 & e_6 & \cdots\} & \cup & S\backslash E & = & S \\
\updownarrow & \updownarrow & \updownarrow & & & \updownarrow & \updownarrow & \updownarrow & & & \updownarrow & & \\
\{d_1 & d_2 & d_3 & \cdots\} & \cup & \{e_1 & e_2 & e_3 & \cdots\} & \cup & S\backslash E & = & D \cup S
\end{array}.$$

(The bijection $S\backslash E \leftrightarrow S\backslash E$ is simply the identity bijection of a set with itself.) Therefore, S and $D \cup S$ are equivalent, and must have the same cardinality, α. But the cardinality of $D \cup S$ is $\aleph_0 + \alpha$. Therefore, $\aleph_0 + \alpha = \alpha$. **QED**

Example 11.9 Using $\aleph_0 = n + \aleph_0$ and the preceding lemma, $c = \aleph_0 + c = (n + \aleph_0) + c = n + (\aleph_0 + c) = n + c$, and we conclude that $n + c = c$ for any natural number n. ■

Example 11.10 What is the cardinality of the infinite set of irrational numbers $\mathbb{I}$? Let its cardinality be α_I. By Lemma 11.2, the cardinality of $\mathbb{Q} \cup \mathbb{I}$ is $\aleph_0 + \alpha_I = \alpha_I$. But $\mathbb{Q} \cup \mathbb{I} = \mathbb{R}$, with cardinality c. Thus, $\alpha_I = c$. ■

A consequence (not proved in this book) of this example is that if we were to choose real numbers at the rate of one per second, but completely at random, from the number line, then there would be zero probability of selecting a rational number, regardless of how long we endured, even for all eternity! Only irrational numbers will turn up, since c is overwhelmingly larger than $\aleph_0$. Since $\mathbb{I}$ is an uncountable blizzard of numbers within $\mathbb{R}$ unrecognized by most people, we might speculate that it takes a conscious, organizing mind to identify and work with one of the rarest species of numbers, the rationals.

We continue with the arithmetic of transfinite numbers. Using Exercise 11.4b below, we will know that $[0, 1]$ is equivalent to $[1, 2]$, and also to $[0, 2]$. Each of these intervals has cardinality c by Exercise 11.11. But $[0, 1] \cup [1, 2] = [0, 2]$. Defining $2c \equiv c + c$ as the cardinality of the union of two sets each of which have cardinality c, we have $2c$ as the cardinality of $[0, 2]$. Therefore, $2c = c$, mimicking the aleph null case. Extending this, since $\bigcup_{j=1}^{n}[j-1, j] = [0, n]$, and $[0, n]$ has cardinality c, we conclude that $nc = c$ for any $n \in \mathbb{N}$. And finally, the entire set $\mathbb{R}$ may be formed by uniting countably many unit intervals: $\bigcup_{j\in\mathbb{Z}}[j-1, j] = \mathbb{R}$. In terms of cardinality, we have $\aleph_0 \cdot c = c$, which parallels Lemma 11.1.

The next, vastly "big idea" we come to is the meaning of $c^2 \equiv c \cdot c$. The easiest way to visualize this is to investigate the cardinality of the set of points inside the unit square, that is, $(0, 1)\times(0, 1)$. In Figure 11.2, for each y value, $0 < y < 1$, the segment between $\langle 0, y\rangle$ to $\langle 1, y\rangle$ (excluding these endpoints) corresponds one-to-one with the interval $(0, 1)$. Thus, the square's interior is the union of uncountably many copies of $(0, 1)$. The cardinality of this union is by definition c^2. At first, Cantor felt sure that this transfinite number was much larger than c.

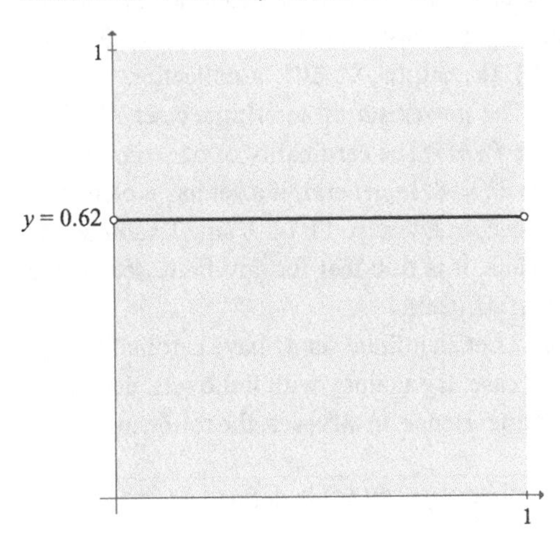

Figure 11.2. The open interval between $\langle 0, 0.62\rangle$ and $\langle 1, 0.62\rangle$ within a unit square. It is one of uncountably many such intervals, which together stack up to form the interior of the square.

Theorem 11.2 $c^2 = c$. (Cantor, 1877)

Proof. It is sufficient to prove that the cardinality of the set of points inside of a unit square is c. Let $\langle x, y\rangle$ be a point in the unit square $(0, 1) \times (0, 1)$, and let $z \in (0, 1)$. Express x and y as decimals, with the requirement that any terminating decimal will be written as the equal non-terminating decimal with endless nines. Now, split the strings of decimals for x and y into segments each of which ends with the first nonzero digit. For example, if $x = 0.35002 = 0.35001999\ldots$, and $y = \frac{\sqrt{2}}{2} = 0.70710678\ldots$, then splitting the string of digits of x yields $3, 5, 001, 9, 9, 9, \ldots$, and for y, we get $7, 07, 1, 06, 7, 8, \ldots$. Next, alternately select segments from x and y: $3, 7, 5, 07, 001, 1, 9, 06, 9, 7, 9, 8, \ldots$, and delete the commas to form the decimal number $z = 0.3750700119069798.\ldots$ This little algorithm creates a function $f(x, y) = z$, $\mathcal{D}_f = (0, 1)\times(0, 1)$, $\mathcal{R}_f = (0, 1)$, because each pair $\langle x, y\rangle$can form only one z-value. Moreover, any $z \in (0, 1)$ with non-terminating decimal form gives rise to a unique $\langle x, y\rangle$ by reversing the algorithm. For example, $z = 0.3086 = 0.3085999\ldots$ yields $3, 08, 5, 9, 9, 9, \ldots$. From this, form $\langle x, y\rangle = \langle 0.3599\ldots, 0.0899\ldots\rangle = \langle 0.36, 0.09\rangle$.

Thus, f is a bijection, $f : (0, 1)\times(0, 1) \leftrightarrow (0, 1)$. We conclude that the unit square is equivalent to the unit interval, and therefore, $c^2 = c$. **QED**

What follows now is the cardinality of the entire plane, $\mathbb{R}\times\mathbb{R} = \mathbb{R}^2$. The plane may be covered by tiling it with a countable collection of unit squares, like an endless linoleum floor (if we wish, we may make this a partition of $\mathbb{R}^2$ by specifying squares like $[0, 1)\times[0, 1)$). The Cartesian plane would therefore have cardinality $\aleph_0 \cdot c^2 = \aleph_0 \cdot c = c$. Consequently, the cardinality of $\mathbb{R}^2$ is still c. This reasoning extends to the set of points of three-dimensional space, $\mathbb{R}^3$. Its cardinality is $c^3 = c \cdot c^2 = c \cdot c = c$ once again. We must admit, then, that boundless space contains *just as many* points as the short interval $(0, 1)$ contains! Surely, we empathize with Cantor when he wrote to fellow mathematician Richard Dedekind about this discovery, "Je le vois, mais je ne le crois pas" ("I see it, but I do not believe it").

11.3 Other Transfinite Cardinal Numbers

At this point, we recognize two transfinite numbers, $\aleph_0$ and c. Cantor began a new line of reasoning by considering the subsets of a given set. For instance, let $A = \{\pi, \heartsuit, ✠\}$. The subsets of A are $\{ \varnothing, \{\pi\}, \{\heartsuit\}, \{✠\}, \{\pi, \heartsuit\}, \{\heartsuit, ✠\}, \{\pi, ✠\}, \{\pi, \heartsuit, ✠\}\}$, a collection of eight sets, notably including the empty set and A itself. The **power set** of an arbitrary set M is the collection of all subsets of M, and we shall denote it $\mathcal{P}(M)$. The cardinality of our example set A is, of course, three, and the cardinality of $\mathcal{P}(A)$ is $2^3 = 8$. In general, if a set has n elements, then its power set has 2^n elements, a good fact to prove, in Exercise 11.14. Using Example 4.2, we can quickly prove that $2^n > n$, for all $n \in \mathbb{N}$. Thus, it is true that for any finite set M, the cardinality of $\mathcal{P}(M)$ is greater than the cardinality of M itself.

Does this inequality remain true for infinite sets? Let an infinite set M have cardinality α. It takes a moment to ponder, but imagine $\mathcal{P}(M)$ in this case. By analogy with finite sets, define the symbol 2^α as the cardinality of $\mathcal{P}(M)$ when M is infinite. Hence, in all cases, the following holds.

> Let α be the cardinality of a set M. The cardinality of $\mathcal{P}(M)$ is defined as 2^α.

We will not let this new concept molder for a second as we state that the cardinality of $\mathcal{P}(\mathbb{N})$ will be $2^{\aleph_0}$, and the cardinality of $\mathcal{P}(\mathbb{R})$ will be 2^c. We can sense the immensity of $\mathcal{P}(\mathbb{R})$ by noting that $\mathbb{R}$, and every interval, open and closed, is a member of $\mathcal{P}(\mathbb{R})$; as well as every subset consisting of exactly one real number (c of them); and every subset with two reals (what is the cardinality of this collection?), three reals, and so on; and every subset that is the union of two disjoint intervals, three disjoint intervals, and so on; and *all* other subsets, with arbitrarily tangled structures!

Theorem 11.3 $2^{\aleph_0} = c$. That is, $\mathcal{P}(\mathbb{N})$ is equivalent to $\mathbb{R}$.

Proof. We appeal to the binary expansion of $x \in [0, 1]$ (an algorithm for converting fractions into binary representation was discussed in Exercise 10.5). Typical binary expansions are $\frac{9}{16} = \frac{1}{2^1} + \frac{0}{2^2} + \frac{0}{2^3} + \frac{1}{2^4} = 0.1001_2$ and $\frac{1}{5} = \frac{0}{2^1} + \frac{0}{2^2} + \frac{1}{2^3} + \frac{1}{2^4} + \frac{0}{2^5} + \frac{0}{2^6} + \cdots = 0.0011001100\ldots_2$, a repeating sequence of two zeros and two ones. To be sure, irrational numbers will generate non-repeating sequences of zeros and ones, just as they generate non-repeating decimal digits in base 10.

The binary digits have a not unusual role in the algorithm that follows. Sequentially number the binary place values after the decimal point in an expansion. If the binary digit of a place value is 1, select the number, and if the digit is 0, do not select the number. The diagram below shows this algorithm applied to $\frac{9}{16}$ and $\frac{1}{5}$. (Do Exercise 11.15 to apply the algorithm to other numbers.)

$\frac{9}{16} =$	0.	1	0	0	1_2		
$\mathbb{N} =$	{	1	2	3	4	5	$\cdots$ }
selected subset	{	1			4 }		

$\frac{1}{5} =$	0.	0	0	1	1	0	0	1	1	0	0	1	$\ldots_2$
$\mathbb{N} =$	{	1	2	3	4	5	6	7	8	9	10	11	$\cdots$ }
selected subset	{			3	4			7	8			11	$\cdots$ }

Notice that the algorithm can be reversed, so that for instance, $\{1, 4\}$ yields the fraction $\frac{9}{16}$, and no other. This means that the algorithm is actually a bijection between $[0, 1]$ and some subsets in $\mathcal{P}(\mathbb{N})$. A slight complication arises because any terminating binary expansion is equal to an expansion with an endless sequence of 1s, as in $\frac{9}{16} = 0.1001_2 = 0.10001111\ldots_2$. But lo! This is just what is needed, because these twin expansions will yield two subsets of $\mathbb{N}$, and both are required.

All the terminating expansions are in one-to-one correspondence with all the *finite* subsets of $\mathbb{N}$ by the algorithm. All of these subsets will be collected into a set $\mathbb{N}_{Fin}$. Interestingly, zero is mapped to the empty set, so $\varnothing \in \mathbb{N}_{Fin}$. Convince yourself that $\mathbb{N}_{Fin}$ has cardinality $\aleph_0$ (or see the paragraph after this proof).

Now, all the real numbers in $(0, 1]$ are either irrational, or non-terminating rational, or else terminating rational, in which case they shall now be expressed with an endless tail of 1s. The algorithm therefore defines a bijection between all $x \in (0, 1]$ and all the *denumerable* subsets of $\mathbb{N}$. Define $\mathbb{N}_{Den}$ as the collection of all denumerable subsets of $\mathbb{N}$. Thus, we have for instance $\frac{1}{5} \leftrightarrow \{3, 4, 7, 8, 11, 12, \ldots\} \in \mathbb{N}_{Den}$, and in its new form, $\frac{9}{16} \leftrightarrow \{1, 5, 6, 7, 8, 9, 10, \ldots\} \in \mathbb{N}_{Den}$.

We have just found that $\mathbb{N}_{Den}$ and $(0, 1]$ are equivalent, both with cardinality c. Furthermore, $\mathbb{N}_{Fin} \cup \mathbb{N}_{Den} = \mathcal{P}(\mathbb{N})$, an interesting partition of this power set. Since the cardinality of $\mathcal{P}(\mathbb{N})$ is $2^{\aleph_0}$, it must be true that $\aleph_0 + c = 2^{\aleph_0}$. But Lemma 11.2 indicates that $\aleph_0 + c = c$, so we conclude that $2^{\aleph_0} = c$, and that the cardinality of $\mathcal{P}(\mathbb{N})$ is c. **QED**

Let's demonstrate that $\aleph_0$ is the cardinality of $\mathbb{N}_{Fin}$ in the proof above. Denote $\varnothing$, the only subset of $\mathbb{N}$ with no elements, by $\mathbb{N}_0$. Let $\mathbb{N}_1 = \{\{1\}, \{2\}, \{3\}, \cdots\} \subset \mathbb{N}_{Fin}$, the collection of singleton subsets of $\mathbb{N}$. $\mathbb{N}_1$ is clearly denumerable. Next, let $\mathbb{N}_2$ be the collection of doublet subsets of $\mathbb{N}$, so that $\mathbb{N}_2 = \{\{1,2\}, \{1,3\}, \{1,4\}, \{2,3\}, \{1,5\}, \{2,4\}, \{1,6\}, \{2,5\}, \cdots\} \subset \mathbb{N}_{Fin}$.

Within $\mathbb{N}_2$, consider $A_n = \{\textit{subsets of } \mathbb{N}_2 : \textit{the sum of the elements of the subset is } n\}$, meaning that $A_3 = \{\{1,2\}\}$, $A_4 = \{\{1,3\}\}$, $A_5 = \{\{1,4\}, \{2,3\}\}$, and so on. Because each A_n is finite and $\mathbb{N}_2 = A_3 \cup A_4 \cup A_5 \cup \cdots$, we discover that $\mathbb{N}_2$ is also denumerable. By a similar argument, we can show that each $\mathbb{N}_j$ is denumerable. Now, $\mathbb{N}_{Fin} = \bigcup_0^\infty \mathbb{N}_j$, which is a countable union of denumerable sets. Consequently, the cardinality of $\mathbb{N}_{Fin}$ is $(\aleph_0)^2 = \aleph_0$.

Since we have seen that $\aleph_0 < c$, the theorem above shows right away that $2^{\aleph_0} > \aleph_0$. This happens to be true for any cardinal number whatsoever, as our next, and very famous, theorem proves.

Theorem 11.4 Cantor's Theorem For any cardinal number α, $2^\alpha > \alpha$. That is, the power set of any set has greater cardinality than that of the set itself.

Proof. The case of a finite set M was proved early in this subsection, concluding with $2^n > n$. Let M now be an infinite set, with cardinality α. Then $\mathcal{P}(M)$ has cardinality 2^α. Assume, on the contrary, that $2^\alpha \le \alpha$. This implies that the subsets of M can be put into one-to-one correspondence with the elements of M, or perhaps with a subset of M. Thus, we shall prove that even $2^\alpha = \alpha$ leads to a contradiction.

Let $x \in M$. The supposed bijection would allow us to write $x \leftrightarrow M_x$, where $M_x \in \mathcal{P}(M)$ is the unique subset of M that is paired with x. It occurs to us that sometimes, $x \in M_x$, and sometimes not. We now focus with great suspicion on the collection of subsets M_x such that $x \notin M_x$. This collection is described as $\{M_x \in \mathcal{P}(M) : x \notin M_x\}$, and we label it Φ. As such, $\Phi \subseteq \mathcal{P}(M)$. The supposed bijection allows us to pinpoint the specific subset of M that contains *all* those x identified by Φ. We name this subset M_z, with the full assurance that there exists some $z \in M$ such that $z \leftrightarrow M_z$. Now, Cantor's famous question, "Is $M_z \in \Phi$?" If indeed $M_z \in \Phi$, then $z \notin M_z$. But $z \in M_z$ must be correct, since M_z is, by definition, the set of all the elements x identified by Φ, and z has been so identified by assuming that $M_z \in \Phi$. We therefore negate our latest assumption, and say that $M_z \notin \Phi$. Unfortunately, this implies that $z \in M_z$, which in turn means that M_z has become a member of Φ! Thus, we find that both of the claims $M_z \in \Phi$ and $M_z \notin \Phi$ lead to self-contradictions. This is checkmate, for where else can M_z be found, except either in Φ, or not in Φ? Therefore, we must repudiate the original assumption, which was that the bijection $x \leftrightarrow M_x$ existed. Our conclusion is that M is not equivalent to $\mathcal{P}(M)$, and that $2^\alpha > \alpha$. **QED**

Using Theorem 11.3 and Cantor's Theorem, we can start with $\aleph_0$ and create an endless chain of increasingly larger cardinals: $\aleph_0 < 2^{\aleph_0} = c < 2^c < 2^{(2^c)} < 2^{(2^{(2^c)})} < \cdots$, with each transfinite number representing the cardinality of the power set of the previous set, as seen below.

SET	$\mathbb{N}$	$\mathbb{R}$	$\mathcal{P}(\mathbb{R})$	$\mathcal{P}(\mathcal{P}(\mathbb{R}))$	$\mathcal{P}(\mathcal{P}(\mathcal{P}(\mathbb{R})))$	$\cdots$
CARDINALITY	$\aleph_0$	c	2^c	$2^{(2^c)}$	$2^{(2^{(2^c)})}$	$\cdots$.

The endless ascension of transfinite cardinal numbers seems like a good place to end our discussion of the infinite. Georg Cantor discovered much more than is discussed here, and his discoveries have fascinated and troubled mathematicians, philosophers, cabalists, and assorted dabblers. But the infinite is properly a *mathematical* concept, for no other human endeavor is capable of navigating far into that abyss.

Note 5: An informal but lucid exposition of the above topics and other fascinating discoveries by Cantor can be found in the wonderful article, "Georg Cantor and the Origins of Transfinite Set Theory," by Joseph W. Dauben, in *Scientific American*, vol. 248, June 1983, pp. 122–131.

A summary of cardinal arithmetic follows. Note that $\mathbb{R}$ and c can be replaced by any uncountable set S with cardinality α, except in $2^{\aleph_0} = c$.

EXPRESSION	*MEANING*	*EXPRESSION*	*MEANING*
$n+m$	The cardinality of the union of two disjoint finite sets is $n+m$.	$\aleph_0 + c = c$	The cardinality of the union of a denumerable set and $\mathbb{R}$ remains c.
$n + \aleph_0 = \aleph_0$	The cardinality of the union of a finite and a denumerable set remains $\aleph_0$.	$n \cdot c = c$	The cardinality of the union of n sets of cardinality c remains c.
$\aleph_0 + \aleph_0 = \aleph_0$	The cardinality of the union of two denumerable sets remains $\aleph_0$.	$\aleph_0 \cdot c = c$	The cardinality of the Cartesian product of a denumerable set and $\mathbb{R}$ remains c.
$n \cdot \aleph_0 = \aleph_0$	The cardinality of the union of n denumerable sets remains $\aleph_0$.	$c^2 = c$	The cardinality of the Cartesian product $\mathbb{R}^2$ remains c.
$(\aleph_0)^2 = \aleph_0$	The cardinality of the Cartesian product of two denumerable sets remains $\aleph_0$.	$c^n = c$	The cardinality of $\mathbb{R}^n$ remains c.
$(\aleph_0)^n = \aleph_0$	The cardinality of the Cartesian product of n denumerable sets remains $\aleph_0$.	2^n	The cardinality of the power set $\mathcal{P}(M)$ of a set M of cardinality n is 2^n.
$\aleph_0 < c$	The cardinality of $\mathbb{N}$ is less than the cardinality of $\mathbb{R}$.	$2^{\aleph_0} = c$	The cardinality of the power set $\mathcal{P}(\mathbb{N})$ is c.
$n + c = c$	The cardinality of the union of a finite set and $\mathbb{R}$ remains c.	$c < 2^c$	The cardinality of $\mathbb{R}$ is less than the cardinality of the power set $\mathcal{P}(\mathbb{R})$.

11.4 Exercises

1. Find a bijection between $\mathbb{N}$ and the set of fractions $\{\frac{n}{5} : n \in \mathbb{Z}\}$, thus proving that this subset of $\mathbb{Q}$ is denumerable.

2. Prove that the set of multiples of three, $\{\pm 3n\}_0^\infty$, is countably infinite by finding a bijection of that set onto $\mathbb{N}$.

3. Any single musical composition must be able to be played in a maximum of, say, 3 days, nonstop. What is your opinion, is the number of distinguishable compositions finite, denumerable, or uncountable? All possible compositions, even noise, must be counted, provided that they can be heard as different by a musician. Rather than counting the possible music scores themselves, which gets too difficult, count the number of digitized songs possible. Hint: The standard for digital recording takes a measurement of the analog wave coming from the artist's soundtrack, 44100 times per second. Each measurement consists of one of $2^{16} = 65536$ voltage levels. Thus, every second of music is converted into a string of 44100 measurements, and each measurement may be any one of 65536 possible voltages.

4. a. Construct a bijection $q : [0, 2\pi) \leftrightarrow [0, 1)$, showing that $[0, 2\pi)$ is infinite. Hint: Draw $[0, 2\pi)$ on the X axis, and $[0, 1)$on the Y axis, and think about a line segment.
   b. Generalize what you did in **a** to a bijection between any nonempty intervals $[a, b]$ and $[c, d]$.

5. Let S be a set with n elements. Prove that no bijection $f : P \leftrightarrow S$ can exist if P is a proper subset of S. Logically speaking, this asserts that a set cannot be both infinite and finite.

6. In the fourth stanza of "Amazing Grace," quoted beneath the chapter's title, the writer lyrically describes how long he will reside in heaven because of salvation by faith. How has he precisely described a countable infinity of years?

7. Prove that the number of points on the curve $y = x^2$ is infinite. Hint: It is sufficient to show that the half-parabola $\{\langle x, x^2\rangle : x \geq 0\}$ is infinite.

8. Prove that the countably infinite union of denumerable sets $\bigcup_{j=1}^{\infty} S_j$ is still a denumerable set, that is, $(\aleph_0)^2 = \aleph_0$, by using the array below. It is important to note that this array exists only by invoking the axiom of choice (see Note 4). The displayed order of each S_i is actually the effect of a bijection specific to each S_i, namely, $f_i(j) = s_{ij}$. But the set of such bijections, B_{S_i}, is uncountable for each S_i, and there are infinitely many S_i. Hence, the axiom of choice is necessary to select one bijection from each B_{S_i} and thence arrive at our array. Hint: Look at Chapter 9.

$$\begin{aligned} S_1 &= \{s_{11} \quad s_{12} \quad s_{13} \quad \cdots\} \\ S_2 &= \{s_{21} \quad s_{22} \quad s_{23} \quad \cdots\} \\ &\cdots \\ S_n &= \{s_{n1} \quad s_{n2} \quad s_{n3} \quad \cdots\} \\ S_{n+1}& \\ &\cdots \end{aligned}$$

9. Prove that the Cartesian product $\mathbb{Z}\times\mathbb{Z}$ has cardinality $\aleph_0$.

10. Prove that the set of prime numbers $\{2, 3, 5, 7, \ldots\}$ has cardinality $\aleph_0$. Hint: Don't search for a bijection between this set and N.

11. Prove that (a, b) and $[a, b]$ have cardinality c. Hint: Don't search for a bijection between them.

12. Prove that the set of transcendental numbers has cardinality c. Hint: The set of transcendental numbers is $\overline{\mathbb{A}}$.

13. A vendor originally sells toy blocks for \$1 each. We saw that an infinitely large box of blocks costs $\$\aleph_0$. Suppose that the cost to manufacture each block is 40¢, and that the vendor advertises a huge sale: the price is 50¢ per block this week only. How much profit will the vendor realize now? Justify your answer.

14. Prove that for a set M with n elements, $\mathcal{P}(M)$ contains 2^n subsets. Hint: Use the binomial theorem (Exercise 4.7) with $a = b = 1$. Explain how that theorem applies.

15. Using the algorithm in the proof of Theorem 11.3,
    a. find the sets produced by $x = 0 = 0.0\ldots_2$, $x = \frac{\pi}{4}$, $x = \frac{11}{32}$, and $x = 1 = 0.1111\ldots_2$. Notice that the terminating form of 1, namely 1.0_2 can't be used in the algorithm.
    b. Given the set of prime numbers $\{2, 3, 5, 7, 11, \ldots\}$, generate the first few digits of the corresponding unique binary expansion, and find an equal decimal for this very peculiar number! What might it be used for?

16. **Subtraction** of transfinite numbers may be defined under the condition that the numbers not represent the same cardinality. Let α and β be the cardinalities of infinite sets S and T, respectively, with $\alpha < \beta$. Then $\beta - \alpha$ is defined as the cardinality of the set $T \setminus S$.
    a. Let $T = [2, 7]$ and $S = \{q \in \mathbb{Q} : 2 \leq q \leq 7\}$. The cardinality of $T \setminus S$ is $c - \aleph_0$. What does this equal?
    b. Let $T = [2, 7]$ and $S = \{q \in \mathbb{Q} : q \geq 6\}$. The cardinality of $T \setminus S$ is $c - \aleph_0$ again. What does this equal?
    c. Prove that $c - \aleph_0 = c$ in general.
    d. Explain why $c - c$ is ambiguous.

17. A very useful theorem in comparing infinite sets is the **Schröder-Bernstein Theorem**. It states that given two sets L and M with the property that L is equivalent to a subset of M, and at the same time M is equivalent to a subset of L, then L and M are equivalent. With a bit of thought, we see that finite sets cannot behave this way if the subsets mentioned are *proper* subsets of L and M. Thus, the impact of the theorem occurs when we apply it to proper subsets of infinite sets L and M. The proof is quite elaborate, and can be found in [33], pp. 17–19 (the theorem is perhaps incorrectly named there, as Cantor did not prove it).

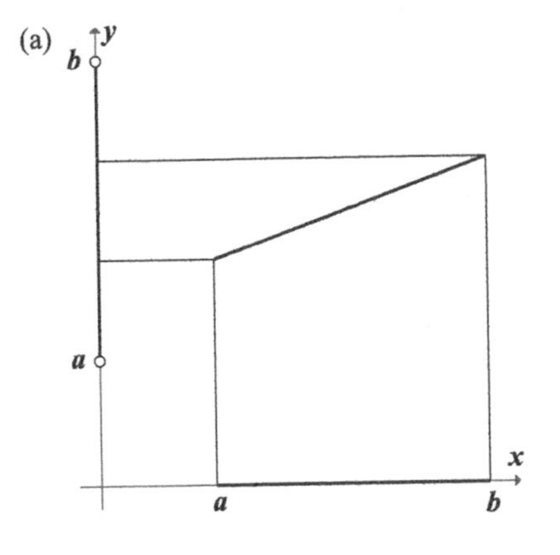

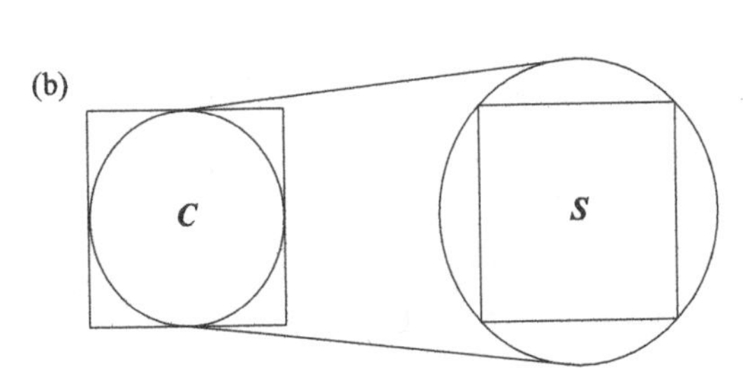

Figure 11.3.

a. Use the Schröder-Bernstein theorem to find an alternative proof to that of Exercise 11.11 that (a, b) and $[a, b]$ have equal cardinality c. Hint: Find a really simple bijection on (a, b) into $[a, b]$, and then use Figure 11.3a to find a bijection on $[a, b]$ into some part of (a, b).

b. Use the Schröder-Bernstein theorem to prove that the set of points S interior to a square is equivalent to the set of points C interior to a circle. Hint: In Figure 11.3b, S and C are shown, with $C \subset S$ on the left, and S is copied at right.

12

The Complete, Ordered Field of Real Numbers

> *... when rational numbers fail us irrational numbers take their place and prove exactly those things which rational numbers could not prove ... we are moved and compelled to assert that they truly are numbers ... On the other hand, other considerations compel us to deny that irrational numbers are numbers at all.*
>
> Michael Stifel (ca. 1486–1567), *Arithmetica Integra* (1544), quoted in Morris Kline, *Mathematical Thought from Ancient to Modern Times*, pg. 251

Up to this point, the set of real numbers $\mathbb{R}$ has been studied without considering its distinguishing characteristic. That is, the set $\mathbb{R}$ is infinite, and it is an ordered field, but other systems have these properties as well. We shall now look at $\mathbb{R}$ as an axiomatic system that extends the system $\mathbb{Q}$. Such an extension is crucial because as simple an equation as $x^2 - 7 = 0$ has no solution in $\mathbb{Q}$. But how do we know that $\sqrt{7} \notin \mathbb{Q}$? We begin with the classic proof that irrational numbers do, in fact, exist.

12.1 Irrational Numbers

The world of Pythagoras of Samos, around 500 B.C., found it very natural to entangle number and geometry with mysticism. They held a sacred tenet that the physical world was explainable strictly by whole numbers and their ratios. This philosophy seemed reasonable, for nothing non-rational—irrational—could be observed or imagined. But the capital theorem of that time, the Pythagorean theorem, contained at its core a consequence that would explode that philosophy. A legend tells of a student of the Pythagorean school who deduced that a diagonal of the unit square, of length $\sqrt{2}$ by their theorem, could not be expressed as a ratio $\frac{a}{b}$ of whole numbers. His reasoning exemplified the deductive thinking that the Pythagoreans were developing.

Lemma 12.1 $\sqrt{2}$ is not a rational number. That is, $\sqrt{2} \notin \mathbb{Q}$.

Proof. Assume on the contrary that $\sqrt{2} = \frac{a}{b}$, and that the ratio is in lowest terms. Then $2 = \frac{a^2}{b^2}$, and $2b^2 = a^2$. This implies that a^2 is an even number. This, in turn, implies that a itself is even (convince yourself of this). It follows that $a = 2m$, for some integer m. Continuing the line of reasoning, $4m^2 = a^2 = 2b^2$. Thus, $2m^2 = b^2$, and b^2 is even. But then b itself is even, and this contradicts the initial assumption that the ratio $\frac{a}{b}$ was in lowest terms. Therefore, no fraction whatsoever can equal $\sqrt{2}$, meaning that $\sqrt{2}$ is irrational. **QED**

This proof must have been at once a triumph of logic, and a defeat for a carefully polished view of the world. Philosophers have been uneasy with mathematicians ever since! The proof of the irrationality of $\sqrt{2}$ can be extended to $\sqrt{p}$, for p any prime number. But a more comprehensive technique uses the following well-known theorem.

Theorem 12.1 The Rational Root Theorem Let $P(x) = p_n x^n + p_{n-1}x^{n-1} + \cdots + p_1 x + p_0$ be a polynomial with integer coefficients, with $p_n \neq 0$ and $p_0 \neq 0$. If $r = \frac{a}{b}$ is a root of $P(x)$, and is in lowest terms, then b is a factor of the lead coefficient p_n, and a is a factor of the constant term p_0.

Proof. Substituting, we have $P(\frac{a}{b}) = p_n(\frac{a}{b})^n + p_{n-1}(\frac{a}{b})^{n-1} + \cdots + p_1(\frac{a}{b}) + p_0 = 0$, so that $p_n a^n + p_{n-1}ba^{n-1} + \cdots + p_1 b^{n-1}a + p_0 b^n = 0$. Expressing this as $p_n a^n = -b(p_{n-1}a^{n-1} + \cdots + p_1 b^{n-2}a + p_0 b^{n-1})$, and remembering that all of these quantities are integers, b must divide the left side of the equation, since it obviously is a factor of the right side. And since b doesn't divide a^n (why not?), then b must be a factor of p_n. Next, we may reexpress $p_n a^n + p_{n-1}ba^{n-1} + \cdots + p_1 b^{n-1}a + p_0 b^n = 0$ as $p_0 b^n = -a(p_n a^{n-1} + p_{n-1}ba^{n-2} + \cdots + p_1 b^{n-1})$, from which we similarly deduce that a must be a factor of p_0 (notice that $a \neq 0$ since $r \neq 0$). **QED**

To see how efficient this theorem is, consider two examples.

Example 12.1 Prove that for any prime number p, $\sqrt{p}$ is irrational. Note that $\sqrt{p}$ is a root of $P(x) = x^2 - p = 0$. If $\sqrt{p}$ were a rational number $\frac{a}{b}$, then the rational root theorem would state that $b = \pm 1$, and $a = \pm p$ or $a = \pm 1$. Thus, $\sqrt{p}$ would equal either $\pm\frac{p}{1}$ or $\pm\frac{1}{1}$, which is false for any prime number. Therefore, $\sqrt{p}$ is irrational (and $\sqrt{p} \notin \mathbb{Q}$). ■

Example 12.2 Prove that $\sqrt{3} + \sqrt{8}$ is irrational. Letting $x = \sqrt{3} + \sqrt{8}$, we have $x - \sqrt{3} = \sqrt{8}$, so that squaring, $x^2 - 2\sqrt{3}x + 3 = 8$. Now $x^2 - 5 = 2\sqrt{3}x$, and squaring again and simplifying, $x^4 - 22x^2 + 25 = 0$. Thus, $P(x) = x^4 - 22x^2 + 25$ has $\sqrt{3} + \sqrt{8}$ as one of its roots (check it!). If this root were rational, then it would equal $\pm\frac{25}{1}$, $\pm\frac{5}{1}$, or $\pm\frac{1}{1}$, according to the rational root theorem. Since none of these choices equals $\sqrt{3} + \sqrt{8}$, this number must be irrational. See Figure 12.1. ■

The set of irrational numbers has already been given the symbol $\mathbb{I}$. We intend to prove that $\mathbb{I} \subset \mathbb{R}$.

12.2 The Axioms for the Real Numbers

A less obvious reason for an extension of the rational number system is that certain infinite series of rational numbers do not have a sum that is rational, for example, $\sum_{n=0}^{\infty} \frac{1}{n!} = e$. In other words, addition is closed in $\mathbb{Q}$ provided that only *finitely* many summands are involved.

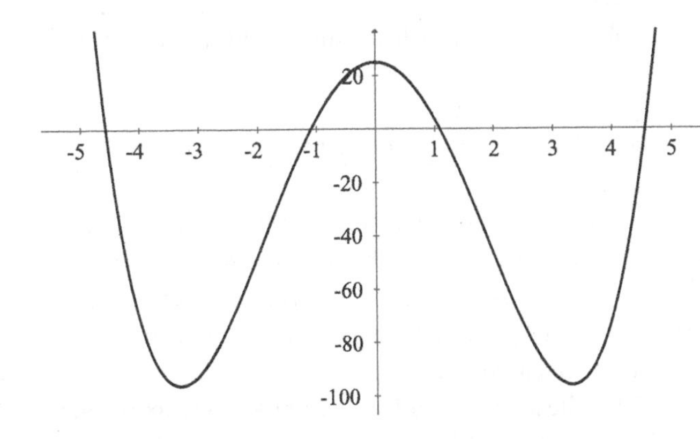

Figure 12.1. The four real roots of $P(x) = x^4 - 22x^2 + 25$ are seen. All of them are irrational by Theorem 12.1, the rational root theorem.

We would like closure for all convergent series. We need a system of axioms that preserves the properties of $\mathbb{Q}$, but encompasses irrational numbers. Because of all this, a new axiom will be added to the system for $\mathbb{Q}$ in Chapter 3, thus forming a new axiomatic system of things called "real numbers," and we intend that $\mathbb{Q} \subset \mathbb{R}$. Axioms inherited from previous axiomatic systems are designated "RQZN," "RQZ," or "RQ."

Create a collection of ordered pairs $\langle A|B \rangle$ of subsets of $\mathbb{Q}$, to be called **real numbers**, along with two operations $+$ and $\times$, that satisfies the list of axioms below. We will make a short investigation of the mysterious sets $A, B \subset \mathbb{Q}$ near the end of this chapter. For ease of notation, real numbers $\langle A|B \rangle$ will be denoted by single letters such as a, b, or x, and the operation $a \times b$ will usually be written ab, or occasionally $a \cdot b$.

RQZN1: For all $a, b \in \mathbb{R}$, $a + b$ and ab are unique elements of $\mathbb{R}$. (Closure and uniqueness)

RQZN2: For all $a, b \in \mathbb{R}$, $a + b = b + a$ and $ab = ba$. (Commutativity)

RQZN3: For all $a, b, c \in \mathbb{R}$, $a + (b + c) = (a + b) + c$ and $a(bc) = (ab)c$. (Associativity)

RQZN4: For all $a, b, c \in \mathbb{R}$, $a(b + c) = ab + ac$. (Distributivity of $\times$ across $+$)

RQZN5: There exists a number $1 \in \mathbb{R}$ such that for any $a \in \mathbb{R}$, $1a = a$. (Existence of a multiplicative identity, named "one")

RQZN6: Let M be a set of natural numbers such that $1 \in M$, and whenever $k \in M$, then $k + 1 \in M$. Then $M = \mathbb{N}$, that is, M contains all the natural numbers. (Mathematical induction)

RQZ7: There exists a number $0 \in \mathbb{R}$, not equal to 1, such that for any $a \in \mathbb{R}$, $0 + a = a$. (Existence of an additive identity, named "zero")

RQZ8: For any $a \in \mathbb{R}$, there exists a unique, corresponding number $-a \in \mathbb{R}$ such that $-a + a = 0$. (Existence of opposites)

RQZ9: There exists a subset $\mathbb{R}^+$ of $\mathbb{R}$ such that (1) for all nonzero $a \in \mathbb{R}$, either $a \in \mathbb{R}^+$ or $-a \in \mathbb{R}^+$ but not both, and (2) for any $a, b \in \mathbb{R}^+$, both $a + b \in \mathbb{R}^+$ and $ab \in \mathbb{R}^+$. (Existence of a positive class)

RQ10: For any $a \in \mathbb{R}$ except 0, there exists a unique, corresponding number $a^{-1} \in \mathbb{R}$ such that $a^{-1}a = 1$. (Existence of reciprocals)

R11: For any nonempty subset $M \subset \mathbb{R}$, if M has an upper bound in $\mathbb{R}$, then M has a least upper bound, or supremum, in $\mathbb{R}$. (Completeness Axiom)

All the properties that were proved for rational numbers continue to hold for the real numbers, since all the axioms for $\mathbb{Q}$ are still affirmed here. All the statements of inequality from Chapter 2 are now based on the following one, which is only slightly modified to fit Axiom RQZ9.

$a < b$ and $b > a$ both mean that $a + x = b$ for some $x \in \mathbb{R}^+$.

The trichotomy property of $\mathbb{R}$ can be established through Axiom RQZ9 in a similar way to how it was derived for the rational numbers in Exercise 3.3a (do this as Exercise 12.2). Thus, we have the partition $\mathbb{R} = \mathbb{R}^- \cup \{0\} \cup \mathbb{R}^+$. Any $x \in \mathbb{R}^+$ is called a **positive real** number. A number is a **negative real** if and only if its opposite is a positive real number. The positive class $\mathbb{R}^+$ is called the **positive reals**, and the set $\mathbb{R}^-$ is the **negative reals**.

Since the first ten axioms are identical to those of system $\mathbb{Q}$, the completeness axiom must be the distinguishing characteristic of the real numbers! Define as **complete** any set for which the completeness axiom is valid.

The set of real numbers $\mathbb{R}$ is a complete, ordered field.

We now look into the terminology of the completeness axiom. A real number u is an **upper bound** for a nonempty set $S \subset \mathbb{R}$ if and only if $s \leq u$ for all $s \in S$. In other words, u is greater than or equal to all members of S. Likewise, a real number l is a **lower bound** for S if and only if $l \leq s$ for all $s \in S$, meaning that l is less than or equal to all members of S. We observe that upper and lower bounds are not unique to a set.

A set $S \subset \mathbb{R}$ is **bounded** if and only if it is a subset of some open interval (l, u) of real numbers, and (l, u) is naturally called a **bounding interval** for S. Any set that isn't bounded is called **unbounded**. As an application, a function f is bounded if and only if its range is a subset of some bounding interval (l, u), and it usually suffices that the range be contained in some closed interval $[-B, B]$ about zero. This requirement is expressed as $|f(x)| \leq B$. A function is unbounded if and only if its range is unbounded.

Example 12.3 Every interval with real endpoints, whether closed, open or otherwise, is obviously bounded. For the bounded interval $S = (0, 5)$, zero or any negative number will serve as a lower bound, and all $x \geq 5$ are perfectly good upper bounds. ■

Example 12.4 Let $S = \{\frac{1}{2^n} : n \in \mathbb{N}\}$. Then S is bounded, and a (rather useless) upper bound for it is $u = 43$. Any negative number would serve as a lower bound, as would $l = 0$. ■

Example 12.5 The empty set is bounded (why?). The union or intersection of any *finite* number of intervals with real endpoints is bounded. Prove this in Exercise 12.3. ■

Example 12.6 $\mathbb{N}$ and $\mathbb{Q}$ are unbounded. $\mathbb{N}$ has all $x \leq 1$ as lower bounds, but has no upper bound. ■

Example 12.7 The sine function is bounded, since for instance, $\mathcal{R}_{\sin} \subset [-3, 3]$ (its domain is unbounded, of course). The function $y = \ln(\frac{1+x}{1-x})$ is unbounded, but has a bounded domain. ■

Example 12.8 Intervals $(-\infty, b)$ and $(-\infty, b]$ have upper bounds, but are unbounded sets. Similarly, (a, ∞) and $[a, \infty)$ have lower bounds, but are unbounded. ■

12.3 The Supremum and Infimum of a Set

In Example 12.3, we see that among all upper bounds of $S = (0, 5)$, the *least* of them is 5. That is, if u is any upper bound for S, then $5 \leq u$. Equivalently, 5 is the least upper bound because any number $x < 5$ will have at least one $s \in S$ such that $s > x$ (indeed, infinitely many members of $(0, 5)$ will exceed such an x). In general:

> The **supremum** of a nonempty set S is a number α such that (1) α is an upper bound of S, and (2) for any number $x < \alpha$, there exists some member $s \in S$ such that $s > x$.

The notation for this is $\sup S = \alpha$, as in $\sup(0, 5) = 5$. We can always say that if u is an upper bound for S, then $\sup S \leq u$. And we have just seen that it is possible that $\sup S \notin S$. Synonyms for "supremum" are **least upper bound** and lowest upper bound.

Continuing with $S = (0, 5)$, we see that among all lower bounds, the *greatest* of them is zero, because any number $x > 0$ will have at least one $s \in S$ such that $s < x$ (indeed, infinitely many members of S will be exceeded by such an x). In general, a **greatest lower bound** or **infimum** for a nonempty set S is a number β such that (1) β is an lower bound of S, and (2) for any number $x > \beta$, there exists some member $s \in S$ such that $s < x$. The notation for this is $\inf S = \beta$, as in $\inf(0, 5) = 0$. We can always say that if l is any lower bound for S, then $\inf S \geq l$. And we have just seen that it is possible that $\inf S \notin S$.

The two previous paragraphs intentionally parallel each other, for the supremum and infimum perform mirror image roles.

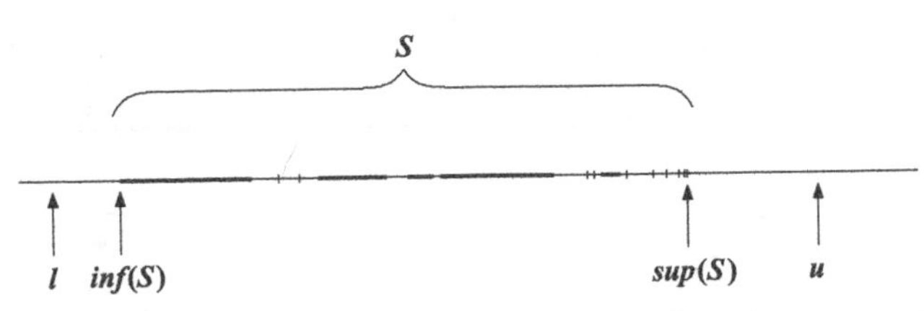

Figure 12.2. An arbitrary, nonempty set S with a typical upper bound u and lower bound l, along with its supremum and infimum.

Example 12.9 For the bounded set $S = [2, 4)$, $\sup S = 4$. And $\inf S = 2$, so that $\inf S$ belongs to S. ■

Example 12.10 For the set $\mathbb{N}$, $\inf \mathbb{N} = 1$, because (1) $\beta = 1$ is a lower bound for $\mathbb{N}$, and (2) for any $x > 1$, for instance, $x = 1.003$, some member of $\mathbb{N}$ (namely 1), is less than x. On the other hand, $\sup \mathbb{N}$ does not exist, since $\mathbb{N}$ has no upper bound. ■

Example 12.11 For the bounded set $S = \{\frac{1}{2^n} : n \in \mathbb{N}\}$, $\sup S = \frac{1}{2}$ because (1) $\alpha = \frac{1}{2}$ is an upper bound for S, and (2) any $x < \alpha$ will have some member of S that is greater than x. Also, $\inf S = 0$, since (1) $\beta = 0$ is a lower bound for S, and (2) any $x > \beta$ will have at least one member $\frac{1}{2^n} \in S$ such that $\frac{1}{2^n} < x$ (actually, infinitely many members of S will be less than such an x). Here, S contains its supremum but not its infimum. ■

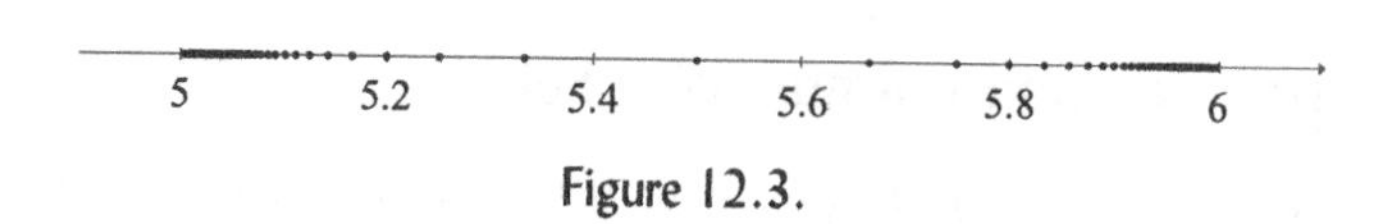

Figure 12.3.

Example 12.12 Let $S = \{5 + \frac{1}{n+1} : n \in \mathbb{N}\} \cup \{6 - \frac{1}{n+1} : n \in \mathbb{N}\}$, seen in Figure 12.3. Then inf $S = 5$ and sup $S = 6$. Although S is bounded, this time, sup $S \notin S$ and inf $S \notin S$. ■

Corollary 12.1 Every nonempty set $S \subset \mathbb{R}$ that has a lower bound in $\mathbb{R}$ has an infimum in $\mathbb{R}$.

Proof. This proof takes advantage of the order of real numbers. Given that l is a lower bound for S, we know that $l \leq s$ for all $s \in S$. Then $-l \geq -s$ for all $s \in S$. A new set appears here, which we shall denote by $-S$, and which contains all the opposites of S: $-S \equiv \{-s : s \in S\}$ (see Figure 12.4). We now see that $-l$ is an upper bound for $-S$. By the completeness axiom, $-S$ must have a supremum, say, $\alpha = \sup(-S)$. Being the supremum, $-s \leq \alpha$ for all $-s \in -S$, so that $-\alpha \leq s$ for all $s \in S$. Thus, $-\alpha$ is a lower bound for S. But if $x > -\alpha$ were a greater lower bound for S, then $-x < \alpha$ would be a lesser upper bound for $-S$, which is impossible. Therefore, $-\alpha$ is the greatest lower bound for S, meaning that inf $S = -\alpha$ exists. **QED**

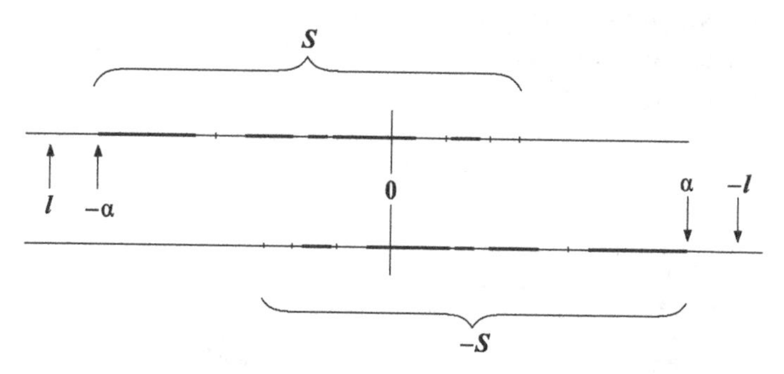

Figure 12.4. An arbitrary set S, some of whose elements are negative, and its infimum $-\alpha$. Also, the set $-S$ and its supremum α.

Lemma 12.2 A nonempty set $S \subset \mathbb{R}$ has at most one supremum. That is, sup S is unique, if it exists.

Proof. Create a proof in Exercise 12.4. **QED**

12.4 The Archimedean Principle

How many milliseconds are there in 50 centuries? Aside from searching for some conversion factors, this is easy to answer (especially with a calculator). But the question hides an assumption so fundamental that our world of experience would seem impossible without it. It is this: sufficiently many milliseconds must eventually surpass 50 centuries! Speaking analytically, is it not obviously true that if $0 < a < b$, then regardless of how large b is, a large enough multiple of a must exceed b? It would seem that all physical processes use this proposition, and that our number system simply reflects this common sense. But alas, we have seen that common sense is an untrustworthy guide for mathematical foundations.

To our surprise, the proposition above about real numbers a and b cannot be proved if we maintain only the axioms for an ordered field (see Exercise 12.11). From a pragmatic point of view, we certainly would want a set of numbers that exhibits all the properties the sciences expect numbers to have. It is the completeness axiom that gives us a new theorem, named after the greatest mathematician, physicist, and engineer of the ancient world, Archimedes of Syracuse (ca. 287 B.C.–212 B.C.). (Archimedes recognized the importance of the theorem, but was not the first to state it.) This theorem will establish the "obvious" proposition above.

Theorem 12.2 The Archimedean Principle Given any $a, b \in \mathbb{R}^+$, there exists $n \in \mathbb{N}$ such that $na > b$.

Proof. To derive a contradiction, suppose $na \leq b$ for all $n \in \mathbb{N}$. Then $S = \{a, 2a, 3a, \cdots\}$ has b as an upper bound. By the completeness axiom, S must have a supremum. Let $\sup S = \alpha$, so that $na \leq \alpha$ for all n. Because $(n+1)a \in S$, it is also true that $(n+1)a \leq \alpha$. This implies that for all n, $na \leq \alpha - a$. But then $\alpha - a$ is an upper bound for S that is strictly less than α! This contradicts the nature of α (remember that α was supposed to be the *least* upper bound), so it follows that for some natural number n, $na > b$. **QED**

Note 1: Aha! Now we have a proof based on the real number axioms that some number n of milliseconds will exceed 50 centuries. Actually, $n = 1.6 \times 10^{14}$ milliseconds is, in fact, more than 50 centuries. Why did such a thing have to be proved?

We are not merely trying to solve a common problem. The Archimedean principle proved the existence of a solution within real numbers to the question "Can a multiple of a exceed b?" This is important since the seemingly obvious answer to the question of existence may not be so obvious in abstract fields (as in Exercise 12.11), or in situations far from common sense.

Suppose that an alien, Mr. Ψ, recognized from youth that the speed of light could not be exceeded, because in his world, relativistic effects were part of life. Then, given that his car can travel at $a = 0.027$ kilometers per second, and the speed of light is $b = 300000$ kps, he would not imagine that a large multiple of a could physically exceed b. For velocities, the Archimedean principle might then be a novelty to him, just as the completeness axiom was to us.

This is to say that Theorem 12.2 proves something without recourse to experience to Mr. Ψ. It is an a priori conclusion, part of Mr. Ψ's intuition, not an a posteriori generalization, as philosopher Immanuel Kant might have said. In this sense, the Archimedean principle is more "real" than the experience of a large number of milliseconds. This principle, the completeness axiom, and things like it bring to light a deeper, more permanent aspect of reality than experience is able to do.

Corollary 12.2 For any $x \in \mathbb{R}^+$, there exists an $n \in \mathbb{N}$ such that $\frac{1}{n} < x$.

Proof. Take $a = 1$ and $b = \frac{1}{x}$ in the Archimedean principle. Then there exists some natural number n such that $n > \frac{1}{x}$, and consequently, $\frac{1}{n} < x$. **QED**

The next corollary's proof should be read carefully to see more about how a supremum behaves.

Corollary 12.3 Let $c \in \mathbb{R}^+$. Then $\sup\{x \in \mathbb{R} : x^2 < c\} = \sqrt{c}$.

Proof. Let $S = \{x \in \mathbb{R} : x^2 < c\} = (-\sqrt{c}, \sqrt{c})$, and we see that it is bounded. By Axiom R11, the supremum of S exists in $\mathbb{R}$; let's call it α. Now, either $\alpha < \sqrt{c}$, $\alpha = \sqrt{c}$, or $\alpha > \sqrt{c}$ by trichotomy. If $\alpha < \sqrt{c}$, then $0 < \sqrt{c} - \alpha$, and by Corollary 12.2, there is a positive number $\frac{1}{n}$ such that $\frac{1}{n} < \sqrt{c} - \alpha$. Thus, $\alpha + \frac{1}{n} < \sqrt{c}$, making $\alpha + \frac{1}{n}$ an element of S that is larger than α. This flatly contradicts the claim that α is an upper bound of S. Next, if $\alpha > \sqrt{c}$, then $0 < \alpha - \sqrt{c}$, and once again, there is a positive number $\frac{1}{m}$ such that $\frac{1}{m} < \alpha - \sqrt{c}$. Then $\sqrt{c} < \alpha - \frac{1}{m} < \alpha$, which means that $\alpha - \frac{1}{m}$ is an upper bound for S that is below α. This contradicts the claim that α is the *least* upper bound of S. Thus, both $\alpha < \sqrt{c}$ and $\alpha > \sqrt{c}$ are false, leaving only $\alpha = \sqrt{c}$. **QED**

We are now in a position to indicate that $\mathbb{R}$ actually contains $\mathbb{Q}$, and has elements that are not in $\mathbb{Q}$. This is the business of the ingenious definition of the reals by means of Dedekind cuts of $\mathbb{Q}$, invented in 1858 by analyst Richard Dedekind (1831–1916). The symbol for this, $\langle A|B\rangle$, appeared just before our list of axioms for the real numbers. A **Dedekind cut** $\langle A|B\rangle$ is an ordered pair of nonempty subsets A and B of $\mathbb{Q}$ such that (1) A and B form a partition of $\mathbb{Q}$, (2) A has no largest element, and (3) whenever $a \in A$ and $b \in B$, then $a < b$.

We distinguish two types of cuts, according to whether B has or hasn't a smallest element. First, for any $r \in \mathbb{Q}$, the partition of $\mathbb{Q}$ given by the intervals $A = \{a \in \mathbb{Q} : a < r\}$ and $B = \{b \in \mathbb{Q} : b \geq r\}$ qualifies $\langle A|B\rangle$ as a Dedekind cut at the rational number r. Since $r = \inf B \in B$, B has a smallest element, and this type of cut will be associated uniquely with rational number r. Very importantly, the type of cut $\langle A|B\rangle$ of $\mathbb{Q}$ where B has no smallest element cannot be associated with a rational number. An example is $A = \{a \in \mathbb{Q} : a < \pi\}$ and $B = \{b \in \mathbb{Q} : b \geq \pi\}$, pictured in Figure 12.5. This $\langle A|B\rangle$ uniquely identifies π since $\pi = \sup A = \inf B$. Thanks to the completeness axiom R11, $\pi \in \mathbb{R}$, and yet, $\pi = \inf B \notin B$, so that this type of cut identifies the other kind of real number—the irrational kind. (Another example appears in the proof of the next theorem.)

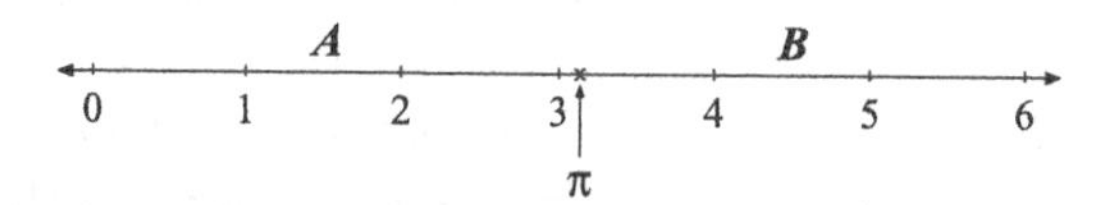

Figure 12.5. The Dedekind cut identifying the irrational number π. The intervals A and B, as defined above, contain only rational numbers.

In order to assert that $\langle A|B\rangle$ is always a real number, even though both A and B are subsets only of rationals, Dedekind carefully defined addition and multiplication of cuts. He showed that the cut at 1, namely $\langle a \in \mathbb{Q} : a < 1 \,|\, b \in \mathbb{Q} : b \geq 1\rangle$, behaved according to RQZ5, and that $\langle a \in \mathbb{Q} : a < 0 \,|\, b \in \mathbb{Q} : b \geq 0\rangle$ satisfied RQZ7. He showed that the collection of all Dedekind cuts actually fulfills all eleven axioms above, so we must treat each $\langle A|B\rangle$ as a real number. The details in all this would remove us from the focus of this chapter, but they just take patience to read through, for instance, in [12], pp. 209–213. For a yet more rigorous and beautifully detailed development of the real numbers, read [35] (Dedekind cuts are the subject of pages 64-69, preceded by a rock-solid foundation of 156 theorems and proofs!)

Note 2: You probably noted that there are two other possible cuts, one in which A has a largest element and B doesn't have a smallest element, and one in which A has a largest element and B has a smallest element. The first one would accomplish the same goal as our definition of a cut, while the second one would not form a partition of $\mathbb{Q}$.

Theorem 12.3 $\mathbb{Q} \subset \mathbb{R}$.

Proof. The collection of both types of Dedekind cuts $\langle A|B \rangle$ described above satisfy RQZN1–R11; they represent the entire set $\mathbb{R}$. Those of the first type identify elements of $\mathbb{Q}$. Thus, $\mathbb{Q} \subseteq \mathbb{R}$. Consider now a prime number p, and the cut $A = \{a \in \mathbb{Q} : a < 0 \ or \ a^2 < p\}$, and $B = \{b \in \mathbb{Q} : b > 0 \ and \ b^2 \geq p\}$. Corollary 12.3 shows that $\sup A = \sqrt{p} = \inf B$ and Lemma 12.2 states that the supremum is unique. Axiom R11 states that $\sqrt{p} \in \mathbb{R}$. But Example 12.1 shows that $\sqrt{p} \notin \mathbb{Q}$, so that $\inf B \notin B$. Hence, for each $\sqrt{p}$, the cut $\langle A|B \rangle$ is of the second type. Therefore, $\mathbb{Q} \subset \mathbb{R}$. **QED**

An immediate consequence of Theorem 12.3 and the two types of cuts is that $\mathbb{R} \backslash \mathbb{Q}$ is exactly the set of irrational numbers $\mathbb{I}$, so that as we also intended to show, $\mathbb{I} \subset \mathbb{R}$. Thus, even if we press into service every member of $\mathbb{Q}$, there will still be uncountably many cuts that don't have rational infima. Exercise 12.5 asks you to give other examples showing that the set of rational numbers $\mathbb{Q}$ does not satisfy the completeness axiom.

This completes our axiomatic construction of the real numbers, using Theorems 3.2, 3.5, and 12.3 to demonstrate that $\mathbb{N} \subset \mathbb{Z} \subset \mathbb{Q} \subset \mathbb{R}$.

12.5 Exercises

1. The rational root theorem can easily show that $\sqrt{3}$ and $\sqrt{8}$ are irrational. Example 12.2 showed that their sum also is irrational. Is the set of irrational numbers $\mathbb{I}$ closed under addition or multiplication?
2. Establish the trichotomy property of $\mathbb{R}$ by using Axiom RQZ9 in a similar way to Exercise 3.3a.
3. Prove that the union or intersection of a finite collection of intervals (disjoint or not) with real endpoints, is always a bounded set (although the union may not be an interval). However, beware of the union of an infinite collection of intervals!
4. Prove that if $\sup S$ exists, then it is unique. Hint: Assume, on the contrary, that S has two suprema.
5. Some Dedekind cuts have already been found that prove that the set of rational numbers $\mathbb{Q}$ doesn't satisfy the completeness axiom. Find several other cuts that do not have rational suprema.

6. Why is it reasonable to define $\sup \varnothing = -\infty$ and $\inf \varnothing = \infty$? Is there any other set A with $\inf A > \sup A$?

7. Evaluate $\sup W$ and $\inf W$ (if they exist) for the set $W = \{\frac{x+|x|}{x} : x \in \mathbb{R}\backslash\{0\}\}$.

8. True or false: if $\sup S = \alpha$ (where S is nonempty), then there are infinitely many members of S in the interval $[x, \alpha]$, where $x < \alpha$.

9. Let $x < y$ and $w < z$ be real numbers. Prove that $x + w < y + z$, but that $xw < yz$ is, in general, false. Is there a condition under which the latter inequality is true?

10. a. Given two sets A and B, define their **sum** as $A + B \equiv \{a + b : a \in A,\ b \in B\}$. Suppose that both $\sup A$ and $\sup B$ exist. Prove that $\sup(A + B)$ exists by proving that $\sup(A + B) = \sup A + \sup B$.
    b. Given two sets A and B, define their **product** as $AB \equiv \{ab : a \in A,\ b \in B\}$ (not to be confused with the Cartesian product $A \times B$). Suppose that both $\sup A$ and $\sup B$ exist. Find a counterexample to show that $\sup(AB) \neq (\sup A)(\sup B)$.

11. We will produce an ordered field that is not Archimedean. Let $\mathbb{J}$ be the set of all reduced rational functions

$$u(x) = \frac{p_n x^n + \cdots + p_1 x + p_0}{q_m x^m + \cdots + q_1 x + q_0},$$

that is, ratios of polynomials with integer coefficients. For instance, $7x^4 - \frac{3}{4}x = \frac{28x^4 - 3x}{4}$ and $\frac{2x}{-x^3+1}$ belong to $\mathbb{J}$, but $\sqrt{x}$ and $\sqrt{3}x^4 + x$ do not. It is true that $\mathbb{J}$ is a field (you may prove this, if you aren't convinced). Define $\mathbb{J}^+$ as the subset of $\mathbb{J}$ where $p_n q_m > 0$. Thus, $7x^4 - \frac{3}{4}x \in \mathbb{J}^+$, but $\frac{2x}{-x^3+1} \notin \mathbb{J}^+$. (This exercise is from [9], pp. 35 and 44.)
    a. Prove that the subset $\mathbb{J}^+$ is closed under addition and multiplication, and that constant polynomial $0 \notin \mathbb{J}^+$. $\mathbb{J}^+$ is then the positive class of $\mathbb{J}$.
    b. Let $\mathbb{J}^-$ be the subset of $\mathbb{J}$ where $p_n q_m < 0$. Show that $0 \notin \mathbb{J}^-$. Show that $\mathbb{J}^-$ is not closed under multiplication.
    c. Show that $\mathbb{J}^- \cap \mathbb{J}^+ = \varnothing$, and that $\mathbb{J} = \mathbb{J}^- \cup \{0\} \cup \mathbb{J}^+$. Parts **a**, **b**, **c** indicate that $\mathbb{J}$ is an ordered field.
    d. To show that the Archimedean principle fails in $\mathbb{J}$, you need only find one element of $\mathbb{J}^+$ that is larger than any given natural number!

    Hint: The Archimedean principle looks like this in quantified form: $\forall u \forall v$(if $u, v \in \mathbb{J}^+$ then $\exists n \in \mathbb{N}$ such that $nv > u$), where u and v are short for rational functions $u(x)$ and $v(x)$, respectively. Its negation, therefore, is $\exists u \exists v$ such that ($u, v \in \mathbb{J}^+$ and $\forall n \in \mathbb{N}$, *not*-($nv > u$)). Follow what it says. For simplicity, take $v(x) = x \in J^+$. Think of $u(x)$ as a curve, and remember that *not*-($nv > u$) means that $nv > u$ is false.

13

Further Properties of Real Numbers

In this chapter, several more key properties of $\mathbb{R}$ will be proved, and we introduce the property of denseness of a set. Our first corollary states what appears to be obvious, that every real number falls between successive integers. The proof, however, isn't obvious!

Corollary 13.1 For any $x \in \mathbb{R}$, there exists an $m \in \mathbb{Z}$ such that $m - 1 \leq x < m$.

Proof. First, consider $x > 0$. In the Archimedean principle, take $a = 1$ and $b = x$, implying that $n > x$ for some $n \in \mathbb{N}$. Defining $T = \{n \in \mathbb{N} : n > x\}$, we see that T is not empty, so by the well-ordering principle, T must have a smallest element m. It follows that $m - 1 \leq x < m$ for some m in $\mathbb{N}$ (and thus in $\mathbb{Z}$).

If $x = 0$, all we need to say is that $m = 1$ satisfies the conclusion.

In case $x < 0$, we have $-x > 0$, whence by the first case, $p - 1 \leq -x < p$ for some $p \in \mathbb{N}$. Define $m = 1 - p$, giving us $-m \leq -x < 1 - m$, where $m \in \mathbb{Z}$. This implies $m - 1 < x \leq m$. If x is a negative noninteger, then this triple inequality is equivalent to $m - 1 \leq x < m$, which is the conclusion. If $x \in \mathbb{Z}^-$ (a negative integer), then the required integer m is simply $m = x + 1$.

Because of trichomy, no other case remains, so the theorem is proved. **QED**

For the next corollary, the relative positions of the various quantities on the number line below will clarify the construction of the proof. It is critical that x and y be separated by a distance greater than 1.

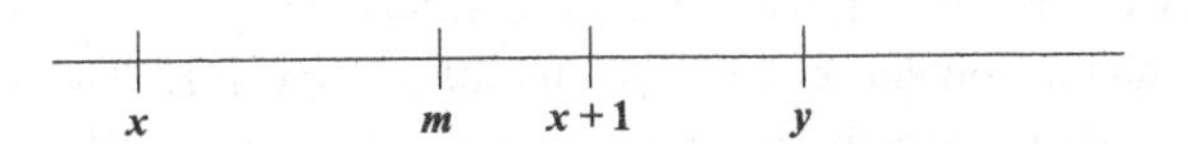

Corollary 13.2 For any $x, y \in \mathbb{R}$ such that $1 < y - x$, there exists an $m \in \mathbb{Z}$ such that $x < m < y$.

Proof. We have $x < x + 1 < y$. Using Corollary 13.1, we find $m \in \mathbb{Z}$ such that $m - 1 \leq x < m$. By successive steps with the triple inequality, $-1 \leq x - m < 0, 0 < m - x \leq 1$, and $x < m \leq x + 1$. Thus, $x < m \leq x + 1 < y$. Therefore, m is the required integer. **QED**

Note 1: Once we admit the decimal expansion of real numbers, the Archimedean principle and its corollaries seem quite obvious. Here are two examples. If the decimal numbers for x and y in Corollary 13.2 differ by more than 1, then just looking at the integer part of the larger of x and y will quickly give us a satisfactory m. And consider Corollary 12.2: let $x = \cot 1.55 \approx 0.021$, and $\frac{1}{0.021} < 48$, thus $\frac{1}{48} < 0.021 \approx \cot 1.55$, apparently giving us $n = 48$ to satisfy the corollary. (This hints at an algorithm for finding n in the corollary.)

However, we are aware that the sense that something is mathematically obvious often hides unforeseen difficulties. ("Hmmm, cot 1.55 is murky..., what if I use better accuracy than two significant digits in approximating it?" Try it!) A more serious objection to using decimals at this point is that many real numbers have non-terminating decimal expansions, and firm foundations for their convergence tacitly use the Archimedean principle. This would create the logical fallacy of circular reasoning, that is, using a conclusion (that decimal expansions converge), as a premise (admitting that the decimal expansion for all real numbers exists).

An important property of some sets of numbers is denseness. Set A is **dense** in set B if between any two elements of B there can be found an element of A. That is, for any $b_1, b_2 \in B$ with $b_1 < b_2$, there exists an $a \in A$ such that $b_1 < a < b_2$. As a special case, set A is **dense in itself** if for any $a_1, a_2 \in A$ with $a_1 < a_2$, we can find $a_3 \in A$ such that $a_1 < a_3 < a_2$.

Example 13.1 Corollary 13.1 proves that $\mathbb{R}$ is dense in $\mathbb{Z}$. On the other hand, Corollary 13.2 indicates that $\mathbb{Z}$ isn't dense in $\mathbb{R}$, since we can easily select reals x and y such that no integer lies between them. ■

The next theorem proves the claim that rational numbers are dense in $\mathbb{R}$. This is amazing, because we have seen that $\mathbb{Q}$ is a vanishingly small subset of $\mathbb{R}$. Our strategy is that regardless of how close two reals x and y are, their difference can be magnified by multiplying it by a positive integer q, and in fact, magnified until the difference is more than 1. But then this interval must contain some integer p.

Theorem 13.1 The set of rational numbers $\mathbb{Q}$ is dense in $\mathbb{R}$. Also, $\mathbb{R}$ is dense in $\mathbb{Q}$.

Proof. Let $x, y \in \mathbb{R}$, with $x < y$. By Corollary 12.2, there exists $q \in \mathbb{N}$ such that $\frac{1}{q} < y - x$, so *(a)* $qx + 1 < qy$. By Corollary 13.1, there exists $p \in \mathbb{Z}$ such that $p - 1 \leq qx < p$. Considering two parts of this inequality, obtain *(b)* $qx < p$, and *(c)* $p \leq qx + 1$. Placing these inequalities in the order *(b)*, *(c)*, and *(a)*, we arrive at $qx < p \leq qx + 1 < qy$. Finally, divide through by q: $x < \frac{p}{q} \leq x + \frac{1}{q} < y$. Thus, a rational number $\frac{p}{q}$ has been found between any two reals $x < y$.

Next, given any two rationals r_1 and r_2, the rational number $\frac{r_1+r_2}{2}$ falls between them (in Exercise 13.2, you will generalize this). But $\frac{r_1+r_2}{2} \in \mathbb{R}$, which means that $\mathbb{R}$ is dense in $\mathbb{Q}$. **QED**

Does an irrational number exist between any two rationals? The answer is "Yes." In other words, $\mathbb{I}$ is dense in $\mathbb{Q}$. Exercise 13.7 below will give hints for a proof.

Corollary 13.3 Let a be a real number such that $0 \leq a < \varepsilon$, where ε is a positive real number (the smaller the ε, the more interesting the situation becomes). Then we conclude that $a = 0$.

Proof. Assume on the contrary, that $a > 0$. Let b be a real number greater than a. By the Archimedean principle, there exists $n \in \mathbb{N}$ such that $na > b$. Then $\frac{b}{n} < a$. By taking $\varepsilon = \frac{b}{n}$, we contradict the premise that $a < \varepsilon$ for *all* positive ε. Thus, $a = 0$. **QED**

Corollary 13.3 has many implications. For instance, it helps us prove continuity by asserting that if $|f(x) - f(b)| < \varepsilon$ for arbitrary $\varepsilon > 0$, then $f(x) = f(b)$ ($|f(x) - f(b)|$ is being used as a).

We end this chapter by introducing two functions. The **floor function**, denoted $\lfloor x \rfloor$ and *floor* (x), may be unambiguously defined as $\lfloor x \rfloor \equiv m - 1$, using m from Corollary 13.1. An easier way to evaluate this function is to find the greatest integer less than or equal to x. This formulation will no doubt be more familiar to you, and older calculus books symbolized this function as $[x]$ or $[[x]]$. For example: $\lfloor 5.9 \rfloor = 5$, $\lfloor 32 \rfloor = 32$, $\lfloor -9 \rfloor = -9$, and $\lfloor -9.3 \rfloor = -10$.

Corollary 13.1 also implies that for any real x, there exists a unique $m \in \mathbb{Z}$ such that $m - 1 < x \leq m$. Using these inequalities, the **ceiling function**, denoted $\lceil x \rceil$ and *ceil* (x), is defined as $\lceil x \rceil \equiv m$. More intuitively, $\lceil x \rceil$ equals the least integer greater than or equal to x. Thus, $\lceil 9.2 \rceil = 10$, $\lceil -6 \rceil = -6$, $\lceil -9.98 \rceil = -9$, and $\lceil -\sqrt{3} \rceil = -1$.

A real number x may be thought of as composed of an integer part added to a fractional (decimal) part $\beta = \beta(x)$. To be precise, β is defined by $x = \lfloor x \rfloor + \beta$, so $\beta = 0$ when x is an integer. From the following examples, we can see why $0 \leq \beta < 1$:

$$
\begin{array}{llll}
\text{If } x = 9.3, & \text{then } x = 9 + 0.3, & \text{and } \lfloor x \rfloor = 9, & \text{so } \beta = 0.3\,. \\
\text{If } x = -3.14, & \text{then } x = -4 + 0.86, & \text{and } \lfloor x \rfloor = -4, & \text{so } \beta = 0.86\,. \\
\text{If } x = -\frac{14}{3}, & \text{then } x = -5 + \frac{1}{3}, & \text{and } \lfloor x \rfloor = -5, & \text{so } \beta = \frac{1}{3}. \\
\text{If } x \in \mathbb{Z}, & \text{then } x = \lfloor x \rfloor + 0, & \text{and } \lfloor x \rfloor = x, & \text{so } \beta = 0.
\end{array}
$$

Exercise 13.3 develops some interesting relationships among the floor and ceiling functions and β.

13.1 Exercises

1. Suppose $x \leq y$, both elements of $\mathbb{R}$. Prove that their arithmetic mean falls between them, that is, that $x \leq \frac{x+y}{2} \leq y$.

2. Suppose $r_1 < r_2$, both elements of $\mathbb{Q}$. Find an expression that will yield an infinite number of distinct rationals between r_1 and r_2. Is $\mathbb{Q}$ dense in itself?

3. This exercise involves the ceiling and floor functions. Recall that $\beta = \beta(x)$ was defined as the nonnegative, fractional (decimal) part of x.
   a. Choose two positive non-integer values of x, two negative non-integer values of x, and two integer values that will demonstrate each of the relationships in the following expressions:
   $$\lfloor x \rfloor = x - \beta, \text{ and } \lceil x \rceil = \begin{cases} x + 1 - \beta & \text{if } x \notin \mathbb{Z} \\ x & \text{if } x \in \mathbb{Z} \end{cases}.$$
   b. Graph the functions *floor* $(x) = \lfloor x \rfloor$ and *ceil* $(x) = \lceil x \rceil$. For the obvious reason, they are called "step functions."

c. Prove that $\lfloor x+m \rfloor = \lfloor x \rfloor + m$ for real x and $m \in \mathbb{Z}$. Show by a counterexample that this cannot be extended to $\lfloor x+y \rfloor = \lfloor x \rfloor + \lfloor y \rfloor$. Is $\lfloor 2x \rfloor = 2\lfloor x \rfloor$? Hint: In $\lfloor x+1 \rfloor$, substitute an equivalent expression in place of x.

d. Prove that $\lfloor x \rfloor = \lceil x \rceil - 1$ for $x \notin \mathbb{Z}$, and $\lfloor x \rfloor = \lceil x \rceil$ otherwise. Are these expressions equivalent to $\lfloor x \rfloor = \lceil x \rceil - \lceil \beta \rceil$, where x is any real?

e. Show that $\lceil x \rceil = -\lfloor -x \rfloor$ for all real numbers.

f. Use the floor and ceiling functions to form a function *round* (x) that will correctly round off a real x to the nearest integer. (If a computer language offers floor and ceiling commands, then this would give you a convenient routine for rounding off.) This brings to memory Exercise 6.10.

4. Use x expressed as $\lfloor x \rfloor + \beta$ to prove Corollary 13.2. Hint: Since $1 < y - x$, then $(\lfloor x \rfloor + \beta) + 1 < y$.

5. Use Corollary 12.2 to prove that for two real numbers $x < y$, there exists a natural number m such that $x < x + \frac{1}{2^m} < y$.

6. If $a < b + \varepsilon$ for all $\varepsilon > 0$, then $a \leq b$. Prove this inference by the methods of contradiction and contraposition.

7. Prove that $\mathbb{I}$ is dense in $\mathbb{Q}$. That is, let $r_1, r_2 \in \mathbb{Q}$, with $r_1 < r_2$, and prove that there exists some $x \in \mathbb{I}$ such that $r_1 < x < r_2$.
Hint: Use Corollary 12.2 to show that there exists $n \in N$ such that $nr_1 + 1 < nr_2$. Select an irrational number between 0 and 1, say, $\frac{1}{\sqrt{2}}$. Form a string of inequalities that includes $nr_1 + \frac{1}{\sqrt{2}} < nr_1 + 1$, which will prove that the irrational number $r_1 + \frac{1}{n\sqrt{2}}$ must be between our two rational numbers. Why are we convinced that $r_1 + \frac{1}{n\sqrt{2}}$ is irrational?

8. a. Recall that the set of terminating decimals $\mathbb{T}$ is a denumerable subset of $\mathbb{Q}$. Show that $\mathbb{T}$ is dense in itself. Show also that (perhaps surprisingly) $\mathbb{T}$ is dense in $\mathbb{R}$. Hint: Let $x < y$ be reals. Then they must differ at some initial decimal place value.

b. Automate the proof that $\mathbb{T}$ is dense in $\mathbb{R}$ by proving that $\frac{\lfloor 10^q x \rfloor + \lceil 10^q y \rceil}{2(10^q)} \in \mathbb{T}$ that lies between two reals $x < y$, where $q = \lceil -\log(y-x) \rceil$ (the common logarithm is necessary). Hint: Exercises 8.1 and 13.3a help.

14

Cluster Points and Related Concepts

Reread Paradox 0.2 at the beginning of this book. We will now develop the analysis that will unravel that enigma. The difficulty arises first of all because an infimum had not been defined, and second because we used the seemingly simple concept of touching intervals. That term shall remain too anthropomorphic to be useful in analysis.

14.1 Neighborhoods

As usual, clear definitions are crucial. A **neighborhood** of a number a is any *open* interval centered at a. That is, $(a - \varepsilon,\ a + \varepsilon)$ is a neighborhood of a whenever $\varepsilon > 0$. Convince yourself that this may be written in set notation as $\{x : |x - a| < \varepsilon, \varepsilon \in \mathbb{R}^+\}$. The distance ε from the center to the endpoints is called the **radius** of the neighborhood. We will also use the symbol $\mathbf{N}_a^\varepsilon$ to denote the neighborhood $(a - \varepsilon,\ a + \varepsilon)$.

A **nested sequence of neighborhoods** of a is a set of neighborhoods $\mathbf{N}_a^{\varepsilon_j}$ all of which are centered at a and each of which contains the next neighborhood $\mathbf{N}_a^{\varepsilon_{j+1}}$. This is nicely written as $\mathbf{N}_a^{\varepsilon_1} \supset \mathbf{N}_a^{\varepsilon_2} \supset \mathbf{N}_a^{\varepsilon_3} \supset \cdots \supset \mathbf{N}_a^{\varepsilon_j} \supset \cdots$, for every $j \in \mathbb{N}$.

14.2 Cluster Points

A real number b is said to be a **cluster point** of a set $S \subset \mathbb{R}$ if and only if every neighborhood of b contains at least one point $s \in S$ different from b. Notice that b itself need not belong to S. The intuitive—but imprecise—sense to start with is that points of S cluster infinitely close to b.

> Note 1: Terms synonymous with cluster point are *accumulation point* and *limit point*. These names have good connotations, too, as Theorem 14.1 will show. Although Karl Weierstrass was the first to understand and use the concept of a cluster point, Georg Cantor was the first name it: a "Grenzpunkt," or limit point.
>
> Since we are working with numbers on the real number line, the words "number" and "point" can be treated as synonyms, with the latter emphasizing the graphic position of the former.

Example 14.1 Every point of the interval $S = (2, 5]$ is a cluster point of S. Especially note that $2 \notin S$ is a cluster point, since every neighborhood $(2 - \varepsilon, 2 + \varepsilon)$ of 2 contains infinitely many points of S different than 2. ■

Example 14.2 The set of cluster points of $S = (-\infty, 5) \cup (5, 6]$ is the entire interval $(-\infty, 6]$. In particular, 5 is a cluster point since every neighborhood $(5 - \varepsilon, 5 + \varepsilon)$ contains elements of S not equal to 5. Notice that $5 \notin S$. Also, because $-\infty$ is not a real number, it is not a cluster point of any set. ■

Example 14.3 Let $S = \{\frac{1}{2^n} : n \in \mathbb{N}\}$. Then S contains no cluster points, since a number $\frac{1}{2^n} \in S$ can be surrounded by a neighborhood such as $(\frac{1}{2^n} - \frac{1}{10^n}, \frac{1}{2^n} + \frac{1}{10^n})$ that contains exactly one point that belongs to S (verify this!). However, S does have a cluster point $b = 0$, since every neighborhood $(-\varepsilon, \varepsilon)$ of 0 contains lots of members of S not equal to 0. ■

Note 2: Saying, "S contains a cluster point b" (that is, $b \in S$) is not the same as saying, "S has a cluster point b." The former implies the latter, but not conversely.

Example 14.4 $\mathbb{Z}$ has no cluster points. On the other hand, every element of $\mathbb{Q}$ is a cluster point of $\mathbb{Q}$. The empty set has no cluster points (why?). ■

Example 14.5 The examples above had either an infinite number of cluster points, or else at most one. It is easy to engineer a set that has, say, three cluster points. Extending the idea in Example 14.3, let $S = \{\frac{1}{n}\} \cup \{2 - \frac{1}{n}\} \cup \{2 + \frac{1}{n}\} \cup \{4 + \frac{1}{n}\}$, where $n \in \mathbb{N}$. Figure 14.1 shows that S has the three cluster points 0, 2, and 4, none of which belong to S. ■

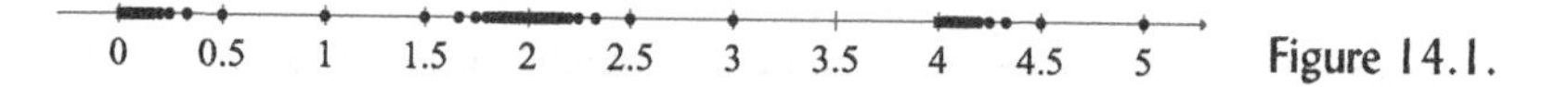

Figure 14.1.

The examples should aid in understanding why the words "cluster" or "accumulation" are appropriate. Cluster points are locations on the real number line near which infinitely many points of S gather. To prove this statement, we use the powerful technique of constructing a nested sequence of neighborhoods. The proof below is algorithmic, so that given a set S, we can construct neighborhoods of the cluster point b. Do Exercise 14.4 for a shorter proof.

Theorem 14.1 Every neighborhood of a cluster point b of a set S contains infinitely many distinct elements of S.

Proof. To describe the algorithm, begin with an arbitrary neighborhood of b given by $\mathbf{N}_b^\varepsilon = (b - \varepsilon, b + \varepsilon)$. Since b is a cluster point, $\mathbf{N}_b^\varepsilon$ contains at least one $s_0 \in S \cap \mathbf{N}_b^\varepsilon$ different from b. Next, find an integer m_1 so large that $\frac{1}{m_1} < |b - s_0|$. Then the neighborhood $A_1 = (b - \frac{1}{m_1}, b + \frac{1}{m_1})$ must contain an $s_1 \in S \cap A_1$ that is different from b, and different from s_0 (verify that $s_1 \neq s_0$). Furthermore, since $\frac{1}{m_1} < \varepsilon$, A_1 is nested inside $\mathbf{N}_b^\varepsilon$: $A_1 \subset \mathbf{N}_b^\varepsilon$. Continuing in this way for $n = 2, 3, \ldots$, find an integer m_n such that $\frac{1}{m_n} < |b - s_{n-1}| < \frac{1}{m_{n-1}}$, and then find a distinct $s_n \in S \cap A_n$ with $s_n \in A_n = (b - \frac{1}{m_n}, b + \frac{1}{m_n})$. Each neighborhood A_n of b is nested as a subset within its predecessor A_{n-1}, and each has a distinct element of S.

In summary, the algorithm produces a nested sequence of neighborhoods $\mathbf{N}_b^\varepsilon \supset A_1 \supset A_2 \supset A_3 \supset \cdots$. They yield infinitely many distinct elements $s_n \in S$, all within $\mathbf{N}_b^\varepsilon$. **QED**

Example 14.6 Form a nested sequence of neighborhoods around the cluster point $b = \pi$ of the interval $S = (-\pi, \pi)$. Begin (quite arbitrarily) with $A_0 = (\pi - 1, \pi + 1) \approx (2.14, 4.14)$, and select (quite randomly) $s_0 = 2.2 \in S \cap A_0$. With a calculator, we find that $\frac{1}{2} < \pi - 2.2$, so that $m_1 = 2$ will do. Next, $A_1 = (\pi - \frac{1}{2}, \pi + \frac{1}{2}) \approx (2.64, 3.64) \subset A_0$. Again, randomly select $s_1 = 3.04 \in S \cap A_1$. Then, $\frac{1}{10} < |\pi - 3.04| < \frac{1}{2}$, so that $m_2 = 10$ will do. And so, $A_2 = (\pi - \frac{1}{10}, \pi + \frac{1}{10}) \approx (3.042, 3.242) \subset A_1$, from which we may select, say, $s_2 = 3.13 \in S \cap A_2$.

The algorithm has now produced the nested sequence $(\pi - 1, \pi + 1) \supset (\pi - \frac{1}{2}, \pi + \frac{1}{2}) \supset (\pi - \frac{1}{10}, \pi + \frac{1}{10})$, with corresponding distinct elements $\{2.2, 3.04, 3.13\} \subset S$, each one closer to π than the previous one. In our mind's eye, we could continue the algorithm without end. ■

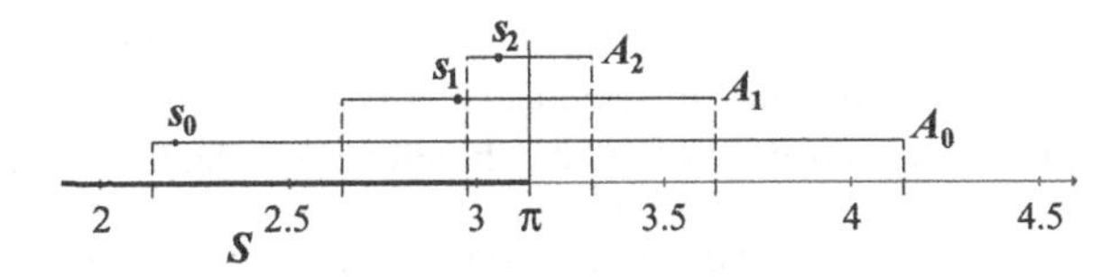

Figure 14.2. The three nested intervals A_0, A_1, and A_2 from Example 14.6 elevated so as to be seen clearly. Also, s_0, s_1, and s_2 are plotted, approaching the vertical segment at π from the left, and always within S.

14.3 Boundary Points

A slight modification of the definition of cluster point yields another interesting idea. A real number b is said to be a **boundary point** of a set $S \subset \mathbb{R}$ if and only if *every* neighborhood of b contains at least one point $s \in S$ and at least one point $x \notin S$. Notice that b itself need not belong to S.

Example 14.7 The interval $S = (2, 5]$ has exactly two boundary points, 2 and 5. Contrast this with Example 14.1. ■

Example 14.8 Let $S = \{\frac{1}{2^n} : n \in \mathbb{N}\}$, as in Example 14.3. Think about $\frac{1}{2^{n+1}} < \frac{1}{2^n} < \frac{1}{2^{n-1}}$ for a moment. For each element $\frac{1}{2^n} \in S$, selecting any $\varepsilon > 0$ allows us to form the neighborhood $\mathrm{N}^{\varepsilon}_{1/2^n}$ that always contains one element of S, namely $\frac{1}{2^n}$ (and some of its neighbors if ε is large enough), as well as infinitely many points not in S. Thus, every element of S is a boundary point. It is important that $0 \notin S$ is also a boundary point, and you should convince yourself that no point outside of $S \cup \{0\}$ can be a boundary point of S. ■

Example 14.9 Every element of $\mathbb{N}$ is a boundary point of $\mathbb{N}$. Every element of $\mathbb{Q}$ is a boundary point of $\mathbb{Q}$. In fact, every element of $\mathbb{R}$ is a boundary point of $\mathbb{Q}$. ■

After numerous examples such as those above, we may notice that if a nonempty set is bounded, then it must have a maximum boundary point. The following theorem reveals that this is true. An equivalent definition of a supremum to that in Chapter 12 will be helpful here.

> For any nonempty set $S \subset \mathbb{R}$, $\alpha = \sup S$ if and only if (1) $s \leq \alpha$ for all $s \in S$, and (2) given any $\varepsilon > 0$, there exists a member $s \in S$ such that $s > \alpha - \varepsilon$.

Theorem 14.2 The supremum of a bounded, nonempty set S is its maximum boundary point, and conversely.

Proof of necessity. Let $\alpha = \sup S$. We will show first that every neighborhood of α contains at least one point $s \in S$ and one point $x \notin S$, thus making α a boundary point. By the two properties above for α, for any $\varepsilon > 0$, there exists a member $s \in S$ such that $\alpha - \varepsilon < s \leq \alpha$. Thus, $(\alpha - \varepsilon, \alpha + \varepsilon)$ is a neighborhood of α that contains some element of S, and definitely many numbers $x \notin S$: all of those $x \in (\alpha, \alpha + \varepsilon)$. Therefore, α is a boundary point of S.

We must now show that no boundary point γ can be greater than α. On the contrary, assume that $\gamma > \alpha$. Then every neighborhood of γ contains at least one point $s' \in S$ and one point $x \notin S$. This leads us to define $\varepsilon = \gamma - \alpha$, which sets up the neighborhood $(\gamma - \varepsilon, \gamma + \varepsilon)$. It's a bad neighborhood, for it contains the point s' that is greater than α. This contradicts the nature of α. We conclude that γ does not exist, and therefore, α is the maximum boundary point.

Proof of sufficiency. Next, we prove the converse statement. Let S have a maximum boundary point α. Thus, every neighborhood $(\alpha - \varepsilon, \alpha + \varepsilon)$ contains at least one point $s \in S$. But no such s can be greater than α, for then s would be a contradictory, larger, maximum boundary point of S. This makes α an upper bound of S. It follows that $s \in (\alpha - \varepsilon, \alpha]$. Moreover, this is true for any $\varepsilon > 0$. By the two properties above, $\alpha = \sup S$. **QED**

Because S may not contain some of its boundary points, S may not contain its supremum, its infimum, or both (see Example 14.7). An analogous, helpful equivalent definition of an infimum appears in Exercise 14.6. That exercise asks for a proof that the infimum of a set is its minimum boundary point.

14.4 Isolated Points

No finite set has a cluster point (verify), and neither do some infinite sets, such as $\mathbb{N}$. They are entirely composed of isolated points. To be precise, an **isolated point** $s \in S \subset \mathbb{R}$ is a point such that some neighborhood $(s - \varepsilon,\ s + \varepsilon)$ contains exactly one point of S, namely s itself.

Example 14.10 The set $S = \mathbb{N} \cup [0, 1]$ contains the set $\{2, 3, 4, \cdots\}$ of isolated points, and no others. ■

Example 14.11 Returning to $S = \{\frac{1}{2^n} : n \in \mathbb{N}\}$, we saw earlier that each member of S can be surrounded by the neighborhood $(\frac{1}{2^n} - \frac{1}{10^n}, \frac{1}{2^n} + \frac{1}{10^n})$, with $\varepsilon = \frac{1}{10^n}$. The *only* member of S within each of these neighborhoods is $\frac{1}{2^n}$ itself (as you verified). Thus, S contains only isolated points. (Nevertheless, recall that S has a cluster point.) ■

14.5 The Maximum and Minimum of a Set

Back in Chapter 4, we asked, "Is there a least member in $\{r \in \mathbb{Q} : 1 < r < 2\}$?" The reason for the negative answer is now apparent: the set does not contain its infimum, 1. Another way to say this is that the set does not have a minimum. To be precise, $\inf S$ is the **minimum** of S if and only if $\inf S$ is finite and $\inf S \in S$. Likewise, $\sup S$ is the **maximum** of S if and only if $\sup S$ is finite and $\sup S \in S$. In view of Theorem 14.2, maximum and minimum boundary points may be used instead of the supremum and infimum, respectively, to define the maximum and minimum of S.

Example 14.12 For the interval $(2, 5]$ in Example 14.7, 5 is the supremum, the maximum boundary point, and the maximum. Also, $\inf(2, 5] = 2$ is its minimum boundary point, but no minimum exists. ■

Example 14.13 The set $S = \{\frac{1}{2^n} : n \in \mathbb{N}\}$ from Example 14.8 contains its maximum boundary point $\frac{1}{2}$, hence that is the set's maximum. Although its minimum boundary point and infimum equal 0, S has no minimum. ■

Example 14.14 The range of the function $f(x) = \frac{\sin x}{x}$ has no maximum, but it has a supremum: $\sup \mathcal{R}_f = 1$. See Exercise 14.7 for the minimum. ■

In the next lemma, the supremum is associated with the maximum cluster point. An analogous lemma for infima appears as Exercise 14.8.

Lemma 14.1 Let S have a supremum, but no maximum element (that is, $\sup S \notin S$). Then $\sup S$ equals the maximum cluster point.

Proof. Let $\alpha = \sup S$. By part (1) of the definition of supremum, and given that no maximum exists, we have that $s < \alpha$ for all $s \in S$. By part (2) of the definition, every neighborhood $(\alpha - \varepsilon, \alpha + \varepsilon)$ contains at least one member $s \in S$, and by what was just said, $s \neq \alpha$. Therefore, α is a cluster point. Now, assume that a greater cluster point γ exists, and let $c = \gamma - \alpha > 0$. Then the neighborhood $(\gamma - c, \gamma + c)$ would contain an $s \in S$ greater than α, which violates part (1) of the definition. Therefore, α is the maximum cluster point. **QED**

We can now put Paradox 0.2 to rest. The intervals involved were $[1, 5]$ and $(5, 6]$. We notice that $\sup[1, 5] = 5 = \inf(5, 6]$. One and the same point, $x = 5$, is doing double duty, which isn't very strange. But $\inf(5, 6] \notin (5, 6]$, so necessarily, $[1, 5] \cap (5, 6] = \varnothing$, which we interpreted as meaning that the intervals do not touch each other. That is reasonable. This is a good opportunity to define $\inf |S - a| \equiv \inf\{|s - a| : s \in S\}$ the **distance** between a number a and a set S. Thus, the distances we are interested in are $d = \inf |[1, 5] - 5| = 0$, and $d = \inf |(5, 6] - 5| = 0$. So, the sets are at zero distance from $x = 5$, but this has nothing to do with "touching" each other. This seemed to be a paradox only because of the vagueness of the literary term "touching," further mixed up by the then unclear nature of suprema and infima.

14.6 Exercises

1. a. Show that every point of a closed interval $[a, b]$ with $a \neq b$ is a cluster point. Hint: Start with the definition of cluster point.
   b. Show that the only boundary points of an interval are the endpoints, regardless of their inclusion in the interval. Hint: First, prove that they are boundary points, and then that no others exist.
2. Prove that no finite, nonempty set F can have a cluster point, but that every point of F is a boundary point.
3. Find any cluster points of $S = \{3 + \frac{\cos(n\pi)}{n} : n \in \mathbb{N}\}$.

4. To prove Theorem 14.1 by another technique, assume on the contrary that there exists some neighborhood $(b - \varepsilon, b + \varepsilon)$ containing only finitely many elements of S. Complete the proof.

5. Show that a nonempty set S will not contain its maximum boundary point if and only if S does not contain its supremum. Hint: Review the exercises for Chapter 1.

6. A helpful equivalent definition of an infimum is: for any nonempty set $S \subset \mathbb{R}$, $\beta = \inf S$ if and only if (1) $s \geq \beta$ for all $s \in S$, and (2) given any $\varepsilon > 0$, there exists a member $s \in S$ such that $s < \beta + \varepsilon$. Prove a theorem analogous to Theorem 14.2: the infimum of a (nonempty) set S is its minimum boundary point, and conversely.

7. Graph the function $f(x) = \frac{\sin x}{x}$ to find $\sup \mathcal{R}_f$. Also, evaluate $\inf \mathcal{R}_f$ to fairly good precision. Does $\mathcal{R}_f$ contain its supremum or its infimum? Hint: Use calculus to find the absolute minimum of the function.

8. Prove a lemma analogous to Lemma 14.1: Let S have an infimum, but no minimum element. Then $\inf S$ equals the minimum cluster point.

9. Consider the curious function $f(x) = \sin(\tan x)$. Graph it, and notice that the roots seem to cluster at odd multiples of $\frac{\pi}{2}$. Explain the behavior. Show that $\frac{\pi}{2}$ is a cluster point of the set of roots of f.

10. Explain why the domain of the function $f(x) = \sqrt{x^3 - x^2}$ has an isolated point.

11. Given two nonempty sets $S \subseteq T \subset \mathbb{R}$ that are bounded, show that $\sup S \leq \sup T$, and $\inf T \leq \inf S$. Hint: First show that there must be some fixed $t \in T$ compared to which every $s \in S$ satisfies $s \leq t$.

15

The Triangle Inequality

The triangle inequality is an extensively used theorem in analysis. In the following work, recall that for any expression X and any nonnegative real a, the single inequality $|X| \leq a$ is equivalent to the triple inequality $-a \leq X \leq a$.

Theorem 15.1 The Triangle Inequality For any $x, y \in \mathbb{R}$, $|x + y| \leq |x| + |y|$.

Proof. Using an equivalent form for absolute value, $|x + y| = \sqrt{(x + y)^2}$, we form a conjecture: is

$$\sqrt{(x + y)^2} \leq |x| + |y| \text{ true?}$$

If so, then squaring both sides (which loses no information since both sides are nonnegative) gives $x^2 + 2xy + y^2 \leq |x|^2 + 2|x||y| + |y|^2$, and then $xy \leq |x||y|$, which is clearly true. Notice that we have not assumed the conclusion in order to prove it; that would be circular reasoning, which can never exist in mathematics. We have simply observed that if the conjecture is true, then $xy \leq |x||y|$, which *is* true. Using the fifth technique of proof from Chapter 2, we will start with the true statement $xy \leq |x||y|$, and reverse our steps. Then $x^2 + 2xy + y^2 \leq x^2 + 2|x||y| + y^2 = |x|^2 + 2|x||y| + |y|^2$ must be true, and then $(x + y)^2 \leq (|x| + |y|)^2$ must be true. Finally, taking the positive square root of both sides,

$$\sqrt{(x + y)^2} \leq |x| + |y|,$$

from which our theorem follows. **QED**

Note 1: Our proof avoided the use of cases: case 1, when both $x, y \geq 0$; case 2, when only one of them is negative; etc. In avoiding cases, we employed better algebraic techniques. We also can see when the inequality becomes an equality (see Exercise 15.1).

Corollary 15.1 $|x| - |y| \leq |x - y|$.

Proof. Consider $|x| = |x - y + y| \leq |x - y| + |y|$, by the triangle inequality. The corollary follows. **QED**

Note 2: You have just seen a tried and true technique from the analyst's arsenal. It is to add and subtract the same quantity to an expression. Keep it in mind!

Note 3: Corollary 15.1 makes an interesting counterpoint to the triangle inequality. A good mnemonic device for $|x + y| \leq |x| + |y|$ is, "If you split a sum by absolute values, you get more." Likewise, $|x| - |y| \leq |x - y|$ can be thought of as, "If you split a difference by absolute values, you get less."

Corollary 15.2 $\big| |x| - |y| \big| \leq |x - y|$.

Proof. Expanding the statement, we will prove that $-|x - y| \leq |x| - |y| \leq |x - y|$. By Corollary 15.1, $|x| - |y| \leq |x - y|$. Multiply this inequality by -1, $-|x - y| \leq |y| - |x|$. Now, since x and y are arbitrary, we may exchange them one for the other: $-|y - x| \leq |x| - |y|$. But $|y - x| = |x - y|$, and putting these details together gives $-|x - y| \leq |x| - |y| \leq |x - y|$ as needed. **QED**

Note 4: Another effective technique was used in proving Corollary 15.2. Exchanging one variable for another (when justified) is simply substitution: $x \longmapsto y$ and $y \longmapsto x$.

A common use of the triangle inequality involves neighborhoods. Consider the neighborhoods $(a - \varepsilon, a + \varepsilon)$ and $(b - \varepsilon, b + \varepsilon)$, where $\varepsilon > 0$ is any radius. If they have a nonempty intersection, then $|b - a| < 2\varepsilon$. Figure 15.1 gives good evidence of this.

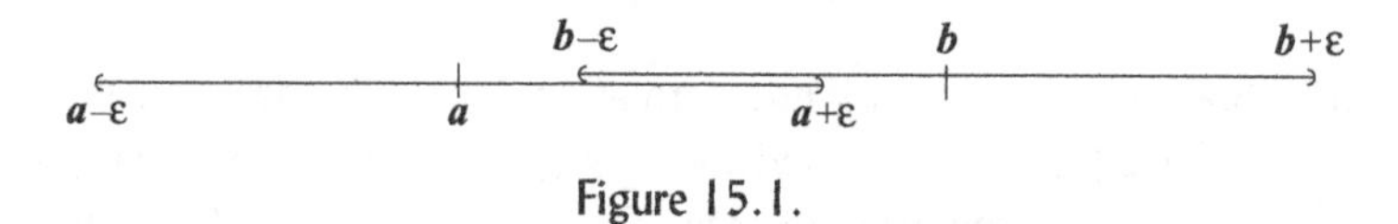

Figure 15.1.

Example 15.1 If the neighborhoods $N_a^\varepsilon = (a - \varepsilon, a + \varepsilon)$ and $N_b^\varepsilon = (b - \varepsilon, b + \varepsilon)$ have $N_a^\varepsilon \cap N_b^\varepsilon \neq \varnothing$, prove that $|b - a| < 2\varepsilon$. We begin by restating the intervals as inequalities: $N_a^\varepsilon = \{x : |x - a| < \varepsilon\}$ and $N_b^\varepsilon = \{x : |x - b| < \varepsilon\}$. Let $z \in N_a^\varepsilon \cap N_b^\varepsilon$. Then simultaneously, $|z - a| < \varepsilon$ and $|z - b| < \varepsilon$. Adding, $|z - a| + |z - b| < 2\varepsilon$. For the purpose of what will follow, we rewrite this as $|z - a| + |b - z| < 2\varepsilon$. By the triangle inequality, we know that $|z - a + b - z| \leq |z - a| + |b - z| < 2\varepsilon$, or $|b - a| < 2\varepsilon$. **QED**

Another common use of the triangle inequality is for putting bounds on functions. Recall from Chapter 12 that given a function g, this instructs us to find (if possible) some number B such that $|g(x)| \leq B$ over its domain, or over an interval. Of course, a bound is not necessarily the maximum value of the function.

Example 15.2 Let $g(x) = 2x^3 - 5x^2 - 9$. Find a bound for this function over $[-3, 4]$.

Since $-3 \leq x \leq 4$, certainly $|x| \leq 4$. Hence, $2|x|^3 \leq 2(4^3) = 128$, and $5|x|^2 \leq 80$. By the triangle inequality, $|2x^3 - 5x^2 - 9| \leq 2|x|^3 + 5|x|^2 + 9 \leq 128 + 80 + 9 = 217$. It follows that $|g(x)| \leq 217$ over $[-3, 4]$. You should verify by graphing that $B = 217$ is actually quite a generous bound.

Here is another approach. Since $-3 \leq x \leq 4$, then $0 \leq 5x^2 \leq 80$, and $-80 \leq -5x^2 \leq 0$. Also, $-27 \leq x^3 \leq 64$, so that $-54 \leq 2x^3 \leq 128$. Adding the needed triple inequalities together,

$-54 - 80 \le 2x^3 - 5x^2 \le 128$, and finally, $-54 - 80 - 9 \le 2x^3 - 5x^2 - 9 \le 128 - 9$. Upon simplifying, we have $-143 \le g(x) \le 119$, which implies that $|g(x)| \le 143$. Thus, we see that bypassing the triangle inequality may give a tighter bound, but at the cost of longer analysis. ■

15.1 Exercises

1. Under what conditions for x and y will equality hold in the triangle inequality?

2. Prove that $|x - y| \le |x| + |y|$. Hint: $x - y = x + (-y)$

3. Prove that $|x| - |y| \le ||x| - |y||$.

4. a. Extend the triangle inequality to $|x + y + z| \le |x| + |y| + |z|$.
   b. Extend the triangle inequality to $\left|\sum_{i=1}^{n} x_i\right| \le \sum_{i=1}^{n} |x_i|$.
   c. The inequality in **b** is true is for any *finite* number of terms. Can we now extend it to $\left|\sum_{i=1}^{\infty} x_i\right| \le \sum_{i=1}^{\infty} |x_i|$? Always be wary of replacing anything finite with the ∞ symbol. An important theorem about absolute convergence will address the question. Look for it in your calculus book.

5. Prove that the converse of the statement in Example 15.1 is true: if neighborhoods $\mathbf{N}_a^{\varepsilon}$ and $\mathbf{N}_b^{\varepsilon}$ have $|b - a| < 2\varepsilon$, then $\mathbf{N}_a^{\varepsilon} \cap \mathbf{N}_b^{\varepsilon} \ne \varnothing$.

6. a. Find a bound for $f(x) = \sqrt{x} - \ln x$ over $[0.1, 5]$.
   b. Find a bound for $g(x) = \sqrt{\sqrt{x} - \ln x}$ over $[4, 10]$.

16

Infinite Sequences

Let us take as a second example the numerical sequence $1, \frac{1}{2}, \frac{1}{3}, \frac{1}{4} \cdots \frac{1}{n}, \frac{1}{n+1}$, *&c....*

Augustin-Louis Cauchy, *Cours d'analyse de l'École royale polytechnique*, 1821, pg. 127

A **sequence** is the range of a function F whose domain is $\mathbb{N}$, or a subset of $\mathbb{N}$. The n**th term** of a sequence is the element a_n associated with subscript $n \in \mathbb{N}$, so that a_n is a shorthand for $F(n)$. Because of the subscripts, a sequence has a fixed order, and will be denoted by angle brackets, as in $\langle \frac{1}{2}, \frac{2}{2^2}, \frac{3}{2^3}, \ldots \rangle$ and $\langle 1, 5, 10, 10, 5, 1 \rangle$. Unlike a set, the terms of a sequence may repeat, since they are, after all, range values. An infinite sequence as a whole is expressed as $\langle a_n \rangle_{n=1}^{\infty}$, or as $\langle a_n \rangle_1^{\infty}$ when the subscripting variable is obvious. With a bit of thought, we see that a sequence may also be defined as a countable set with a specified order. Thus, it differs from a set in two ways: (1) a sequence can't have an uncountable number of elements, and (2) the elements cannot be rearranged since they must appear in the order prescribed by the subscripts.

Note 1. Since the coordinates that locate points in space must be in order, we have also used angle brackets for them, as in $\langle 5, 3 \rangle \in \mathbb{R}^2$ and $\langle 5, 3, 8 \rangle \in \mathbb{R}^3$.

Example 16.1 The infinite sequence of fractions above can be written as $\langle a_n \rangle_1^{\infty}$, where $a_n = \frac{n}{2^n}$ and $n \in \mathbb{N}$. The function $F(n) = \frac{n}{2^n}$ evaluates term a_n, so that $\mathcal{R}_F \subset \mathbb{R}$. We won't often need to be explicit about F; writing $a_n = \frac{n}{2^n}$ is enough. The form $\langle \frac{n}{2^n} \rangle_1^{\infty}$ is much more brief. This isn't just a useful shorthand; it also brings forth the internal structure of the sequence. Thus, for instance, the difference between successive terms is $a_{n+1} - a_n = \frac{n+1}{2^{n+1}} - \frac{n}{2^n} = \frac{1-n}{2^{n+1}}$. ■

Example 16.2 The sequence of **binomial coefficients** $\binom{m}{n}$ or ${}_mC_n$ is a finite sequence of natural numbers given by the function $F(n) = \frac{m!}{n!(m-n)!} \equiv \binom{m}{n}$, where $m \in \mathbb{N} \cup \{0\}$ is fixed, and $n \in \{0, 1, 2, \ldots, m\}$ is the domain of F. Selecting $m = 5$, we have $\langle \binom{5}{n} \rangle_0^5 = \langle 1, 5, 10, 10, 5, 1 \rangle$. Recall that the sequence of binomial coefficients is central to the binomial theorem, as can be seen in the expansion $(x + y)^5 = \sum_{n=0}^{5} \binom{5}{n} x^{5-n} y^n = x^5 + 5x^4y + 10x^3y^2 + 10x^2y^3 + 5xy^4 + y^5$. ■

Example 16.3 The set $\mathbb{Q}^+$ is denumerable according to Theorem 9.1. This implies that we could write $\mathbb{Q}^+$ as the infinite sequence $\langle \frac{1}{1}, \frac{1}{2}, \frac{2}{1}, \frac{1}{3}, \ldots \rangle$. In fact, we were able to write all of $\mathbb{Q}$ as an infinite sequence. ■

Example 16.4 Sequences need not be composed of numbers. A sequence of functions can be created as follows: let $a_n = f_n(x) = \frac{n}{x^2+1}$. Explicitly, this gives us $\langle f_n(x) \rangle_{n=1}^{\infty} = \langle \frac{1}{x^2+1}, \frac{2}{x^2+1}, \frac{3}{x^2+1}, \ldots \rangle$. Graph $f_n(x)$ for $n = 1, 2, 3$ to see what this sequence begins to look like. If someone insists, the function generating this sequence would be $F(n) = \frac{n}{x^2+1}$, with $\mathcal{R}_F = \{\frac{n}{x^2+1} : n \in \mathbb{N}\}$. Here x is just a dummy variable, so $F(n) = \frac{n}{t^2+1}$ will give exactly the same sequence. ■

Example 16.5 Example 5.1 defined $S_n = \{\text{points of the circle } x^2 + y^2 = (\frac{n}{n+1})^2\}$, which is an uncountable set of points for each n. Let's modify this to $C = \langle \text{circles } x^2 + y^2 = (\frac{n}{n+1})^2 \rangle_{n=1}^{\infty}$, so that the terms of the sequence are the circles themselves. Thus, the range of function C forms an infinite sequence $\langle c_n \rangle_1^{\infty}$ of *circles*, $\mathcal{R}_C = \{\text{circles with center at the origin and radii } \frac{n}{n+1} : n \in \mathbb{N}\}$, which may be seen in Figure 5.2 of Chapter 5. For example, $C(10) = c_{10}$ is the circle $x^2 + y^2 = (\frac{10}{11})^2$. ■

Example 16.6 The very famous **Fibonacci sequence** is $F = \langle 1, 1, 2, 3, 5, 8, 13, \ldots \rangle$. Its definition is recursive, that is, later terms are defined by means of earlier terms. Here is the definition: $F_n = F_{n-1} + F_{n-2}$, with $F_1 = F_2 = 1$. Consequently, for instance, $F_5 = F_4 + F_3 = (F_3 + F_2) + (F_2 + F_1) = 3F_2 + 2F_1 = 5$. See Exercise 16.5. ■

For now, we will concentrate on real-valued sequences, that is, functions $F : \mathbb{N} \to \mathbb{R}$ (or sometimes $F : \mathbb{N} \cup \{0\} \to \mathbb{R}$) as in Example 16.1. Treating a sequence of real numbers as a function has the advantage that it can be graphed as points $\langle n, a_n \rangle$ in $\mathbb{R}^2$. The examples below display some of the many ways infinite sequences behave.

Example 16.7 The behaviors of $\langle \sqrt{n} \rangle_1^{\infty}$, $\langle \frac{3n-1}{5n} \rangle_1^{\infty}$, and $\langle \frac{n}{2^n} \rangle_1^{\infty}$ can be compared easily by graphing them. First, they are graphed as points in $\mathbb{R}$, then as points in $\mathbb{R}^2$ in Figures 16.1 to 16.3; each graph has its advantages. ■

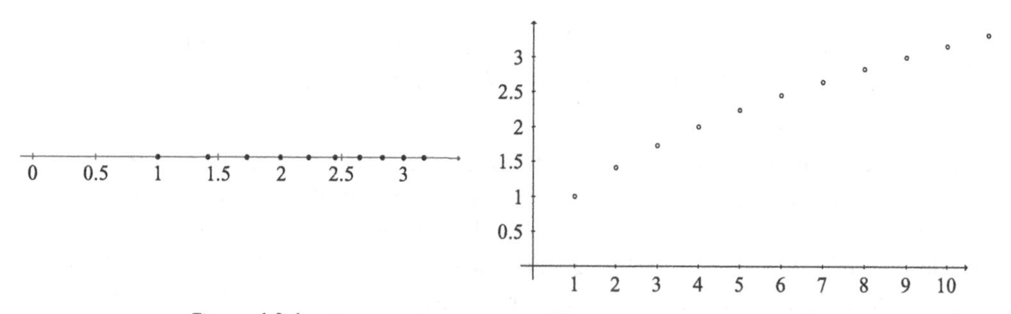

Figure 16.1. The infinite sequence $\langle \sqrt{n} \rangle_1^{\infty}$ graphed in $\mathbb{R}$ and in $\mathbb{R}^2$.

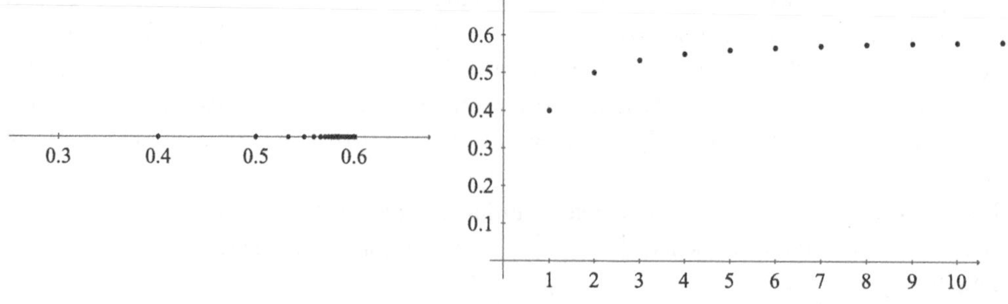

Figure 16.2. The infinite sequence $\langle \frac{3n-1}{5n} \rangle_1^{\infty}$ graphed in $\mathbb{R}$ and in $\mathbb{R}^2$.

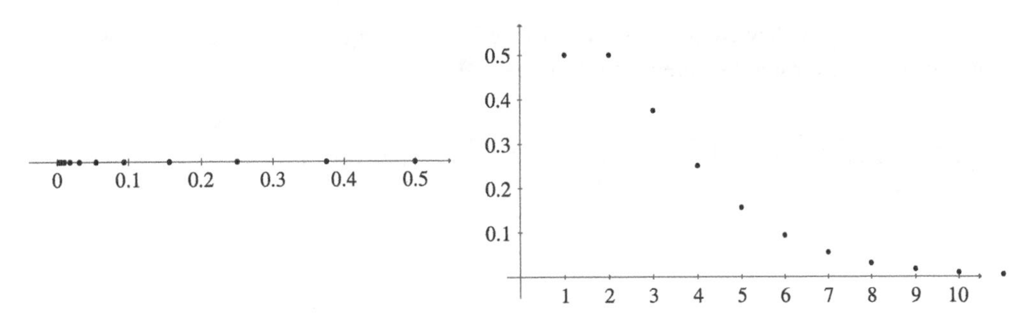

Figure 16.3. The infinite sequence $\langle \frac{n}{2^n} \rangle_1^\infty$ graphed in $\mathbb{R}$ and in $\mathbb{R}^2$.

Example 16.8 The sequence defined by the function below selects some of the points of the set S pictured in Example 14.5 (why is the definition of S there not a function?)

$$a_n = \begin{cases} \frac{1}{n} & \text{when } n = 4k - 3 \\ 2 - \frac{1}{n} & \text{when } n = 4k - 2 \\ 2 + \frac{1}{n} & \text{when } n = 4k - 1 \\ 4 + \frac{1}{n} & \text{when } n = 4k \end{cases},$$

where $k \in \mathbb{N}$.

The sequence loses its mystery once we begin selecting values of n. For instance, if $n = 10$, then the only possible choice is the row with $10 = 4k - 2$, since k must be a natural number (namely, $k = 3$). Thus, $a_{10} = 2 - \frac{1}{10} = \frac{19}{10}$. The graph of S in Example 14.5 is comparable to the graph in Figure 16.4, which skips some points of S but shows the behavior near cluster points. The three cluster points appear as horizontal asymptotes of the graphed sequence. ■

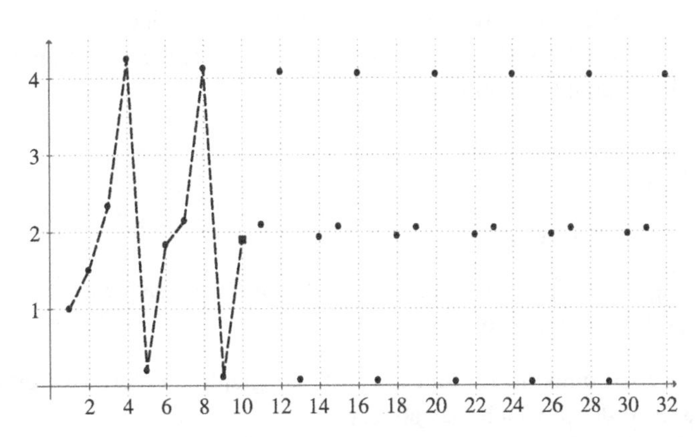

Figure 16.4. The sequence above is graphed as $\langle n, a_n \rangle$ according to the four-part rule given there. The dotted line shows the sequence's order of points. The term a_{10} appears at $\langle 10, \frac{19}{10} \rangle$ as a black square.

Example 16.9 The sequence given by **Stirling's formula** has $a_n = e^{-n} n^n \sqrt{2\pi n}$. This remarkable formula approximates the sequence of factorials $\langle n! \rangle_1^\infty$ with decreasing relative error as

$n \to \infty$. The table below compares some terms of the two sequences. Exercise 16.4 will ask you to investigate further this interesting formula. ■

n	1	2	3	4	5	6
$\langle e^{-n} n^n \sqrt{2\pi n} \rangle$	0.9221...	1.919...	5.836...	23.50...	118.0...	710.0...
$\langle n! \rangle$	1	2	6	24	120	720

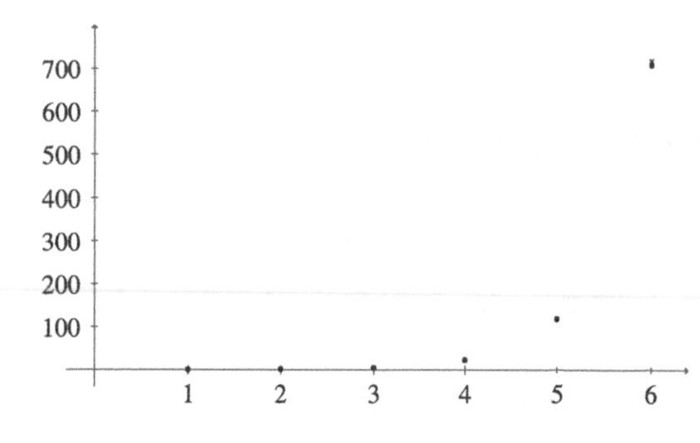

Figure 16.5. The Stirling's formula sequence is graphed using + symbols, and the factorial sequence is graphed using ∘ symbols. The approximation is so good that only one absolute difference is distinctly visible at this scale.

Note 2: There are many proofs of the fact that Stirling's formula approximates $n!$ with ever-increasing precision, and none is easy. Perhaps the shortest is "An Elementary Proof of Stirling's Formula," by P. Diaconis and D. Freedman, in *The American Mathematical Monthly*, vol. 93, no. 2 (Feb. 1986), pp. 123–125.

It seems traditional in analysis for the word "elementary" to be used in a loose way. That article's proof is elementary only in comparison to the proofs mentioned in its references. Nevertheless, it does bring the demonstration down to the level of intermediate analysis (quite a feat!).

A sequence $\langle a_n \rangle_1^\infty$ of real numbers is **bounded** if there exists some positive constant B such that $|a_n| < B$ for all terms a_n. If this is not the case, then $\langle a_n \rangle_1^\infty$ is **unbounded**. For instance, $\langle \frac{3n-1}{5n} \rangle_1^\infty$ and $\langle \frac{n}{2^n} \rangle_1^\infty$ in Example 16.7 are bounded sequences, as is the sequence in Example 16.8. The sequences in Example 16.9 are unbounded. Since sequences of real numbers are specialized sets of real numbers, the definition here conforms to the definition of a bounded set in Chapter 12, as logic would demand.

16.1 Exercises

1. Why is $\mathbb{R}$ not an infinite sequence?

2. The set $S = \{\frac{1}{2^n} : n \in \mathbb{N}\}$ from Example 14.3 becomes the sequence $\langle \frac{1}{2^n} \rangle_1^\infty$. Graph it as $\langle n, \frac{1}{2^n} \rangle$ to see it in two dimensions.

3. Plot the first twelve points $\langle n, a_n \rangle$ that appear in the graph of Figure 16.4 by hand on graph paper, noting the order in which they appear.

4. a. It seems that the Stirling's formula sequence makes a large error at $a_6 \approx 710.08$ compared to $6!$. The relative error of x compared to y is defined as $\frac{x-y}{y}$. Calculate the relative errors of Stirling's formula values to factorial values in the table of Example 16.9. In fact, the sequence of relative errors of the two sequences can be shown to approach zero.

b. Most calculators will not evaluate 100000!. However, take the common logarithm of both sides of $n! \approx e^{-n}n^n\sqrt{2\pi n}$, and substitute $n = 10^5$. How many digits does 100000! have? What is its approximate value?

5. For each of the following sequences, find a function that produces the general term a_n.
   a. $\langle 1, -1, \frac{1}{2}, \frac{-1}{6}, \frac{1}{24}, \frac{-1}{120}, \ldots\rangle$
   b. $\langle \frac{1}{3}, \frac{3}{5}, \frac{5}{7}, \frac{7}{9}, \ldots\rangle$
   c. $\langle 1, 4, 27, 256, 3125, \ldots\rangle$
   d. $\langle 1, 3, 6, 10, 15, 21, \ldots\rangle$
   e. $\langle 1, 0, 1, 0, 1, \ldots\rangle$ (can you find more than one function?)
   f. $\langle 2,\ 2{\cdot}4,\ 2{\cdot}4{\cdot}6,\ 2{\cdot}4{\cdot}6{\cdot}8, \ldots\rangle$

6. a. If a sequence $\langle a_n\rangle_1^\infty$ of positive terms is bounded, must the sequence $\langle \frac{1}{a_n}\rangle_1^\infty$ be bounded? If so, prove it; if not, find a counterexample.
   b. If $\langle a_n\rangle_1^\infty$ is bounded, must the sequence $\langle 1 - a_n\rangle_1^\infty$ be bounded? If so, prove it; if not, find a counterexample.

7. a. For a sequence $\langle a_n\rangle_1^\infty$, suppose that the ratios $|\frac{a_{n+1}}{a_n}| \leq B$, where $B > 1$. Find an example to prove that this is not a sufficient condition for the boundedness of $\langle a_n\rangle_1^\infty$.
   b. Now, suppose that $|\frac{a_{n+1}}{a_n}| \leq B < 1$. Prove that $\langle a_n\rangle_1^\infty$ must be bounded. Hint: $|a_{n+1}| \leq B|a_n|$.

17

Limits of Sequences

One says that a magnitude is the limit of another magnitude when the second may approach the first more closely than a given quantity, as small as one would desire it to be, however, never allowing the approaching magnitude to surpass the magnitude to which it approaches; so that the difference between such a magnitude and its limit is absolutely unassignable.

Jean le Rond d'Alembert, from his article "Limite" in the *Encyclopédie*, 1765.

Establishing the definition of a limit was one of the toughest intellectual struggles in the development of mathematics. Its roots begin with the extraordinary approach by Archimedes to approximating π (see [1], pp. 64–67). Implicit in his geometric analysis is that successively better estimates of π could be calculated, π being the unknown limiting value of his algorithm. Newton's definition of the derivative of a function as the quantity remaining after vanishingly small quantities were disregarded captured the sense of a limit, but his procedure was deservedly criticized by the philosopher Bishop George Berkeley. The first firm idea of a limit perhaps belongs to the mathematician Jean le Rond d'Alembert (1717–1783) (see the quote above). But it is Augustin-Louis Cauchy who deserves the credit for algebraically explaining that a limit is a number that a function approaches. Here, we apply the definition of limit to infinite sequences, as Karl Weierstrass would have written it in the mid-1800s.

> The **limit of a sequence** $\langle a_n \rangle_1^\infty$—if it exists—is a number L with the property that
> [Definition 1] every neighborhood $(L-\varepsilon,\ L+\varepsilon)$ contains all but a finite number of terms of $\langle a_n \rangle_1^\infty$,
> or
> [Definition 2] for any $\varepsilon > 0$, there exists a corresponding number N_ε such that for all terms with subscripts $n > N_\varepsilon$, we have $|a_n - L| < \varepsilon$.

Whenever L exists, we will say that sequence $\langle a_n \rangle_1^\infty$ **converges** to L, and write $\lim_{n\to\infty} a_n = L$, or simply $\langle a_n \rangle_1^\infty \to L$.

Comparing the two definitions, note that "for all terms with subscripts $n > N_\varepsilon$, we have $|a_n - L| < \varepsilon$" is logically equivalent to "$(L-\varepsilon, L+\varepsilon)$ contains all but a finite number of terms of $\langle a_n \rangle_1^\infty$" (verify the equivalence). And the phrase "for any $\varepsilon > 0$" in the second definition has the same role as "every neighborhood" in the first. Definition 2 goes one step further in requiring us to find a function $N(\varepsilon) \to N_\varepsilon$ that outputs a unique N_ε for each $\varepsilon > 0$ we input.

Note 1: In Definition 1, the phrase "all but a finite number of terms" is not the same as "infinitely many terms." For an infinite sequence, the former implies the latter, since $\aleph_0 - n = \aleph_0$. But the latter does not imply the former, for extracting infinitely many terms from the sequence may still leave behind infinitely many terms. For these reasons, in Definition 2, all terms after term a_{N_ε} are required, not just infinitely many terms after a_{N_ε}.

Let's apply the two definitions to prove that $\langle \frac{3n-1}{5n} \rangle_1^\infty$ does have a limit. Using a calculator, it seems clear that as n gets large, $\langle \frac{3n-1}{5n} \rangle_1^\infty \to \frac{3}{5}$. Figure 17.1 certainly hints that $L = \frac{3}{5}$ (this sequence appeared in Example 16.7).

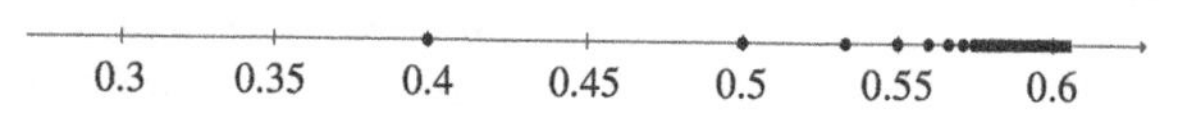

Figure 17.1. Terms $a_n = \frac{3n-1}{5n}$ starting at $a_1 = \frac{2}{5}$ and increasing toward the cluster point $\frac{3}{5}$.

Example 17.1 To prove that $\lim_{n\to\infty} \frac{3n-1}{5n} = \frac{3}{5}$, we may apply Definition 1. We must show that every neighborhood $(\frac{3}{5} - \varepsilon, \frac{3}{5} + \varepsilon)$ contains all but a finite number of terms of $\langle \frac{3n-1}{5n} \rangle_1^\infty$. This in turn means that $\frac{3}{5} - \varepsilon < \frac{3n-1}{5n} < \frac{3}{5} + \varepsilon$ must be true for all but a few terms of the sequence. Continuing, $-\varepsilon < \frac{3n-1}{5n} - \frac{3}{5} < \varepsilon$, and combining fractions, $-\varepsilon < -\frac{1}{5n} < \varepsilon$. Regardless of how small ε is, Corollary 12.2 assures us that an n does exist that fulfills this triple inequality. The challenge remains: *every* neighborhood $(\frac{3}{5} - \varepsilon, \frac{3}{5} + \varepsilon)$, even minute neighborhoods, must contain all but a few terms. Continuing, $-\varepsilon < -\frac{1}{5n} < \varepsilon$ implies that $n > \frac{1}{5\varepsilon} > -n$. Since ε is positive, we evidently need just this part: $n > \frac{1}{5\varepsilon}$.

The challenge has now been answered. Given any (even a microscopic) neighborhood $(\frac{3}{5} - \varepsilon, \frac{3}{5} + \varepsilon)$ defined by ε, all but a finite set of subscripts n are greater than $\frac{1}{5\varepsilon}$. It will then follow that every such term $a_n = \frac{3n-1}{5n}$ will fall inside the neighborhood of $\frac{3}{5}$. Therefore, $\lim_{n\to\infty} \frac{3n-1}{5n} = \frac{3}{5}$. ■

To demonstrate the analysis just done, let's actually select a tiny neighborhood of $\frac{3}{5}$, say $(\frac{3}{5} - 0.001, \frac{3}{5} + 0.001) = (0.599, 0.601)$ (keep an eye on Figure 17.1). Thus, we have chosen $\varepsilon = 0.001$. We may now identify the actual term of $\langle \frac{3n-1}{5n} \rangle_1^\infty$ after which all the terms of the sequence will fall inside the neighborhood. This was done algebraically in the example above. As derived there, we require $n > \frac{1}{5(0.001)} = 200$. With some excitement, we evaluate $a_{201} = \frac{3(201)-1}{5(201)} = 0.5990049\ldots \in (0.599, 0.601)$. Success! Just as essential, the analysis also proves that *all* terms after a_{200} will be members of the selected interval. As counterpoints, observe that $a_{200} = 0.599 \notin (0.599, 0.601)$, and that the finite set of terms that did not fall into the interval was $\{a_1, \ldots, a_{200}\}$.

Looking at Figure 17.1, it becomes clear that the neighborhoods $(\frac{3}{5} - \varepsilon, \frac{3}{5} + \varepsilon)$ can be thought of as tolerance intervals (as in manufacturing specifications) within which we want the terms of the sequence to fall. Thus, in the check above, we set the tolerance at $\varepsilon = 0.001$. We then found that the first 200 terms of the sequence did not meet the tolerance, but that every term after that met (and surpassed) it. It is as if a manufacturing robot steadily learned to increase its precision, until finally, starting at item number 201, the robot got its dimensions right, and then improved forever after (of course, that's a fairy tale ending, as anyone in industry knows!).

Example 17.2 Use Definition 2 of limit to prove that $\lim_{n\to\infty} \frac{3n-1}{5n} = \frac{3}{5}$. We must prove that for any $\varepsilon > 0$, there exists a corresponding number N_ε such that whenever the subscript $n > N_\varepsilon$, then $|\frac{3n-1}{5n} - \frac{3}{5}| < \varepsilon$. We are searching for the function $N(\varepsilon)$ that yields N_ε with the needed properties, for any $\varepsilon > 0$.

The abiding curiosity of this analysis is that we usually work backwards from the consequent, namely $|\frac{3n-1}{5n} - \frac{3}{5}| < \varepsilon$, to discover the function $N(\varepsilon)$. Thus, combining the fractions within the absolute value, $|-\frac{1}{5n}| < \varepsilon$, so that $|\frac{1}{5n}| < \varepsilon$. Since n is positive, we use $\frac{1}{5n} < \varepsilon$. Solving for n, $\frac{1}{5\varepsilon} < n$. This now yields a function $N(\varepsilon)$ that specifies the number N_ε, namely, $N(\varepsilon) = \frac{1}{5\varepsilon} = N_\varepsilon$.

To prove that $N(\varepsilon)$ does in fact satisfy the consequent $|\frac{3n-1}{5n} - \frac{3}{5}| < \varepsilon$ whenever $n > N_\varepsilon$, we unroll the steps of the derivation just made. Thus, with $N_\varepsilon = \frac{1}{5\varepsilon}$, whenever $n > N_\varepsilon$ it follows that $n > \frac{1}{5\varepsilon}$. Then, $\frac{1}{5n} < \varepsilon$ and $|\frac{1}{5n}| < \varepsilon$. This brings us back to $|\frac{3n-1}{5n} - \frac{3}{5}| < \varepsilon$.

Finally, we assert the definition as proof that $\frac{3}{5}$ is the limit L. For any $\varepsilon > 0$, there exists a corresponding number $N_\varepsilon = \frac{1}{5\varepsilon}$ such that whenever $n > N_\varepsilon$, then $|\frac{3n-1}{5n} - \frac{3}{5}| < \varepsilon$. Hence, $\lim_{n\to\infty} \frac{3n-1}{5n} = \frac{3}{5}$. ■

As was done after Example 17.1, we make a comforting partial check. Select $\varepsilon = 0.0123$. According to our analysis, $N(0.0123) = N_{0.0123} = \frac{1}{5(0.0123)} = 16.2601\ldots$ (see Figure 17.2). Thus, if subscript $n \geq 17$, then $|\frac{3n-1}{5n} - \frac{3}{5}| < 0.0123$. At $n = 17$, we get $|\frac{3(17)-1}{5(17)} - \frac{3}{5}| = 0.01176\ldots$, which is less than 0.0123. Success! Moreover, the analysis in the proof shows that $|\frac{3n-1}{5n} - \frac{3}{5}| < 0.0123$ for *all* $n \geq 17$. Exercise 17.1 will ask you to do the check when $\varepsilon = 0.001$, and compare it to Example 17.1.

D'Alembert's definition (beneath the chapter's title) was a great step forward, but it has some deficiencies. Two shortcomings are easy to spot. First, d'Alembert implies that the limit is a number that the second, varying, magnitude cannot equal. While this is true for the sequence above, it isn't correct in general. Second, he seems to confine himself to a varying magnitude

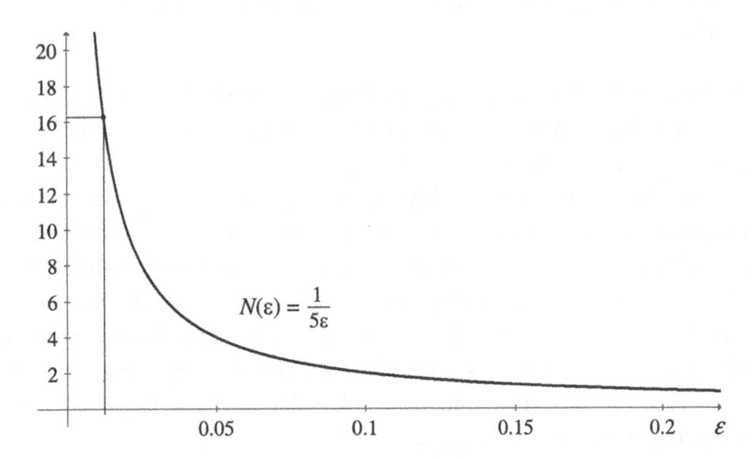

Figure 17.2. Graph of the function $N(\varepsilon) = \frac{1}{5\varepsilon}$. As the tolerance ε decreases, ever increasing ordinates N_ε are needed to insure that the tolerance is met by terms a_n. The check point $\langle 0.0123, 16.26\ldots\rangle$ is shown. For any $\varepsilon \geq 0.2$, n may remain equal to 1 since $N_\varepsilon \leq 1$.

that steadily increases towards the limit. The next example shows a sequence that oscillates towards its limit.

Example 17.3 Prove that $\langle \frac{(-1)^n}{e^n} \rangle_1^\infty \to 0$. The first few terms (always evaluate the first few terms), reveal the oscillating nature: $\langle \frac{-1}{e}, \frac{1}{e^2}, \frac{-1}{e^3}, \ldots \rangle$. We must show that for any $\varepsilon > 0$, there exists a corresponding number N_ε such that whenever $n > N_\varepsilon$, then $|\frac{(-1)^n}{e^n} - 0| < \varepsilon$. Beginning with the last inequality, simplify it to $\frac{1}{e^n} < \varepsilon$. Solving for n, we arrive at $\ln \frac{1}{\varepsilon} < n$ (after verifying with your calculus text that if $0 < x < y$, then $\ln x < \ln y$). Therefore, taking $N_\varepsilon = \ln \frac{1}{\varepsilon}$ gives us a correspondence such that whenever $n > N_\varepsilon$, then $|\frac{(-1)^n}{e^n} - 0| < \varepsilon$. It follows that $\lim_{n\to\infty} \frac{(-1)^n}{e^n} = 0$.

The graph of $\langle \frac{(-1)^n}{e^n} \rangle_1^\infty$ in $\mathbb{R}^2$ reveals much. In Figure 17.3, one sees that even-numbered terms decrease toward 0, while odd-numbered terms increase toward 0. Also included is the $\varepsilon = 0.04$ neighborhood of the limit. We see that every term starting at $a_4 = e^{-4}$ falls within the neighborhood. Prove this statement as Exercise 17.3. ■

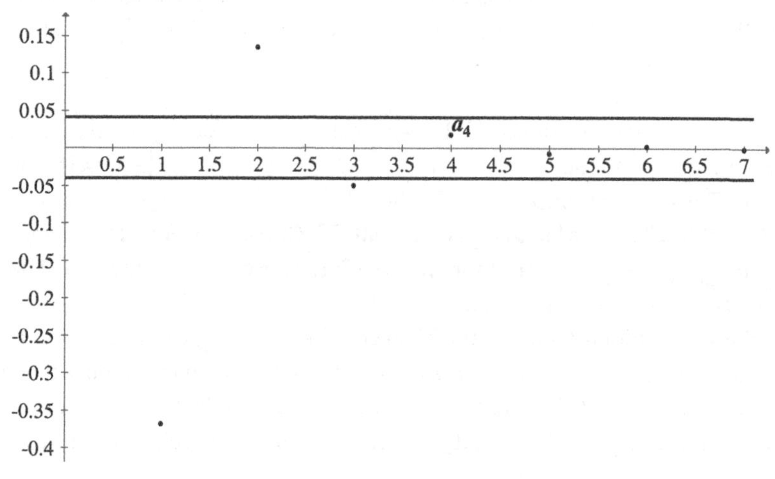

Figure 17.3. Graph of $\langle (-1)^n e^{-n} \rangle_1^\infty$ in $\mathbb{R}^2$. An $\varepsilon = 0.04$ neighborhood of the limit, namely $(-0.04, 0.04)$, appears as a thin channel.

Example 17.4 Prove that the sequence $\langle \frac{n}{2^n} \rangle_1^\infty$ pictured in Example 16.7 has $\lim_{n\to\infty} \frac{n}{2^n} = 0$. We must once again show that for any $\varepsilon > 0$, there exists a corresponding N_ε such that for all terms with subscripts $n > N_\varepsilon$, it must be true that $|\frac{n}{2^n} - 0| < \varepsilon$.

Beginning with the concluding inequality, we simplify it to $\frac{n}{2^n} < \varepsilon$. Now a difficult problem arises: we cannot solve for n, as we did in the previous examples. However, the definition does not require us to find the very first a_n that falls within the selected neighborhood $(-\varepsilon, \varepsilon)$. We need only find *some* term after which every term will fall within $(-\varepsilon, \varepsilon)$. This is our way out.

It is easy to show that $\frac{1}{2^n} < \frac{1}{n^2}$ for all $n \geq 5$. Create a short proof by induction that $n^2 < 2^n$ using the anchor case $n = 5$ (or review Example 4.2), and arrive at the desired inequality. It follows that $\frac{n}{2^n} < \frac{n}{n^2} = \frac{1}{n}$. Thus, if we find an n no less than 5 and so large that $\frac{1}{n} < \varepsilon$, then we will have identified a_n, because clearly, $a_n = \frac{n}{2^n} < \frac{1}{n} < \varepsilon$.

Hence, instead of attempting to solve $\frac{n}{2^n} < \varepsilon$ for n, we just solve $\frac{1}{n} < \varepsilon$ for n, giving $\frac{1}{\varepsilon} < n$. And now, $N_\varepsilon = \frac{1}{\varepsilon}$ is surely large enough to guarantee that $|\frac{n}{2^n} - 0| < \varepsilon$ for all $n > N_\varepsilon$, provided that $N_\varepsilon \geq 5$.

To not admit n less than 5, we finally state that given any $\varepsilon > 0$, if we take $N_\varepsilon = \max\{\frac{1}{\varepsilon}, 5\}$, it follows that $|\frac{n}{2^n} - 0| < \varepsilon$ for all $n > N_\varepsilon$. This means that $\lim_{n\to\infty} \frac{n}{2^n} = 0$. ■

Once again, randomly select a tolerance, such as $\varepsilon = 0.017$. Then $\frac{1}{0.017} = 58.823\cdots = N_\varepsilon$. Thus, every term starting at $a_{59} = \frac{59}{2^{59}} \approx 1.02 \times 10^{-16}$ is certainly within the neighborhood $(-0.017, 0.017)$. But isn't this a bit overboard? Surely terms well before a_{59} are already within the selected neighborhood. The great point here is that it doesn't matter! We reemphasize that it is not necessary to find the first term within $(-0.017, 0.017)$. (In fact, it's impossible to do this algebraically in the present example.) We only needed to be sure that all terms of the sequence after the 58th term are in $(-0.017, 0.017)$. See Exercise 17.4.

Example 17.5 Prove that $\langle\frac{5n^4+87n^3}{4n^4+7n+8}\rangle_1^\infty \to \frac{5}{4}$. This is trickier than Example 17.2, but similar, because the terms in both are rational functions of n. By Definition 2, for any $\varepsilon > 0$, we must find an N_ε such that

$$\left|\frac{5n^4 + 87n^3}{4n^4 + 7n + 8} - \frac{5}{4}\right| < \varepsilon$$

is always true when $n > N_\varepsilon$.

Again, working with $|\frac{5n^4+87n^3}{4n^4+7n+8} - \frac{5}{4}| < \varepsilon$, we simplify:

$$\left|\frac{5n^4 + 87n^3}{4n^4 + 7n + 8} - \frac{5}{4}\right| = \frac{348n^3 - 35n - 40}{4(4n^4 + 7n + 8)} < \varepsilon.$$

It's a monstrous task to solve this for n. Instead, increase the magnitude of the fraction by conveniently increasing the numerator, or decreasing the denominator (keeping it positive), or both:

$$\frac{348n^3 - 35n - 40}{4(4n^4 + 7n + 8)} \le \frac{400n^3}{4(4n^4)} = \frac{25}{n}.$$

Simplified by this technique, if $\frac{25}{n} < \varepsilon$, then certainly $|\frac{5n^4+87n^3}{4n^4+7n+8} - \frac{5}{4}| < \varepsilon$. Now, $\frac{25}{n} < \varepsilon$ implies $n > \frac{25}{\varepsilon} = N_\varepsilon$. Therefore, given any $\varepsilon > 0$, $N_\varepsilon = \frac{25}{\varepsilon}$ is large enough to guarantee that $|\frac{5n^4+87n^3}{4n^4+7n+8} - \frac{5}{4}| < \varepsilon$ for all $n > N_\varepsilon$. ■

Note 2: The technique of adjusting the size of a positive fraction is heavily used in analysis. The details are as follows. Let $0 < \frac{a}{b}$, let u be positive, and let $0 < v < b$. The technique uses these conclusions: $\frac{a}{b} < \frac{a+u}{b}$, and $\frac{a}{b} < \frac{a}{b-v}$, and together, $\frac{a}{b} < \frac{a+u}{b-v}$. Likewise, $\frac{a}{b} > \frac{a-u}{b}$, $\frac{a}{b} > \frac{a}{b+v}$, and together, $\frac{a}{b} > \frac{a-u}{b+v}$.

The next theorem is fundamental to the study of limits. Two proofs appear; the first uses Definition 1, and the second uses Definition 2.

Theorem 17.1 If a sequence $\langle a_n\rangle_1^\infty$ has a limit L, then it is unique. That is, a sequence cannot have more than one limit.

First Proof. Assume, on the contrary, that $\lim_{n\to\infty} a_n = L$, $\lim_{n\to\infty} a_n = M$, and $L \ne M$. Take $\varepsilon = \frac{1}{2}|M - L| > 0$. Then, since $\lim_{n\to\infty} a_n = L$, all but a finite number of terms must lie in the neighborhood $(L - \varepsilon, L + \varepsilon)$. Now, $(M - \varepsilon, M + \varepsilon) \cap (L - \varepsilon, L + \varepsilon) = \varnothing$, and this leaves M out in the cold, so to speak, for in the neighborhood $(M - \varepsilon, M + \varepsilon)$ there can only be a *finite* number of terms of the sequence, consisting of some (and maybe not even all) of the terms outside of $(L - \varepsilon, L + \varepsilon)$. This contradicts the assumed nature of M, and therefore distinct limits are impossible. **QED**

Second Proof. Let $\lim_{n\to\infty} a_n = L$ and $\lim_{n\to\infty} a_n = M$. (This is not a proof by contradiction, since we won't assume that $L \neq M$.) Then for any $\varepsilon > 0$, there exists an N_ε so large that both $|a_n - L| < \frac{\varepsilon}{2}$ and $|a_n - M| < \frac{\varepsilon}{2}$. Adding, $|a_n - L| + |a_n - M| < \varepsilon$, which is equivalent to $|L - a_n| + |a_n - M| < \varepsilon$. Now, apply the triangle inequality: $|(L - a_n) + (a_n - M)| \leq |L - a_n| + |a_n - M| < \varepsilon$. Thus, $|L - M| < \varepsilon$. This states that $|L - M|$ is less than any positive quantity ε. By Corollary 13.3, $|L - M| = 0$, and therefore, $L = M$. **QED**

17.1 Limits as Cluster Points

Consider Figure 17.1 once again. The sequence $\langle \frac{3n-1}{5n} \rangle_1^\infty$ converges to $\frac{3}{5}$, and considered as a set of points $S = \{\frac{3n-1}{5n}\}_1^\infty$, the number $x = \frac{3}{5}$ is its one and only cluster point. Recall that a cluster point is a location on the real number line near which infinitely many points of a set gather, so it seems that all limits are cluster points. This is almost correct. As a counterexample, the sequence $\langle 3, 2, 1, 1, 1, 1, 1, \ldots \rangle$ has a limit of 1, but considered as a set, this sequence collapses to $\{3, 2, 1\}$, which has no cluster points at all. With this in mind, we formulate the next theorem.

Theorem 17.2 If a sequence $\langle a_n \rangle_1^\infty$ with infinitely many, but not necessarily all, distinct terms has the limit L, then the set $A = \{a_n\}_{n=1}^\infty$ has L as its only cluster point.

Proof. The strategy follows the logic. We must first show that the definition of a limit implies the definition of a cluster point (Chapter 14). By the premise of convergence to L, every neighborhood $\mathbf{N}_L^\varepsilon$ contains all but a finite number of terms of $\langle a_n \rangle_1^\infty$.

Assume that even one of the neighborhoods, say $\mathbf{N}_L^{\varepsilon*}$, were to have every term a_n within it equal to L. Then only a finite number of terms may differ from L (those outside of the neighborhood). This would contradict the other premise, that $\langle a_n \rangle_1^\infty$ has infinitely many distinct terms. Hence, every neighborhood of L contains at least one point $a_n \in A$ different from L. Therefore, L is a cluster point of A. Now that existence has been proved, you must demonstrate uniqueness (that is, that no other cluster point exists), in Exercise 17.11. **QED**

17.2 Exercises

1. Make another check similar to the one after Example 17.2, but take $\varepsilon = 0.001$. Compare your results with the check after Example 17.1.

2. For the sequence in Example 17.2, what integer N_ε will guarantee that all terms a_n will be within $\varepsilon = 10^{-5}$ of the limit? What is the value of that first term that falls within that tolerance?

3. For the sequence in Example 17.3, prove that given $\varepsilon = 0.04$, all terms starting at $n = 4$ satisfy $|\frac{(-1)^n}{e^n} - 0| < \varepsilon$. Does this prove that the limit of the sequence is 0?

4. After the proof in Example 17.4, we saw that $\varepsilon = 0.017$ gave us an enormously tiny $a_{59} = \frac{59}{2^{59}}$, well within the neighborhood $(-0.017, 0.017)$. Where in the proof did we allow such extreme subscripts to be considered?

5. Suppose that $\langle a_n \rangle_1^\infty$, with no negative terms, has a limit L. Prove that $L \geq 0$. Hint: Try a proof by contrapositive.

6. Let $\langle a_n \rangle_1^\infty$ have no negative terms. Prove that if $\langle a_n \rangle_1^\infty \to 0$, then $\langle \sqrt{a_n} \rangle_1^\infty \to 0$, and conversely.

7. Prove that the following limits are correct (use either of the two definitions):
   a. $\lim_{n\to\infty} \frac{5n-37}{3n+1} = \frac{5}{3}$ b. $\lim_{n\to\infty} \frac{5n^2-3n}{3n^2+1} = \frac{5}{3}$ c. $\lim_{n\to\infty} \frac{\cos(n\pi)}{n} = 0$
   d. $\lim_{n\to\infty} \frac{12n^4-5n^3}{3n^4+\sqrt{n}} = 4$ e. $\lim_{n\to\infty} \frac{1}{2\sqrt[3]{n}} = 0$ f. $\lim_{n\to\infty} \frac{2^n}{n!} = 0$
   Hint: As long as $n \geq 4$, $\frac{2^n}{n!} = \frac{2}{1} \cdot \frac{2}{2} \cdot (\frac{2}{3} \cdot \cdots \cdot \frac{2}{n-1}) \cdot \frac{2}{n} < \frac{2}{1} \cdot \frac{2}{2} \cdot \frac{2}{n}$, since all the fractions in parentheses are less than one.

8. Generalize Exercise 17.7f to a proof that $\lim_{n\to\infty} \frac{b^n}{n!} = 0$ for any real b. Hint: Start with positive b, which need not be an integer, and natural number $n \geq b + 2$, and refine the proof of that exercise.

9. Prove that $\langle \frac{\cos n}{n} \rangle_1^\infty$ converges. You must make a hypothesis about what the limit is, before you can begin the proof.

10. Prove that an isolated point of a sequence can never be the limit of the sequence.

11. In Theorem 17.2, prove that another cluster point $M \neq L$ cannot exist.

12. State the converse of Theorem 17.2, and find a counterexample to prove that it is false. Hint: The minor premise "infinitely many distinct elements" is a premise in the converse statement as well. Form a sequence that visits all neighborhoods of its cluster point, but always has some outlier points not in such neighborhoods.

18

Divergence: The Non-Existence of a Limit

A word of warning is in order, however: Do not confuse infinite symbols with numbers, and write such nonsense as $|a_n - \infty| < \varepsilon$ when dealing with an infinite limit!

John M. H. Olmsted [18], pg. 32

It takes some logical care to negate our definitions for the existence of a limit, and thereby obtain statements that deny the existence of a limit. Using skills from Chapter 1, we conclude the following.

> A real number L is not the limit of a sequence $\langle a_n \rangle_1^\infty$ if and only if:
> [Negated Definition 1] there exists some neighborhood $(L - \varepsilon^*, L + \varepsilon^*)$ that does not contain—that excludes—an infinite number of terms of $\langle a_n \rangle_1^\infty$,
> or
> [Negated Definition 2] for some $\varepsilon^* > 0$, infinitely many terms a_n exist such that $|a_n - L| \geq \varepsilon^*$.

When every real number L fails to be the limit of sequence $\langle a_n \rangle_1^\infty$, we say that $\langle a_n \rangle_1^\infty$ is a **divergent sequence**, and that the sequence **diverges**.

Example 18.1 It is worth remembering that graphs often give deceptive evidence for the existence of a limit. Based on the graph of $\langle \frac{1}{2\sqrt[3]{n}} \rangle_1^\infty$ seen below, this sequence seems to converge to perhaps $L = 0.2$.

To prove that this is false, we shall apply Negated Definition 1. Notice that $a_{512} = \frac{1}{2\sqrt[3]{512}} = 0.0625$, for instance, is far below 0.2. Let's choose $\varepsilon^* = 0.1$, yielding the neighborhood $(0.1, 0.3)$ around L. We want to determine which terms of the sequence fall within this neighborhood. Hence, we solve $\frac{1}{10} < \frac{1}{2\sqrt[3]{n}} < \frac{3}{10}$. Taking reciprocals, $\frac{10}{3} < 2\sqrt[3]{n} < 10$, and then, $\frac{1000}{216} < n < 125$. We now see that only the *finite* collection of terms $a_5 = \frac{1}{2\sqrt[3]{5}}$ through $a_{124} = \frac{1}{2\sqrt[3]{124}}$ lie in the neighborhood $(0.1, 0.3)$. In other words, $(0.1, 0.3)$ excludes infinitely many terms. Consequently, $L = 0.2$ cannot be the limit of the sequence. (You proved that the limit was zero in Exercise 17.7e.) ■

Example 18.2 Prove that $\langle \frac{n^2}{n+5} \rangle_1^\infty$ diverges. We will use Negated Definition 2 this time. We must prove there exists an $\varepsilon^* > 0$ such that infinitely many terms of the sequence satisfy $|\frac{n^2}{n+5} - L| \geq \varepsilon^*$, regardless of the value of L. We pick a convenient tolerance, say, $\varepsilon^* = 1$. Then

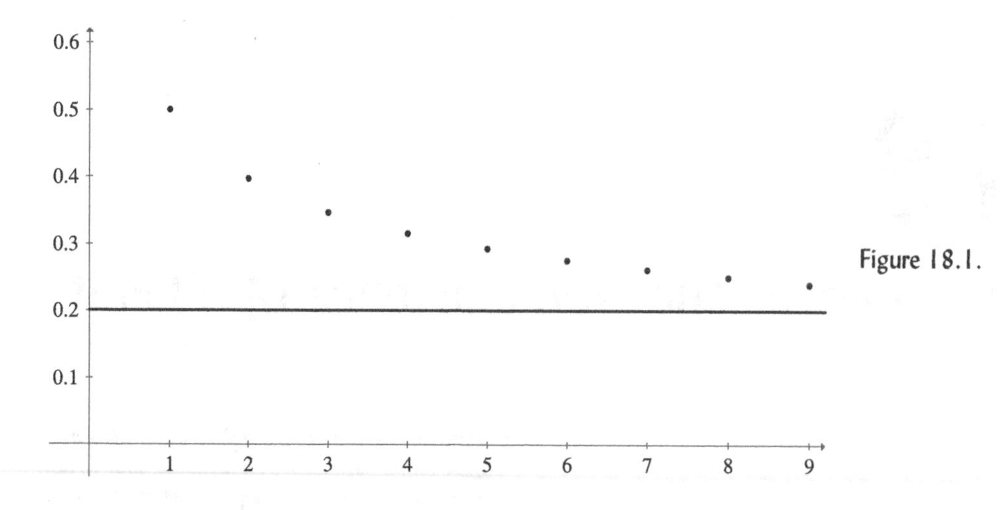

Figure 18.1.

$|\frac{n^2}{n+5} - L| \geq 1$ implies that either $\frac{n^2}{n+5} - L \geq 1$ or $\frac{n^2}{n+5} - L \leq -1$. For our purposes, showing that $\frac{n^2}{n+5} - L \geq 1$ will be sufficient. Thus, we need to find n so large that $\frac{n^2}{n+5} \geq 1 + L$, regardless of L. We could solve for n, but let's make our job a bit easier. As long as $n \geq 6$, we have $\frac{n^2}{n+5} > \frac{n^2}{2n} = \frac{n}{2}$ (verify!). Thus, for any L, we may take $\frac{n}{2} \geq 1 + L$ (that is, $n \geq 2(1 + L)$), and $n \geq 6$, both of which are satisfied by taking $n \geq \max\{2(1 + L), 6\}$.

Gathering it all together, we have shown that for $\varepsilon^* = 1$, if we choose $n \geq \max\{2(1 + L), 6\}$, then $|a_n - L| \geq 1$ for infinitely many terms, regardless of the value of L. As a result, $\langle \frac{n^2}{n+5} \rangle_1^\infty$ converges nowhere; that is to say, it diverges. (Now, do Exercise 18.1.) ■

Note 1: Example 18.2 became more difficult than the first example because the first example was testing only a particular value $L = 0.2$ for non-compliance.

Example 18.3 Prove that $\langle (-1)^n \frac{n}{2n+1} \rangle_1^\infty$ diverges. We must show that no real L satisfies the definition of a limit. In Figure 18.2, the sequence of even-numbered terms seems to increase towards $\frac{1}{2}$, while the odd-numbered terms seem to decrease towards $-\frac{1}{2}$. This leads to a strategy. Consider for the moment only odd-numbered terms: $n = 2k - 1$, $k \in \mathbb{N}$. We must prove that

$$\left\langle (-1)^{2k-1} \frac{2k-1}{2(2k-1)+1} \right\rangle_{k=1}^{\infty} \to -\frac{1}{2}.$$

The terms simplify to $\langle -\frac{2k-1}{4k-1} \rangle_1^\infty$. To prove that this sequence converges to $-\frac{1}{2}$, we need to show that for any $\varepsilon > 0$, there is a corresponding N_ε such that if $k > N_\varepsilon$, then $| -\frac{2k-1}{4k-1} + \frac{1}{2}| < \varepsilon$. Once again, simplify:

$$\left| -\frac{2k-1}{4k-1} + \frac{1}{2} \right| = \frac{1}{8k-2} < \varepsilon.$$

Considering that $\frac{1}{8k-2} < \frac{1}{k}$ for all $k \in \mathbb{N}$, let us require that $\frac{1}{k} < \varepsilon$. Then $k > \frac{1}{\varepsilon}$, and we set $N_\varepsilon = \frac{1}{\varepsilon}$. We may reverse our steps and now assert that for any $\varepsilon > 0$, if $k > N_\varepsilon = \frac{1}{\varepsilon}$, then

$$\left| (-1)^{2k-1} \frac{2k-1}{2(2k-1)+1} + \frac{1}{2} \right| = \left| -\frac{2k-1}{4k-1} + \frac{1}{2} \right| = \frac{1}{8k-2} < \frac{1}{k} < \varepsilon.$$

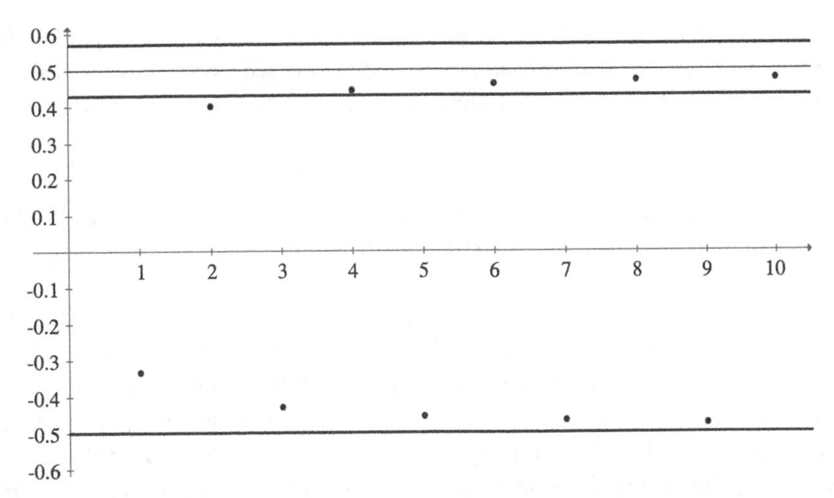

Figure 18.2. The sequence $\langle(-1)^n \frac{n}{2n+1}\rangle_1^\infty$ showing even-numbered terms increasing towards 0.5, and odd-numbered terms decreasing towards -0.5. A neighborhood of $L = 0.5$ is seen to exclude infinitely many terms. Thus, 0.5 cannot be the limit.

Consequently,

$$\lim_{k\to\infty}(-1)^{2k-1}\frac{2(2k-1)+1}{2k-1} = \lim_{odd\ n\to\infty}(-1)^n\frac{2n+1}{n} = -\frac{1}{2}.$$

Next, consider only even-numbered terms, that is, $n = 2k$. We must prove that $\langle(-1)^{2k}\frac{2k}{2(2k)+1}\rangle_1^\infty \to \frac{1}{2}$. The analysis is similar to what was just done, and is left as Exercise 18.4.

Now, back to our objective. We must disprove that every real number L is a limit of $\langle(-1)^n\frac{n}{2n+1}\rangle_1^\infty$. If $L > \frac{1}{2}$, then set $\varepsilon = L - \frac{1}{2}$, that is, ε is the distance between L and $\frac{1}{2}$. Then the neighborhood $(L-\varepsilon, L+\varepsilon) = (\frac{1}{2}, \frac{1}{2}+2\varepsilon)$ will exclude infinitely many of terms of the sequence—it actually contains none! By Negated Definition 1, no $L > \frac{1}{2}$ may be a limit. Next, $L = \frac{1}{2}$ cannot be a limit, for a neighborhood such as (0.43, 0.57), drawn in Figure 18.2, excludes an infinity of odd-numbered terms of the sequence, namely those clustered near $-\frac{1}{2}$. Next, if $-\frac{1}{2} < L < \frac{1}{2}$, we are able once again to form a neighborhood of L that contains only a finite number of terms a_n, thus excluding infinitely many terms. Next, $L = -\frac{1}{2}$ cannot be a limit, as in the case $L = \frac{1}{2}$. And finally, if $L < -\frac{1}{2}$, we can again construct a neighborhood of L that is void of terms a_n (thereby excluding infinitely many terms).

Thus, no real number L serves as a limit, and $\langle(-1)^n\frac{n}{2n+1}\rangle_1^\infty$ diverges. ■

The previous example provided practice with definitions and with algebra, but it was long. Are there more efficient ways of showing divergence? Yes! The next example shows a shorter proof, and other ways will appear later.

Example 18.4 We prove again that $\langle(-1)^n\frac{n}{2n+1}\rangle_1^\infty$ diverges. Assume on the contrary that it converges to a limit L. By Theorem 17.2, L is the unique cluster point of this set of sequential points. It isn't difficult to show that $\{(-1)^n\frac{n}{2n+1} : n = 2k,\ k \in \mathbb{N}\}$ has the cluster point $L = \frac{1}{2}$. On the other hand, $\{(-1)^n\frac{n}{2n+1} : n = 2k-1,\ k \in \mathbb{N}\}$ has the cluster point $M = -\frac{1}{2}$, which is a contradiction. Therefore, $\langle(-1)^n\frac{n}{2n+1}\rangle_1^\infty$ has no limit, which is to say, it diverges. ■

We know that every term of $\langle n\rangle_1^\infty$ is an isolated point, and that this sequence diverges. But all terms of $\langle 2^{-n}\rangle_1^\infty$ are also isolated points, and yet the sequence converges to zero. The following lemma describes how the points of a sequence must fall to prevent it from converging to a limit.

Lemma 18.1 Given a sequence $\langle a_n\rangle_1^\infty$, if there exists a fixed $\varepsilon^* > 0$ for which infinitely many terms are isolated points, then the sequence diverges.

Proof. Let us denote the infinite subset of isolated points of $\langle a_n\rangle_1^\infty$ by $\{p_k\}_{k=1}^\infty$. Thus, the premises give us infinitely many neighborhoods $(p_k - \varepsilon^*,\ p_k + \varepsilon^*)$ containing exactly one term of $\langle a_n\rangle_1^\infty$ in each. Then any two of the terms must be at least ε^* distance apart. But since $\{p_k\}_{k=1}^\infty$ is infinite, the positions of points p_k stretch out over an unbounded interval (we are applying the Archimedean principle). Therefore, for any claimed limit L, Negated Definition 2 applies: for the given $\varepsilon^* > 0$, infinitely many terms p_k of $\langle a_n\rangle_1^\infty$ exist such that $|p_k - L| \geq \varepsilon^*$. **QED**

Example 18.5 For $\langle nb\rangle_{n=1}^\infty$, where b is any nonzero constant, all terms are separated by distance $|b|$. Thus, we choose $\varepsilon^* = |b|$, and Lemma 18.1 tells us that $\langle nb\rangle_{n=1}^\infty$ diverges. ■

Example 18.6 Let

$$a_n = \begin{cases} \dfrac{1}{n} & \text{if } n \text{ is even} \\ n & \text{if } n \text{ is odd} \end{cases}.$$

The sequence $\langle a_n\rangle_1^\infty$ has a cluster point at 0, but since the odd-numbered terms are isolated points with, say, $\varepsilon^* = 1$, the sequence has no limit. ■

Example 18.7 Review Example 14.11. The sequence $\langle 2^{-n}\rangle_1^\infty$ has nothing but isolated points, and yet its limit is 0. Does this sequence violate Lemma 18.1? Exercise 18.6 asks you to explain the disparity. ■

18.1 Divergence to $\pm\infty$

Look at Negated Definition 1 at the beginning of this chapter once again. We read that L is not the limit of $\langle a_n\rangle_1^\infty$ if there exists some neighborhood $(L - \varepsilon^*, L + \varepsilon^*)$ that excludes an infinite number of terms of $\langle a_n\rangle_1^\infty$. This will definitely be the case if the sequence is unbounded, but we use a stronger requirement here.

Suppose that given any positive number b, all terms after a certain a_n are greater than b. This implies that $\langle a_n\rangle_1^\infty$ has no upper bound, but more than that, it implies that any open interval (b, ∞) will contain all but a finite number of terms of the sequence. This seems to conform with the requirement for a convergent sequence, although now the sequence is not converging toward any real number. Nevertheless, this idea prompts us to define a **neighborhood of** $+\infty$ as any open interval (b, ∞). We may say that this type of sequence will have $+\infty$ as a limit, where we are now extending the meaning of the word "limit." To be precise:

The sequence $\langle a_n\rangle_1^\infty$ has **limit** $+\infty$ if and only if corresponding to any positive number b, there exists some subscript N_b such that $a_n > b$ for all terms with subscripts $n \geq N_b$.

We will express this as $\lim_{n\to\infty} a_n = +\infty$ or $\langle a_n\rangle_1^\infty \to +\infty$. We say that the sequence **diverges to** $+\infty$, and that it **increases without bound**.

Suppose now that given any positive number c, all terms after a certain a_{N_c} are less than $-c$. Similar considerations as above lead us to define a **neighborhood of** $-\infty$ as any open interval $(-\infty, -c)$. The corresponding definition is:

> The sequence $\langle a_n\rangle_1^\infty$ has **limit** $-\infty$ if and only if corresponding to any positive number c, there exists some subscript N_c such that $a_n < -c$ for all terms with subscripts $n \geq N_c$.

We will express this as $\lim_{n\to\infty} a_n = -\infty$ or $\langle a_n\rangle_1^\infty \to -\infty$. We say that the sequence **diverges to** $-\infty$ and that it **decreases without bound**.

Note 2: We are creating an interesting collection of nested subsets. Among the divergent sequences, unbounded sequences form a subset. Within this subset is found the collection of sequences that diverge to $+\infty$. Within this collection, even more specialized are the sequences that *monotonically* diverge to $+\infty$. They have the additional requirement that $a_{n+1} \geq a_n$ for all n. And within this collection, yet more specialized are the sequences that diverge to $+\infty$ *strictly monotonically*, where $a_{n+1} > a_n$ for all n.

Similarly, sequences that monotonically diverge to $-\infty$ have the extra requirement that $a_{n+1} \leq a_n$ (and $a_{n+1} < a_n$ for strict monotonicity).

Example 18.8 $\mathbb{N}$ expressed as the sequence $\langle n\rangle_1^\infty$ has limit $+\infty$. Also, $\langle -n^3\rangle_1^\infty \to -\infty$. And $\lim_{n\to\infty} \sqrt{n} = +\infty$, according to your work in Exercise 18.7 below. ■

Example 18.9 Let

$$a_n = \begin{cases} n-1 & \text{if } n \text{ is even} \\ n+1 & \text{if } n \text{ is odd} \end{cases}.$$

Graph the sequence to see its interesting behavior. To simplify the analysis, we see that $a_n \geq n-1$ for all n.

Now, given any $b > 0$, let $N_b = \lceil b\rceil + 1$. Thus, for all $n \geq N_b$, we see that $a_n \geq n-1 \geq N_b - 1 = \lceil b\rceil \geq b$. Hence, $\langle a_n\rangle_1^\infty$ has limit $+\infty$. As a check, select a large number b, evaluate N_b, and then check that indeed $a_n \geq b$ for all $n \geq N_b$. We write $\langle a_n\rangle_1^\infty \to +\infty$. ■

Example 18.10 Let $q > 1$ be a constant. Prove that $\langle q^n\rangle_1^\infty$ has limit $+\infty$. We make a typical application of Bernoulli's inequality $(1+p)^n \geq 1+np$, where $p \geq -1$. Select p so that $q = 1+p$. Then $q^n \geq 1 + n(q-1) > n(q-1)$ for any $n \in \mathbb{N}$. This implies that $\lim_{n\to\infty} q^n = +\infty$.

The definition may be applied more closely. Since $q-1$ is positive, the Archimedean principle states that there exists an $N_b \in \mathbb{N}$ such that $N_b(q-1) > b$, for any positive b. If $n \geq N_b$, it's easy to show by induction that $q^n \geq q^{N_b}$. Hence, for all $n \geq N_b$, $q^n \geq q^{N_b} > N_b(q-1) > b$ for any positive b. This means that $\lim_{n\to\infty} q^n = +\infty$. (Do Exercise 18.9.) ■

Example 18.11 The sequence given by Stirling's formula in Example 16.9 diverges to $+\infty$. ■

18.2 Oscillating Sequences

The most obvious oscillating sequence may very well be $\langle(-1)^n\rangle_1^\infty$. The sequence $\langle\cos n\rangle_1^\infty$ also oscillates, as is seen in Example 19.6. The sequence $\langle 2, 1, 4, 3, 6, 5, \cdots\rangle$ has a graph that looks like the peaks of a mountain range (verify), but curiously, it does not qualify as an oscillating sequence. Clearly, we need a clear definition!

An **oscillating sequence** is a sequence that diverges, but does not diverge to either $+\infty$ or $-\infty$. By virtue of this definition, we have partitioned the set of all divergent sequences of real numbers into the subset of sequences that diverge to $+\infty$, the subset of those that diverge to $-\infty$, and the subset of those that oscillate.

Example 18.12 The sequence $\langle n + (-1)^n n\rangle_1^\infty = \langle 0, 4, 0, 8, 0, \cdots\rangle$ does not diverge to $+\infty$ (why?), and it certainly does not diverge to $-\infty$. Neither does it converge, since the subsequence of even terms diverges to $+\infty$. Therefore, this sequence oscillates, and it is also an unbounded sequence. ■

Example 18.13 The sequence $\langle(-1)^n \frac{2n+1}{n}\rangle_1^\infty = \langle -\frac{3}{1}, \frac{5}{2}, -\frac{7}{3}, \cdots\rangle$ in Figure 18.3 contains the reciprocals of the sequence in Example 18.3. Both sequences oscillate between their respective cluster points. ■

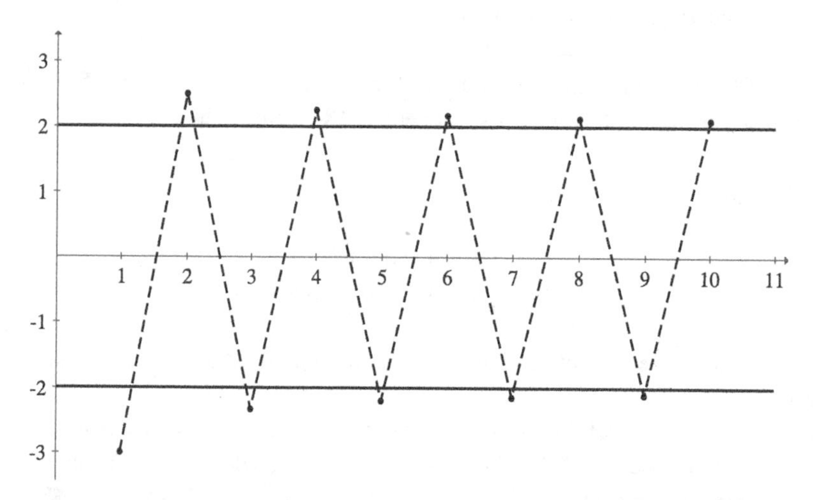

Figure 18.3. The sequence $\langle(-1)^n \frac{2n+1}{n}\rangle_1^\infty$ oscillates between its cluster points 2 and -2.

Example 18.14 The sequence $\langle 2, 1, 4, 3, 6, 5, \cdots\rangle$ was found in Example 18.9 to diverge to $+\infty$. It "oscillates" in the common sense of the word, but it is not an oscillating sequence. ■

18.3 Exercises

1. In Example 18.2, using $\varepsilon^* = 1$ specifically, we found that $|\frac{n^2}{n+5} - L| \geq 1$ whenever $n \geq \max\{2(1 + L), 6\}$. Select $L = 55$, $L = 3.3$, $L = 0$, and $L = -6$, and verify the statement in each case.

2. a. The terms of the sequence $\langle n + \frac{1}{n} \rangle_1^\infty$ soon become nearly one unit apart. Show that the sequence diverges. Hint: Find an expression for the difference of any two successive terms of this sequence. Then, use Lemma 18.1.
   b. Observe that $\frac{n^2+1}{n} = n + \frac{1}{n}$ by division. In Example 18.2, a long division gives $\frac{n^2}{n+5} = n - 5 + \frac{25}{n+5}$, which, as in **a**, shows that later terms are about one unit apart. Hence, apply the technique from **a** for a faster proof that $\langle \frac{n^2}{n+5} \rangle_1^\infty$ diverges.

3. Show that the sequence $\langle \cos(n\pi) \rangle_1^\infty$ diverges.

4. Complete the proof in Example 18.3 by proving that $\lim\limits_{\text{even } n \to \infty} \langle (-1)^n \frac{n}{2n+1} \rangle_1^\infty = \frac{1}{2}$. Check your analysis by selecting $\varepsilon = 0.003$, and showing that for every even natural number $n > N_\varepsilon$, the inequality $|(-1)^n \frac{n}{2n+1} - \frac{1}{2}| < 0.003$ is true.

5. Mark each of the following as true or false, and if false, draw a graph of a counterexample sequence.
   a. A sequence with no lower bound will have $-\infty$ as a limit.
   b. If, given any real number U, there is at least one term $a_n > U$, then the sequence diverges to $+\infty$.
   c. If a sequence has terms all of which decrease with no lower bound, then the sequence diverges to $-\infty$.
   d. If a sequence diverges to $+\infty$, then its terms satisfy $a_{n+1} \geq a_n$.
   e. If a sequence oscillates, then it is bounded.
   f. If two sequences oscillate, then their sum also oscillates.

6. All terms of $\langle 2^{-n} \rangle_1^\infty$ are isolated points, and yet, $\lim\limits_{n\to\infty} 2^{-n} = 0$, in apparent contradiction to Lemma 18.1. Clear the matter up. Hint: See Example 14.11.

7. Prove that $\lim\limits_{n\to\infty} \sqrt{n} = +\infty$. Hint: Begin by writing the definition.

8. Prove that $\lim\limits_{n\to\infty} \sqrt[n]{n!} = +\infty$. Hint: Use Exercise 2.8.

9. In the inequalities $q^n \geq N_b(q-1) > b$ in Example 18.10, suppose $q = 1.1$ and we want $\langle q^n \rangle_1^\infty$ to exceed 7.2. What is the value of $N_{7.2}$? Check that $1.1^{N_{7.2}} \geq N_{7.2}(1.1 - 1) > 7.2$. Since there are integers n such that $1.1^{N_{7.2}} > 1.1^n > 7.2$, is the value $N_{7.2}$ too large?

10. The sequence $\langle \cos n \rangle_1^\infty$ can be proved to diverge. (Example 19.6 graphs the sequence. A convincing argument for divergence can be made after trying Exercise 20.5.) Assuming divergence, prove that $\langle \cos n \rangle_1^\infty$ is an oscillating sequence.

19

Four Great Theorems in Real Analysis

Certain sequences, when graphed, look to be either going uphill or going downhill. We define each type individually. A **monotone increasing sequence** $\langle a_n \rangle_1^\infty$ has successive terms that do not decrease, that is, $a_{n+1} \geq a_n$ for all n. A **monotone decreasing sequence** $\langle a_n \rangle_1^\infty$ has successive terms that do not increase, that is, $a_{n+1} \leq a_n$ for all n. When we wish to talk about these two classes of sequences collectively, we shall call them the set of **monotone sequences**. Also, we speak of **eventually monotone sequences**. These are sequences that become monotone after some term a_M, the idea being that the first few terms constitute only the finite front end of the sequence (and in analysis, even a billion trillion terms is still "the first few"). It is the infinitely long tail starting at a_M that is always interesting.

Monotone sequences play an important role in analysis mainly due to the following important theorem.

Theorem 19.1 The Monotone Convergence Theorem Every bounded, monotone sequence $\langle a_n \rangle_1^\infty$ converges.

Proof. Case I: the bounded sequence $\langle a_n \rangle_1^\infty$ is monotone increasing. Since $\langle a_n \rangle_1^\infty$ is bounded, it has a supremum $\sup\langle a_n \rangle_1^\infty = \alpha$ by the completeness axiom. By the definition of supremum, (1) $a_n \leq \alpha$ for all terms of the sequence, and (2) for any number $x < \alpha$, some term a_n is larger than x, so that $x < a_n \leq \alpha$.

Our goal is to prove that the supremum turns out to be the limit of the sequence. To do this, the monotonicity property $a_{n+1} \geq a_n$ is necessary. Given any $\varepsilon > 0$, property (2) implies that $\alpha - \varepsilon < a_n \leq \alpha$ for some term a_n. Let us associate that term to ε by renaming its subscript N_ε. Thus, $\alpha - \varepsilon < a_{N_\varepsilon} \leq \alpha$. By monotonicity, $a_{N_\varepsilon} \leq a_n$ for all $n > N_\varepsilon$. We are certain now that given any $\varepsilon > 0$, $a_n \leq \alpha$ for all terms with $n > N_\varepsilon$. This means that regardless of how small ε is, all but a finite number of terms a_n of the sequence fall in the neighborhood $(\alpha - \varepsilon, \alpha + \varepsilon)$. Therefore, the sequence converges to α. We conclude that $\lim_{n\to\infty} a_n = \alpha = \sup\langle a_n \rangle_1^\infty$.

Case II: the bounded sequence $\langle a_n \rangle_1^\infty$ is monotone decreasing. The conclusion is that $\langle a_n \rangle_1^\infty$ converges to its infimum: $\lim_{n\to\infty} a_n = \inf\langle a_n \rangle_1^\infty$. This is left as Exercise 19.2. **QED**

Note 1: The monotone convergence theorem begins by treating $\{a_n\}_1^\infty$ as a bounded set of points, which must have a supremum. Then, the theorem uses the monotonicity of the sequence $\langle a_n \rangle_1^\infty$ to prove that the limit exists.

Example 19.1 The usefulness of the monotone convergence theorem is illustrated here. A recursively defined sequence has $a_1 = 1$, with succeeding terms arising according to $a_{n+1} = \sqrt{1 + a_n}$. We wonder if $\langle a_n \rangle_1^\infty$ converges. Explicitly, we have

$$a_n = \underbrace{\sqrt{1 + \sqrt{1 + \cdots + \sqrt{1 + \sqrt{1 + 1}}}}}_{n \text{ 1s}},$$

which helps only for calculating the values of terms.

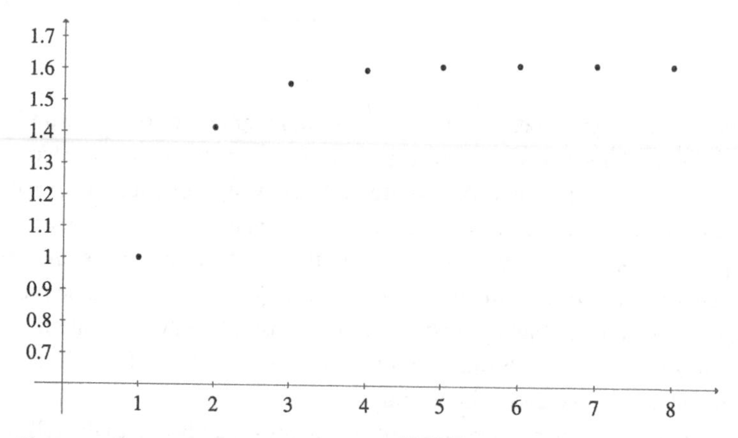

Figure 19.1. The recursive sequence $a_1 = 1$, $a_{n+1} = \sqrt{1 + a_n}$, plotted in $\mathbb{R}^2$, is monotone increasing and shows evidence of convergence.

First, we will use induction to prove that $\langle a_n \rangle_1^\infty$ is monotone increasing. As the anchor for the induction, we have $a_2 = \sqrt{2} > a_1$. Next, assume $a_n > a_{n-1}$. From $a_{n+1}^2 = 1 + a_n$ and the induction hypothesis, we find that

$$a_{n+1} = \frac{1 + a_n}{a_{n+1}} > \frac{1 + a_{n-1}}{a_{n+1}} = \frac{a_n^2}{a_{n+1}}.$$

Thus, $a_{n+1}^2 > a_n^2$, so $a_{n+1} > a_n$, and the induction is complete.

Second, we prove that $\langle a_n \rangle_1^\infty$is bounded above. From $a_{n-1} = a_n^2 - 1 = (a_n - 1)(a_n + 1)$, conclude that $a_n - 1 = \frac{a_{n-1}}{a_n + 1} < 1$. Thus $a_n < 2$ for all $n \in \mathbb{N}$, and our sequence is bounded above by 2. Its convergence follows from Theorem 19.1. The graph in Figure 19.1 hints at convergence (but of course, can't prove it). See Exercise 19.3 for the surprising value of the limit. ■

The next theorem is related to Theorem 14.1. Figure 19.2 illustrates what will be going on.

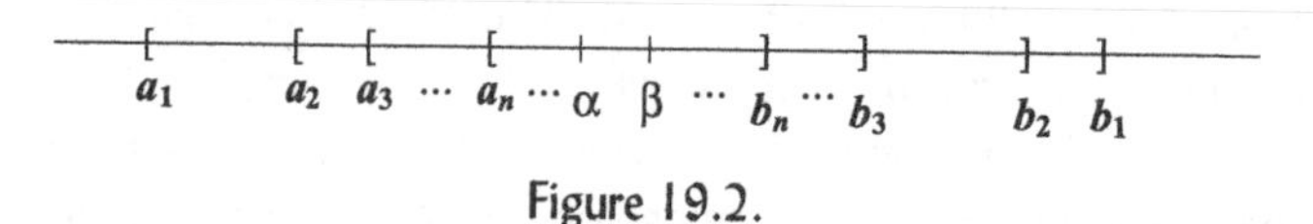

Figure 19.2.

Theorem 19.2 The Nested Intervals Theorem For each $n \in \mathbb{N}$, let $I_n = [a_n, b_n]$ be a sequence of closed, nested intervals, $[a_1, b_1] \supset [a_2, b_2] \supset \cdots \supset [a_n, b_n] \supset \cdots$. Then $\bigcap_{n=1}^\infty I_n$ is closed and not empty.

Proof. By the construction of the intervals, we have for any $n \in \mathbb{N}$, $a_{n+1} \geq a_n$, and $b_{n+1} \leq b_n$. This implies the interesting scenario in Figure 19.2 that for any two subscripts m, n whatsoever, $a_m \leq b_n$ (for example, $a_{54} \leq b_7$). So, the sequence $\langle a_n \rangle_1^\infty$ is monotone increasing and bounded above by b_1 (among other choices). Therefore, by the monotone convergence theorem, this sequence has a limit $\alpha = \sup\langle a_n \rangle$. Likewise, $\langle b_n \rangle_1^\infty$ is monotone decreasing and bounded below by a_1, so it has a limit $\beta = \inf\langle b_n \rangle$.

We claim that $\alpha \leq \beta$. Suppose, on the contrary, that $\beta < \alpha$. Then there exists a natural number M such that for all $m \geq M$, $\beta < a_m \leq \alpha$ (remember $\langle a_m \rangle_{m=1}^\infty$ is monotone increasing towards α). Using our observation above, $\beta < a_M \leq b_n$ for all $n \in \mathbb{N}$. But this contradicts the fact that β is the infimum of $\langle b_n \rangle_1^\infty$. Therefore, $\alpha \leq \beta$ was correct.

We have now shown that $a_n \leq \alpha \leq \beta \leq b_n$ for all $n \in \mathbb{N}$. This implies that $[\alpha, \beta] \subset I_n$ for all $n \in \mathbb{N}$. Thus, $\bigcap_{n=1}^\infty I_n = [\alpha, \beta]$, and this interval always contains at least one real number. If it happens that $\alpha = \beta$, then $\bigcap_{n=1}^\infty I_n$ contains exactly one real number. **QED**

Note 2: Theorem 19.2 makes a bridge between the geometry of the real number line and the decimal representation of real numbers. Admitting for the moment that every real number α has a decimal representation, say, $\alpha = m.d_1d_2d_3\ldots$ (where d_1 is the tenths-place digit of α, and so on), then the nested intervals theorem gives us a way to locate α on the real number line. First, we know that $\alpha \in [m, m+1]$, where $m = \lfloor \alpha \rfloor$. Next,

$$\alpha \in \left[m + \frac{d_1}{10}, m + \frac{d_1+1}{10}\right] \subset [m, m+1].$$

Next,

$$\alpha \in \left[m + \frac{d_1}{10} + \frac{d_2}{100}, m + \frac{d_1}{10} + \frac{d_2+1}{100}\right] \subset \left[m + \frac{d_1}{10}, m + \frac{d_1+1}{10}\right].$$

Continue successively in this way to form a nested sequence of intervals whose intersection contains only α. We have homed in on the location of α. You may use this algorithm to locate π^2, for example.

The converse is interesting. If we place a point γ on the real number line, it is possible to find its decimal representation. Since $\gamma \in [m, m+1)$ for some integer m, then $m = \lfloor \gamma \rfloor$. Next, divide $[m, m+1)$ into 10 equal subintervals, and there must exist a unique number $d_1 \in \{0, 1, \ldots, 9\}$ such that

$$\gamma \in \left[m + \frac{d_1}{10}, m + \frac{d_1+1}{10}\right) \subset [m, m+1).$$

Thus, so far, $\gamma = m.d_1$. Similarly,

$$\gamma \in \left[m + \frac{d_1}{10} + \frac{d_2}{100}, m + \frac{d_1}{10} + \frac{d_2+1}{100}\right) \subset \left[m + \frac{d_1}{10}, m + \frac{d_1+1}{10}\right),$$

so d_2 becomes the hundredths-place digit of γ. As many decimal digits as one wishes can be determined in this way. Since the successive interval lengths shrink towards zero, Theorem 19.2 guarantees that γ is the unique real number represented by all these digits.

Another, more elaborate, application of this theorem appears in the last exercise of Appendix D, where you meet continued fractions.

19.1 Subsequences

A **subsequence** of $\langle a_n\rangle_1^\infty$ is any sequence arising by extracting terms from $\langle a_n\rangle_1^\infty$ without changing their order of appearance. Thus, for instance, the sequence of primes $\langle 2, 3, 5, 7, 11, \ldots\rangle$ is a subsequence of $\mathbb{N}$, and $\langle 1\rangle_1^\infty$ is the subsequence of $\langle \cos(n\pi)\rangle_1^\infty$ found by selecting only the even-numbered terms. Example 16.8, to be reviewed shortly, is most conveniently seen as being formed from four subsequences.

To be precise, take any *strictly increasing* function $N : \mathbb{N} \to \mathbb{N}$ expressed as $n_k = N(k)$, and replace the subscripts of a sequence $\langle a_n\rangle_1^\infty$ this way: $\langle a_{n_k}\rangle_{k=1}^\infty$. Then $\langle a_{n_k}\rangle_{k=1}^\infty$ is a subsequence of $\langle a_n\rangle_1^\infty$. Now n is no longer an independent subscript, but has become the output of an increasing, integer-valued, function of k. This definition will be detailed in the following examples. Natural numbers k are in the position of a **sub-subscript** in a_{n_k}.

Example 19.2 Given $A = \langle \frac{(-1)^n}{n}\rangle_1^\infty = \langle -1, \frac{1}{2}, \frac{-1}{3}, \frac{1}{4}, \ldots\rangle$, the strictly increasing function $n = N(k) = n_k = 2k$ will generate the subsequence $B = \langle \frac{(-1)^{2k}}{2k}\rangle_1^\infty = \langle \frac{1}{2k}\rangle_1^\infty = \langle \frac{1}{2}, \frac{1}{4}, \ldots\rangle$ of even-numbered terms. We may express this as $b_k = a_{2k}$ with $k \in \mathbb{N}$. Thus, the kth term of B equals the $2k$th term of A. We also write $a_{n_k} = a_{2k} = b_k = \frac{1}{2k}$, as seen in the following table.

k		1		2		3		4	$\cdots$
$n_k = 2k$		2		4		6		8	$\cdots$
$b_k = a_{n_k} = \frac{1}{2k}$		$\frac{1}{2}$		$\frac{1}{4}$		$\frac{1}{6}$		$\frac{1}{8}$	$\cdots$
		$\Updownarrow$		$\Updownarrow$		$\Updownarrow$		$\Updownarrow$	
$a_n = \frac{(-1)^n}{n}$	-1	$\frac{1}{2}$	$\frac{-1}{3}$	$\frac{1}{4}$	$\frac{-1}{5}$	$\frac{1}{6}$	$\frac{-1}{7}$	$\frac{1}{8}$	$\cdots$
n	1	2	3	4	5	6	7	8	$\cdots$

■

Example 19.3 Let $B = \langle \frac{1}{2k}\rangle_1^\infty$, and choose the strictly increasing function $N(\ell) = 4^\ell$, $\ell \in \mathbb{N}$, which we may write as $k_\ell = 4^\ell$. This extracts the subsequence $C = \langle c_\ell\rangle_1^\infty = \langle b_{k_\ell}\rangle_{\ell=1}^\infty = \langle b_{4^\ell}\rangle_1^\infty$. That is, $b_{4^\ell} = \frac{1}{2(4^\ell)} = \frac{1}{2^{2\ell+1}} = c_\ell$. In the following table, the ℓth term of C equals the 4^ℓth term of B.

ℓ				1			2		3	$\cdots$
$k_\ell = 4^\ell$				4			16		64	$\cdots$
$c_\ell = b_{4^\ell} = \frac{1}{2^{2\ell+1}}$				$\frac{1}{8}$			$\frac{1}{32}$		$\frac{1}{128}$	$\cdots$
				$\Updownarrow$			$\Updownarrow$		$\Updownarrow$	
$b_k = \frac{1}{2k}$	$\frac{1}{2}$	$\frac{1}{4}$	$\frac{1}{6}$	$\frac{1}{8}$	$\frac{1}{10}$	$\cdots$	$\frac{1}{32}$	$\cdots$	$\frac{1}{128}$	$\cdots$
k	1	2	3	4	5	$\cdots$	16	$\cdots$	64	$\cdots$

■

Example 19.4 Since the denominators of C (above) increase much more rapidly than those of B, we see that C is a very sparse subsequence of B. On the other hand, $\langle \frac{1}{2k}\rangle_3^\infty$ is another subsequence of B that isn't sparse at all, omitting just two terms, and it is extracted from B by the function $k = N(\ell) = \ell + 2$, $\ell \in \mathbb{N}$. ■

The theoretical underpinning of Examples 19.2, 19.3, and 19.4 is the composition of functions. In Example 19.2, let $F(n) = a_n = \frac{(-1)^n}{n}$ be the function that generates A, and we considered $n = N(k) = 2k$. Then $(F \circ N)(k) = F(N(k)) = F(2k) = a_{2k} = \frac{1}{2k} = b_k$. In other words, the function that generates subsequence B is $F \circ N$. The only requirement for such a composition to define a subsequence is that the function N be strictly increasing, per the definition of a subsequence. Continue this investigation in Exercise 19.6.

It's a bit surprising that anything substantial can be said about all sequences. The next theorem is, therefore, a surprise.

Theorem 19.3 The Monotone Subsequence Theorem Every sequence $\langle a_n \rangle_1^\infty$ has a monotone subsequence.

Proof. (This proof is by Hugh Thurston, "A Simple Proof That Every Sequence Has a Monotone Subsequence," in *Mathematics Magazine*, vol. 67, no. 5 (Dec. 1994), pg. 344.)

Given $\langle a_n \rangle_1^\infty$, consider its tail starting at subscript k ($k \geq 1$): $a_k, a_{k+1}, a_{k+2}, \ldots$. If the tail is found to have no greatest term, then it contains an increasing, and consequently monotone, subsequence of $\langle a_n \rangle_1^\infty$, confirming the theorem in this case.

If there is no such tail, let a_p be a maximum of $a_1, a_2, a_3, \ldots$, let a_q be a maximum of the tail $a_{p+1}, a_{p+2}, \ldots$, let a_r be a maximum of the tail $a_{q+1}, a_{q+2}, \ldots$, and so on. Then $a_p \geq a_q \geq a_r \geq \cdots$ becomes a monotone decreasing subsequence, confirming the theorem once again. **QED**

Example 19.5 Consider the sequence in Example 16.8. Its graph shows evidence of four monotone subsequences. They are all easy to see, since the full sequence was created with them in mind. One of them is $a_n = 2 - \frac{1}{n}$ when $n = 4k - 2$. We now express this subsequence as $a_{n_k} = 2 - \frac{1}{4k-2}$, $k \in \mathbb{N}$. Hence, this is a monotone increasing subsequence. Do Exercise 19.7 at this time. ■

Example 19.6 The sequence $\langle \cos n \rangle_0^\infty$ has points scattered (but not uniformly distributed) throughout $(-1, 1]$, as can be seen in Figure 19.3.

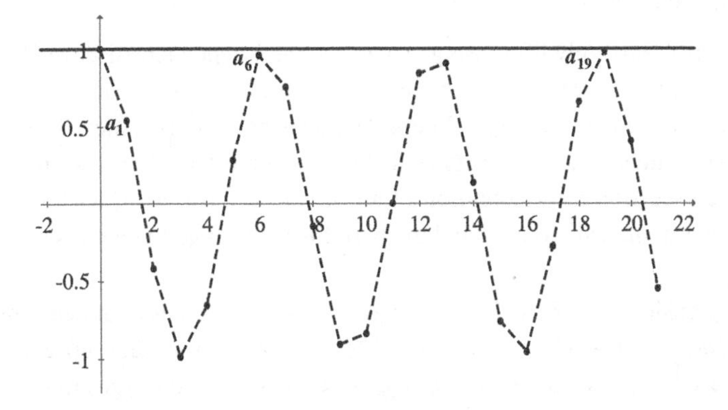

Figure 19.3.

This time, no obvious monotone subsequence can be seen. We apply Thurston's procedure in the proof of Theorem 19.3. The greatest term in $\langle \cos n \rangle_0^\infty$ is $a_0 = 1$ (prove that $\cos n \neq 1$ for any $n > 0$). Furthermore, the tail starting at a_1 has no greatest term (we won't prove this, but a strong argument for it is in the solution to Exercise 20.5). We may begin with $a_1 = \cos 1 \approx 0.5403$, and then select $a_6 = \cos 6 \approx 0.9602$, followed by $a_{19} = \cos 19 \approx 0.9887$.

Although we have no formula for the general term, the subsequence $\langle a_1, a_6, a_{19}, \ldots \rangle$ begins a monotone increasing subsequence. (It is by no means the only one!) ■

On a number line, draw the interval [2, 7]. Confining yourself to that interval, place many, many points, representing terms a_n of a sequence. If all of your points are distinct, then it's unavoidable: you will be creating at least one cluster point. The famous theorem that follows perfects this observation. It singles out bounded sequences, in contrast to Theorem 19.3, but derives a much stronger conclusion. It is named after Bernhard Bolzano, who first investigated its ideas, and the great analyst Karl Weierstrass (1815–1897), who developed five versions of the theorem between 1865 and 1886.

Theorem 19.4 The Bolzano-Weierstrass Theorem for Sequences Every bounded sequence $\langle a_n \rangle_1^\infty$ has a convergent subsequence.

First Proof. Let $\langle a_n \rangle_1^\infty$ be bounded within a closed interval $[x_1, y_1]$, with midpoint m_1. Then, either $[x_1, m_1]$ or $[m_1, y_1]$ must have infinitely many points of the sequence. Without loss of generality, suppose it is $[x_1, m_1]$, with midpoint m_2. Now, either $[x_1, m_2]$ or $[m_2, m_1]$ must contain infinitely many points of the original sequence. Let's suppose it is $[m_2, m_1]$, with midpoint m_3. By this process, a sequence of closed, nested intervals is formed, and by Theorem 19.2, their intersection is some interval $[\alpha, \beta]$. But additionally, each interval is exactly half as long as the previous one. Thus, the length of $[\alpha, \beta]$ must be zero, by Corollary 13.3. Therefore, $\alpha = \beta$.

In the first interval $[x_1, y_1]$, select one term a_{n_1} from the sequence. In the second interval $[x_1, m_1]$, select a second term a_{n_2} (it may or may not equal a_{n_1}), and so on. The subsequence $\langle a_{n_k} \rangle_1^\infty$ that arises is convergent, with limit α (convince yourself of this), and we are done. **QED**

There is a second, elegant proof of the Bolzano-Weierstrass theorem for sequences that you will discover in Exercise 19.8.

Note 3: There is a roughly fifty-year gap between the independent discoveries of the theorem by Bolzano and Weierstrass. Bolzano, a Czechoslovakian priest, had discussed many ideas about analysis and the infinite, including Galileo's paradox (Paradox 0.3), decades before mainstream mathematicians recognized them as important. We saw in Chapter 11 that Cantor had used Bolzano's definition of an infinite set. Boyer [3] writes, "Bolzano was a 'voice crying in the wilderness,' and many of his results had to be rediscovered later."

Example 19.7 Consider $\langle a_n \rangle_1^\infty$, with $a_n = 10 + (-1)^n \frac{n}{n+1}$, which is bounded. The subsequence of even terms, $a_{2n} = 10 + \frac{2n}{2n+1}$ converges to 11, the supremum of this subsequence (by which theorem?). The subsequence of odd terms, $a_{2n-1} = 10 - \frac{2n-1}{2n}$ converges to its infimum, 9. We see that a bounded sequence can have subsequences with different limits. Figure 19.4 shows this behavior. ■

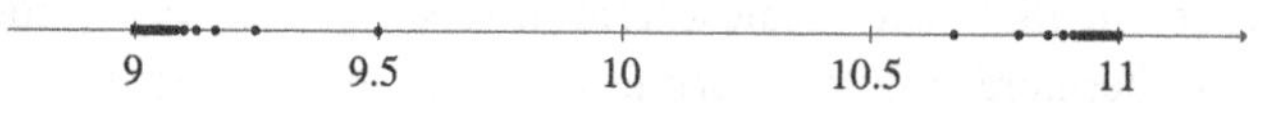

Figure 19.4. A sequence $\langle a_n \rangle_1^\infty$ with two subsequences converging to separate limits. The subsequential limits are cluster points of the original sequence.

Collect the set of limits of all convergent subsequences of a sequence. This is called the sequence's **set of subsequential limits**. In Example 19.7, the set of subsequential limits is $\{9, 11\}$.

Lemma 19.1 Every cluster point of a sequence is a subsequential limit of that sequence.

Proof. For a sequence $\langle a_n \rangle_1^\infty$, let α be one of its cluster points. Then any neighborhood $(\alpha - \varepsilon, \alpha + \varepsilon)$ will contain infinitely many points of the sequence. By the algorithmic proof of Theorem 14.1, we construct a subsequence of terms a_{n_k} that has limit α. Thus, α is a subsequential limit of $\langle a_n \rangle_1^\infty$. **QED**

Do Exercise 19.14 at this time.

Lemma 19.2 A sequence $\langle a_n \rangle_1^\infty$ converges to a limit L if and only if all of its subsequences converge to L.

Proof of Necessity. Suppose $\langle a_n \rangle_1^\infty \to L$, and consider a subsequence $\langle a_{n_k} \rangle_{k=1}^\infty$. We might as well write the definition of a limit at L, for it's almost certain that we will need it: for any $\varepsilon > 0$, $|a_n - L| < \varepsilon$ for all $n > N_\varepsilon$. Recall that $n_k = N(k)$ is an increasing function, with only natural numbers in its domain and range. This implies that sub-subscript k of a subsequence always lags behind the term number n_k of the sequence, that is, $k \le n_k$ for all $k \in \mathbb{N}$. Now we apply the definition of a limit, using k: we know that for any $\varepsilon > 0$, it is true that $|a_k - L| < \varepsilon$ for all $k > N_\varepsilon$. It follows that $|a_{n_k} - L| < \varepsilon$, because $n_k \ge k$. Hence, any and all subsequences converge to L.

Proof of Sufficiency. To prove the converse, let all subsequences $\langle a_{n_k} \rangle_{k=1}^\infty$ converge to a unique L. The proof is almost done! See Exercise 19.9. **QED**

Example 19.8 Let r be a constant such that $-1 < r < 1$, and suppose that $\langle r^n \rangle_1^\infty$ converges to L. A new way to evaluate L is to focus on a convenient subsequence, say, the one with $n_k = k + 3$. Thus, we look at $\langle r^{k+3} \rangle_1^\infty = \langle r^4, r^5, \ldots \rangle$. By Lemma 19.2, this subsequence converges to the same limit. In other words, $L = \lim_{n\to\infty} r^n = \lim_{k\to\infty} r^{k+3} = r^3 \lim_{k\to\infty} r^k = r^3 L$. When $r \neq 0$, this implies that $L = 0$. (What if $r = 0$?)

Alas, this clever argument rests upon the presumed existence of the limit L. Theorem 20.6 will prove that L exists in its fuller treatment of this important sequence. ■

19.2 Exercises

1. Describe the class of sequences that are both monotone increasing and monotone decreasing.
2. Prove Case II of the monotone convergence theorem.
3. The sequence $a_1 = 1$, $a_{n+1} = \sqrt{1 + a_n}$ was shown in Example 19.1 to converge, so let $\lim_{n\to\infty} a_n = \varphi$. Then obviously $\lim_{n\to\infty} a_{n+1} = \varphi$. Take limits of both sides of $a_{n+1}^2 = 1 + a_n$, and find a quadratic equation in φ. Solve it and reject the negative solution (why?). Do you recognize the famous constant φ that this sequence converges to?

4. What problem would arise in the nested intervals theorem if the closed intervals were replaced by open intervals?

5. Find $\bigcap_{n=1}^{\infty}[-\frac{n^2+n+1}{2n^2}, \frac{n^2+n+1}{2n^2}]$. Hint: Show that $\langle\frac{n^2+n+1}{2n^2}\rangle_1^\infty$ is monotone decreasing towards a limit.

6. In Examples 19.2 and 19.3, we saw $A = \langle\frac{(-1)^n}{n}\rangle_1^\infty$, $B = \langle\frac{1}{2k}\rangle_1^\infty$, and $C = \langle\frac{1}{2^{2\ell+1}}\rangle_1^\infty$. Writing the first few terms of each, note that C is a subsequence of B, which is a subsequence of A. Therefore, there must be a function that extracts C directly from A. Find it by forming compositions.

7. Example 19.5 investigated the sequence from Example 16.8,

$$a_n = \begin{cases} \frac{1}{n} & \text{when } n = 4k-3 \\ 2-\frac{1}{n} & \text{when } n = 4k-2 \\ 2+\frac{1}{n} & \text{when } n = 4k-1 \\ 4+\frac{1}{n} & \text{when } n = 4k \end{cases},$$

where $k \in \mathbb{N}$, and found an explicit expression for the clearly visible monotone increasing subsequence. Find expressions for the other three subsequences.

8. Use some other theorems in this chapter to make a second proof of the Bolzano-Weierstrass theorem for sequences.

9. Finish the proof of Lemma 19.2 with one more statement.

10. Prove the only if part of Lemma 19.2 by contradiction. Hint: Let $\langle a_n\rangle_1^\infty$ converge to a limit L, and assume on the contrary that two particular subsequences have unequal subsequential limits: $\lim_{k\to\infty} a_{n_k} = L \neq M = \lim_{i\to\infty} a_{n_i}$.

11. Recall Example 18.4, which gave a shorter proof that $\langle(-1)^n\frac{n}{2n+1}\rangle_1^\infty$ diverges. State the contrapositive of Lemma 19.2 and use it to prove for a third time that the sequence diverges.

12. a. Find two monotone subsequences in $\langle n^{(-1)^n}\rangle_1^\infty$.
    b. In Example 19.7, find another convergent subsequence besides the two discussed there, and find a divergent subsequence. Hint: For the divergent sequence, you need to find an endless supply of terms that approaches 9, and another endless supply of terms that approaches 11.
    c. Find the first few terms of the monotone subsequence selected by Thurston's procedure in the monotone subsequence theorem from the sequence $\langle(1+\sin n)\frac{n}{5+n^2}\rangle_1^\infty$ graphed in Figure 19.5. (Incidentally, a_{11} is nearly zero, indicating that $\sin 11 \approx -1$. Why is this approximation so good?)

13. Suppose the sequence $\langle a_n\rangle_1^\infty$ is bounded within the interval $[2, 7]$. For the sequence not to have any convergent subsequence, infinitely many terms a_n would have to be at some minimum distance from every other term. Draw $[2, 7]$ and try to locate infinitely many points a_n under this restriction. In other words, try to circumvent the Bolzano-Weierstrass theorem for sequences!

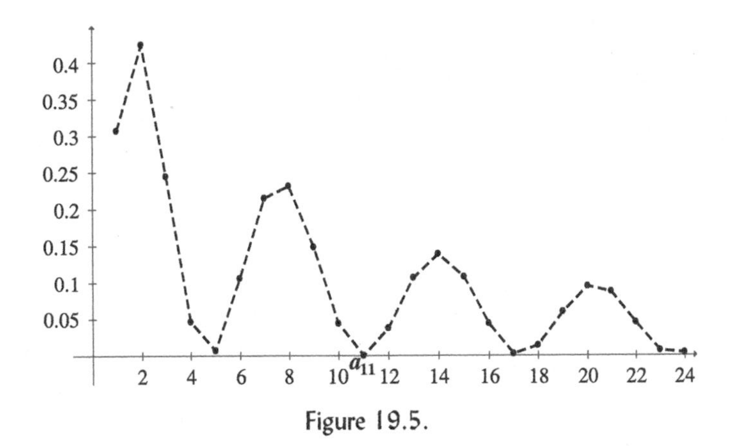

Figure 19.5.

14. Because of sequences that are eventually constant, Lemma 19.1 has only a partial converse. Prove that if $\langle a_n \rangle_1^\infty$ has a subsequential limit α, and $a_n = \alpha$ for only a *finite* number of terms, then α is a cluster point of the sequence. Compare this with Theorem 17.2. Hint: Let $\varepsilon > 0$ be given, and there is a corresponding N_ε such that with regard to the subsequence, for all $n_k > N_\varepsilon$, we have $|a_{n_k} - \alpha| < \varepsilon$.

15. Prove that if a sequence diverges to $+\infty$, then every subsequence does the same.

20

Limit Theorems for Sequences

When the successive values attributed to a variable approach indefinitely to a fixed value, in such a way that they end by differing from it by as little as one wishes, this last is called the limit *of all the others.*

Augustin-Louis Cauchy, *Cours d'analyse de l'Écoleroyale polytechnique*, 1821, pg. 4

We have purposely delayed using the standard theorems about limits of sequences $\langle a_n \rangle_1^\infty$ in order to investigate some deeper properties. At some point, which depends upon one's zeal for δ-ε proofs, it becomes tiresome to prove convergence or divergence directly from the definitions. The following theorems streamline the process of finding limits, but first they must be proved, using the definitions—of course!

Theorem 20.1 Let c be a constant, and let $\lim\limits_{n\to\infty} a_n = L$. Then $\lim\limits_{n\to\infty} (ca_n) = c \lim\limits_{n\to\infty} a_n = cL$.

Proof. We must prove that given any $\varepsilon > 0$, there exists an N_ε such that for all subscripts $n > N_\varepsilon$, it is true that $|ca_n - cL| < \varepsilon$. If $c = 0$, then N_ε is arbitrary, and the inequality and the theorem are easily true. Working with the inequality when $c \neq 0$, we must show that $|c||a_n - L| < \varepsilon$, or $|a_n - L| < \frac{\varepsilon}{|c|}$. But this is true from the premise that $\langle a_n \rangle_1^\infty$ converges; we just find an N_ε so large that $|a_n - L| < \frac{\varepsilon}{|c|}$.

Retracing our steps when $c \neq 0$, for any $\varepsilon > 0$, we determineN_ε so large that for all $n > N_\varepsilon$, $|a_n - L| < \frac{\varepsilon}{|c|}$. Then $|ca_n - cL| < \varepsilon$, and we conclude that the limit of $\langle ca_n \rangle_1^\infty$ is cL. **QED**

Example 20.1 Use the fact that $\lim_{n\to\infty} \frac{3n^2+7}{4n^2+n} = \frac{3}{4}$ to prove that $\lim_{n\to\infty} 4\left(\frac{3n^2+7}{4n^2+n}\right) = 3$. In this case, $c = 4$, so we need to display an N_ε so large that $\left|\frac{3n^2+7}{4n^2+n} - \frac{3}{4}\right| < \frac{\varepsilon}{4}$ for all $n > N_\varepsilon$ (following the proof above). We observe that this is little more than a restatement of the definition of $\lim_{n\to\infty} \frac{3n^2+7}{4n^2+n} = \frac{3}{4}$, with ε replaced by $\frac{\varepsilon}{4}$. Accordingly,

$$\left|\frac{3n^2+7}{4n^2+n} - \frac{3}{4}\right| = \left|\frac{3n-28}{16n^2+4n}\right| < \frac{3n}{16n^2} = \frac{3}{16n},$$

so we wish to make $\frac{3}{16n} < \frac{\varepsilon}{4}$, implying that $n > \frac{3}{4\varepsilon}$. This leads us to set $N_\varepsilon = \frac{3}{4\varepsilon}$. Hence, for all $n > \frac{3}{4\varepsilon}$, $\left|\frac{3n^2+7}{4n^2+n} - \frac{3}{4}\right| < \frac{\varepsilon}{4}$. We now multiply by 4, $\left|4\left(\frac{3n^2+7}{4n^2+n}\right) - 3\right| < \varepsilon$, demonstrating that $\lim_{n\to\infty} 4\left(\frac{3n^2+7}{4n^2+n}\right) = 3$.

As expected, smaller values of ε necessitate larger values of N_ε in order that values of $a_n = 4\left(\frac{3n^2+7}{4n^2+n}\right)$ fall within the ε tolerance of $L = 3$. As a partial check, take $\varepsilon = 0.01$. Then $N_{0.01} = 75$. We now know that for all $n > 75$, we will have $\left|\frac{3n^2+7}{4n^2+n} - \frac{3}{4}\right| < \frac{0.01}{4} = 0.0025$. Multiplying by 4 gives $\left|4\left(\frac{3n^2+7}{4n^2+n}\right) - 3\right| < 0.01$. At $n = 76$, for instance, $\left|4\left(\frac{3(76^2)+7}{4(76^2)+76}\right) - 3\right| \approx 0.0086$, which is indeed less than the 0.01 tolerance. ∎

Theorem 20.2 Let $\lim_{n\to\infty} a_n = L$ and $\lim_{n\to\infty} b_n = M$ be the limits of two convergent sequences. Then

$$\lim_{n\to\infty} (a_n + b_n) = \lim_{n\to\infty} a_n + \lim_{n\to\infty} b_n = L + M,$$

and

$$\lim_{n\to\infty} (a_n - b_n) = \lim_{n\to\infty} a_n - \lim_{n\to\infty} b_n = L - M.$$

Proof. Given any $\varepsilon > 0$, by virtue of the convergence of the two sequences there is an N_ε so large that for all $n > N_\varepsilon$, $|a_n - L| < \frac{\varepsilon}{2}$ and simultaneously $|b_n - M| < \frac{\varepsilon}{2}$. Then $|a_n - L| + |b_n - M| < \frac{\varepsilon}{2} + \frac{\varepsilon}{2} = \varepsilon$. Our goal is to prove that $|(a_n + b_n) - (L + M)| < \varepsilon$. Applying the triangle inequality, $|(a_n + b_n) - (L + M)| = |(a_n - L) + (b_n - M)| \le |a_n - L| + |b_n - M| < \varepsilon$ for all $n > N_\varepsilon$. Therefore, $\lim_{n\to\infty} (a_n + b_n) = L + M = \lim_{n\to\infty} a_n + \lim_{n\to\infty} b_n$.

For the difference, use what we have just proved and Theorem 20.1:

$$\begin{aligned}\lim_{n\to\infty} (a_n - b_n) &= \lim_{n\to\infty} (a_n + (-b_n)) = \lim_{n\to\infty} a_n + \lim_{n\to\infty} (-b_n)\\ &= \lim_{n\to\infty} a_n - \lim_{n\to\infty} b_n = L - M. \quad \textbf{QED}\end{aligned}$$

Lemma 20.1 All convergent sequences are bounded sets (and hence bounded sequences).

Proof. Let $\lim_{n\to\infty} a_n = L$ for a convergent sequence $\langle a_n \rangle_1^\infty$. This means that for any ε, the neighborhood $(L - \varepsilon, L + \varepsilon)$ contains all but a finite number of terms of the sequence, in particular the neighborhood $(L - 1, L + 1)$. If no terms are excluded from this neighborhood, then the sequence is bounded by the interval $(L - 1, L + 1)$, and we are done. Another way to write this conclusion is that $|a_n| \le C$ for all $n \in \mathbb{N}$, where $C = \max\{|L - 1|, |L + 1|\}$ is a constant.

If some terms (even a trillion!) are excluded from the neighborhood, finish the proof as Exercise 20.2. **QED**

Theorem 20.3 Let $\lim_{n\to\infty} a_n = L$ and $\lim_{n\to\infty} b_n = M$ be the limits of two convergent sequences. Then

$$\lim_{n\to\infty} (a_n b_n) = \left(\lim_{n\to\infty} a_n\right)\left(\lim_{n\to\infty} b_n\right) = LM.$$

Proof. Given any $\varepsilon > 0$, we must prove that $|a_n b_n - LM| < \varepsilon$ for large enough n. The triangle inequality is applied in

$$\begin{aligned} |a_n b_n - LM| &= |a_n b_n - a_n M + a_n M - LM| = |a_n(b_n - M) + M(a_n - L)| \\ &\le |a_n(b_n - M)| + |M(a_n - L)| = |a_n||b_n - M| + |M||a_n - L|. \end{aligned} \tag{1}$$

The trick (employed every so often) of adding and subtracting the same quantity is actually the inspired finale of a search for absolute differences that may be made as small as we wish. Mimicking the proof of Theorem 20.2, there exists an N_ε so large that $|a_n - L| < \varepsilon$ and $|b_n - M| < \varepsilon$ when $n > N_\varepsilon$. But still we are in a delicate position, for while $|M||a_n - L|$ can be made as small as we wish, $|a_n||b_n - M|$ depends on possibly very large values $|a_n|$, so that we are not sure that $|a_n b_n - LM|$ can be forced to be arbitrarily small (remember that we want $|a_n b_n - LM| < \varepsilon$).

Fortunately, Lemma 20.1 guarantees that $(a_n)_1^\infty$ is a bounded set. This means that $|a_n| < C$ for all terms of $(a_n)_1^\infty$, where C is a positive constant. We can now proceed, using (1): $|a_n b_n - LM| \le |a_n||b_n - M| + |M||a_n - L| < C\varepsilon + |M|\varepsilon$. The proof is essentially done, since $C\varepsilon + |M|\varepsilon$ can be made as small as desired by varying *only* ε.

However, the definition formally demands that $|a_n b_n - LM| < \varepsilon$, and not what was just obtained. We refine our analysis slightly. Because of the convergence of the two sequences, we take N_ε so large that when $n > N_\varepsilon$, $|M||a_n - L| < \frac{\varepsilon}{2}$ and simultaneously $C|b_n - M| < \frac{\varepsilon}{2}$ (the first inequality is trivially fulfilled if $M = 0$). It then follows from *(1)* that

$$|a_n b_n - LM| \le |a_n||b_n - M| + |M||a_n - L| < C|b_n - M| + |M||a_n - L| < \frac{\varepsilon}{2} + \frac{\varepsilon}{2} = \varepsilon,$$

just as the definition demands.

We conclude that $\lim_{n\to\infty}(a_n b_n) = LM = (\lim_{n\to\infty} a_n)(\lim_{n\to\infty} b_n)$ **QED**

Note 1: Authors may use slightly modified versions of the proof above. For example, some write that there exists an N_ε such that when $n > N_\varepsilon$, $|a_n - L| < \frac{\varepsilon}{2C}$ and a usually different N'_ε such that when $n > N'_\varepsilon$, $|b_n - M| < \frac{\varepsilon}{2|M|}$ (provided $M \neq 0$). Then, for all $n > \max\{N_\varepsilon, N'_\varepsilon\}$, we have $|a_n b_n - LM| < C\frac{\varepsilon}{2C} + |M|\frac{\varepsilon}{2|M|} = \varepsilon$ as needed. This may cause students to wonder about the sudden appearance of things like $\frac{\varepsilon}{2|M|}$. The reason is that cancelling common factors leads to ε alone.

Furthermore, to include the case $M = 0$, (but to obscure the proof even more) we may pick N_ε so large that $|b_n - M| < \frac{\varepsilon}{2(|M|+1)}$ instead of $\frac{\varepsilon}{2|M|}$.

Corollary 20.1 If $\lim_{n\to\infty} a_n = L$, then $\lim_{n\to\infty} |a_n| = |L|$.

Proof. Create a proof as Exercise 20.3. **QED**

Lemma 20.2 Let $(b_n)_1^\infty$ be a sequence of nonzero numbers with limit $M \neq 0$. Then

$$\lim_{n\to\infty}\left(\frac{1}{b_n}\right) = \frac{1}{\lim\limits_{n\to\infty} b_n} = \frac{1}{M}.$$

Proof. We must prove that for any $\varepsilon > 0$, $|\frac{1}{b_n} - \frac{1}{M}| < \varepsilon$ for sufficiently large n. Starting as usual with the last inequality,

$$\left|\frac{1}{b_n} - \frac{1}{M}\right| = \left|\left(\frac{M - b_n}{M}\right)\frac{1}{b_n}\right| = \frac{|b_n - M|}{|M|}\frac{1}{|b_n|}.$$

Since $\langle b_n\rangle_1^\infty \to M$, the factor $|\frac{b_n - M}{M}|$ can be made as small as we wish. However, the other factor, $\frac{1}{|b_n|}$, may be very large. Our proof focuses on this factor for a moment.

By Corollary 20.1, $\lim_{n\to\infty} |b_n| = |M| > 0$. Hence, only a finite number of terms of $\langle |b_n|\rangle_1^\infty$ are excluded from the neighborhood $\left(\frac{1}{2}|M|, \frac{3}{2}|M|\right)$ centered at $|M|$. This means that for some N^*, the endless tail $\langle |b_n|\rangle_{n=N^*}^\infty$ must fall inside our neighborhood, implying that $\frac{1}{2}|M| < |b_n|$ for all $n \geq N^*$. Consequently, $\frac{1}{|b_n|} \leq \frac{2}{|M|}$ for all terms of $\langle |b_n|\rangle_{n=N^*}^\infty$. We have discovered that $\frac{1}{|b_n|}$ may be large, but not unboundedly large, as n increases.

Returning to the main argument, we know that for any $\varepsilon > 0$ and for all $n > N_\varepsilon$, $|b_n - M| < \varepsilon$. Thus, $|\frac{1}{b_n} - \frac{1}{M}| = \frac{|b_n - M|}{|M|}\frac{1}{|b_n|} < \frac{\varepsilon}{|M|}\frac{2}{|M|} = \frac{2\varepsilon}{|M|^2}$, provided that $n \geq N^*$ as well. We are essentially done, since $\frac{2\varepsilon}{|M|^2}$ may be made as small as we wish by varying only ε. But someone may request that we follow the definition formally. Hence, we choose N_ε large enough so that for all $n > \max\{N^*, N_\varepsilon\}$, $|b_n - M| < \frac{1}{2}|M|^2\varepsilon$. It follows that

$$\left|\frac{1}{b_n} - \frac{1}{M}\right| = \frac{|b_n - M|}{|M|}\frac{1}{|b_n|} < \frac{\frac{1}{2}|M|^2\varepsilon}{|M|}\frac{2}{|M|} = \varepsilon,$$

as requested. The lemma follows. **QED**

The sequence $\langle |b_n|\rangle_1^\infty$ was positive, but more importantly, it did not contain terms that were *arbitrarily* close to zero. Such a sequence is said to be bounded away from zero. In general, a sequence $\langle a_n\rangle_1^\infty$ is **bounded away** from a constant C if we can find another constant ξ such that $C < \xi \leq a_n$ (or $a_n \leq \xi < C$ when the terms are less than C) for all $n \in \mathbb{N}$. In the proof just finished, the neighborhood's radius of $\frac{1}{2}|M|$ is almost arbitrary. It's necessary only that it be less than $|M|$, so that we can certify that the tail end of $\langle |b_n|\rangle_1^\infty$ is bounded away from zero. Consider Exercise 20.4 now.

Theorem 20.4 Let $\lim_{n\to\infty} a_n = L$, and $\lim_{n\to\infty} b_n = M \neq 0$, where no b_n is zero. Then

$$\lim_{n\to\infty}\left(\frac{a_n}{b_n}\right) = \frac{\lim\limits_{n\to\infty} a_n}{\lim\limits_{n\to\infty} b_n} = \frac{L}{M}.$$

Proof. This is left as Exercise 20.6. **QED**

Lemma 20.2 cannot tell us that the sequence $\langle\frac{1}{n}\rangle_1^\infty$ converges, since the denominators generate the divergent sequence $\langle n\rangle_1^\infty$. The following lemma gives the result we want.

Lemma 20.3 $\lim_{n\to\infty} \frac{1}{n} = 0$.

Proof. We must prove that for any $\varepsilon > 0$, $|\frac{1}{n} - 0| < \varepsilon$ for sufficiently large n. By Corollary 12.2, $\frac{1}{n} < \varepsilon$ for some natural number n, regardless of how small ε is. Thus, the limit is zero. **QED**

Example 20.2 Reestablish the result from Example 17.5 that

$$\lim_{n\to\infty}\frac{5n^4+87n^3}{4n^4+7n+8}=\frac{5}{4}.$$

In order to use Theorem 20.4, we use some standard algebra:

$$\frac{5n^4+87n^3}{4n^4+7n+8}=\frac{5+\frac{87}{n}}{4+\frac{7}{n^3}+\frac{8}{n^4}}.$$

Now, using the theorems and lemmas above,

$$\lim_{n\to\infty}\frac{5+\frac{87}{n}}{4+\frac{7}{n^3}+\frac{8}{n^4}}=\frac{\lim_{n\to\infty}\left(5+\frac{87}{n}\right)}{\lim_{n\to\infty}\left(4+\frac{7}{n^3}+\frac{8}{n^4}\right)}=\frac{5+0}{4+0+0}=\frac{5}{4}.\quad\blacksquare$$

The technique of rationalizing the numerator appears in the proof of the next theorem.

Theorem 20.5 Suppose that $\langle a_n\rangle_1^\infty$ has no negative terms, and that $\lim_{n\to\infty} a_n = L$. Then

$$\lim_{n\to\infty}\sqrt{a_n}=\sqrt{\lim_{n\to\infty}a_n}=\sqrt{L}.$$

Proof. As a first step, it is necessary to show that $L \geq 0$. This was done in Exercise 17.5. It will soon be seen that we must also prove as a separate case that $\langle a_n\rangle_1^\infty \to 0$ implies $\langle\sqrt{a_n}\rangle_1^\infty \to 0$. This was done in Exercise 17.6. Hence, we may proceed with $L > 0$ as a premise.

We must show that for any $\varepsilon > 0$, there is a corresponding number N_ε such that for all $n > N_\varepsilon$, $|\sqrt{a_n}-\sqrt{L}| < \varepsilon$. At this point, we rationalize the numerator:

$$|\sqrt{a_n}-\sqrt{L}|=\left|\frac{(\sqrt{a_n}-\sqrt{L})(\sqrt{a_n}+\sqrt{L})}{\sqrt{a_n}+\sqrt{L}}\right|=\left|\frac{a_n-L}{\sqrt{a_n}+\sqrt{L}}\right|=\frac{|a_n-L|}{\sqrt{a_n}+\sqrt{L}}.$$

The premise already asserts that for any $\varepsilon > 0$, there is indeed a number N_ε such that for all $n > N_\varepsilon$, $|a_n - L| < \varepsilon$. Dividing by $\sqrt{a_n}+\sqrt{L}$, we have $\frac{|a_n-L|}{\sqrt{a_n}+\sqrt{L}} < \frac{\varepsilon}{\sqrt{a_n}+\sqrt{L}} \leq \frac{\varepsilon}{\sqrt{L}}$. No embarrassing division by zero can occur in these steps (why?).

Gathering our statements, we have shown that when $L > 0$, for any $\varepsilon > 0$ and for all $n > N_\varepsilon$, $|\sqrt{a_n}-\sqrt{L}| < \frac{\varepsilon}{\sqrt{L}}$. This is sufficient to conclude that $\sqrt{L}$ is the limit. But as in the previous proofs, we instead choose N_ε such that for all

$$n > N_\varepsilon,\ |a_n-L| < \varepsilon\sqrt{L}.\ \text{Then}\ |\sqrt{a_n}-\sqrt{L}| < \frac{\varepsilon\sqrt{L}}{\sqrt{L}}=\varepsilon,$$

as formally required. Consequently, $\lim_{n\to\infty}\langle\sqrt{a_n}\rangle_1^\infty=\sqrt{L}$. **QED**

The next theorem is easy to see, and its proof reviews some past concepts.

Theorem 20.6 Let r be a constant. Then $\lim_{n\to\infty} r^n = 0$ if and only if $|r| < 1$.

Proof of sufficiency. Given that $|r| < 1$, $|r|^{n+1} < |r|^n$ for all $n \in \mathbb{N}$. Thus, $\left(|r|^n\right)_1^\infty$ is bounded by $[0, 1]$, and is monotone decreasing. Hence, the monotone convergence theorem indicates that $\left(|r|^n\right)_1^\infty$ converges to its infimum, β. Using the technique in Example 19.8,

$$\beta = \lim_{n\to\infty} |r|^n = \lim_{k\to\infty} |r|^{k+1} = |r| \lim_{k\to\infty} |r|^k = |r|\beta.$$

Since $|r| < 1$, this implies that $\beta = 0$.

Because $\lim_{n\to\infty} |r|^n = 0$, every neighborhood $(-\varepsilon, \varepsilon)$ of zero contains all but a finite set of terms of $\langle |r|^n \rangle_1^\infty$ (the set typically changes depending on ε). This implies that $0 \le |r^n| < \varepsilon$, and hence, $-\varepsilon < r^n < \varepsilon$ except for a finite set of terms. This finally implies that $\left(r^n\right)_1^\infty \to 0$.

Proof of necessity. The converse statement is "If $\lim_{n\to\infty} r^n = 0$ then $|r| < 1$,"and we will prove its contrapositive. Thus, let $|r| \ge 1$. Since we are working with $|r|^n$, Bernoulli's inequality applies nicely: $(1+p)^n \ge 1 + np$ for all $n \in \mathbb{N}$, and we will need $p \ge 0$. Let $|r| = 1 + p$, giving us $|r|^n \ge 1 + n\left(|r| - 1\right)$. In the case that $|r| > 1$, this shows that $|r|^n$ may be made larger than any given number, because of the Archimedean principle. Therefore, $\left(r^n\right)_1^\infty$ must diverge. When $|r| = 1$, $|r|^n = 1$ for every $n \in \mathbb{N}$, which implies that $\lim_{n\to\infty} r^n = 1$. In either case, $\lim_{n\to\infty} r^n \ne 0$, and the contrapositive is proved. **QED**

Theorem 20.7 The Pinching Theorem for Sequences Let $\left(a_n\right)_1^\infty$, $\left(b_n\right)_1^\infty$, and $\left(c_n\right)_1^\infty$ be three sequences such that $a_n \le b_n \le c_n$ for all $n \in \mathbb{N}$. Also, let $\lim_{n\to\infty} a_n = \lim_{n\to\infty} c_n = L$. Then $\left(b_n\right)_1^\infty$ also converges, and to the same limit as the others.

Proof. The first premise leads us to

$$a_n - L \le b_n - L \le c_n - L \text{ for all } n \in \mathbb{N}. \qquad \text{(a)}$$

The other premise implies that for any $\varepsilon > 0$, there is an N_ε so large that both $|a_n - L| < \varepsilon$ and $|c_n - L| < \varepsilon$ when $n > N_\varepsilon$. These mean that

$$-\varepsilon < a_n - L < \varepsilon \qquad \text{(b)}$$

and

$$-\varepsilon < c_n - L < \varepsilon \text{ for all } n > N_\varepsilon. \qquad \text{(c)}$$

Putting parts of inequalities (b), (a) and (c) together (in that order), we obtain $-\varepsilon < b_n - L < \varepsilon$ for all $n > N_\varepsilon$. This means that $\lim_{n\to\infty} b_n = L$. **QED**

Example 20.3 Find the limit of $\left(\frac{\cos n}{n}\right)_1^\infty$. The pinching theorem for sequences not only proves that the limit exists, it evaluates the limit. Since $-1 \le \cos n \le 1$, we have $\frac{-1}{n} \le \frac{\cos n}{n} \le \frac{1}{n}$. Identify $a_n = \frac{-1}{n}$, $b_n = \frac{\cos n}{n}$, and $c_n = \frac{1}{n}$, and note that $L = \lim_{n\to\infty} a_n = \lim_{n\to\infty} c_n = 0$ by Lemma 20.3. Therefore, $\lim_{n\to\infty} \frac{\cos n}{n} = 0$. ■

Example 20.4 Find $\lim_{n\to\infty} \frac{1}{\sqrt[3]{n}}$. Our strategy uses $\sqrt[4]{n} \le \sqrt[3]{n} \le \sqrt{n}$ for all $n \in \mathbb{N}$. Thus, $\frac{1}{\sqrt{n}} \le \frac{1}{\sqrt[3]{n}} \le \frac{1}{\sqrt[4]{n}}$. By using Theorem 20.5 and Lemma 20.3, we derive

$$\lim_{n\to\infty} \frac{1}{\sqrt{n}} = \sqrt{\lim_{n\to\infty} \frac{1}{n}} = 0,$$

and

$$\lim_{n\to\infty} \frac{1}{\sqrt[4]{n}} = \sqrt{\sqrt{\lim_{n\to\infty} \frac{1}{n}}} = 0.$$

Hence, the pinching theorem for sequences tells us that $\lim_{n\to\infty} \frac{1}{\sqrt[3]{n}} = 0$ as well. ∎

20.1 Exercises

1. Use the information above and in the previous chapter to evaluate the following limits.
   a. $\lim_{n\to\infty} \frac{7}{4-3n}$ b. $\lim_{n\to\infty} \sqrt{\frac{9n-n^3}{5+n^2-6n^3}}$ (think about $n = 1$ and $n = 2$)
   c. $\lim_{n\to\infty} \sqrt[3]{\frac{5|n|}{2n-7}}$ d. $\lim_{n\to\infty} \left(\sqrt{n+1} - \sqrt{n}\right)$
2. a. Finish the proof of Lemma 20.1.
   b. Show that the converse of Lemma 20.1 is false.
3. a. Prove Corollary 20.1.
   b. State and disprove the converse of Corollary 20.1.
4. Prove or disprove: If a sequence of positive terms $\langle a_n \rangle_1^\infty$ is bounded away from zero, then it must have a minimum term a_M. That is, $0 < a_M \le a_n$ for all n.
5. Write a convincing argument (or a proof) for or against this statement: $\langle \cos n \rangle_1^\infty$ is bounded away from one. Hint: $\cos(2kp)$ approaches 1 when p approximates π more and more closely.
6. Prove Theorem 20.4.
7. Suppose we are proving Theorem 20.6, and we already know that $\langle |r|^n \rangle_1^\infty \to 0$. If the following argument is valid, fill in some key reasons. If it isn't valid, why not?

   $$0 = \lim_{n\to\infty} |r|^n = \lim_{n\to\infty} \left(\sqrt{r^2}\right)^n = \lim_{n\to\infty} \sqrt{r^{2n}} = \sqrt{\lim_{n\to\infty} r^{2n}}.$$

   This implies that

   $$0 = \lim_{n\to\infty} r^{2n} = \lim_{n\to\infty} \left(r^n\right)^2 = \left(\lim_{n\to\infty} r^n\right)^2.$$

   Hence, $0 = \lim_{n\to\infty} r^n$. **QED?**
8. Prove that for a sequence $\langle a_n \rangle_1^\infty$ of positive numbers, $\lim_{n\to\infty} a_n = 0$ if and only if $\lim_{n\to\infty} \frac{1}{a_n} = +\infty$. What can be said about the sequence $\langle -a_n \rangle_1^\infty$ of negative numbers?
9. a. Prove the following: if $r \ge 1$, then $\lim_{n\to\infty} r^{1/n} = 1$. Hint: As in the proof of Theorem 20.6, use Bernoulli's inequality, letting $1 + p = r^{1/n}$. Derive that $r^{1/n} \le \frac{n+r-1}{n}$.

b. Extend the result of **a** into a theorem: $\lim_{n\to\infty} r^{1/n} = 1$ for all $r > 0$. (For more interesting results, see Exercise 35.10.) Hint: For $0 < r < 1$, let $s = \frac{1}{r}$.

10. Make a Venn diagram that has $\mathscr{L} = \{\textit{real number sequences}\}$ as the universal set, and that shows the proper intersections of the following subsets of $\mathscr{L}$: $C = \{\textit{convergent sequences}\}$, $D = \{\textit{divergent sequences}\}$, $M = \{\textit{monotone sequences}\}$, $B = \{\textit{bounded sequences}\}$, $D_{+\infty} = \{\textit{sequences divergent to} + \infty\}$, $D_{-\infty} = \{\textit{sequences divergent to} - \infty\}$, $O = \{\textit{oscillating sequences}\}$. Once you are sure that the diagram is correct, print the impressive design onto a T-shirt.

21

Cauchy Sequences and the Cauchy Convergence Criterion

On one occasion when Laplace and several other notables were present, Lagrange pointed to young Cauchy in his [corner office] and said, "You see that little young man? Well! He will supplant all of us in so far as we are mathematicians."

E. T. Bell [2], pg. 274.

Augustin-Louis Cauchy(1789–1857) is a towering figure in mathematical analysis. We know that for a sequence to have a limit L, some tail end of it must lie within any neighborhood of L that we choose. Cauchy observed that this means that eventually any two terms a_n and a_m must be as close together as we please. This bit of intuitive thinking is solidified in his definition (a definition foreseen by Bernhard Bolzano, and perhaps by others before him).

> A **Cauchy sequence** is a sequence $\langle a_n \rangle_1^\infty$ with the property that given any $\varepsilon > 0$, there exists an $N_\varepsilon > 0$ such that $|a_n - a_m| < \varepsilon$ whenever we choose the two subscripts $m, n \geq N_\varepsilon$.

Note 1: The subscripts m and n must be independent of each other, and *both* must be at least N_ε. Once in a while, the alternative $m, n > N_\varepsilon$ will be more convenient. Without loss of generality, we could use $n \geq m \geq N_\varepsilon$, or similar statements such as $m > n \geq N_\varepsilon$ for convenience. It is also possible to use $m = n + p$, with $p \in \mathbb{N}$, and $n \geq N_\varepsilon$. (Warnings are found in Exercise 21.6.) Also, N_ε may be a natural number, but need not be.

The inequality $|a_n - a_m| < \varepsilon$ shows that pairs of terms a_n and a_m will be found squeezed together as tightly as we wish. The key question is, "If they are squeezed together this way, are they then also squeezing against a limit?" In other words, is this clustering sufficient to indicate the existence of a limit for the sequence? We shall now see that it's not only sufficient, but also necessary!

Theorem 21.1 The Cauchy Convergence Criterion A sequence of real numbers is convergent if and only if it is a Cauchy sequence.

Proof of necessity. Given a sequence $\langle a_n \rangle_1^\infty$ that converges to the limit L, then for any $\varepsilon > 0$, there exists an N_ε so large that for all $n \geq N_\varepsilon$, we are sure that $|a_n - L| < \frac{\varepsilon}{2}$. Then for any $m \geq n$, it is also true that $|a_m - L| < \frac{\varepsilon}{2}$. Hence, for any $m \geq n \geq N_\varepsilon$, the following holds, with the help of the triangle inequality:

$$|a_n - a_m| = |(a_n - L) + (L - a_m)| \leq |a_n - L| + |L - a_m| < \frac{\varepsilon}{2} + \frac{\varepsilon}{2} = \varepsilon.$$

And so, $\langle a_n \rangle_1^\infty$ is indeed a Cauchy sequence.

Proof of sufficiency. To prove sufficiency, we begin with a Cauchy sequence $\langle a_n \rangle_1^\infty$. First we prove that it is bounded. Select a convenient ε, such as $\varepsilon = 1$, and find an integer N_1 so that Cauchy's definition implies that for all $m \geq n = N_1$, we have $|a_m - a_{N_1}| < 1$. Here, we have determined subscript $n = N_1$, which fixes the term a_{N_1}. Then, $|a_m| - |a_{N_1}| \leq |a_m - a_{N_1}| < 1$, and $|a_m| < |a_{N_1}| + 1$. This means that the tail $\langle a_n \rangle_{n=N_1}^\infty$ beginning at term a_{N_1} is bounded! It is now easy to find a bound for the whole sequence $\langle a_n \rangle_1^\infty$. A hint is in Exercise 21.1.

We will now prove that the sequence converges. By the Bolzano-Weierstrass theorem for sequences, $\langle a_n \rangle_1^\infty$ will have a convergent subsequence $\langle a_{n_k} \rangle_{k=1}^\infty$, and we will call its limit β. Given any $\varepsilon > 0$, there is a corresponding N_ε so large that for all the subsequence's subscripts $n_k \geq N_\varepsilon$, we guarantee that $|a_{n_k} - \beta| < \frac{\varepsilon}{2}$.

Furthermore, because $\langle a_n \rangle_1^\infty$ is a Cauchy sequence, and using the same ε, there exists an M_ε such that for all $m, n_k \geq M_\varepsilon$, we have $|a_m - a_{n_k}| < \frac{\varepsilon}{2}$. We link this with the previous paragraph by choosing the larger of N_ε and M_ε, and make the following statement: if $m, n_k \geq \max\{M_\varepsilon, N_\varepsilon\}$, then

$$|a_m - \beta| = |(a_m - a_{n_k}) + (a_{n_k} - \beta)| \leq |a_m - a_{n_k}| + |a_{n_k} - \beta| < \frac{\varepsilon}{2} + \frac{\varepsilon}{2} = \varepsilon.$$

Since ε is arbitrary, we have proved that the Cauchy sequence $\langle a_n \rangle_1^\infty$ converges (to β). **QED**

Note 2: The Cauchy convergence criterion is a necessary and sufficient theorem for sequences of real numbers. The sufficiency ("if") part of the criterion is a consequence of the completeness axiom of the real numbers (see Exercise 21.7). Because every Cauchy sequence in $\mathbb{R}$ converges to a real number, $\mathbb{R}$ is called a *complete space* (review the discussion after the axioms for $\mathbb{R}$ in Chapter 12).

It is possible to study sequences of terms other than real numbers, or sequences that may exist in spaces with novel definitions of distance. In such strange realms, the completeness axiom may not be in force, and it may happen that a Cauchy sequence does not converge. These ideas will be investigated more fully in Exercise 21.8 and in Chapter 53.

Recall that in order to use the definition of convergence of a sequence, we needed to explicitly state what that limit was, since the definition used that value. The Cauchy convergence criterion guarantees convergence if we can just establish the sequence to be a Cauchy sequence. No knowledge of the limit itself is needed. (This was also a strength of the monotone convergence theorem.)

Example 21.1 We prove that $\left\langle\frac{e^n}{\pi^n}\right\rangle_1^\infty$ is a Cauchy sequence. Take $m \geq n$, so that $\left(\frac{e}{\pi}\right)^m \leq \left(\frac{e}{\pi}\right)^n$ (verify), and consider

$$|a_n - a_m| = \left|\left(\frac{e}{\pi}\right)^n - \left(\frac{e}{\pi}\right)^m\right| < \left(\frac{e}{\pi}\right)^n.$$

We must show that $\left(\frac{e}{\pi}\right)^n < \varepsilon$ whenever $n > N_\varepsilon$. Bernoulli's inequality is helpful here, in the form $b^n \geq 1 + n(b-1)$ with $b \geq 1$ (and $n \in \mathbb{N}$ as usual). Then $\frac{1}{b^n} \leq \frac{1}{1+n(b-1)}$, and we choose $b = \frac{\pi}{e}$. This yields

$$\left(\frac{e}{\pi}\right)^n \leq \frac{1}{1+n\left(\frac{\pi}{e}-1\right)} < \frac{1}{n\left(\frac{\pi}{e}-1\right)} = \frac{e}{n(\pi - e)} < \frac{7}{n}.$$

It follows that for any $\varepsilon > 0$, if we set $N_\varepsilon = \frac{7}{\varepsilon}$, then for all $n \geq N_\varepsilon$, it is true that $\left(\frac{e}{\pi}\right)^n < \frac{7}{n} \leq \varepsilon$.

Unrolling our steps, we are certain that to any $\varepsilon > 0$, there corresponds the number $N_\varepsilon = \frac{7}{\varepsilon}$ with the property that for all

$$m \geq n \geq N_\varepsilon, |a_n - a_m| < \left(\frac{e}{\pi}\right)^n < \frac{7}{n} \leq \varepsilon.$$

Hence, $\left\langle\frac{e^n}{\pi^n}\right\rangle_1^\infty$ is a Cauchy sequence (and furthermore, it converges; but we knew this by Theorem 20.6).

Other approaches to proving that $\left\langle\frac{e^n}{\pi^n}\right\rangle_1^\infty$ is a Cauchy sequence are yours to find in Exercise 21.2. ■

Example 21.2 We show that every decimal $0.d_1d_2d_3\ldots d_n\ldots$, where $d_i \in \{0, 1, 2, \ldots, 9\}$ are the digits, actually converges (hence, to a unique real number). Set up the sequence of truncated decimal approximations $\langle a_n\rangle_1^\infty = \langle 0.d_1,\ 0.d_1d_2,\ 0.d_1d_2d_3, \ldots\rangle$ taken from $0.d_1d_2d_3\ldots d_n\ldots$, and ask, "Is $\langle a_n\rangle_1^\infty$ a Cauchy sequence?" We must show that for any $\varepsilon > 0$, there exists an N_ε such that for any

$$m > n \geq N_\varepsilon, |a_m - a_n| = \frac{d_{n+1}}{10^{n+1}} + \frac{d_{n+2}}{10^{n+2}} + \cdots + \frac{d_m}{10^m} < \varepsilon.$$

Since $0 \leq d_i \leq 9$ is always the case, we have

$$\frac{d_{n+1}}{10^{n+1}} + \cdots + \frac{d_m}{10^m} \leq \frac{9}{10^{n+1}} + \cdots + \frac{9}{10^m} = 9\sum_{k=n+1}^{m}\frac{1}{10^k}.$$

The summation is a finite geometric series of terms that has the sum

$$\sum_{k=n+1}^{m}\frac{1}{10^k} = \frac{\left(\frac{1}{10}\right)^{n+1} - \left(\frac{1}{10}\right)^{m+1}}{1-\frac{1}{10}} = \frac{\left(\frac{1}{10}\right)^n - \left(\frac{1}{10}\right)^m}{9}$$

(applying Exercise 4.9). Hence,

$$\frac{d_{n+1}}{10^{n+1}} + \frac{d_{n+2}}{10^{n+2}} + \cdots + \frac{d_m}{10^m} \leq \left(\frac{1}{10}\right)^n - \left(\frac{1}{10}\right)^m < \left(\frac{1}{10}\right)^n.$$

It follows that if $\frac{1}{10^n} < \varepsilon$, then certainly $|a_m - a_n| < \varepsilon$.

Therefore, given any $\varepsilon > 0$, choose $N_\varepsilon = \max\left\{\frac{-\ln\varepsilon}{\ln 10}, 1\right\}$. We may now assert that for any

$$m > n \geq N_\varepsilon, |a_m - a_n| < \frac{1}{10^n} < \varepsilon$$

is true. Consequently, $\langle a_n \rangle_1^\infty = \langle 0.d_1,\ 0.d_1d_2, 0.d_1d_2d_3, \ldots \rangle$ is a Cauchy sequence, and it must converge to a real number by the Cauchy convergence criterion. **QED**

The question of the decimal representation of reals was considered in Note 2 of Chapter 19, and will be looked at again in Chapter 28 from another point of view.

21.1 Exercises

1. Finish the part of the proof of the Cauchy convergence criterion that establishes a bound for the whole sequence $\langle a_n \rangle_1^\infty$. Hint: Use $\max\{|a_n|\}$ for $1 \le n < N_1$, which is the finite front end of the sequence.

2. a. Reconsider Example 21.1 with $a_n = \frac{e^n}{\pi^n}$. Fill in the blanks and complete an alternative proof: Proving that $\left(\frac{e}{\pi}\right)^m \le \left(\frac{e}{\pi}\right)^n$ when $m \ge n$ amounts to proving that this sequence is ________ ________. Since it is also bounded below by ______, the sequence ______.
   Hence, by the ________ ________ ________, it is a Cauchy sequence.
   b. Staying with Example 21.1, there are other approaches to finding a suitable N_ε. We may simply solve for n in $\left(\frac{e}{\pi}\right)^n < \varepsilon$. Verify that this is equivalent to $n > \frac{\ln \varepsilon}{\ln \frac{e}{\pi}}$. Fill in the details about finding N_ε, and finish this alternative proof that $\left\langle \frac{e^n}{\pi^n} \right\rangle_1^\infty$ is a Cauchy sequence. (If $0 < \varepsilon < 1$, which is the interesting case, then $\frac{\ln \varepsilon}{\ln \frac{e}{\pi}} > 0$, but if $\varepsilon > 1$, then $\frac{\ln \varepsilon}{\ln \frac{e}{\pi}} < 0$. What must be said?)
   Can N_ε be determined in an even simpler manner?

3. Use the definition of a Cauchy sequence to prove that the limit of $\left\langle \frac{n!}{n^n} \right\rangle_1^\infty$ is zero. Hints: With the strategy of Example 21.1, you will need to show that $\frac{n!}{n^n} > \frac{m!}{m^m}$ whenever $m > n$ by proving that $\frac{m^m}{n^n} > \frac{m!}{n!}$ whenever $m > n$. Soon after that, you will have to show that for large enough n it is true that $\frac{n!}{n^n} < \varepsilon$. Alas, for a given (small) ε, this inequality doesn't give much guidance for selecting a corresponding N_ε. But prove and use this: $\frac{n!}{n^n} \le \frac{1}{n}$, for all n.

4. Prove that a sequence $\langle a_n \rangle_1^\infty$ converges if and only if given any $\varepsilon > 0$, there exists an $N_\varepsilon \in \mathbb{N}$ such that if $n > N_\varepsilon$ then $|a_n - a_{N_\varepsilon}| < \varepsilon$.

5. Prove that if $\langle a_n \rangle_1^\infty$ is a sequence such that $|a_{n+1} - a_n| < 2^{-n}$, then $\langle a_n \rangle_1^\infty$ is a Cauchy sequence (and as a consequence, convergent). Fill in details in the following sequence of steps.
   (a) Extend $|a_{n+1} - a_n| < 2^{-n}$ to $|a_{n+2} - a_{n+1}| < 2^{-(n+1)}, \ldots \quad , |a_{n+p} - a_{n+p-1}| < 2^{-(n+p-1)}$, and show that $\sum_{k=1}^{p} |a_{n+k} - a_{n+k-1}| < \sum_{k=1}^{p} \frac{1}{2^{n+k-1}}$ for any $p \in \mathbb{N}$.
   (b) Show by induction on p (or otherwise) that $\sum_{k=1}^{p} \frac{1}{2^{n+k-1}} < \frac{1}{2^{n-1}}$.
   (c) Show that $\frac{1}{2^{n-1}} \le \frac{1}{n}$ for $n \in \mathbb{N}$.
   (d) Using the triangle inequality, show that $|a_{n+p} - a_n| < \frac{1}{n}$ for any $p \in \mathbb{N}$.
   (e) Finish the proof by letting $m = n + p$ (remember, p is arbitrary).
   Carefully contrast this exercise with the next one.

6. a. Show that linking m and n in the definition of Cauchy convergence by setting $m = n + 1$ gives a necessary but not sufficient condition for convergence of a sequence $\langle a_n \rangle_1^\infty$. That is, can you find a sequence such that given any $\varepsilon > 0$, $|a_{n+1} - a_n| < \varepsilon$ for all $n \ge N_\varepsilon$, and even so, the sequence does not converge? (Meaning: this condition is not sufficient.) To finish,

show that if a sequence converges, then it is true that given any $\varepsilon > 0$, $|a_n - a_{n+1}| < \varepsilon$ for all $n \geq N_\varepsilon$. (Meaning: this condition is necessary.) Hint: Consider a_n given by $a_1 = 1$, $a_2 = 1 + \frac{1}{2} = \frac{3}{2}$, $a_3 = 1 + \frac{1}{2} + \frac{1}{3} = \frac{11}{6}$, $a_4 = \frac{25}{12}, \ldots$ and recall a property of the harmonic series.

b. Show that even linking m and n in the definition of Cauchy convergence by setting $m = n + p$, where p is a *fixed* natural number, once again gives a necessary but not sufficient condition for convergence of a sequence. Hint: Consider $\{1, 2, 2\frac{1}{2}, 3, 3\frac{1}{3}, 3\frac{2}{3}, 4, 4\frac{1}{4}, 4\frac{2}{4}, 4\frac{3}{4}, 5, 5\frac{1}{5}, 5\frac{2}{5}, 5\frac{3}{5}, 5\frac{4}{5}, 6, \cdots\}$ (from [18], pg. 66). The sequence is obviously unbounded. But pick say, $p = 7$, and then convince yourself that $|a_{n+7} - a_n|$ can be as small as we wish, when n is large enough. In fact, suppose $|a_{n+7} - a_n| < \varepsilon = 0.002$, for instance. Find a corresponding $N_{0.002}$.

7. Where in the proof of sufficiency of the Cauchy convergence criterion is the completeness axiom of real numbers needed?

8. Define $\mathbb{O}$ to be the massively huge set of all open intervals on the number line: $\mathbb{O} = \{(a, b) : a, b \in \mathbb{R},\ a < b\}$. Imagine now sequences $\langle s_n \rangle_1^\infty$ not of real numbers but strictly of elements of $\mathbb{O}$, and consider this sequence: $s_n = \bigcap_{k=1}^{n} \left(1 - \frac{1}{k}, 4 + \frac{1}{k}\right) \in \mathbb{O}$. That is, $s_1 = (0, 5)$, $s_2 = \left(\frac{1}{2}, \frac{9}{2}\right)$, and $s_n = (1 - \frac{1}{n}, 4 + \frac{1}{n})$ in general.

a. Show that $\lim_{n\to\infty} s_n = [1, 4]$, a closed interval. Since we are confined to the space of open intervals, $\langle s_n \rangle_1^\infty$ does not converge, due to the important technicality that its limit is outside of $\mathbb{O}$.

b. Define a distance D between two open intervals (a, b) and (c, d) by $D = |c - a| + |d - b|$. We can now tell how far apart two intervals are from each other. How far apart are $(1, 2)$ and $(3, 8)$? How far apart are s_2 and s_5, members of $\mathbb{O}$? What is the distance between arbitrary terms s_n and s_m of $\mathbb{O}$?

c. Given any $\varepsilon > 0$, find an N_ε such that if $m, n \geq N_\varepsilon$, then the distance D between s_n and s_m will be less than ε. Conclude that our $\langle s_n \rangle_1^\infty$ is a Cauchy sequence that does not follow the the "if" part of the Cauchy convergence criterion within its space $\mathbb{O}$.

Note 3: A caution: distance can't be defined any way we like. A metric is any function of two arguments that follows certain rules which create for us an analytic concept of distance. The formula for the distance between two points that you learned in analytic geometry satisfies the rules of a metric. So does the D function here. We will study metric functions in Chapter 52, and this particular one in Example 52.5.

Philosophically speaking, it was a great shock to find out that distance was not necessarily something like the length of a thread stretched out between points A and B in space. That is, the concept of distance could not be used in Immanuel Kant's "synthetic *a priori*" category of judgements by which mankind creates experience. Distance could be redefined so as to create a different universe that would nevertheless be comprehensible.

And it isn't just philosophy! In the early 1900s, Albert Einstein's friend Hermann Minkowski revealed a special distance called the space-time metric to measure relativistic distances between events. As you can imagine, his metric involved time and the speed of light.

9. Take a look at the fascinating expressions called continued fractions in Appendix D, and a slightly more elaborate application of the Cauchy convergence criterion.

22

The Limit Superior and Limit Inferior of a Sequence

We have seen that some sequences, even if bounded, do not have a limit. This is because they have more than one cluster point, or equivalently, because at least two subsequences converge to different limits. Such a case is graphed in Figure 22.1,

$$\text{where } a_n = \begin{cases} 4 + \dfrac{1}{n} & \text{if } n \text{ is odd} \\ 3 - \dfrac{1}{n} & \text{if } n \text{ is even} \end{cases}.$$

The two explicit subsequences are fairly evident, converging to the two subsequential limits $\lim_{n\to\infty} a_{2n-1} = 4$ and $\lim_{n\to\infty} a_{2n} = 3$. Note that $\sup\langle a_n\rangle_1^\infty = 5$. Even though $\langle a_n\rangle_1^\infty$ has no limit, we would like to be able to identify its maximum and minimum subsequential limits.

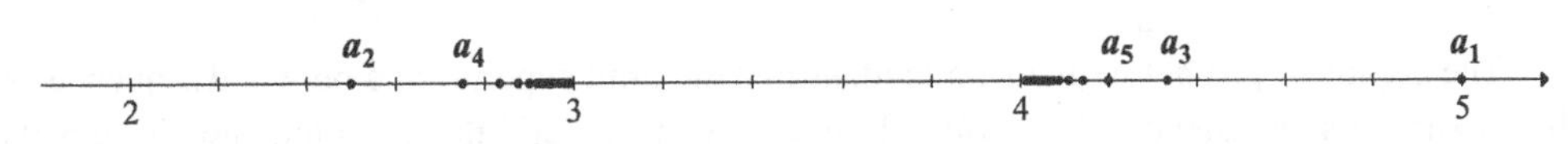

Figure 22.1. $\langle a_n\rangle_1^\infty$ with $a_n = 4 + \frac{1}{n}$ for odd n, and $a_n = 3 - \frac{1}{n}$ for even n.

To find the maximum subsequential limit, we begin to cut offsome of the largest terms of $\langle a_n\rangle_1^\infty$. In Figure 22.1, let us delete a_1, a_3, and a_5. The remaining part of the sequence, namely $\langle a_2, a_4, a_6, a_7, a_8, a_9, \dots\rangle$, has a supremum that is closer to 4 than originally. In fact, $\sup\langle a_2, a_4, a_6, a_7, a_8, a_9, \dots\rangle = a_7 = 4 + \frac{1}{7}$.

We now notice that concentrating on the odd-numbered terms is unnecessary. It's simpler to remove successive terms, one by one, starting at a_1. As we continue deleting terms, we create a sequence of suprema. Symbolically, we have $\langle \sup\langle a_n\rangle_1^\infty, \sup\langle a_n\rangle_2^\infty, \sup\langle a_n\rangle_3^\infty, \dots\rangle = \langle \sup\langle a_n\rangle_{n=k}^\infty\rangle_{k=1}^\infty$. In the case of Figure 22.1, the resulting sequence is $\langle a_1, a_1, a_3, a_3, a_5, a_5, \dots\rangle$, with terms given by

$$\sup\langle a_n\rangle_{n=k}^\infty = \begin{cases} 4 + \dfrac{1}{k} & \text{if } k \text{ is odd} \\ 4 + \dfrac{1}{k-1} & \text{if } k \text{ is even} \end{cases}.$$

As k increases, $\sup\langle a_n\rangle_{n=k}^{\infty}$ decreases, yielding a monotone decreasing sequence of suprema. The limit of the new sequence is the maximum subsequential limit of $\langle a_n\rangle_1^{\infty}$, namely, 4. That was our goal!

These ideas must now be put on firmer ground, and we should consider how general our observations were. First of all, it is true *in general* that for a bounded sequence $\langle a_n\rangle_1^{\infty}$, the resulting sequence of suprema $\langle \sup\langle a_n\rangle_{n=k}^{\infty}\rangle_{k=1}^{\infty}$ is monotone decreasing. This becomes part of our first lemma.

Lemma 22.1 For any bounded sequence $\langle a_n\rangle_1^{\infty}$, the sequence of suprema $\langle \sup\langle a_n\rangle_{n=k}^{\infty}\rangle_{k=1}^{\infty}$ is monotone decreasing. Moreover, the limit of $\langle \sup\langle a_n\rangle_{n=k}^{\infty}\rangle_{k=1}^{\infty}$, symbolically $\lim\limits_{k\to\infty} \sup\langle a_n\rangle_{n=k}^{\infty}$, exists.

Proof. By definition, $\langle a_n\rangle_1^{\infty} \supset \langle a_n\rangle_2^{\infty} \supset \langle a_n\rangle_3^{\infty} \supset \cdots$, and applying Exercise 14.11, we arrive at $\sup\langle a_n\rangle_1^{\infty} \geq \sup\langle a_n\rangle_2^{\infty} \geq \sup\langle a_n\rangle_3^{\infty} \geq \cdots$.

Since the original sequence $\langle a_n\rangle_1^{\infty}$ is bounded, $\inf\langle a_n\rangle_1^{\infty}$ must exist. But $\inf\langle a_n\rangle_1^{\infty} \leq \sup\langle a_n\rangle_k^{\infty}$ for any $k \in \mathbb{N}$, as this says that the infimum of a sequence cannot be greater than the supremum of a subsequence lying within it. We see then that $\langle \sup\langle a_n\rangle_{n=k}^{\infty}\rangle_{k=1}^{\infty}$ fulfills the premises of the monotone convergence theorem (Theorem 19.1), and therefore converges. In other words, the limit of $\langle \sup\langle a_n\rangle_1^{\infty}, \sup\langle a_n\rangle_2^{\infty}, \sup\langle a_n\rangle_3^{\infty}, \dots\rangle$ exists for bounded sequences. **QED**

Thus, we have identified $\lim\limits_{k\to\infty} \sup\langle a_n\rangle_{n=k}^{\infty}$. Let's name this new quantity.

The **limit superior** of a bounded sequence $\langle a_n\rangle_1^{\infty}$, written $\limsup\langle a_n\rangle_1^{\infty}$ (pronounced "lim soup"), is $\lim\limits_{k\to\infty} \sup\langle a_n\rangle_{n=k}^{\infty}$.

The argument just before the lemma identified the limit superior of a bounded sequence as the maximum subsequential limit. We will now prove in general that the limit superior equals the maximum subsequential limit of the original sequence.

Theorem 22.1 Let $\langle a_n\rangle_1^{\infty}$ be a bounded sequence. Then $\limsup\langle a_n\rangle_1^{\infty}$ equals the maximum subsequential limit of $\langle a_n\rangle_1^{\infty}$.

Proof. By the Bolzano-Weierstrass theorem for sequences (Theorem 19.4), $\langle a_n\rangle_1^{\infty}$ has at least one convergent subsequence. Let L be the maximum subsequential limit (see Note 1 regarding a detail). Focusing on a subsequence $\langle a_{n_i}\rangle_{i=1}^{\infty}$ that converges to L, we know that for any $\varepsilon > 0$, all but a finite number of terms of $\langle a_{n_i}\rangle_{i=1}^{\infty}$ fall within $(L-\varepsilon, L+\varepsilon)$. Furthermore, since L is the maximum subsequential limit, there can be only finitely many terms a_n of the original sequence (and perhaps none) such that $a_n \geq L+\varepsilon$. Among these terms, let's select the one with the largest subscript and call it a_μ (and if there is none, let $\mu = 0$, since a_0 does not exist). It follows that $a_{\mu+1} < L+\varepsilon$, and from this we infer that $\sup\langle a_n\rangle_{n=\mu+1}^{\infty} < L+\varepsilon$. But the sequence of suprema is monotone decreasing and convergent (Lemma 22.1), so $\lim\limits_{k\to\infty} \sup\langle a_n\rangle_{n=k}^{\infty} \leq \sup\langle a_n\rangle_{n=\mu+1}^{\infty} < L+\varepsilon$.

Next, for any $\varepsilon > 0$, there are infinitely many terms of the sequence $\langle a_n \rangle_1^\infty$ such that $a_n > L - \varepsilon$, (since by the observations above, infinitely many terms belonging to the *subsequence* must satisfy this). Thus, beginning at any such term a_ω, $\sup \langle a_n \rangle_{n=\omega}^\infty > L - \varepsilon$, and this implies that $\lim\limits_{\omega \to \infty} \sup \langle a_n \rangle_{n=\omega}^\infty \geq L$.

The final inequalities in the last two paragraphs tell us that for any $\varepsilon > 0$, $L \leq \limsup \langle a_n \rangle_1^\infty < L + \varepsilon$. Therefore, $\limsup \langle a_n \rangle_1^\infty = L$. **QED**

Note 1: It is possible for a bounded sequence to have not just one or two, but infinitely many subsequential limits. If this is the case, then we would obtain the supremum of the set of subsequential limits, which then is itself a cluster point of the original sequence. But then this supremum becomes a subsequential limit, in fact, the maximum one. The proof of the theorem therefore continues as is.

Example 22.1 In the example illustrated in Figure 22.1, there were terms greater than the limit superior. Let's consider the sequence $\langle \arctan n \rangle_1^\infty$, all of whose terms are less than its limit, $\frac{\pi}{2}$. Hence, the maximum—and only—subsequential limit is the limit itself. Consequently, we expect $\limsup \langle \arctan n \rangle_1^\infty = \lim\limits_{n \to \infty} \arctan n = \frac{\pi}{2}$. The graph in Figure 22.2 shows the situation.

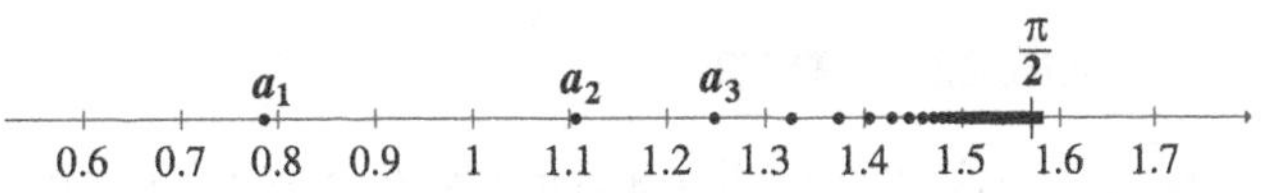

a_1 a_2 a_3 $\frac{\pi}{2}$

0.6 0.7 0.8 0.9 1 1.1 1.2 1.3 1.4 1.5 1.6 1.7

Figure 22.2. $\langle \arctan n \rangle_1^\infty$.

In this case, $\langle \langle a_n \rangle_{n=k}^\infty \rangle_{k=1}^\infty = \langle \langle \arctan n \rangle_1^\infty, \langle \arctan n \rangle_2^\infty, \langle \arctan n \rangle_3^\infty, \ldots \rangle$, which is a sequence of subsequences constructed by deleting successively $a_1, a_2, a_3, \ldots$, the least remaining term of the immediately preceding subsequence. This leads us to $\langle \sup \langle \arctan n \rangle_{n=k}^\infty \rangle_{k=1}^\infty = \langle \frac{\pi}{2} \rangle_{k=1}^\infty$, a constant sequence. None of these suprema belong to the original sequence. Nevertheless, $\lim\limits_{k \to \infty} \sup \langle \arctan n \rangle_{n=k}^\infty = \limsup \langle \arctan n \rangle_1^\infty = \frac{\pi}{2}$, as expected. ■

Let's return to the sequence pictured in Figure 22.1,

$$a_n = \begin{cases} 4 + \dfrac{1}{n} & \text{if } n \text{ is odd} \\ 3 - \dfrac{1}{n} & \text{if } n \text{ is even} \end{cases}.$$

In a process like a mirror imageof what was done there, we may begin to cut off some of the least terms, a_2, a_4, and so on, and investigate the *infima* of the resulting subsequences. Once again, it's easier to cut off each successive term, starting at a_1, and consider $\langle \inf \langle a_n \rangle_{n=k}^\infty \rangle_{k=1}^\infty$. The infima form the sequence $\langle a_2, a_2, a_4, a_4, a_6, a_6, \ldots \rangle$. Guiding ourselves by Figure 22.1, we see that the sequence of infima is monotone increasing, and its limit is 3. This identifies the minimum subsequential limit, and we write $\lim\limits_{k \to \infty} \inf \langle a_n \rangle_{n=k}^\infty = 3$. The corresponding definition follows.

The **limit inferior** of a bounded sequence $\langle a_n \rangle_1^\infty$, written $\liminf \langle a_n \rangle_1^\infty$ (pronounced "lim inf"), is $\lim\limits_{k \to \infty} \inf \langle a_n \rangle_{n=k}^\infty$.

Thus, $\liminf \langle a_n \rangle_1^\infty = 3$ for the sequence in Figure 22.1.

Twins to Lemma 22.1 and Theorem 22.1 are needed at this point.

Lemma 22.2 For any bounded sequence $\langle a_n\rangle_1^\infty$, the sequence of infima $\langle \inf\langle a_n\rangle_{n=k}^\infty\rangle_{k=1}^\infty$ is monotone increasing. Moreover, $\lim_{k\to\infty} \inf\langle a_n\rangle_{n=k}^\infty$, the limit inferior of the sequence, exists.

Proof. By definition, $\langle a_n\rangle_1^\infty \supset \langle a_n\rangle_2^\infty \supset \langle a_n\rangle_3^\infty \supset \cdots$, and applying Exercise 14.11, we arrive at $\inf\langle a_n\rangle_1^\infty \le \inf\langle a_n\rangle_2^\infty \le \inf\langle a_n\rangle_3^\infty \le \cdots$.

Since the original sequence $\langle a_n\rangle_1^\infty$ is bounded, $\sup\langle a_n\rangle_1^\infty$ must exist. But $\sup\langle a_n\rangle_1^\infty \ge \inf\langle a_n\rangle_k^\infty$ for any $k \in \mathbb{N}$ (why?). We see that $\langle \inf\langle a_n\rangle_{n=k}^\infty\rangle_{k=1}^\infty$ fulfills the premises of the Monotone Convergence Theorem (Theorem 19.1), and therefore must converge. In other words, the limit of $\langle \inf\langle a_n\rangle_1^\infty, \inf\langle a_n\rangle_2^\infty, \inf\langle a_n\rangle_3^\infty, \ldots\rangle$ always exists for bounded sequences. This is the number $\lim_{k\to\infty} \inf\langle a_n\rangle_{n=k}^\infty = \liminf\langle a_n\rangle_1^\infty$. **QED**

Theorem 22.2 Let $\langle a_n\rangle_1^\infty$ be any bounded sequence. Then $\liminf\langle a_n\rangle_1^\infty$ is the minimum subsequential limit of $\langle a_n\rangle_1^\infty$.

Proof. This is left as Exercise 22.6. **QED**

Example 22.2 In Example 22.1, we see $\lim_{n\to\infty} \arctan n = \frac{\pi}{2}$. Since $\frac{\pi}{2}$ is the only subsequential limit in town, our latest theorem immediately tells us that $\liminf\langle \arctan n\rangle_1^\infty = \frac{\pi}{2}$. But (for the sake of understanding) look at this through Lemma 22.2. We start with the same sequence as in Example 22.1: $\langle\langle a_n\rangle_{n=k}^\infty\rangle_{k=1}^\infty = \langle\langle \arctan n\rangle_1^\infty, \langle \arctan n\rangle_2^\infty, \langle \arctan n\rangle_3^\infty, \ldots\rangle$. The resulting sequence of infima is $\langle \frac{\pi}{4}, \arctan 2, \arctan 3, \ldots\rangle$, which is monotone increasing, as stated in the lemma. Therefore, $\lim_{k\to\infty} \inf\langle \arctan n\rangle_{n=k}^\infty = \liminf\langle \arctan n\rangle_1^\infty = \frac{\pi}{2}$. ∎

Example 22.3 Consider the sequence $\left\langle \frac{n^2-10n+25}{n^2-6n+10}\right\rangle_1^\infty$ whose terms are graphed in Figure 22.3 as points in $\mathbb{R}^2$. From previous chapters, we know that $\lim_{n\to\infty} a_n = 1$ (a horizontal dashed line in the graph). As can be seen, the sequence becomes monotone increasing beginning at $a_5 = 0$, while its first few terms have disorganized values. We will see that the first few terms cannot have any effect on the limits superior and inferior.

The supremum of the entire sequence equals the maximum term, $a_2 = \frac{9}{2}$, but our procedure requires us to successively cut off a_1 and determine $\sup\langle a_n\rangle_{n=2}^\infty$, then cut off a_2 and determine $\sup\langle a_n\rangle_{n=3}^\infty$, and so on. Hence, the sequence of suprema is $\langle \sup\langle a_n\rangle_{n=k}^\infty\rangle_{k=1}^\infty = \langle \frac{9}{2}, \frac{9}{2}, 4, 1, 1, 1, \ldots\rangle$. We see just three distinct suprema, drawn as dashed lines, and the limit of this sequence is 1. The conclusion is that $\limsup\left\langle \frac{n^2-10n+25}{n^2-6n+10}\right\rangle_1^\infty = 1$.

The numbers are quite different as we evaluate the limit inferior. The infimum of the entire sequence equals the minimum term, $a_5 = 0$; but our procedure requires us to successively cut off a_1, determine $\inf\langle a_n\rangle_{n=2}^\infty$, then cut off a_2, determine $\inf\langle a_n\rangle_{n=3}^\infty$, and so on.... Hence, the sequence of infima is $\langle \inf\langle a_n\rangle_{n=k}^\infty\rangle_{k=1}^\infty = \langle 0, 0, 0, 0, 0, \frac{1}{10}, \frac{4}{17}, \frac{9}{26}, \ldots\rangle$. It is a monotone increasing sequence, shown with solid lines in Figure 22.3, and its limit is 1. Consequently, $\liminf\left\langle \frac{n^2-10n+25}{n^2-6n+10}\right\rangle_1^\infty = 1$. ∎

The last property that we shall prove is becoming apparent in the examples above, and will be reinforced in Exercise 22.3: the limits superior and inferior of a sequence are equal if and only if the sequence converges to a limit.

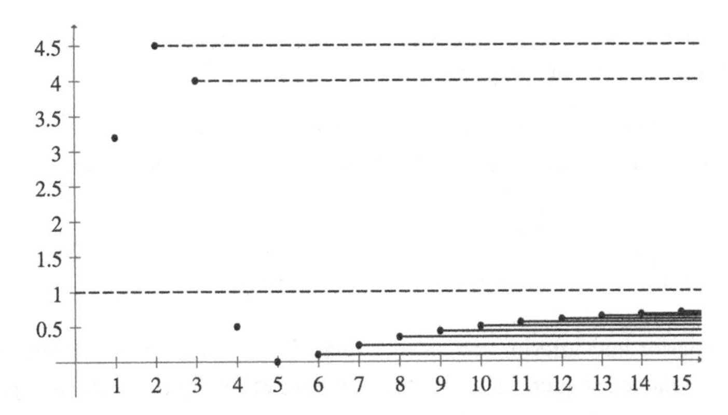

Figure 22.3. The sequence $\langle a_n \rangle_1^\infty = \left\langle \frac{n^2-10n+25}{n^2-6n+10} \right\rangle_1^\infty$ in $\mathbb{R}^2$. It is eventually monotone increasing towards the limit 1. The sequence of suprema $\left\langle \frac{9}{2}, \frac{9}{2}, 4, 1, 1, 1, \ldots \right\rangle$ is represented by exactly three lines (dashed). The sequence of infima $\left\langle 0, 0, 0, 0, 0, \frac{1}{10}, \frac{4}{17}, \frac{9}{26}, \ldots \right\rangle$ must be represented by infinitely many lines, some of them are shown (solid lines).

Theorem 22.3 Let $\langle a_n \rangle_1^\infty$ be any bounded sequence. Then $\liminf \langle a_n \rangle_1^\infty = \limsup \langle a_n \rangle_1^\infty$ if and only if $\lim\limits_{n\to\infty} a_n$ exists, and in those cases, all three limits are equal.

Proof of sufficiency. Let $\lim\limits_{n\to\infty} a_n = L$. Then all subsequences converge to L. In particular, $\inf \langle a_n \rangle_{n=k}^\infty = L$ for every $k \in \mathbb{N}$. Hence, $\lim\limits_{k\to\infty} \inf \langle a_n \rangle_{n=k}^\infty = L$. Likewise, $\sup \langle a_n \rangle_{n=k}^\infty = L$ for every $k \in \mathbb{N}$. Hence, $\lim\limits_{k\to\infty} \sup \langle a_n \rangle_{n=k}^\infty = L$, and we conclude that $\liminf \langle a_n \rangle_1^\infty = \limsup \langle a_n \rangle_1^\infty = \lim\limits_{n\to\infty} a_n$.

Proof of necessity. We are given that $\liminf \langle a_n \rangle_1^\infty = \limsup \langle a_n \rangle_1^\infty = Q$. Concentrating first on the limit superior, $Q = \lim\limits_{k\to\infty} \sup \langle a_n \rangle_{n=k}^\infty$ tells us that for any $\varepsilon > 0$, there exists a supremum $\sup \langle a_n \rangle_{n=k}^\infty$ identified by subscript k_ε such that $\sup \langle a_n \rangle_{n=k_\varepsilon}^\infty < Q + \varepsilon$. On the lower side of the sequence, $Q = \liminf \langle a_n \rangle_1^\infty = \lim\limits_{k\to\infty} \inf \langle a_n \rangle_{n=k}^\infty$, which says that there exists an infimum $\inf \langle a_n \rangle_{n=\ell_\varepsilon}^\infty$ such that $Q - \varepsilon < \inf \langle a_n \rangle_{n=\ell_\varepsilon}^\infty$. We shouldn't expect the identifying subscripts k_ε and ℓ_ε to be equal for the same ε.

Thus, so far we know that $Q - \varepsilon < \inf \langle a_n \rangle_{n=\ell_\varepsilon}^\infty \le \sup \langle a_n \rangle_{n=k_\varepsilon}^\infty < Q + \varepsilon$. Choose the larger of k_ε and ℓ_ε and call it ξ. This implies that all terms of the sequence beginning with a_ξ fall between $Q - \varepsilon$ and $Q + \varepsilon$. Hence, for any $\varepsilon > 0$, all but a finite number of terms of $\langle a_n \rangle_1^\infty$ fall within the interval $(Q - \varepsilon, Q + \varepsilon)$. This means that $\lim\limits_{n\to\infty} a_n$ exists and equals Q. Consequently, $\liminf \langle a_n \rangle_1^\infty = \limsup \langle a_n \rangle_1^\infty = \lim\limits_{n\to\infty} a_n$. **QED**

22.1 Exercises

1. Given the sequence $\langle 2.1, 1.1, 2.01, 1.01, 2.001, 1.001, \ldots \rangle$, write the six terms indicated by $\langle \sup \langle a_n \rangle_{n=k}^\infty \rangle_{k=1}^6$, and then find $\limsup \langle a_n \rangle_1^\infty$ by observation. Explain why the sequence diverges. Hint: You are looking for the decreasing sequence $\langle \sup \langle a_n \rangle_1^\infty, \sup \langle a_n \rangle_2^\infty, \sup \langle a_n \rangle_3^\infty, \sup \langle a_n \rangle_4^\infty, \sup \langle a_n \rangle_5^\infty, \sup \langle a_n \rangle_6^\infty \rangle$.

2. Using the sequence in Exercise 22.1, write the six terms indicated by $\langle \inf \langle a_n \rangle_{n=k}^{\infty} \rangle_{k=1}^{6}$, and then find $\liminf \langle a_n \rangle_1^{\infty}$ by observation.

3. Use the graphs that you draw of the following bounded sequences to evaluate their limits superior and limits inferior. Note which two of them converge to a limit.
   a. $\langle 3^{(-1)^n} \rangle_1^{\infty}$
   b. $\left\langle \dfrac{1}{n} \right\rangle_1^{\infty}$
   c. $\left\langle \dfrac{1}{n} \sin \dfrac{n\pi}{2} + 1 \right\rangle_1^{\infty}$
   d. $\langle \cos n \rangle_0^{\infty}$ (see Example 19.6)

4. The limits superior and inferior are closely related to cluster points of sets (Chapter 14). Show that for bounded sequences $\langle a_n \rangle_1^{\infty}$ that are not eventually constant (we are considering $\langle a_n \rangle_1^{\infty}$ as an infinite set of points), $\limsup \langle a_n \rangle_1^{\infty}$ locates the maximum cluster point, and $\liminf \langle a_n \rangle_1^{\infty}$ locates the minimum cluster point. Hint: Recall Exercise 19.14.

5. a. Simplify $\langle \sup \langle a_n \rangle_{n=k}^{4} \rangle_{k=1}^{4}$. Hint: You're only dealing with four terms.
   b. Let $\langle a_n \rangle_1^{\infty}$ be a bounded, monotone increasing sequence. Simplify $\langle \sup \langle a_n \rangle_{n=k}^{\infty} \rangle_{k=1}^{4}$.
   c. Let $\langle a_n \rangle_1^{\infty}$ be a bounded, monotone decreasing sequence. Simplify $\langle \sup \langle a_n \rangle_{n=k}^{\infty} \rangle_{k=1}^{4}$.
   d. Let $\langle a_n \rangle_1^{\infty}$ be a bounded, monotone decreasing sequence. Simplify $\langle \inf \langle a_n \rangle_{n=k}^{\infty} \rangle_{k=1}^{4}$.
   e. Let $\langle a_n \rangle_1^{\infty}$ be a bounded, monotone increasing sequence. Simplify $\langle \inf \langle a_n \rangle_{n=k}^{\infty} \rangle_{k=1}^{4}$.

6. Prove Theorem 22.2 in a parallel fashion to Theorem 22.1.

7. In all the examples of this chapter, it seems that $\limsup \langle a_n \rangle_1^{\infty}$, which is $\lim_{k \to \infty} \sup \langle a_n \rangle_{n=k}^{\infty}$, is also equal to $\inf_k (\sup \langle a_n \rangle_{n=k}^{\infty})$, that is, the final limit operation can be replaced by the infimum operation. Prove that this is true in general. Likewise, it is true that $\liminf \langle a_n \rangle_1^{\infty} = \lim_{k \to \infty} \inf \langle a_n \rangle_{n=k}^{\infty} = \sup_k (\inf \langle a_n \rangle_{n=k}^{\infty})$.

23

Limits of Functions

The limit of a function f is related to the limit of a sequence. One distinction is that the latter's limit (if it exists) is found exclusively as $n \to \infty$, whereas the limit of a function can exist in a neighborhood of a real number a, as well as when $x \to \infty$. The concept of a limit replaced the "vanishingly small ratio" of quantities that produced a derivative as Leibniz and Newton had envisioned. The modern definition of limit first appeared in the writings of Cauchy (see Chapter 36 for more details). The algebraic form here was perhaps first written by a contemporary of Karl Weierstrass, the analyst Eduard Heine (1821–1881), in 1872:

> A function f has a **limit** L at a real number a if and only if given any $\varepsilon > 0$, we can find a corresponding $\delta > 0$ such that if $0 < |x - a| < \delta$ (with $x \in \mathcal{D}_f$), then $|f(x) - L| < \varepsilon$. In such a case, we write $\lim_{x \to a} f(x) = L$.

To write the definition in terms of neighborhoods, we first define a **deleted neighborhood** of a number a as a neighborhood of a with its midpoint a excluded from it. Three analytical notations for this are $(a - \delta, a) \cup (a, a + \delta)$, $(a - \delta, a + \delta)\backslash\{a\}$, and $\{x : 0 < |x - a| < \delta\}$, where the positive number δ is the "radius" of the neighborhood. (The last form is used in the definition.) Since none of these forms is concise, let us create the symbol $\mathbf{D}_a^\delta$ to represent the deleted neighborhood of a with radius δ. We can now formulate an equivalent definition of the limit of a function.

> A function f has a **limit** L at a real number a if and only if given any $\mathbf{N}_L^\varepsilon$, we can find a corresponding $\mathbf{D}_a^\delta$ such that if $x \in \mathbf{D}_a^\delta \cap \mathcal{D}_f$, then $f(x) \in \mathbf{N}_L^\varepsilon$. In such a case, we write $\lim_{x \to a} f(x) = L$.

The definition of a limit does not require a to belong to the domain of f. However, it is important that $x \neq a$, hence the introduction of deleted neighborhoods. Also, it will be important to understand that the value L of the limit—when it exists—is independent of the value $f(a)$, even when *it* exists.

Note 1: When both the limit and the function exist at a, and are equal, we have the enormously important case of continuity of the function at a. This will be studied in the next chapter.

Example 23.1 Prove that $\lim_{x\to 0}|x| = 0$ (the domain is $\mathbb{R}$ here). Trying out the first definition, we must verify that given any $\varepsilon > 0$, we can find a corresponding $\delta > 0$ such that if $0 < |x - 0| < \delta$, then $||x| - 0| < \varepsilon$. Simplifying, we wish to be certain that $|x| < \varepsilon$ whenever $0 < |x| < \delta$. This will be the case if we set δ equal to ε. Thus, the limit $L = 0$. The absolute value function has limit 0 at $a = 0$.

Looking at the second definition, we must verify that given any neighborhood of $L = 0$, namely $(-\varepsilon, \varepsilon)$, we can find a corresponding $\mathbf{D}_0^\delta = (-\delta, \delta)$ such that if $x \in \mathbf{D}_0^\delta$, then $|x| \in (-\varepsilon, \varepsilon)$. If we set δ equal to ε, then $x \in (-\delta, \delta)$ is equivalent to $|x| \in [0, \delta)$, which is contained in $(-\varepsilon, \varepsilon)$. Therefore, the absolute value function has the limit $L = 0$ at $a = 0$. The moral is that the two definitions always shake logical hands. ■

Example 23.2 The function in Figure 23.1 is said to have a **removable** or **gap discontinuity** at $a = 2$; it is drawn as a small circle at $\langle 2, 5\rangle$ on the curve. But remember, $\langle 2, 10\rangle \in f$ and not $\langle 2, 5\rangle$. Nevertheless, we shall prove that $\lim_{x\to 2} f(x) = 5$.

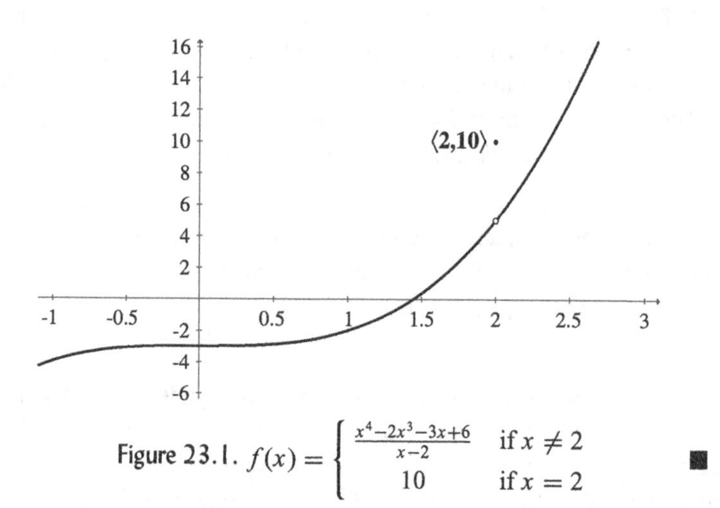

Figure 23.1. $f(x) = \begin{cases} \frac{x^4-2x^3-3x+6}{x-2} & \text{if } x \neq 2 \\ 10 & \text{if } x = 2 \end{cases}$ ■

Is the ordinate of the gap actually $y = 5$? Direct substitution does not give an answer. However, the numerator of the fraction factors as $(x^3 - 3)(x - 2)$ (incidentally, how does one determine that without a calculator?). Thus, for all $x \neq 2$, $f(x) = x^3 - 3$. Since for all x near 2, $f(x)$ is near 5, it seems reasonable that the gap is located at $\langle 2, 5\rangle$. The sense of nearness is made precise by the definition of the limit, and the limit will indeed be the ordinate of the gap.

According to the second definition, we are to show that any time we pick abscissas within $\mathbf{D}_2^\delta$, the function will always yield ordinates within a neighborhood of radius ε of 5. Or, by the first definition, we are to show that given any $\varepsilon > 0$, we can find a corresponding $\delta > 0$ such that if $0 < |x - 2| < \delta$, then $|(x^3 - 3) - 5| < \varepsilon$.

Preparation. Starting with the last inequality, we get $|x^3 - 8| = |x - 2||x^2 + 2x + 4| < \varepsilon$. We'll need the inequality

$$|x - 2||x^2 + 2x + 4| < \varepsilon. \tag{1}$$

Now, the factor $|x-2|$ can be as small as we wish (but not zero) because it is less than δ, but the factor $|x^2+2x+4|$ must not be exceedingly large, in order that the product in (1) be less than our selected ε. We must therefore put a bound on the size of the factor $|x^2+2x+4|$.

Expanding $|x-2|<\delta$, we get $2-\delta<x<2+\delta$. If we temporarily require $\delta<1$, then we restrict x to the neighborhood $1<x<3$. And then, $1<x^2<9$ and $2<2x<6$, so that $7<x^2+2x+4<19$. (You may verify this by graphing $y=x^2+2x+4$ over the interval $1<x<3$.) We now have a bound on the size of $|x^2+2x+4|$, namely,

$$|x^2+2x+4|<19 \text{ if } \delta<1. \tag{2}$$

Using $|x-2|<\delta$, (2) yields $|x-2||x^2+2x+4|<19\delta$. Comparing this with (1), we realize that we should set $19\delta=\varepsilon$, or $\delta=\frac{\varepsilon}{19}$, provided someone gives us an ε small enough so that $\delta<1$. In other words, if $\varepsilon<19$, then $\delta=\frac{\varepsilon}{19}$ is our corresponding δ. What if $\varepsilon\geq 19$? We will consider that case shortly.

Proof. Given any $\varepsilon>0$, first check that $\varepsilon<19$. Suppose it is. Then the corresponding $\delta=\frac{\varepsilon}{19}<1$. Now, select x such that $0<|x-2|<\delta$, that is, x is inside $\mathbf{D}_2^{\delta}$. Then our bound $|x^2+2x+4|<19$ takes effect. Then (1) is true because $|x-2||x^2+2x+4|<19\delta=\varepsilon$. But then, $|f(x)-5|=|x^3-8|=|x-2||x^2+2x+4|<\varepsilon$. That is, all function values fall within a neighborhood of radius ε of $y=5$.

What if $\varepsilon\geq 19$? Then it is only necessary to fix $\delta=1$. For, if once again $0<|x-2|<\delta=1$, then (2) will give us $|x^2+2x+4|\leq 19$. Then $|f(x)-5|=|x-2|\,|x^2+2x+4|<19\delta=19\leq\varepsilon$. Again, we have landed within the ε-neighborhood of $y=5$.

Joining the two cases together, we find that given any $\varepsilon>0$, the corresponding $\delta=\min\{\frac{\varepsilon}{19},1\}$. It follows that if $0<|x-2|<\delta$, then $|f(x)-5|<\varepsilon$. Therefore, $\lim_{x\to 2} f(x)=5$.

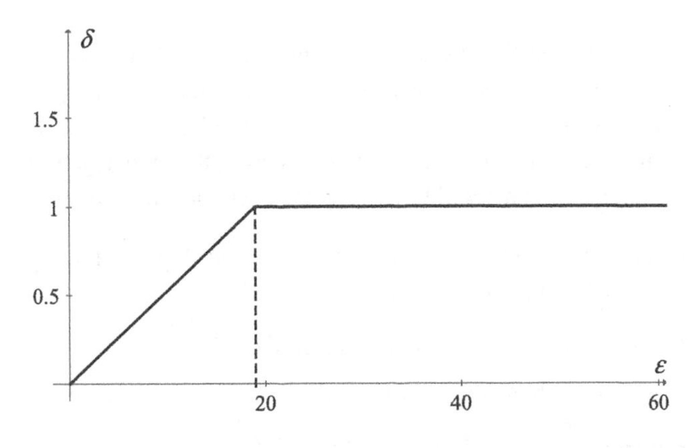

Figure 23.2. Graph of the ε-to-δ correspondence $\delta=\min\{\frac{\varepsilon}{19},1\}$.

$f(2)$ could have been defined as any value other than 10, or left undefined, but $\lim_{x\to 2} f(x)=5$ is independent of such considerations. ■

Note 2: Discontinuity of a function will be defined in the next chapter. For now, let terms such as "removable discontinuity" be indications of peculiar graphic behavior.

Example 23.3 Figure 23.3 shows the graph of the important function $s(\theta)=\frac{\sin\theta}{\theta}$ with $\mathcal{D}_s=\mathbb{R}\backslash\{0\}$. It displays a removable discontinuity at $\theta=0$. We will determine $\lim_{\theta\to 0} s(\theta)$ in the last theorem of this chapter. For now, investigate the limit by evaluating $s(0.05)$ and $s(-0.003)$, for instance. It's easy to guess what the limit should be, and the graph supports your guess. But a guess it must remain, until an analytical proof is provided. ■

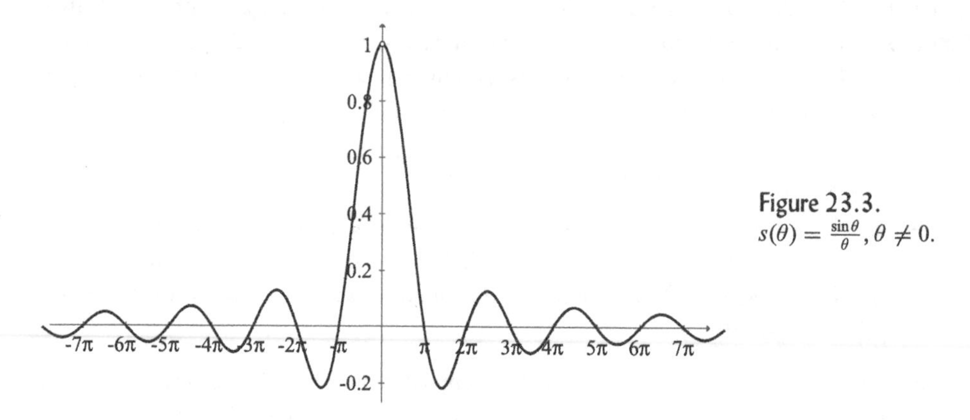

Figure 23.3.
$s(\theta) = \frac{\sin\theta}{\theta}, \theta \neq 0.$

23.1 Non-Existence of a Limit

It is important to understand the negation of the definition of limit at a point. In writing it, recall that the negation of the universal "for any" is the particular "for some," and vice versa. Leaving the details to Note 3, we state:

A function f **does not have a limit** L at a real number $x = a$ if and only if for some $\varepsilon^* > 0$, and for any $\delta > 0$, there exists at least one $x_\delta \in \mathcal{D}_f$ that while satisfying $0 < |x_\delta - a| < \delta$, nevertheless yields $|f(x_\delta) - L| \geq \varepsilon^*$.

The notation x_δ is meant to remind us that the abscissa that ruins the claim that L is the limit depends upon our choice of δ. To put it a little less precisely, L cannot be the limit if $f(x_\delta)$ refuses to remain in the ε^*-neighborhood of L. The negation neatly disposes of any particular L as a candidate for the limit. If we want to show that f has no limit whatsoever, then the negation must be applied to an *arbitrary* real number L. This will be done in the examples that follow.

Note 3: The negation of the existence of a limit L is carried out in quantified logic following a plan similar to Example 1.11. First, negate the definition's statement:

$$\textit{not-}(\forall\varepsilon, (\text{if } \varepsilon > 0, \text{then } (\exists\delta > 0 \text{ such that } \forall x, (\text{if } x \in \mathcal{D}_f \text{ and } 0 < |x - a| < \delta, \text{then} |f(x) - L| < \varepsilon)))).$$

Next, negate the first universal quantifier:

$$\exists\varepsilon > 0 \text{ such that } \textit{not-}(\exists\delta > 0 \text{ such that } \forall x, (\text{if } x \in \mathcal{D}_f \text{ and } 0 < |x - a| < \delta, \text{then } |f(x) - L| < \varepsilon)).$$

Next, negate the existential quantifier:

$$\exists\varepsilon > 0 \text{ such that } \forall\delta, (\text{if } \delta > 0, \text{then } \textit{not-}(\forall x, (\text{if } x \in \mathcal{D}_f \text{ and } 0 < |x - a| < \delta, \text{then } |f(x) - L| < \varepsilon))).$$

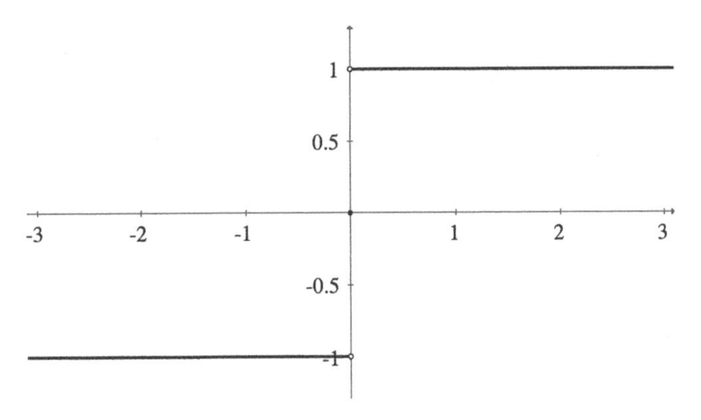

Figure 23.4.

$$\operatorname{sgn} x = \begin{cases} 1 & \text{if } x > 0 \\ 0 & \text{if } x = 0 \\ -1 & \text{if } x < 0 \end{cases}.$$

Now, negate the last universal quantifier:

$$\exists \varepsilon > 0 \text{ such that } \forall \delta, (\text{if } \delta > 0, \text{then } (\exists x \in \mathcal{D}_f$$
$$\text{and } 0 < |x - a| < \delta \text{ and } |\, f(x) - L| \geq \varepsilon)).$$

The defining statement above just plumps up this conclusion.

The **signum function** is defined and graphed in Figure 23.4. Its output indicates the parity or "sign" of a nonzero number. Its graph shows a **jump discontinuity** at $x = 0$, so no limit should exist there. (Perforce, this is not a removable discontinuity.)

Example 23.4 Prove that $\lim_{x \to 0} \operatorname{sgn} x$ does not exist. We must present an $\varepsilon^* > 0$ such that for any L and for any $\delta > 0$, there exists at least one x satisfying $0 < |x_\delta - 0| < \delta$ and such that, alas, $|\operatorname{sgn} x_\delta - L| \geq \varepsilon^*$. That is, while remaining inside the deleted δ-neighborhood of 0, the signum function will throw us outside the ε^*-neighborhood of L. Notice the size of the jump. Let's therefore pick $\varepsilon^* = \frac{1}{2}$.

Take any $\delta > 0$, form $\mathbf{D}_0^\delta$, and suppose L is any positive claimed limit. Select first a negative $x_\delta \in \mathbf{D}_0^\delta$. Then, $|\operatorname{sgn} x_\delta - L| = |-1 - L| = 1 + L > 1 > \varepsilon^*$. The function has forced us outside of the ε^*-neighborhood. So, suppose L is any negative claimed limit. Thinking ahead, select a positive $x_\delta \in \mathbf{D}_0^\delta$. Then, $|\operatorname{sgn} x_\delta - L| = 1 - L > 1 > \varepsilon^*$. Again, we are outside the ε^*-neighborhood.

As a last resort, we suppose $L = 0$. Select a nonzero $x_\delta \in \mathbf{D}_0^\delta$. Then, $|\operatorname{sgn} x_\delta - L| = |\pm 1 - 0| = 1 > \varepsilon^*$. Hence, we can always find at least one x_δ such that $\operatorname{sgn} x_\delta$ falls outside of the $\varepsilon^* = \frac{1}{2}$ neighborhood of L. Therefore, no limit exists for the signum function at $a = 0$. ■

Example 23.5 Prove that $\lim_{x \to 0} e^{1/x}$ does not exist. The graph of the function $g(x) = e^{1/x}$, whose domain is $\mathbb{R} \backslash \{0\}$, is in Figure 23.5. It shows an **infinite jump discontinuity** at $a = 0$, so no limit should exist there.

As in Example 23.4, we must present an $\varepsilon^* > 0$ such that for any L and for any $\delta > 0$, there exists some x_δ satisfying $0 < |x_\delta - 0| < \delta$ such that (perhaps unfortunately), $|g(x_\delta) - L| \geq \varepsilon^*$. Looking at the graph, let's take $\varepsilon^* = 1$.

Suppose L is any claimed limit. We will need the inequality $e^u > u$ (see Exercise 23.8). Thus, $e^{1/x} - L > \frac{1}{x} - L$. For any $\mathbf{D}_0^\delta$, we can always find some x_δ in this deleted neighborhood—we will actually need x_δ to be positive—such that $\frac{1}{x_\delta} > L + 1$. This leads us to $e^{1/x_\delta} - L > \frac{1}{x_\delta} - L > 1 = \varepsilon^*$. Hence, $|e^{1/x_\delta} - L| > \varepsilon^*$ for some x_δ in any $\mathbf{D}_0^\delta$. This proves that $g(x) = e^{1/x}$ has no limit at $a = 0$.

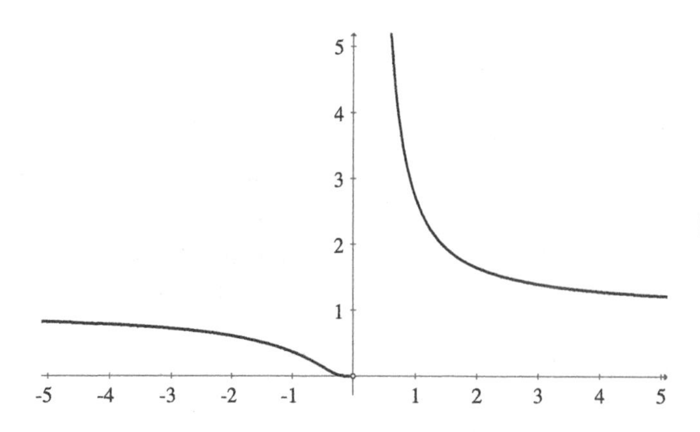

Figure 23.5. $g(x) = e^{1/x}, x \neq 0.$

To see the numerical machinery working here, suppose that $L = 10.4$. For any $\mathbf{D}_0^\delta$, take x_δ to be positive, in $\mathbf{D}_0^\delta$, and such that $\frac{1}{x} > 10.4 + 1 = 11.4$. That is, take $x < \frac{1}{11.4} \approx 0.0877$ and in $\mathbf{D}_0^\delta$. We may consider $\mathbf{D}_0^{0.05}$ for instance, and select $x_{0.05} = 0.049$ for instance. Then $|e^{1/0.049} - 10.4| \approx 7.3 \times 10^8$, far greater than $\varepsilon^* = 1$. Since 10.4 was terribly off the mark, perhaps $L = 999$ instead. Again, for any $\mathbf{D}_0^\delta$, take x_δ to be positive, within $\mathbf{D}_0^\delta$, and such that $x < \frac{1}{1000} = 10^{-3}$. Consider $\mathbf{D}_0^{0.2}$ for instance, and select $x_{0.2} = 0.0009 \in \mathbf{D}_0^{0.2}$ for instance. Now $|e^{1/0.0009}| - 999|$ is astronomically greater than ε^*. We can see that g may send us outside of $(L-1, L+1)$ regardless of the value of L. ■

As the graph in Figure 23.5 shows, the reason that $g(x) = e^{1/x}$ fails to have a limit at $a = 0$ is that the function's range is unbounded over every neighborhood $\mathbf{N}_0^\delta$ (bounded and unbounded functions were mentioned briefly in Chapter 12 in connection with unbounded sets). The analytical definitions for these behaviors are:

A function f is **bounded** over a neighborhood $\mathbf{N}_a^\delta$ if and only if there exists some $B > 0$ such that for all $x \in \mathcal{D}_f \cap \mathbf{N}_a^\delta$, $|f(x)| \leq B$.	A function f is **unbounded** over every neighborhood $\mathbf{N}_a^\delta$ if and only if for any (think "small") $\mathbf{N}_a^\delta$ and any $B > 0$, there exists at least one $x_B \in \mathcal{D}_f \cap \mathbf{N}_a^\delta$ that yields $|f(x_B)| \geq B$.

The first definition identifies a bounded function (in its entirety) when the neighborhood becomes all $\mathbb{R}$. The second definition allows, but doesn't require, that $x = a$ (the first definition is trivially true at any $a \in \mathcal{D}_f$). $\lim_{x \to a} f(x)$ does not exist when f is unbounded at $x = a$. As in previous definitions, the subscript B indicates that x_B usually depends on the value of B. Under the special circumstances that f is unbounded at $x = a$ because $f(x_B) \geq B$ for all $x_B \in \mathcal{D}_f \cap \mathbf{N}_a^\delta$, we symbolically write that $\lim_{x \to a} f(x) = \infty$. Similarly, if $f(x_B) \leq -B$ for all such x_B, we write $\lim_{x \to a} f(x) = -\infty$.

23.2 Other Types of Limits

The function in Figure 23.4 has no limit at $a = 0$, but still, the two "sides" of $\operatorname{sgn} x$ do have limits. In Figure 23.5, the part of $g(x) = e^{1/x}$ in Quadrant II should have a limit at $a = 0$. It makes sense therefore to state the following definition.

A function f will have a **limit** L **from the left** of $x = a$ if and only if given any $\varepsilon > 0$, we can find a corresponding $\delta > 0$ such that if $x \in (a - \delta, a) \cap \mathcal{D}_f$, then $|f(x) - L| < \varepsilon$.

In such a case, we write $\lim_{x \to a-} f(x) = L$.

Considering the side of a function to the right of a, we state a corresponding definition.

Function f will have a **limit L from the right** of $x = a$ if and only if given any $\varepsilon > 0$, we can find a corresponding $\delta > 0$ such that if $x \in (a, a + \delta) \cap \mathcal{D}_f$, then $| f(x) - L| < \varepsilon$.

In this case, we write $\lim_{x \to a+} f(x) = L$. Exercises 23.9, 23.13, and 23.14 investigate the limits from the left and right. Collectively, we call $\lim_{x \to a-} f(x)$ and $\lim_{x \to a+} f(x)$ **one-sided limits.**

Looking once more at g in Figure 23.5, we see what may be a horizontal asymptote at $y = 1$. And in Figure 23.4, do the lines $y = \pm 1$ qualify as horizontal asymptotes of $\operatorname{sgn} x$? The following definitions will settle the issue.

A function f will have a **limit L at $+\infty$** if and only if given any $\varepsilon > 0$, we can find a corresponding $N_\varepsilon > 0$ such that for all $x > N_\varepsilon$ (with $x \in \mathcal{D}_f$), then $|f(x) - L| < \varepsilon$. In such a case, we write

$$\lim_{x \to \infty} f(x) = L.$$

A function f will have a **limit L at $-\infty$** if and only if given any $\varepsilon > 0$, we can find a corresponding $N_\varepsilon > 0$ such that for all $x < -N_\varepsilon$ (with $x \in \mathcal{D}_f$), then $|f(x) - L| < \varepsilon$. In such a case, we write

$$\lim_{x \to -\infty} f(x) = L.$$

One may say that f has the limit L in a neighborhood of $+\infty$ or $-\infty$, correspondingly. Graphically, the existence of a limit L at $+\infty$ or $-\infty$ appears as a **horizontal asymptote** at $y = L$.

Example 23.6 To show that $y = \operatorname{sgn} x$ does have horizontal asymptotes at $y = \pm 1$, note that $| \operatorname{sgn} x - 1| = 0$ whenever x is positive. To put this in terms of the definition, let $\varepsilon > 0$ be given, and take $N_\varepsilon = \frac{1}{\varepsilon}$. Then for all $x > N_\varepsilon$, we find that $0 = | \operatorname{sgn} x - 1| < \varepsilon$. Thus, $\lim_{x \to \infty} \operatorname{sgn} x = 1$, and $y = 1$ is a horizontal asymptote of the function. A parallel argument will show that $\lim_{x \to -\infty} \operatorname{sgn} x = -1$, and $y = -1$ is the other horizontal asymptote. This is an example of a function that equals its horizontal asymptotes infinitely many times. ■

Our first theorem relates the limit of a sequence of function values $\langle f(x_n)\rangle_1^\infty$ to the function's limit itself.

Theorem 23.1 Sequential Criteria for Limits We are given a function $f : \mathbb{R} \to \mathbb{R}$. All sequences $\langle x_n\rangle_1^\infty$ mentioned here will belong to the domain of f.

A The function f has a limit at $x = a$, let's say $\lim_{x \to a} f(x) = L$, if and only if for every sequence such that $x_n \to a$ and $x_n \neq a$, we have $\lim_{n \to \infty} f(x_n) = L$.

B The function f has a limit at $+\infty$, say $\lim_{x \to \infty} f(x) = L$, if and only if for every sequence such that $x_n \to \infty$, we have $\lim_{n \to \infty} f(x_n) = L$.

C The function f has a limit at $-\infty$, say $\lim_{x \to -\infty} f(x) = L$, if and only if for every sequence such that $x_n \to -\infty$, we have $\lim_{n \to -\infty} f(x_n) = L$.

Proof of sufficiency. **A** We will prove the contrapositive of the sufficiency statement: if f does not have limit L in any neighborhood of $x = a$, then there exists a sequence such that $x_n \to a$, with $x_n \neq a$, and $\langle f(x_n)\rangle_1^\infty$ does not have limit L.

By what we have learned above, there does exist an $\varepsilon^* > 0$ such that regardless of how small $\delta > 0$ may be, we can find a number $x \in \mathcal{D}_f$ such that $0 < |x - a| < \delta$ and $|f(x) - L| \geq \varepsilon^*$. In particular, $0 < |x - a| < \frac{1}{n}$ and $|f(x) - L| \geq \varepsilon^*$ can be made simultaneously true for each $n \in \mathbb{N}$ by finding an x corresponding to each n, and calling it x_n. This creates a sequence $\langle x_n\rangle_1^\infty$ such that $x_n \to a$, $x_n \neq a$, and yet $|f(x_n) - L| \geq \varepsilon^*$. This in turn indicates that $\langle f(x_n)\rangle_1^\infty$ does not have limit L at $x = a$, and we have shown the contrapositive to be true.

B To prove the contrapositive in this case, let it be given that f does not have limit L in any neighborhood of $+\infty$. This means that there exists an $\varepsilon^* > 0$ such that regardless of how large $n \in \mathbb{N}$ is, we can find some $x^* > n$ (and in $\mathcal{D}_f$) such that $|f(x^*) - L| \geq \varepsilon^*$. Now, for each natural number n, denote one of the numbers x^* (or maybe the only such one) by x_n. This creates a sequence $\langle x_n\rangle_1^\infty$ such that $x_n \to \infty$, and yet $|f(x_n) - L| \geq \varepsilon^*$. This implies that $\langle f(x_n)\rangle_1^\infty$ does not have limit L at $+\infty$, thus proving the contrapositive.

C An argument parallel to the previous one shows the contrapositive to be true in this case.

Proof of necessity. **A** We are given that $\lim_{x\to a} f(x) = L$. That is, given any $\varepsilon > 0$, there does exist a $\delta > 0$ such that $0 < |x - a| < \delta$ implies $|f(x) - L| < \varepsilon$. Let $\langle x_n\rangle_1^\infty$ be a sequence with $x_n \to a$ and $x_n \neq a$. Then we also know that corresponding to our δ, there exists an N_ε such that for all subscripts $n > N_\varepsilon$, $0 < |x_n - a| < \delta$. This implies that $|f(x_n) - L| < \varepsilon$ for all $n > N_\varepsilon$, which means that $\lim_{n\to\infty} f(x_n) = L$.

B Given that $\lim_{x\to\infty} f(x) = L$, we know that for any $\varepsilon > 0$, there does exist an $N_\varepsilon > 0$ such that for all $x > N_\varepsilon$, $|f(x) - L| < \varepsilon$. Let $\langle x_n\rangle_1^\infty$ be a sequence with $x_n \to \infty$ and $x_n \neq a$. Then we infer that for any $\varepsilon > 0$ and for all $x_n > N_\varepsilon$, $|f(x_n) - L| < \varepsilon$. This means that $\lim_{n\to\infty} f(x_n) = L$.

C An argument parallel to the previous one shows the statement to be true in this case. **QED**

The theorem just proved may be teamed up with the pinching theorem for sequences (Theorem 20.7) to provide a corresponding pinching theorem for functions, to be proved as a corollary in Exercise 23.12.

Example 23.7 Return to the function f in Example 23.2, and use a sequential criterion to prove once again that $\lim_{x\to 2} f(x) = 5$. Let $\langle x_n\rangle_1^\infty$ be any sequence such that $x_n \to 2$ and $x_n \neq 2$. We need to show that $\lim_{n\to\infty} f(x_n) = 5$. By the same algebra as in that example, $\lim_{n\to\infty} f(x_n) = \lim_{n\to\infty}(x_n^3 - 3)$ (notice how the algebra emphasized that $x_n \neq 2$), so we must show that $\lim_{n\to\infty}(x_n^3 - 3) = 5$. This is easy to do, thanks to several theorems about limits for sequences in Chapter 20. Thus, $\lim_{n\to\infty}(x_n^3 - 3) = 2^3 - 3 = 5$. Since $\langle x_n\rangle_1^\infty$ was arbitrary except for the requirements of Theorem 23.1A, we have proved that $\lim_{x\to 2} f(x) = 5$. ■

Example 23.8 Show that $g(x) = e^{1/x}$ has $\lim_{x\to\infty} g(x) = \lim_{x\to-\infty} g(x) = 1$; Figure 23.5 supports this. In Exercise 35.5f, you will prove that if $u > v$, then $e^u > e^v$, a fact that we will need soon.

Review Exercise 20.9a and its proof. Since $e > 1$, the exercise informs us that $\lim_{n\to\infty} e^{1/n} = 1$. But this is the limit of just one sequence, $\langle e^{1/n}\rangle_1^\infty$.

The proof that the function $y = e^{1/x}$ has limit 1 at $+\infty$ is next. Using the properties of real numbers in Chapter 13, for any real $x \geq 1$, $n = \lfloor x \rfloor$ has the property that $0 < n \leq x < n+1$. Clearly, as $x \to \infty$, so does n. We are also sure that $\frac{1}{n+1} < \frac{1}{x} \leq \frac{1}{n}$, implying that $e^{1/(n+1)} < e^{1/x} \leq e^{1/n}$. Now, each term of an *arbitrary* sequence $\langle x_k\rangle_1^\infty$ that diverges to $+\infty$ will satisfy $e^{1/(n+1)} < e^{1/x_k} \leq e^{1/n}$ for an appropriate $n = n_k$. Thus, $\lim_{k\to\infty} e^{1/x_k} = \lim_{n\to\infty} e^{1/n} = 1$, and Theorem 23.1B lets us conclude that $\lim_{x\to\infty} e^{1/x} = \lim_{n\to\infty} e^{1/n} = 1$.

Once you prove that $\lim_{x\to-\infty} g(x) = 1$ in Exercise 23.10, we will be certain that $y = 1$ is the only horizontal asymptote of this function. ■

Lemma 23.1 $\lim_{\theta\to 0} \sin\theta = 0$ and $\lim_{\theta\to 0} \cos\theta = 1$.

Proof. Begin with the observation that any chord of a circle has a shorter length than that of the minor arc associated with it. Therefore, in the unit circle in Figure 23.6, and for any $\theta \in \left(0, \frac{\pi}{2}\right)$, we see that $0 < 2\sin\theta < 2\theta$; thus $0 < \sin\theta < \theta$. By the pinching theorem for functions (see Exercise 23.12), $\lim_{\theta\to 0+} \sin\theta = 0$. Since $\sin(-\theta) = -\sin\theta$ (the sine is an odd function), we also have $\lim_{\theta\to 0-} \sin\theta = 0$. Hence, by the result of Exercise 23.9, $\lim_{\theta\to 0} \sin\theta = 0$. In Exercise 23.16, show furthermore that $\lim_{\theta\to 0} \cos\theta = 1$. **QED**

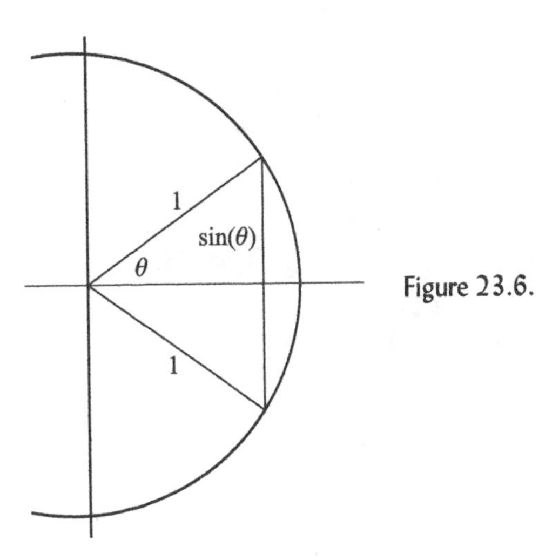

Figure 23.6.

Theorem 23.2 $\lim_{\theta\to 0} \frac{\sin\theta}{\theta} = 1$ (see Figure 23.3).

Proof. Inscribe a regular n-sided polygon in a unit circle. In Figure 23.7, AC and CD are two sides of the polygon. Segment AB is half of a side, with length $\sin\frac{\pi}{n}$. For $n \geq 3$, let the central angle θ be such that $\frac{\pi}{n+1} < \theta \leq \frac{\pi}{n}$. This, and the fact that the sine function is strictly increasing over $\left(0, \frac{\pi}{2}\right)$, give us the inequalities

$$\frac{\sin\frac{\pi}{n+1}}{\frac{\pi}{n}} \leq \frac{\sin\frac{\pi}{n+1}}{\theta} < \frac{\sin\theta}{\theta} < \frac{\sin\theta}{\frac{\pi}{n+1}} \leq \frac{\sin\frac{\pi}{n}}{\frac{\pi}{n+1}}. \tag{3}$$

We will now assume that the circumference of a circle can be approximated to as high a precision as desired by increasing n, the number of sides in our polygon. Thus, the polygon's perimeter, which is $2n \sin \frac{\pi}{n}$ according to Figure 23.7, approaches the circle's circumference of 2π as n increases. In the language of limits, we may write

$$\lim_{n\to\infty} \frac{2n \sin \frac{\pi}{n}}{2\pi} = 1, \quad \text{or} \quad \lim_{n\to\infty} \frac{\sin \frac{\pi}{n}}{\frac{\pi}{n}} = 1.$$

Knowing this, the right side of (3) yields

$$\lim_{n\to\infty} \frac{\sin \frac{\pi}{n}}{\frac{\pi}{n+1}} = \lim_{n\to\infty} \left(\frac{n+1}{n} \frac{\sin \frac{\pi}{n}}{\frac{\pi}{n}} \right) = 1,$$

and on the left side, $\lim_{n\to\infty} \frac{\sin \frac{\pi}{n+1}}{\frac{\pi}{n}} = \lim_{n\to\infty} \left(\frac{n}{n+1} \frac{\sin \frac{\pi}{n+1}}{\frac{\pi}{n+1}} \right) = 1.$

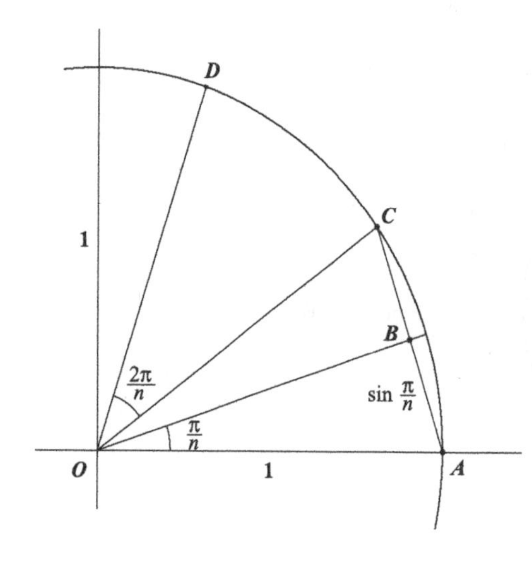

Figure 23.7.

Since each term of an arbitrary sequence $\langle \theta_k \rangle_1^\infty$ that converges to zero from the right will satisfy (3) for an appropriate $n = n_k$, we have that

$$\lim_{k\to\infty} \frac{\sin \theta_k}{\theta_k} = \lim_{n_k\to\infty} \frac{\sin \frac{\pi}{n_k}}{\frac{\pi}{n_k}} = 1.$$

Theorem 23.1A applies now, telling us that $\lim_{\theta\to 0+} \frac{\sin\theta}{\theta} = 1$.

As the last detail, since

$$\lim_{\theta \to 0-} \frac{\sin\theta}{\theta} = \lim_{\theta \to 0+} \frac{\sin(-\theta)}{-\theta} = \lim_{\theta \to 0+} \frac{\sin\theta}{\theta} = 1,$$

we conclude that $\lim_{\theta \to 0} \frac{\sin\theta}{\theta} = 1$. (See also Exercise 23.17.) **QED**

Note 4: The proof just finished relies critically on the definition of the arclength of a curve as the limit of the total length of secants properly drawn to the curve. Fortunately, a circle is a smooth enough curve that the inscribed regular polygon provides the needed secants. It was Archimedes who first used this idea as a geometric method to approximate π, in his brilliant treatise, *On the Measurement of the Circle.* A captivating overview of Archimedes and his work is found in Chapter 6 of [1]. The common alternative proof of Theorem 23.2 found in calculus texts relies on areas, an equally geometric idea.

A strictly analytical proof of Theorem 23.2 would be quite abstract. The sine function would first be defined as the solution to a differential equation, or as a function defined through an integral. Then, analytical properties of the function—properties that correspond to the well-known trigonometric identities—would be derived. The derivative of the sine would be found, and our limit would be proved, perhaps by l'Hospital's rule (Appendix E). You may study these approaches in [9], pp. 222–223, or in [13], pp. 230–235.

Historically, analysis moved away from proofs based on geometry and towards analytic and axiomatic methods long ago. Thus, Newton's (and physics') greatest work, *Philosophiae naturalis principia mathematica* ("Mathematical Principles of Natural Philosophy," 1687), is decidedly geometric (see Chapter 36 for a quote). Joseph Louis Lagrange's *Théorie des fonctions analytiques* ("Theory of Analytical Functions," 1797) begins to base calculus upon algebra, and bears no diagrams at all, let alone geometric ones! It is the same with Cauchy's lengthy *Cours d'analyse de l'École royale polytechnique* ("Course of Analysis of the Royal Polytechnic School," 1821). That cautious tradition continues, as in Walter Rudin's modern text [37] (1976).

A likely reason for this trend is that geometric figures may lure us to fallacious conclusions, as in Paradox 0.5. Another example of this is the "proof" that all triangles are isosceles, in [44], pp. 203–204, a collection by Ian Stewart of 180 delightful curiosities.

23.3 Exercises

1. a. Prove that for the parabola $f(x) = 6 - x^2$, $\lim_{x\to 4} f(x) = -10$. Start this proof by writing: "We must verify that given any _____, we can find a corresponding _____ such that if ____________ $< \delta$, then ____________ $< \varepsilon$." Now, finish the proof.
   b. Make a partial check of **a** using first $\varepsilon = 0.002$ and then $\varepsilon = 145.9$.

2. Graph $y = \text{sgn}(\sin x)$ and just by inspection, decide at which points this function has a limit.

3. Review the graph of $y = \lfloor x \rfloor$ in Exercise 13.3b. Prove that the function does not have a limit at any $a \in \mathbb{Z}$.

4. Recall from Chapter 12 that a function $f : \mathcal{D}_f \to \mathcal{R}_f$ is bounded if the range $\mathcal{R}_f$ is a bounded set. Show that if f is not bounded in any $\mathbf{D}_a^\delta$, then it cannot have a limit at a. Hint: Use the contrapositive statement.

5. Observe the graph of the ε-to-δ correspondence $\delta = \min\left\{\frac{\varepsilon}{19}, 1\right\}$ in Figure 23.2. Why must the correspondence in the definition of limit be a positive function? Must it be a function?

6. Find two functions f and g with domains containing a deleted neighborhood of a, such that $\lim_{x\to a} f(x)$ exists, $\lim_{x\to a} g(x)$ does not exist, but $\lim_{x\to a}(f(x)g(x))$ does exit.

7. Convince yourself that it is not possible to find two functions f and g with domains containing a deleted neighborhood of a such that $\lim_{x\to a} f(x)$ exists, $\lim_{x\to a} g(x)$ does not exist, and $\lim_{x\to a}(f(x)+g(x))$ exits. (A rigorous proof will be done in a later chapter.)

8. Supply the missing steps in the proofs of the following chain of statements that demonstrate that $e^u > u$ for all real u. The case of negative u is trivially true. (i) By Bernoulli's inequality, $e^n \geq 1 + n(e-1)$, for any $n \in \mathbb{N}\cup\{0\}$. (ii) $e^n \geq 1 + n$ for any $n \in \mathbb{N}\cup\{0\}$. (iii) Given any $u \geq 0$, we can find an integer $N \geq 0$ such that $N + 1 > u \geq N$. (iv) $e^u \geq e^N \geq 1 + N > u$.

9. Prove that $\lim_{x\to a} f(x)$ exists if and only if both $\lim_{x\to a-} f(x)$ and $\lim_{x\to a+} f(x)$ exist and are equal. Hint: The second definition works nicely.

10. Prove that $\lim_{x\to -\infty} g(x) = 1$ in Example 23.8.

11. Prove that $\lim_{x\to\infty} \frac{\sin x}{x} = 0$ (see Figure 23.3).

12. Use the sequential criteria for limits (Theorem 23.1) and the pinching theorem for sequences (Theorem 20.7) to prove the following corollary, which we may call the **Pinching Theorem for Functions**: Let f, g, and h be functions such that over a common domain containing a deleted neighborhood of a, such that $f(x) \leq g(x) \leq h(x)$, and $\lim_{x\to a} f(x) = \lim_{x\to a} h(x) = L$. Then $\lim_{x\to a} g(x) = L$ also.

13. Show that the signum function has $\lim_{x\to 0-} \operatorname{sgn} x = -1$ and $\lim_{x\to 0+} \operatorname{sgn} x = 1$.

14. Show that the function in Figure 23.5 has $\lim_{x\to 0-} e^{1/x} = 0$, but that $\lim_{x\to 0+} e^{1/x}$ does not exist. Hint: Use Exercises 23.8 and 23.12.

15. Prove that $\lim_{x\to 0-} \frac{e^{1/x}}{x} = 0$. Hint: Let $w = -x$, and use l'Hospital's rule (Appendix E).

16. Show that $\lim_{\theta\to 0} \cos\theta = 1$, to finish Lemma 23.1.

17. In Figure 23.7, the maximum distance between arc AC and chord AC is $1 - \cos\frac{\pi}{n}$ (verify). Thus, every other point on the arc has a perpendicular distance to the chord that is less than $1 - \cos\frac{\pi}{n}$. Now, let the number of sides in the inscribed polygon increase. As $n \to \infty$, all such distances approach zero, and the arc itself approaches the chord. But one may argue that this is obvious and says nothing since both the arc and its chord shrink towards zero length.

    Archimedes' famous strategy in Note 4 can be made more plausible if instead of fixing the circle and shortening the lengths of the polygon's sides, we fix a chord and increase the radii of circles subtending this chord. The length of the arc should then have a limit equal to the length of the chord. To that end,

(i) Draw a circle of radius r with center at a point O in Quadrant I that passes through $P = \langle 0, 0\rangle$ and $Q = \langle 2a, 0\rangle$, where a is a positive constant. Let $C = \langle a, 0\rangle$.

(ii) Show that the length of arc PQ is $s = 2r\theta = 2r\sin^{-1}\left(\frac{a}{r}\right)$, where θ is the measure of angle POC.

(iii) Find $\lim_{r\to\infty} 2r\sin^{-1}\left(\frac{a}{r}\right)$. Discuss what this means in terms of Archimedes' strategy for estimating π.

24

Continuity and Discontinuity

Now, if to every x there corresponds a single, finite y in such a way that, as x continuously passes through the interval from a to b, $y = f(x)$ also gradually changes, then y is called a continuous function of x in this interval.

Peter Gustav Lejeune-Dirichlet, 1837

Continuity is one of the fundamental concepts in real analysis. If someone asks what is meant by a continuous function, we may draw pictures of unbroken functions. But if we suppose that continuous functions must be smooth, then we have overstated the requirement, because the function in Figure 24.1 is not smooth, but is nevertheless continuous.

The sense of unbrokenness observed by Dirichlet (quoted above) leads to an almost comprehensive definition of continuity—in fact, it was Leonhard Euler's working definition: a function is continuous if and only if its graph can be drawn without lifting pencil from paper. This rule efficiently weeds out the functions in Figures 23.4 and 23.5 of Chapter 23. The functions in Figures 24.1 and 24.2 follow this rule for continuity, but is the function in Figure 24.3 continuous, considering that 0^0 has been replaced by 1? And can the very curious function in Figure 24.4 be drawn in the required way? This function exhibits an unbounded frequency of oscillation in any neighborhood of zero, but the amplitude of the oscillations diminishes towards zero. We investigate all these functions in this chapter.

It may come as a surprise that all sequences are continuous functions, even though nothing looks more broken up than their graphs. So, it seems that words such as "unbroken" and "smooth" are full of connotations that make good mental images, but fuzzy analysis. The precise definition follows (recall Example 1.10).

> A function f is **continuous at a point** $x = a$ in its domain if and only if given any $\varepsilon > 0$, we can find a corresponding $\delta > 0$ such that if $x \in \mathcal{D}_f$ and $|x - a| < \delta$, then $|f(x) - f(a)| < \varepsilon$.

In terms of neighborhoods, this definition reads:

> A function f is **continuous at a point** $x = a$ in its domain if and only if given any $\varepsilon > 0$, we can find a corresponding $\delta > 0$ such that if $x \in \mathbf{N}_a^{\delta} \cap \mathcal{D}_f$, then $f(x) \in \mathbf{N}_{f(a)}^{\varepsilon}$.

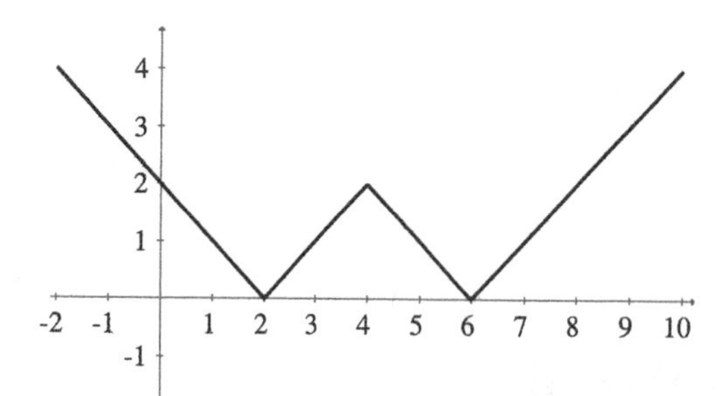

Figure 24.1. $y = ||x - 4| - 2|$

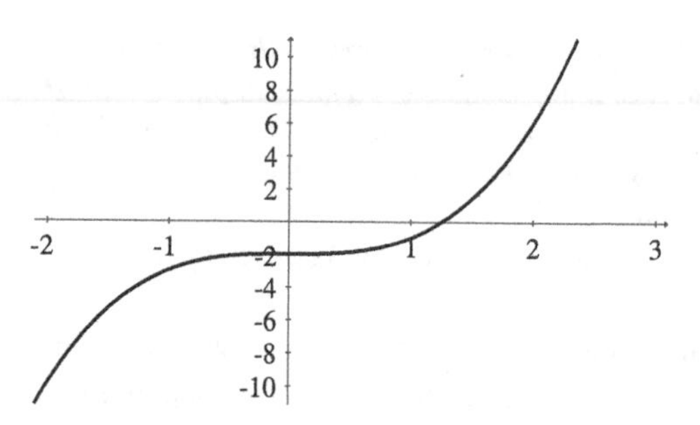

Figure 24.2. $y = x^3 - 3$

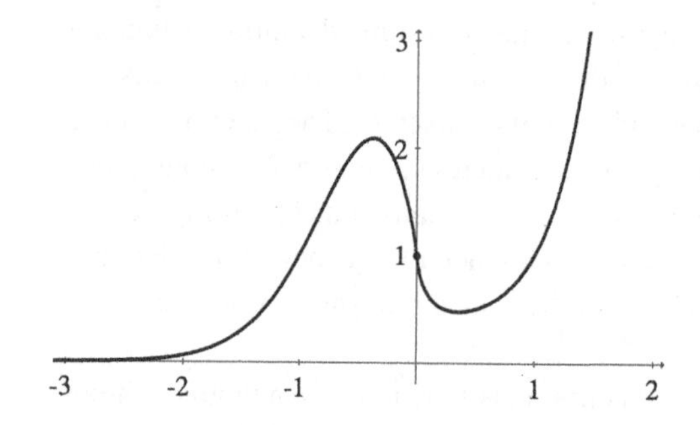

Figure 24.3. $y = \begin{cases} (x^2)^x & \textit{if } x \neq 0 \\ 1 & \textit{if } x = 0 \end{cases}$

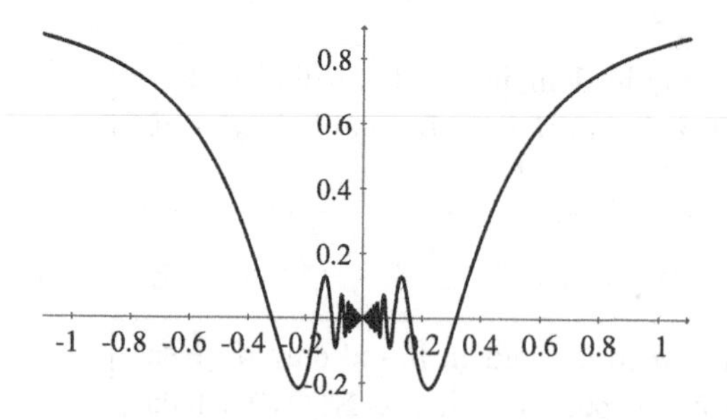

Figure 24.4. $y = \begin{cases} x \sin(\frac{1}{x}) & \textit{if } x \neq 0 \\ 0 & \textit{if } x = 0 \end{cases}$

It will become apparent that the value of δ in the definitions may not depend on just the given ε, but also on a. That is, δ is actually a function $\delta(\varepsilon, a)$.

We speak of continuity as a **local** property of a function since it occurs (or doesn't occur) at each point of the function, and we need not investigate the function as a whole. However, if we manage to prove that a function is continuous at every point in a set S, then it is said that f is **continuous over** or **on** S. In the case that $S = \mathcal{D}_f$, then f is **continuous over its domain** (or just "continuous"), and if $S = \mathbb{R}$, then f is **continuous everywhere**.

Theorem 24.1 A function $f : \mathbb{R} \to \mathbb{R}$ is continuous at a point $x = a$ if and only if three conditions hold: (1) $f(a)$ exists, (2) $\lim_{x\to a} f(x)$ exists, and (3) the real numbers in conditions (1) and (2) are equal. In short, $f(a) = \lim_{x\to a} f(x)$.

Proof of necessity. Let $f(x)$ be continuous at $x = a$. Then a must belong to the domain of f, meaning that $f(a)$ exists. The conditional "if $|x - a| < \delta$ and $x \in \mathcal{D}_f$, then $|f(x) - f(a)| < \varepsilon$" implies "if $0 < |x - a| < \delta$ and $x \in \mathcal{D}_f$, then $|f(x) - L| < \varepsilon$" when we select $L = f(a)$. This means that the limit exists, and equals $f(a)$.

Proof of sufficiency. The three conditions mean that $\lim_{x\to a} f(x) = f(a)$, which implies that if $0 < |x - a| < \delta$ (with $x \in \mathcal{D}_f$), then $|f(x) - f(a)| < \varepsilon$. Since x may equal a, the requirement that $0 < |x - a|$ is dropped. And when $x = a$, $|f(x) - f(a)| = 0 < \varepsilon$ is always fulfilled. Thus, $f(x)$ is continuous at a. **QED**

An important interpretation of Theorem 24.1 is that the continuity of f allows us to exchange the limit operation with the operations performed by f. That is, $f(\lim_{x\to a} x) = \lim_{x\to a} f(x)$.

Example 24.1 Use the definition to prove that all linear functions $f(x) = mx + b$ are continuous everywhere.

Let $a \in \mathbb{R}$. We must verify that given any $\varepsilon > 0$, we can find a corresponding $\delta > 0$ such that if $|x - a| < \delta$, then $|f(x) - f(a)| = |mx + b - (ma + b)| < \varepsilon$. Begin work with the last inequality: $|mx + b - (ma + b)| = |m||x - a| < \varepsilon$. Since m is a constant, $|m||x - a|$ can be made smaller than ε if $|x - a|$ can be made small. But we assert that x is in a δ-neighborhood of a, that is, $|x - a| < \delta$. Thus, $|m||x - a| < |m|\delta$. Since we want to conclude that $|m||x - a| < \varepsilon$, we realize that the ε-to-δ correspondence should come from $\varepsilon = |m|\delta$, or $\delta = \frac{\varepsilon}{|m|}$ when $m \neq 0$.

For $m \neq 0$, we have that given any $\varepsilon > 0$, $\delta = \frac{\varepsilon}{|m|}$ will guarantee that if $|x - a| < \delta$, then $|mx + b - (ma + b)| = |m||x - a| < |m|\delta = \varepsilon$. Therefore, all non-constant linear functions $f(x) = mx + b$ are continuous over $\mathbb{R}$.

If $m = 0$, then $|m||x - a| = 0 < \varepsilon$ no matter what δ-neighborhood of a one chooses. Hence, all constant functions are continuous over $\mathbb{R}$.

It follows that all linear functions are continuous everywhere. **QED**

Example 24.2 Prove that $g(x) = \frac{7}{x}$ is continuous at $a = 4$.

Preparation. Here, $g(a) = \frac{7}{4}$. We must verify that given any $\varepsilon > 0$, we can find a corresponding $\delta > 0$ such that if $|x - 4| < \delta$, then $|\frac{7}{x} - \frac{7}{4}| < \varepsilon$. We note right away that the δ-neighborhood $4 - \delta < x < 4 + \delta$ contains $x = 0$ if $\delta > 4$. So we will first consider cases that yield, say, $\delta < 1$. Start with the inequality $|\frac{7}{x} - \frac{7}{4}| = |\frac{28-7x}{4x}| = \frac{7|x-4|}{|4x|}$. For this fraction to be less than even tiny ε's, not only must the numerator be small, but also the denominator must *not* be small. The numerator will be controlled by $|x-4|<\delta$. In order to control the denominator, we need only to keep it bounded away from zero. This means that we must be able to display a positive number that is always less than $|4x|$.

Return to $4-\delta<x<4+\delta$ with $\delta<1$. Then we can say that $3<|x|<5$. Then also $12<|4x|<20$, and here we see that $|4x|$ is bounded away from zero. We are therefore justified in taking reciprocals, $\frac{1}{20}<\frac{1}{|4x|}<\frac{1}{12}$. This tells us that $\frac{1}{|4x|}$ is less than $\frac{1}{12}$ but not too small, always provided that $\delta<1$.

Now we can be sure that

$$\left|\frac{7}{x}-\frac{7}{4}\right|=\frac{7|x-4|}{|4x|}<\frac{7\delta}{|4x|}<\frac{7\delta}{12}.$$

We realize that provided $\delta<1$, we should set $\frac{7\delta}{12}=\varepsilon$, so that $\delta=\frac{12\varepsilon}{7}$. Thus, given any positive $\varepsilon<\frac{7}{12}$, we have a corresponding $\delta=\frac{12\varepsilon}{7}<1$, which guarantees that

$$|\frac{7}{x}-\frac{7}{4}|<\frac{7\delta}{12}=\varepsilon.$$

What if $\varepsilon\geq\frac{7}{12}$ is given? In that case, we can hold $\delta=1$. This works because going through the above inequalities again, we find that

$$\frac{1}{20}\leq\frac{1}{|4x|}\leq\frac{1}{12}.$$

This implies that

$$\left|\frac{7}{x}-\frac{7}{4}\right|=\frac{7|x-4|}{|4x|}\leq\frac{7|x-4|}{12}<\frac{7\delta}{12}=\frac{7}{12}\leq\varepsilon,$$

meaning that once again,

$$\left|\frac{7}{x}-\frac{7}{4}\right|<\varepsilon.$$

Proof. We have now fulfilled the definition. Given any $\varepsilon>0$, we have found a corresponding δ, specifically $\delta=\min\{\frac{12\varepsilon}{7},1\}$. If $\varepsilon<\frac{7}{12}$, then $\delta=\frac{12\varepsilon}{7}$, and we are sure that whenever $|x-4|<\delta$, then

$$|g(x)-g(4)|=\left|\frac{7}{x}-\frac{7}{4}\right|=\frac{7|x-4|}{|4x|}<\frac{7\delta}{|4x|}<\frac{7\delta}{12}=\varepsilon.$$

If $\varepsilon\geq\frac{7}{12}$, we have $\delta=1$, in which case

$$|g(x)-g(4)|=\frac{7|x-4|}{|4x|}\leq\frac{7|x-4|}{12}<\frac{7\delta}{12}=\frac{7}{12}\leq\varepsilon.$$

Therefore, $g(x)=\frac{7}{x}$ is continuous at $x=4$. ■

The next example will emphasize that although $\lim_{x\to a}f(x)$ is independent of the function's value $f(a)$, continuity demands their equality.

Example 24.3 The polynomial $p(x)=x^3-3$ (Figure 24.2) is continuous at $a=2$ by Theorem 24.1, since $p(2)=\lim_{x\to 2}p(x)=5$. In Example 23.2, we encountered the function

$$f(x)=\begin{cases}\frac{x^4-2x^3-3x+6}{x-2} & \text{if } x\neq 2\\ 10 & \text{if } x=2\end{cases}.$$

and saw that $f(x) = p(x)$ whenever $x \neq 2$. In other words, $f(x) = p(x)$ in every deleted neighborhood $\mathbf{D}_2^\delta$. This implies that $\lim_{x\to 2} f(x) = \lim_{x\to 2} p(x)$. However, $f(2)$ is independently (peevishly?) set at 10, and this destroys the continuity of $f(x)$, Theorem 24.1 being an if and only if statement. ■

Example 24.4 Prove that $f(x) = \sqrt{x}$ is continuous at any number a of its domain.

Preparation. We must verify that given any $\varepsilon > 0$, we can find a corresponding $\delta > 0$ such that if $|x - a| < \delta$ (with $x \in \mathcal{D}_f$), then $|\sqrt{x} - \sqrt{a}| < \varepsilon$.

In the special case that $a = 0$, we must show that if $0 \leq x < \delta$, then $\sqrt{x} < \varepsilon$. This is satisfied when the ε-to-δ correspondence is $\delta = \varepsilon^2$. Thus, we will consider $a > 0$ in what follows.

In order to work with $|\sqrt{x} - \sqrt{a}| < \varepsilon$, the classic procedure is to rationalize the numerator,

$$|\sqrt{x} - \sqrt{a}| = |\frac{(\sqrt{x} - \sqrt{a})(\sqrt{x} + \sqrt{a})}{\sqrt{x} + \sqrt{a}}| = \frac{|x - a|}{\sqrt{x} + \sqrt{a}} < \frac{\delta}{\sqrt{x} + \sqrt{a}} \leq \frac{\delta}{\sqrt{a}}.$$

Now we realize that we may set $\varepsilon = \frac{\delta}{\sqrt{a}}$, so $\delta = \varepsilon\sqrt{a}$. (A possible objection here is that for values of a near zero and large values of ε, the neighborhood $(a - \delta, a + \delta)$ contains negative abscissas, but this is countered by remembering that $x \in \mathcal{D}_f$.)

Proof. For $a > 0$, given any $\varepsilon > 0$, the corresponding δ is $\delta = \varepsilon\sqrt{a}$. This correspondence gives us $|\sqrt{x} - \sqrt{a}| = \frac{|x-a|}{\sqrt{x}+\sqrt{a}} < \frac{\delta}{\sqrt{a}} = \varepsilon$ whenever $|x - a| < \delta$. Returning to the first detail, if $a = 0$, we use $\delta = \varepsilon^2$.

In terms of neighborhoods, the proof reads: given any $\varepsilon > 0$, whenever $x \in \mathcal{D}_f$ is in the neighborhood $\mathbf{N}_a^\delta$ given by

$$\delta = \begin{cases} \varepsilon\sqrt{a} & \text{if } a > 0 \\ \varepsilon^2 & \text{if } a = 0 \end{cases},$$

we are certain that $\sqrt{x}$ is in the ε-neighborhood $\mathbf{N}_{\sqrt{a}}^\varepsilon = (\sqrt{a} - \varepsilon, \sqrt{a} + \varepsilon)$. Therefore, $f(x) = \sqrt{x}$ is a continuous function over its domain.

In Chapter 27 we will investigate $f(x) = \sqrt{x}$ more closely and discover that the dependence of δ on a can actually be sidestepped. ■

Example 24.5 Show that $s(x) = 3|x + 6|$ is continuous at every real number a. We must verify that given any $\varepsilon > 0$, we can find a corresponding $\delta > 0$ such that if $x \in \mathcal{D}_s = \mathbb{R}$ and $|x - a| < \delta$, then $|s(x) - s(a)| < \varepsilon$.

The last inequality demands that $|3|x + 6| - 3|a + 6|| = 3||x + 6| - |6 + a|| < \varepsilon$. Now, by Corollary 15.2, $3||x + 6| - |6 + a|| \leq 3|(x + 6) - (6 + a)| = 3|x - a| < 3\delta$. This indicates that a useful ε-to-δ correspondence is $\delta = \frac{\varepsilon}{3}$.

Thus, whenever $|x - a| < \delta$, we retrace our steps to arrive at $|3|x + 6| - 3|6 + a|| \leq 3|x - a| < 3\delta = \varepsilon$. The definition of continuity is now satisfied everywhere. ■

Example 24.6 Show that

$$u(x) = \begin{cases} |x + 6| & \text{if } x > 3 \\ x^2 & \text{if } x \leq 3 \end{cases}$$

is continuous at $a = 3$. Let's use Theorem 24.1. First, $u(3) = 9$ exists. Second, $\lim_{x \to 3+} |x + 6| = 9$ and $\lim_{x \to 3-} x^2 = 9$. This implies that $\lim_{x \to 3} u(x) = 9$ independently of $u(3)$. Thirdly, $u(3) = \lim_{x \to 3} u(x)$. Consequently, $u(x)$ is continuous at $a = 3$. ■

24.1 Discontinuity

A function that is discontinuous (not continuous) at a point is said to have a discontinuity at that point. Thus, discontinuity is a local property of a function. The logical negation of the definition of continuity is as follows (details appear in Example 1.11).

A function f is **discontinuous at a point** $x = a$ if and only if there exits an $\varepsilon^* > 0$ such that for any $\delta > 0$, we can find at least one $x \in \mathcal{D}_f$ such that $|x - a| < \delta$, and also $|f(x) - f(a)| < \varepsilon^*$ is false.

The inequality $|f(x) - f(a)| < \varepsilon^*$ may be falsified in two ways. The first way, when $a \in \mathcal{D}_f$, is that $|f(x) - f(a)| \geq \varepsilon^*$ will have to be shown to be true. The second way, when $a \notin \mathcal{D}_f$, is that $|f(x) - f(a)| < \varepsilon^*$ is automatically false because $f(a)$ doesn't exist. It follows that a function cannot be continuous outside of its domain. You should rewrite the definition of discontinuity in terms of neighborhoods as Exercise 24.12.

Perhaps the easiest way to determine discontinuity is to negate the statements in Theorem 24.1.

Corollary 24.1 A function $f : \mathbb{R} \to \mathbb{R}$ is discontinuous at a point $x = a$ if and only if any of the following are true: (1) $f(a)$ does not exist (that is, $a \notin \mathcal{D}_f$), (2) $\lim_{x \to a} f(x)$ does not exist, or (3) $\lim_{x \to a} f(x) \neq f(a)$.

Proof. Theorem 24.1 has the form $F \equiv (A \text{ and } B \text{ and } C)$. This is equivalent to $not\text{-}F \equiv (not\text{-}A$ or $not\text{-}B$ or $not\text{-}C)$. **QED**

Example 24.7 The signum function $y = \operatorname{sgn} x$ of the previous chapter is discontinuous at $a = 0$ and nowhere else. This is because statement (2) of the corollary is true at that point (of course, it follows that (3) must also be true). ■

Example 24.8 Show that

$$f(x) = \begin{cases} |x + 6| & \text{if } x \neq 3 \\ 10 & \text{if } x = 3 \end{cases}$$

is discontinuous at $a = 3$. Corollary 24.1 is the key. Notice that $\lim_{x \to 3} |x + 6| = 9$ independently of $f(3)$, but $f(3) = 10$. Hence, $\lim_{x \to 3} f(x) \neq f(3)$, and it follows that f is discontinuous at $a = 3$. This is a removable discontinuity, since redefining $f(3)$ to equal 9 patches the gap. ■

Example 24.9 Name the function in Figure 24.3 as

$$Y(x) = \begin{cases} (x^2)^x & \text{if } x \neq 0 \\ 1 & \text{if } x = 0 \end{cases}.$$

Perhaps defining $Y(0) = 0^0 \equiv 1$ will allow us to draw Y without lifting pencil from paper. Prove in Exercise 24.10 that Y is, in fact, continuous with this definition. (If Y were not defined at zero, then the point $\langle 0, 1\rangle$ would be a removable discontinuity.) ■

It's worthwhile to name the several types of discontinuity that a function can exhibit, just for graphic emphasis. The discontinuity in each example occurs at $a = 0$.

TYPE OF DISCONTINUITY	*DEFINITION*	*EXAMPLE*
Removable Gap	$f(a)$ doesn't exist, but $\lim_{x\to a} f(x)$ exists	$y = \frac{\sin x}{x}$
Removable Gap	$f(a)$ and $\lim_{x\to a} f(x)$ exist, but they aren't equal.	$y = \lfloor \cos x \rfloor$
Gap at Infinity	$\lim_{x\to a-} f(x) = \infty$ and $\lim_{x\to a+} f(x) = \infty$, or $\lim_{x\to a-} f(x) = -\infty$ and $\lim_{x\to a+} f(x) = -\infty$.	$y = \lvert\frac{1}{x}\rvert$
Finite Jump	Both one-sided limits exist but aren't equal.	$y = \operatorname{sgn} x$
Infinite Jump	One one-sided limit exists and the other is $\pm\infty$.	$y = e^{1/x}$
Infinite Jump	$\lim_{x\to a-} f(x) = \infty$ and $\lim_{x\to a+} f(x) = -\infty$, or $\lim_{x\to a-} f(x) = -\infty$ and $\lim_{x\to a+} f(x) = \infty$.	$y = \cot x$
Bounded Oscillatory	Neither one-sided limit exists, but f is bounded in a neighborhood of a.	$y = \sin\frac{1}{x}$
Unbounded Oscillatory	Neither one-sided limit exists and neither is $\pm\infty$, but f is unbounded in all neighborhoods of a.	$y = \frac{1}{x}\sin\frac{1}{x}$

Monotone functions are interesting in this context because they are severely restricted with regard to discontinuity, as we now see.

Lemma 24.1 Let $f : \mathbb{R} \to \mathbb{R}$ be monotone over an interval $A \subseteq \mathcal{D}_f$. If f is discontinuous, then only finite jump discontinuities are possible.

Proof. First of all, any interval A over which f is constant does not bother us, since f is continuous there. Thus, we focus on strictly monotone, discontinuous functions over A, and for now, let f be strictly increasing. Reviewing the table above, we must prove that the only type of discontinuity at $x = a$ is where the one-sided limits $\lim_{x\to a-} f(x)$ and $\lim_{x\to a+} f(x)$ exist, but are not equal. Note that $f(a)$ exists. Figure 24.1 in Chapter 48 shows a monotone function with the discontinuities that will be proved to exist.

Suppose a is not an endpoint of A. The set $T = \{f(x) : x < a\}$ is nonempty and bounded above by $f(a)$, and therefore has a supremum, L. Thus, L has the property that for any $\varepsilon > 0$, there exists an $f(x) \in T$ such that $f(x) \geq L - \varepsilon$, which in this case implies $|f(x) - L| \leq \varepsilon$. Set T may just as well be described as $T = \{f(a - \delta) : \delta > 0\}$. We now have that for any $\varepsilon > 0$, there exists a $\delta > 0$ such that $f(a - \delta) \in T$ and $|f(a - \delta) - L| \leq \varepsilon$. This is equivalent to $\lim_{x \to a-} f(x) = L$. Moreover, since $f(x) < f(a)$ for $f(x) \in T$, $\lim_{x \to a-} f(x) \leq f(a)$.

Similarly, the set $S = \{f(x) : x > a\}$ has an infimum, U. By reasoning as above (which you should do), we find that for any $\varepsilon > 0$, there exists a $\delta > 0$ such that $f(a + \delta) \in S$ and $|f(a + \delta) - U| \leq \varepsilon$. This is equivalent to $\lim_{x \to a+} f(x) = U$, and since $f(x) > f(a)$ for $f(x) \in S$, $\lim_{x \to a+} f(x) \geq f(a)$.

So now we have $\lim_{x \to a-} f(x) \leq f(a) \leq \lim_{x \to a+} f(x)$. We have ruled out equality of the three quantities, since that would imply continuity at a. Thus, we are left with three possible types of jump discontinuity for a strictly increasing, discontinuous function over an interval: (1) $\lim_{x \to a-} f(x) = f(a) < \lim_{x \to a+} f(x)$, (2) $\lim_{x \to a-} f(x) < f(a) < \lim_{x \to a+} f(x)$, or (3) $\lim_{x \to a-} f(x) < f(a) = \lim_{x \to a+} f(x)$. If a is the greater (or only) endpoint of A, then only the analysis for set T applies, and we arrive at $\lim_{x \to a-} f(x) < f(a)$. If a is the lesser (or only) endpoint of A, then consider only set S, and we have $\lim_{x \to a+} f(x) > f(a)$. The case where A has no endpoints has already been covered. And finally, if f is strictly decreasing, we may apply this proof to $-f$. The conclusion will be that only jump discontinuities (1), (2), and (3) can exist, with their inequalities reversed. **QED**

Because Lemma 24.1 allows only finite jump discontinuities, monotone functions cannot suffer from uncountably many discontinuities.

Theorem 24.2 A monotone function over an interval can only have a countable number of discontinuities.

Proof. We begin with a monotone function f over $[a, b] \subseteq \mathcal{D}_f$. Let $D = |f(b) - f(a)|$. If $D = 0$, then f is constant over $[a, b]$, and we are done. Hence $D > 0$, and without loss of generality, we may assume that $D = 1$. By Lemma 24.1, only finite jump discontinuities are possible. Suppose that a jump discontinuity occurs at $t \in (a, b)$, and let $d = |\lim_{x \to t+} f(x) - \lim_{x \to t-} f(x)|$ be the jump distance. (If $t = a$ or $t = b$, then the jump distances are $d = |\lim_{x \to a+} f(x) - f(a)|$ and $d = |f(b) - \lim_{x \to b-} f(x)|$, respectively.) It is always the case that $0 < d \leq 1$. But if $d = 1$, then there is no vertical space left for a second discontinuity. If $\frac{1}{2} \leq d \leq 1$ for all jump distances, then there can be no more than two jumps in $[a, b]$. If $\frac{1}{3} \leq d \leq 1$ for all jump distances, then there are no more than three jumps in $[a, b]$. In general, for each $n \in \mathbb{N}$, if $d = \frac{1}{n}$ is the minimum jump distance, then there are at most n jump discontinuities in $[a, b]$. Hence, to each member of the denumerable set $\{\frac{1}{n}\}_1^\infty$ there corresponds a finite number of jump discontinuities. Therefore, by Lemma 11.1, the collection of all discontinuities is at most countably infinite.

When the interval of interest $A \subseteq \mathcal{D}_f$ is either not closed or not bounded, f may not be bounded anymore, that is, D may not exist. In Exercise 24.13, you are asked to extend the proof to cover bounded intervals other than $[a, b]$. Exercise 24.14 covers the case of unbounded intervals. **QED**

24.2 Exercises

1. For each of the functions **a** through **m**, check the box if the stated condition of Theorem 24.1 is true for that function at the given value of a.

	Function	Comments	$f(a)$ exists?	$\lim_{x\to a} f(x)$ exists?	Continuous at $x = a$?
a.	$f(x) = \begin{cases} x^2 & \text{if } x \geq 0 \\ -x^2 & \text{if } x < 0 \end{cases}$	piecewise function $a = 0 \in \mathcal{D}_f$	✓	✓	✓
b.	$f(x) = \begin{cases} x^2 & \text{if } x > 0 \\ -x^2 & \text{if } x < 0 \end{cases}$	gap at $a = 0$			
c.	$f(x) = \begin{cases} x^2 & \text{if } x \geq 0 \\ x^2 + 1 & \text{if } x < 0 \end{cases}$	finite jump at $a = 0$			
d.	$f(x) = \begin{cases} x^2 & \text{if } x \neq 3 \\ \varnothing & \text{if } x = 3 \end{cases}$	$a = 3$ deleted from $\mathcal{D}_f$, gap at $a = 3$			
e.	$f(x) = \frac{x(x^2-1)}{x+1}$	gap at $a = -1$			
f.	$f(x) = \begin{cases} \operatorname{sgn} x & \text{if } x \neq 0 \\ 5 & \text{if } x = 0 \end{cases}$	finite jump at $a = 5$			
g.	$f(x) = \begin{cases} \frac{1}{x-8} & \text{if } x \neq 8 \\ 2 & \text{if } x = 8 \end{cases}$	infinite jump at $a = 8$			
h.	$f(x) = \begin{cases} \frac{x^2-64}{x-8} & \text{if } x \neq 8 \\ 12 & \text{if } x = 8 \end{cases}$	gap at $a = 8$			
i.	$f(x) = \begin{cases} \frac{x^2-64}{x-8} & \text{if } x \neq 8 \\ 16 & \text{if } x = 8 \end{cases}$	$a = 8 \in \mathcal{D}_f$			
j.	$f(x) = \begin{cases} \tan x & \text{if } x \neq \pi/2 \\ 0 & \text{if } x = \pi/2 \end{cases}$	infinite jump at $a = \pi/2$			
k.	$f(x) = \ln\lvert x + 1\rvert$	unbounded at $a = -1$			
l.	$f(x) = \begin{cases} x^x & \text{if } x > 0 \\ 1 & \text{if } x = 0 \end{cases}$	$a = 0 \in \mathcal{D}_f$			
m.	$f(x) = \cos(x + \lvert x\rvert)$	$a = 0 \in \mathcal{D}_f$			

2. Go through the steps as in Example 24.1 to prove that $y = \frac{1}{\pi}x - \sin 2$ is continuous at $a = \sqrt{6}$.

3. a. Show that Theorem 24.1 is equivalent to: a function f is continuous at $x = a$ if and only if $\lim_{h\to 0} f(a + h) = f(a)$.

b. Prove that $g(x) = \sin x$ is continuous over its entire domain by using the version of Theorem 24.1 in **a**. Hint: You will need to verify that $\lim_{h\to 0} \sin h = 0$ and $\lim_{h\to 0} \cos h = 1$.
c. Prove that $g(x) = \cos x$ is continuous over its entire domain.

4. a. Prove that $h(x) = \frac{x}{x-6}$ is continuous at $a = 12$. Graph the function in a neighborhood of $a = 12$. Start the proof by writing: "We must verify that given any ____, we can find a corresponding ____ such that if $x \in \mathcal{D}_f$ and ________ $< \delta$, then ______________ $< \varepsilon$." Hint: The work is similar to Example 24.2.
b. Partially check your proof in **a** by showing that the δ corresponding to $\varepsilon = 0.0054$ does in fact fulfill the definition of continuity.
c. We routinely say that $h(x) = \frac{x}{x-6}$ is a discontinuous function. Is it true that $h(x) = \frac{x}{x-6}$ is continuous over its domain, even though it has a vertical asymptote at $x = 6$?

5. a. Prove that $w(x) = ||x-4|-2|$ is continuous at $a = 4$ (graph in Figure 24.1). Hint: The graph is a helpful guide. Smaller εs require neighborhoods with $\delta < 2$, which imply $w(x) = 2 - |x-4|$. Why do bigger εs just need $\delta = 2$?
b. Partially check your proof in **a** by showing that the δ corresponding to $\varepsilon = 0.0054$ does in fact fulfill the definition of continuity.

6. Prove that

$$\operatorname{sinc}(x) = \begin{cases} \frac{\sin x}{x} & \text{if } x \neq 0 \\ 1 & \text{if } x = 0 \end{cases}$$

is continuous at $a = 0$. Could $\operatorname{sinc}(x)$ remain continuous if we choose $\operatorname{sinc}(0) \neq 1$? Hint: The function $\operatorname{sinc}(x)$ is almost the same as $s(\theta) = \frac{\sin\theta}{\theta}$ from the previous chapter.

7. a. Prove that graph in Figure 24.4, $f(x) = x \sin\frac{1}{x}$ with $f(0) = 0$, is continuous at $a = 0$. This furnishes us with an example of a function that fails Euler's rough-and-ready test for continuity. Could $f(x)$ remain continuous if we choose $f(0) \neq 0$? Hint: $|\sin\frac{1}{x}| \leq 1$.
b. Partially check your proof in **a** by showing that the δ corresponding to $\varepsilon = 0.0054$ does in fact fulfill the definition of continuity.

8. Generalize Example 24.2 by proving that $g(x) = \frac{7}{x}$ is continuous at any $a \neq 0$. Of course, δ depends on ε, but it turns out that for this function, the dependence of δ on a is unavoidable (in contrast to other examples in this chapter). Hint: To keep things manageable, work with positive a first. You will have to use $|x-a| < \delta$ in order to keep x bounded away from zero, so that $\frac{1}{|ax|}$ remains bounded too. One suggestion is to take $\delta < \frac{1}{2}a$ somewhere along your proof.

9. Suppose that a belongs to the domain of f. We shall call $\langle a, f(a)\rangle$ a **type 1 isolated point** of the function in $\mathbb{R}^2$ if a is not a cluster point of the domain, so that a small circle (a circular neighborhood!) can be drawn around $\langle a, f(a)\rangle$ within which no other points of f exist.
a. Graph $g(x) = (x-1)\sqrt{x-2}$ by hand, and show that it has exactly one type 1 isolated point. Hint: Don't graph it by calculator.
b. Prove that $g(x)$ is continuous at $a = 1$. Hint: The definition says, "... whenever $x \in N_1^{\delta} \cap D_g$, then $g(x) \in N_{g(1)}^{\varepsilon}$." What if $\delta = \frac{1}{2}$, for example?
c. Prove that in general, a function is continuous at any type 1 isolated point of its domain.
d. Prove therefore that all sequences $\langle a_n \rangle_1^{\infty}$ are continuous functions.

10. Prove that

$$Y(x)\begin{cases}(x^2)^x & \text{if } x \neq 0\\ 1 & \text{if } x = 0\end{cases}$$

in Figure 24.3 is continuous at $a = 0$. Hint: Simplify things by noting that $Y(-x) = \frac{1}{Y(x)}$. Calculus techniques are worthwhile to find the limit at $x = 0$.

11. Suppose that $x = a$ belongs to the domain of f. In comparison to Exercise 24.9, we shall call $\langle a, f(a)\rangle$ a **type 2 isolated point** of the function in $\mathbb{R}^2$ if a is a cluster point of the domain and if a small circle (a circular neighborhood again) can be drawn around $\langle a, f(a)\rangle$ within which no other points of f exist.
    a. Graph

    $$H(x) = \begin{cases}-x+4 & \text{if } x > 2\\ 1 & \text{if } x = 2\\ -x & \text{if } x < 2\end{cases}$$

    and show that it has exactly one type 2 isolated point. Use this fact and Corollary 24.1 to prove that H is discontinuous at that point.
    b. Graph the curious function $g(x) = \lfloor \sin x \rfloor$ and show that it has a type 2 isolated point at $a = \frac{\pi}{2}$.
    c. Prove that g in **b** is discontinuous at $\frac{\pi}{2}$, but for a different reason than H in **a**. Hint: Prove that $\lim_{x\to\pi/2} \lfloor \sin x \rfloor \neq \lfloor \sin \frac{\pi}{2} \rfloor$ and use Corollary 24.1.
    d. Prove that in general, a function is discontinuous at any type 2 isolated point of its domain.

12. Rewrite the definition of discontinuity at a point in terms of neighborhoods.

13. Continuing the proof of Theorem 24.2, suppose that f is bounded over the interval of interest $A \subseteq \mathcal{D}_f$, and that A is any non-closed, bounded interval, namely (a, b), $(a, b]$, or $[a, b)$. Prove that f will still have the property of Theorem 24.2. Hint: Modify f so that it conforms to the theorem's proof.

14. a. Continuing the proof of Theorem 24.2, suppose that f is unbounded over the interval of interest $A \subseteq \mathcal{D}_f$, where A is any one of (a, b), $[a, b)$, $(a, b]$, $(-\infty, b)$, $(-\infty, b]$, $[a, \infty)$, (a, ∞), or $(-\infty, \infty)$. Then the proof's D does not exist, but it is nevertheless impossible for any jump distance d to be infinite at some $t_0 \in A$. Prove this.
    b. Finish the proof of Theorem 24.2 by proving the case for monotone increasing, discontinuous, unbounded functions f over intervals of interest (a, b) or $(-\infty, b)$, where $\lim_{x\to b-} f(x) = \infty$. All other cases may be proved similarly.

25

The Sequential Criterion for Continuity

It is useful to combine the theory of sequences with the definition of continuity to arrive at the main theorem of this chapter. The idea here is to consider any sequence $\langle x_n \rangle_1^\infty$ of points on the X axis such that $\lim_{n\to\infty} x_n = a$. Now consider their images, $\langle f(x_n) \rangle_1^\infty$. If these images converge to $f(a)$ on the Y axis, then f must be continuous at a. The key ingredient here will be Theorem 23.1, the sequential criteria for limits (of course!).

The functions in Figures 25.1 to 25.4 exhibit behaviors that may be analyzed by sequential criteria. The function in Figure 25.1 has an unbounded frequency of oscillation in any neighborhood of $a = 0$. The function in Figure 25.2 show the same behavior at points where $\tan x$ does not exist. The function in Figure 25.3 is **Dirichlet's function**; what we see there can only be a rough sketch of the actual function. Equally strange is the **ruler function** (or **Thomae's function**) in Figure 25.4, with a pattern of points that crowds relentlessly towards the X axis.

Theorem 25.1 The Sequential Criterion for Continuity A function $f : \mathbb{R} \to \mathbb{R}$ is continuous at $x = a$ if and only if for any sequence $\langle x_n \rangle_1^\infty \subseteq \mathcal{D}_f$ that converges to a, it is true that $\langle f(x_n) \rangle_1^\infty \to f(a)$. That is, given $\lim_{n\to\infty} x_n = a$, the function f is continuous at a if and only if $f(\lim_{n\to\infty} x_n) = \lim_{n\to\infty} f(x_n)$.

Proof of necessity. (Compare this proof with that of Theorem 23.1A; the only essential change is to replace L by $f(a)$.) Let f be continuous at $x = a$ and let $\langle x_n \rangle_1^\infty \subseteq \mathcal{D}_f$ be such that $\lim_{n\to\infty} x_n = a$. We take any $\varepsilon > 0$. Because f is continuous at a, there exists a corresponding $\delta > 0$ such that whenever $x \in \mathcal{D}_f$ and $|x - a| < \delta$, then we are certain that $|f(x) - f(a)| < \varepsilon$. We now substitute the terms x_n of the sequence into the continuity assertion. Since $\lim_{n\to\infty} x_n = a$, there exists a number N_ε such that for all $n > N_\varepsilon$, $|x_n - a| < \delta$. This implies that $|f(x_n) - f(a)| < \varepsilon$. And this, in turn, implies that the sequence of ordinates $\langle f(x_n) \rangle_1^\infty \to f(a)$.

Written another way, $f(a) = f(\lim_{n\to\infty} x_n) = \lim_{n\to\infty} f(x_n)$.

Proof of sufficiency. (Again, compare this with the proof of Theorem 23.1A.) We won't prove that if $\langle f(x_n) \rangle_1^\infty \to f(a)$ for every sequence $\langle x_n \rangle_1^\infty \subseteq \mathcal{D}_f$ that converges to a, then f is continuous at a. Instead, we prove its contrapositive, and assume that f is not continuous at $x = a$. Since $f(a)$ exists, the assumption implies that $\lim_{x\to a} f(x)$ does not exist. This means that there exists some $\varepsilon^* > 0$ such that for any $\delta > 0$, there is at least one $x_\delta \in \mathcal{D}_f$ in the neighborhood $|x - a| < \delta$ such that $|f(x_\delta) - f(a)| \geq \varepsilon^*$ (see Chapter 23, "Non-Existence of a Limit").

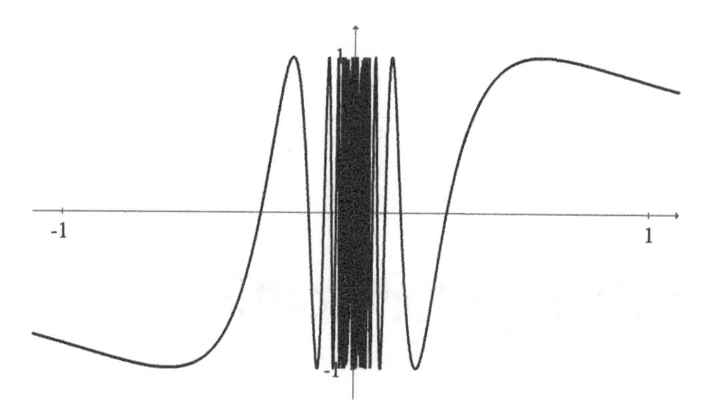

Figure 25.1. $y = \sin \frac{1}{x}$

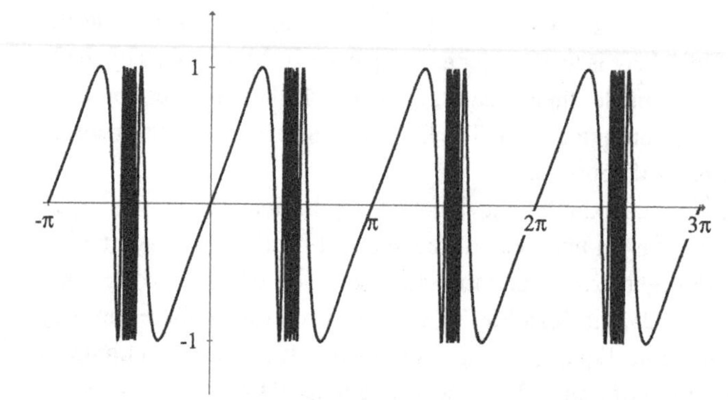

Figure 25.2. $y = \sin(\tan x)$

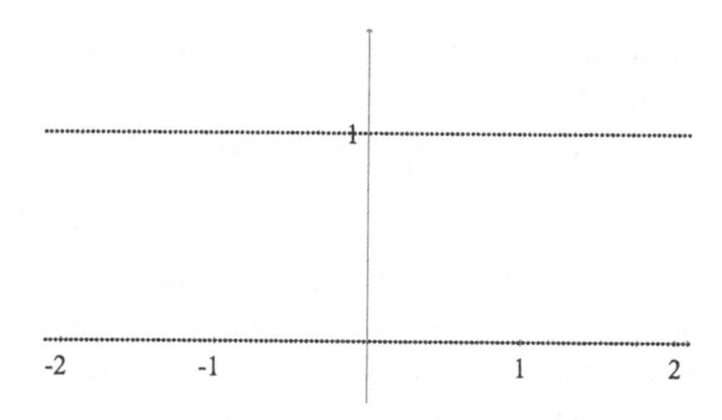

Figure 25.3. $y = \begin{cases} 1 & \text{if } x \text{ is rational.} \\ 0 & \text{if } x \text{ is irrational.} \end{cases}$

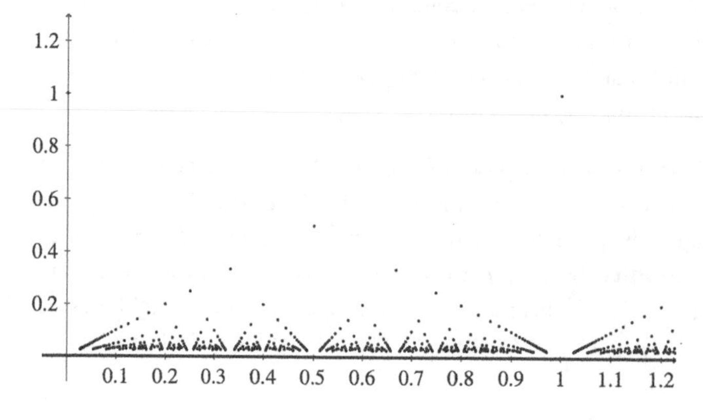

Figure 25.4.
$y = \begin{cases} \frac{1}{|q|} & \text{if } x = \frac{p}{q}\text{, reduced.} \\ 0 & \text{if } x \text{ is irrational.} \end{cases}$

In particular, select $\delta = \frac{1}{n}$. For each $n \in \mathbb{N}$, what was just stated guarantees some $x_n \in \mathcal{D}_f$ satisfying $|x_n - a| < \frac{1}{n}$ and such that $|f(x_n) - f(a)| \geq \varepsilon^*$. This creates a sequence $\langle x_n \rangle_1^\infty \to a$ whose images $\langle f(x_n) \rangle_1^\infty \not\to f(a)$. Hence, the contrapositive statement is proved, and therefore, f must be continuous at $x = a$. **QED**

Example 25.1 Show that

$$f(x) = \begin{cases} |x+6| & \text{if } x > 3 \\ x^2 & \text{if } x \leq 3 \end{cases}$$

is continuous at $a = 3$. Let $\langle x_n \rangle_1^\infty$ be any sequence that converges to 3. We must show that $f(3) = 9 = f(\lim_{n\to\infty} x_n) = \lim_{n\to\infty} f(x_n)$. Since the ordinates of f are evaluated by two rules, we expect both $|x+6|$ and x^2 to be in a neighborhood of 9 when x_n is in a neighborhood of 3. Thus, we will use the facts that $\lim_{n\to\infty} |x_n + 6| = 9 = \lim_{n\to\infty} x_n^2$ when $x_n \to 3$, but we must pay attention to details for a moment. If $x_n \leq 3$, then $f(x_n) = x^2$, but if $x_n > 3$, then $f(x_n) = |x+6|$. We therefore segregate the terms of $\langle x_n \rangle_1^\infty$: a subsequence $\langle x_{n_k} \rangle_{k=1}^\infty$ with $x_{n_k} \leq 3$ shall be denoted $\langle x_k^- \rangle_1^\infty$, and a subsequence $\langle x_{n_k} \rangle_{k=1}^\infty$ with $x_{n_k} > 3$ shall be denoted $\langle x_k^+ \rangle_1^\infty$. Now, we must show that $\lim_{k\to\infty} |x_k^+ + 6| = 9 = \lim_{k\to\infty} (x_k^-)^2$ when $x_n \to 3$. Since the absolute value function and the squaring function are both continuous, we have $\lim_{k\to\infty} |x_k^+ + 6| = |\lim_{k\to\infty} x_k^+ + 6| = 9$ and $\lim_{k\to\infty} (x_k^-)^2 = (\lim_{k\to\infty} x_k^-)^2 = 9$.

Theorem 25.1 now lets us conclude that this piecewise function is continuous at $a = 3$. ■

Discontinuity at a point may be demonstrated by negating Theorem 25.1, which entails finding a sequence that does not obey $\langle f(x_n) \rangle_1^\infty \to f(a)$. The next examples show such behavior.

Example 25.2 Consider $f(x) = \sin(\tan x)$ in Figure 25.2 (and in Exercise 14.9). Clearly, $a = \frac{\pi}{2} \notin \mathcal{D}_f$, and there are infinitely many oscillations in any neighborhood of $\frac{\pi}{2}$. Let's define $f(\frac{\pi}{2}) = b$, where b is any real number we like (or dislike). To prove that f is discontinuous at $\frac{\pi}{2}$ regardless of the choice of b, we seek a sequence $\langle x_n \rangle_1^\infty \to \frac{\pi}{2}$ that yields $\sin(\tan x_n) = \pm 1$. Visually, we are looking for the abscissas of the maxima and minima of the function near $a = \frac{\pi}{2}$. This sequence will prove the function to be discontinuous at $\frac{\pi}{2}$ by Theorem 25.1.

Thus, we may take $\tan x_n = \frac{\pi}{2} + n\pi$, for $n \in \mathbb{N}$, giving us $x_n = \arctan(\frac{\pi}{2} + n\pi)$. This interesting sequence increases towards $\frac{\pi}{2}$ (verify!), and furthermore, $f(x_n) = \sin(\tan x_n) = (-1)^n$. Hence we have a case in which $\lim_{n\to\infty} x_n = \frac{\pi}{2}$, but $f(\lim_{n\to\infty} x_n) = b \neq \lim_{n\to\infty} f(x_n)$, because $\lim_{n\to\infty} f(x_n)$ does not exist. Consequently, this function is not continuous at $a = \frac{\pi}{2}$, and the discontinuity cannot be removed by defining $f(\frac{\pi}{2})$ as any real number. ■

Example 25.3 Figure 25.3 displays Dirichlet's function

$$D(x) = \begin{cases} 1 & \text{if } x \in \mathbb{Q} \\ 0 & \text{if } x \notin \mathbb{Q} \end{cases}.$$

This function would actually appear as two solid horizontal lines due to the density of rational and irrational numbers in $\mathbb{R}$. The analyst and number theorist Peter Lejeune-Dirichlet invented it to demonstrate that functions are capable of being discontinuous at *every* real number!

To prove this, first let a be a rational number. Then there exists a sequence of irrational numbers $\langle \xi_n \rangle_1^\infty \to a$ (by reason of the density of the irrationals). Consequently, $1 = D(a) = D(\lim_{n\to\infty} \xi_n) \neq \lim_{n\to\infty} D(\xi_n) = 0$. Hence, the function is discontinuous at all rational numbers. Next, let α be an irrational number. Then there exists a sequence of rational numbers $\langle r_n\rangle_1^\infty \to \alpha$ (the rationals are dense). But now, $1 = \lim_{n\to\infty} D(r_n) \neq D(\lim_{n\to\infty} r_n) = D(\alpha) = 0$. Hence, the function is discontinuous at all irrational numbers as well. ■

Example 25.4 The function in Figure 25.4 is often called the ruler function for a reason that becomes clear when vertical segments are drawn from each point down to the X axis. First, we show that

$$R(x) = \begin{cases} \frac{1}{|q|} & \text{if } x = \frac{p}{q} \text{ in lowest terms} \\ 0 & \text{if } x \text{ is irrational} \end{cases}$$

is discontinuous at every rational number $\frac{p}{q}$ reduced to lowest terms. Consider a sequence of irrational numbers $\langle \xi_n\rangle_1^\infty \to \frac{p}{q}$. Then $\langle R(\xi_n)\rangle_1^\infty \to 0$, while $R(\frac{p}{q}) \neq 0$, implying that R is discontinuous at every $\frac{p}{q}$.

To discover the behavior of the function at an irrational number α, let $\varepsilon > 0$ be given, and take $\delta = \varepsilon$. Form a sequence $\langle x_n\rangle_1^\infty \to \alpha$. If x_n is irrational, then $R(x_n) = 0 \in \mathbf{N}_0^\varepsilon$. If x_n is rational, consider the neighborhood $(\alpha - \delta, \alpha + \delta) = \mathbf{N}_\alpha^\delta$, and let $x_n = \frac{p_n}{q_n} \in \mathbf{N}_\alpha^\delta$. Then $R(\frac{p_n}{q_n}) = \frac{1}{|q_n|} \notin \mathbf{N}_0^\varepsilon$ for only a *finite* number of terms x_n, namely those which satisfy $\frac{1}{|q_n|} \geq \varepsilon$ (that is, those terms with denominators $|q_n| \leq \frac{1}{\varepsilon}$). Hence, for any $\varepsilon > 0$, only a finite number of terms of $\langle R(x_n)\rangle_1^\infty$ fall outside of $\mathbf{N}_0^\varepsilon$ even though $\mathbf{N}_\alpha^\delta$ always contains the infinite tail of $\langle x_n\rangle_1^\infty$. In short, for any $\langle x_n\rangle_1^\infty \to \alpha$, we have found that $\langle R(x_n)\rangle_1^\infty \to R(\alpha) = 0$, according to Definition 1 of Chapter 17. Theorem 25.1 now indicates that R is continuous at α. Therefore, the ruler function is continuous at irrational numbers, and only at irrational numbers—very strange indeed! ■

An **additive function** g is a function that has the property $g(x + y) = g(x) + g(y)$. For instance, $g(x) = 7x$ has it, since $g(x + y) = 7(x + y) = 7x + 7y = g(x) + g(y)$. That was easy, but our mathematical curiosity raises the question of whether there are other additive functions besides the linear $y = mx$ ones. As often happens, a simple question like this is the invitation to a great field of mathematical study—in this case, functional analysis. The following theorem employs sequences of rational numbers to prove a surprising result. The theorem has passed through the alembic of many analysts, beginning with Cauchy, and continuing with Gaston Darboux (1842–1917) in 1880. This version is found in "A Characterization of Conditional Probability" by Paul Teller and Arthur Fine, *Mathematics Magazine*, vol. 48 (Nov. 1975), pp. 267–268.

Theorem 25.2 The only nonnegative additive function $g : \mathbb{R} \to \mathbb{R}$ over a domain $[0, a]$ is the linear function $g(x) = mx$, where we establish the constant $m = \frac{g(a)}{a}$ ($g(a)$ can be any nonnegative real number we wish).

Proof. First, we show that g is strictly increasing, that is, if $0 \leq x < y \leq a$, then $g(x) < g(y)$. To see this, let $y - x = \xi \in (0, a]$. Under the premises that g is additive and nonnegative, we have $g(y) = g(x + \xi) = g(x) + g(\xi) > g(x)$.

Next, $g(0) = g(0 + 0) = 2g(0)$, implying that $g(0) = 0$. Also, $g(a) = \frac{g(a)}{a} \cdot a = ma$. Thus, so far, we have found that $g(x) = mx$ when $x = 0$ and when $x = a$.

Next, consider any division of $(0, a]$ into q equal parts $\frac{a}{q}$, where $q \in \mathbb{N}$. Because g is additive, we have

$$g(a) = g\left(q \cdot \frac{a}{q}\right) = \underbrace{g\left(\frac{a}{q}\right) + \cdots + g\left(\frac{a}{q}\right)}_{q \text{ terms}} = q \cdot g\left(\frac{a}{q}\right),$$

so that $g(\frac{a}{q}) = \frac{g(a)}{q} = m\frac{a}{q}$. We will now extend this to any fractional (rational) part of a that is in $(0, a]$, namely $\frac{p}{q}a$, where $p \in \{1, 2, \cdots, q\}$. Once again by additivity, we see that

$$g\left(p \cdot \frac{a}{q}\right) = \underbrace{g\left(\frac{a}{q}\right) + \cdots + g\left(\frac{a}{q}\right)}_{p \text{ terms}} = p \cdot g\left(\frac{a}{q}\right) = m\frac{pa}{q}.$$

We conclude that $g(x) = mx$ when $x \in \{0, \frac{1}{q}a, \frac{2}{q}a, \cdots, a\}$, or when x is any fractional part of a. Now, at last, consider any real number $x \in (0, a)$. Since $\mathbb{Q}$ is dense in $\mathbb{R}$, we can always find sequences $\langle r_n \rangle_1^\infty$ and $\langle s_n \rangle_1^\infty$ of fractional parts of a that converge to x and such that $r_n \le x \le s_n$ for all $n \in \mathbb{N}$. But g is strictly increasing, implying that $g(r_n) \le g(x) \le g(s_n)$. And by the work above, $g(r_n) = mr_n$ and $g(s_n) = ms_n$. Thus, $mr_n \le g(x) \le ms_n$, which implies by the pinching theorem for sequences that $g(x) = mx$ for all real x in $(0, a)$. The theorem now follows. **QED**

25.1 Exercises

1. Prove that the function in Figure 25.1, $y = \sin\frac{1}{x}$, is discontinuous at $x = 0$, and that it is not a removable discontinuity. Hint: Review Example 25.2.

2. Discuss the continuity of

$$f(x) = \begin{cases} \sin\frac{1}{x} + \operatorname{sgn} x & \text{if } x \neq 0 \\ 0 & \text{if } x = 0 \end{cases},$$

   especially at $x = 0$.

3. Create a function h that is discontinuous at every number of its domain, but such that $|h(x)|$ is continuous everywhere. Hint: Modify Dirichlet's function.

4. a. Where (if anywhere) is the function

$$w(x) = \begin{cases} 1 - x & \text{if } x \text{ is rational} \\ x & \text{if } x \text{ is irrational} \end{cases}$$

   over domain $[0, 1]$ continuous? (Graph it!)

   b. Show that w has an inverse function. Show also that w is not a monotonic function. Thus, we have an example of a function that is neither strictly increasing nor strictly decreasing, but nonetheless has an inverse. (This shows that the converse of Theorem 7.2 is false.)

5. From the point of view of physics, suppose a baseball oscillates with position (in kilometers) at time t given by $s(t) = 10\sin(\frac{1}{5-t})$, $t \ge 0$ (in seconds). Graph this function over $[0, 5]$. Its velocity is given by the derivative $s'(t)$. Show that the velocity becomes undefined at $t = 5$ seconds. Evaluate $|s'(4.9942)|$ to show that the ball will be oscillating violently, with a speed

of almost 3×10^5 km/sec (the speed of light). Well before that, however, the nightmarish forces pushing the baseball would have smashed it to smithereens.

6. Modify the ruler function in Example 25.4 to

$$f(x) = \begin{cases} \frac{1}{|q|} & \text{if } x = \frac{p}{q} \text{ in lowest terms} \\ 1 & \text{if } x \text{ is irrational} \end{cases}.$$

Prove that f is discontinuous everywhere.

7. Prove that if f and g are continuous over a common domain and $f(r) = g(r)$ for all rationals r, then $f(x) = g(x)$ for all x in the domain. (Suppose that f expresses a model for a physical process as an equation, and g is the result of a different, untested, model for the process, and a researcher can manage to show that $f(r) = g(r)$ only for rational numbers r. Assuming (reasonably) the continuity of the physical process, he/she may then conclude that f and g are functionally the same model. This is thought-provoking, isn't it?)

26

Theorems About Continuous Functions

A continuous function is an especially well-behaved kind of function; the theorems in this chapter will state some of the important properties that make such functions so "nice." We often say, without fear of confusion, that a property about continuous functions is a property about continuity. For instance, one nice property is that the arithmetic operations of functions are closed with regard to continuity. Our first theorem states this precisely.

Theorem 26.1 Let the functions $f, g : \mathbb{R} \to \mathbb{R}$ be continuous at $x = a$. Then the following functions are continuous at $x = a$:

A $f + g$
B $f - g$
C fg and
D $\frac{f}{g}$, provided $g(a) \neq 0$.

Proof. Each part could be proved directly from the definition of continuity, with great toil (try Exercise 26.2). However, we have the powerful Theorem 24.1 available. We are given that $\lim_{x\to a} f(x) = f\left(\lim_{x\to a} x\right) = f(a)$, and likewise for g. All the limits are understood to be taken in neighborhoods of a such that $(a - \delta, a + \delta) \subset \mathcal{D}_f \cap \mathcal{D}_g$.

A First, $(f + g)(a) = f(a) + g(a)$ exists. In the following chain of equalities, the one with the check mark is justified by Theorems 25.1 and 20.2:

$$\begin{aligned}(f+g)(a) &= (f+g)\left(\lim_{x\to a} x\right) = (f+g)(a)\\ &= f(a) + g(a) = f\left(\lim_{x\to a} x\right) + g\left(\lim_{x\to a} x\right)\\ &= \lim_{x\to a} f(x) + \lim_{x\to a} g(x) \overset{\checkmark}{=} \lim_{x\to a}(f(x) + g(x)) = \lim_{x\to a}(f+g)(x).\end{aligned}$$

By Theorem 24.1, we conclude that $f + g$ is continuous at a.

B Same proof as for **A**, with the additions changed to subtractions.

C First, $(fg)(a) = f(a) \cdot g(a)$ exists. A proof similar to the one in **A** may be written, with addition replaced by multiplication, and with the use of Theorem 20.3.

D First, $(\frac{f}{g})(a) = \frac{f(a)}{g(a)}$ exists provided $g(a) \neq 0$. A proof similar to the one in **A** may be written, with addition replaced by division, and with the use of Theorem 20.4. **QED**

Theorem 26.2 Let the function $g : \mathbb{R} \to \mathbb{R}$ be continuous at $x = a$, and let the function $f : \mathbb{R} \to \mathbb{R}$ be continuous at $b = g(a)$. Then the composition $f \circ g$ is continuous at $x = a$.

Proof. Applying Theorem 24.1, we must prove that $\lim_{x\to a}(f \circ g)(x) = (f \circ g)(\lim_{x\to a} x) = (f \circ g)(a)$. First of all, $(f \circ g)(a) = f(g(a)) = f(b)$ exists. Let $y = g(x)$. In the following chain of equalities, the ones with the check marks are due to the continuity of the two functions:

$$\lim_{x\to a}(f \circ g)(x) = \lim_{x\to a} f(g(x)) = \lim_{y\to b} f(y) \overset{\checkmark}{=} f\left(\lim_{x\to a} g(x)\right)$$
$$\overset{\checkmark}{=} f\left(g\left(\lim_{x\to a} x\right)\right) = (f \circ g)\left(\lim_{x\to a} x\right) = (f \circ g)(a). \quad (1)$$

Hence, $\lim_{x\to a}(f \circ g)(x) = (f \circ g)(a)$, and the composition is continuous at a. **QED**

Example 26.1 To see an application of each equality in *(1)*, let $g(x) = \frac{7}{x}$ and $w(x) = ||x - 4| - 2|$. We will show that $w \circ g$ is continuous at $a = 4$. In Chapter 24, g was shown to be continuous at $a = 4$, and w can be shown to be continuous at $g(4) = \frac{7}{4} = b$ (by work similar to Exercise 24.5a there). Follow the chain of equalities *(1)*, with $a = 4$:

$$\lim_{x\to 4}(w \circ g)(x) = \lim_{x\to 4}\left|\left|\frac{7}{x} - 4\right| - 2\right| = \lim_{y\to 7/4}||y - 4| - 2| = \left|\left|\left(\lim_{y\to 7/4} y\right) - 4\right| - 2\right|$$
$$= \left|\left|\left(\lim_{x\to 4}\frac{7}{x}\right) - 4\right| - 2\right| = \left|\left|\left(\frac{7}{\lim_{x\to 4} x}\right) - 4\right| - 2\right| = w\left(g\left(\lim_{x\to 4} x\right)\right)$$
$$= (w \circ g)\left(\lim_{x\to 4} x\right) = (w \circ g)(4) = \left|\left|\frac{7}{4} - 4\right| - 2\right| = \frac{1}{4}.$$

Through a rather microscopic exposition, we have shown that $\lim_{x\to 4}(w \circ g)(x) = (w \circ g)(4)$, implying continuity at $a = 4$. ■

Reconsidering the proof of **D** in Theorem 26.1, it's worth noting that since g is continuous at a and nonzero there, there is a neighborhood of a where $g(x) \neq 0$. The sense of connectedness that continuous functions possess leads us to a lemma having to do with neighborhoods over which they are strictly positive or negative.

Lemma 26.1 Let $f : \mathbb{R} \to \mathbb{R}$ be continuous over an interval $(a, b) \subseteq \mathcal{D}_f$.

A Let $f(c) > 0$ for some $c \in (a, b)$. Then there exists a neighborhood of c throughout which $f(x) > 0$.

B Let $f(c) < 0$ for some $c \in (a, b)$. Then there exists a neighborhood of c throughout which $f(x) < 0$.

Proof. **A** Given such a function f, pick $\varepsilon = \frac{1}{2}f(c)$. By continuity, there corresponds a neighborhood $(c - \delta, c + \delta)$ about c such that whenever x is in it, we have $|f(x) - f(c)| < \varepsilon = \frac{1}{2}f(c)$. Thus, $\frac{1}{2}f(c) < f(x) < \frac{3}{2}f(c)$, which means that $f(x) > 0$ throughout the neighborhood.

B Prove this part as Exercise 26.3. **QED**

The next theorem is an important consequence of continuity on an interval. It will tell us that if a function is continuous on a closed, bounded interval, and if it has at least one positive ordinate and one negative ordinate, then it must cross the X axis at least once. Thus, this theorem gives sufficient conditions for the existence of a root of the function.

Theorem 26.3 Let $f : \mathbb{R} \to \mathbb{R}$ be continuous over an interval $[a, b] \subseteq \mathcal{D}_f$, and let $f(a)$ have the opposite sign of $f(b)$. Then there exits a number $c \in (a, b)$ such that $f(c) = 0$. That is, f has a root between a and b.

Proof. Our first case is that $f(a)$ is negative and $f(b)$ is positive. Define $A = \{x \in [a, b] : f(x) < 0\}$. A is not empty, since $a \in A$, and A has an upper bound, namely b. Thus, the completeness axiom guarantees the existence of a least upper bound c. That is, $c = \sup(A)$.

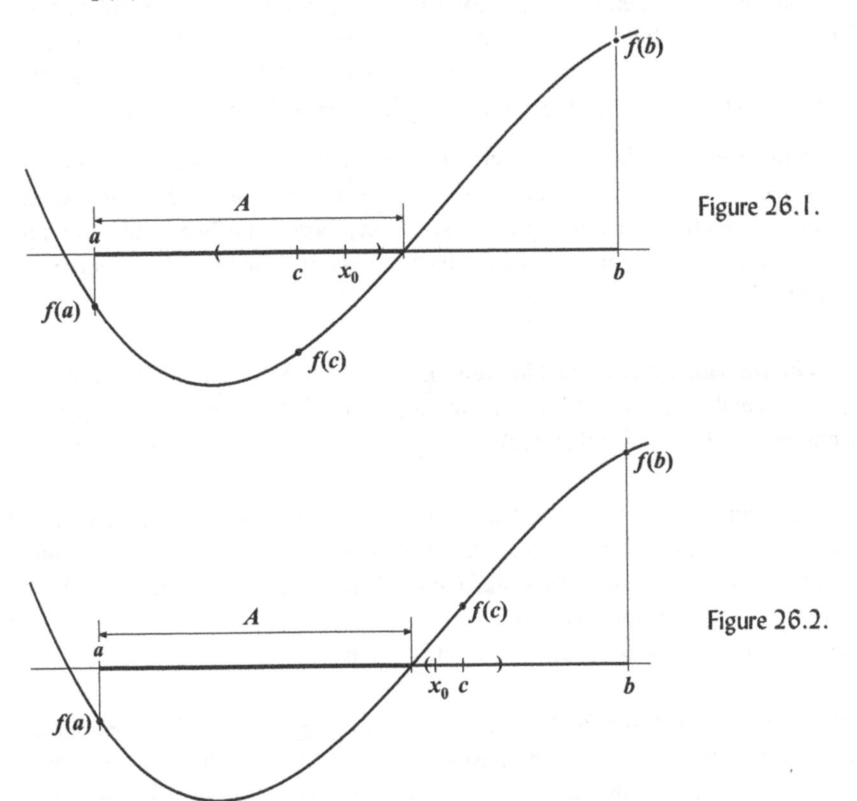

Figure 26.1.

Figure 26.2.

Suppose $f(c) < 0$ (see Figure 26.1). By Lemma 26.1B, there must be a neighborhood $(c - \delta, c + \delta)$ about c throughout which $f(x)$ is negative. Select a number $x_0 \in (c, c + \delta)$. Then $x_0 \in A$ and $x_0 > c$. But this contradicts the property that c is an upper bound for A. Hence, $f(c) \not< 0$.

Suppose instead that $f(c) > 0$ (see Figure 26.2). By Lemma 26.1A, there must be a neighborhood $(c - \delta, c + \delta)$ throughout which $f(x)$ is positive. Select a number $x_0 \in (c - \delta, c)$.

Then $x_0 < c$ and there are no members of A greater than x_0. This would make x_0 an upper bound of A that is *less* than $\sup(A)$. Again, this is a contradiction, so that $f(c) \not> 0$.

Because of trichotomy, we are driven to the conclusion that $f(c) = 0$.

The second case is that $f(a) > 0$, and $f(b) < 0$, and we can apply the above proof to the function $-f$ and conclude once more that a root exists in (a, b). **QED**

Note 1: The root theorem gives us an efficient algorithm for finding the roots of functions when it isn't possible to solve $f(x) = 0$ for x. It is called the *Bisection Method*, for reasons that will become clear.

Suppose that we have a continuous function over $[a, b]$, with $f(a) < 0$, and $f(b) > 0$. Immediately check the sign of f at the midpoint, that is, the sign of $f(\frac{a+b}{2})$. If $f(\frac{a+b}{2}) > 0$, then the root theorem states that a root must exist in the half-interval $(a, \frac{a+b}{2})$. On the other hand, if $f(\frac{a+b}{2}) < 0$, then a root must exist in the other half-interval $(\frac{a+b}{2}, b)$. If $f(\frac{a+b}{2}) = 0$, then our search is over and we feel very fortunate!

Begin the second iteration of the algorithm by checking f at the midpoint of the half-interval that contains the root, either the midpoint $\frac{3a+b}{2} \in (a, \frac{a+b}{2})$ or the midpoint $\frac{a+3b}{2} \in (\frac{a+b}{2}, b)$. The bisection that follows now yields a quarter-interval that must contain a root, and so on. After n iterations, the subinterval containing a root will have length $(b - a)/2^n$, making the bisection method quite efficient for computers. Try it in Exercise 26.5.

The next theorem generalizes the root theorem. It confirms our intuitive sense that continuous functions have no gaps or jumps in their range. More precisely, it states that continuous functions over closed, bounded intervals must intersect every horizontal line within certain bounds. This is sometimes called the intermediate value property of a function. The theorem is attributed to Bernhard Bolzano.

Theorem 26.4 The Intermediate Value Theorem Let $f : \mathbb{R} \to \mathbb{R}$ be continuous over an interval $[a, b] \subseteq \mathcal{D}_f$, and let $f(a) \neq f(b)$. Then for any y_0 that is between $f(a)$ and $f(b)$, we will find a number $c \in (a, b)$ such that $f(c) = y_0$.

Proof. As the first case, suppose $f(a) < y_0 < f(b)$. Define $g(x) = f(x) - y_0$ over $[a, b]$. Then $g(a) = f(a) - y_0 < 0$, while $g(b) = f(b) - y_0 > 0$. Also, $g(x)$ is continuous over $[a, b]$. Consequently, the root theorem applies, and tells us that there exists a root $c \in (a, b)$ for $g(x)$. That is, $0 = g(c) = f(c) - y_0$ and therefore, $f(c) = y_0$.

Do Exercise 26.6 for the case that $f(a) > y_0 > f(b)$. **QED**

Example 26.2 By Theorems 26.1 and 26.2, the function $q(x) = \sum_{n=1}^{4} |x - n|$ is continuous everywhere. Graph it over $[0, 7] \subset \mathcal{D}_q$, and consider only $x \in [2, 7]$, so that $q(2) = 4$ and $q(7) = 18$. We choose, quite randomly, $y_0 = 11 \in (4, 18)$. By the intermediate value theorem, there must exist some $c \in (2, 7)$ such that $q(c) = y_0 = 11$. There's no easy algebraic way to solve $11 = \sum_{n=1}^{4} |c - n|$, but the graph hints at $c = 5.25$, and substitution verifies that $q(5.25) = 11$. Continue the investigation of q in Exercise 26.10. ■

Draw several pictures of continuous functions $f : \mathbb{R} \to \mathbb{R}$ and it becomes increasingly clear that f must have a maximum and a minimum in any $[a, b] \subseteq \mathcal{D}_f$. No theorem so far has said this, but now we introduce a lemma that will lead us to such a theorem.

Lemma 26.2 Let $f : \mathbb{R} \to \mathbb{R}$ be continuous over an interval $[a, b] \subseteq \mathcal{D}_f$. Then f is bounded over the interval. That is, there exists a positive number B such that $|f(x)| \leq B$ for all $x \in [a, b]$. In short, continuous functions over closed, bounded intervals are bounded.

Proof. To derive a contradiction, suppose that f is as given, but is unbounded over $[a, b]$. This implies that for any $n \in \mathbb{N}$, we can find an $x_n \in [a, b]$ such that $f(x_n) \geq n$. This creates a bounded sequence $\langle x_n \rangle_1^\infty$. By the Bolzano-Weierstrass theorem for sequences, there will therefore be a convergent subsequence $\langle x_{n_k} \rangle_{k=1}^\infty$, with a limit that we may call L, and $L \in [a, b]$. Since f is continuous, $\lim_{k\to\infty} f(x_{n_k}) = f(L)$. This implies that the sequence of ordinates $\langle f(x_{n_k}) \rangle_{k=1}^\infty$, having limit L, is bounded. However, we have seen that sequence $\langle f(x_n) \rangle_1^\infty$ is unbounded, so that every one of its subsequences must be unbounded also. Our assumption must therefore be false, and we conclude that f is bounded over $[a, b]$. **QED**

Theorem 26.5 The Extreme Value Theorem Let $f : \mathbb{R} \to \mathbb{R}$ be continuous over an interval $[a, b] \subseteq \mathcal{D}_f$. Then f attains both a minimum m and a maximum M (its extrema) over the interval. To be precise, there exist points $\langle x_m, m \rangle$ and $\langle x_M, M \rangle$ belonging to f, with $x_m, x_M \in [a, b]$, and such that $f(x_m) = m \leq f(x) \leq M = f(x_M)$, for all $x \in [a, b]$. Concisely put, $m, M \in f([a, b])$.

Proof. Let f be as required. By the lemma above, f is bounded over $[a, b]$. Thus, f must have a least upper bound, and we shall call it M. That is, $M = \sup f([a, b])$, and we have established $f(x) \leq M$ for all $x \in [a, b]$.

But is M the image of some $x_M \in \mathcal{D}_f$? Because M is the supremum, we can find a number $y_n \in f([a, b])$ such that $M - \frac{1}{n} < y_n < M$ for any $n \in \mathbb{N}$. This creates a sequence $\langle f(x_n) \rangle_1^\infty$ with $f(x_n) = y_n$. Moreover, $\lim_{n\to\infty} f(x_n) = M$ by the pinching theorem for sequences. Belonging to $[a, b]$, $\langle x_n \rangle_1^\infty$ is bounded, and therefore has a convergent subsequence $\langle x_{n_k} \rangle_{k=1}^\infty$ (the Bolzano-Weierstrass theorem again) with a limit that we shall denote x_M, and $x_M \in [a, b]$. Because f is continuous, we have $\lim_{k\to\infty} f(x_{n_k}) = f\left(\lim_{k\to\infty} x_{n_k}\right) = f(x_M)$. But $\langle f(x_{n_k}) \rangle_{k=1}^\infty$ is a subsequence of the convergent sequence $\langle f(x_n) \rangle_1^\infty$, implying a common limit: $f(x_M) = \lim_{k\to\infty} f(x_{n_k}) = \lim_{n\to\infty} f(x_n) = M$. We have therefore found the claimed point $\langle x_M, M \rangle \in f$ with $x_M \in [a, b]$. Equivalently, f attains a maximum value M in $[a, b]$.

In Exercise 26.8, you will be asked to prove that f also attains a minimum value m. **QED**

Notice how important the sequential criterion for continuity (Theorem 25.1) was in the two previous proofs.

Note 2: The premises of the extreme value theorem include that the interval be closed and bounded. This is necessary, in that a function continuous over so much as a half-closed interval may fail to have an extremum, as the next example will show. Continuity is the other necessary condition. Recalling Example 23.3, $y = \frac{\sin x}{x}$ does not have a maximum over $[-1, 1]$, even given that it's a bounded function.

The theorem does not locate the extreme values; that is, it is an existence theorem. Furthermore, nothing is implied about the uniqueness of the extrema.

Example 26.3 We know by Theorem 26.1D that $f(x) = \frac{x}{x^2+1}$ is continuous everywhere, and in particular over $[0, 2]$. The minimum is $f(0) = 0$ at the endpoint $x_m = 0$, and the maximum is at $x_M = 1$, not an endpoint (verify!). If f is restricted to $(0, 2]$, then no minimum value can be attained, and the extreme value theorem doesn't apply. ■

Finally, we consider a useful property of some very specialized functions, those that are continuous and strictly monotone. It seems pretty obvious that if a continuous function has an inverse, then the inverse is continuous also. But the proof isn't obvious at all.

Theorem 26.6 Let $f : \mathbb{R} \to \mathbb{R}$ be a strictly monotone (hence, one-to-one), continuous function over any type of interval $A \subseteq \mathcal{D}_f$. Then its inverse function f^{-1} is also continuous over $f(A) \subseteq \mathcal{D}_{f^{-1}}$.

Proof. We are to prove the continuity of f^{-1} at a point $y_0 \in f(A)$. Thus, given any $\varepsilon_1 > 0$, we must be able to find a corresponding $\delta > 0$ such that if $|y - y_0| < \delta$ is a neighborhood of y_0 in $f(A) \subseteq \mathcal{D}_{f^{-1}}$, then $|f^{-1}(y) - f^{-1}(y_0)| < \varepsilon_1$. We have identified $f^{-1}(y_0) = x_0$, a point in the interior of A (see Figure 26.3). Now, for larger ε_1s, the neighborhood $(f^{-1}(y_0) - \varepsilon_1, f^{-1}(y_0) + \varepsilon_1) = (x_0 - \varepsilon_1, x_0 + \varepsilon_1)$ may not be a subset of $\mathcal{R}_{f^{-1}} = \mathcal{D}_f$. To remove this annoyance, let a be the lesser endpoint of A, and b be its greater endpoint (a may be $-\infty$ and b may be ∞). Take $\varepsilon \equiv \min\{x_0 - a, b - x_0, \varepsilon_1\} \le \varepsilon_1$. Then $(x_0 - \varepsilon, x_0 + \varepsilon) = \mathbf{N}^{\varepsilon}_{x_0} \subseteq A$ in every case, and we shall continue with ε.

Consider first the case of strictly increasing f. Begin with the neighborhood $\mathbf{N}^{\varepsilon}_{x_0} = (x_0 - \varepsilon, x_0 + \varepsilon)$ just defined. Then $f(x_0 - \varepsilon) < f(x) < f(x_0 + \varepsilon)$ for every $x \in \mathbf{N}^{\varepsilon}_{x_0}$, and we infer that the set $U = \{f(x) : x \in \mathbf{N}^{\varepsilon}_{x_0}\} = f(\mathbf{N}^{\varepsilon}_{x_0})$ is an open interval $U = (f(x_0 - \varepsilon), f(x_0 + \varepsilon))$ by Exercise 26.11c (the continuity of f is required there). Also, $U \subseteq f(A)$, by Theorem 6.1C. Since $y_0 = f(x_0)$, y_0 belongs to U. From the interior of U we can therefore create a neighborhood of y_0, namely $y_0 - \delta < y < y_0 + \delta$, where $\delta = \frac{1}{2}\min\{y_0 - f(x_0 - \varepsilon), f(x_0 + \varepsilon) - y_0\}$, similar to the situation above. Symbolically, we now have $\mathbf{N}^{\delta}_{y_0} \subset U \subseteq f(A)$; $\mathbf{N}^{\delta}_{y_0}$ is the highlighted neighborhood in Figure 26.3 (the $\frac{1}{2}$ factor is just for visual clarity). Then, by Theorem 6.2C, $f^{-1}(\mathbf{N}^{\delta}_{y_0}) \subseteq f^{-1}(U)$. Next, by Theorem 7.2, f^{-1} is also strictly increasing, so we know that $f^{-1}(U) \subseteq (x_0 - \varepsilon, x_0 + \varepsilon)$ (we can't say that $f^{-1}(U)$ is an interval, because the continuity of f^{-1} is still in question). Hence, $f^{-1}(\mathbf{N}^{\delta}_{y_0}) \subseteq (x_0 - \varepsilon, x_0 + \varepsilon) \subseteq A$.

Now, $\mathbf{N}^{\delta}_{y_0}$ is exactly the set of points satisfying $|y - y_0| < \delta$, and $(x_0 - \varepsilon, x_0 + \varepsilon)$ is written as $\{x : |f^{-1}(y) - f^{-1}(y_0)| < \varepsilon\}$. What has now been discovered is that given any $\varepsilon_1 > 0$, there

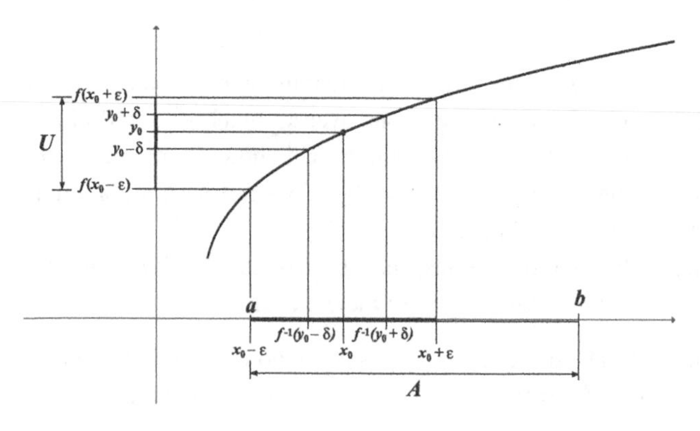

Figure 26.3. Here, f and f^{-1} are strictly increasing functions. $\mathbf{N}^{\delta}_{y_0}$ is highlighted. $f^{-1}(\mathbf{N}^{\delta}_{y_0})$ is seen as the interval $(f^{-1}(y_0 - \delta), f^{-1}(y_0 + \delta))$ in $\mathcal{R}_{f^{-1}} = \mathcal{D}_f$. The interval contains x_0, but it is not, in general, a neighborhood of x_0.

is indeed a corresponding $\delta > 0$ such that if $|y - y_0| < \delta$ in $f(A)$, then $|f^{-1}(y) - f^{-1}(y_0)| < \varepsilon \leq \varepsilon_1$ in A. This is the statement of continuity of f^{-1} at any y_0 in the interior of $f(A)$.

If it happens that y_0 is an endpoint of $f(A)$, the argument need be modified only slightly. Last of all, you are asked to supply the proof for the case of strictly decreasing f in Exercise 26.12. **QED**

Example 26.4 Modifying the domain of the function in the previous example to $[-1, 1]$, we shall rename it $h(x) = \frac{x}{x^2+1}$. It is now strictly increasing, with $\mathcal{R}_h = \left[-\frac{1}{2}, \frac{1}{2}\right]$. Since h inherits the continuity of the original, we are certain by Theorem 26.6 that it has a continuous inverse h^{-1} over $\mathcal{D}_{h^{-1}} = \left[-\frac{1}{2}, \frac{1}{2}\right]$, given by

$$h^{-1}(y) = \begin{cases} \dfrac{1 - \sqrt{1 - 4y^2}}{2y} & \text{if } y \in \left[-\dfrac{1}{2}, \dfrac{1}{2}\right] \setminus \{0\} \\ 0 & \text{if } y = 0 \end{cases}.$$

You should verify both algebraically and graphically that this is the inverse, including the fact that $\lim_{y \to 0} h^{-1}(y) = 0$, and that it too is strictly increasing. ■

26.1 Exercises

1. Use Theorems 26.1 and 26.2 and examples and exercises in Chapter 24 to explain why the following functions are continuous over their entire domains.
   a. $H(x) = \sqrt{x} + 3|x + 8|$ b. $K(x) = \frac{1}{x^2} \sin(\frac{1}{x})$ c. $F(x) = \sin x$
2. a. What is the definition of continuity for the function $f + g$ at $x = a$? It's a good workout to use it to prove that if both f and g are continuous at a, then so is $f + g$. Try it! (Compare this proof with that of Theorem 20.2.) Hints: Given any $\varepsilon > 0$, we must find a $\delta > 0$ such that if $|x - a| < \delta$, then $|(f + g)(x) - (f + g)(a)| < \varepsilon$, and as usual, x belongs to $\mathcal{D}_f \cap \mathcal{D}_g$. But we are given only the separate continuity conditions of f and g. Hence, given any $\varepsilon > 0$, expect there to be different corresponding δs, a δ_1 for f, and a δ_2 for g. How can a single δ be found for $f + g$?
   b. (This exercise requires the proof in **2a.**) What is wrong with the following argument? Let $u(x) = \frac{1}{2}x$ and $v(x) = 2x$. To prove that $u + v$ is continuous, we choose $\delta = \varepsilon$. Then given any $\varepsilon > 0$, whenever $|x - a| < \delta$, then $|u(x) + v(x) - \frac{5}{2}a| < \varepsilon$. Thus $u + v$ is continuous at any $x = a$.
3. Prove Lemma 26.1B. Then, create a function that has $f(2) = -1$, and yet is positive over a deleted neighborhood of $c = 2$. Were you able to make it continuous?
4. How does the intermediate value theorem generalize the root theorem, or put another way, how does it make the root theorem a corollary?
5. Write a short program (on a programmable calculator, for example), or a flowchart, that employs the bisection method, and test it by finding the unique root of $u(x) = \arcsin(\frac{x}{2}) - (x^2 - x - 1 + 0.3\pi)$ over $[0, 2]$. The value of the root is quite famous! Hint: For any continuous function u, the root theorem applies to $[a, b]$ whenever $u(a) \cdot u(b) < 0$. See also Exercise 19.3.

6. Prove the intermediate value theorem for the case that $f(a) > y_0 > f(b)$.

7. Example 26.3 showed that a closed, bounded interval was a necessary premise for the extreme value theorem. Create a function that shows that a closed, bounded interval is also necessary for the intermediate value theorem to hold.

8. Prove the second part of the extreme value theorem, that (with the notation of the theorem), $\langle x_m, m\rangle \in f$.

9. a. Theorem 26.1 says in part that if f and g are continuous at $x = a$, then $f + g$ is also. Create a counterexample to show that the converse is false.
   b. Theorem 26.1 also says that if f and g are continuous at $x = a$, then fg is also. Create a counterexample to show that the converse is false.

10. What does the intermediate value theorem say if we select $y_0 = 4$ in Example 26.2? Notice how the graph of $q(x) = \sum_{n=1}^{4} |x - n|$ seems roughly parabolic. Find a connection to the graph of a parabola with vertex at $\langle 2.5, 4\rangle$.

11. a. Let f (which need not be one-to-one) be continuous over $[a, b]$ in its domain ($[a, b] \subseteq \mathcal{D}_f$). Prove that the image $f([a, b])$ of a closed, bounded interval is also a closed, bounded interval (in the range of course). Does this still hold if f is constant? Hint: Show that $f([a, b]) = [m, M] \subseteq \mathcal{R}_f$.
    b. Find a fairly simple continuous function f that acts as a counterexample to the claim that the image $f((a, b))$ of an open interval is an open interval.
    c. Show that if f is continuous and monotone (not necessarily strictly monotone) over (a, b), then the image $f((a, b))$ of an open interval is an open interval (which may have one or both endpoints at infinity). Note that this result still holds when f isn't continuous at a or b; it may not even exit at either endpoint.

12. Finish the proof of Theorem 26.6, for the case of strictly decreasing f.

13. Consider the branches of the hyperbola $x^2 - y^2 = 1$ situated in the interior of Quadrants I and III. They form a bijective function with domain $(-\infty, -1) \cup (1, \infty)$. Can Theorem 26.6 be applied to this piecewise-defined function? Explain.

27

Uniform Continuity

Continuous functions play a central role in real analysis, but sometimes a stronger form of continuity is necessary. In Example 24.1, we found that every linear function $f(x) = mx + b$ is continuous at any point $x = a$. More than that, for slopes $m \neq 0$, the ε-to-δ correspondence $\delta = \frac{\varepsilon}{|m|}$ was independent of a. This means that given any positive ε, a single, corresponding, δ-neighborhood of a, namely $(a - \frac{\varepsilon}{|m|}, a + \frac{\varepsilon}{|m|})$, would yield abscissas that satisfied $|mx + b - (ma + b)| < \varepsilon$ at any location a on the X axis. This is the effect when δ is a function of *only* ε.

We now state the definition of this stronger, uniform, type of continuity.

A function f is **uniformly continuous on a set** $A \subseteq \mathcal{D}_f$ if and only if given any $\varepsilon > 0$, we can find a corresponding $\delta_\varepsilon > 0$ such that whenever $|x_1 - x_2| < \delta_\varepsilon$ with $x_1, x_2 \in A$, then $|f(x_1) - f(x_2)| < \varepsilon$. (The notation δ_ε means that δ depends only upon ε.)

First, this says that the nearness of ordinates, $|f(x_1) - f(x_2)| < \varepsilon$, over A can be controlled by a single specification on the nearness of abscissas over A, $|x_1 - x_2| < \delta_\varepsilon$. Second, we see that uniform continuity occurs over a set, not at a point. This is sometimes expressed by saying that uniform continuity is a **global** property of a function, not a local or pointwise property. Third, the definition implies that if f is uniformly continuous on A, then it certainly is continuous there. The converse is false, as Example 27.8 will demonstrate.

Example 27.1 In the introductory paragraph above, we found that all non-constant linear functions are uniformly continuous over all $\mathbb{R}$, with $\delta = \delta_\varepsilon = \frac{\varepsilon}{|m|}$. (Show that linear functions with $m = 0$ also have this property, as Exercise 27.1.) It seems reasonable that $g(x) = |x|$ should also be uniformly continuous over $\mathbb{R}$. A quick way to show this is to restate the function as

$$g(x) = \begin{cases} x & \text{if } x \geq 0 \\ -x & \text{if } x < 0 \end{cases}.$$

Since $m = \pm 1$ for this piecewise linear function, we choose $\delta_\varepsilon = \frac{\varepsilon}{|\pm 1|} = \varepsilon$. Now, given any $\varepsilon > 0$, if $|x_1 - x_2| < \delta_\varepsilon = \varepsilon$, then $|g(x_1) - g(x_2)| = ||x_1| - |x_2|| \leq |x_1 - x_2| < \varepsilon$, using Corollary 15.2. The absolute value function is therefore uniformly continuous over $\mathbb{R}$. ■

Example 27.2 Prove that $f(x) = \sqrt{x}$ is uniformly continuous over the interval $[1, \infty)$. For any

$$x_1, x_2 \in [1, \infty), \left|\sqrt{x_1} - \sqrt{x_2}\right| = \left|\frac{x_1 - x_2}{\sqrt{x_1} + \sqrt{x_2}}\right| < |x_1 - x_2|$$

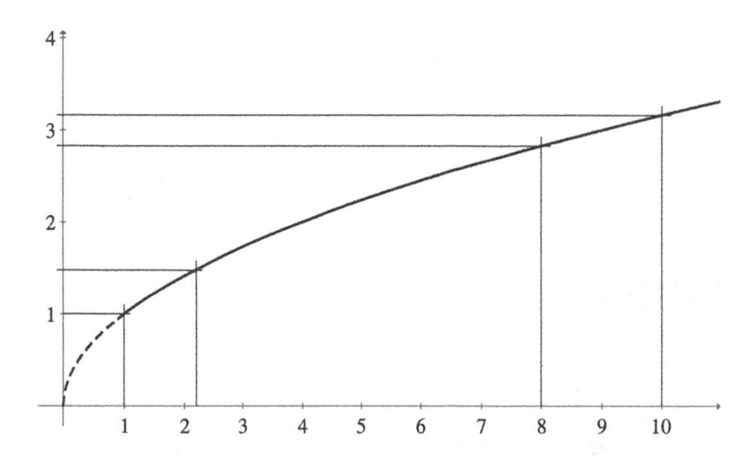

Figure 27.1. Over $[1, \infty)$, every ε-channel drawn from the Y axis is narrower than the corresponding δ-channel rising up from the X axis. Thus, we may set $\delta_\varepsilon = \varepsilon$ to fulfill the definition of uniform continuity over $[1, \infty)$.

since $\sqrt{x_1} + \sqrt{x_2} > 1$. Hence, given any $\varepsilon > 0$, choose $\delta_\varepsilon = \varepsilon$. It follows that whenever $|x_1 - x_2| < \delta_\varepsilon$, then $\left|\sqrt{x_1} - \sqrt{x_2}\right| < \varepsilon$, with δ_ε completely independent of the values of x_1 and x_2. This implies uniform continuity of f over the given interval.

From the graph in Figure 27.1 of this function over $[1, \infty)$, we see that the secant slopes are bounded (in fact, we proved that

$$\frac{\Delta y}{\Delta x} = \frac{\sqrt{x_1} - \sqrt{x_2}}{x_1 - x_2} = \frac{\left|\sqrt{x_1} - \sqrt{x_2}\right|}{|x_1 - x_2|} < 1).$$

This will turn out to be a sufficient condition for uniform continuity. ■

Example 27.3 Show that $f(x) = \sqrt{x}$ is uniformly continuous over the interval $[0, 1]$, this despite unbounded secant slopes near the origin. Take $x_1, x_2 \in [0, 1]$. We want to show that if $|x_1 - x_2| < \delta_\varepsilon$, then $\left|\sqrt{x_1} - \sqrt{x_2}\right| < \varepsilon$. Without loss of generality, we may assume that $x_2 \leq x_1$. Prove the following interesting inequality as Exercise 27.2: $0 \leq \sqrt{x_1} - \sqrt{x_2} \leq \sqrt{x_1 - x_2}$. This gives us a clue that $\delta_\varepsilon = \varepsilon^2$ may work. Pursuing this idea, consider that for any $\varepsilon > 0$, if $|x_1 - x_2| < \delta_\varepsilon = \varepsilon^2$, then $\sqrt{|x_1 - x_2|} < \varepsilon$, which implies $\left|\sqrt{x_1} - \sqrt{x_2}\right| \leq \left|\sqrt{x_1 - x_2}\right| = \sqrt{|x_1 - x_2|} < \varepsilon$. Hence, $f(x) = \sqrt{x}$ is uniformly continuous over $[0, 1]$.

To get a more concrete sense of the proof, select some values of ε, x_1, and x_2, and test the inequality $\left|\sqrt{x_1} - \sqrt{x_2}\right| < \varepsilon$:

Pick ε	$\delta_\varepsilon = \varepsilon^2$	Pick x_1	Pick x_2	Is $\lvert x_1 - x_2\rvert < \delta_\varepsilon$?	Is $\lvert\sqrt{x_1} - \sqrt{x_2}\rvert < \varepsilon$?
0.1	0.01	0.9	0.891	$0.009 < 0.01$ ✓	$\lvert\sqrt{0.9} - \sqrt{0.891}\rvert$ $\approx 0.0048 < 0.1$ ✓
0.1	0.01	0.0099	0	$0.00999 < 0.01$ ✓	$\lvert\sqrt{0.0099} - \sqrt{0}\rvert$ $\approx 0.0995 < 0.1$ ✓
0.001	10^{-6}	0.001	0.0009991	$9 \times 10^{-7} < 10^{-6}$✓	$\lvert\sqrt{0.001} - \sqrt{0.0009991}\rvert$ $\approx 1 \times 10^{-5} < 0.001$ ✓

■

The three examples above demonstrate that proving uniform continuity may involve some tricky algebra. Thus, a welcome property of continuous functions appears in our first theorem.

Note 1: Cauchy at first did not recognize the distinction between regular continuity and uniform continuity. In his defense, we should remember that he was working at a time when the definition of continuity itself was just coming into focus.

Theorem 27.1 If a function $f : \mathbb{R} \to \mathbb{R}$ is continuous over a closed and bounded interval, then it is uniformly continuous there.

The proof will be deferred to the end of this chapter, since we will need the negation of uniform continuity in order to prove the theorem's contrapositive: if a function f is not uniformly continuous over $[a, b]$, then it is not continuous over $[a, b]$.

Example 27.4 Using Theorem 27.1, $y = \cos x$ is uniformly continuous over any interval $[a, b]$, and $y = \tan x$ is uniformly continuous over any $[a, b] \subset (-\frac{\pi}{2}, \frac{\pi}{2})$. Shortly, we will be able to prove that $y = \cos x$ is uniformly continuous over all $\mathbb{R}$, whereas $y = \tan x$ certainly is not. ■

The next theorem helps us to patch together a uniformly continuous function from sections of it that were previously known to be uniformly continuous.

Theorem 27.2 Suppose I and J are intervals of any type, closed or not, and bounded or not, and suppose that $I \cap J \neq \varnothing$. Let $f : \mathbb{R} \to \mathbb{R}$ be uniformly continuous on I and J, both subsets of $\mathcal{D}_f$. Then f is uniformly continuous on $I \cup J$.

Proof. Let $\varepsilon > 0$ be given. When both $x_1, x_2 \in I$, there exists a $\delta_{I,\varepsilon} > 0$ depending only upon ε, such that the definition of uniform continuity is satisfied over I. When both $x_1, x_2 \in J$, there exists a $\delta_{J,\varepsilon} > 0$ depending only upon the same ε, such that the definition of uniform continuity is satisfied over J.

In case $x_1 \in I$ and $x_2 \in J$, we can still say that for any $x_3 \in I \cap J$, whenever $|x_1 - x_3| < \delta_{I,\varepsilon}$ then $|f(x_1) - f(x_3)| < \varepsilon$, and whenever $|x_2 - x_3| < \delta_{J,\varepsilon}$ then $|f(x_2) - f(x_3)| < \varepsilon$. But then we have that

$$|x_1 - x_2| = |x_1 - x_3 + x_3 - x_2| \le |x_1 - x_3| + |x_3 - x_2| < \delta_{I,\varepsilon} + \delta_{J,\varepsilon}.$$

Whenever x_1 and x_2 are this close together, it follows that

$$|f(x_1) - f(x_2)| = |f(x_1) - f(x_3) + f(x_3) - f(x_2)| \le |f(x_1) - f(x_3)| + |f(x_3) - f(x_2)| < 2\varepsilon.$$

To collect the three cases together, it is sufficient to use $\delta_\varepsilon = \min\{\delta_{I,\varepsilon}, \delta_{J,\varepsilon}, \delta_{I,\varepsilon} + \delta_{J,\varepsilon}\} = \min\{\delta_{I,\varepsilon}, \delta_{J,\varepsilon}\}$, which is a quantity independent of all values in $I \cup J$ and dependent solely upon ε. We now find that for any $x, y \in I \cup J$, whenever $|x - y| < \delta_\varepsilon$, then $|f(x) - f(y)| < 2\varepsilon$. Consequently, f is uniformly continuous over $I \cup J$. (A detail of form is discussed in Exercise 27.9.) **QED**

Do Exercise 27.4 now.

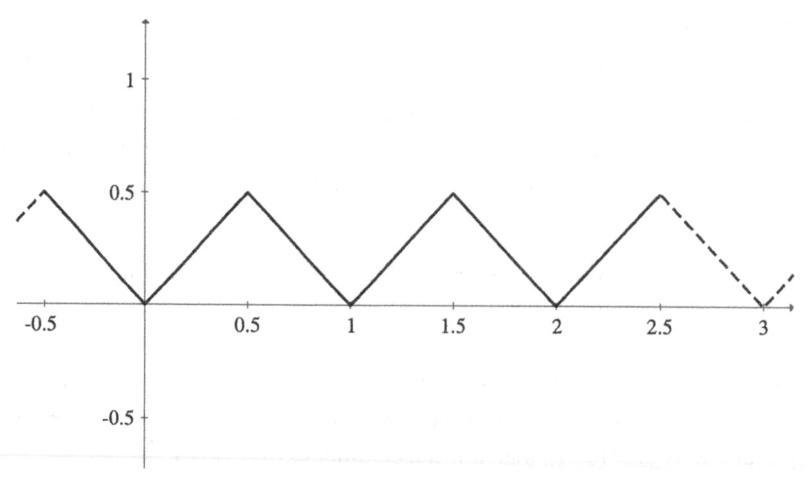

Figure 27.2. $Q(x) = |x - \lfloor x + \frac{1}{2} \rfloor|$ is uniformly continuous over $[-0.5, 2.5]$.

Example 27.5 Consider $q(x) = |x|$ with $\mathcal{D}_q = \left[-\frac{1}{2}, \frac{1}{2}\right]$. We know that q is uniformly continuous over its domain. We can create the sawtooth function Q in Figure 27.2 by duplicating q over successive intervals, in this way: $Q(x) = \left|x - \left\lfloor x + \frac{1}{2} \right\rfloor\right|$, with $\mathcal{D}_Q = \mathbb{R}$. The easiest way to show that the sawtooth function is uniformly continuous over, let's say, $\left[-\frac{1}{2}, \frac{5}{2}\right]$, is to think about the union of $q_0(x) = |x|$ over $\left[-\frac{1}{2}, \frac{1}{2}\right]$, $q_1(x) = |x - 1|$ over $\left[\frac{1}{2}, \frac{3}{2}\right]$, and $q_2(x) = |x - 2|$ over $\left[\frac{3}{2}, \frac{5}{2}\right]$. Each q_i is uniformly continuous over its respective interval, and each interval is not disjoint from its successor. Hence, Q is uniformly continuous over $\bigcup_{i=0}^{2} \left[i - \frac{1}{2}, i + \frac{1}{2}\right] = \left[-\frac{1}{2}, \frac{5}{2}\right]$, by the theorem just proved.

A warning: Theorem 27.2 can extend the uniform continuity of Q to any interval $\left[-\frac{n}{2}, \frac{n}{2}\right]$, but this does *not* imply uniform continuity over all $\mathbb{R}$. To look into this matter, see Exercise 27.7. ■

Our next theorem presents a very useful test for uniform continuity. It involves the secant slopes m_{sec} of a function, and thus may be applied nicely to function Q in the previous example.

Theorem 27.3 If a function $f : \mathbb{R} \to \mathbb{R}$ has bounded secant slopes over a set $A \subseteq \mathcal{D}_f$, then it is uniformly continuous over A. That is, if $|m_{\text{sec}}| = \left|\frac{f(x_1) - f(x_2)}{x_1 - x_2}\right| \leq M$, where M is a positive constant and x_1 and x_2 are any distinct values in A, then f is uniformly continuous over A.

Proof. We will show that the definition of uniformly continuity is satisfied. Given any $\varepsilon > 0$, we must find a corresponding $\delta_\varepsilon > 0$ depending upon only ε, such that if $|x_1 - x_2| < \delta_\varepsilon$ (and of course, $x_1, x_2 \in A$), then $|f(x_1) - f(x_2)| < \varepsilon$. Our major premise tells us that $|f(x_1) - f(x_2)| \leq M|x_1 - x_2|$ for all distinct $x_1, x_2 \in A$. We now have that $M|x_1 - x_2| < M\delta_\varepsilon$. Finish the proof by finding the specific ε-to-δ_ε correspondence that will fulfill the definition of uniform continuity, as Exercise 27.8. **QED**

Example 27.6 The sawtooth function Q in Figure 27.2 is uniformly continuous over all $\mathbb{R}$. To show that this is so, focus on the individual triangular sections. Convince yourself that all secant slopes over [0, 1] satisfy $|m_{\text{sec}}| \le 1$. We observe the same behavior over $\left[\frac{1}{2}, \frac{3}{2}\right]$. Now, Q is a periodic function, meaning that, for instance, its ordinates over [1, 2] are the same as those over [0, 1], and those over $\left[\frac{3}{2}, \frac{5}{2}\right]$ are the same as over $\left[\frac{1}{2}, \frac{3}{2}\right]$, and so on in both directions of the X axis. Therefore, for any two distinct points at most one unit apart, that is, $0 < |x_1 - x_2| \le 1$, we have $|m_{\text{sec}}| \le 1$. On the other hand, if $|x_1 - x_2| > 1$, then since $|Q(x_1) - Q(x_2)| \le \frac{1}{2}$, we find that $|m_{\text{sec}}| = \frac{|Q(x_1)-Q(x_2)|}{|x_1-x_2|} < \frac{1}{2}$. So it is always true that $|m_{\text{sec}}| \le 1$. Consequently, we may use $M = 1$ in Theorem 27.3, and conclude that $Q(x)$ is uniformly continuous over $\mathbb{R}$. ■

Example 27.7 To prove that $y = \cos x$ is uniformly continuous over all $\mathbb{R}$, we would like to prove that the secant slopes $\left|\frac{\cos x_1 - \cos x_2}{x_1 - x_2}\right|$ are bounded—and the graph of the cosine function provides evidence! To that end, we search for a helpful trigonometric identity: $\cos x_1 - \cos x_2 = -2 \sin \frac{x_1+x_2}{2} \sin \frac{x_1-x_2}{2}$. Then

$$\left|\frac{\cos x_1 - \cos x_2}{x_1 - x_2}\right| = \left|\sin \frac{x_1 + x_2}{2}\left(\frac{\sin \frac{1}{2}(x_1 - x_2)}{\frac{1}{2}(x_1 - x_2)}\right)\right| < \left|\sin \frac{x_1 + x_2}{2}\right| \le 1,$$

because $\left|\frac{\sin\theta}{\theta}\right| < 1$ for all nonzero θ. Consequently, $\left|\frac{\cos x_1 - \cos x_2}{x_1 - x_2}\right| < 1$, and the secant slopes are bounded by $M = 1$. Theorem 27.3 now indicates that $y = \cos x$ is uniformly continuous over all $\mathbb{R}$. ■

Bounded secant slopes is a sufficient, but not necessary, condition for uniform continuity. We have encountered a simple function in Example 27.3 that is uniformly continuous over [0, 1], but which doesn't have bounded secant slopes there.

Note 2: A function such as $y = \cos x$ that has bounded secant slopes over its domain is said to satisfy the *Lipshitz condition* $|f(x_1) - f(x_2)| \le M|x_1 - x_2|$, where M is a positive constant. Thus, Theorem 27.3 states that the Lipshitz condition is sufficient to guarantee uniform continuity. Functions which satisfy the Lipshitz condition with $M \le 1$ are called *contraction mappings*.

27.1 Negation of Uniform Continuity

It was useful to understand the negation of continuity in Chapter 24, and we now present the negation of uniform continuity.

A function f is **not uniformly continuous** on a set $A \subseteq \mathcal{D}_f$ if and only if there exists an $\varepsilon^* > 0$ such that for any $\delta > 0$ (regardless of how small), there are distinct numbers $x_1, x_2 \in A$ such that $|x_1 - x_2| < \delta$, but $|f(x_1) - f(x_2)| \ge \varepsilon^*$.

Note 3: Since all uniformly continuous functions on $A \subseteq \mathcal{D}_f$ are continuous there, the negation above encompasses nonuniformly continuous functions on A along with functions discontinuous on A.

Using the principles of logic in Chapter 1, we outline the negation of uniform continuity as follows.

Negate the definition of uniform continuity:

$$not\text{-}\{\forall\varepsilon, \{\text{ if } \varepsilon > 0, \text{ then } [\exists\delta_\varepsilon > 0 \text{ such that } \forall x_1, x_2 \in A,$$
$$(\text{if } |x_1 - x_2| < \delta, \text{ then } |f(x_1) - f(x_2)| < \varepsilon)]\}\}.$$

The negation works its way through the nested parentheses, δ_ε is replaced by δ, and the final form is:

$$\exists\varepsilon > 0 \text{ such that } \{\forall\delta, [\text{if } \delta > 0, \text{ then } \exists x_1, x_2 \in A$$
$$\text{such that } (|x_1 - x_2| < \delta \text{ and } |f(x_1) - f(x_2)| \geq \varepsilon)]\}.$$

Marking the existing ε as ε^* and fleshing out the statement gives us the definition above. Note that the numbers x_1 and x_2 will usually depend on the choice of δ.

Example 27.8 The simple parabola $g(x) = x^2$ is not uniformly continuous over its domain $\mathbb{R}$. Because of the parabola's symmetry, we need only consider $x \in \mathbb{R}^+$. Quite arbitrarily, select $\varepsilon^* = 1$. We must show that $|x_1^2 - x_2^2| \geq 1$ for some positive, distinct x_1 and x_2 such that $|x_1 - x_2| < \delta$, regardless of how small δ is. We should be able to do this by picking x_1 and x_2 quite close together, but really large.

Sometimes it's easier to work with the single value x_1 and link x_2 to it in some way. Thus, if we set $x_2 = x_1 + \frac{\delta}{2}$, then $|x_1 - x_2| < \delta$. Under this condition, it is necessary to show only that $|x_1^2 - (x_1 + \frac{\delta}{2})^2| \geq 1$ for some large enough x_1. Expanding and simplifying, $|x_1^2 - (x_1 + \frac{\delta}{2})^2| = |x_1\delta + \frac{1}{4}\delta^2| = \delta\,|x_1 + \frac{1}{4}\delta| > \delta x_1$. And $\delta x_1 \geq 1$ provided that we take $x_1 \geq \frac{1}{\delta}$.

We conclude that for $\varepsilon^* = 1$ and for any $\delta > 0$, choosing $x_1 \geq \frac{1}{\delta}$ and $x_2 = x_1 + \frac{\delta}{2}$ will yield $|x_1^2 - x_2^2| > \delta x_1 \geq \varepsilon^*$, even though we are certain that $0 < |x_1 - x_2| < \delta$. Consequently, $g(x) = x^2$ is not uniformly continuous over $\mathbb{R}^+$, and this implies that g cannot be uniformly continuous over $\mathbb{R}$.

To get a firm sense of the inequalities in the proof, it is always good to test them with several values of δ and x_1. Here are some typical results.

δ	Pick $x_1 \geq \frac{1}{\delta}$	$x_2 = x_1 + \frac{\delta}{2}$	Is $\lvert x_1^2 - x_2^2\rvert \geq 1 = \varepsilon^*$?
3	$x_1 = \frac{1}{2}$	$x_2 = 2$	$\lvert\frac{1}{4} - 4\rvert = 3.75 \geq 1$ ✓
0.1	$x_1 = 10$	$x_2 = 10.05$	$\lvert 10^2 - 10.05^2\rvert = 1.0025 \geq 1$ ✓
10^{-6}	$x_1 = 2 \times 10^6$	$x_2 = 2 \times 10^6 + 5 \times 10^{-7}$	$\lvert(2 \times 10^6)^2 - (2 \times 10^6 + 5 \times 10^{-7})^2\rvert \approx 2 \geq 1$ ✓

Thus, we see that $|g(x_1) - g(x_2)|$ can be kept greater than 1 even if x_2 is quite near to x_1, by taking a large enough value for x_1. ■

Finally, here is the promised proof of Theorem 27.1.

Proof of Theorem 27.1. We shall prove the contrapositive: if a function f is not uniformly continuous over $[a, b]$, then it is not continuous over $[a, b]$. Applying the negation of uniform continuity, we are given some $\varepsilon^* > 0$ such that there exists sequences of points $w_n, x_n \in [a, b]$ such that for every subscript n, we may place them as close together as we please by setting

$|w_n - x_n| < \frac{1}{n}$, but such that $|f(w_n) - f(x_n)| \geq \varepsilon^*$. Since closed intervals contain all of their cluster points (Exercise 14.1a), and since every bounded sequence has a convergent subsequence (the Bolzano-Weierstrass theorem, Theorem 19.4), $\langle x_n \rangle_1^\infty$ will have a subsequence $\langle x_{n_k} \rangle_{k=1}^\infty$ that converges to some point $\xi \in [a, b]$. The corresponding subsequence $\langle w_{n_k} \rangle_{k=1}^\infty$ will also converge to ξ, since the parent sequences obey $|w_n - x_n| < \frac{1}{n}$.

We now consider the possibility that both (a) $|f(\xi) - f(x_{n_k})| < \frac{\varepsilon^*}{2}$ and (b) $|f(\xi) - f(w_{n_k})| < \frac{\varepsilon^*}{2}$ are true for sufficiently large k. Together, they imply that $|f(w_{n_k}) - f(x_{n_k})| < \varepsilon^*$ (verify!). But this inequality blatantly contradicts the premise that $|f(w_n) - f(x_n)| \geq \varepsilon^*$ for all $n \in \mathbb{N}$. It follows that at least one of statements (a) and (b) must be false. But if either of them is false, then f fails the requirements for continuity at $\xi \in [a, b]$ as stated in Theorem 25.1, the sequential criterion for continuity.

The contrapositive of Theorem 27.1 is valid, and consequently, the theorem itself. **QED**

27.2 Exercises

1. In the introduction, we saw that linear functions with nonzero slopes were uniformly continuous. Show by the definition that the constant functions $f(x) = b$ are uniformly continuous over $\mathbb{R}$. Therefore, all linear functions are uniformly continuous.

2. Prove that for any nonnegative real numbers $x_2 \leq x_1$, $\sqrt{x_1} - \sqrt{x_2} \leq \sqrt{x_1 - x_2}$.

3. Write the details of a proof of uniform continuity from the definition for the following functions. Include a short check table such as the one in Example 27.3.
   a. $f(x) = -|3x|$ over $\mathbb{R}$.
   b. $h(x) = x + |x|$ over $\mathbb{R}$.
   c. Use either Theorem 27.2 or 27.3 to prove uniform continuity in **a** and **b**.

4. Example 24.4 showed that the square root function was continuous over its domain. Prove that $f(x) = \sqrt{x}$ is actually uniformly continuous over its domain, even though it is unbounded, and even though it has a vertical tangent within its domain.

5. Prove that $u(x) = \frac{1}{x}$ is uniformly continuous over $[1, \infty)$. Of course, Theorem 27.1 does not apply! Hint: Consider secant slopes.

6. Prove that $u(x) = \frac{1}{x}$ is continuous over $(0, 1]$, but not uniformly continuous there. Follow this outline:
   (i) For continuity, we need not reinvent the wheel. Instead, use Theorem 26.1.
   (ii) Write the negation of uniform continuity with $\varepsilon^* = 1$, as applied to function u. Because $x_1, x_2 \in (0, 1]$, values $\delta \geq 1$ are not relevant!
   (iii) As in Example 27.8, choose $x_2 = x_1 + \frac{\delta}{2}$. Derive that $\left|\frac{1}{x_1} - \frac{1}{x_2}\right| = \frac{\delta}{2x_1^2 + \delta x_1}$.
   (iv) Find a function of δ with domain $\delta \in (0, 1)$ by which you can determine an $x_1 \in (0, 1]$ that will guarantee that $\frac{\delta}{2x_1^2 + \delta x_1} > 1$. That is, find some $f(\delta) = x_1$ such that $f(\delta) \in (0, 1]$ and $\delta > 2f^2(\delta) + \delta f(\delta)$.
   Hint: We want $\delta > 2x_1^2 + \delta x_1$. Given $0 < \delta < 1$, try $f(\delta) = x_1 = \frac{\delta}{3}$ for example.
   (v) Rewrite the negation of uniform continuity as an assertion, using the information that you found in (iii) and (iv).

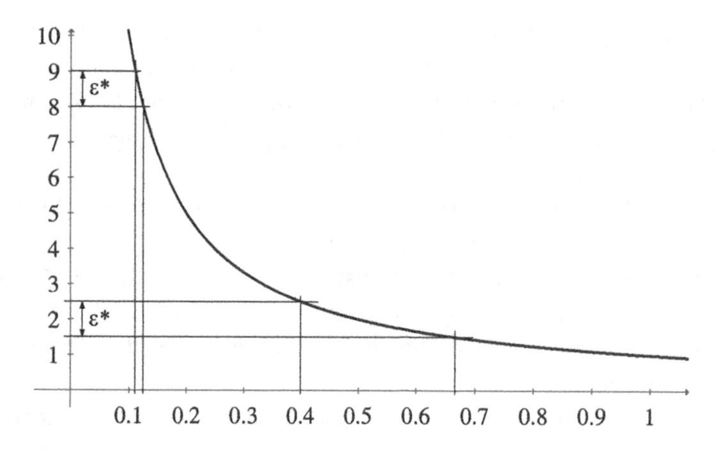

Figure 27.3.

A graph is always useful. The ε-channels shown in Figure 27.3 have widths $\varepsilon^* = 1$, and are caused by finding x_1 and $x_2 = x_1 + \frac{\delta}{2}$ so close together that $\left|\frac{1}{x_1} - \frac{1}{x_2}\right| = \varepsilon^*$.

7. What is wrong with the following argument that $y = x^2$ is uniformly continuous over its entire domain, contrary to Example 27.8?

   By Theorem 27.1, $y = x^2$ is uniformly continuous over $[0, 1]$, and in general, over $[n-1, n]$, for all n. By Theorem 27.2, $y = x^2$ is uniformly continuous over $\bigcup_{n=1}^{\infty} [n-1, n] = [0, \infty)$. By symmetry, $y = x^2$ is uniformly continuous over all $\mathbb{R}$. **QED!?**

8. Finish the proof of Theorem 27.3.

9. A detail in the proof of Theorem 27.2 my be questioned: we found that if $|x - y| < \delta_\varepsilon$, then $|f(x) - f(y)| < 2\varepsilon$. But the definition needs $|f(x) - f(y)| < \varepsilon$. Either justify the form in the proof, or adjust it to fit the definition.

10. Suppose f is continuous over (a, b), $\lim_{x \to a+} f(x) = y_0$, and $\lim_{x \to b-} f(x) = y_1$ exist. Prove that defining f^* as

$$f^*(x) \equiv \begin{cases} y_0 & \text{if } x = a \\ f(x) & \text{if } x \in (a, b) \\ y_1 & \text{if } x = b \end{cases}$$

    makes f^* uniformly continuous over $[a, b]$. Apply this lemma to
    a. $g(x) = x \sin \frac{1}{x}$ over $(0, 2)$ b. $h(x) = \frac{\ln x}{x-1}$ over $(\frac{1}{2}, 1)$

11. a. Prove that if f and g are uniformly continuous over A, then $f + g$ is uniformly continuous over A.

    b. We know that $f(x) = x$ is uniformly continuous over $\mathbb{R}$. But alas, the product of two uniformly continuous functions need not be uniformly continuous, a counterexample being $f^2(x) = x^2$ in Example 27.8. Look again at the uniformly continuous sawtooth function $Q(x) = \left|x - \lfloor x + \frac{1}{2} \rfloor\right|$ of Example 27.5. Show that the product fQ is not uniformly continuous. Hints: Draw the graph of the product function $xQ(x)$ (which we may name the "sharkstooth function"), and show that it is an odd function, so you only need to prove that it is not uniformly continuous over R^+. Do this by noting that $xQ(x) = x(x-k)$ over $x \in \left[k, k+\frac{1}{2}\right)$, where k is any natural number. To make things easier, choose $x_1 = k$ and link x_2 to x_1 so that $x_2 \in \left[k, k+\frac{1}{2}\right)$.

c. The cause of the problem in b is that $f(x) = x$ is not bounded. Prove that if both f and g are uniformly continuous and *bounded* over A, then fg is uniformly continuous over A. Hint: Reconsider the technique used in the proof of continuity of a product of functions.

d. Prove that if f and g are uniformly continuous and bounded over A, and g is bounded away from zero, then $\frac{f}{g}$ is uniformly continuous over A.

28

Infinite Series of Constants

$\frac{\pi}{2\sqrt{2}} = 1 + \frac{1}{3} - \frac{1}{5} - \frac{1}{7} + \frac{1}{9} + \frac{1}{11} - \frac{1}{13} - \frac{1}{15} + \cdots$ *etc.*
Joseph Fourier, *Analytical Theory of Heat*, 1822, Article 181.

The theory that has been developed for infinite sequences of real numbers now forms the foundation for studying infinite series. Given an infinite sequence of real numbers $\langle a_n \rangle_1^\infty$, the indicated sum $a_1 + a_2 + \cdots + a_n + \cdots$ of the terms of the sequence is an **infinite series**. We will abbreviate this as $\sum_{n=1}^{\infty} a_n$. In this case, however, indicating a sum doesn't guarantee that the sum exists. This is because Axiom RQZN1 only implies closure of the sum of a finite set of real numbers, so $\sum_{n=1}^{\infty} a_n$ takes us into uncharted territory.

To determine when an infinite series has a total, we consider the sums of successively longer portions of the front end of the series, called its **partial sums**, as follows:

$$s_1 = a_1 = \sum_{n=1}^{1} a_n$$

$$s_2 = a_1 + a_2 = \sum_{n=1}^{2} a_n = s_1 + a_2$$

$$s_3 = a_1 + a_2 + a_3 = \sum_{n=1}^{3} a_n = s_2 + a_3$$

$$\cdots$$

$$s_m = a_1 + a_2 + \cdots + a_m = \sum_{n=1}^{m} a_n = s_{m-1} + a_m$$

$$s_{m+1} = s_m + a_{m+1}$$

$$s_{m+2} = s_{m+1} + a_{m+2}$$

$$\cdots .$$

Thus, we see that the mth partial sum is explicitly stated as $s_m = \sum_{n=1}^{m} a_n$, or recursively as $s_m = s_{m-1} + a_m$ for $m \geq 2$.

The convergence of the series is now linked to the convergence of the *sequence* of partial sums $\langle s_1, s_2, s_3, \ldots \rangle$:

> An infinite series $\sum_{n=1}^{\infty} a_n$ **converges** to a real number A if and only if the sequence of partial sums $\langle s_m \rangle_1^{\infty}$ converges to A.
>
> We also say that $\sum_{n=1}^{\infty} a_n$ is a **convergent series**, and that its **sum** is A. This is written as $\lim_{m\to\infty} \sum_{n=1}^{m} a_n = A$, or as $\sum_{n=1}^{\infty} a_n = A$.

An immediate consequence of the definition is that if an infinite series converges, then its sum is unique. This follows from the fact that the limit of a sequence (of partial sums in this case) is unique, confirmed by Theorem 17.1.

Whenever $\langle s_1, s_2, s_3, \ldots \rangle$ diverges, the series $\sum_{n=1}^{\infty} a_n$ is said to **diverge**, and it is a **divergent series**. If and only if $s_m \to +\infty$, we write $\sum_{n=1}^{\infty} a_n \to +\infty$. Likewise, $\sum_{n=1}^{\infty} a_n \to -\infty$ if and only if $s_m \to -\infty$. And finally, $\sum_{n=1}^{\infty} a_n$ diverges by oscillation if and only if $\langle s_m \rangle_1^{\infty}$ diverges by oscillation.

Example 28.1 The **harmonic series** is the series $\sum_{n=1}^{\infty} \frac{1}{n} = 1 + \frac{1}{2} + \frac{1}{3} + \cdots$, which will be shown to diverge in the next chapter, and the **alternating harmonic series** is

$$\sum_{n=1}^{\infty} \frac{(-1)^{n-1}}{n} = 1 - \frac{1}{2} + \frac{1}{3} - \cdots,$$

which will be shown to converge. The latter series may also be written $\sum_{m=0}^{\infty} \frac{(-1)^m}{m+1}$ by the substitution $n = m + 1$. Thus, it is quite possible for the first subscript to be something other than 1. ■

Example 28.2 The series $\sum_{n=0}^{\infty} 0.1^n = 1 + 0.1 + 0.1^2 + 0.1^3 + \cdots$, is nicely abbreviated as $1.111\ldots$, and this brings to mind the convergence of non-terminating decimals, which was discussed in Example 21.2. This topic will be revisited later in this chapter. ■

Example 28.3 An **arithmetic series** has the form $\sum_{n=0}^{\infty}(qn + r)$, where q and r are constants, so that the difference between successive terms $a_{n+1} - a_n = q$ is constant. Thus, $\sum_{n=0}^{\infty}(\frac{3}{2}n + 1) = 1 + \frac{5}{2} + 4 + \frac{11}{2} + \cdots$ is such a series. ■

Example 28.4 A more intricately defined alternating series is

$$\sum_{n=0}^{\infty} \frac{(-1)^n}{(2n)!}\left(\frac{\pi}{4}\right)^{2n} = \frac{1}{1!} - \frac{1}{2!}\left(\frac{\pi}{4}\right)^2 + \frac{1}{4!}\left(\frac{\pi}{4}\right)^4 - \frac{1}{6!}\left(\frac{\pi}{4}\right)^6 + \cdots,$$

which happens to converge to $\cos\frac{\pi}{4} = \frac{\sqrt{2}}{2}$. Here, $a_{2n-1} = 0$ for all $n \in \mathbb{N}$, so that only even-numbered terms appear. You may recognize this as an instance of the Maclaurin series for the cosine function. ■

Applying the associative and commutative axioms to only a finite number of terms of a series is equivalent to working with just the front end of the series, which means that the endless tail of the series is left untouched. The following lemma assures us that the series is essentially unaffected.

Lemma 28.1 Rearranging (commuting) a *finite* number of terms in an infinite series, or inserting or removing a *finite* number of parentheses (associating), will not change its sum if the series is convergent, nor will it change divergence to convergence.

Proof. Let $\sum_{n=1}^{\infty} a_n$ be the series that is to be altered in any of the ways stated. Since there is only a finite number of alterations, there will be a term a_M with maximum subscript M that is affected either by having been switched with a term before it, or by being included within parentheses, or by being freed from parentheses. It follows that the partial sum $s_M = \sum_{n=1}^{M} a_n$ of the unaltered series must equal S_M, the Mth partial sum of the altered series. Then $s_{M+1} = s_M + a_{M+1} = S_{M+1}$, and in general, $s_{M+p} = s_{M+p-1} + a_{M+p} = S_{M+p}$, with $p \in \mathbb{N}$. Hence, starting at $M+1$, the tails of partial sum sequences $\langle s_m \rangle_{m=M+1}^{\infty}$ and $\langle S_m \rangle_{m=M+1}^{\infty}$ are identical. This implies that the full sequences either diverge or converge, and if they converge, the limits are equal. **QED**

Our first theorem gives us a sense of how carefully one must proceed when dealing with infinitely numerous groupings of a series. The theorem extends Axiom RQZN3 (associativity), but only under certain conditions.

Theorem 28.1 If an infinite series *converges*, then inserting infinitely many parentheses so as to group the terms does not affect the sum of the series. Equivalently, if a series with grouped terms diverges, then removing infinitely many parentheses doesn't halt divergence.

Proof. Let $\sum_{n=1}^{\infty} a_n = A$ be a convergent series. Thus, its sequence of partial sums $\langle s_n \rangle_1^{\infty} \to A$. Insert infinitely many parentheses, anywhere, shown as

$$a_1 + a_2 + \cdots + a_{n_1} + (a_{n_1+1} + a_{n_1+2} + \cdots + a_{n_2}) + a_{n_2+1} + a_{n_2+2} + \cdots + a_{n_3}$$
$$+ (a_{n_3+1} + a_{n_3+2} + \cdots + a_{n_4}) + \cdots .$$

This effectively groups all the terms, for we then have

$$(a_1 + a_2 + \cdots + a_{n_1}) + (a_{n_1+1} + a_{n_1+2} + \cdots + a_{n_2}) + (a_{n_2+1} + a_{n_2+2} + \cdots + a_{n_3})$$
$$+ \cdots + (a_{n_k+1} + a_{n_k+2} + \cdots + a_{n_{k+1}}) + \cdots .$$

We now identify the sums within the parentheses as follows,

$$b_0 = a_1 + a_2 + \cdots + a_{n_1}$$
$$b_1 = a_{n_1+1} + a_{n_1+2} + \cdots + a_{n_2}$$
$$\cdots$$
$$b_k = a_{n_k+1} + a_{n_k+2} + \cdots + a_{n_{k+1}}$$
$$\cdots ,$$

and make a new series $\sum_{k=0}^{\infty} b_k$. This series in turn produces a new sequence of partial sums $\langle s_{n_k} \rangle_{k=0}^{\infty}$ where $s_{n_0} = b_0 = \sum_{j=1}^{n_1} a_j$, $s_{n_1} = b_0 + b_1 = \sum_{j=1}^{n_2} a_j$, $\ldots$, $s_{n_k} = \sum_{j=0}^{k} b_j = \sum_{j=1}^{n_{k+1}} a_j$, and so on.

Now we observe that the new sequence of partial sums $\langle s_{n_k} \rangle_{k=0}^{\infty}$ is nothing other than a subsequence of the original $\langle s_n \rangle_1^{\infty}$. Lemma 19.2 tells us that the subsequence must converge to

A. Therefore, $\sum_{k=0}^{\infty} b_k = A$, which means that the inserted parentheses had no effect other than grouping.

The second statement in the theorem follows from the contrapositive of the first statement (verify this in Exercise 28.14). **QED**

Example 28.5 Given that the alternating harmonic series converges to a number A, the following grouping does not change the sum:

$$A = (1) + \left(-\frac{1}{2}+\frac{1}{3}\right) + \left(-\frac{1}{4}+\frac{1}{5}-\frac{1}{6}\right) + \left(\frac{1}{7}-\frac{1}{8}+\frac{1}{9}-\frac{1}{10}\right)$$
$$+ \left(\frac{1}{11}-\frac{1}{12}+\frac{1}{13}-\frac{1}{14}+\frac{1}{15}\right) + \cdots,$$

simplified as

$$A = 1 - \frac{1}{6} - \frac{13}{60} + \frac{73}{2520} + \frac{4789}{60060} - \cdots.$$

Nor can any other grouping affect the sum. ■

Example 28.6 The situation is very different when a series diverges. The classic example appears in Paradox 0.1, where we now see that $\sum_{n=0}^{\infty}(-1)^n = 1 - 1 + 1 - 1 + \cdots$ is divergent, since its partial sums form the divergent sequence $\langle 1, 0, 1, 0, \ldots \rangle$. However, inserting parentheses as $(1-1)+(1-1)+\cdots$ produces the series $0+0+\cdots$, which of course converges to zero. Hence, inserting parentheses can have the most drastic effect on divergent series. See Exercise 28.2. ■

Note 1: We can now dispose of Paradox 0.1. It arose because the associative axiom was erroneously assumed to apply to an infinite set of summands. Leibniz and his contemporaries can't be faulted severely, since the axiomatization of $\mathbb{R}$ was still far in their future.

Example 28.7 A striking consequence of the previous example is that the removal of infinitely many parentheses from a convergent series may destroy the convergence. Just reverse the train of reasoning in that example. We know that $(1-1)+(1-1)+\cdots$ converges to zero, but removing all the parentheses yields a divergent series. ■

Example 28.8 A special class of series is called **telescoping**, for the reason shown in this example. Consider

$$\sum_{n=1}^{\infty} \frac{1}{n(n+1)} = \frac{1}{2} + \frac{1}{6} + \frac{1}{12} + \cdots,$$

whose terms can be separated into partial fractions: $\frac{1}{n(n+1)} = \frac{1}{n} - \frac{1}{n+1}$. Thus, we have the equivalent grouped series (a)

$$\sum_{n=1}^{\infty}\left(\frac{1}{n}-\frac{1}{n+1}\right) = \left(1-\frac{1}{2}\right)+\left(\frac{1}{2}-\frac{1}{3}\right)+\left(\frac{1}{3}-\frac{1}{4}\right)+\cdots.$$

At this point, we would like very much to treat this as $1 + (-\frac{1}{2}+\frac{1}{2}) + (-\frac{1}{3}+\frac{1}{3}) + \cdots$, in effect telescoping the entire series (a) down to its initial term 1. But to obtain this, notice that an infinite set of parentheses must be removed from (a), leaving us with (b) $1 - \frac{1}{2} + \frac{1}{2} - \frac{1}{3} + \frac{1}{3} - \cdots$. In light of Example 28.7, we have to be cautious. Therefore, we calculate the partial sums of (b), which are $\langle 1, \frac{1}{2}, 1, \frac{2}{3}, 1, \frac{3}{4}, \ldots \rangle$, and this sequence converges to 1 (verify). Now we may apply Theorem 28.1, and we are confident that

$$1 - \frac{1}{2} + \frac{1}{2} - \frac{1}{3} + \frac{1}{3} - \cdots = \left(1 - \frac{1}{2}\right) + \left(\frac{1}{2} - \frac{1}{3}\right) + \left(\frac{1}{3} - \frac{1}{4}\right) + \cdots .$$

But by the same theorem,

$$1 - \frac{1}{2} + \frac{1}{2} - \frac{1}{3} + \frac{1}{3} - \cdots = 1 + \left(-\frac{1}{2} + \frac{1}{2}\right) + \left(-\frac{1}{3} + \frac{1}{3}\right) + \cdots ,$$

which telescopes (converges instantly) to a sum of 1. In conclusion,

$$\sum_{n=1}^{\infty} \frac{1}{n(n+1)} = \sum_{n=1}^{\infty} \left(\frac{1}{n} - \frac{1}{n+1}\right) = 1 + \sum_{n=2}^{\infty} \left(\frac{1}{n} - \frac{1}{n}\right) = 1.$$

The moral of this example is that any regrouping of infinitely many terms must be justified. ■

Theorem 28.2 The term-by-term addition of two convergent series yields a series that converges to the sum of the individual series. That is, if

$$\sum_{n=1}^{\infty} a_n = A \text{ and } \sum_{n=1}^{\infty} b_n = B, \text{ then } \sum_{n=1}^{\infty} (a_n + b_n) = \sum_{n=1}^{\infty} a_n + \sum_{n=1}^{\infty} b_n = A + B.$$

Proof. We are given that the partial sums $\langle s_n \rangle_1^\infty$ of $\sum_{n=1}^\infty a_n$ converge to A, $\langle s_1, s_2, \ldots, s_n, \ldots \rangle \to A$, and that the partial sums $\langle S_n \rangle_1^\infty$ of $\sum_{n=1}^\infty b_n$ converge to B, $\langle S_1, S_2, \ldots, S_n, \ldots \rangle \to B$. A direct application of Theorem 20.2 tells us that $\langle s_1 + S_1, s_2 + S_2, \ldots, s_n + S_n, \ldots \rangle \to A + B$. Each term $s_n + S_n$ of this sequence can be rearranged as $(a_1 + a_2 + \cdots + a_n) + (b_1 + b_2 + \cdots + b_n) = (a_1 + b_1) + (a_2 + b_2) + \cdots + (a_n + b_n)$. Thus, $\sum_{k=1}^n (a_k + b_k) = s_n + S_n$. Since we know that $\sum_{n=1}^\infty a_n + \sum_{n=1}^\infty b_n = A + B$, it follows that $\sum_{k=1}^\infty (a_k + b_k) = A + B$. **QED**

Note 2: For finite series, the convenience of writing $\sum_{n=1}^m (a_n + b_n) = \sum_{n=1}^m a_n + \sum_{n=1}^m b_n$ amounts to repeated application of the associative and commutative axioms, resulting in grouping (or ungrouping) the finite number of terms in $\langle a_n \rangle_1^m$ and $\langle b_n \rangle_1^m$. Theorem 28.2 allows us to do the same with the *infinity* of terms in $\langle a_n \rangle_1^\infty$ and $\langle b_n \rangle_1^\infty$, provided that the series are individually convergent.

See Exercise 28.6 for the effects of adding two divergent series and of adding a convergent series to a divergent series.

Theorem 28.3 The Distributive Axiom can be extended to apply to a convergent infinite series. That is, if c is a constant, and if $\sum_{n=1}^\infty a_n = A$, then $\sum_{n=1}^\infty c a_n = c \sum_{n=1}^\infty a_n = cA$.

Proof. Provide a proof as Exercise 28.7. **QED**

Perhaps the simplest *necessary* condition for convergence of a series is that the terms approach zero. This is the logical content of the following theorem.

Theorem 28.4 If the sequence of terms $\langle a_n \rangle_1^\infty \not\to 0$, then the series $\sum_{n=1}^{\infty} a_n$ will diverge.

First Proof. We prove the contrapositive statement which, by the way, is often the form in which this theorem is stated: If the series $\sum_{n=1}^{\infty} a_n$ converges, then the sequence of terms $\langle a_n \rangle_1^\infty \to 0$.

We are given, then, that the series converges to, let's say, A. Thus, regarding its partial sums, $\lim_{n\to\infty} s_n = A$. For each $n \in \mathbb{N}$, $a_{n+1} = s_{n+1} - s_n$. Taking limits, $\lim_{n\to\infty} a_{n+1} = \lim_{n\to\infty} s_{n+1} - \lim_{n\to\infty} s_n = 0$. This implies that $\lim_{n\to\infty} a_n = 0$.

Second Proof. Again using the contrapositive statement, $\lim_{n\to\infty} s_n = A$ means that for any $\varepsilon > 0$, there exists an N_ε such that for all $n > N_\varepsilon$, $|s_n - A| < \frac{\varepsilon}{2}$. This also implies that $|s_{n+1} - A| < \frac{\varepsilon}{2}$. We know that $|a_{n+1}| = |s_{n+1} - s_n|$. But by the triangle inequality, $|s_{n+1} - s_n| = |s_{n+1} - A + A - s_n| \le |s_{n+1} - A| + |s_n - A| < \varepsilon$. Hence, $|a_{n+1}| < \varepsilon$ for all $n > N_\varepsilon$, which implies that $\langle a_n \rangle_1^\infty \to 0$. **QED**

Reflecting a bit on the two proofs, we see that they are essentially the same, with the second proof employing the definition of the limit of a sequence explicitly.

Example 28.9 The series $\sum_{n=1}^{\infty} \frac{n+2}{4n-1}$ diverges, because $\langle \frac{n+2}{4n-1} \rangle_1^\infty \to \frac{1}{4}$. ■

The harmonic series serves as a reminder that $\langle a_n \rangle_1^\infty \to 0$ is *not* a sufficient condition for convergence of a series, since $a_n = \frac{1}{n} \to 0$, and yet $\sum_{n=1}^{\infty} \frac{1}{n}$ diverges, as will be proved in Corollary 29.1.

28.1 Geometric Series

A **geometric series** has the form $a + ar + ar^2 + \cdots = \sum_{n=0}^{\infty} ar^n$, where a and r are constants, and r is the **common ratio** of the series. (Set $0^0 \equiv 1$ for convenience in this context.) "Common ratio" is an appropriate name, for the ratio between successive terms is constant: $\frac{ar^{n+1}}{ar^n} = r$, this assuming neither a nor r are zero. A key fact making geometric series friendly is that a simple formula for their partial sums is available. We begin with a fairly well-known polynomial identity, $(1-r)(1+r+r^2+\cdots+r^m) = 1 - r^{m+1}$. Multiplying both sides by a and using summation notation, we get $a(1-r)\sum_{n=0}^{m} r^n = a(1-r^{m+1})$. Provided that $r \neq 1$, this yields the formula for a partial sum of a geometric series: $s_m = \sum_{n=0}^{m} ar^n = a\frac{1-r^{m+1}}{1-r}$. Note that this is equivalent to the result of Exercise 4.9.

The following theorem is of great importance since it settles the question of convergence for the entire class of geometric series.

Theorem 28.5 Let a be a nonzero constant. The geometric series $\sum_{n=0}^{\infty} ar^n$ converges if and only if the common ratio satisfies $|r| < 1$, in which case $\sum_{n=0}^{\infty} ar^n = \frac{a}{1-r}$. (If $a = 0$, then the series converges to zero regardless of r.)

Proof of sufficiency. Suppose that $|r| < 1$. For the series to converge,

$$\lim_{m\to\infty} s_m = \lim_{m\to\infty} a\frac{1-r^{m+1}}{1-r} = \frac{a}{1-r}\lim_{m\to\infty}(1-r^{m+1})$$

must exist. By Theorem 20.6, if $|r| < 1$, then $\lim_{m\to\infty} r^{m+1} = 0$. Thus, $|r| < 1$ is a sufficient condition for convergence, in which case $\lim_{m\to\infty} s_m = \frac{a}{1-r}$. In summary, if $|r| < 1$, then $\sum_{n=0}^{\infty} ar^n = \frac{a}{1-r}$. (This conclusion remains true even when a or r is zero.)

Proof of necessity. In the case that $|r| \geq 1$, $\langle r^n\rangle_0^\infty$ diverges (Theorem 20.6 again), which forces the sequence of partial sums $\sum_{n=0}^{m} ar^n$ to diverge as well, since $a \neq 0$. Thus, $\sum_{n=0}^{\infty} ar^n$ diverges. (We have proved the contrapositive of the necessity statement.) **QED**

Example 28.10 Find the sum of $\sum_{n=1}^{\infty} \frac{4^{n-3}+(-5)^{n+2}}{7^n}$. Begin by investigating $\sum_{n=1}^{\infty} \frac{4^{n-3}}{7^n}$ and $\sum_{n=1}^{\infty} \frac{(-5)^{n+2}}{7^n}$ separately. The first one is a convergent geometric series with $r = \frac{4}{7} < 1$, found this way: $\sum_{n=1}^{\infty} \frac{4^{n-3}}{7^n} = 4^{-3} \sum_{n=1}^{\infty} (\frac{4}{7})^n$. Noting that the series begins at $n = 1$, we adjust Theorem 28.5 to get $\sum_{n=1}^{\infty} ar^n = \frac{ar}{1-r}$. Hence, the first sum is

$$\frac{4^{-3}(\frac{4}{7})}{1-\frac{4}{7}} = \frac{1}{48}.$$

Likewise, $\sum_{n=1}^{\infty} \frac{(-5)^{n+2}}{7^n} = (-5)^2 \sum_{n=1}^{\infty} (\frac{-5}{7})^n$. Since $|-\frac{5}{7}| < 1$ the sum is

$$\frac{(-5)^2(-\frac{5}{7})}{1+\frac{5}{7}} = -\frac{125}{12}.$$

Since both series converge, their sum $\sum_{n=1}^{\infty} \frac{4^{n-3}+(-5)^{n+2}}{7^n}$ converges (by Theorem 28.2) to $\frac{1}{48} - \frac{125}{12} = -\frac{499}{48}$.

The series with common ratio $r = -\frac{5}{7}$ is another example of an alternating series:

$$\sum_{n=1}^{\infty} \left(\frac{-5}{7}\right)^n = -\frac{5}{7} + \frac{25}{49} - \frac{125}{343} + \cdots. \quad \blacksquare$$

Example 28.11 An avid but slightly addled coffee drinker begins with a full cup of java and sips $\frac{1}{3}$ of it, pausing to savor the aroma. He then sips $\frac{1}{3}$ of the remaining coffee, pausing again. He continues, always drinking $\frac{1}{3}$ of what remains. If this strange behavior is continued forever, how much coffee, if any, is left? (Just as strangely, assume that no evaporation or other losses occur, and that the fluid is continuous, not composed of discrete molecules.)

Interestingly, the amounts remaining form a sequence, and the amount drunk forms a related series. After the first sip, $\frac{2}{3}$ of a cup remains; after the second sip, $\frac{2}{3}(\frac{2}{3}) = \frac{4}{9}$ remains, and in general, after n sips, $(\frac{2}{3})^n$ remains. And $\langle(\frac{2}{3})^n\rangle_1^\infty \to 0$, implying that the cup is emptied after an eternity of pauses. The amounts sipped add up as $\frac{1}{3}(1) + \frac{1}{3}(\frac{2}{3}) + \frac{1}{3}(\frac{4}{9}) + \cdots$. We recognize this as a convergent geometric series, since $r = \frac{2}{3}$. Using summation notation, we have

$$\sum_{n=0}^{\infty} \frac{1}{3}\left(\frac{2}{3}\right)^n = \frac{1}{3}\sum_{n=0}^{\infty}\left(\frac{2}{3}\right)^n = \frac{1}{3}\frac{1}{(1-\frac{2}{3})} = 1.$$

This is as it should be, for when no coffee remains, it's all been sipped.

This scenario is very similar to Zeno's conception in Paradox 0.4, and yet, no paradox arises.... ■

Note 3: Paradox 0.4 will now fall. It arises in the application of mathematics to dynamics. Ancient Greek physicists mostly limited themselves to statics, probably due to their poor understanding of motion. The notion of speed as the ratio of change in position, Δs, to change in time, Δt, was not admissible to them since distance and time were incommensurable quantities. That is, a length could not be a multiple of a time increment, in contrast to different lengths, one of which was always a multiple of the other (even if the ratio was irrational). Thus, without the vital ratio that defines speed, Zeno would imagine successive positions in space occupied by Achilles and the tortoise, and quite disconnectedly, he would imagine the corresponding times of occupation by the two characters.

And so, Zeno sees Achilles covering the sequence of space increments $\langle 100, 0.1, 0.0001, \ldots \rangle$ meters, in the sequence of time increments $\langle 10, 0.01, 0.00001, \ldots \rangle$ seconds. The paradox could even then have been resolved, since the sequences produce geometric series which converge respectively to the location at which Achilles meets the tortoise, and to the instant when that happens. But unfortunately, convergence to a limit would not be explored until Archimedes' powerful mind arrived. Thus, all Zeno could reasonably do was to consider the successive distances between the two racers. Since this is also an infinite sequence of nonzero terms, a monstrous paradox resulted!

How astonished the old philosopher would have been had you been there to show him that $100 + 0.1 + 0.0001 + \cdots = 100\frac{100}{999}$ meters, and that $10 + 0.01 + 0.00001 + \cdots = 10\frac{10}{999}$ seconds. Alas, over 2000 years would elapse before mathematicians would truly understand what those ellipsis marks meant. By then, Galileo (building upon Oresme and others) would simply state $v_{Achilles} = \frac{\Delta s}{\Delta t} = 10\,\text{m}/\text{s}$, so $s_{Achilles} = 10t$ m, and $v_{tortoise} = \frac{\Delta s}{\Delta t} = 0.01\,\text{m}/\text{s}$, so that $s_{tortoise} = 100 + 0.01t$ m. Equating the distances traveled, $10t = 100 + 0.01t$, he finds the instant when the warrior catches up to the reptile: $t = 10\frac{10}{999}$ seconds, in complete agreement with the geometric series' sum. How simple in hindsight, yet how brilliant the triumph for dynamics and mathematics.

Example 28.12 In Figure 28.1, the central rays are each 30° apart. Beginning at point $P = \langle 1, 0 \rangle$, a segment PQ_1 perpendicular to ray OQ_1 is drawn. From that intersection, a second segment Q_1Q_2 perpendicular to the next ray is drawn. Continue this geometric construction and a lovely segmented spiral is formed. Is the spiral infinitely long, and if not, how long is it?

Our first segment is a leg PQ_1 of right triangle PQ_1O, with length sin 30°. Leg OQ_1 has length cos 30°, and this becomes the hypotenuse of the second triangle, Q_1Q_2O. Leg Q_1Q_2 is

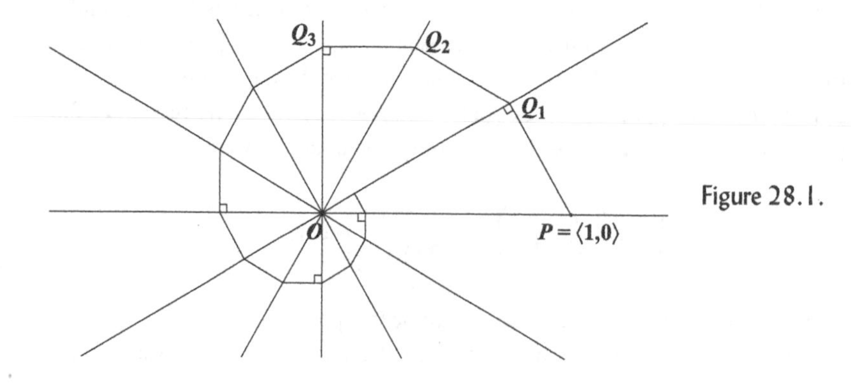

Figure 28.1.

our second segment, with length $\sin 30^\circ \cos 30^\circ$ (supply a reason). Next, leg Q_2Q_3 is our third segment, with length $\sin 30^\circ \cos^2 30^\circ$. The sum of all the segments creates a geometric series $\sum_{n=0}^{\infty} \sin 30^\circ \cos^n 30^\circ = \frac{1}{2}\sum_{n=0}^{\infty}(\frac{\sqrt{3}}{2})^n$. Since $r = \frac{\sqrt{3}}{2} < 1$, the series converges. Consequently, its sum is $\frac{1}{2}\frac{1}{(1-\sqrt{3}/2)} = 2 + \sqrt{3} \approx 3.732$, the length of the entire spiral! ■

28.2 Decimal Representation of Real Numbers

We have alluded to and used the decimal representation of real numbers fairly often, without establishing what such a representation actually is. First of all, a nonnegative terminating decimal representation is by definition a finite series of the form $a_0 + \sum_{n=1}^{m} d_n 10^{-n}$ where $a_0 \in \mathbb{N} \cup \{0\}$, and $\langle d_n \rangle_{n=1}^{m}$ contains only the decimal digits $d_n \in \{0, 1, 2, \ldots, 9\}$. There is no question about convergence in this case, and it isn't hard to see that without exception, $a_0 + \sum_{n=1}^{m} d_n 10^{-n} = \frac{p}{q} \in \mathbb{Q}$. Negative terminating decimal representations enter the picture at this point in a natural way.

By a **non-terminating decimal representation** we shall mean the infinite series $a_0 + \sum_{n=1}^{\infty} d_n 10^{-n}$ or its opposite, where $a_0 \in \mathbb{N} \cup \{0\}$, and $\langle d_n \rangle_{n=1}^{\infty}$ contains only the digits $d_n \in \{0, 1, 2, \ldots, 9\}$. If, after a certain digit, nothing but zeros follow, then the non-terminating representation is replaced by its equal terminating form, and we needn't be concerned about it. In any other case, the representation becomes an infinite series, which requires us to investigate convergence. Moreover, we shall see that such a series of rational numbers may converge to an irrational sum, for $\mathbb{Q}$ does not support R11, the completeness axiom. The convergence of this representation was proved once before, in Example 21.2, by applying the Cauchy convergence criterion. Let's look at this representation again.

Theorem 28.6 Any decimal representation converges to a unique real number.

Proof. As we have just seen, convergence is not an issue with terminating representations. For non-terminating representations, the sequence of partial sums $\langle \sum_{k=1}^{n} d_k 10^{-k} \rangle_{k=1}^{\infty}$ is monotone increasing and bounded above by 1, since $\sum_{k=1}^{\infty} 9(10^{-k}) = 1$ (verify!). The monotone convergence theorem then tells us that the representation converges to its real, unique, nonnegative, supremum. Hence, $a_0 + \sum_{n=1}^{\infty} d_n 10^{-n}$ converges to a unique real number. **QED**

With Theorem 28.6 in hand, the algorithm described in Note 2 of Chapter 19 allows us to establish the position of every real number on a number line. The converse statement, that every real number has a decimal representation, can be demonstrated by selecting a point of the real number line, and reversing the algorithm in that note. Be precise about how this is done in Exercise 28.9. The representation is not unique when it terminates, for example, $\frac{27}{4} = 6.75000\ldots = 6.74999\ldots$, but this doesn't present a problem.

Note 4: Aside from the tradition begun by Leonardo of Pisa, there is nothing special about the decimal, or base ten, representation of real numbers. Today, many people are familiar with binary (base two) representation, first investigated by Leibniz. Recall the algorithm for converting a fraction into its binary representation in Exercise 10.5. Ternary, or base three, expansions are discussed in Chapter 55.

Do you see that the geometric series $\sum_{k=0}^{\infty} \frac{a_k 10^{-1}+b_k 10^{-2}+c_k 10^{-3}}{10^{3k}}$, where a_k, b_k, and c_k are decimal digits, can be interpreted as an expansion in base 1000? Every block of three decimal digits $a_k b_k c_k$ from 000 to 999 becomes a digit in this bloated number system! Thus, π would appear as $003.141, 592, 653, \ldots$, where you are seeing only four digits in base 1000.

Different, and intriguing, number-base representations are discussed in "Deconstructing Bases: Fair, Fitting, and Fast Bases," by Thomas Q. Sibley in *Mathematics Magazine*, vol. 76, no. 5 (Dec. 2003), pp. 380–385.

Example 28.13 Example 28.2 introduces

$$\sum_{n=0}^{\infty} 0.1^n = 1 + 0.1 + 0.1^2 + 0.1^3 + \cdots = 1.111\ldots.$$

By Theorem 28.6, this represents a real number. Since $\sum_{n=0}^{\infty} 0.1^n$ is a geometric series with $r = 0.1$, its sum is $\frac{1}{1-0.1} = \frac{10}{9}$. This is the reason that we say $1.111\ldots = \frac{10}{9}$. ■

Recall that a decimal representation with a block of digits that eventually appears and repeats forever, as in $8.42\overline{905}$, is claimed to be a rational number. The following algorithm—which we shall call the xi algorithm because it sounds exotic—proves that the claim is true, and depends completely upon the positional notation of decimals. Let $\alpha = 8.42\overline{905}$. Then

$$\begin{aligned} 1000\alpha &= 8429.05905905\ldots \\ \underline{-\ \alpha} &= \underline{-\ 8.42905905\ldots} \\ 999\alpha &= 8420.63, \end{aligned}$$

from which we derive that $\alpha = \frac{842063}{99900} \in \mathbb{Q}$. The key step in this algorithm was to multiply α by 1000. In Exercise 28.8, generalize the xi algorithm so that it will handle any decimal representation of this type.

The following theorem could use the xi algorithm to prove the sufficiency part, but it seems worthwhile to see what a decimal representation with repeating blocks of digits looks like as an infinite series.

Theorem 28.7 A real number α is rational if and only if it has either a terminating decimal representation or a non-terminating representation in which eventually a block of b digits repeats endlessly ($b \in \mathbb{N}$). Hence, irrational numbers must have non-terminating, non-repeating decimal representations.

Proof of sufficiency. (We will consider only real numbers $\alpha \geq 0$, since the argument will then easily extend to negatives.) The case of a terminating representation is fairly obvious:

$$\alpha = a_0 + \sum_{n=1}^{m} d_n 10^{-n} = \frac{p}{q} \in \mathbb{Q}.$$

Next, suppose that α has a representation in which, after m digits, there appears a repeating block of b digits, $b \geq 1$. We may express the first block as

$$10^{-m}\left(\frac{a_1}{10}+\frac{a_2}{10^2}+\cdots+\frac{a_b}{10^b}\right).$$

As a series, then,

$$\alpha = a_0 + \sum_{n=1}^{m} d_n 10^{-n} + 10^{-m}\sum_{k=0}^{\infty}\frac{a_1 10^{-1} + a_2 10^{-2} + \cdots + a_b 10^{-b}}{10^{kb}},$$

(notice that the first block corresponds to the term $k = 0$).

Is α a rational number? The front part, $a_0 + \sum_{n=1}^{m} d_n 10^{-n}$, is rational, and we will prove that $\sum_{k=0}^{\infty}\frac{a_1 10^{-1}+a_2 10^{-2}+\cdots+a_b 10^{-b}}{10^{kb}}$ is also rational. Reexpressing this, we have

$$(a_1 10^{-1} + a_2 10^{-2} + \cdots + a_b 10^{-b})\sum_{k=0}^{\infty}(10^{-b})^k.$$

We now recognize a convergent geometric series since $r = 10^{-b} < 1$. Hence,

$$\sum_{k=0}^{\infty}(10^{-b})^k = \frac{1}{1-10^{-b}} = \frac{10^b}{10^b - 1}.$$

Therefore,

$$\alpha = a_0 + \sum_{n=1}^{m} d_n 10^{-n} + 10^{-m}(a_1 10^{-1} + a_2 10^{-2} + \cdots + a_b 10^{-b})\left(\frac{10^b}{10^b-1}\right),$$

which is a rational number (admittedly, an awkward one to express!).

Proof of necessity. (Again without loss of generality, consider only nonnegative real numbers.) Let $\alpha = \frac{p}{q} \geq 0$. We have seen from Exercise 8.1 that the denominator of α is of the form $2^m 5^n$ if and only if α has a terminating decimal representation. If the denominator is not of that form, then we perform a long division, $p \div q$. The only possible remainders at each step of the division belong to the set $\{1, 2, 3, \ldots, q-1\}$. Since the division algorithm does not end, one of the remainders must eventually reappear. At this point, the division algorithm repeats itself, thus yielding a repeated block of digits in the quotient. The algorithm will continue to yield the same block of digits successively and forever.

Thus, $\alpha \in \mathbb{Q}$ if and only if it has either of the stated decimal representations. It also follows that irrational numbers will have neither of the stated representations. **QED**

Note 5: Not having repeated blocks of digits does not necessarily imply that irrational numbers must display a random jumble of digits. Some, such as π, seem to have a random succession of digits. But look at the Liouville number in Chapter 8 to see an irrational number with a strict pattern in its digits. Another irrational with a clear pattern is the very curious Champernowne's constant $0.12345678910111213\ldots$.

Example 28.14 The xi algorithm found that $8.42\overline{905} = \frac{842063}{99900}$. Express this decimal as the type of series in the preceding proof. We have $a_0 = 8$, $m = 2$ initial digits, and $b = 3$, the length of the repeating block. Thus,

$$8.42\overline{905} = 8 + (4 \times 10^{-1} + 2 \times 10^{-2}) + 10^{-2} \sum_{k=0}^{\infty} \frac{9 \times 10^{-1} + 0 \times 10^{-2} + 5 \times 10^{-3}}{10^{3k}}. \quad \blacksquare$$

Trying to extend Theorem 28.2, we wonder if assuming the convergence of two series $\sum_{n=1}^{\infty} a_n$ and $\sum_{n=1}^{\infty} b_n$ is sufficient to guarantee the convergence of $\sum_{n=1}^{\infty} a_n b_n$, the term-by-term product. Unfortunately, this is in general not sufficient. Chapter 33 will give requirements that will guarantee the convergence of the series of products.

Note 6: Much more detailed than Theorem 28.7 is this theorem by one of the greatest mathematicians of all time, Carl Friedrich Gauss (1777–1855):

> Let $\frac{p}{q} \in \mathbb{Q}$, where $0 < p < q$, and the fraction is reduced. Factorize the denominator as $q = 2^i 5^j M$, where M has no factors 2 or 5. Then the decimal expansion of $\frac{p}{q}$ will begin with $m = \max\{i, j\}$ digits after the decimal point, followed by a block of b endlessly repeated digits, where b is the smallest natural number such that $10^b - 1$ is a multiple of M (that is, b solves $10^b - 1 = nM$ for some $n \in \mathbb{N}$).

Applying Gauss's theorem to the fraction $\frac{2009}{3400}$, we first factorize $q = 3400 = 2^3 5^2 \cdot 17$, so $M = 17$. Then $\frac{2009}{3400}$ will have $m = \max\{2, 3\} = 3$ initial digits after the decimal point, followed by a block of $b = 16$ repeating digits, since 16 is the smallest exponent $b \in \mathbb{N}$ that solves $10^b - 1 = 17n$. The actual decimal: $\frac{2009}{3400} = 0.590\,\overline{8823529411764705}$.

Even as a teenager, Gauss was discovering groundbreaking theorems. This small theorem appears in Gauss's first publication, the monumental *Disquisitiones arithmeticae* of 1801, which systematized the entire branch of mathematics called number theory. Mathematical rigor is at its highest in all of his publications. As was common for such men, Gauss contributed greatly to other fields of science. He was one of the inventors of the telegraph, for instance (with his friend, physicist Wilhelm Weber). Read the article "The Discovery of Ceres: How Gauss Became Famous," by Donald Teets and Karen Whitehead, in *Mathematics Magazine*, vol. 72, no. 2 (April 1999), pp. 83–93, to get a sense of Gauss's extraordinary mathematical acuity, and to learn about one of his contributions to astronomy.

28.3 Exercises

1. a. Verify that the fourth partial sum s_3 of the arithmetic series in Example 28.3 is $s_3 = 1 + \frac{5}{2} + 4 + \frac{11}{2} = \frac{3+1}{2}(a_0 + a_3)$ (why is 3 the subscript?).
   b. Prove that the partial sum $\sum_{n=0}^{m}(qn + r)$ of the arithmetic series in Example 28.3 is given in closed form by $s_m = \frac{m+1}{2}(a_0 + a_m)$. Note that the first partial sum is $s_0 = a_0 = r$. Hint: You may need the formula in Exercise 4.2a.
   c. Prove that all arithmetic series diverge, except if $q = r = 0$.
2. Something is wrong with this line of reasoning, which has Example 28.6 as a counterexample: A divergent series remains divergent when any number of parentheses are inserted, since this always creates a subsequence $\langle s_{n_k} \rangle_{k=1}^{\infty}$ of partial sums, and if the original partial sum sequence $\langle s_m \rangle_1^{\infty}$ already diverges, then so must $\langle s_{n_k} \rangle_{k=1}^{\infty}$. Where is the loophole?

3. Let $p \in \mathbb{N}$ be a constant. Verify by checking partial sums that making the change of subscript $\sum_{m=1+p}^{\infty} a_m$ (equivalent to $\sum_{n=1}^{\infty} a_{n+p}$) does not disrupt the convergence or divergence of the original series $\sum_{n=1}^{\infty} a_n$. Hint: Fill in the table

n	1	2	$\cdots$	$p-1$	p	$p+1$	$p+2$	$\cdots$
a_n	a_1	a_2	$\cdots$					$\cdots$
a_m								$\cdots$.

4. Find the sum $\sum_{n=0}^{\infty} \frac{(0.3)^{n+1}-4^{n-2}}{(-5)^{n-1}}$ as a rational number.

5. In Example 28.12, suppose that the plane is divided into n equal slices, each with central angle $\theta = \frac{2\pi}{n}$. As θ decreases, spirals have increasingly more segments, but each segment shortens in length. Determine what happens to the total length of a segmented spiral as $\theta \to 0$.

6. a. Prove the corollary that if $\sum_{n=1}^{\infty} a_n$ converges and $\sum_{n=1}^{\infty} b_n$ diverges, then $\sum_{n=1}^{\infty}(a_n + b_n)$ diverges. Hint: Do not try to use $\sum_{n=1}^{\infty} a_n + \sum_{n=1}^{\infty} b_n = \sum_{n=1}^{\infty}(a_n + b_n)$ (why?).
   b. Adding two divergent series does not guarantee a divergent result! For an uninteresting example, suppose that $\sum_{n=1}^{\infty} a_n$ diverges. Then so must $\sum_{n=1}^{\infty}(-a_n)$. But their sum converges: $\sum_{n=1}^{\infty}(a_n - a_n) = 0$. An example that isn't so trivial uses the following two divergent series: $\sum_{n=1}^{\infty} \frac{1}{n}$ and $\sum_{n=1}^{\infty} \frac{-1}{n+1}$ (we will prove the first one divergent in the next chapter). Convince yourself that the second one is divergent, and find out why $\sum_{n=1}^{\infty}(\frac{1}{n} + \frac{-1}{n+1})$ converges.
   c. Find a counterexample that shows that the converse of Theorem 28.2 is false. In other words, knowing that $\sum_{n=1}^{\infty}(a_n + b_n)$ converges does not allow us to conclude that $\sum_{n=1}^{\infty} a_n$ and $\sum_{n=1}^{\infty} b_n$ converge individually.

7. Prove Theorem 28.3, the distributive law for infinite series.

8. Generalize the xi algorithm so that it will find the rational equivalent for a decimal representation with a repeating block that is b digits long, $b = 1, 2, 3, \ldots$.

9. Prove that when a decimal representation terminates, that is, $\alpha = a_0.d_1d_2\ldots d_m$ with $d_m \neq 0$, then there is a non-terminating representation that is also equal to α, namely $\alpha = a_0.d_1d_2\ldots q999\ldots$, where $q = d_m - 1$.

10. Demonstrate the partial converse of Theorem 28.6: given a real number α on the number line, it has a decimal representation. The difficulty is that if α is has a terminating representation, then it has two representations (see the previous exercise), but if α can't be expressed as terminating, then the representation is unique! Hint: Use Note 2 of Chapter 19. To make things easier, consider just positive α.

11. Let the partial sums of $\sum_{n=1}^{\infty} a_n$ be bounded, but not necessarily convergent. Show that you can insert parentheses in the series so as to make it convergent. This gives a general setting for Example 28.6. Hint: Review the Bolzano-Weierstrass theorem for sequences (Theorem 19.4).

12. Prove that if a series of nonzero terms $\sum_{n=1}^{\infty} a_n$ converges, then $\sum_{n=1}^{\infty} \frac{1}{a_n}$ diverges, but not conversely. Hint: Use Exercise 20.8.

13. Does $\sum_{n=1}^{\infty} \sin n$ converge? Why or why not?

14. Let Cu = *an ungrouped series converges*, I = *insert infinitely many parentheses*, Cg = *the grouped series converges*. Then Theorem 28.1 symbolically says, "If (Cu and I), then Cg." Verify that the contrapositive, if *not*-Cg then *not*-(Cu and I), may be rewritten as "(I and *not*-Cg) implies *not*-Cu," which is the theorem's second statement.

29

Series with Positive Terms

$$\frac{\pi^2}{6} = \frac{1}{1^2} + \frac{1}{2^2} + \frac{1}{3^2} + \frac{1}{4^2} + \cdots$$

Leonhard Euler discovers the sum in 1734.

In order to study infinite series further, it is wise to concentrate first on series with positive (actually, nonnegative) terms. The lemma below should be compared with Theorem 28.1, which dealt with grouping and ungrouping without reference to nonnegative terms.

Lemma 29.1 If an infinite series of nonnegative terms converges, then inserting infinitely many parentheses (grouping) or removing infinitely many parentheses (ungrouping) will not change its sum, nor will grouping or ungrouping change divergence of this type of series to convergence. In short, an infinite series of nonnegative terms is unaffected by grouping or ungrouping.

Proof. Suppose that a series with nonnegative terms converges, meaning that its sequence of partial sums $\langle s_n \rangle_1^\infty$ converges. Then Theorem 28.1 applies, and the grouped series converges to the same limit as the original series. Thus, inserting parentheses does not disturb the sum.

Conversely, suppose that a grouped series of nonnegative terms converges. Write this series as $(a_1 + a_2 + \cdots + a_{n_1}) + (a_{n_1+1} + a_{n_1+2} + \cdots + a_{n_2}) + \cdots + (a_{n_k+1} + \cdots + a_{n_{k+1}}) + \cdots$, with its corresponding sequence of grouped partial sums $\langle s_{n_k} \rangle_{k=1}^\infty$. The premise implies that this is a monotone increasing sequence that converges to a limit s. Removing all the parentheses produces the sequence of partial sums $S = \langle s_1, s_2, \ldots, s_{n_1}, s_{n_1+1}, s_{n_1+2}, \ldots, s_{n_2}, s_{n_2+1}, \ldots, s_{n_3}, \ldots, s_{n_k}, \ldots \rangle$ which is, in fact, the sequence of partial sums $\langle s_n \rangle_1^\infty$ of the ungrouped series. S is, of course, monotone increasing, but it contains the subsequence $\langle s_{n_k} \rangle_{k=1}^\infty$. Therefore, S has the same limit, s. This shows that removing parentheses does not disturb the sum.

The remaining statements in the lemma will be addressed in Exercise 29.7g. **QED**

Our first, and powerful, theorem allows us to compare a series of known convergence to one whose convergence we wish to determine.

Theorem 29.1 The Comparison Test Let two series $\sum_{n=1}^{\infty} a_n$ and $\sum_{n=1}^{\infty} b_n$ be such that all terms compare term by term as $0 \leq a_n \leq b_n$. Then (1) if $\sum_{n=1}^{\infty} b_n$ converges, then $\sum_{n=1}^{\infty} a_n$ converges, and (2) if $\sum_{n=1}^{\infty} a_n$ diverges, then $\sum_{n=1}^{\infty} b_n$ diverges.

Proof. To prove (1), given that $\sum_{n=1}^{\infty} b_n$ converges, consider its sequence of partial sums $\langle B_n \rangle_1^{\infty} = \langle b_1, b_1 + b_2, \ldots \rangle$. This is a convergent sequence (why?), with limit B, say, and it is monotone increasing, implying that $B_n \leq B$ for all $n \in \mathbb{N}$. We compare it term by term with the corresponding sequence $\langle A_n \rangle_1^{\infty}$ of partial sums of $\sum_{n=1}^{\infty} a_n$, and we find that $a_1 \leq b_1$, and $a_1 + a_2 \leq b_1 + b_2$, and so on. In other words, $A_n \leq B_n$ for all $n \in \mathbb{N}$. The sequence of partial sums of $\sum_{n=1}^{\infty} a_n$ is monotone increasing, and it is bounded above by B since $A_n \leq B_n \leq B$. Hence, the monotone convergence theorem of Chapter 19 guarantees that $\langle A_n \rangle_{n=1}^{\infty}$ converges to its own supremum. This means that the original series $\sum_{n=1}^{\infty} a_n$ will converge.

Proving (2) comes down to one key word. Exercise 29.6 asks you to prove this part. **QED**

Note 1: The series in the comparison test need not satisfy $0 \leq a_n \leq b_n$ starting at $n = 1$. As long as the inequalities are satisfied for the tail end starting at some integer $M \geq 2$, we can apply the theorem to the series starting at a_M and b_M, respectively.

Corollary 29.1 The harmonic series $\sum_{n=1}^{\infty} \frac{1}{n}$ diverges.

Proof. (The gist of this proof is by Nicole Oresme, from the 1300s!) Lemma 29.1 allows us to insert parentheses at will. We will group the first two terms, after which we will form groups of 2, 4, 8, ..., 2^k, ... terms, as follows.

Number of terms	2	2^1	2^2	2^3	$\cdots$	2^k	$\cdots$
$\sum_{n=1}^{\infty} \frac{1}{n} =$	$(1 + \frac{1}{2})$	$+ (\frac{1}{3} + \frac{1}{4})$	$+ (\frac{1}{5} + \frac{1}{6} + \frac{1}{7} + \frac{1}{8})$	$+ (\frac{1}{9} + \cdots + \frac{1}{16})$	$+ \cdots +$	$(\frac{1}{2^k+1} + \cdots + \frac{1}{2^{k+1}})$	$+ \cdots$
Compared to	$\frac{1}{2}$	$+ \frac{2}{4}$	$+ \frac{4}{8}$	$+ \frac{8}{16}$	$+ \cdots +$	$\frac{2^k}{2^{k+1}}$	$+ \cdots.$

Let the grouped terms sequence be denoted as $\langle u_n \rangle_1^{\infty}$, so that for instance, $u_3 = \frac{1}{5} + \frac{1}{6} + \frac{1}{7} + \frac{1}{8}$. We will compare $\sum_{n=1}^{\infty} u_n$ to the series $\sum_{n=1}^{\infty} \frac{1}{2}$. Term by term, the comparison shows $\frac{1}{2} \leq u_n$. Since the series $\sum_{n=1}^{\infty} \frac{1}{2} \to +\infty$, we have $\sum_{n=1}^{\infty} u_n \to +\infty$ by the comparison test. Hence, by Lemma 29.1, $\sum_{n=1}^{\infty} \frac{1}{n} \to +\infty$. **QED**

(Some amazing facts about the harmonic series will be uncovered in Exercise 34.7.)

An important criterion for the convergence of series with nonnegative terms appears in the next theorem.

Theorem 29.2 Let $\sum_{n=1}^{\infty} a_n$ be a series of nonnegative terms. Then $\sum_{n=1}^{\infty} a_n$ converges if and only if the sequence of partial sums $\langle s_n \rangle_1^{\infty}$ is bounded. In fact, $\sum_{n=1}^{\infty} a_n = \sup \langle s_n \rangle_1^{\infty}$.

Proof of sufficiency. First, suppose that $\langle s_n\rangle_1^\infty$ is bounded. Since every term $a_n \geq 0$ and $s_{n+1} = s_n + a_{n+1}$, the sequence of partial sums is monotone increasing. Hence, the monotone convergence theorem (Theorem 19.1) tells us that $\lim_{n\to\infty} s_n = \sup\langle s_n\rangle_1^\infty$. By definition, $\lim_{n\to\infty} s_n = \sum_{n=1}^\infty a_n$, and we conclude that the series converges to $\sup\langle s_n\rangle_1^\infty$.

Proof of necessity. Let $\sum_{n=1}^\infty a_n$ converge, say to A. This means that $\lim_{n\to\infty} s_n = A$. Then Lemma 20.1 tells us that $\langle s_n\rangle_1^\infty$ must be bounded. **QED**

The next test uses integration, an idea that emerges when we consider $\sum_{n=1}^\infty a_n$ as a sum of rectangles of dimensions 1 by a_n. It is said that a function $f(x)$ **generates** the sequence $\langle a_n\rangle_1^\infty$ when $\mathbb{N} \subseteq \mathcal{D}_f$ and $f(n) = a_n$ for each $n \in \mathbb{N}$.

Theorem 29.3 The Integral Test Let f be a positive, decreasing function that generates the sequence $\langle a_n\rangle_1^\infty$. Then the series $\sum_{n=1}^\infty a_n$ converges if and only if the improper integral $\int_1^\infty f(x)dx$ converges. (If they do converge, we wouldn't expect the values to be equal, although that is possible.)

Proof. Consider first of all the finite area under f and over the interval $[n, n+1]$, where n is a natural number. Since f is decreasing, the areas of the two rectangles in Figure 29.1 compare to the area found by the integral as $a_{n+1} \leq \int_n^{n+1} f(x)dx \leq a_n$. These inequalities hold for any $n \in \mathbb{N}$. Let M be a natural number greater than one. If we now add the terms of the series from $n = 1$ up to $n = M - 1$, we have

$$\sum_{n=1}^{M-1} a_{n+1} \leq \sum_{n=1}^{M-1} \int_n^{n+1} f(x)dx \leq \sum_{n=1}^{M-1} a_n.$$

The expression in the middle of the triple inequality is

$$\int_1^2 f(x)dx + \int_2^3 f(x)dx + \cdots + \int_{M-1}^M f(x)dx = \int_1^M f(x)dx.$$

Thus, we now have

$$\sum_{n=1}^{M-1} a_{n+1} \leq \int_1^M f(x)dx \leq \sum_{n=1}^{M-1} a_n, \text{ for any } M \geq 2. \tag{1}$$

Focusing just on

$$\sum_{n=1}^{M-1} a_{n+1} \leq \int_1^M f(x)dx,$$

let $M \to \infty$ and suppose that

$$\lim_{M\to\infty} \int_1^M f(x)dx = \int_1^\infty f(x)dx = I$$

is a finite area. Then every partial sum $\sum_{n=1}^{M-1} a_{n+1} \leq I$. So, the sequence of partial sums is monotone increasing and bounded by I. Consequently, the series converges. We have proved that convergence of the improper integral implies convergence of the series.

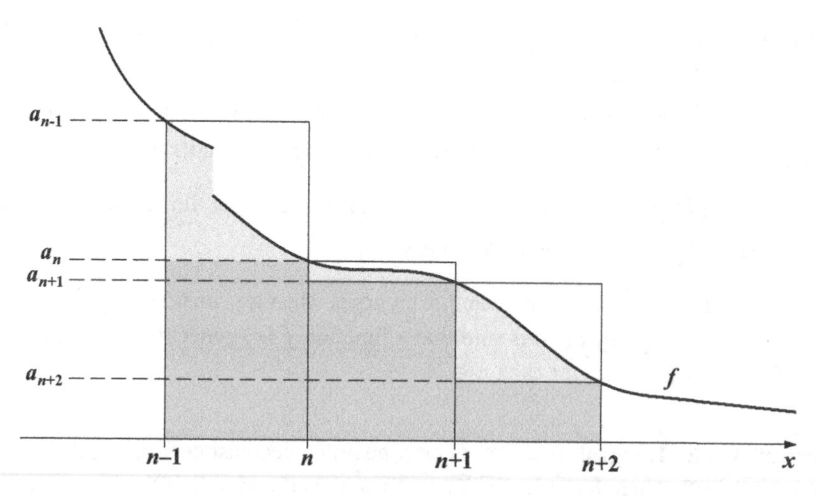

Figure 29.1. The areas of the two rectangles over base $[n, n+1]$ show that $a_{n+1} \leq \int_n^{n+1} f(x)dx \leq a_n$. Even if f is discontinuous at a point, as seen within $[n-1, n]$, $a_n \leq \int_{n-1}^{n} f(x)dx \leq a_{n-1}$holds.

Now, in (1) focus just on $\int_1^M f(x)dx \leq \sum_{n=1}^{M-1} a_n$. Again, let $M \to \infty$ but this time suppose that $\sum_{n=1}^{\infty} a_n = A$ is a convergent series. This implies that the monotone increasing sequence of areas $\langle \int_1^M f(x)dx \rangle_{M=2}^{\infty}$ is bounded by A. Hence, the integral converges. We have proved that convergence of the series implies convergence of the improper integral.

The statements of convergence of the integral and the series are therefore logically equivalent. **QED**

Note 2: In the integral test, function f need only decrease after some point x_0, as long as it generates the series from that point on. As seen above, f may even have a number of jump discontinuities. We don't even have to evaluate the improper integral. All that is necessary is to be certain that $\int_1^{\infty} f(x)dx$ is a finite number.

Also, because of the equivalence of convergence of the integral and the series, divergence of either one implies divergence of the other.

Example 29.1 Determine what values of p will cause the p**-series** $\sum_{n=1}^{\infty} \frac{1}{n^p}$ to converge and to diverge. We see that $f(x) = \frac{1}{x^p} > 0$ generates the terms of the series, and that f is decreasing over $[1, \infty)$ provided $p > 0$. We therefore proceed to evaluate $\int_1^{\infty} \frac{1}{x^p}dx$, if possible. This brings us to

$$\lim_{c\to\infty} \left(\frac{x^{-p+1}}{-p+1} \Big|_1^c \right) = \frac{1}{1-p} \lim_{c\to\infty} (c^{1-p} - 1).$$

If $p > 1$, then the limit exists, the integral converges, and the series converges by the integral test.

If $0 < p < 1$, then $\lim_{c\to\infty} (c^{1-p} - 1) \to +\infty$, the integral diverges, and the series diverges. If $p = 1$, the antiderivative is meaningless, but we know that in this case, $\sum_{n=1}^{\infty} \frac{1}{n}$ is the divergent harmonic series. And if $p \leq 0$, then the integral test can't be used, but $\langle \frac{1}{n^p} \rangle_{n=1}^{\infty} \not\to 0$ as n increases, so the series diverges.

Collecting all our observations, the p-series $\sum_{n=1}^{\infty} \frac{1}{n^p}$ converges if $p > 1$, and diverges if $p \leq 1$. ■

A free lunch comes with the integral test! If $\sum_{n=1}^{\infty} a_n$ has sum A, then we can use it to provide us with an estimate of how close we are to A when we can add only a finite number of terms, say $\sum_{n=1}^{M} a_n$. This is important, for example, when a computer program evaluates series. The program should have some internal guarantee that a 12-decimal output is accurate, when perhaps hundreds of terms have been added.

Corollary 29.2 Let $f(x)$ generate the terms of the positive series $\sum_{n=1}^{\infty} a_n$ as required by the integral test. If $\sum_{n=1}^{\infty} a_n = A$, then the Mth partial sum $s_M = \sum_{n=1}^{M} a_n$ differs from A according to the estimates $\int_{M+1}^{\infty} f(x)dx \leq A - s_M \leq \int_{M}^{\infty} f(x)dx$.

Proof. From the proof of the integral test, we have that $a_{n+1} \leq \int_{n}^{n+1} f(x)dx \leq a_n$ for any $n \in \mathbb{N}$. Applying this to the case $n - 1$ gives

$$\int_{n}^{n+1} f(x)dx \leq a_n \leq \int_{n-1}^{n} f(x)dx$$

(see this in Figure 29.1). Now, taking any $M \in \mathbb{N}$, add together terms from $M + 1$ to some integer

$$N - 1 > M + 1 : \sum_{n=M+1}^{N-1} \int_{n}^{n+1} f(x)dx \leq \sum_{n=M+1}^{N-1} a_n \leq \sum_{n=M+1}^{N-1} \int_{n-1}^{n} f(x)dx,$$

giving

$$\int_{M+1}^{N} f(x)dx \leq \sum_{k=M+1}^{N-1} a_n \leq \int_{M}^{N-1} f(x)dx.$$

Next, let N increase without bound:

$$\int_{M+1}^{\infty} f(x)dx \leq \sum_{k=M+1}^{\infty} a_n \leq \int_{M}^{\infty} f(x)dx.$$

The integrals converge by the integral test. But

$$\sum_{k=M+1}^{\infty} a_n = \sum_{k=1}^{\infty} a_n - \sum_{k=1}^{M} a_n = A - s_M,$$

so we conclude that

$$\int_{M+1}^{\infty} f(x)dx \leq A - s_M \leq \int_{M}^{\infty} f(x)dx. \quad \textbf{QED}$$

Example 29.2 Let's try the corollary on a geometric series whose sum we actually know,

$$\sum_{n=1}^{\infty} 0.96^n = \frac{0.96}{1 - 0.96} = 24 = A.$$

The function that generates the terms is $f(x) = 0.96^x$, which is decreasing over $[0, \infty)$. Suppose we add $M = 10$ terms. We are thus working with

$$\int_{11}^{\infty} 0.96^x dx \leq A - s_{10} \leq \int_{10}^{\infty} 0.96^x dx.$$

Using a bit of calculus, the integrals evaluate as

$$\int_{11}^{\infty} 0.96^x\,dx = -\frac{0.96^{11}}{\ln 0.96} \approx 15.635$$

and

$$\int_{10}^{\infty} 0.96^x\,dx = -\frac{0.96^{10}}{\ln 0.96} \approx 16.286,$$

so that $15.635 \le A - s_{10} \le 16.286$.

Knowing this much already tells us that the 10-term partial sum gives us poor precision for the value of A. Suppose that it was not possible to evaluate the sum A. Then with a calculator, $\sum_{n=1}^{10} 0.96^n = s_{10} \approx 8.044$. Thus, $15.635 \le A - 8.044 \le 16.286$, and hence, $23.679 \le A \le 24.330$. This is an estimate for A, but as expected, it isn't very precise. ■

Example 29.3 Use Corollary 29.2 to estimate the sum of $\sum_{n=1}^{\infty} \frac{1}{n^2+1}$, if it converges. The function $f(x) = \frac{1}{x^2+1}$ generates the terms of the series, and satisfies the requirements of the integral test. Since

$$\int_{1}^{\infty} \frac{1}{x^2+1}dx = \frac{\pi}{4},$$

it follows that the series converges, let's say, to B. Corollary 29.2 indicates that

$$\int_{M+1}^{\infty} \frac{1}{x^2+1}dx + s_M \le B \le \int_{M}^{\infty} \frac{1}{x^2+1}dx + s_M. \tag{2}$$

Select $M = 100$ because $a_{100} = \frac{1}{10001}$, which affects the fourth decimal place. Evaluating by calculator, $s_{100} = \sum_{n=1}^{100} \frac{1}{n^2+1} \approx 1.0667242$, and by calculus,

$$\int_{101}^{\infty} \frac{1}{x^2+1}dx = \arctan(\frac{1}{101}) \approx 0.0099007,$$

and

$$\int_{100}^{\infty} \frac{1}{x^2+1}dx = \arctan(\frac{1}{100}) \approx 0.0099997.$$

Substituting into (2) gives $1.07662 \le B \le 1.07673$, which shows agreement to three decimal places, rounded. Hence, we can confidently state that $\sum_{n=1}^{\infty} \frac{1}{n^2+1} \approx 1.077$. ■

The next test for convergence of series is often effective when terms involve powers or factorials, since it employs the ratios of successive terms. It was discovered by Jean Le Rond d'Alembert (see Note 4).

Theorem 29.4 The Ratio Test Let $\sum_{n=1}^{\infty} a_n$ be a series of positive terms, and form the ratio of successive terms $\frac{a_{n+1}}{a_n}$. Consider $\rho = \lim_{n\to\infty} \frac{a_{n+1}}{a_n}$. Then

(1) if $0 \le \rho < 1$, then $\sum_{n=1}^{\infty} a_n$ converges,
(2) if $1 < \rho \le +\infty$, then $\sum_{n=1}^{\infty} a_n$ diverges, and
(3) if $\rho = 1$ or does not exist because $(\frac{a_{n+1}}{a_n})_1^{\infty}$ oscillates, then $\sum_{n=1}^{\infty} a_n$ may either converge or diverge, so the test concludes nothing!

Proof. Let $r_n = \frac{a_{n+1}}{a_n}$, so that $\rho = \lim_{n\to\infty} r_n$. It is impossible for ρ to be negative, since all $r_n > 0$.

To prove the first case, let $0 \le \rho < 1$. This means that there exists an r and an $N \in \mathbb{N}$ such that for all $n \ge N, r_n \le r < 1$. (The existence of r with these properties is a delicate point; see Note 3.) Thus, $a_{n+1} = r_n a_n \le r a_n$ for all $n \ge N$. In particular, $a_{N+1} \le r a_N$. This implies that $a_{N+2} \le r a_{N+1} \le r^2 a_N$, and by induction, $a_{N+m} \le r^m a_N$ for all $m \in \mathbb{N}$.

The last inequality allows us to compare $\sum_{m=1}^{\infty} a_{N+m}$ with $\sum_{m=1}^{\infty} r^m a_N = a_N \sum_{m=1}^{\infty} r^m$. We have established that $|r| < 1$; hence, the geometric series $a_N \sum_{m=1}^{\infty} r^m$ converges. It follows by the comparison test that $\sum_{m=1}^{\infty} a_{N+m}$ converges also. Since this series is the tail end of the given $\sum_{n=1}^{\infty} a_n$, the given series converges when $0 \le \rho < 1$.

To prove (2), let $\rho > 1$. Hence, there exists a subscript N such that for all $n \ge N, 1 < r \le r_n$. By parallel reasoning to the first case, $a_{N+m} \ge r^m a_N$, for all $m \in \mathbb{N}$. This leads us to a comparison with the *divergent* geometric series $a_N \sum_{m=1}^{\infty} r^m$. Therefore, the tail end $\sum_{m=1}^{\infty} a_{N+m}$ diverges by the comparison test. And so, the given series $\sum_{n=1}^{\infty} a_n$ diverges when $1 < \rho \le +\infty$.

With regard to case (3), in Exercise 29.4 you are asked to create four series with positive terms: two for which $\rho = \lim_{n\to\infty} \frac{a_{n+1}}{a_n} = 1$, but one converges and the other diverges, and another two for which ρ fails to exist, but again one converges and the other diverges. It follows that the test is useless for testing convergence in case (3). **QED**

Note 3: Let's look more closely at the proof of case (1) of the ratio test, and suppose that we are not dealing with a constant series. If no such r existed, then $\langle r_n \rangle_1^{\infty}$ could have a limit of exactly 1, contradicting $\rho < 1$.

The premise that ρ exists implies that only a finite number of terms are greater than any constant r that is itself greater than ρ. For instance, if $\rho = 0.88$ and we select $r = 0.90$, then all but a finite number of terms of $\langle r_n \rangle_1^{\infty}$ fall inside of a neighborhood, say (0.86, 0.90), of ρ. The finite collection of outlier terms need not even be successive. They could be, say, $r_3 = 0.96, r_{10} = 5.34$, and $r_{47} = 0.80$.

What we are observing here is that $\rho = \limsup \langle r_n \rangle_1^{\infty}$, and this would actually formulate a stronger version of the ratio test, in which the regular limit of $\langle r_n \rangle_1^{\infty}$ need not exist (review Chapter 22).

Example 29.4 Test the series $\sum_{n=1}^{\infty} \frac{n!}{10^n}$ for convergence. We find that

$$\frac{a_{n+1}}{a_n} = \frac{(n+1)!}{10^{n+1}} \frac{10^n}{n!} = \frac{n+1}{10}.$$

Hence, $\rho = \lim_{n\to\infty} \frac{a_{n+1}}{a_n} = +\infty$, and the series diverges. Note that this isn't apparent when we look at the first few terms,

$$\frac{1}{10} + \frac{2}{100} + \frac{6}{1000} + \frac{24}{10000} + \cdots .$$

The moral is to not draw hasty conclusions from the first few terms of a series. ■

Example 29.5 Test the series $\sum_{n=1}^{\infty} \frac{\sqrt{n}+5}{\sqrt[3]{n}+3}$ for convergence. The ratio test uses

$$\rho = \lim_{n\to\infty} \left(\frac{\sqrt{n+1}+5}{\sqrt[3]{n+1}+3} \cdot \frac{\sqrt[3]{n}+3}{\sqrt{n}+5} \right) = \left(\lim_{n\to\infty} \frac{\sqrt{n+1}+5}{\sqrt{n}+5} \right) \left(\lim_{n\to\infty} \frac{\sqrt[3]{n}+3}{\sqrt[3]{n+1}+3} \right) = 1.$$

Thus, the test is inconclusive.

Next, we try the common procedure of rationalizing the numerator of $a_n = \frac{\sqrt{n}+5}{\sqrt[3]{n}+3}$, and get

$$a_n = \frac{n-25}{(\sqrt[3]{n}+3)(\sqrt{n}-5)} = \frac{1-\frac{25}{n}}{(\frac{1}{n^{1/6}}+\frac{3}{n^{1/2}})(1-\frac{5}{n^{1/2}})} \not\to 0$$

as $n \to \infty$. Hence, the series diverges.

We can also use the comparison test. Finding an expression smaller than a_n is the tricky part, but we are aided by a small clue: $\frac{\sqrt{n}}{\sqrt[3]{n}} = \sqrt[6]{n}$. Now, verify that $\sqrt[6]{n} - 1 < \frac{\sqrt{n}+5}{\sqrt[3]{n}+3}$ for all $n \in \mathbb{N}$. Since $\sum_{n=1}^{\infty}(\sqrt[6]{n}-1)$ diverges, so must the series that it is being compared to. ■

Example 29.6 If we carelessly apply the ratio test to the harmonic series, we may conclude that it converges, since $\frac{a_{n+1}}{a_n} = \frac{n}{n+1} < 1$ for all $n \in \mathbb{N}$. The oversight is that affirming $\frac{a_{n+1}}{a_n} < 1$ for all n is not sufficient. The theorem is clear: the *limit* of ratios $\frac{a_{n+1}}{a_n}$ must be less than 1. ■

Example 29.7 Let M be a fixed natural number. An interesting series with polynomial denominators is

$$\sum_{n=1}^{\infty} \frac{1}{n(n+1)(n+2)\cdots(n+M)}. \tag{3}$$

For instance, if $M = 3$, we get

$$\sum_{n=1}^{\infty} \frac{1}{n(n+1)(n+2)(n+3)} = \frac{1}{24} + \frac{1}{120} + \frac{1}{360} + \cdots + \frac{1}{n(n+1)(n+2)(n+3)} + \cdots.$$

See Exercise 29.5 for application of the ratio test and the comparison test to (3). But better than that, we are able to find a simple formula for the sum. To find the sum (3), we reexpress the nth term as

$$\begin{aligned}&\frac{1}{n(n+1)(n+2)\cdots(n+M)}\\ &\quad= \frac{1}{M}\left(\frac{1}{n(n+1)(n+2)\cdots(n+M-1)} - \frac{1}{(n+1)(n+2)\cdots(n+M)}\right).\end{aligned}$$

Then each partial sum of (3) telescopes:

$$\begin{aligned}S_m &= \sum_{n=1}^{m} \frac{1}{n(n+1)(n+2)\cdots(n+M)}\\ &= \frac{1}{M}\sum_{n=1}^{m}\left(\frac{1}{n(n+1)(n+2)\cdots(n+M-1)} - \frac{1}{(n+1)(n+2)\cdots(n+M)}\right)\\ &= \frac{1}{M}\left(\frac{1}{M!} - \frac{1}{(m+1)(m+2)\cdots(m+M)}\right).\end{aligned}$$

Since this is a closed-form expression for S_m, (3) must equal $\lim_{m\to\infty} S_m = \frac{1}{M}\left(\frac{1}{M!}\right)$. Thus, when $M = 3$, we discover that

$$\frac{1}{24} + \frac{1}{120} + \frac{1}{360} + \cdots = \frac{1}{3}\left(\frac{1}{3!}\right) = \frac{1}{18}. \quad ■$$

Note 4: The surname "Le Rond d'Alembert" is not that of the man's parents. There had been an affair between an artillery officer, the Chevalier Destouches, and Madame de Tencin, a profligate Parisian socialite. And so, a foundling was brought in from the steps of St.-Jean-Le-Rond, a tiny chapel in Paris, one cold day late in 1717. Since his mother repudiated him, the young man adopted the church's name, for it had lovingly adopted him in his youth. Destouches, in contrast, reclaimed his honor by providing for Le Rond's education and livelihood. The origin of the other surname, d'Alembert, is not clear.

Jean Le Rond d'Alembert became a mathematician, physicist, and philosopher of great influence in turbulent 18th century France. Among other achievements, he extended the applications of Newton's mechanics, collaborated with Euler and others to develop the study of differential equations, and became the editor of all scientific articles for the *Encyclopédie*, a massive work that was intended to present all of human knowledge from the point of view of the French Enlightenment. He wrote the first satisfactory definition of a limit for an article in that work, as we saw in Chapter 17. And, in a treatise of mathematics named the *Opuscules mathématiques* (eight volumes, written between 1761 and 1780!), d'Alembert gave us his ratio test for convergence.

D'Alembert never reconciled with his mother. Perhaps her most selfless act towards him, therefore, was that she let him live, rather than die of exposure, as could easily have been done in those days. One cannot but wonder how many unborn mathematicians were never offered even that smallest cup of mercy.

29.1 Exercises

1. Determine whether each series in the following diverse collection converges or diverges, using the theorems in this and the previous chapter. You should be able to determine the exact sum of the series with asterisks (some with the help of your calculus text). Some series can be proved to converge or diverge by more than one method, which makes great practice!

a. $\sum_{n=1}^{\infty} e^{-2n}$ *

b. $\sum_{n=2}^{\infty} \frac{1}{n^2-1}$ *

c. $\sum_{n=1}^{\infty} \frac{\sqrt{n}}{n+10}$

d. $\sum_{n=0}^{\infty} \frac{8n^4}{n!}$

e. $\sum_{n=1}^{\infty} \frac{n+6}{\sqrt{n}+30}$

f. $\frac{1}{\ln 2} + \frac{1}{\ln 3} + \frac{1}{\ln 4} + \frac{1}{\ln 5} + \cdots$

g. $\sum_{n=1}^{\infty} \frac{1}{\sqrt{n!}}$

h. $\sum_{n=0}^{\infty} (\cos n - \sin n)$

i. $\sum_{n=1}^{\infty} \frac{n^2}{\pi^n}$

j. $\frac{1}{1\cdot 3} + \frac{1}{2\cdot 4} + \frac{1}{3\cdot 5} + \frac{1}{4\cdot 6} + \cdots$ *

k. $\sum_{n=1}^{\infty} \frac{1}{n^3+3n^2+2n}$ *

l. $8 + \frac{8}{\sqrt[3]{2}} + \frac{8}{\sqrt[3]{3}} + \frac{8}{\sqrt[3]{4}} + \cdots$

m. $\sum_{n=1}^{\infty} \frac{(n!)^2}{(2n)!}$

n. $\sum_{n=0}^{\infty} \frac{n^{12.7}}{1.03^n}$

o. $\sum_{n=2}^{\infty} \frac{1}{n \ln n}$

p. $\sum_{n=0}^{\infty} \frac{(e^n)^2}{n!}$ *

q. $\sum_{n=0}^{\infty} \frac{e^{(n^2)}}{n!}$

r. $1 + \frac{1}{4} + \frac{1}{9} + \frac{1}{16} + \cdots$

s. $\sum_{n=1}^{\infty} \frac{1}{n}\left(\frac{2}{3}\right)^n$ *

t. $\sum_{n=0}^{\infty} \frac{(\sqrt{2})^n}{n!}$ *

u. $\frac{1}{3} + \frac{1\cdot 2}{3\cdot 5} + \frac{1\cdot 2\cdot 3}{3\cdot 5\cdot 7} + \frac{1\cdot 2\cdot 3\cdot 4}{3\cdot 5\cdot 7\cdot 9} + \cdots$

v. $\sum_{n=0}^{\infty} \frac{n!+1}{(n+3)!}$

2. Regarding Example 29.4,
   a. At what n do the terms $a_n = \frac{n!}{10^n}$ begin to increase in value?
   b. Prove the more general case: $\sum_{n=1}^{\infty} \frac{n!}{b^n}$ diverges for any positive constant b.

3. We know that $\langle n! \rangle_1^\infty$ increases extremely fast, but the coupled exponential sequence $\langle n^n \rangle_1^\infty$ increases even faster. Use the ratio test to show that the latter sequence increases so fast, in fact, that $\sum_{n=1}^{\infty} \frac{n!}{n^n}$ converges.

4. a. To show that the ratio test is inconclusive when $\rho = 1$, find a series with $\rho = 1$ that you know is divergent, and find another series with $\rho = 1$ that is convergent. Hint: What do you remember about p-series?
   b. To show that the ratio test is inconclusive when ρ does not exist due to oscillation of the ratio of terms $\langle \frac{a_{n+1}}{a_n} \rangle_1^\infty$, construct a series that converges and a series that diverges under that condition. Hint: For a convergent series, use $2 + \frac{2}{2^2} + \frac{3}{2^1} + \frac{4}{2^4} + \frac{5}{2^3} + \frac{6}{2^6} + \frac{7}{2^5} + \cdots$, and observe the nature of the oscillation of $\langle \frac{a_{n+1}}{a_n} \rangle_1^\infty$. A divergent series with oscillating $\langle \frac{a_{n+1}}{a_n} \rangle_1^\infty$ is not hard to invent.

5. a. Evaluate $\sum_{n=1}^{\infty} \frac{(n-1)!}{(n+4)!}$.
   b. Evaluate $\sum_{n=1}^{\infty} \frac{1}{(n+1)(n+2)(n+3)(n+4)}$.
   c. Does the ratio test prove convergence for (3) in Example 29.7? Does the comparison test prove convergence?

6. Finish proving Theorem 29.1, the comparison test.

7. Theorem 28.1 and Lemma 29.1 proved two statements each (but the lemma goes further). They are compared below, where we let Σ be any series whatsoever, and Φ be any series of positive terms:

Statements	Theorem 28.1	Statements	Lemma 29.1
S	If Σ converges, then grouped Σ converges.	Q	If Φ converges, then grouped Φ converges.
P	If grouped Σ diverges, then Σ diverges.	R	If grouped Φ converges, then Φ converges.

   a. What logical relationship do statements S and P have in Theorem 28.1?
   b. What about Q and R in Lemma 29.1?
   Answer the following as true or false, based on Theorem 28.1, Lemma 29.1, and the examples in Chapters 28 and this chapter:
   c. If grouped Φ diverges, then Φ diverges.
   d. If Φ diverges, then grouped Φ diverges.
   e. If grouped Σ converges, then Σ converges.
   f. If Σ diverges, then grouped Σ diverges.
   g. Lemma 29.1 may be rewritten as: Φ converges if and only if grouped Φ converges, or equivalently, Φ diverges if and only if grouped Φ diverges. The proof there covered the implications in statements Q and R above. Finish the proof. In short, the convergence behavior of a series of *nonnegative* terms is immune to insertion or removal of parentheses.

8. Devise an alternative proof of Theorem 28.6, based on the comparison test.

9. Up to now, the only convergent series whose sums we have actually evaluated are the geometric and the telescoping series. Quite a number of techniques can be found in the mathematical literature for summing specific types of infinite series (besides Taylor series,

which will be studied in later chapters). Here is a very nice one, "Evaluating the Sum of the Series $\sum_{k=1}^{\infty} \frac{k^j}{M^k}$," by Alan Gorfin in *The College Mathematics Journal*, vol. 20, no. 4 (Sept. 1989), pp. 329–331 (we will change j to q and k to n for consistency).

In the series $T_q = \sum_{n=1}^{\infty} \frac{n^q}{M^n}$, q is a nonnegative integer constant and M is a real number constant: $|M| > 1$. If $q = 0$, then $T_0 = \sum_{n=1}^{\infty} \frac{1}{M^n} = \frac{1}{M-1}$ is a plain geometric series. We shall evaluate $\sum_{n=1}^{\infty} \frac{n^q}{M^n}$ exactly, for any $q \in \mathbb{N} \cup \{0\}$.

To evaluate $T_1 = \sum_{n=1}^{\infty} \frac{n}{M^n}$, consider

$$MT_1 = M\sum_{n=1}^{\infty} \frac{n}{M^n} = \sum_{n=1}^{\infty} \frac{n}{M^{n-1}} = \sum_{n=0}^{\infty} \frac{n+1}{M^n},$$

which equals

$$\sum_{n=0}^{\infty} \frac{1}{M^n} + \sum_{n=0}^{\infty} \frac{n}{M^n}$$

by Theorem 28.2. Hence, $MT_1 = \frac{M}{M-1} + T_1$, which has the unique solution $T_1 = \frac{M}{(M-1)^2}$. This is a brand-new formula:

$$\sum_{n=1}^{\infty} \frac{n}{M^n} = \frac{1}{M} + \frac{2}{M^2} + \frac{3}{M^3} + \cdots = \frac{M}{(M-1)^2}.$$

Next, to evaluate

$$T_2 = \sum_{n=1}^{\infty} \frac{n^2}{M^n},$$

try the same technique:

$$\begin{aligned} MT_2 &= \sum_{n=1}^{\infty} \frac{n^2}{M^{n-1}} = \sum_{n=0}^{\infty} \frac{(n+1)^2}{M^n} = \sum_{n=0}^{\infty} \frac{n^2}{M^n} + 2\sum_{n=0}^{\infty} \frac{n}{M^n} + \sum_{n=0}^{\infty} \frac{1}{M^n} \\ &= T_2 + 2T_1 + \frac{M}{M-1}. \end{aligned}$$

Solving

$$MT_2 = T_2 + \frac{2M}{(M-1)^2} + \frac{M}{M-1}$$

yields another new formula:

$$T_2 = \sum_{n=1}^{\infty} \frac{n^2}{M^n} = \frac{M^2 + M}{(M-1)^3}.$$

The technique can be applied to any $T_q = \sum_{n=1}^{\infty} \frac{n^q}{M^n}$, where we must expand the binomial in the numerator of $MT_q = \sum_{n=0}^{\infty} \frac{(n+1)^q}{M^n}$. The general result is

$$T_q = \sum_{n=1}^{\infty} \frac{n^q}{M^n} = \frac{1 + \binom{q}{1}T_{q-1} + \binom{q}{2}T_{q-2} + \cdots + \binom{q}{q-1}T_1 + T_0}{M-1}.$$

Even though it is a recursive form, this is the exact sum of such series! (We will return to this type of series in Exercises 44.9 and 44.12.) To experience the versatility of this formula, find the sums of:

a. $\sum_{n=1}^{\infty} \frac{n}{2(5^n)}$

b. $\sum_{n=1}^{\infty} \frac{n^2}{\pi^n}$ (this is the series in Exercise 29.1i above)

c. $\sum_{n=0}^{\infty} \frac{n^3-3n-2}{9^n}$

d. $\sum_{n=1}^{\infty}(-1)^n \frac{n^2}{3^n}$ Hint: Use $\frac{n^2}{(-3)^n}$.

30

Further Tests for Series with Positive Terms

The tests for convergence in the previous chapter go a long way, but this subject is vast and deep. For instance, we sometimes speak of a "more refined test." This means that a certain test gives conclusions for all the series that another test can decide, and also gives conclusions for still other series. Cauchy discovered the following test, which does indeed give conclusions for some series that the ratio test can't handle.

Theorem 30.1 The Root Test Let $\sum_{n=1}^{\infty} a_n$ be a series of nonnegative terms, and consider $\sigma = \lim_{n\to\infty} \sqrt[n]{a_n}$. Then

(1) if $0 \le \sigma < 1$, then $\sum_{n=1}^{\infty} a_n$ converges,
(2) if $1 < \sigma \le +\infty$, then $\sum_{n=1}^{\infty} a_n$ diverges, and
(3) if $\sigma = 1$ or does not exist because $\langle \sqrt[n]{a_n} \rangle_1^{\infty}$ oscillates, then $\sum_{n=1}^{\infty} a_n$ may either converge or diverge, so the test concludes nothing.

Proof. Let $r_n = \sqrt[n]{a_n}$, so that $\sigma = \lim_{n\to\infty} r_n$. To prove the first case, let $0 \le \sigma < 1$. This means that there exists r and $N \in \mathbb{N}$ such that for all subscripts $n \ge N$, $\sqrt[n]{a_n} \le r < 1$. (The existence of r is discussed in the footnote after the proof of the ratio test in the previous chapter.) Consequently, $a_n \le r^n < 1$ for all $n \ge N$. We may therefore compare $\sum_{n=N}^{\infty} a_n$ with the convergent geometric series $\sum_{n=N}^{\infty} r^n$. By the comparison test, $\sum_{n=N}^{\infty} a_n$ converges also. Hence, $\sum_{n=1}^{\infty} a_n$ must converge.

To prove (2), let $\sigma > 1$. Hence, there exists a subscript N such that for all $n \ge N$, $1 < r \le \sqrt[n]{a_n}$. This implies that $a_n > 1$ for all $n \ge N$, which in turn implies that $a_n \not\to 0$. Theorem 28.4 now indicates that $\sum_{n=1}^{\infty} a_n$ diverges.

To prove (3), we need to display series that converge and diverge under the premise that $\sigma = 1$ (Exercise 30.2), and under the premise that $\langle \sqrt[n]{a_n} \rangle_1^{\infty}$ oscillates (Exercises 30.3c,d). **QED**

Example 30.1 Use the root test to show that $\sum_{n=0}^{\infty} e^{-2n}$ converges. Since $\sqrt[n]{a_n} = \sqrt[n]{e^{-2n}} = e^{-2}$, we have $\sigma = \lim_{n\to\infty} e^{-2} = e^{-2} < 1$, and (1) holds. This is to be expected, for $\sum_{n=0}^{\infty} e^{-2n}$ is a convergent geometric series. In the same way, the root test gives a handy proof that a geometric series $\sum_{n=0}^{\infty} r^n$ converges if $0 < r < 1$, and diverges if $r > 1$. The reason is that for these series,

$\sigma = r$ (verify!). When $r = \pm 1$, the test fails, but it's easy to prove divergence in these two cases. ∎

In the next application of the root test, the $\lim_{n\to\infty} \sqrt[n]{n}$ is needed. Finding it is not a trivial problem (in calculus courses, the limit is derived by using l'Hospital's rule). In Exercise 30.10, show that $\lim_{n\to\infty} \sqrt[n]{n} = 1$ by theorems not dependent on calculus.

Example 30.2 Is the root test more refined than the ratio test? Well, perhaps it sips tea while wearing silk gloves, but we need a concrete example. In Exercise 29.4b, we encountered the peculiar series $2 + \frac{2}{2^2} + \frac{3}{2^1} + \frac{4}{2^4} + \frac{5}{2^3} + \frac{6}{2^6} + \frac{7}{2^5} + \cdots$ whose convergence could not be decided by the ratio test. To proceed with the root test, we need the general term

$$\sqrt[n]{a_n} = \begin{cases} \sqrt[n]{n2^{-n}} = \frac{\sqrt[n]{n}}{2} & \text{if } n \text{ is even} \\ \sqrt[n]{n2^{2-n}} = \frac{\sqrt[n]{4}\sqrt[n]{n}}{2} & \text{if } n \text{ is odd} \end{cases}.$$

Both $\lim_{n\to\infty} \frac{\sqrt[n]{n}}{2}$ and $\lim_{n\to\infty} \frac{\sqrt[n]{4}\sqrt[n]{n}}{2}$ equal $\frac{1}{2}$, and therefore, $\sigma = \lim_{n\to\infty} \sqrt[n]{a_n} = \frac{1}{2} < 1$. We can conclude by the root test that the series converges to a sum.

On the other hand, it is true that every series for which the ratio test gives a conclusion is a series for which the root test yields a conclusion (see [13], pg. 86). Hence, the root test is the more refined test of the two. ∎

The limit comparison test is closely related to the comparison test, and we consider it next.

Theorem 30.2 The Limit Comparison Test Let $\sum_{n=1}^{\infty} a_n$ and $\sum_{n=1}^{\infty} b_n$ be series of positive terms, and determine $\rho = \lim_{n\to\infty} \frac{a_n}{b_n}$. Then

(1) if $\rho > 0$, then $\sum_{n=1}^{\infty} a_n$ and $\sum_{n=1}^{\infty} b_n$ both converge or both diverge,
(2) if $\rho = 0$ and if $\sum_{n=1}^{\infty} b_n$ converges, then $\sum_{n=1}^{\infty} a_n$ converges also.
(3) if $\rho = +\infty$ and if $\sum_{n=1}^{\infty} b_n$ diverges, then $\sum_{n=1}^{\infty} a_n$ diverges also.

Proof. In (1), since $\rho > 0$, we know that the ratio $\frac{a_n}{b_n}$ is bounded away from zero. Thus, consider the interval $(\frac{1}{t}, t)$, where t is a positive constant large enough that $\rho \in (\frac{1}{t}, t)$. Because of convergence, $\frac{a_n}{b_n} \in (\frac{1}{t}, t)$ for all but a finite number of ratios. That is, there exists some N such that for all subscripts $n \geq N$, $\frac{1}{t} < \frac{a_n}{b_n} < t$, and so, $\frac{1}{t} b_n < a_n < t b_n$.

Focusing on $\frac{1}{t} b_n < a_n$, suppose $\sum_{n=1}^{\infty} a_n$ converges. Then by the comparison test, $\sum_{n=N}^{\infty} \frac{1}{t} b_n$ converges also, and thus $\sum_{n=1}^{\infty} b_n$ converges. Furthermore, looking at $a_n < t b_n$, if $\sum_{n=1}^{\infty} a_n$ diverges, then $\sum_{n=N}^{\infty} t b_n$ diverges, and thus $\sum_{n=1}^{\infty} b_n$ diverges. By parallel reasoning, when $\sum_{n=1}^{\infty} b_n$ converges or diverges, then $\sum_{n=1}^{\infty} a_n$ behaves in the same way.

In (2), since $\rho = 0$, we know that there exists some N such that for all subscripts $n \geq N$, $0 < \frac{a_n}{b_n} < 1$, and so, $0 < a_n < b_n$. It follows by the comparison test that if $\sum_{n=1}^{\infty} b_n$ converges, then $\sum_{n=1}^{\infty} a_n$ converges, but not conversely.

In (3), $\rho = +\infty$ implies that $\lim_{n\to\infty} \frac{b_n}{a_n} = 0$. The conclusion drawn from (2) by exchanging the two series is that if $\sum_{n=1}^{\infty} a_n$ converges, then $\sum_{n=1}^{\infty} b_n$ converges. The contrapositive is what we want: if $\rho = +\infty$ and if $\sum_{n=1}^{\infty} b_n$ diverges, then $\sum_{n=1}^{\infty} a_n$ diverges. **QED**

Example 30.3 Use the limit comparison test to decide if $\sum_{n=1}^{\infty} \frac{25n^3+500n^2-7}{5n^4-2n+3}$ converges. To proceed, we must make a conjecture: we suspect that the series diverges, since for large n, the terms are roughly equal to $\frac{5}{n}$. This is found by forming the ratio $\frac{25n^3}{5n^4}$ of the dominant terms in the numerator and denominator. This is sometimes expressed by saying that $\frac{25n^3+500n^2-7}{5n^4-2n+3}$ is of the same order of magnitude as $\frac{5}{n}$ (see Note 1). Thus, use the limit comparison test with $b_n = \frac{5}{n}$, knowing beforehand that $\sum_{n=1}^{\infty} \frac{5}{n}$ diverges. Then

$$\rho = \lim_{n\to\infty} \left(\frac{25n^3 + 500n^2 - 7}{5n^4 - 2n + 3} \cdot \frac{n}{5} \right) = 1.$$

It follows that the given series diverges as well. ■

Note 1: The term "same order of magnitude" used above means that $f(n) = \frac{25n^3+500n^2-7}{5n^4-2n+3}$ is bounded within two positive constant multiples of $g(n) = \frac{5}{n}$ as $n \to \infty$. You should verify that $g(n) \le f(n)$ for all $n \in \mathbb{N}$. Furthermore, verify, at least by graphing, that for $n \ge 21$, $\frac{5}{n} < f(n) < 2 \cdot \frac{5}{n}$, meaning that $g(n) \le f(n) \le 2g(n)$ for large n.

In a more general setting, a positive function f is of the *same order of magnitude* as a positive function g if positive constants K and M exist such that $Kg(x) \le f(x) \le Mg(x)$ as $x \to \infty$ (or as the case may be, $x \to a$). Graphically, the two constants allow g to form border curves above and below f (although the borders may be quite far from f in general). In the present example, we found that $K = 1$ and $M = 2$ do the job. Because taking reciprocals of the (positive) quantities in $Kg(x) \le f(x) \le Mg(x)$ brings us to $\frac{1}{M} f(x) \le g(x) \le \frac{1}{K} f(x)$ (verify), we can also say that g is of the same order of magnitude as f. Hence, we simply say that f and g are of the same order of magnitude. These definitions carry over to the sequences that f and g generate, as in Example 30.3.

As another example, $f(x) = x^2 + 50x - 3$ is of the same order of magnitude as $g(x) = 4x^2$. This is true since as $x \to \infty$, $\frac{1}{4}g(x) \approx f(x)$, so that $K = \frac{1}{5}$ will do as the first constant, and $M = \frac{1}{3}$ works as the second constant (other choices also work). You should verify that $\frac{1}{5}(4x^2) \le x^2 + 50x - 3 \le \frac{1}{3}(4x^2)$ for sufficiently large x. Graphically, $y = \frac{1}{5}(4x^2)$ makes a border beneath $f(x)$, while $y = \frac{1}{3}(4x^2)$ makes a border above $f(x)$.

The reciprocals of f and g in Example 30.3 are of the same order of magnitude. This is so since, for instance,

$$\frac{1}{2}\left(\frac{n}{5}\right) \le \frac{5n^4 - 2n + 3}{25n^3 + 500n^2 - 7} \le 1\left(\frac{n}{5}\right) \quad \text{as } n \to \infty.$$

That is, we have found two positive constants, $K = \frac{1}{2}$ and $M = 1$, that allow $1/g$ to form border curves about $1/f$.

A convenient theorem to use is that positive functions f and g are of the same order of magnitude if and only if $\lim_{x\to\infty} \frac{f(x)}{g(x)}$ is a positive real number (likewise for $\lim_{x\to a} \frac{f(x)}{g(x)}$ in the other case).

Example 30.4 Test the series $\sum_{n=1}^{\infty} \frac{\ln n}{n^2}$ for convergence. After one or two guesses for the limit comparison test, we try $b_n = \frac{1}{n^{3/2}}$. Then $\rho = \lim_{n\to\infty} \frac{\ln n}{n^{1/2}} = 0$ (prove this in Exercise 30.11). Since the p-series $\sum_{n=1}^{\infty} \frac{1}{n^{3/2}}$ converges, the limit comparison test concludes that the given series converges. (By the way, we have shown that $f(n) = \ln n$ is not of the same order of magnitude as $g(n) = \sqrt{n}$.) ■

Cauchy also devised the following curious convergence test.

Theorem 30.3 The Cauchy Condensation Test Let $\langle a_n \rangle_1^\infty$ be a monotone decreasing sequence of positive terms. Then $\sum_{n=1}^{\infty} a_n$ converges if and only if $\sum_{n=1}^{\infty} 2^n a_{2^n}$ converges.

Proof of sufficiency. Let $\sum_{n=1}^{\infty} 2^n a_{2^n}$ be a convergent series, and for convenience let $b_n = 2^n a_{2^n}$. Starting at a_2, insert parentheses to group two, four, ..., 2^n, terms:

$$(a_2 + a_3) + (a_4 + a_5 + a_6 + a_7) + (a_8 + \cdots + a_{15}) + \cdots .$$

From monotonicity, we have $a_2 + a_3 \le 2a_2 = b_1$, $a_4 + a_5 + a_6 + a_7 \le 4a_4 = b_2$, and in general, $\sum_{n=2^k}^{2^{k+1}-1} a_n \le 2^k a_{2^k} = b_k$, where $k \in \mathbb{N}$. In the following expression, the double summation expresses the sum of terms that are grouped as shown above:

$$\sum_{k=1}^{\infty} \left(\sum_{n=2^k}^{2^{k+1}-1} a_n \right) \le \sum_{k=1}^{\infty} b_k.$$

The comparison test now tells us that since $\sum_{k=1}^{\infty} b_k$ converges, so must $\sum_{k=1}^{\infty} (\sum_{n=2^k}^{2^{k+1}-1} a_n)$. Removing parentheses, we find that $\sum_{n=2}^{\infty} a_n$ converges; hence clearly $\sum_{n=1}^{\infty} a_n$ converges as well.

Proof of necessity. Now, let $\sum_{n=1}^{\infty} a_n$ converge. Finish the proof by showing that $\sum_{n=1}^{\infty} 2^n a_{2^n}$ converges as Exercise 30.6. **QED**

Example 30.5 Determine the values of p which force the p-series $\sum_{n=1}^{\infty} \frac{1}{n^p}$ to converge. This was done earlier by means of the integral test. Let's compare that test with the condensation test. We see that $\langle n^{-p} \rangle_{n=1}^{\infty}$ is monotone decreasing for all $p > 0$. The p-series converges if and only if

$$\sum_{n=1}^{\infty} 2^n a_{2^n} = \sum_{n=1}^{\infty} 2^n \frac{1}{(2^n)^p} = \sum_{n=1}^{\infty} 2^{n(1-p)}$$

converges. Since p is constant, this series is geometric, with common ratio $r = 2^{1-p}$ (verify). For convergence we require $2^{1-p} < 1$, which implies that $p > 1$; and for divergence, $0 < p \le 1$. The condensation test certainly streamlined the p-series analysis! ■

Example 30.6 Does the series

$$1 + \frac{\sqrt{2}}{2^2} + \frac{\sqrt[3]{3}}{3^2} + \frac{\sqrt[4]{4}}{4^2} + \cdots = \sum_{n=1}^{\infty} \frac{n^{1/n}}{n^2}$$

converge? To determine this by the condensation test, first verify that the general term $a_n = \frac{1}{n^{2-1/n}}$ is monotone decreasing. Next,

$$\sum_{n=1}^{\infty} 2^n a_{2^n} = \sum_{n=1}^{\infty} 2^n \frac{(2^n)^{1/2^n}}{(2^n)^2} = \sum_{n=1}^{\infty} \frac{1}{2^{n-n/2^n}}.$$

The form of the exponent suggests using the comparison test, but against what series? A first attempt might be $\sum_{n=1}^{\infty} \frac{1}{2^n}$, since we know that it converges. However, verify that $\frac{1}{2^{n-n/2^n}} < \frac{1}{2^n}$ is incorrect. That attempt leads us to $\frac{1}{2^{n-n/2^n}} < \frac{2}{2^n}$, true for all $n \in \mathbb{N}$. Now, since $\sum_{n=1}^{\infty} \frac{2}{2^n}$ converges nicely, so will $\sum_{n=1}^{\infty} \frac{1}{2^{n-n/2^n}}$. Therefore, by the condensation test, $\sum_{n=1}^{\infty} \frac{n^{1/n}}{n^2}$ will converge. (Try Exercise 30.7 now.) ■

Yet more refined than the ratio and root tests is the next one, which is often used when those tests are inconclusive. It was discovered by analyst Joseph Ludwig Raabe (1801–1851), the first professor of mathematics at the Polytechnikum in Zurich, Switzerland, an institution that fostered the careers of an illustrious list of mathematicians.

Theorem 30.4 Raabe's Test Let $\sum_{n=1}^{\infty} a_n$ be a series of positive terms, and consider $\mu = \lim_{n\to\infty} n(1 - \frac{a_{n+1}}{a_n})$. Then

(1) if $1 < \mu \le +\infty$, then $\sum_{n=1}^{\infty} a_n$ converges,
(2) if $-\infty \le \mu < 1$, then $\sum_{n=1}^{\infty} a_n$ diverges, and
(3) if $\mu = 1$, or does not exist because $\langle n(1 - \frac{a_{n+1}}{a_n})\rangle_1^{\infty}$ oscillates, then $\sum_{n=1}^{\infty} a_n$ may either converge or diverge, so the test concludes nothing.

Proof. To prove (1), suppose that $1 < \mu$. Let b be any number such that $1 < b < \mu$. Then there exists N (which depends on the choice of b) such that for all subscripts $n \ge N$, we have $b < n(1 - \frac{a_{n+1}}{a_n})$, implying $na_{n+1} < na_n - ba_n$ and $na_{n+1} < (n-1)a_n - (b-1)a_n$. Since $b > 1$, we arrive at

$$0 < (b-1)a_n < (n-1)a_n - na_{n+1}. \tag{1}$$

The last difference in *(1)* shows us that $\langle na_{n+1}\rangle_{n=N}^{\infty}$ is monotone decreasing for $n \ge N$.

Next, we are going to show that the partial sums $\sum_{n=N}^{K} a_n$ are bounded. At successive values of n, $n \in \{N, N+1, N+2, \ldots, K\}$, *(1)* evaluates as follows:

$$\begin{array}{c|c} N & (b-1)a_N < (N-1)a_N - Na_{N+1} \\ N+1 & (b-1)a_{N+1} < Na_{N+1} - (N+1)a_{N+2} \\ N+2 & (b-1)a_{N+2} < (N+1)a_{N+2} - (N+2)a_{N+3}\,. \\ \cdots & \cdots \\ K & (b-1)a_K < (K-1)a_K - Ka_{K+1} \end{array}$$

Adding the inequalities in the right-hand column gives $(b-1)\sum_{n=N}^{K} a_n < (N-1)a_N - Ka_{K+1}$. What is interesting about this is that $(N-1)a_N - Ka_{K+1}$ is positive and less than the constant $(N-1)a_N$. Accordingly, $(b-1)\sum_{n=N}^{K} a_n$ and $\sum_{n=N}^{K} a_n$ itself are bounded partial sums for every $K \ge N$. But this is sufficient to prove that $\sum_{n=N}^{\infty} a_n$ converges, by the monotone convergence theorem. It follows that the entire series $\sum_{n=1}^{\infty} a_n$ converges, so (1) is true.

To prove (2), this time let b be such that $\mu < b < 1$. Then we find a specific N such that $n(1 - \frac{a_{n+1}}{a_n}) < b$ for all $n \ge N$. With a bit of algebra, we get $(n-b)a_n < na_{n+1}$. By our choice of b, this implies that $(n-1)a_n < na_{n+1}$ for all $n \ge N$, which means that $\langle na_{n+1}\rangle_{n=N}^{\infty}$ is monotone increasing. As such, for all $n \ge N$, $na_{n+1} \ge Na_{N+1} = C$, a constant. Hence, $a_{n+1} \ge \frac{C}{n}$, and every partial sum compares as

$$\sum_{n=N}^{K} a_{n+1} \ge \sum_{n=N}^{K} \frac{C}{n} = C \sum_{n=N}^{K} \frac{1}{n}.$$

We recognize $\sum_{n=N}^{K} \frac{1}{n}$ as part of a partial sum of the harmonic series. Since $\lim_{K\to\infty} \sum_{n=N}^{K} \frac{1}{n} = +\infty$, by comparison $\lim_{K\to\infty} \sum_{n=N}^{K} a_{n+1} = +\infty$. This means that $\sum_{n=1}^{\infty} a_n$ diverges as (2) posits.

Case (3) can be demonstrated as in the previous tests by exhibiting series that converge and diverge and have $\mu = 1$, or have an oscillating $\langle n(1 - \frac{a_{n+1}}{a_n})\rangle_1^\infty$. The simplest instance of divergence happens with the harmonic series, where $\mu = \lim_{n\to\infty} n(1 - \frac{n}{n+1}) = 1$ (verify). In Exercise 30.13, you must prove that a series having $\mu = 1$ converges.

The series $\sum_{n=2}^{\infty} \frac{1}{n^2+(-1)^n n}$ produces

$$n\left(1 - \frac{a_{n+1}}{a_n}\right) = n\frac{2n+1-(-1)^n(2n+1)}{(n+1)^2-(-1)^n(n+1)} = \begin{cases} \dfrac{4n^2+2n}{(n+1)^2+(n+1)} & \text{if } n \text{ is odd} \\ 0 & \text{if } n \text{ is even} \end{cases}.$$

Consequently,

$$n\left(1 - \frac{a_{n+1}}{a_n}\right) \to \begin{cases} 4 & \text{if } n \text{ is odd} \\ 0 & \text{if } n \text{ is even} \end{cases},$$

which means that the limit μ fails to exist by oscillation. Furthermore,

$$\frac{1}{n^2+(-1)^n n} < \frac{1}{(n-1)^2}$$

for $n \geq 2$, so that by comparison with $\sum_{n=2}^{\infty} \frac{1}{(n-1)^2}$, the given series converges. Thus, a series has been found to converge even though μ in Raabe's test does not exist by oscillation.

In Exercise 30.14, you must prove that a series with no limit μ because of oscillation diverges. Once Exercises 30.13 and 30.14 are completed, this theorem is proved. **QED**

Example 30.7 Show that the series $\frac{2}{9} + \frac{2\cdot 5}{9\cdot 12} + \frac{2\cdot 5\cdot 8}{9\cdot 12\cdot 15} + \cdots$ converges. Of course, we must first find an expression for the general term a_n. Work out the details to arrive at

$$a_n = \frac{(2)(5)\cdots(3n-7)(3n-4)(3n-1)}{(9)(12)\cdots(3n)(3n+3)(3n+6)},$$

where there are exactly n factors in the numerator and denominator. Then

$$\frac{a_{n+1}}{a_n} = \frac{3n+2}{3n+9},$$

and we evaluate

$$\lim_{n\to\infty} n\left(1 - \frac{3n+2}{3n+9}\right) = \frac{7}{3} = \mu.$$

Since $\mu > 1$, the given series converges by Raabe's test. The ratio test gives $\rho = \lim_{n\to\infty} \frac{a_{n+1}}{a_n} = 1$, and the root test gives $\sigma = \lim_{n\to\infty} \sqrt[n]{a_n} = 1$ (see Note 2), so that Raabe's test seems to go beyond their reach (we will not prove that Raabe's test is a refinement of the root test). ■

Note 2: In the example, rewrite a_n as

$$a_{m/3} = \frac{(m-1)(m-4)(m-7)\cdots(m-(3n-2))}{(m+6)(m+3)(m)\cdots(m-(3n-9))},$$

where $m = 3n$. This curious substitution lets the numerator and denominator of a_n appear as polynomials in m which expand as $m^n + N(m)$ and $m^n + D(m)$ respectively, where $N(m)$ and $D(m)$ are of degree less than n. Thus,

$$\sigma = \lim_{n\to\infty} \sqrt[n]{a_n} = \lim_{m\to\infty} (a_{m/3})^{3/m} = \left(\lim_{m\to\infty} \sqrt[m]{\frac{m^n + N(m)}{m^n + D(m)}}\right)^3 = 1.$$

30.1 Exercises

1. Use the root test to prove that $\sum_{n=1}^{\infty} \frac{3^n}{n!}$ converges. Hint: You will need Exercise 18.8.
2. Find a divergent series and a convergent series for which $\sigma = 1$ in the root test, demonstrating one possibility in (3) of the theorem.
3. a. Does the root test prove convergence of the series

$$\sum_{n=1}^{\infty} \frac{1}{2^n(3-(-1)^n)} = \frac{1}{8} + \frac{1}{8} + \frac{1}{32} + \frac{1}{32} + \cdots?$$

   (The ratio test fails on this one.)

   b. Why does the root test not prove anything about

$$\sum_{n=1}^{\infty} \frac{1}{5-(-1)^n} = \frac{1}{6} + \frac{1}{4} + \frac{1}{6} + \cdots?$$

   (The ratio test *must* fail on this one.) Show that it diverges using some other means.

   c. Why does the root test not prove anything about

$$\sum_{n=1}^{\infty} (\frac{1}{3-(-1)^n})^n = \frac{1}{4} + \frac{1}{4} + \frac{1}{64} + \frac{1}{16} + \frac{1}{1024} + \cdots?$$

   (The ratio test fails again.) Show that it converges using some other means.

   d. Why does the root test not prove divergence of

$$\sum_{n=1}^{\infty} \left(\frac{1}{2-(-1)^n}\right)^{\frac{n}{2}} = \frac{\sqrt{3}}{3} + 1 + \frac{\sqrt{3}}{9} + 1 + \frac{\sqrt{3}}{27} + 1 + \cdots?$$

   (Did we mention the ratio test fails?) Show that it diverges using some other means.

4. Can we use the limit comparison test to say that if $\rho = 0$, then if $\sum_{n=1}^{\infty} a_n$ diverges, then $\sum_{n=1}^{\infty} b_n$ diverges?
5. Use the limit comparison test to determine the convergence of

$$\sum_{n=1}^{\infty} \frac{\sqrt[3]{n^5 + n^4}}{3n^3 + 50n}.$$

Hint: The dominant term in the numerator is $\sqrt[3]{n^5}$.

6. Finish the proof of Cauchy's condensation test by proving that if $\sum_{n=1}^{\infty} a_n$ converges, then $\sum_{n=1}^{\infty} 2^n a_{2^n}$ converges. Hint: Adjust slightly the technique in the first part of the proof of the theorem.

7. Determine whether the series $1 + \frac{\sqrt{2}}{2} + \frac{\sqrt[3]{3}}{3} + \frac{\sqrt[4]{4}}{4} + \cdots = \sum_{n=1}^{\infty} \frac{n^{1/n}}{n}$ converges, using the condensation test.

8. a. Prove that $\sum_{n=3}^{\infty} \frac{1}{n \ln n}$ diverges by the condensation test. Notice that because $\frac{1}{n \ln n} < \frac{1}{n}$ for $n \geq 3$, this series diverges even more slowly than the harmonic series!
   b. Show that $\sum_{n=3}^{\infty} \frac{(\ln n)^2}{n^2}$ converges, even though $\frac{(\ln n)^2}{n^2} > \frac{1}{n^2}$ for $n \geq 3$.

9. Test the series $\sum_{n=0}^{\infty} \frac{n!+1}{(n+2)!}$ for convergence.

10. To complete the proof of convergence in Example 30.2, it's necessary to prove that $\lim_{n\to\infty} \sqrt[n]{n} = 1$. Do so. Hint: Use the arithmetic mean-geometric mean inequality, and $n = \sqrt{n} \cdot \sqrt{n}$.

11. Use the previous exercise to help prove that $\lim_{n\to\infty} \frac{\ln n}{n} = 0$. You may assume that the natural logarithm function is continuous. Go further by proving that for any positive constant M, $\lim_{n\to\infty} \frac{\ln n}{n^M} = 0$. This result is fascinating, for it shows that eventually, the logarithmic function increases more slowly (is of a lesser order of magnitude) than any positive power. Thus, for instance, $\lim_{n\to\infty} \frac{\ln n}{\sqrt[10]{n}} = 0$.

12. a. Test the series $\frac{1}{8} + \frac{1\cdot 2}{8\cdot 9} + \frac{1\cdot 2\cdot 3}{8\cdot 9\cdot 10} + \cdots$ for convergence by the ratio test, the root test, and Raabe's test. Hint: For the root test, you should arrive at

$$\sigma = \lim_{n\to\infty} \sqrt[n]{\frac{7!n!}{(n+7)!}},$$

which is easy to evaluate using $\lim_{n\to\infty} \sqrt[n]{n+k} = 1$.
   b. We have used the p-series before to compare the methods of various tests for convergence. When p is any *integer*, apply Raabe's test to $\sum_{n=1}^{\infty} \frac{1}{n^p}$ and see how it arrives at the conclusions about p-series.

13. To continue the demonstrations in (3) in Raabe's test, first prove by some other test that $\sum_{n=2}^{\infty} \frac{1}{n(\ln n)^2}$ converges. To show that $\mu = \lim_{n\to\infty} n(1 - \frac{a_{n+1}}{a_n}) = 1$, verify the steps saying "should" in what follows.

First,

$$\frac{a_{n+1}}{a_n} = \frac{n \ln^2 n}{(n+1)\ln^2(n+1)},$$

where we use the shorthand $\ln^2 n$ for $(\ln n)^2$.

Next, if it exists,

$$\mu = \lim_{n\to\infty} n\left(1 - \frac{a_{n+1}}{a_n}\right) \text{ should equal } \lim_{n\to\infty} \frac{n^2(\ln^2(n+1) - \ln^2 n) + n\ln^2(n+1)}{(n+1)\ln^2(n+1)}.$$

This should equal

$$1+\lim_{n\to\infty}\frac{n^2(\ln(n+1)-\ln n)(\ln(n+1)+\ln n)}{(n+1)\ln^2(n+1)}.$$

This should equal

$$1+\lim_{n\to\infty}\frac{n^2\ln(\frac{n+1}{n})}{(n+1)\ln(n+1)}\cdot\lim_{n\to\infty}\frac{\ln(n+1)+\ln n}{\ln(n+1)}.$$

And this should equal

$$1+\frac{\lim_{n\to\infty}\ln(1+\frac{1}{n})^n}{\lim_{n\to\infty}\left(\frac{n+1}{n}\ln(n+1)\right)}\cdot\lim_{n\to\infty}\left(1+\frac{\ln n}{\ln(n+1)}\right).$$

The three limits are easy to find. Simplifying should yield $\mu=1$ (nice workout!).

14. For the series

$$\sum_{n=2}^{\infty}\frac{1}{n+(-1)^n\sqrt{n}},$$

verify the chain of equalities:

$$n\left(1-\frac{a_{n+1}}{a_n}\right)=n\left(1-\frac{n+(-1)^n\sqrt{n}}{n+1-(-1)^n\sqrt{n+1}}\right)=n\left(\frac{1-(-1)^n(\sqrt{n+1}+\sqrt{n})}{n+1-(-1)^n\sqrt{n+1}}\right)$$
$$=\frac{1-(-1)^n(\sqrt{n+1}+\sqrt{n})}{\frac{n+1}{n}-(-1)^n\frac{\sqrt{n+1}}{n}}.$$

In the last expression, the denominator has a limit; what is it? The numerator oscillates as n increases. Hence, $\mu=\lim_{n\to\infty}n(1-\frac{a_{n+1}}{a_n})$ fails to exist by oscillation. Furthermore, the series itself diverges by the limit comparison test, as you should now show. What have you demonstrated with regard to the proof of Raabe's test?

31

Series with Negative Terms

$$\frac{\pi}{4} = \frac{1}{1} - \frac{1}{3} + \frac{1}{5} - \frac{1}{7} + \frac{1}{9} - \cdots$$

James Gregory almost certainly discovers the sum in 1671;
Gottfried Leibniz certainly discovers it in 1674.

The series in this chapter will contain an infinite number of positive and negative terms. The positive and negative signs often come in a prescribed pattern or order, but they may occur with no order at all. The simplest possible pattern is to alternate positive and negative signs, and we investigate this first.

31.1 Alternating Series

An important class of infinite series is that of **alternating series**, defined as $\sum_{n=1}^{\infty}(-1)^{n+1}a_n = a_1 - a_2 + a_3 - a_4 + \cdots$, where all a_n are positive (having $a_n > 0$ certifies that the signs actually alternate). For convenience, we may also use $\sum_{n=0}^{\infty}(-1)^n a_n = a_0 - a_1 + a_2 - a_3 + \cdots$.

One type of alternating series that has already been studied, in Chapter 28, is the alternating geometric series $\sum_{n=0}^{\infty} ar^n$, when $r < 0$. It was proved there that $\sum_{n=0}^{\infty} ar^n = a - a|r| + a|r|^2 - a|r|^3 + \cdots$ converges if and only if $|r| < 1$ (put $0^0 \equiv 1$ when $r = 0$).

The fundamental test for convergence of alternating series is the following one, named after its discoverer.

Theorem 31.1 Leibniz's Alternating Series Test An alternating series $\sum_{n=1}^{\infty}(-1)^{n+1}a_n$ will converge if (1) the sequence $\langle a_n\rangle_1^{\infty}$ of positive numbers is monotone decreasing, and (2) $\lim_{n\to\infty} a_n = 0$.

Proof. We investigate first the sequence of partial sums with an even number of terms, $S_{2n} = (a_1 - a_2) + (a_3 - a_4) + \cdots + (a_{2n-1} - a_{2n})$. Since each difference in parenthesis is positive, the sequence of even partial sums $\langle S_{2n}\rangle_1^{\infty}$ is monotone increasing. But also, $S_{2n} = a_1 - (a_2 - a_3) - \cdots - (a_{2n-2} - a_{2n-1}) - a_{2n}$ for $n \in \mathbb{N}$, implying that $S_{2n} < a_1 = S_1$, which tells us that the sequence is bounded above by S_1. Therefore, by the monotone convergence theorem, $\langle S_{2n}\rangle_1^{\infty}$ converges as a sequence to a number S.

For similar reasons, the sequence of partial sums with an odd number of terms, $S_{2n-1} = a_1 - (a_2 - a_3) - \cdots - (a_{2n-2} - a_{2n-1})$ is monotone decreasing. Also, $S_{2n-1} = (a_1 - a_2) + (a_3 - a_4) + \cdots + (a_{2n-3} - a_{2n-2}) + a_{2n-1} > a_1 - a_2 = S_2$ for $n \in \mathbb{N}$, so that $\langle S_{2n-1}\rangle_1^{\infty}$ is

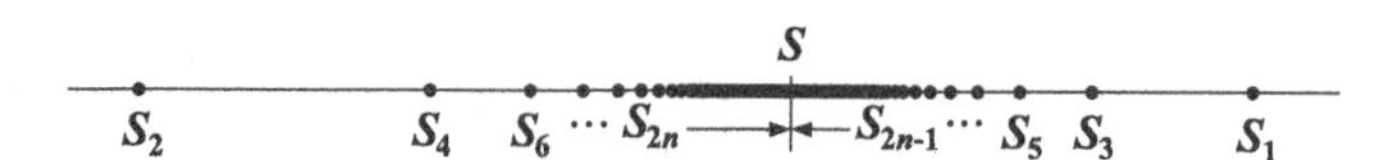

Figure 31.1. Even and odd partial sums of a convergent alternating series.

bounded below by S_2. Thus, $\langle S_{2n-1}\rangle_1^\infty$ also converges as a sequence. From what is suggested by Figure 31.1, we need to show that $\lim_{n\to\infty} S_{2n-1} = S$ (since $S = \lim_{n\to\infty} S_{2n}$).

At this point, we know that for any $\varepsilon > 0$, there exists a corresponding N_ε such that for all $n > N_\varepsilon$, it is true that $|S_{2n} - S| < \frac{\varepsilon}{2}$. We have not used premise (2) above, which would never have been inserted needlessly by as brilliant a mathematician as Leibniz. We now use (2) to assert that there is an N'_ε such that for all $n > N'_\varepsilon$, $a_{2n} < \frac{\varepsilon}{2}$. Then both epsilon inequalities must hold whenever $n > M_\varepsilon = \max\{N_\varepsilon, N'_\varepsilon\}$. We will need the identity $S_{2n-1} = S_{2n} + a_{2n}$. Now, by the triangle inequality,

$$|S_{2n-1} - S| = |S_{2n} + a_{2n} - S| \le |S_{2n} - S| + |a_{2n}| < \frac{\varepsilon}{2} + \frac{\varepsilon}{2} = \varepsilon$$

for all $n > M_\varepsilon$. This proves that $\lim_{n\to\infty} S_{2n-1} = S$, and with $\lim_{n\to\infty} S_{2n} = S$, it follows that the alternating series $\sum_{n=1}^{\infty}(-1)^{n+1}a_n$ converges (to S). **QED**

Example 31.1 Does the alternating p-series $\sum_{n=1}^{\infty}(-1)^{n+1}\frac{1}{n^p} = 1 - \frac{1}{2^p} + \frac{1}{3^p} - \cdots$ converge for any values of p? We observe that $\langle \frac{1}{n^p}\rangle_{n=1}^{\infty}$ is monotone decreasing *and* that $\lim_{n\to\infty}\frac{1}{n^p} = 0$ whenever $p > 0$. Therefore, Leibniz's alternating series test guarantees convergence if $p > 0$.

When $p = 1$, this generates the alternating harmonic series, which therefore converges. You have perhaps found in calculus that its sum is $\ln 2$. Exercise 31.8 develops a proof independent of calculus.

If $p \le 0$, then $a_n = \frac{1}{n^p} \not\to 0$, and the alternating p-series diverges as it oscillates (see Figure 31.2). ■

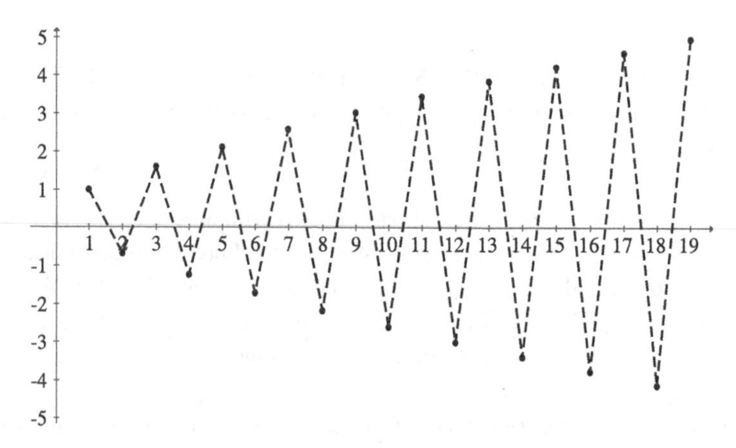

Figure 31.2. The partial sums $S_n = \sum_{k=1}^{n}(-1)^{k+1}\frac{1}{k^{-3/4}}$ of an alternating p-series with $p = \frac{-3}{4}$ show oscillation that never damps down. The series therefore diverges.

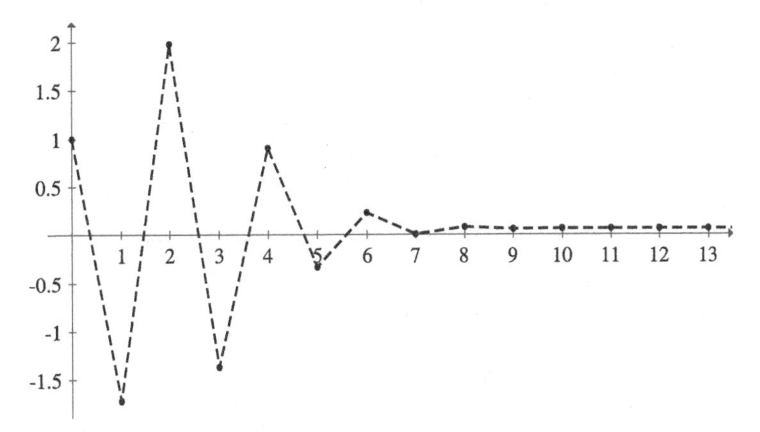

Figure 31.3. Partial sums $S_n = \sum_{k=0}^{n} \frac{(-e)^k}{k!}$ showing a rapid convergence to the sum $e^{-e} \approx 0.066$.

Example 31.2 Does the series $\sum_{n=0}^{\infty}(-1)^n \frac{e^n}{n!}$ converge? Sometimes, decreasing monotonicity can be checked by proving that $\frac{a_{n+1}}{a_n} < 1$. Here,

$$\frac{a_{n+1}}{a_n} = \frac{e^{n+1}}{(n+1)!}\frac{n!}{e^n} = \frac{e}{n+1},$$

which is less than 1 for all $n \geq 2$. That $\lim_{n\to\infty} \frac{e^n}{n!} = 0$ has been previously established (Exercise 17.8). Thus, the alternating series test indicates that the series converges. (The Maclaurin series for $y = e^x$ yields the actual sum, $\sum_{n=0}^{\infty}(-1)^n \frac{e^n}{n!} = e^{-e} \approx 0.066$.)

The graph in Figure 31.3 of the sequence of partial sums $S_n = \sum_{k=0}^{n}(-1)^k \frac{e^k}{k!}$ shows the typical behavior of alternating series that converge: they alternately over- and underestimate the sum. Here, $\langle S_n \rangle_0^\infty$ converges quickly thanks to the factorial in the denominator. ■

Example 31.3 Does the **logarithmic series**

$$\sum_{n=1}^{\infty}(-1)^n \frac{\ln n}{n} = \frac{\ln 2}{2} - \frac{\ln 3}{3} + \frac{\ln 4}{4} - \cdots$$

converge? To use the alternating series test, note that monotonicity begins at $\frac{\ln 3}{3} > \frac{\ln 4}{4}$. First, we must show that

$$a_n = \frac{\ln n}{n} > \frac{\ln(n+1)}{n+1} = a_{n+1},$$

or $\ln n^{n+1} > \ln(n+1)^n$, or finally, that $n > (1 + \frac{1}{n})^n$ (verify the algebra!). To prove the last inequality, you should recall the definition $\lim_{n\to\infty}(1 + \frac{1}{n})^n = e$, and it will be proved in Chapter 34 that $e > (1 + \frac{1}{n})^n$ for all $n \in \mathbb{N}$. Hence, $n > e > (1 + \frac{1}{n})^n$ when $n \geq 3$, implying that $a_n > a_{n+1}$ for $n \geq 3$, as we wanted to show. Second, $\lim_{n\to\infty} \frac{\ln n}{n} = 0$ has been proved in Exercise 30.11. Consequently, $\sum_{n=1}^{\infty}(-1)^n \frac{\ln n}{n}$ converges.

As in the previous example, the sequence of partial sums $\langle S_n \rangle_1^\infty$ over- and underestimates the sum. This time, the convergence is so slow that we can't use the graph to estimate the sum with much precision. ■

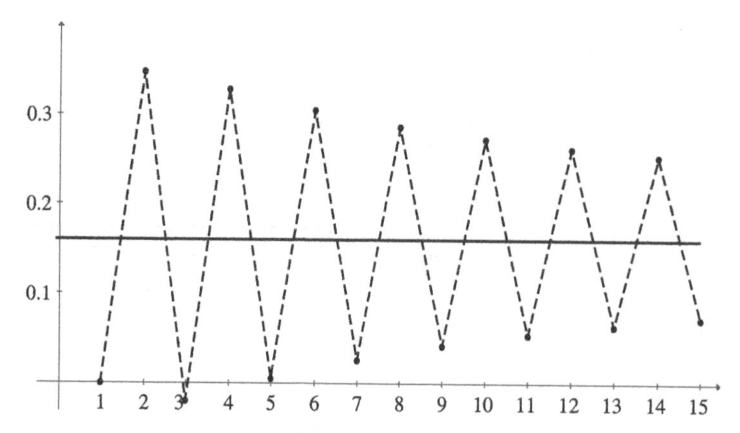

Figure 31.4. Partial sums $S_n = \sum_{k=1}^{n}(-1)^k \frac{\ln k}{k}$ showing a slow convergence to a sum of approximately 0.16.

The graphs in Figures 31.3 and 31.4 show that partial sums S_n of convergent alternating series approximate their respective sums with less and less absolute error. The following corollary to Leibniz's alternating series test shows this observation to be true always.

Corollary 31.1 If an alternating series converges, $\sum_{n=1}^{\infty}(-1)^{n+1}a_n = S$, and its nth partial sum is S_n, then $|S - S_n| \leq a_{n+1}$. That is, the error at S_n is at worst equal to the absolute value of the first unused term.

Proof. The proof of the alternating series test indicates that $S_{2n-1} \geq S_{2n+1} \geq S$ and $S_{2n} \leq S$. These inequalities yield two results. First, $0 \leq S_{2n-1} - S \leq S_{2n-1} - S_{2n} = a_{2n}$, so that $|S_{2n-1} - S| \leq a_{2n}$. Second, $0 \leq S - S_{2n} \leq S_{2n+1} - S_{2n} = a_{2n+1}$, so that $|S - S_{2n}| \leq a_{2n+1}$. Hence, it is always true that $|S - S_j| \leq a_{j+1}$, regardless of whether subscript j is even $(j = 2n)$ or odd $(j = 2n - 1)$. This, with a different letter, is the corollary's conclusion. **QED**

Example 31.4 How much of an error is gleaned using S_{10} in the series $\sum_{n=0}^{\infty}(-1)^n \frac{e^n}{n!}$ from Example 31.2? Since this series begins at $n = 0$, avoid subscript errors by reexpressing it as

$$\sum_{n=1}^{\infty}(-1)^{n-1}\frac{e^{n-1}}{(n-1)!}$$

in order to conform to Corollary 31.1. Then, $|S - S_{10}| \leq a_{11} = \frac{e^{10}}{10!} \approx 0.00607$.

Furthermore, the sum S must lie within the interval $[S_{2n}, S_{2n+1}]$ (see Figure 31.1). With a calculator,

$$S_{10} = \sum_{n=1}^{10}(-1)^{n-1}\frac{e^{n-1}}{(n-1)!} \approx 0.06114,$$

and $S_{11} = S_{10} + a_{11} \approx 0.06721$. Therefore, $S \in [0.06114, 0.06721]$.

It happens that we know the sum (see Example 31.2): $S = e^{-e} \approx 0.06599$. This allows us to verify the error stated in the corollary. We have $|S - S_{10}| \approx |0.06599 - 0.06114| = 0.00485$, which indeed checks out as less than a_{11}. ■

31.2 Absolute and Conditional Convergence

Another classification of infinite series is that of absolute convergence versus conditional convergence. These are mutually exclusive subsets of the set of convergent series, as the following definitions show.

A series $\sum_{n=1}^{\infty} a_n$ **converges absolutely** or is **absolutely convergent** if and only if the series $\sum_{n=1}^{\infty} |a_n|$ of absolute values converges.

A series **converges conditionally** or is **conditionally convergent** if and only if it converges, but is not absolutely convergent, that is, $\sum_{n=1}^{\infty} |a_n|$ diverges.

In this context, the given series may be alternating, but not necessarily: the positive and negative signs may come in any order, even randomly. The definition states that the conditionally convergent series form a subset of all convergent series, disjoint from the set of absolutely convergent series. After Theorem 31.2, we will know that absolutely convergent series also form a subset of the convergent series.

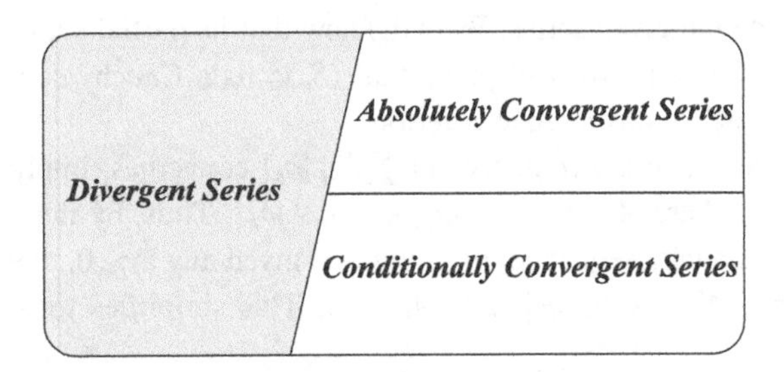

Figure 31.5. The set of all infinite series is usefully partitioned into three subsets. Note that all convergent series must either be absolutely or conditionally convergent, but never both.

Example 31.5 We saw in Example 31.1 that the alternating harmonic series $1 - \frac{1}{2} + \frac{1}{3} - \frac{1}{4} + \cdots$ converges by Theorem 31.1. But it is not absolutely convergent, for we know that $1 + \frac{1}{2} + \frac{1}{3} + \frac{1}{4} + \cdots$ diverges. Hence, using Figure 31.5, the alternating harmonic series is conditionally convergent. ■

Example 31.6 Every convergent geometric series is absolutely convergent. In other words, there are no conditionally convergent geometric series. For example, $\sum_{n=0}^{\infty}(-0.75)^n$ is absolutely convergent, since $\sum_{n=0}^{\infty} |-0.75|^n = 4$. Of course, the original series cannot converge to the same sum: $\sum_{n=0}^{\infty}(-0.75)^n = \frac{4}{7}$. See Exercise 31.3. ■

Example 31.7 The series $\sum_{n=1}^{\infty} \frac{\cos n}{n^2} = (0.540\ldots) + (-0.104\ldots) + (-0.109\ldots) + \cdots$ has an unorganized pattern of signs. But since

$$\sum_{n=1}^{\infty} \left|\frac{\cos n}{n^2}\right| \leq \sum_{n=1}^{\infty} \frac{1}{n^2},$$

and this p-series converges, it follows that $\sum_{n=1}^{\infty} \frac{\cos n}{n^2}$ converges absolutely. ■

We soon realize that if a series has no negative terms to begin with, then absolute convergence is identical to convergence. By virtue of the definition of absolute convergence,

All the convergence tests in the previous three chapters are also tests for absolute convergence.

One strategy for handling a series with infinitely many negative and positive terms is to first test it for absolute convergence. After that, we would go further and test the original series for convergence, with signs intact. Fortunately, the following key theorem saves us half the work, for it asserts that if we can prove the absolute convergence of a series, then the series in its *original* form automatically converges. We won't need to use Theorem 31.1 on an alternating series if we can at the start prove that it converges absolutely.

Theorem 31.2 If a series is absolutely convergent, then it is a convergent series in its original form.

Proof. Let $\sum_{k=1}^{\infty} a_k$ be the given series. We will show that its partial sums $S_n = \sum_{k=1}^{n} a_k$ have a limit. In order to show this, we will prove that $\langle S_n \rangle_1^{\infty}$ is a Cauchy sequence and appeal to Theorem 21.1, the Cauchy convergence criterion.

The premise is absolute convergence, so $\sum_{k=1}^{\infty} |a_k|$ converges, implying that $\langle A_n \rangle_1^{\infty}$ is a convergent sequence, where $A_n = |a_1| + |a_2| + \cdots + |a_n|$. Thus, by the Cauchy convergence criterion, $\langle A_n \rangle_1^{\infty}$ is a Cauchy sequence, meaning that given any $\varepsilon > 0$, there exists an N_ε such that for any $m > n \geq N_\varepsilon$, we have $|A_m - A_n| < \varepsilon$. This simplifies to $A_m - A_n < \varepsilon$ by our choice of m and n.

Returning to $\langle S_n \rangle_1^{\infty}$, observe that

$$\begin{aligned} |S_m - S_n| &= \left| \sum_{k=1}^{m} a_k - \sum_{k=1}^{n} a_k \right| = |a_m + a_{m-1} + \cdots + a_{n+1}| \\ &\leq |a_m| + |a_{m-1}| + \cdots + |a_{n+1}| = A_m - A_n < \varepsilon, \end{aligned}$$

for any $m > n \geq N_\varepsilon$. Therefore, $\langle S_n \rangle_1^{\infty}$ must be a Cauchy sequence, and by the Cauchy convergence criterion, $\langle S_n \rangle_1^{\infty}$ will converge. It follows by definition that $\sum_{k=1}^{\infty} a_k$ converges. **QED**

Example 31.8 In Exercise 29.1s, you proved that $\sum_{n=1}^{\infty} \frac{1}{n}\left(\frac{2}{3}\right)^n$ converged by the ratio test. Now, since all of its terms are positive, view the series as the absolutely convergent counterpart of $\sum_{n=1}^{\infty} (-1)^{n+1} \frac{1}{n}\left(\frac{2}{3}\right)^n$. By the theorem just proved, this alternating series converges. ■

Example 31.9 In Example 31.7, we found that $\sum_{n=1}^{\infty} \frac{\cos n}{n^2}$ converges absolutely by the comparison test. By the theorem just proved, the series itself is guaranteed to converge. ■

What happens when the monotonicity premise of Leibniz's alternating series test is dismissed? Does the series in question diverge? Not necessarily, for that is the illicit converse fallacy! In fact, the series may either converge or diverge. The following examples demonstrate this, and show common practices and worries when dealing with alternating series in general. They are found in "On A Convergence Test for Alternating Series," by R. Lariviere, in *Mathematics Magazine*, vol. 29 (Nov.–Dec. 1955), pg. 88.

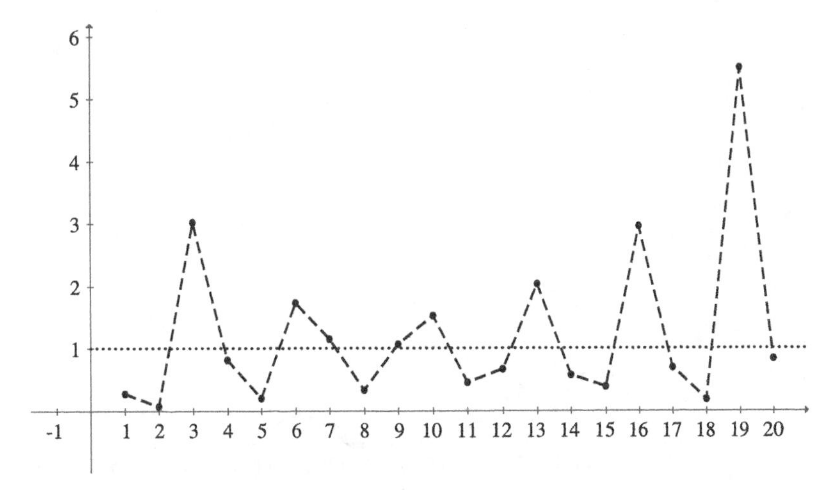

Figure 31.6. The sequence of ratios $\langle \frac{n^2|\cos 1 + \sin 1 \cot n|}{(n+1)^2} \rangle_1^\infty$.

Example 31.10 The alternating series $\sum_{n=1}^{\infty}(-1)^{n+1}\frac{|\sin n|}{n^2}$ (where all $a_n = \frac{|\sin n|}{n^2} \neq 0$) satisfies premise (2) of Theorem 31.1, but fails the monotonicity premise (1). To show this, we form

$$\frac{a_{n+1}}{a_n} = \frac{n^2}{(n+1)^2}\frac{|\sin(n+1)|}{|\sin n|} = \frac{n^2}{(n+1)^2}|\cos 1 + \sin 1 \cot n|.$$

Thus, $\frac{a_{n+1}}{a_n} > 1$ whenever n approximates a multiple of π, but $\frac{a_{n+1}}{a_n} < 1$ when $\cot n \approx 0$. These inequalities will happen infinitely many times, implying that $\langle \frac{|\sin n|}{n^2} \rangle_1^\infty$ will never become monotone. The haphazard graph of the sequence of ratios $\langle \frac{n^2}{(n+1)^2}|\cos 1 + \sin 1 \cot n| \rangle_1^\infty$ in Figure 31.6 supports this conclusion.

Yet, $\sum_{n=1}^{\infty}\frac{|\sin n|}{n^2}$ is convergent (by what test?). Hence, by Theorem 31.2, $\sum_{n=1}^{\infty}(-1)^{n+1}\frac{|\sin n|}{n^2}$ is convergent. ■

Example 31.11 The elaborate-looking alternating series

$$\sum_{n=1}^{\infty}\frac{(-1)^{n+1}}{n}\sqrt{2}^{(-1)^{n+1}} = \sqrt{2} - \frac{1}{2\sqrt{2}} + \frac{\sqrt{2}}{3} - \frac{1}{4\sqrt{2}} + \frac{\sqrt{2}}{5} - \cdots$$

also satisfies premise (2) of the alternating series test (show this in Exercise 31.4). But a little algebra shows that $\frac{a_{n+1}}{a_n} = (\frac{n}{n+1})2^{(-1)^n}$. Consequently, when n is even, $\frac{a_{n+1}}{a_n} > 1$, and when n is odd, $\frac{a_{n+1}}{a_n} < 1$. Thus,

$$\left\langle \frac{1}{n}\sqrt{2}^{(-1)^{n+1}} \right\rangle_1^\infty$$

is not monotone decreasing, and premise (1) of the alternating series test is not the case again. This sequence has a pretty graph, seen in Figure 31.7.

To prove that the series diverges, rewrite it as

$$\frac{\sqrt{2}}{1} - \frac{\sqrt{2}}{4} + \frac{\sqrt{2}}{3} - \frac{\sqrt{2}}{8} + \frac{\sqrt{2}}{5} - \frac{\sqrt{2}}{12} + \cdots + \frac{\sqrt{2}}{2n+1} - \frac{\sqrt{2}}{4n+4} + \cdots$$

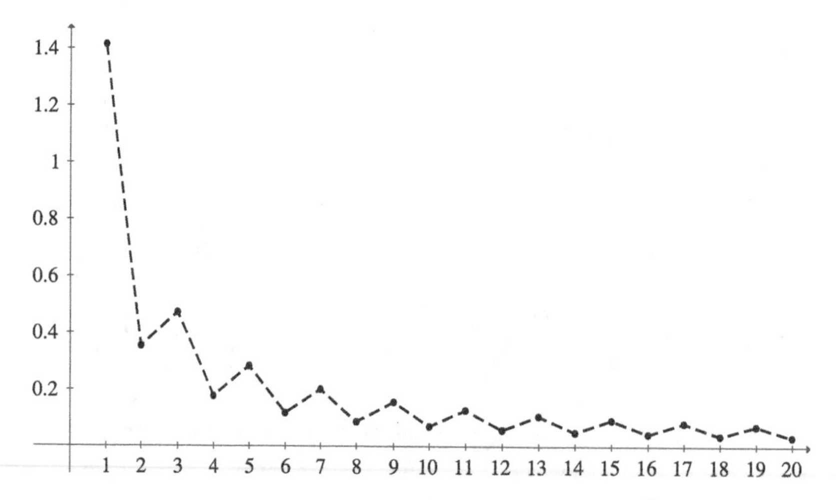

Figure 31.7. Sequence of absolute terms $\langle \frac{1}{n}\sqrt{2}^{(-1)^{n+1}} \rangle_1^\infty$.

where $n \in \mathbb{N} \cup \{0\}$. We would like to insert parentheses and group pairs of terms, but recalling Example 28.6, we are warned that grouping may transform a divergent series into a convergent one. Fortunately, if a grouped series diverges, then removing parentheses cannot make it converge, by Theorem 28.1. Thus, we will show that

$$\left(\frac{\sqrt{2}}{1} - \frac{\sqrt{2}}{4}\right) + \left(\frac{\sqrt{2}}{3} - \frac{\sqrt{2}}{8}\right) + \left(\frac{\sqrt{2}}{5} - \frac{\sqrt{2}}{12}\right) + \cdots + \left(\frac{\sqrt{2}}{2n+1} - \frac{\sqrt{2}}{4n+4}\right) + \cdots$$

diverges. This grouped series is expressed as

$$\sum_{n=0}^{\infty}\left(\frac{\sqrt{2}}{2n+1} - \frac{\sqrt{2}}{4n+4}\right) = \frac{\sqrt{2}}{4}\sum_{n=0}^{\infty}\frac{2n+3}{(2n+1)(n+1)},$$

and the integral test (among others) indicates divergence. It follows that the original, ungrouped series diverges. ■

Examples 31.10 and 31.11 have shown us that the monotonicity premise (1) of Theorem 31.1 is necessary, and we have used some interesting techniques to determine convergence and divergence.

31.3 Exercises

1. Determine whether the following series converge absolutely or conditionally. Find the sums for the series marked with asterisks.

   a. $\sum_{n=0}^{\infty} \frac{(-1)^n}{2n+2}$ *

   b. $\sum_{n=0}^{\infty} (-\frac{\pi}{4})^n$ *

   c. $\sum_{n=2}^{\infty} \frac{(-1)^n}{\ln n}$

   d. $\sum_{n=1}^{\infty} (-1)^n \frac{\ln n}{n}$ (the logarithmic series from Example 31.3)

   e. $\sum_{n=1}^{\infty} \frac{(-1)^{n+1}}{n^3+3n^2+2n}$

   f. $\sum_{n=1}^{\infty} \frac{(-1)^n}{n(n+1)}$ *

   g. $\sum_{n=1}^{\infty} \frac{(-1)^{n-1}}{\sqrt{n}}$

h. $\sum_{n=1}^{\infty} \frac{\operatorname{sgn}(\sin n^2)}{n^{3/2}}$ (evaluate the first eight or nine terms)
i. Any alternating p-series with $0 < p \leq 1$.

2. Make a graph of the partial sums of Exercise 31.1b similar to the one in Example 31.3. Also, graph a horizontal line at the sum that you found for the series. Does the series converge quickly?

3. Prove this lemma: if a series $\sum_{n=1}^{\infty} a_n$ with positive and negative terms converges absolutely, then $\sum_{n=1}^{\infty} a_n \leq |\sum_{n=1}^{\infty} a_n| \leq \sum_{n=1}^{\infty} |a_n|$. Hint: Compare the partial sums.

4. In Example 31.11, prove that premise (2) of the alternating series test is true:

$$\lim_{n\to\infty} \frac{\sqrt{2}^{(-1)^{n+1}}}{n} = 0.$$

5. For the series in Exercise 31.1d, find the maximum error made in estimating the actual sum by the sum of the first 100 terms. Find an interval in which the sum must lie.

6. a. Show that the alternating series test cannot be applied to $\sum_{n=2}^{\infty} \frac{(-1)^n}{\sqrt{n}+(-1)^n}$. However, use another test to prove that the series is not absolutely convergent. Hence, if it converges, it will be conditionally convergent.
b. Assume that $\sum_{n=2}^{\infty} \frac{(-1)^n}{\sqrt{n}+(-1)^n}$ converges (conditionally, by a). Since $\sum_{n=2}^{\infty} \frac{(-1)^n}{\sqrt{n}}$ converges, then by Theorem 28.2, the difference

$$\sum_{n=2}^{\infty} \frac{(-1)^n}{\sqrt{n}} - \sum_{n=2}^{\infty} \frac{(-1)^n}{\sqrt{n}+(-1)^n}$$

must equal the convergent series

$$\sum_{n=2}^{\infty} \left(\frac{(-1)^n}{\sqrt{n}} - \frac{(-1)^n}{\sqrt{n}+(-1)^n} \right) = \sum_{n=2}^{\infty} \frac{1}{n+(-1)^n\sqrt{n}}$$

(verify). But this series *diverges*, as proved in Exercise 30.13. Explain.

7. Prove that $\sum_{n=1}^{\infty} \frac{n^2 3^n}{(-5)^n}$ converges, and find its sum. Hint: See Exercise 29.9.

8. This exercise uses the interesting article "The Alternating Harmonic Series" by Leonard Gillman, in *The College Mathematics Journal*, vol. 33, no. 2 (March 2002), pp. 143–145.

To evaluate the sum of the alternating harmonic series without resorting to power series in calculus, prove the several steps in what follows. We will need a famous number called Euler's constant, γ, to be defined as $\gamma \equiv \lim_{n\to\infty}(H_n - \ln n)$ in Chapter 34, where $H_n = \sum_{k=1}^{n} \frac{1}{k}$ are the partial sums of the harmonic series.
(a) Explain why $\gamma = \lim_{n\to\infty}(H_{2n} - \ln 2n)$.
(b) Let $A_n = \sum_{k=1}^{n}(-1)^{k-1}\frac{1}{k}$ be the partial sums of the alternating harmonic series. Notice that, for example,

$$A_6 = 1 - \frac{1}{2} + \frac{1}{3} - \frac{1}{4} + \frac{1}{5} - \frac{1}{6} = 1 + \frac{1}{2} + \frac{1}{3} + \frac{1}{4} + \frac{1}{5} + \frac{1}{6} - \left(1 + \frac{1}{2} + \frac{1}{3}\right)$$
$$= H_6 - H_3.$$

This is true in general! To be precise, prove by induction that $A_{2n} = H_{2n} - H_n$ for all $n \in \mathbb{N}$.

(c) Since we now know that $A_{2n} = (H_{2n} - \ln 2n) - (H_n - \ln n) + \ln 2n - \ln n$, show that $\lim_{n \to \infty} A_{2n} = \ln 2$.

(d) Finally, prove that $\lim_{n \to \infty} A_n = \ln 2$.

Note 1: Another very nice demonstration that the alternating harmonic series converges to $\ln 2$ is based on the fact that $\int_1^2 \frac{1}{x} dx = \ln 2$. It is "A Geometric Proof of the Formula for $\ln 2$" by Frank Kost, in *Mathematics Magazine*, vol. 44, no. 1 (Jan. 1977), pp. 37–38.

32

Rearrangements of Series

The divergent series are the invention of the devil, and it is a shame to base on them any demonstration whatsoever. By using them, one may draw any conclusion he pleases and that is why these series have produced so many fallacies and paradoxes.

Niels Henrik Abel, 1828

This chapter considers the effects of rearranging or commuting the terms of an infinite series. In Chapter 28, it was noted that the commutative and associative axioms of $\mathbb{R}$ were not written to include infinite series. Up to now, only Theorems 28.1, 28.2, and Lemma 29.1 have given us light in this new direction. Let's repeat parts of them here for convenience. Theorem 28.1: the sum of a convergent series is not affected by grouping of terms. Theorem 28.2: if

$$\sum_{n=1}^{\infty} a_n = A \quad \text{and} \quad \sum_{n=1}^{\infty} b_n = B, \text{ then } \sum_{n=1}^{\infty} (a_n + b_n) = \sum_{n=1}^{\infty} a_n + \sum_{n=1}^{\infty} b_n = A + B,$$

which is a rearrangement and regrouping of a denumerable number of terms when we look at the conclusion in detail:

$$(a_1 + b_1) + (a_2 + b_2) + \cdots + (a_n + b_n) + \cdots = (a_1 + a_2 + \cdots + a_n + \cdots) \\ + (b_1 + b_2 + \cdots + b_n + \cdots).$$

Lemma 29.1: the sum of a convergent series of nonnegative terms is not affected by grouping or ungrouping of its terms.

The classic problem with rearrangements is demonstrated by the alternating harmonic series. We know by Leibniz's alternating series test that the series has a sum, evaluated in Exercise 31.8:

$$\sum_{n=1}^{\infty} \frac{(-1)^{n+1}}{n} = 1 - \frac{1}{2} + \frac{1}{3} - \frac{1}{4} + \frac{1}{5} - \frac{1}{6} + \frac{1}{7} - \frac{1}{8} + \cdots = \ln 2 \approx 0.693.$$

Do some innocent-looking rearranging so that two negative terms follow one positive term:

$$1-\frac{1}{2}-\frac{1}{4}+\frac{1}{3}-\frac{1}{6}-\frac{1}{8}+\frac{1}{5}-\frac{1}{10}-\frac{1}{12}+\cdots$$
$$=\left(1-\frac{1}{2}\right)-\frac{1}{4}+\left(\frac{1}{3}-\frac{1}{6}\right)-\frac{1}{8}+\left(\frac{1}{5}-\frac{1}{10}\right)-\frac{1}{12}+\cdots.$$

Assuming that the left-hand side converges, it must equal the right-hand side by Theorem 28.1. Simplifying the parentheses and using Theorem 28.3,

$$\frac{1}{2}-\frac{1}{4}+\frac{1}{6}-\frac{1}{8}+\frac{1}{10}-\frac{1}{12}+\cdots=\frac{1}{2}\left(1-\frac{1}{2}+\frac{1}{3}-\frac{1}{4}+\frac{1}{5}-\frac{1}{6}+\cdots\right)=\frac{1}{2}\ln 2.$$

This is absurd, for it shows us that rearranging the alternating harmonic series produces half its original sum! How can a convergent series transmogrify itself in such a terrible manner?

The cause of the problem is the mode of convergence of the series, namely, conditional convergence. We now develop some new ideas that apply not just to alternating series, but to *any* series $\sum_{n=1}^{\infty} a_n$. Define

$$p_n \equiv \frac{1}{2}(a_n+|a_n|)=\max\{0,a_n\} \text{ and } q_n \equiv \frac{1}{2}(a_n-|a_n|)=\min\{0,a_n\}.$$

Thus, for each $n \in \mathbb{N}$, $p_n \geq 0$ and $q_n \leq 0$, and at least one of them equals zero (verify). The two defining equations lead to (a) $a_n = p_n + q_n$ and (b) $|a_n| = p_n - q_n$. We now form three series closely connected with the given one: the **nonnegative part** $\sum_{n=1}^{\infty} p_n$, the **nonpositive part** $\sum_{n=1}^{\infty} q_n$, and the series of absolute values $\sum_{n=1}^{\infty} |a_n|$.

Example 32.1 For the alternating harmonic series $\sum_{n=1}^{\infty} \frac{(-1)^{n+1}}{n}$, we have, of course,

$$\langle a_n\rangle_1^{\infty}=\left\langle 1,\frac{-1}{2},\frac{1}{3},\frac{-1}{4},\frac{1}{5},\frac{-1}{6},\ldots\right\rangle.$$

And now,

$$\langle p_n\rangle_1^{\infty}=\left\langle 1,0,\frac{1}{3},0,\frac{1}{5},\ldots\right\rangle,\ \langle q_n\rangle_1^{\infty}=\left\langle 0,\frac{-1}{2},0,\frac{-1}{4},0,\frac{-1}{6},\ldots\right\rangle,$$

and

$$\langle |a_n|\rangle_1^{\infty}=\left\langle 1,\frac{1}{2},\frac{1}{3},\frac{1}{4},\frac{1}{5},\frac{1}{6},\ldots\right\rangle.$$

Consequently, the nonnegative part of the alternating harmonic series is

$$\sum_{n=1}^{\infty} p_n = 1+0+\frac{1}{3}+0+\frac{1}{5}+\cdots,$$

while the nonpositive part is $\sum_{n=1}^{\infty} q_n = 0-\frac{1}{2}-0-\frac{1}{4}-0-\frac{1}{6}-\cdots$. The third series is $\sum_{n=1}^{\infty}|a_n| = 1+\frac{1}{2}+\frac{1}{3}+\frac{1}{4}+\frac{1}{5}+\cdots$. ■

We investigate the partial sums of the four series, namely

$$S_n = a_1 + a_2 + \cdots + a_n$$
$$P_n = p_1 + p_2 + \cdots + p_n$$
$$Q_n = q_1 + q_2 + \cdots + q_n$$
$$A_n = |a_1| + |a_2| + \cdots + |a_n|.$$

Directly from (a) and (b), we infer that $S_n = P_n + Q_n$ and $A_n = P_n - Q_n$, for all $n \in \mathbb{N}$. These formulations lead us to an easily-proved lemma.

Lemma 32.1 **A** If any two of the four series $\sum_{n=1}^{\infty} a_n$, $\sum_{n=1}^{\infty} p_n$, $\sum_{n=1}^{\infty} q_n$, and $\sum_{n=1}^{\infty} |a_n|$, converge, then all of them converge.
B If $\sum_{n=1}^{\infty} a_n$ converges absolutely, then both $\sum_{n=1}^{\infty} p_n$ and $\sum_{n=1}^{\infty} q_n$ converge.
C If $\sum_{n=1}^{\infty} a_n$ converges conditionally, then both $\sum_{n=1}^{\infty} p_n$ and $\sum_{n=1}^{\infty} q_n$ diverge.

Proof. **A** Begin by assuming $\langle P_n \rangle_1^{\infty}$and $\langle Q_n \rangle_1^{\infty}$ converge, and use Theorem 20.2. Complete the proof of all three parts as Exercise 32.3. **QED**

Example 32.2 In the previous example involving the alternating harmonic series, both

$$\sum_{n=1}^{\infty} p_n = \sum_{n=1}^{\infty} \frac{1}{2n-1} \quad \text{and} \quad \sum_{n=1}^{\infty} q_n = \sum_{n=1}^{\infty} \frac{-1}{2n}$$

diverge, as Lemma 32.1C asserts. ■

The third statement in the lemma holds the resolution to the problem of rearrangement. The following amazing theorem is by Bernhard Riemann.

Theorem 32.1 Let $\sum_{n=1}^{\infty} a_n$ be a conditionally convergent series. For any real number B, there exists a rearrangement of the series that will converge to B. Also, there exists a rearrangement that will make the series diverge.

Proof. By Lemma 32.1, the nonnegative part $\sum_{n=1}^{\infty} p_n$ and nonpositive part $\sum_{n=1}^{\infty} q_n$ diverge. Since these series are monotone, they diverge to $+\infty$ and $-\infty$, respectively. We are given a real number B. The following pointed algorithm will let us construct a rearrangement of $\sum_{n=1}^{\infty} a_n$ that converges to B.

Begin by selecting the least number of terms N_1 from the nonnegative part so that $\sum_{n=1}^{N_1} p_n > B$. This inequality is certainly attainable, for we have seen that $\sum_{n=1}^{\infty} p_n$ increases without bound (sometimes, we may need only the $N_1 = 1$ term to surpass B). Next, select the least number M_1 of terms from the nonpositive part so that $\sum_{n=1}^{N_1} p_n + \sum_{n=1}^{M_1} q_n < B$. This inequality is also attainable, because $\sum_{n=1}^{\infty} q_n$ decreases without bound. Continue with the next N_2 terms (the fewest possible) from the nonnegative part so that

$$\sum_{n=1}^{N_1} p_n + \sum_{n=1}^{M_1} q_n + \sum_{n=N_1+1}^{N_2} p_n > B,$$

followed by the least M_2 such that

$$\sum_{n=1}^{N_1} p_n + \sum_{n=1}^{M_1} q_n + \sum_{n=N_1+1}^{N_2} p_n + \sum_{n=M_1+1}^{M_2} q_n < B,$$

and so on.

In our symbols, this algorithm is forming the rearrangement with partial sums

$$\begin{aligned}&\big(P_{N_1},\ P_{N_1} + Q_{M_1},\ P_{N_2} + Q_{M_1},\ P_{N_2} + Q_{M_2},\ P_{N_3} + Q_{M_2},\ P_{N_3} + Q_{M_3}, \ldots,\\ &P_{N_i} + Q_{M_{i-1}},\ P_{N_i} + Q_{M_i}, \ldots\big)_{i=1}^{\infty}\end{aligned} \tag{1}$$

(define $Q_{M_0} = 0$ for notation's sake). Furthermore, it also stipulates that $P_{N_1} - p_{N_1} < B < P_{N_1}$, $P_{N_1} + Q_{M_1} < B < P_{N_1} + Q_{M_1} - q_{M_1}$, $P_{N_2} - p_{N_2} + Q_{M_1} < B < P_{N_2} + Q_{M_1}$, and so forth. In general, for even $i \in \mathbb{N}$, $P_{N_i} + Q_{M_i} < B < P_{N_i} + Q_{M_i} - q_{M_i}$, which implies that $0 < B - (P_{N_i} + Q_{M_i}) < -q_{M_i}$. And for odd $i \in \mathbb{N}$, $P_{N_i} - p_{N_i} + Q_{M_i} < B < P_{N_i} + Q_{M_i}$, which implies that $-p_{N_i} < B - (P_{N_i} + Q_{M_i}) < 0$. Due to the convergence of $\sum_{n=1}^{\infty} a_n$, we know that both $p_n \to 0$ and $q_n \to 0$ (Theorem 28.4). Hence, $\lim_{i\to\infty} (P_{N_i} + Q_{M_i}) = B$.

The terms $P_{N_i} + Q_{M_i}$ are the even-numbered terms of (1), and we now wonder about the odd-numbered terms $P_{N_i} + Q_{M_{i-1}}$. Thinking about how Q_{M_i} is derived from $Q_{M_{i-1}}$, we have $P_{N_i} + Q_{M_{i-1}} = P_{N_i} + Q_{M_i} - \sum_{n=M_{i-1}+1}^{M_i} q_n$, from which follows

$$\lim_{i\to\infty} (P_{N_i} + Q_{M_{i-1}}) = \lim_{i\to\infty} (P_{N_i} + Q_{M_i}) - \lim_{i\to\infty} \left(\sum_{n=M_{i-1}+1}^{M_i} q_n \right) = B - 0 = B.$$

Therefore, the rearrangement of $\sum_{n=1}^{\infty} a_n$ in (1) will converge to the arbitrarily given number B.

To finish the proof, you are asked in Exercise 32.5 to develop a rearrangement of $\sum_{n=1}^{\infty} a_n$ that will diverge. **QED**

Note 1: The theorem just proved needs a prescribed number B, and then tells us how to rearrange any conditionally convergent series to sum to B. Unfortunately, given a rearrangement of a conditionally convergent series, the theorem can't tell us anything in general about a sum B, or if convergence even occurs.

Example 32.3 Find a rearrangement of the alternating harmonic series that converges to $B = 1.5$. Using Example 32.1, begin with

$$1 + \frac{1}{3} + \frac{1}{5} = \sum_{n=1}^{5} p_n = P_5 > 1.5$$

(and $1 + \frac{1}{3} < 1.5$), so that $N_1 = 5$. Next, $P_5 - \frac{1}{2} < 1.5$, so we use $Q_2 = -\frac{1}{2}$, implying that $M_1 = 2$, since $\frac{1}{2}$ is the first nonpositive term to push the sum below the 1.5 threshold. Thus, $P_5 + Q_2 < 1.5$. Next,

$$P_5 + Q_2 + \frac{1}{7} + \frac{1}{9} + \frac{1}{11} + \frac{1}{13} + \frac{1}{15} = P_{15} + Q_2 > 1.5,$$

which indicates that $N_2 = 15$. Once more, $P_{15} + Q_2 - \frac{1}{4} = P_{15} + Q_4 < 1.5$, so $M_2 = 4$. Riemann's algorithm is producing the rearrangement

$$\left(1 + \frac{1}{3} + \frac{1}{5}\right) - \frac{1}{2} + \left(\frac{1}{7} + \frac{1}{9} + \frac{1}{11} + \frac{1}{13} + \frac{1}{15}\right) - \frac{1}{4} + \cdots,$$

with the sequence of partial sums $\langle P_5,\ P_5 + Q_2,\ P_{15} + Q_2,\ P_{15} + Q_4, \cdots\rangle$. Continuing without end, this rearrangement will converge to 1.5. ■

Note 2: As opposed to the situation in Note 1, any rearrangement of the alternating harmonic series made by consistently taking m positive terms followed by n negative terms, does have a known sum. That sum is $\ln\left(2\sqrt{\frac{m}{n}}\right)$.

For example,

$$\frac{1}{1} - \frac{1}{2} - \frac{1}{4} - \frac{1}{6} - \frac{1}{8} + \frac{1}{3} - \frac{1}{10} - \frac{1}{12} - \frac{1}{14} - \frac{1}{16} + \frac{1}{5} - \cdots$$

consistently takes $m = 1$ positive terms followed by $n = 4$ negative terms. Then this rearrangement of the alternating harmonic series converges to $\ln\left(2\sqrt{\frac{1}{4}}\right) = 0$. Quite amazing, Professor Riemann!

For the proof of this theorem and other related facts, read "On Rearrangements of the Alternating Harmonic Series" by F. Brown, L. O. Cannon, J. Elich, and D. G. Wright, in *The College Mathematics Journal*, vol. 16, no. 2 (March 1985), pp. 135–138.

Roughly speaking, Theorem 32.1 tells us that conditionally convergent series converge due to a delicate balance of positive and negative terms. A rearrangement of infinitely many terms may alter the balance and affect the sum, or even completely tip the scales into divergence. In contrast to this behavior, absolutely convergent series are totally robust in that no rearrangement whatsoever will succeed in changing their sums. The next theorem, discovered by Dirichlet, proves that proposition.

Theorem 32.2 Let $\sum_{n=1}^{\infty} a_n$ be an absolutely convergent series, with sum A. Then every rearrangement of the series will also have sum A, and the rearrangements are absolutely convergent as well.

Proof. The proof begins with the special case of a nonnegative series $\sum_{n=1}^{\infty} d_n$ that converges to A. Let $D_n = \sum_{j=1}^{n} d_j$ be its nth partial sum. Let $\sum_{n=1}^{\infty} b_n$ be any rearrangement of $\sum_{n=1}^{\infty} d_n$, and let $B_n = \sum_{j=1}^{n} b_j$ be its nth partial sum. Now, each B_n contains a finite number of terms of $\sum_{n=1}^{\infty} d_n$. It follows that $B_n \le D_m$ for some D_m that contains all the terms in B_n. Since $\langle D_m\rangle_1^{\infty}$ is monotone increasing with limit A, we have that $B_n \le A$ for all $n \in \mathbb{N}$. Thus, $\langle B_n\rangle_1^{\infty}$ is a bounded, monotone increasing sequence; therefore it must have a limit B (by what theorem?), and $B \le A$. Furthermore, and clever indeed, $\sum_{n=1}^{\infty} d_n$ is a rearrangement of $\sum_{n=1}^{\infty} b_n$, which allows us to rewrite the argument and arrive at $A \le B$. Consequently, $A = B$ for any rearrangement of a convergent nonnegative series $\sum_{n=1}^{\infty} d_n$.

Now to the general case of any absolutely convergent series $\sum_{n=1}^{\infty} a_n$. By Lemma 32.1B, the nonnegative and nonpositive parts of the series converge, allowing us to write $\sum_{n=1}^{\infty} p_n = P$ and $\sum_{n=1}^{\infty} q_n = Q$. Suppose $\sum_{n=1}^{\infty} b_n$ is any rearrangement of $\sum_{n=1}^{\infty} a_n$. It follows that the nonnegative part of $\sum_{n=1}^{\infty} b_n$, let's write it as $\sum_{n=1}^{\infty} b_n^{[+]}$, is a rearrangement of $\sum_{n=1}^{\infty} p_n$. For the same reason, the nonpositive part of $\sum_{n=1}^{\infty} b_n$, written as $\sum_{n=1}^{\infty} b_n^{[-]}$, is a rearrangement of $\sum_{n=1}^{\infty} q_n$. Now, the special case just proved applies to $\sum_{n=1}^{\infty} b_n^{[+]}$ and $-\sum_{n=1}^{\infty} b_n^{[-]}$. It implies that

$$\sum_{n=1}^{\infty} b_n^{[+]} = P \quad \text{and} \quad -\sum_{n=1}^{\infty} b_n^{[-]} = -Q.$$

We have shown that

$$\sum_{n=1}^{\infty} b_n = \sum_{n=1}^{\infty} b_n^{[+]} + \sum_{n=1}^{\infty} b_n^{[-]} = P + Q = A = \sum_{n=1}^{\infty} a_n.$$

Therefore, any rearrangement of the absolutely convergent series $\sum_{n=1}^{\infty} a_n$ has no effect at all on the sum of the series.

Lemma 32.1A shows us that $\sum_{n=1}^{\infty} |b_n|$ also converges, which means that the rearrangement $\sum_{n=1}^{\infty} b_n$ is absolutely convergent, as the theorem states. **QED**

Example 32.4 It is easy to show that $\sum_{n=0}^{\infty} \frac{(-2)^n}{n!} = e^{-2}$ is absolutely convergent. We may rearrange the series to display three successive positive terms followed by two successive negative terms, without end. We will have

$$1 + \frac{4}{2!} + \frac{16}{4!} - \frac{2}{1!} - \frac{8}{3!} + \frac{64}{6!} + \frac{256}{8!} + \frac{1024}{10!} - \cdots.$$

Then by Theorem 32.2, this series also converges to e^{-2}, although the rate of convergence is much slower than that of the original series. ■

32.1 Exercises

1. Try another trick with the alternating harmonic series:

$$\begin{aligned} 2\ln 2 &= 2\left(1 - \frac{1}{2} + \frac{1}{3} - \frac{1}{4} + \frac{1}{5} - \frac{1}{6} + \frac{1}{7} - \frac{1}{8} + \cdots\right) \\ &= 2 - 1 + \frac{2}{3} - \frac{1}{2} + \frac{2}{5} - \frac{1}{3} + \frac{2}{7} - \frac{1}{4} + \cdots \\ &= (2-1) - \frac{1}{2} + \left(\frac{2}{3} - \frac{1}{3}\right) - \frac{1}{4} + \left(\frac{2}{5} - \frac{1}{5}\right) - \frac{1}{6} + \cdots. \end{aligned}$$

Simplify, and arrive at another astonishing conclusion.

2. a. $\sum_{n=2}^{\infty} \frac{(-1)^n}{\ln n}$ was shown to converge conditionally in Exercise 31.1c. Conclude something about the series $\sum_{n=1}^{\infty} \frac{1}{\ln(2n)}$ and $\sum_{n=1}^{\infty} \frac{-1}{\ln(2n+1)}$.

   b. The logarithmic series $\sum_{n=1}^{\infty} (-1)^n \frac{\ln n}{n}$ was shown to converge conditionally in Exercise 31.1d (its sum is known). What can be said about $\sum_{n=1}^{\infty} \frac{-\ln(2n-1)}{2n-1}$?

3. Prove Lemma 32.1.

4. Extend Example 32.3 by finding the next two terms in the sequence

$$\langle P_5,\ P_5 + Q_2,\ P_{15} + Q_2,\ P_{15} + Q_4, \cdots \rangle .$$

5. Finish the proof of Theorem 32.1 by describing a rearrangement of a conditionally convergent series $\sum_{n=1}^{\infty} a_n$ that must diverge. And now, use your scheme to form a rearrangement of the alternating harmonic series that forces it to diverge to $+\infty$.

6. Are there divergent series that have rearrangements that converge? If you suspect that it is so, construct such a series. If you think this is impossible, prove it.

33

Products of Series

$\frac{\pi}{4} = \cos y - \frac{1}{3}\cos 3y + \frac{1}{5}\cos 5y - \frac{1}{7}\cos 7y + \frac{1}{9}\cos 9y-$ etc.... we give the variable y a value included between $-\frac{1}{2}\pi$ and $+\frac{1}{2}\pi$. It would be easy to prove that this series is always convergent, that is to say that writing instead of y any number whatever... we approach more and more to a fixed value, so that the difference of this value from the sum of the calculated terms becomes less than any assignable magnitude.

Joseph Fourier, "*Analytical Theory of Heat*," 1822, Article 177.

How would one attempt to multiply two infinite series, as in $(\sum_{n=0}^{\infty} a_n)(\sum_{n=0}^{\infty} b_n)$? We will consider two possibilities. The first will remind us of polynomial multiplication; the second will remind us of the inner product of two vectors. In either case, the convergence of the product must be addressed.

33.1 Cauchy Products

Let's look at the product of two polynomials, and generalize in some way. Multiplying two polynomials of equal degree, let's say the third, produces an intriguing pattern:

$$\begin{aligned}(a_0 + a_1x + a_2x^2 + a_3x^3)(b_0 + b_1x + b_2x^2 + b_3x^3) = \; & a_0b_0x^0 \\ & + a_0b_1x \; + a_1b_0x \\ & + a_0b_2x^2 + a_1b_1x^2 + a_2b_0x^2 \\ & + a_0b_3x^3 + a_1b_2x^3 + a_2b_1x^3 + a_3b_0x^3,\end{aligned}$$

plus six more terms of higher degree that won't concern us. We notice that the terms seen here are in the form $a_kb_{n-k}x^n$, where $0 \le k \le n$. Thus, the coefficient of x^n is $\sum_{k=0}^{n} a_kb_{n-k}$, for $n \le 3$. It is not hard to imagine that the coefficients will have this form for any n, where n is the lesser of the degrees of the two polynomials. Next, the leap: suppose n is infinite. This is, of course, an error in thinking, for n cannot be $\aleph_0$. However, a precise definition supplied by Cauchy puts us back on track. The **Cauchy product** of two series, which we denote as $\sum_{n=0}^{\infty} c_n$, is defined as follows.

$$\left(\sum_{n=0}^{\infty} a_n\right)\left(\sum_{n=0}^{\infty} b_n\right) = \sum_{n=0}^{\infty} c_n$$

where $c_n = a_0b_n + a_1b_{n-1} + \cdots + a_{n-1}b_1 + a_nb_0$, and $n \in \mathbb{N}\cup\{0\}$.

Using summation notation, $c_n = \sum_{k=0}^{n} a_k b_{n-k}$. In expanded form, we have the Cauchy product

$$
\begin{aligned}
&(a_0 + a_1 + \cdots + a_n + \cdots)(b_0 + b_1 + \cdots + b_n + \cdots) \\
&= c_0 + c_1 + c_2 + c_3 + \cdots + c_n + \cdots \\
&= \underbrace{a_0 b_0}_{c_0} + \underbrace{(a_0 b_1 + a_1 b_0)}_{c_1} + \underbrace{(a_0 b_2 + a_1 b_1 + a_2 b_0)}_{c_2} \\
&+ \underbrace{(a_0 b_3 + a_1 b_2 + a_2 b_1 + a_3 b_0)}_{c_3} + \cdots + \underbrace{\sum_{k=0}^{n} a_k b_{n-k}}_{c_n} + \cdots .
\end{aligned}
$$

Note 1: Cauchy's definition mimics the product of two power series, and for good reasons, as will be seen in Chapter 43:

$$
\begin{aligned}
&\left(\sum_{n=0}^{\infty} a_n x^n\right)\left(\sum_{n=0}^{\infty} b_n x^n\right) = (a_0 + a_1 x + a_2 x^2 + \cdots + a_n x^n + \cdots)(b_0 + b_1 x + b_2 x^2 \\
&+ \cdots + b_n x^n + \cdots) = a_0 b_0 + (a_0 b_1 + a_1 b_0)x + (a_0 b_2 + a_1 b_1 + a_2 b_0)x^2 + \cdots \\
&+ \left(\sum_{k=0}^{n} a_k b_{n-k}\right) x^n + \cdots = \sum_{n=0}^{\infty} c_n x^n .
\end{aligned}
$$

In the spirit of 17th century analysis, such a multiplication would have been done with few worries about convergence. In those days, analysis was sometimes performed simply by analogy to common algebra (today, this is called *formal manipulation*). In this case, the product of infinitely many terms was seen as just an extension of the product of two polynomials! It was perhaps assumed that if the two factors $\sum_{n=0}^{\infty} a_n$ and $\sum_{n=0}^{\infty} b_n$ individually converged, then so would the product series.

Alas, it is not true.... A counterexample: let $a_n = b_n = \frac{(-1)^n}{\sqrt{n+1}}$. Then the series $\sum_{k=0}^{n} a_n$ and $\sum_{k=0}^{n} b_n$ converge, and

$$
c_n = \sum_{k=0}^{n} a_k b_{n-k} = \sum_{k=0}^{n} \frac{(-1)^k}{\sqrt{k+1}} \frac{(-1)^{n-k}}{\sqrt{n-k+1}} = (-1)^n \sum_{k=0}^{n} \frac{1}{\sqrt{k+1}\sqrt{n-k+1}} .
$$

By showing that $|c_n| \geq 1$, you can prove that $\sum_{n=0}^{\infty} c_n$ diverges.

Example 33.1 Find the Cauchy product

$$
\left(\sum_{n=0}^{\infty} \frac{2^n}{3^n}\right)\left(\sum_{n=0}^{\infty} \frac{(-1)^n}{5^n}\right).
$$

We identify $a_n = (\frac{2}{3})^n$ and $b_n = (\frac{-1}{5})^n$. Then

$$
c_n = \sum_{k=0}^{n} a_k b_{n-k} = \sum_{k=0}^{n} \left(\frac{2}{3}\right)^k \left(\frac{-1}{5}\right)^{n-k} = \sum_{k=0}^{n} \left(\frac{2}{3}(-5)\right)^k \left(\frac{-1}{5}\right)^n = \left(\frac{-1}{5}\right)^n \sum_{k=0}^{n} \left(\frac{-10}{3}\right)^k .
$$

It follows that the Cauchy product is

$$\sum_{n=0}^{\infty} c_n = 1 - \frac{1}{5}\left(1 - \frac{10}{3}\right) + \frac{1}{25}\left(1 - \frac{10}{3} + \frac{100}{9}\right) - \frac{1}{125}\left(1 - \frac{10}{3} + \frac{100}{9} - \frac{1000}{27}\right) + \cdots.$$

Being convergent geometric series,

$$\sum_{n=0}^{\infty} \frac{2^n}{3^n} = 3 \quad \text{and} \quad \sum_{n=0}^{\infty} \frac{(-1)^n}{5^n} = \frac{5}{6}, \quad \text{giving us} \quad \left(\sum_{n=0}^{\infty} \frac{2^n}{3^n}\right)\left(\sum_{n=0}^{\infty} \frac{(-1)^n}{5^n}\right) = \frac{5}{2}.$$

But does the Cauchy product converge, and if so, is its sum $\frac{5}{2}$? The answers will arrive soon. ■

The examples below show that you have already encountered Cauchy products.

Example 33.2 In a typical calculus exercise, we are asked to find the Taylor series for the function $f(x) = e^x \ln x$ about $x = 1$. (We will carefully study Taylor series beginning in Chapter 45.) Finding the general coefficient $\frac{f^{(n)}(1)}{n!}$ is almost painful. Instead, we turn to the product of the individual Taylor series for e^x and $\ln x$:

$$\left(e + e(x-1) + \frac{e}{2!}(x-1)^2 + \frac{e}{3!}(x-1)^3 + \cdots\right)$$
$$\times \left((x-1) - \frac{(x-1)^2}{2} + \frac{(x-1)^3}{3} - \frac{(x-1)^4}{4} + \cdots\right). \tag{1}$$

You have undoubtedly multiplied series such as these as if they were awfully long polynomials, by formal manipulation. The resulting product, in powers of $x - 1$, is the required Taylor series,

$$e(x-1) + \frac{1}{2}e(x-1)^2 + \frac{1}{3}e(x-1)^3 + 0e(x-1)^4 + \frac{3}{40}e(x-1)^5 - \frac{7}{144}e(x-1)^6 + \cdots. \tag{2}$$

This is none other than the Cauchy product of the Taylor series in (1)! Note that term $b_0 = \ln 1 = 0$ in the second series of (1) has been left out.

Now, for instance, substitute $x = \frac{1}{3}$ in (1). We wish to determine the Cauchy product

$$\left(e - \frac{2}{3}e + \frac{2}{9}e - \frac{4}{81}e + \cdots\right)\left(-\frac{2}{3} - \frac{2}{9} - \frac{8}{81} - \frac{4}{81} - \cdots\right). \tag{3}$$

If these were the two series given us to begin with, we of course wouldn't have the guidance provided by the exponents of $x - 1$ in order to multiply. Instead, we rely on Cauchy's definition:

$$\left(e - \frac{2e}{3} + \frac{2e}{9} - \frac{4e}{81} + \cdots\right)\left(-\frac{2}{3} - \frac{2}{9} - \frac{8}{81} - \frac{4}{81} - \cdots\right) = e\cdot\frac{-2}{3} + \left(e\cdot\frac{-2}{9} + \frac{-2e}{3}\cdot\frac{-2}{3}\right)$$
$$+ \left(e\cdot\frac{-8}{81} + \frac{-2e}{3}\cdot\frac{-2}{9} + \frac{2e}{9}\cdot\frac{-2}{3}\right) + \left(e\cdot\frac{-4}{81} + \frac{-2e}{3}\cdot\frac{-8}{81} + \frac{2e}{9}\cdot\frac{-2}{9} + \frac{-4e}{81}\cdot\frac{-2}{3}\right) + \cdots$$
$$= -\frac{2}{3}e + \frac{2}{9}e - \frac{8}{81}e + 0 - \cdots.$$

This is identical to the result that we obtain if we substitute $x = \frac{1}{3}$ into (2). See Exercise 33.3 regarding the effect of including term $b_0 = 0$ in (1), and the convergence of this product. ■

Example 33.3 Find the Maclaurin series for the function $f(x) = e^x \tan^{-1} x$ at $x = \frac{1}{2}$. Finding the general coefficient $\frac{f^{(n)}(0)}{n!}$ is daunting. Thus, we find the product of the individual Maclaurin series, $e^x = \sum_{n=0}^{\infty} \frac{x^n}{n!}$ and $\tan^{-1} x = \sum_{n=0}^{\infty} (-1)^n \frac{x^{2n+1}}{2n+1}$:

$$\left(1 + x + \frac{1}{2!}x^2 + \frac{1}{3!}x^3 + \frac{1}{4!}x^4 + \frac{1}{5!}x^5 + \cdots\right)\left(x - \frac{1}{3}x^3 + \frac{1}{5}x^5 - \frac{1}{7}x^7 + \cdots\right)$$

$$= x + x^2 + \frac{1}{6}x^3 - \frac{1}{6}x^4 + \frac{3}{40}x^5 + \cdots. \tag{4}$$

Once again, this product is the Cauchy product of the two Maclaurin series. When $x = \frac{1}{2}$, the factors become

$$\left(\sum_{n=0}^{\infty} \frac{1}{2^n n!}\right)\left(\sum_{n=0}^{\infty} \frac{(-1)^n}{2^{2n+1}(2n+1)}\right),$$

which yields the Cauchy product

$$\left(1 + \frac{1}{2} + \frac{1}{2^2 2!} + \frac{1}{2^3 3!} + \frac{1}{2^4 4!} + \frac{1}{2^5 5!} + \cdots\right)\left(\frac{1}{2} - \frac{1}{2^3 3} + \frac{1}{2^5 5} - \frac{1}{2^7 7} + \cdots\right)$$

$$= 1 \cdot \frac{1}{2} + \left(1 \cdot \frac{-1}{2^3 3} + \frac{1}{2} \cdot \frac{1}{2}\right) + \left(1 \cdot \frac{1}{2^5 5} + \frac{1}{2} \cdot \frac{-1}{2^3 3} + \frac{1}{2^2 2!} \cdot \frac{1}{2}\right) + \cdots. \tag{5}$$

This simplifies to $\frac{1}{2} + \frac{5}{24} + \frac{23}{480} + \cdots$, but this is *not* the result of substituting $x = \frac{1}{2}$ into the right side of (4). Is something wrong? Why is this procedure different from that of Example 33.2?

Observe that the inverse tangent, being an odd function, generates a Maclaurin series with strictly odd powers. Hence, for the Cauchy product to develop in parallel with (4), we should write

$$\left(1 + x + \frac{1}{2!}x^2 + \frac{1}{3!}x^3 + \frac{1}{4!}x^4 + \frac{1}{5!}x^5 + \cdots\right)$$
$$\left(0 + x + 0x^2 - \frac{1}{3}x^3 + 0x^4 + \frac{1}{5}x^5 + 0x^6 - \frac{1}{7}x^7 + \cdots\right).$$

(That is, we explicitly write $b_{2n} = 0x^{2n}$ for $n \in \mathbb{N} \cup \{0\}$.) Upon substituting $x = \frac{1}{2}$, we obtain

$$\left(1 + \frac{1}{2} + \frac{1}{2^2 2!} + \frac{1}{2^3 3!} + \frac{1}{2^4 4!} + \frac{1}{2^5 5!} + \cdots\right)\left(0 + \frac{1}{2} + 0 - \frac{1}{2^3 3} + 0 + \frac{1}{2^5 5} + 0 - \cdots\right),$$

giving the Cauchy product

$$1 \cdot 0 + \left(1 \cdot \frac{1}{2} + \frac{1}{2} \cdot 0\right) + \left(1 \cdot 0 + \frac{1}{2} \cdot \frac{1}{2} + \frac{1}{2^2 2!} \cdot 0\right) + \left(1 \cdot \frac{-1}{2^3 3} + \frac{1}{2} \cdot 0 + \frac{1}{2^2 2!} \cdot \frac{1}{2} + \frac{1}{2^3 3!} \cdot 0\right) + \cdots.$$

This simplifies to $\frac{1}{2} + \frac{1}{4} + \frac{1}{48} - \frac{1}{96} + \cdots$, which *is* the result of substituting $x = \frac{1}{2}$ into the right side of (4). On second thought, we may consider the Cauchy product (5) on its own merits, with $a_n = \frac{1}{2^n n!}$ and $b_n = \frac{(-1)^n}{2^{2n+1}(2n+1)}$, yielding $\frac{1}{2} + \frac{5}{24} + \frac{23}{480} + \cdots$ for $n \in \mathbb{N} \cup \{0\}$ (as we have seen). The series $\frac{1}{2} + \frac{5}{24} + \frac{23}{480} + \cdots$ and the previous one, $\frac{1}{2} + \frac{1}{4} + \frac{1}{48} - \frac{1}{96} + \cdots$ will converge to $f\left(\frac{1}{2}\right) = e^{1/2} \tan^{-1} \frac{1}{2} = 0.7644\ldots$ because $x = \frac{1}{2}$ lies within the common interval of convergence of the two Maclaurin series, and because of the uniqueness of power series expansions. (We will study these ideas in Chapters 43 to 46.) ■

Our first theorem, by the analyst and algebraist Franz Mertens (1840–1927), answers the question about conditions under which a Cauchy product will converge, independently of the study of power series.

Theorem 33.1 Mertens' Theorem Suppose $\sum_{n=0}^{\infty} a_n = A$ is absolutely convergent and $\sum_{n=0}^{\infty} b_n = B$ is convergent (either conditionally or absolutely). Then the Cauchy product $\sum_{n=0}^{\infty} c_n$ converges to $C = AB$.

Proof. We consider first the special case that $B = 0$, so we must prove that $C = 0$. Establish the partial sums $B_n = \sum_{k=0}^{n} b_k$ and $C_n = \sum_{k=0}^{n} a_k b_{n-k}$. Using the premise that $\sum_{n=0}^{\infty} a_n$ is absolutely convergent, let $A' = \sum_{n=0}^{\infty} |a_n|$, and let its partial sums be $A'_n = \sum_{k=0}^{n} |a_k|$. Clearly, $A' \geq 0$, but if $A' = 0$, then $A = 0$, and the proof is finished (fill in the details in Exercise 33.4a; it's not because $C = 0 \times 0$!) Thus, we proceed with the added premise that $A' > 0$.

It's not hard to demonstrate that $C_n = a_0 B_n + a_1 B_{n-1} + \cdots + a_n B_0$, for any $n \in \mathbb{N} \cup \{0\}$ (do this in Exercise 33.4b).

Since $\sum_{n=0}^{\infty} b_n = 0$, we know that $\lim_{n \to \infty} B_n = 0$. Therefore, given any $\varepsilon > 0$, we are able to choose N_ε so large that for any $n \geq N_\varepsilon$, $|B_n| < \frac{\varepsilon}{2A'}$ holds. Based on this, we derive a sequence of inequalities: $|a_{n-N_\varepsilon} B_{N_\varepsilon}| \leq |a_{n-N_\varepsilon}| \frac{\varepsilon}{2A'}$, $|a_{n-N_\varepsilon-1} B_{N_\varepsilon+1}| \leq |a_{n-N_\varepsilon-1}| \frac{\varepsilon}{2A'}$, and so on, until $|a_0 B_n| \leq |a_0| \frac{\varepsilon}{2A'}$.

Apply the triangle inequality:

$$|a_0 B_n + \cdots + a_{n-N_\varepsilon-1} B_{N_\varepsilon+1} + a_{n-N_\varepsilon} B_{N_\varepsilon}| \leq |a_0 B_n| + \cdots + |a_{n-N_\varepsilon-1} B_{N_\varepsilon+1}|$$
$$+ |a_{n-N_\varepsilon} B_{N_\varepsilon}| \leq |a_0| \frac{\varepsilon}{2A'} + \cdots + |a_{n-N_\varepsilon-1}| \frac{\varepsilon}{2A'} + |a_{n-N_\varepsilon}| \frac{\varepsilon}{2A'} = A'_{n-N_\varepsilon} \frac{\varepsilon}{2A'} \leq \frac{\varepsilon}{2},$$

in view of the sequence of inequalities just derived. Thus, we have shown that $|a_0 B_n + \cdots + a_{n-N_\varepsilon-1} B_{N_\varepsilon+1} + a_{n-N_\varepsilon} B_{N_\varepsilon}| \leq \frac{\varepsilon}{2}$, for any $n \geq N_\varepsilon$.

Next, use the triangle inequality again:

$$|C_n| = |a_0 B_n + a_1 B_{n-1} + \cdots + a_n B_0| \leq |\, a_0 B_n + a_1 B_{n-1} + \cdots + a_{n-N_\varepsilon} B_{N_\varepsilon}|$$
$$+ |a_{n-(N_\varepsilon-1)} B_{N_\varepsilon-1} + a_{n-(N_\varepsilon-2)} B_{N_\varepsilon-2} + \cdots + a_n B_0| \leq \frac{\varepsilon}{2} + |a_{n-(N_\varepsilon-1)} B_{N_\varepsilon-1}$$
$$+ a_{n-(N_\varepsilon-2)} B_{N_\varepsilon-2} + \cdots + a_n B_0| \leq \frac{\varepsilon}{2} + |a_{n-(N_\varepsilon-1)} B_{N_\varepsilon-1}| + \cdots + |a_n B_0|.$$

Furthermore, $\langle B_n \rangle_0^\infty$ is a bounded sequence, so let $|B_n| < d$ for all n. Continuing, $\frac{\varepsilon}{2} + |a_{n-(N_\varepsilon-1)} B_{N_\varepsilon-1}| + \cdots + |a_n B_0| < \frac{\varepsilon}{2} + (|a_{n-(N_\varepsilon-1)}| + \cdots + |a_n|)d = \frac{\varepsilon}{2} + (A'_n - A'_{n-N_\varepsilon})d$.

Collecting these inequalities, we arrive at

$$|C_n| < \frac{\varepsilon}{2} + (A'_n - A'_{n-N_\varepsilon})d. \tag{6}$$

Still more, since $\langle A'_n \rangle_0^\infty$ converges, it is a Cauchy sequence. The Cauchy convergence criterion (Theorem 21.1) guarantees that for the ε being used, there in fact exists an M_ε large enough that

$$A'_n - A'_{n-N_\varepsilon} = |A'_n - A'_{n-N_\varepsilon}| < \frac{\varepsilon}{2d} \tag{7}$$

whenever n and $n - N_\varepsilon$ are no less than M_ε

From (6) and (7), it follows that given any $\varepsilon > 0$, there exists a $Q_\varepsilon = N_\varepsilon + M_\varepsilon$ such that for all $n > Q_\varepsilon$, we have $|C_n| < \frac{\varepsilon}{2} + \frac{\varepsilon}{2d}d = \varepsilon$. In other words, $\lim_{n\to\infty} C_n = 0$, which tells us that $C = \sum_{n=0}^{\infty} c_n = 0$.Finally, for the case that $B \neq 0$, define $h_0 \equiv b_0 - B$ and $h_n \equiv b_n$ for $n \in \mathbb{N}$. Then $\sum_{n=0}^{\infty} h_n = 0$. In Exercise 33.4c, explain how this little "trick" allows us to use the results of the first case to finish the proof. **QED**

Example 33.4 Write the first three terms of the Cauchy product $(\sum_{n=1}^{\infty} \frac{1}{n^2})(\sum_{n=1}^{\infty} \frac{1}{(n+3)!})$, and evaluate the sum of the infinite series. We could evaluate terms c_1, c_2, and c_3 directly from the definition, but in the spirit of Example 33.2, multiply the polynomials

$$\left(\sum_{n=1}^{3} \frac{1}{n^2}x^n\right)\left(\sum_{n=1}^{3} \frac{1}{(n+3)!}x^n\right) = \frac{1}{24}x^2 + \frac{3}{160}x^3 + \frac{7}{864}x^4 + \frac{11}{8640}x^5 + \frac{1}{6480}x^6.$$

This product guarantees the correct Cauchy product coefficients for terms up to x^4 but not higher (why?). Hence, letting $x = 1$ gives us the required first three terms:

$$\frac{1}{24} + \frac{3}{160} + \frac{7}{864}.$$

Since

$$\sum_{n=1}^{\infty} \frac{1}{n^2} = \frac{\pi^2}{6} \quad \text{and} \quad \sum_{n=1}^{\infty} \frac{1}{(n+3)!} = \sum_{n=4}^{\infty} \frac{1}{n!} = e - \frac{8}{3}$$

(from other sources), and both series are absolutely convergent, Mertens' theorem states that the Cauchy product equals $\frac{\pi^2}{6}(e - \frac{8}{3})$. ■

33.2 Inner Product Series

At the end of Chapter 28, we briefly wondered if given two convergent series $\sum_{n=1}^{\infty} a_n$ and $\sum_{n=1}^{\infty} b_n$, we could be sure that the **term by term product** or **inner product series** $\sum_{n=1}^{\infty} a_n b_n = a_1 b_1 + a_2 b_2 + \cdots$ was convergent. This represents quite a different way of thinking about a product of series, although it is the counterpart to the way we add two series.

A counterexample now shows that the convergence of two series isn't sufficient for the convergence of their inner product series. The convergent series $\sum_{n=1}^{\infty} \frac{(-1)^n}{\sqrt{n}}$, when multiplied term by term with itself, gives $\sum_{n=1}^{\infty} \frac{1}{n}$, which we know diverges. We therefore take a step back and consider the weaker premise of two *nonnegative* series that converge. It is not difficult to show that with this condition, the inner product series $\sum_{n=1}^{\infty} a_n b_n$ will converge. Prove this fact in Exercise 33.7.

The following lemma proves an algebraic identity for real numbers discovered by a tragically short-lived genius of mathematics, Niels Henrik Abel (1802–1829).

Lemma 33.1 Abel's Lemma Let $\langle a_n \rangle_1^{\infty}$ and $\langle b_n \rangle_1^{\infty}$ be sequences, and let the partial sums of the second sequence be $B_k = \sum_{j=1}^{k} b_j$. Then the following is an algebraic identity for all $m \geq n \geq 1$:

$$\sum_{k=n}^{m} a_k b_k = a_{m+1} B_m - a_n B_{n-1} + \sum_{k=n}^{m} (a_k - a_{k+1}) B_k \quad \text{(with } B_0 \equiv 0\text{)}.$$

Proof. We have $b_k = B_k - B_{k-1}$ for all $k \in \mathbb{N}$, so that the first summation becomes

$$\sum_{k=n}^{m} a_k b_k = \sum_{k=n}^{m} a_k(B_k - B_{k-1}) = \sum_{k=n}^{m} a_k B_k - \sum_{k=n}^{m} a_k B_{k-1} = \sum_{k=n}^{m} a_k B_k - \sum_{k=n-1}^{m-1} a_{k+1} B_k.$$

Watching the limits of summation closely, we continue:

$$\sum_{k=n}^{m} a_k B_k - \sum_{k=n-1}^{m-1} a_{k+1} B_k = \sum_{k=n}^{m} a_k B_k - \sum_{k=n}^{m-1} a_{k+1} B_k - a_n B_{n-1}$$

$$= \sum_{k=n}^{m} a_k B_k - \sum_{k=n}^{m} a_{k+1} B_k - a_n B_{n-1} + a_{m+1} B_m = a_{m+1} B_m - a_n B_{n-1} + \sum_{k=n}^{m} (a_k - a_{k+1}) B_k.$$

QED

Our first test for convergence of inner product series can be seen as a test for conditional convergence of series in general, and it can also be considered as a refinement of Leibniz's alternating series test, which can be proved as a corollary of it.

Theorem 33.2 Dirichlet's Test Let $\langle a_n \rangle_1^\infty$ be a monotone decreasing sequence such that $\lim_{n\to\infty} a_n = 0$ and let $\sum_{n=1}^{\infty} b_n$ be a series whose sequence of partial sums is bounded. Then $\sum_{n=1}^{\infty} a_n b_n$ converges.

Proof. Letting $B_n = \sum_{k=1}^{n} b_k$ (with $B_0 \equiv 0$), we are told that a number d exists such that $|B_n| < d$ for all $n \in \mathbb{N}$. Considering just a finite number of terms of the inner product series, we apply Abel's lemma: for all

$$m \geq n \geq 1, \sum_{k=n}^{m} a_k b_k = a_{m+1} B_m - a_n B_{n-1} + \sum_{k=n}^{m} (a_k - a_{k+1}) B_k.$$

Using the triangle inequality twice,

$$\left|\sum_{k=n}^{m} a_k b_k\right| \leq |a_{m+1} B_m| + |a_n B_{n-1}| + \left|\sum_{k=n}^{m} (a_k - a_{k+1}) B_k\right| \leq |a_{m+1} B_m|$$

$$+ |a_n B_{n-1}| + \sum_{k=n}^{m} |(a_k - a_{k+1}) B_k|.$$

Now, substitute our bound d:

$$|a_{m+1} B_m| + |a_n B_{n-1}| + \sum_{k=n}^{m} |(a_k - a_{k+1}) B_k| \leq |a_{m+1}| d + |a_n| d + \sum_{k=n}^{m} (|a_k - a_{k+1}| d)$$

$$= \left(|a_{m+1}| + |a_n| + \sum_{k=n}^{m} |a_k - a_{k+1}|\right) d = \left(a_{m+1} + a_n + \sum_{k=n}^{m} (a_k - a_{k+1})\right) d,$$

by the given properties of $\langle a_n \rangle_1^\infty$. But $\sum_{k=n}^{m} (a_k - a_{k+1}) = a_n - a_{m+1}$. Thus,

$$(a_{m+1} + a_n + \sum_{k=n}^{m} (a_k - a_{k+1})) d = 2 a_n d.$$

Because the limit of $\langle a_n \rangle_1^\infty$ is 0, a_n can be diminished at will. Applying these deductions to our proof, given any $\varepsilon > 0$, it is certain that we can find an N_ε large enough so that for all $n \geq N_\varepsilon$ and $m \geq n$, $|\sum_{k=n}^{m} a_k b_k| \leq 2a_n d < \varepsilon$. We have now discovered that $|\sum_{k=n}^{m} a_k b_k| < \varepsilon$, for any $m \geq n \geq N_\varepsilon$. Rewriting,

$$\left|\sum_{k=n}^{m} a_k b_k\right| = \left|\sum_{k=1}^{m} a_k b_k - \sum_{k=1}^{n} a_k b_k\right| = |p_m - p_n| < \varepsilon,$$

where p_m and p_n are partial sums of the inner product series. This means that the partial sums form a Cauchy sequence, and by the Cauchy convergence criterion, we conclude that $\sum_{n=1}^{\infty} a_n b_n$ converges. **QED**

Example 33.5 Does $\sum_{n=1}^{\infty} \frac{n^{-1/3}+n^{-1/2}}{2^n+n}$ converge? We are now able to consider this as the inner product series of

$$\sum_{n=1}^{\infty}(n^{-1/3} + n^{-1/2}) \quad \text{and} \quad \sum_{n=1}^{\infty} \frac{1}{2^n + n}.$$

The first series diverges (verify); nevertheless, $\langle \frac{1}{n^{1/3}} + \frac{1}{n^{1/2}} \rangle_1^\infty$ is monotone decreasing towards zero. The second series converges (verify), implying that its partial sums are bounded. Thus, Dirichlet's test applies, and the inner product series $\sum_{n=1}^{\infty} \frac{n^{-1/3}+n^{-1/2}}{2^n+n}$ converges. ■

Example 33.6 The series $\sum_{n=1}^{\infty} \frac{\sin n}{n}$ resists our previous tests for convergence (try them!). We have two choices for applying Dirichlet's test. The first is to let $\langle \sin n \rangle_1^\infty$ be the sequence, and $\sum_{n=1}^{\infty} \frac{1}{n}$ be the series. This is useless, since $\sin n \not\to 0$, and just as bad, the partial sums of the harmonic series are not bounded. The second choice is to think instead about $\langle \frac{1}{n} \rangle_1^\infty$ and $\sum_{n=1}^{\infty} \sin n$. Now, $\langle \frac{1}{n} \rangle_1^\infty$ is monotone decreasing towards zero, so we hope to prove that $\sum_{n=1}^{\infty} \sin n$ has bounded partial sums. Unfortunately, you showed in Exercise 28.13 that $\sum_{n=1}^{\infty} \sin n$ diverges. An interesting trigonometric identity provides a closed-form expression for the partial sums: $\sum_{k=1}^{n} \sin k = \frac{1}{2} \sin n + \frac{1}{2}(\cot \frac{1}{2})(1 - \cos n)$. Prove this statement as Exercise 33.8.Is $\langle |\sum_{k=1}^{n} \sin k| \rangle_{n=1}^\infty$ bounded, as required by Dirichlet's test? We continue, $|\sum_{k=1}^{n} \sin k| \leq \frac{1}{2}|\sin n| + \frac{1}{2}(\cot \frac{1}{2})|1 - \cos n| \leq \frac{1}{2} + \cot \frac{1}{2} \approx 2.33$. Thus, the question is answered, "yes." Consequently, $\sum_{n=1}^{\infty} \frac{\sin n}{n}$ converges. **QED**

Note 2: The example has demonstrated that the inner product series of two divergent series may converge. That is, both $\sum_{n=1}^{\infty} \frac{1}{n}$ and $\sum_{n=1}^{\infty} \sin n$ diverge, and yet $\sum_{n=1}^{\infty} \frac{\sin n}{n}$ converges.
A counterintuitive fact about the series is that

$$\sum_{n=1}^{\infty} \frac{\sin n}{n} = \sum_{n=1}^{\infty} \left(\frac{\sin n}{n}\right)^2. \text{ Their sum is } \frac{\pi - 1}{2}.$$

The last theorem that we will discuss modifies the premises of Dirichlet's test, allowing different sequences $\langle a_n \rangle_1^\infty$, but restricting us to convergent $\sum_{n=1}^{\infty} b_n$. We get a glimpse of Abel's insight, for he proved this test without the benefit of the previous one.

Corollary 33.1 Abel's Test Let $\langle a_n \rangle_1^\infty$ be a convergent, monotone sequence, and let $\sum_{n=1}^{\infty} b_n$ be a convergent series. Then $\sum_{n=1}^{\infty} a_n b_n$ converges.

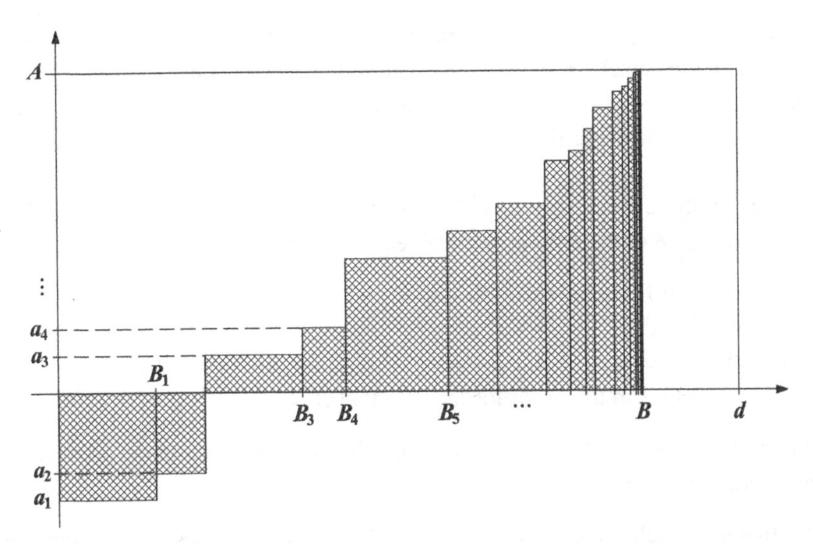

Figure 33.1. $\sum_{n=1}^{\infty} a_n b_n$ converges by Abel's test. For clarity, all b_n are positive here. We see partial sums $B_n = B_{n-1} + b_n$, and $\sum_{n=1}^{\infty} b_n = B$. Here, $\langle a_n \rangle_1^{\infty}$ is monotone increasing with limit A. The sum of successive rectangle areas $a_n b_n$ is eventually monotone increasing and is less than Ad, where d is any upper bound for $\langle B_n \rangle_1^{\infty}$.

Proof. Let $\lim_{n\to\infty} a_n = A$. In the case that $\langle a_n \rangle_1^{\infty}$ is monotone increasing, let $v_n = A - a_n$. Then $\langle v_n \rangle_1^{\infty}$ is monotone decreasing towards zero. We may therefore apply Dirichlet's test to the series $\sum_{n=1}^{\infty} v_n b_n$, concluding that it converges. Thus, $\sum_{n=1}^{\infty}(A - a_n)b_n$ converges, and since $\sum_{n=1}^{\infty} Ab_n$ converges, we conclude that

$$\sum_{n=1}^{\infty} a_n b_n = \sum_{n=1}^{\infty} Ab_n - \sum_{n=1}^{\infty}(A - a_n)b_n$$

converges also. If it happens that $\langle a_n \rangle_1^{\infty}$ is monotone decreasing, let $v_n = a_n - A$. Once again, $\langle v_n \rangle_1^{\infty}$ is monotone decreasing towards zero, and a parallel argument lets us conclude that $\sum_{n=1}^{\infty} a_n b_n$ converges. **QED**

Example 33.7 Does $\sum_{n=1}^{\infty}(\frac{1}{n^2})(1 + \frac{1}{n})^n$ converge? As an inner product series, we notice that $\sum_{n=1}^{\infty} \frac{1}{n^2}$ is a convergent series (verify), and $\langle (1 + \frac{1}{n})^n \rangle_1^{\infty}$ is a convergent, monotone sequence (its limit is e, as we will prove in the next chapter). By Abel's test, the series $1(2) + \frac{1}{4}(\frac{3}{2})^2 + \frac{1}{9}(\frac{4}{3})^3 + \frac{1}{16}(\frac{5}{4})^4 + \cdots$ converges. ■

33.3 Exercises

1. Check whether Mertens' theorem applies to the Cauchy products below, and evaluate whichever sums you can.

   a. $\left(\sum_{n=0}^{\infty} \frac{1}{(n+2)(n+3)}\right)\left(\sum_{n=0}^{\infty} \frac{1}{\pi^n}\right)$ b. $\left(\sum_{n=1}^{\infty} \frac{(-1)^n}{n}\right)\left(\sum_{n=1}^{\infty} \frac{(-1)^n}{(n+3)!}\right)$

2. Prove that the Cauchy product in Example 33.1 converges, and find the sum.

3. a. Insert the term $b_0 = 0$ in (1) of Example 33.2, and then substitute $x = \frac{1}{3}$. Is there any change in the Cauchy product? Hint: Multiply part of (1) as $(e + ey + \frac{e}{2!}y^2 + \frac{e}{3!}y^3 + \frac{e}{4!}y^4 + \frac{e}{5!}y^5)(z + y - \frac{1}{2}y^2 + \frac{1}{3}y^3 - \frac{1}{4}y^4 + \frac{1}{5}y^5)$ where $y = x - 1$ for convenience and z is a dummy variable that later is set to zero. Observe the role of z. Compare with Example 33.3.
   b. Prove that the Cauchy product (3) in Example 33.2 converges. Hint: Find the intervals of convergence of the two series involved.

4. a. In the proof of Mertens' theorem, explain why we say, "if $A' = 0$, then $A = 0$, and the proof is finished."
   b. Prove that $C_n = a_0 B_n + a_1 B_{n-1} + \cdots + a_n B_0$ for all $n \in \mathbb{N} \cup \{0\}$.
   c. Fill in the details of the proof for the case that $B \neq 0$.

5. Could we prove a theorem similar to Mertens' theorem, but without the premise of absolute convergence? Hint: See Note 1.

6. Prove that the Cauchy product of two absolutely convergent series is absolutely convergent.

7. Prove by Mertens' theorem that if $\sum_{n=0}^{\infty} a_n$ and $\sum_{n=0}^{\infty} b_n$ are convergent series of nonnegative terms, then their inner product series converges. Also, show that neither Dirichlet's test nor Abel's test can be used with these premises. Hint: The comparison test can be applied, but requires an observation about c_n.

8. Prove the trigonometric identity in Example 33.6 by setting $\theta = 1$ in the multiple angle identity $\sin(k\theta) = 2\sin((k-1)\theta)\cos\theta - \sin((k-2)\theta)$, and then considering $\sum_{k=1}^{n+1} \sin k = 2\cos 1 \sum_{k=1}^{n+1} \sin(k-1) - \sum_{k=1}^{n+1} \sin(k-2)$.

9. Plot the values of the first 30 partial sums of $\sum_{n=1}^{\infty} \frac{\sin n}{n}$ from Example 33.6. Use the graph to estimate the sum.

10. a. Generalize Example 33.6 by showing that $\sum_{n=1}^{\infty} a_n \sin n$ converges provided that $\langle a_n \rangle_1^{\infty}$ decreases monotonically to a limit of zero.
    b. Generalize **a** by showing that $\sum_{n=1}^{\infty} a_n \sin(n\theta)$ converges for any fixed, real θ, provided that $\langle a_n \rangle_1^{\infty}$ decreases monotonically to a limit of zero. Hint: In the multiple angle identity $\sin(k\theta) = 2\sin((k-1)\theta)\cos\theta - \sin((k-2)\theta)$, do not assign a particular value to θ (in contrast to Exercise 33.8).

11. Use Dirichlet's test to prove Leibniz's alternating series test as a corollary.

12. Prove that $1 + \frac{1}{2} - \frac{1}{3} - \frac{1}{4} + \frac{1}{5} + \frac{1}{6} - \cdots$ converges, by Dirichlet's test. As an aside, find a convenient expression for the nth term of this series.

13. Test the series $\frac{1}{3} + \frac{1}{5\sqrt{3}} + \frac{1}{3^2\sqrt{3^2}} + \frac{1}{5^2\sqrt{3^3}} + \frac{1}{3^3\sqrt{3^4}} + \frac{1}{5^3\sqrt{3^5}} + \frac{1}{3^4\sqrt{3^6}} + \frac{1}{5^4\sqrt{3^7}} + \cdots$ for convergence. Hint: Use $\langle(\sqrt{3})^{-n}\rangle_0^{\infty}$.

14. A gambler has a probability p of winning a certain game every time it is repeated (a trial). It is understood that $0 \leq p \leq 1$, and $q = 1 - p$ is the probability of losing at each trial. The gambler compulsively continues to bet until he wins, somehow possessing unlimited funding and time. At each trial, he decides to bet $\sqrt[n]{d}$ dollars, so the first bet is \$$d$, and if he loses, then the second bet is \$$\sqrt{d}$, and so on. If he wins, his payoff is \$$\frac{6d}{5}$, and then he

quits. From probability theory, the net expected amount of money won or lost (the expected value of the game) is $p(\frac{6d}{5})\sum_{n=1}^{\infty} q^{n-1} - \sum_{n=1}^{\infty} q^n \sqrt[n]{d}$ dollars. Use Abel's test to show that $\sum_{n=1}^{\infty} q^n \sqrt[n]{d}$ converges.

Hence, the gambler can't lose infinitely much money. To the nearest cent, what is the expected amount won or lost if $d = \$100$ and $p = 0.25$? If five trials were played, and he wins on the fifth trial, how much did the gambler win or lose? Based on this, would James Bond be attracted to the game?

15. The great mathematical physicist Joseph Fourier is quoted at the start of the chapter as stating that $f(y) = \sum_{n=0}^{\infty} \frac{(-1)^n}{2n+1} \cos((2n+1)y)$ converges for any y in $(-\pi/2, \pi/2)$. Fourier series in general have the form $\frac{a_0}{2} + \sum_{n=1}^{\infty}(a_n \cos nx + b_n \sin nx)$. These series are highly important in any science that deals with periodic phenomena, especially physics. It is the job of Fourier analysis to show that f does sum, as Fourier claims, to $\frac{\pi}{4}$ over the given open interval. (Chapter 9 of [15] provides a fast introduction to Fourier series.)

Prove Fourier's assertion of convergence. Note that the series for f is not alternating since the cosine factor does not have constant sign. To apply Dirichlet's test, you will need the following identity, where $y \neq \frac{k\pi}{2}$ and $\varphi = 2y$:

$$\begin{aligned}
&\sum_{n=0}^{N}(-1)^n \cos((2n+1)y) \\
&= \left(\frac{1}{4\sin y}\right)(\sin\varphi + \sin((2\lfloor N/2\rfloor + 1)\varphi) - \sin((2N - 2\lfloor N/2\rfloor)\varphi)) \\
&\quad + \left(\frac{1}{4\cos y}\right)(1 - \cos\varphi + \cos((2\lfloor N/2\rfloor + 1)\varphi) - \cos((2N - 2\lfloor N/2\rfloor)\varphi)).
\end{aligned}$$

The typical partial sum $\sum_{n=0}^{8} \frac{(-1)^n}{2n+1} \cos((2n+1)y)$ has a very interesting graph, best appreciated over the interval $[-0.5\pi, 2.5\pi]$. Use the highest resolution your calculator can provide.

(The identity crucial to this exercise was found using some messy trigonometry, and **Lagrange's trigonometric identities**:

$$\sum_{n=0}^{N} \sin(n\theta) = \frac{1}{2}\cot\frac{\theta}{2} - \frac{\cos(2N+1)\frac{\theta}{2}}{2\sin\frac{\theta}{2}}$$

and

$$\sum_{n=0}^{N} \cos(n\theta) = \frac{1}{2} + \frac{\sin(2N+1)\frac{\theta}{2}}{2\sin\frac{\theta}{2}}, \theta \neq 2k\pi.)$$

34

The Numbers e and γ

Ponamus autem brevitatis gratia pro numero hoc 2,718281828459 &c. constanter litteram e, quae ergo denotabit basin Logarithmorum naturalium . . . (Moreover, for the sake of brevity, let us put for this number 2.718281828459 etc. consistently the letter e, which therefore shall denote the base of natural logarithms . . .)

Leonhard Euler, *Introductio in analysin infinitorum*, 1748, Chapter VII, pg. 90.

This chapter will first analytically define and investigate the famous number e. Following that, we will define a real number γ known as Euler's constant, and then approximate its value.

34.1 $e = 2.718281828459045235 36\ldots$

As a historical point of interest, e makes its first, fleeting appearance in a 1618 translation of *Mirifici logarithmorum canonis descriptio* (A Description of the Wonderful Rule of Logarithms), wherein the cleric and mathematician John Napier (1550–1617) introduced logarithms to the world. The role of e as the base of the exponential function $y = e^x$ arose from the investigation by Leibniz in 1684 of solutions to the differential equation $y = y'$. He asked, "What nonzero function has its slope equal to its ordinate at every point?" (The question almost asks itself in Leibniz's study of the derivative.) Leonhard Euler was the first to use "e" in a manuscript around 1727, *Meditatio in experimenta explosione tormentorum nuper instituta* (A Study of Recently Undertaken Experiments on the Firing of Cannons). Subsequent work by Euler disclosed more and more of this number's marvelous properties.

Theorem 34.1 The sequence $\left\langle \left(1 + \frac{1}{n}\right)^n \right\rangle_1^\infty$ converges, defining e.

Proof. Since n is an integer, a direct proof is suggested by applying the binomial theorem to $\left(1 + \frac{1}{n}\right)^n$. Expanding (and recalling that the expansion has $n + 1$ terms),

$$\left(1 + \frac{1}{n}\right)^n = 1 + \frac{n}{1!}\left(\frac{1}{n}\right) + \frac{n(n-1)}{2!}\left(\frac{1}{n}\right)^2 + \frac{n(n-1)(n-2)}{3!}\left(\frac{1}{n}\right)^3 + \cdots$$

$$+ \frac{n(n-1)\cdots(2)}{(n-1)!}\left(\frac{1}{n}\right)^{n-1} + \frac{n(n-1)\cdots(2)(1)}{n!}\left(\frac{1}{n}\right)^n \tag{1}$$

$$= 1 + \underbrace{\begin{array}{c} 1 + \frac{1}{2!}\left(1-\frac{1}{n}\right) + \frac{1}{3!}\left(1-\frac{1}{n}\right)\left(1-\frac{2}{n}\right) + \cdots \\ + \frac{1}{(n-1)!}\left[\left(1-\frac{1}{n}\right)\cdots\left(1-\frac{n-2}{n}\right)\right] + \frac{1}{n!}\left[\left(1-\frac{1}{n}\right)\cdots\left(1-\frac{n-1}{n}\right)\right] \end{array}}_{n \text{ terms}} \quad (2)$$

$$\le 1 + 1 + \frac{1}{2!} + \frac{1}{3!} + \cdots + \frac{1}{n!} \le 1 + \underbrace{1 + \frac{1}{2^1} + \frac{1}{2^2} + \cdots + \frac{1}{2^{n-1}}}_{n \text{ terms}}$$

$$= 1 + \sum_{k=0}^{n-1} \frac{1}{2^k} = 3 - \frac{1}{2^{n-1}} < 3 \quad \text{for all } n \in \mathbb{N}. \quad (3)$$

This means that $\left\langle\left(1+\frac{1}{n}\right)^n\right\rangle_1^\infty$ is a bounded sequence. (As Exercise 34.1, justify the equalities and inequalities in (1) through (3).)

Next, we would like to show that $\left\langle\left(1+\frac{1}{n}\right)^n\right\rangle_1^\infty$ is monotone increasing (why?). Reconsidering (2), we compare the expansion of $\left(1+\frac{1}{n}\right)^n$ with that of $\left(1+\frac{1}{n+1}\right)^{n+1}$:

$\left(1+\frac{1}{n}\right)^n$	$\left(1+\frac{1}{n+1}\right)^{n+1}$
$1+$	$1+$
$1+$	$1+$
$\frac{1}{2!}\left(1-\frac{1}{n}\right)+$	$\frac{1}{2!}\left(1-\frac{1}{n+1}\right)+$
$\frac{1}{3!}\left(1-\frac{1}{n}\right)\left(1-\frac{2}{n}\right)+$	$\frac{1}{3!}\left(1-\frac{1}{n+1}\right)\left(1-\frac{2}{n+1}\right)+$
$\cdots+$	$\cdots+$
$\frac{1}{(n-1)!}\left[\left(1-\frac{1}{n}\right)\cdots\left(1-\frac{n-2}{n}\right)\right]+$	$\frac{1}{(n-1)!}\left[\left(1-\frac{1}{n+1}\right)\cdots\left(1-\frac{n-2}{n+1}\right)\right]+$
$\frac{1}{n!}\left[\left(1-\frac{1}{n}\right)\cdots\left(1-\frac{n-1}{n}\right)\right].$	$\frac{1}{n!}\left[\left(1-\frac{1}{n+1}\right)\cdots\left(1-\frac{n-1}{n+1}\right)\right]+$
	$\frac{1}{(n+1)^{n+1}}.$

One by one (except for the 1s), the terms in the series at left are less than the matching terms in the series at right, and on top of that, the latter has an extra positive term. Hence, $\left(1+\frac{1}{n+1}\right)^{n+1} > \left(1+\frac{1}{n}\right)^n$ for all $n \in \mathbb{N}$.

Therefore, $\left\langle\left(1+\frac{1}{n}\right)^n\right\rangle_1^\infty$ is indeed a monotone increasing sequence (see Figure 34.1). Since we proved that it is also bounded, the monotone convergence theorem allows us to conclude that the sequence converges. This means that $\lim_{n\to\infty}\left(1+\frac{1}{n}\right)^n$ exists, and the limit is universally denoted e. **QED**

Next, we show that a famous infinite series converges to e.

Theorem 34.2 $\sum_{k=0}^\infty \frac{1}{k!} = \frac{1}{0!} + \frac{1}{1!} + \frac{1}{2!} + \frac{1}{3!} + \cdots$ converges, and its sum is e.

Proof. The partial sum $s_n = \sum_{k=0}^n \frac{1}{k!}$ appears in (3). It follows that $\left(1+\frac{1}{n}\right)^n \le s_n$ for all $n \in \mathbb{N}$.

Using a calculator, we find that for any natural number n we pick, $s_n < \left(1+\frac{1}{n}\right)^{n+1}$, so this seems to be provable. To that end, we compare the $n+1$ terms in

$$1 + 1 + \frac{1}{2!} + \frac{1}{3!} + \cdots + \frac{1}{(n-1)!} + \frac{1}{n!} \quad (5)$$

with the corresponding terms in

$$\left(1+\frac{1}{n}\right)^{n+1} = \left(\frac{n+1}{n}\right)\left(1+\frac{1}{n}\right)^{n}.$$

Expanding this, we get

$$\frac{n+1}{n} + \frac{n+1}{n} + \frac{(n+1)(n)(n-1)}{2!n^3} + \frac{(n+1)(n)(n-1)(n-2)}{3!n^4} + \cdots$$
$$+ \frac{(n+1)(n)(n-1)\cdots(3)}{(n-2)!n^{n-1}} + \frac{(n+1)(n)}{n^n} + \frac{n+1}{n^{n+1}}. \tag{6}$$

The first two terms of (6) are clearly greater than the first two terms of (5); their excess is $\frac{2(n+1)}{n} - 2 = \frac{2}{n}$. But from that point on, the terms of (6) are *less* than those of (5), for instance,

$$\frac{(n+1)(n)(n-1)}{2!n^3} < \frac{1}{2!}$$

($n \geq 3$, of course). This leads us to the observation that in order for $s_n < \left(1+\frac{1}{n}\right)^{n+1}$, beginning with

$$\frac{1}{2!} - \frac{(n+1)(n)(n-1)}{2!n^3}$$

the total excess of the terms in (5) over those in (6) must be less than $\frac{2}{n}$. Proving this demands some lengthy analysis, which is left to Appendix B.

With that result behind us, we now have

$$\left(1+\frac{1}{n}\right)^{n} \leq s_n < \left(1+\frac{1}{n}\right)^{n+1}$$

for all $n \in \mathbb{N}$ (Figure 34.1 illustrates this). But also,

$$\lim_{n\to\infty}\left(1+\frac{1}{n}\right)^{n} = \lim_{n\to\infty}\left(1+\frac{1}{n}\right)^{n+1} = e$$

by Theorem 34.1. Consequently, the pinching theorem for sequences tells us that $\lim_{n\to\infty} s_n = \sum_{k=0}^{\infty} \frac{1}{k!} = e$. **QED**

Note 1: In calculus texts, $\sum_{k=0}^{\infty} \frac{1}{k!} = e$ is easily shown by using the Maclaurin series for $f(x) = e^x$. This demonstrates the power of calculus. However, many concepts (such as the series representation of a function) need to be established before Maclaurin series can enter the picture.

Since factorials increase terribly fast, the partial sums of $\sum_{k=0}^{\infty} \frac{1}{k!}$ converge very quickly to e, implying that the partial sums can be used to estimate e to high precision. The error made by using the Mth partial sum to underestimate e is

$$e - s_M = \sum_{k=M+1}^{\infty} \frac{1}{k!} = \frac{1}{(M+1)!} + \frac{1}{(M+2)!} + \frac{1}{(M+3)!} + \cdots$$
$$= \frac{1}{M!}\left(\frac{1}{M+1} + \frac{1}{(M+1)(M+2)} + \cdots\right) < \frac{1}{M!}\left(\frac{1}{M+1} + \frac{1}{(M+1)^2} + \cdots\right),$$

and the last expression in parentheses is a convergent geometric series since $M \geq 1$. Hence,

$$e - s_M < \frac{1}{M!}\left(\frac{\frac{1}{M+1}}{1-\frac{1}{M+1}}\right) = \frac{1}{M!}\left(\frac{1}{M}\right) \tag{7}$$

is a bound on the error $e - s_M$ when we stop at the Mth partial sum.

Example 34.1 How precise is $s_9 = \frac{1}{0!} + \frac{1}{1!} + \frac{1}{2!} + \cdots + \frac{1}{9!}$ as an estimate for e? We have $e - s_9 < \frac{1}{9!}\left(\frac{1}{9}\right) \approx 3 \times 10^{-7}$, so that $\sum_{k=0}^{9} \frac{1}{k!} = 2.7182815\ldots$ is correct to six decimal places. ■

Corollary 34.1 e is irrational.

Proof. Assume, on the contrary, that $e = \frac{a}{b}$, a rational number with $a, b \in \mathbb{N}$. Then $b!e$ is a natural number. Since $b!s_b = b!\left(1 + 1 + \frac{1}{2!} + \frac{1}{3!} + \cdots + \frac{1}{b!}\right)$ is also in $\mathbb{N}$, it follows that $b!e - b!s_b$ is also in $\mathbb{N}$ (verify). However, using our estimate (7), $0 < b!\,(e - s_b) < \frac{1}{b}$. This is immediately a contradiction, since no integer exists in $(0, 1)$. Therefore, e is irrational. **QED**

Note 2: In Note 2 of Chapter 8, we mentioned that e is not just irrational, but is transcendental. The proof of this is quite a bit more intricate than the proof you have just read, but it's still accessible as an undergraduate analysis topic, and may be found in the reference in that note.

A final lemma gives us some important properties used in this and the next chapter. Its proof uses some of the analysis of Problem 12 in *100 Great Problems of Elementary Mathematics: Their History and Solution*, by Heinrich Dörrie, Dover, 1965.

Lemma 34.1 For any $n \in \mathbb{N}$, $\left(1 + \frac{1}{n}\right)^n < e < \left(1 + \frac{1}{n}\right)^{n+1}$.

Proof. As Exercise 34.3c, prove the left-hand inequality using what we have found so far in this chapter.

To prove the right-hand inequality, we will first show that $\left(\left(1 + \frac{1}{n}\right)^{n+1}\right)_1^\infty$ is monotone decreasing (see Figure 34.1). We will need some delicate algebra based on the arithmetic mean—geometric mean inequality $\sqrt[m]{a_1 a_2 \cdots a_m} \leq \frac{1}{m}\sum_{i=1}^{m} a_i$, true for any collection $\{a_i\}_1^m$ of positive real numbers (see Example 4.3). In this inequality, substitute $r \in \mathbb{Q}^+$ for exactly ℓ of the factors a_i, and let the remaining $m - \ell$ factors be 1. We therefore have

$$\sqrt[m]{\underbrace{1 \cdot \cdots \cdot 1}_{m-\ell \text{ factors}}} \cdot \sqrt[m]{\underbrace{r \cdot \cdots \cdot r}_{\ell \text{ factors}}} = \sqrt[m]{r^\ell} \leq \frac{1}{m}(\ell r + m - \ell).$$

Equivalently, with $0 < \frac{\ell}{m} \leq 1$,

$$r^{\ell/m} \leq 1 + \frac{\ell}{m}(r - 1). \tag{8}$$

Next, it is always possible to express $\frac{\ell}{m} = \frac{1+p}{1+q}$ (not necessarily reduced), where $p \leq q$ are both natural numbers. Now, let

$$r = 1 - \frac{1}{1+p} = \frac{p}{1+p}$$

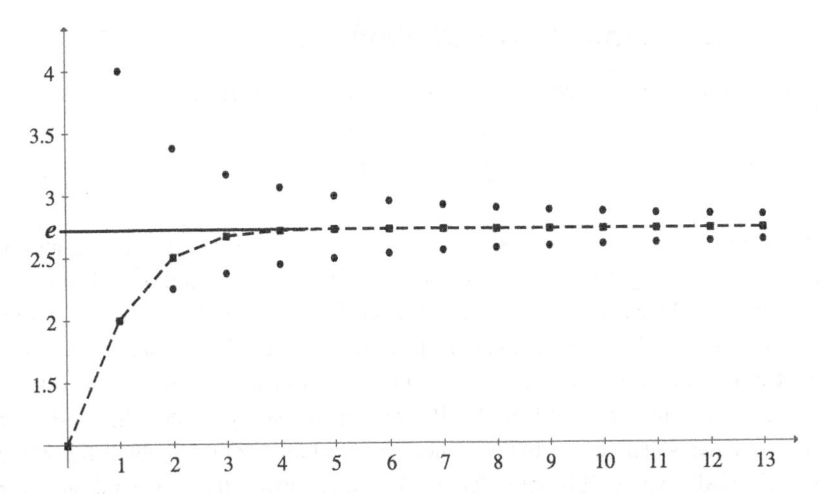

Figure 34.1. The monotone sequences $\langle(1+\frac{1}{n})^n\rangle_1^\infty$ and $\langle(1+\frac{1}{n})^{n+1}\rangle_1^\infty$ respectively increase and decrease slowly towards $y = e$. The partial sums $\sum_{k=0}^{n}\frac{1}{k!}$ (in square marks) fall between the two sequences, as proved in Theorem 34.2. Their convergence towards e is very fast.

(so $r \in \mathbb{Q}^+$ still holds). Substituting into (8), we get

$$\left(\frac{p}{1+p}\right)^{(1+p)/(1+q)} \le 1 + \left(\frac{1+p}{1+q}\right)\left(\frac{-1}{1+p}\right) = \frac{q}{1+q}.$$

Consequently,

$$\left(\frac{p}{1+p}\right)^{1+p} \le \left(\frac{q}{1+q}\right)^{1+q}.$$

Taking reciprocals of the fractions on both sides, we finally arrive at

$$\left(1+\frac{1}{p}\right)^{1+p} \ge \left(1+\frac{1}{q}\right)^{1+q}$$

whenever $p \le q$. This inequality proves that $\langle(1+\frac{1}{n})^{n+1}\rangle_1^\infty$ is indeed monotone decreasing.

Since $\lim_{n\to\infty}(1+\frac{1}{n})^{n+1} = e$, and e is irrational, we conclude that $e < (1+\frac{1}{n})^{n+1}$ for all $n \in \mathbb{N}$. **QED**

Note 3: Although Lemma 34.1 is important in analysis, it is quite impractical for computation. For instance, using $n = 1000$ yields only two correct decimal places for e. But some clever modifications of the expression $(1+\frac{1}{n})^n$ will produce rapidly converging sequences, as in the article "Novel Series-based Approximations to e" by John A. Knox and Harlan J. Brothers, in *The College Mathematics Journal*, vol. 30, no. 4 (September 1999), pp. 269–275.

34.2 $\gamma = 0.5772156649015328606 0\ldots$

The following classic sequence was introduced by Leonhard Euler in 1735:

$$\left\langle \sum_{k=1}^{n} \frac{1}{k} - \ln n \right\rangle_{n=1}^{\infty}.$$

The general term is $c_n = \frac{1}{1} + \frac{1}{2} + \cdots + \frac{1}{n} - \ln n$. This sequence is intriguing because the harmonic series slowly diverges to $+\infty$, but each partial sum $\sum_{k=1}^{n} \frac{1}{k}$ is diminished by $\ln n$. Are these subtractions sufficient to keep the series from diverging? Euler not only proved it was so, but in his typical style, later calculated the limit to 16 decimal places! He wrote about this constant in the article *De progressionibus harmonicis observationes* (Observations about the Harmonic Series), in a journal from St. Petersburg, Russia, published in 1740. Today, the limit of the sequence is denoted γ, and is named **Euler's constant** (also, the Euler-Mascheroni constant). Incidentally, you will become famous if you can prove that γ is irrational, for this is unknown even to this day.

Theorem 34.3 $\left\langle \sum_{k=1}^{n} \frac{1}{k} - \ln n \right\rangle_{n=1}^{\infty}$ converges (to Euler's constant γ).

Proof. (Since γ involves the natural logarithm function, we can expect the proof to use the properties of logarithms.)

By Lemma 34.1, $\left(1 + \frac{1}{n}\right)^{n+1} > e$. Hence, $\ln\left(1 + \frac{1}{n}\right) > \frac{1}{n+1}$. Now, with regard to the terms $c_n = \sum_{k=1}^{n} \frac{1}{k} - \ln n$ of this sequence, we find that

$$c_n - c_{n+1} = \ln(n+1) - \ln n - \frac{1}{n+1} = \ln\left(1 + \frac{1}{n}\right) - \frac{1}{n+1} > 0.$$

Therefore, $c_n > c_{n+1}$ for all $n \in \mathbb{N}$, meaning that $\left\langle \sum_{k=1}^{n} \frac{1}{k} - \ln n \right\rangle_{n=1}^{\infty}$ is monotone decreasing.

It remains for us to demonstrate that our sequence is bounded below—by zero, as we now prove. We use induction to prove that $c_n > 0$ for all $n \in \mathbb{N}$. Actually, we will use the stronger statement $\sum_{k=1}^{n-1} \frac{1}{k} - \ln n > 0$. Accordingly, our anchor is case $n = 2$: $\sum_{k=1}^{1} \frac{1}{k} - \ln 2 > 0$. We assume as the induction hypothesis (case n) that $\sum_{k=1}^{n-1} \frac{1}{k} > \ln n$.

Again by Lemma 34.1, $e > \left(1 + \frac{1}{n}\right)^n$. This leads directly to $\frac{1}{n} > \ln\left(\frac{n+1}{n}\right)$. Using this and the induction hypothesis (as we must), we now have

$$\frac{1}{n} + \sum_{k=1}^{n-1} \frac{1}{k} > \frac{1}{n} + \ln n > \ln\left(\frac{n+1}{n}\right) + \ln n = \ln(n+1).$$

Consequently, $\sum_{k=1}^{n} \frac{1}{k} > \ln(n+1)$, which is the case $n + 1$.

Through induction, we have confirmed that $c_n > 0$ for all $n \geq 2$. Since $\left\langle \sum_{k=1}^{n} \frac{1}{k} - \ln n \right\rangle_{n=1}^{\infty}$ is both monotone decreasing and bounded below, it converges. The limit is denoted γ. **QED**

How might we evaluate Euler's constant γ to reasonable precision? First, do Exercise 34.6a to get a sense of how slowly the defining sequence converges. It is therefore surprising that the very similar sequence

$$\left\langle \sum_{k=1}^{n} \frac{1}{k} - \ln(n + 0.5) \right\rangle_{n=1}^{\infty}$$

converges fairly rapidly to γ. Let $b_n = \sum_{k=1}^{n} \frac{1}{k} - \ln(n+0.5)$ for convenience. Then, at $n = 50$, $b_{50} = \sum_{k=1}^{50} \frac{1}{k} - \ln 50.5 \approx 0.5772$, which is γ rounded to four decimal places. In comparison, it takes about 20,000 terms of Euler's sequence to achieve such precision! It turns out that the new sequence has an error bound that is inversely proportional to the square of n (compare with Exercise 34.6b):

$$\frac{1}{24(n+1)^2} < b_n - \gamma < \frac{1}{24n^2}. \tag{9}$$

Note 4: The sequence described above is from "A Geometric Look at Sequences that Converge to Euler's Constant," by Duane W. DeTemple, in *The College Mathematics Journal*, vol. 37, no. 2 (March 2006), pp. 128–131.

Example 34.2 Presuming that we do not know the value of γ beforehand, we are nevertheless able to state that with $n = 200$ terms in (9), the error bounds in the new estimate are

$$\frac{1}{24(201)^2} < b_{200} - \gamma < \frac{1}{24(200)^2}.$$

That is, the difference $b_{200} - \gamma$ is between 1.03×10^{-6} and 1.04×10^{-6}.

With a bit of algebra, (9) yields

$$b_{200} - \frac{1}{24(200)^2} < \gamma < b_{200} - \frac{1}{24(201)^2}.$$

With a calculator, we get

$$b_{200} = \sum_{k=1}^{200} \frac{1}{k} - \ln 200.5 = 0.577216701\ldots.$$

Then

$$b_{200} - \frac{1}{24(200)^2} = 0.577215659\ldots,$$

while

$$b_{200} - \frac{1}{24(201)^2} = 0.577215670\ldots.$$

Thus, seven-place agreement is achieved. In comparison, barely two places are correct in $c_{200} = \sum_{k=1}^{200} \frac{1}{k} - \ln 200 = 0.5797\ldots$. ■

34.3 Exercises

1. a. Carefully apply the binomial theorem and derive (1) and (2).
   b. Prove the inequalities and the equality in (3).

2. How many decimal places will be correct if we use the 30th partial sum s_{30} of the series for e?

3. Prove:
   a. $\lim_{n\to\infty} \left(\frac{n}{n+1}\right)^n = \frac{1}{e}$. Hint: $\frac{n}{n+1} = \frac{1}{1+\frac{1}{n}}$.

b. $\lim_{n\to\infty}\left(1+\frac{1}{bn}\right)^n = \sqrt[b]{e}$, where $b \in \mathbb{N}$. Hint: Let $m = bn \in \mathbb{N}$.
c. $\left(1+\frac{1}{n}\right)^n < e$ for all $n \in \mathbb{N}$.

4. Use the sequence $b_n = \sum_{k=1}^{n}\frac{1}{k} - \ln(n+0.5)$ with $n = 500$ terms and the inequality $b_n - \frac{1}{24n^2} < \gamma < b_n - \frac{1}{24(n+1)^2}$ mentioned in Example 34.2 to estimate γ, and compare your estimate to the approximation in the subtitle.

5. Prove that $\lim_{n\to\infty}\left(\sum_{k=1}^{n}\frac{1}{k} - \ln(n+0.5)\right) = \gamma$.

6. a. Use a calculator to evaluate the 100th, 200th (it's in Example 34.2), and 300th term of the defining sequence for Euler's constant γ. Does the sequence seem to converge rapidly?
   b. The article "Euler's Constant," by Robert M. Young in *The Mathematical Gazette*, vol. 75 (June 1991), pg. 187, shows that $\frac{1}{2(n+1)} < c_n - \gamma < \frac{1}{2n}$. This means that c_n has error bounds proportional to just the first power of n. Use this and c_{200} to estimate γ by algebra similar that in Example 34.2 and Exercise 34.4.

7. a. Corollary 29.1 proves that the harmonic series diverges, but it doesn't hint at how slowly it does so. We can turn the tables on Theorem 34.3 and use γ to estimate the number of terms that the harmonic series needs to finally surpass a partial sum of, say, 100. Given that $\gamma = 0.5772156\ldots = \lim_{n\to\infty}\left(\sum_{k=1}^{n}\frac{1}{k} - \ln n\right)$, we estimate that $0.577216 + \ln n \approx \sum_{k=1}^{n}\frac{1}{k}$ for large n. Solve $0.577216 + \ln n = 100$ to find approximately the number of terms in that partial sum (the answer to this always amazes me).
   b. Thus, the harmonic series diverges at a maddeningly slow pace. On the other side of things, we have seen in Exercise 34.6a that some series *converge* so slowly that a calculator has a hard time estimating the sum. Consider the harmonic-like series

$$1+\frac{1}{2}+\frac{1}{3}+\cdots+\frac{1}{8}+\frac{1}{10}+\frac{1}{11}+\cdots+\frac{1}{18}+\frac{1}{20}+\cdots$$

   where no denominator containing a 9 appears. It actually converges, but an ice age might pass before one gets anything fruitful! The article "Thinning Out the Harmonic Series" by G. Hossein Behforooz in *Mathematics Magazine*, vol. 68, no. 4 (Oct. 1995), pp. 289–293, demonstrates this. Try programming a computer to output several partial sums, and compare them to 22.921, the approximate sum.

8. A very different proof of Theorem 34.3 can be created if we introduce the integral $\int_1^t \frac{1}{x}dx$ as the definition of the natural logarithm function $y = \ln t$. (Of course, this presupposes the definition of the Riemann integral.)
   a. Begin with the graph of $f(x) = \frac{1}{x}$ in Figure 34.2. Considering areas, prove that $\frac{1}{n+1} < \ln(n+1) - \ln n < \frac{1}{n}$ for any $n \in \mathbb{N}$. What fundamental theorem from calculus did you employ?

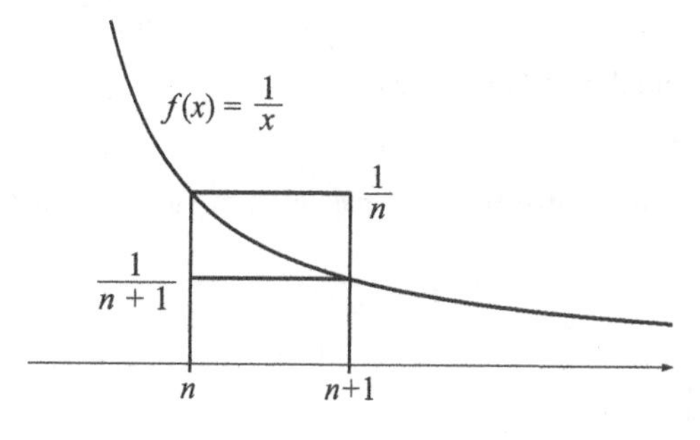

Figure 34.2.

b. Use **a** to prove that

$$c_n - c_{n+1} = \ln(n+1) - \ln n - \frac{1}{n+1} > 0$$

for any $n \in \mathbb{N}$. Show that $\left(\sum_{k=1}^{n} \frac{1}{k} - \ln n\right)_{n=1}^{\infty}$ is monotone decreasing.

c. Using **a** again, sum to get

$$\sum_{k=1}^{n-1} \int_k^{k+1} \frac{1}{x} dx < \sum_{k=1}^{n-1} \frac{1}{k} < \sum_{k=1}^{n} \frac{1}{k}$$

(k is a dummy integer). Express the sum $\sum_{k=1}^{n-1} \int_k^{k+1} \frac{1}{x} dx$ as one integral, free of k.

d. Use **c** to conclude that $c_n = \sum_{k=1}^{n} \frac{1}{k} - \ln n > 0$ for all $n \in \mathbb{N}$. This means that the sequence is bounded below by zero. What theorem now implies that $\left(\sum_{k=1}^{n} \frac{1}{k} - \ln n\right)_{n=1}^{\infty}$ converges?

9. As in Exercise 34.8, admit the properties of the natural logarithm function $y = \ln x$, including that it is the inverse of $y = e^x$. Then a very short proof of Theorem 34.1 can be devised. Hint: Begin with $\frac{1}{n+1} < \ln(n+1) - \ln n < \frac{1}{n}$ from Exercise 34.8a.

35

The Functions exp x and ln x

35.1 The Exponential Function $y = e^x = \exp x$

Anyone may define the function $\exp x = e^x$ in any (even a silly) fashion, but the merit of the following definition is that it allows e^x to have all the properties of the exponential function so crucial to calculus and the sciences (to be sure, other definitions exist that do the same). From Theorem 34.1, $e = \lim_{n\to\infty}(1+\frac{1}{n})^n$, so that for a real number x, $e^x \equiv (\lim_{n\to\infty}(1+\frac{1}{n})^n)^x$. This defines the **exponential function**, also written as $\exp x$, with domain $\mathcal{D}_{\exp} = \mathbb{R}$. Next, considering that for every $n \in \mathbb{N}$, $((1+\frac{1}{n})^n)^x = (1+\frac{1}{n})^{nx}$, we have $e^x = \lim_{n\to\infty}((1+\frac{1}{n})^{nx})$. This will lead to important properties of $y = e^x$ below and in Exercise 35.5.

Lemma 35.1 $\lim_{x\to 0} e^x = 1$.

Proof. From Lemma 34.1, for any fixed $n \in \mathbb{N}$, $(1+\frac{1}{n})^n < e < (1+\frac{1}{n})^{n+1}$, hence for $x > 0$, $(1+\frac{1}{n})^{nx} < e^x < (1+\frac{1}{n})^{(n+1)x}$. Therefore, by the pinching theorem for functions, $\lim_{x\to 0+} e^x = 1$. Next, $\lim_{x\to 0-} e^x = \lim_{x\to 0+} e^{-x} = \frac{1}{\lim_{x\to 0+} e^x} = 1$. The limits from the right and left of zero are equal, so the lemma is proved. **QED**

Theorem 35.1 The function $y = e^x = \exp x$ is continuous over its entire domain, $\mathbb{R}$.

Proof. According to Theorem 24.1, we must prove that $\lim_{x\to a} e^x = e^a$, for any $a \in \mathbb{R}$. Writing $a = (x-a)+a$, and $u = x-a$, we may show equivalently that $\lim_{u\to 0} e^{u+a} = e^a$. But this is now easy to prove, for $\lim_{u\to 0} e^{u+a} = e^a \lim_{u\to 0} e^u = e^a(1) = e^a$, by the previous lemma and Exercise 35.5d. **QED**

Back in Exercise 23.8, you showed that $x < e^x$ for all real x. Now we show a much stronger inequality.

Theorem 35.2 For all real x, $1+x \leq e^x$.

Proof. We use Bernoulli's inequality: $1+mp \leq (1+p)^m$ for any real $p \geq -1$ and $m \in \mathbb{N}\cup\{0\}$. Substitute $p = \frac{1}{n}$, where p is selected so that $n \in \mathbb{N}$, giving us $1+\frac{m}{n} \leq (1+\frac{1}{n})^m$ for any $\frac{m}{n} \geq 0$. In Lemma 34.1, we found that $(1+\frac{1}{n})^n < e$ for all $n \in \mathbb{N}$, from which we easily derive that $(1+\frac{1}{n})^m \le e^{m/n}$. Consequently, $1+\frac{m}{n} \le e^{m/n}$ for all nonnegative rational numbers (with equality when $m=0$).

The case for negative rational numbers begins with $1+\frac{m}{n} \le (1+\frac{1}{n})^m$, where m is as above and $p=\frac{1}{n}$, but p is selected so that n is any *negative* integer (verify that this can be done). Furthermore, $(1+\frac{-1}{|n|})^{|n|} < e^{-1}$, as you must show in Exercise 35.1. Hence, for any negative integer n,

$$1+\frac{m}{n} \le \left(1+\frac{1}{n}\right)^m = \left(1+\frac{-1}{|n|}\right)^m \le (e^{-1})^{m/|n|} = e^{m/n},$$

where $\frac{m}{n} \le 0$. At this point, then, we have proved that $1+r \le e^r$ for all $r \in \mathbb{Q}$.

Our final step is the transition from rationals to real numbers. Let $x \in \mathbb{R}$ be arbitrary, and let $\langle r_n \rangle_1^\infty$ be any sequence of rational numbers converging to x. Then $\lim_{n\to\infty} e^{r_n} = e^x$ by the continuity of the exponential function. Since $1+r_n \le e^{r_n}$ for all $n \in \mathbb{N}$ by the previous paragraph, we have $1+x = \lim_{n\to\infty}(1+r_n) \le \lim_{n\to\infty} e^{r_n} = e^x$. We conclude that $1+x \le e^x$ for all real numbers x. **QED**

Note 1: We could attempt to prove the theorem by noting that $\frac{d^2e^x}{dx^2} = e^x > 0$ for all real x. Thus, the exponential function is always concave up, meaning that it lies above all its tangent lines. Since $y = 1+x$ is tangent to $y=e^x$ at the origin, we quickly conclude that $1+x \le e^x$. Unfortunately, this demonstration falls into the trap of circular reasoning (a form of begging the question), since $1+x \le e^x$ must be used to prove that $\frac{de^x}{dx} = e^x$ in the first place. See Theorem 37.5.

In Cauchy's amazing textbook on analysis, the *Cours d'analyse de l'École royale polytechnique*, he establishes by very elaborate non-calculus means the famous series $e^x = 1 + x + \frac{x^2}{2!} + \frac{x^3}{3!} + \cdots$. Then, on page 442, he says, "To the previous theorems we will add the following, from which one may derive many important consequences." The theorem that follows is that $1+x < e^x$ for all x, (he apparently disregards $x=0$). To prove this, Cauchy notes that we only need to concern ourselves with the case $1+x>0$, since e^x is always positive. He rewrites the series as $1+x+\frac{x^2}{2!}(1+\frac{x}{3}) + \frac{x^4}{4!}(1+\frac{x}{5}) + \cdots$. Behold! Starting with $\frac{x^2}{2!}(1+\frac{x}{3})$, every term is positive, and he concludes that $1+x<e^x$. Isn't that elegant?

35.2 The Logarithmic Function $y = \ln x$

The exponential function $y = \exp x$ is strictly increasing (proved in Exercise 35.5f), with domain $\mathcal{D}_{\exp} = \mathbb{R}$ and range $\mathcal{R}_{\exp} = \mathbb{R}^+$. Because of its monotonicity, the function is a bijection between its domain and range, and it must therefore have an inverse function (see Theorem 7.2). There is no algebraic method involving the four arithmetic operations, powers, and roots, that can solve $y=e^x$ for x, so we create a new function that explicitly does this, $x = \ln y$, the **logarithmic function**. Then $\mathcal{D}_{\ln} = \mathbb{R}^+$ and $\mathcal{R}_{\ln} = \mathbb{R}$, because the exchange of domain and range is part of the definition of an inverse function. Practicing the notation of Chapter 7, $\langle 0,1\rangle \in \exp$ if and only if $\langle 1,0\rangle \in \ln$, which, in plain language, states that $\ln 1 = 0$.

Also immediately available from the theory in Chapter 7 are the compositions $x = \ln(\exp x)$ and $y = \exp(\ln y)$, where $y>0$. These may be more familiar in the form $x = \ln(e^x)$ and $y = e^{\ln y}$. These relationships are the basis for the well-known laws of logarithms.

Theorem 35.3 The logarithmic function $y = \ln x$, has the following properties, for any $a, b \in \mathbb{R}^+$.

A The logarithmic function is strictly increasing, that is, if $a > b$, then $\ln a > \ln b$.
B $\ln(ab) = \ln a + \ln b$.
C $\ln(\frac{a}{b}) = \ln a - \ln b$.
D $\ln a^s = s \ln a$ for any real number s.

Proof. **A** This follows from the strictly increasing property of the exponential function (Exercise 35.5f) and Theorem 7.2.

B Note that $e^{\ln a + \ln b} = e^{\ln a} e^{\ln b} = ab$ by the compositions mentioned above. For the same reason, $e^{\ln(ab)} = ab$. Thus, $e^{\ln(ab)} = e^{\ln a + \ln b}$. But the exponential function is bijective, which implies that $\ln(ab) = \ln a + \ln b$.
C This is proved in a parallel way, as Exercise 35.6.
D We know that $a^s = e^{\ln a^s}$. Also, $e^{s \ln a} = (e^{\ln a})^s = a^s$. Hence, $e^{\ln a^s} = e^{s \ln a}$, and being bijective, $\ln a^s = s \ln a$. **QED**

Note 2: The proof of part D is easier for weak eyes to see when written as $a^s = \exp(\ln a^s)$, and $\exp(s \ln a) = (\exp \ln a)^s = a^s$, hence, $\ln a^s = s \ln a$. The point is that you should be comfortable using either notation for the exponential function!

Corollary 35.1 $\ln y \leq y - 1$ for all $y \in \mathbb{R}^+$.

Proof. Supply a proof as Exercise 35.8. **QED**

Corollary 35.2 The logarithmic function is continuous over its domain $\mathbb{R}^+$.

Proof. This follows from Theorems 35.1 and 26.6. **QED**

Note 3: For a marvelously detailed, and very original, analytic development of logarithms, see Chapter 2 of [34] by the eminent analyst Edmund Landau (1877–1938).

The best short account of the discovery of logarithms is "A Brief History of Logarithms" by R. C. Pierce, Jr. in *The Two-Year College Mathematics Journal*, vol. 8, no. 1 (Jan. 1977), pp. 22–26.

The exponential function $y = e^x$ can be generalized to any base $b > 0$ as follows: $b^x \equiv e^{x \ln b}$, where $x \in \mathbb{R}$. We may now define any power b^a as a limit: $b^a = e^{a \ln b} = \lim_{n \to \infty}(1 + \frac{a \ln b}{n})^n$ (see Exercise 35.5b). This presents a comprehensive analytical basis for any expression of the form b^a, where $b \in \mathbb{R}^+$ and $a \in \mathbb{R}$. For instance, the number $\pi^{\sqrt{5}}$ should be the limit of $\langle 3^2, 3.1^{2.2}, 3.14^{2.23}, 3.141^{2.236}, \ldots \rangle$, based on the decimal expansions of π and $\sqrt{5}$. On the up side, the convergence of the sequence is easy to prove (is it monotone increasing and bounded?). On the down side, we are facing a new situation with no theorems to help us: $\lim_{\langle a,b \rangle \to \langle \sqrt{5}, \pi \rangle}(b^a)$, where $\langle a, b \rangle$ is a variable *point* approaching $\langle \sqrt{5}, \pi \rangle$ along any *path* in $\mathbb{R}^2$. Our answer to this two-variable difficulty is that

$$\pi^{\sqrt{5}} = \lim_{n\to\infty}\left(1 + \frac{\sqrt{5}\ln\pi}{n}\right)^n.$$

That is, defining the terms

$$a_n = \left(1 + \frac{\sqrt{5}\ln\pi}{n}\right)^n, n \in \mathbb{N},$$

gives us a sequence $\langle a_n\rangle_1^\infty$ amenable to all of our theorems, and its limit is actually $\pi^{\sqrt{5}}$.

We can, in fact, prove all the laws of exponents in this most general context. For example, prove that $(b^x)(b^y) = b^{x+y}$ in Exercise 35.9.

35.3 Exercises

1. Prove that $(1 + \frac{-1}{n})^n < e^{-1}$, where n is a natural number. Hint: In Lemma 34.1, use $n = m - 1$.

2. Lemma 34.1 does not prove that $(1 + \frac{1}{u})^u \leq e$ for all real $u \geq 1$, but only for $n \in \mathbb{N}$. This inequality can now be proved. Please do so!

3. For $x \geq -2$, prove that $(\frac{x}{2} + 1)^2 \leq e^x$.

4. Use one of the tests of convergence to prove that $\sum_{n=0}^{\infty} \frac{t^n}{n!}$ converges for any fixed real number t (stipulate that $0^0 \equiv 1$ so that future details stay consistent). In particular, what happens when $t = 1$?

5. a. Prove that $\lim_{x\to\infty}(1 - \frac{1}{x})^x = e^{-1}$ Hint: We can't change variables to $w = -x$ since the definition does not entertain that direction of travel. Use

$$1 + \tfrac{-1}{x} = \frac{1}{1 + \frac{1}{x-1}},$$

and let $w = x - 1$, so that now, $w \to +\infty$.

   b. Prove that $\lim_{x\to\infty}(1 + \frac{a}{x})^x = e^a$ for any real a. (This generalizes **a**.) Hint: Let $w = \frac{x}{a}$ when $a > 0$. Apply **a** when a is negative.

   c. Use the definition of e^x or any of its properties proved thus far to prove that $e^{-a} = \frac{1}{e^a}$ for any real a, so that the function e^x as defined satisfies this important property of exponents.

   d. Prove that $e^{y+x} = e^y e^x$ which, in view of **b**, is true for all real y and x (yet another expected property).

   e. Observe from the definition of e^x that $e^x > 1$ whenever $x > 0$. Now show that $e^x > 0$ for all real x.

   f. Finally, prove that $e^y > e^x$ for any two real numbers $y > x$. Thus, the function $\exp x = e^x$ is strictly increasing throughout its domain (again, as expected). Hint: Let $y = x + d$.

6. Finish the proof of Theorem 35.3 by showing that $\ln(\frac{a}{b}) = \ln a - \ln b$ for any $a, b \in \mathbb{R}$.

7. Prove that if $0 < w < 1$, then $\ln w < 0$, and if $w > 1$, then $\ln w > 0$.

8. Prove Corollary 35.1. Draw the graphs of $y = \ln x$ and $y = x - 1$ to help understand the corollary. Hint: What theorem does this seem be a corollary of?

9. Prove that $(b^x)(b^y) = b^{x+y}$ using the definitions $b^x = \lim_{n\to\infty}(1 + \frac{x \ln b}{n})^n$, $b^y = \lim_{n\to\infty}(1 + \frac{y \ln b}{n})^n$, and $b^{x+y} = \lim_{n\to\infty}(1 + \frac{(x+y)\ln b}{n})^n$. Hint: Eventually, you may need l'Hospital's rule.

10. In Exercise 20.9b, we proved that $\lim_{n\to\infty} r^{1/n} = 1$ for $r > 0$. Further interesting results can be found now as a consequence of Theorem 35.2 in the form $1 + \frac{\ln r}{n} \le e^{\ln r/n} = r^{1/n}$ for all real $r > 0$ and $n \in \mathbb{N}$. In Exercise 20.9a, you showed that $1 \le r^{1/n} \le \frac{n+r-1}{n}$, for $r \ge 1$.
    a. Carefully investigate how Bernoulli's inequality applies for $0 < r < 1$, proving that $1 + \frac{\ln r}{n} \le r^{1/n} \le \frac{n+r-1}{n} < 1$.
    b. Over the domain $[1, \infty)$, with continuous x replacing the discrete n, graph the three functions $y_1 = 1 + \frac{\ln r}{x}$, $y_2 = r^{1/x}$, and $y_3 = \frac{x+r-1}{x}$ first with $r = 0.3$ and next with $r = 3.5$ to see what the pretty inequalities $1 + \frac{\ln r}{x} \le r^{1/x} \le \frac{x+r-1}{x}$ look like. Do you see that $\lim_{x\to\infty} r^{1/x} = 1$ (as in Exercise 20.9b)?
    c. Logarithms may be entirely avoided by using the algebraic substitution $1 - \frac{1}{rn} \le e^{-1/rn}$ in Theorem 35.2. First, for $r > 0$, show that $e^{-1/r} < r$. Hence, show that $1 - \frac{1}{rn} < r^{1/n}$. We now find that $1 - \frac{1}{rn} < r^{1/n} \le \frac{n+r-1}{n}$, with a new lower bounding function to graph: $y_4 = 1 - \frac{1}{rx}$. Once again, $\lim_{x\to\infty} r^{1/x} = 1$. Fascinating!

11. In the study of probability, the binomial and the Poisson probability distributions are widely used. In the binomial distribution, the probability of r successes in n trials, when the probability of a success in any one trial is p, is $P_{bino}(r) = \binom{n}{r} p^r (1-p)^{n-r}$. In the Poisson distribution, the probability of r occurrences of an event with a mean rate λ of occurrence, is $P_{Pois}(r) = \frac{\lambda^r}{r!} e^{-\lambda}$. It is important that as $n \to \infty$ and $p \to 0$ but $\lambda = np$ remains constant, Poisson probabilities approximate binomial probabilities better and better.
    a. See how well P_{bino} and P_{Pois} agree at $r = 1$ when $n = 200$ and $p = 0.001$.
    b. To prove the claimed approximation, begin with the binomial probability

$$\binom{n}{r} p^r (1-p)^{n-r} = \frac{n!}{r!(n-r)!} \left(\frac{\lambda}{n}\right)^r \left(1 - \frac{\lambda}{n}\right)^{n-r}$$
$$= \left(\frac{n!}{n^r (n-r)!}\right) \left(\frac{\lambda^r}{r!}\right) \left(1 - \frac{\lambda}{n}\right)^n \left(1 - \frac{\lambda}{n}\right)^{-r}.$$

Find the limit of the four factors on the right as $n \to \infty$.

36
The Derivative

It is easy to obtain the derived function $y' = \frac{dy}{dx}$ when one takes for y one of the simple functions ... for $y = \sin x$, $\frac{\Delta y}{\Delta x} = \frac{\sin \frac{1}{2} i}{\frac{1}{2} i} \cos(x + \frac{1}{2} i)$, [so that] $y' = \frac{dy}{dx} = \cos x = \sin(x + \frac{\pi}{2})$

Augustin-Louis Cauchy, *Leçons sur le calcul différentiel*, 1829, pp. 289–290.

One of the pregnant problems circulating about the mathematical world of the first half of the 17th century was that of finding the tangent to a curve at a given point. Great mathematicians of the time such as René Descartes (1596–1650), Pierre de Fermat (1601–1665), Gilles de Roberval (1602–1675), John Wallis (1616–1703), and James Gregory (1638–1675) developed special cases and constructions that yielded tangent lines. Newton's mentor Isaac Barrow (1630–1677) came close to being a codiscoverer of calculus through his study of this problem. As a curiosity, you may enjoy finding the equation of the tangent line to a parabola without using any calculus, in Exercise 36.1.

It fell upon Isaac Newton in his *annus mirabilis* 1666, his miracle year, to create the key concept of the ratio of vanishingly small increments that yield a tangent slope. Both he and Gottfried Leibniz had a firm grasp of this general method for tangents by 1675, Leibniz using the symbol $\frac{dy}{dx}$ for the first time in a manuscript dated November 11, 1675.

Note 1: Here is an excerpt from the first section of Newton's foundational treatise on physics (called natural philosophy in those days), *Philosophiae naturalis principia mathematica*, published in 1687. Instantaneous velocity—Newton calls it "ultimate velocity"—is discussed in a manner that seems vague to us today. But notice the idea of a limit throughout the passage. Newton was grappling for the first time with what we now blithely write as $v = \frac{ds}{dt} = \lim_{t_2 \to t_1} \frac{s_2 - s_1}{t_2 - t_1}$. (A fine account of the mathematics of that exciting century appears in Chapter VI of [8].)

> ... by the ultimate velocity is meant that with which the body is moved, neither before it arrives at its last place and the motion ceases, nor after, but at the very instant it arrives; that is, that velocity with which the body arrives at its last place, and with which the motion ceases. And in like manner, by the ultimate ratio of evanescent quantities is to be understood the ratio of the quantities not before they vanish, nor afterwards, but with which they vanish. In like manner the first ratio of nascent quantities is that with which they begin to be.... There is a limit which the velocity

at the end of the motion may attain, but not exceed. This is the ultimate velocity. And there is the like limit in all quantities and proportions that begin and cease to be. And since such limits are certain and definite, to determine the same is a problem strictly geometrical.

Book I, section I, last scholium. Translated from the Latin by Andrew Motte, 1729.

We can judge how far the concept of the derivative evolved in one and a half centuries by comparing Newton's passage above with Cauchy's statements in his 1829 *Leçons sur le calcul différentiel*, "Lessons on Differential Calculus." He writes on page 288 that $\frac{\Delta y}{\Delta x} = \frac{f(x+i)-f(x)}{i}$ (where $i = \Delta x$) may approach a limit as both the numerator and denominator tend to zero, a limit which, when it exists, depends on x. Cauchy reminds us that this new function, producing the limits of the given ratio, has been named the *derived function*. We are standing upon firm ground now! If anything remained mysterious in his statement, it was the issue about i approaching zero. And that issue would be settled not long afterward by Karl Weierstrass's school of analysis. Thus, we have the following, modern statement. (Recall that the δ-neighborhood $(x - \delta, x + \delta)$ of a point x is denoted by $\mathbf{N}_x^\delta$.)

> A function f has a **derivative** or is **differentiable** at a point $x \in \mathbf{N}_x^\delta \subseteq \mathcal{D}_f$ if and only if $\lim_{h \to 0} \frac{f(x+h)-f(x)}{h}$ exists.

This limit is a function variously written as f', $\frac{df}{dx}$, or Df, and when emphasis on the independent variable is desired, by $f'(x)$, $\frac{df(x)}{dx}$, or $D_x f$.

If we write $x = c$, $x_1 = c + h$, and $\Delta x = h = x_1 - c$, then we have the equivalent expressions

$$f'(c) = \left.\frac{df(x)}{dx}\right|_c = \lim_{x_1 \to c} \frac{f(x_1) - f(c)}{x_1 - c} = \lim_{\Delta x \to 0} \frac{f(c + \Delta x) - f(c)}{\Delta x}.$$

The emphasis here is on finding the derivative at a specific value $c \in \mathbf{N}_c^\delta \subseteq \mathcal{D}_f$, and on the derivative as the limit of the slopes of secants $\frac{f(x_1)-f(c)}{x_1-c}$ drawn through the fixed point $\langle c, f(c)\rangle$ and any other point $\langle x_1, f(x_1)\rangle$, as $x_1 \to c$.

Note 2: It is necessary that $x \in \mathbf{N}_x^\delta \subseteq \mathcal{D}_f$ since we must be able to evaluate $f(x + h)$ for all values of $x + h$ in a neighborhood of x. This will soon be modified for one-sided derivatives. Extending the concept, the derivative could be defined at a cluster point $x \in \mathcal{D}_f$.

The first thing to note is that the limit involved in finding a derivative is not a limit for f, but rather for the divided difference function $Q_x(h) = \frac{f(x+h)-f(x)}{h}$. This is a function of both x and h, with $\{x, x + h\} \subset \mathbf{N}_x^\delta \subseteq \mathcal{D}_f$(the notation $Q_x(h)$ emphasizes this). Q_x has the sole purpose of yielding secant slopes between the given point $\langle x, f(x)\rangle$ and the variable point $\langle x + h, f(x + h)\rangle$. This implies that if $f'(x)$ exists and we choose a value of h close to zero, then the value of $Q_x(h)$ will approximate the value of $f'(x)$. The dependency on h may be removed either by fixing h or by taking the limit (if possible) as $h \to 0$. Let us try these ideas on a well-known function.

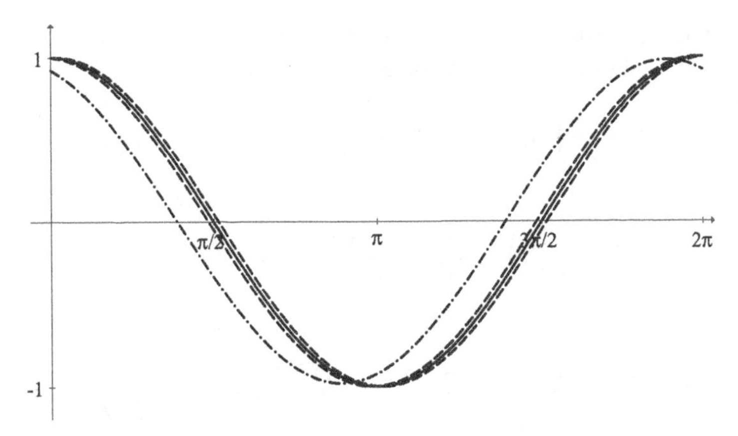

Figure 36.1. The graph of $f'(x) = \cos x$, shown in solid line, is close by $Q_x(0.1)$ on its left and $Q_x(-0.1)$ on its right, shown in broken lines. $Q_x(0.7)$ is also graphed, in the dash-dot line.

Example 36.1 Given $f(x) = \sin x$ and $h = 0.1$, compare

$$Q_x(h) = \frac{f(x+h) - f(x)}{h}$$

with the exact $f'(x) = \cos x$ at $x = \frac{\pi}{3}$, and in general for any x.

We have

$$Q_x(0.1) = \frac{\sin(x + 0.1) - \sin x}{0.1},$$

which brings us to the secant slope $Q_{\pi/3}(0.1) = 10(\sin(\frac{\pi}{3} + 0.1) - \sin\frac{\pi}{3}) \approx 0.456$, in rough agreement with $\cos\frac{\pi}{3} = \frac{1}{2}$. Next, we extend the comparison to other xs besides $\frac{\pi}{3}$.

Early graphing calculators (pre-1990, practically neolithic) did not have a command for graphing $f'(x)$. Suppose that a student did not know that $f'(x) = \cos x$. In those days, he or she could approximate the derivative by $Q_x(0.1) = 10(\sin(x + 0.1) - \sin x)$, a function of x alone. The graph of $Q_x(0.1)$ appears in Figure 36.1. It is shifted just slightly to the left of the graph of $y = \cos x$, thus showing fairly close agreement. Likewise, the graph of $Q_x(-0.1) = -10(\sin(x - 0.1) - \sin x)$ is symmetrically displaced to the right of $y = \cos x$. On the other hand, the figure shows that raising h to 0.7 makes $Q_x(0.7)$ an awful approximation for the cosine.

Figure 36.1 strongly suggests that

$$\lim_{h\to 0} Q_x(h) = \lim_{h\to 0}\frac{\sin(x+h) - \sin x}{h} = \cos x. \quad \blacksquare$$

Example 36.2 Prove that $\frac{d\sin x}{dx} = \cos x$. (After studying the previous example we expect this result!) The analysis here is by Cauchy, and he may have been the first to prove this important formula rigorously. With $y = \sin x$, he must demonstrate that $\lim_{h\to 0}\frac{\sin(x+h)-\sin x}{h}$ exists (we use h instead of his i). He begins with a standard trigonometric identity, $\sin\alpha\cos\beta = \frac{1}{2}\sin(\alpha+\beta) + \frac{1}{2}\sin(\alpha-\beta)$. Substituting $\alpha = \frac{1}{2}h$ and $\beta = x + \frac{1}{2}h$, he has

$$\sin\left(\frac{1}{2}h\right)\cos\left(x+\frac{1}{2}h\right) = \frac{1}{2}\sin(x+h) + \frac{1}{2}\sin(-x),$$

so

$$2\sin\left(\frac{1}{2}h\right)\cos\left(x+\frac{1}{2}h\right) = \sin(x+h) - \sin x.$$

Consequently,

$$\frac{\Delta y}{\Delta x} = \frac{\sin(x+h)-\sin x}{h} = \frac{\sin\left(\frac{1}{2}h\right)}{\frac{1}{2}h}\cos\left(x+\frac{1}{2}h\right).$$

Cauchy had previously shown, and we know by Theorem 23.2, that $\lim_{\theta\to 0}\frac{\sin\theta}{\theta} = 1$. It follows that

$$\frac{d\sin x}{dx} = \lim_{h\to 0}\frac{\sin(x+h)-\sin x}{h} = \lim_{h\to 0}\left(\frac{\sin(\frac{1}{2}h)}{\frac{1}{2}h}\cos(x+\frac{1}{2}h)\right) = 1\cdot\cos(x+0) = \cos x$$

(he just states the conclusion). **QED**

Example 36.3 Show that $g(x) = \sqrt[3]{x}$ is differentiable everywhere except at $x = 0$. The definition indicates that

$$\frac{d\sqrt[3]{x}}{dx} = \lim_{h\to 0}\frac{\sqrt[3]{x+h}-\sqrt[3]{x}}{h} = \lim_{h\to 0}\frac{(\sqrt[3]{x+h}-\sqrt[3]{x})((\sqrt[3]{x+h})^2+\sqrt[3]{x+h}\sqrt[3]{x}+(\sqrt[3]{x})^2)}{h((\sqrt[3]{x+h})^2+\sqrt[3]{x+h}\sqrt[3]{x}+(\sqrt[3]{x})^2)}$$

$$= \lim_{h\to 0}\frac{x+h-x}{h((\sqrt[3]{x+h})^2+\sqrt[3]{x+h}\sqrt[3]{x}+(\sqrt[3]{x})^2)},$$

where the numerator has been rationalized by using the identity $(A-B)(A^2+AB+B^2) = A^3 - B^3$. Continuing, we arrive at

$$\lim_{h\to 0}\frac{1}{\sqrt[3]{(x+h)^2}+\sqrt[3]{x+h}\sqrt[3]{x}+\sqrt[3]{x^2}} = \frac{1}{3\sqrt[3]{x^2}},$$

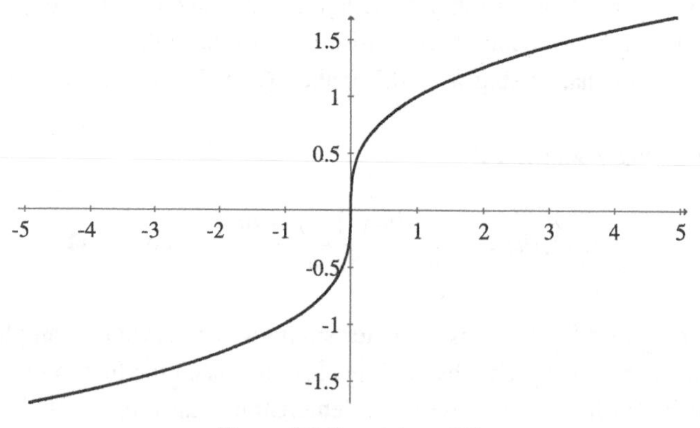

Figure 36.2. $u(x) = \sqrt[3]{x}$

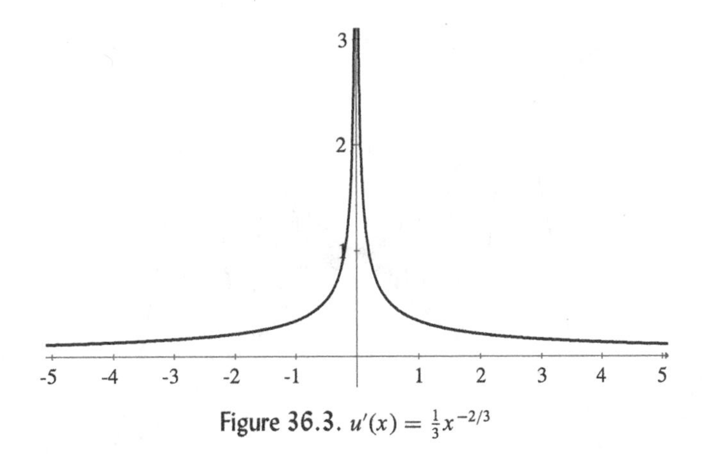

Figure 36.3. $u'(x) = \frac{1}{3}x^{-2/3}$

provided $x \neq 0$. This is the undoubtedly familiar result that $\frac{d\sqrt[3]{x}}{dx} = \frac{1}{3}x^{-2/3}$. Lastly, at $x = 0$,

$$\left.\frac{d\sqrt[3]{x}}{dx}\right|_{x=0} = \lim_{h\to 0} \frac{\sqrt[3]{0+h} - \sqrt[3]{0}}{h} = \lim_{h\to 0} \frac{1}{h^{2/3}},$$

which does not exist.

The graph of $g(x) = \sqrt[3]{x}$ shows the vertical tangent at $\langle 0, 0\rangle$ as the cause of non-differentiability there. The graph of $g'(x)$ vividly shows the discontinuity of the derivative at $\langle 0, 0\rangle$. ■

A second thing to note is that for $Q_x(h) = \frac{f(x+h)-f(x)}{h}$ to have a real-valued limit, it is necessary (but not sufficient) for the numerator to have a zero limit as $h \to 0$. This brings us to the main theorem that connects differentiability and continuity.

Theorem 36.1 If a function f is differentiable at x, then it is also continuous there.

Proof. Given that $f'(x)$ exists, we have

$$\begin{aligned}\lim_{h\to 0}(f(x+h) - f(x)) &= \lim_{h\to 0}\left(h \cdot \frac{f(x+h) - f(x)}{h}\right)\\ &= \left(\lim_{h\to 0} h\right)\left(\lim_{h\to 0} \frac{f(x+h) - f(x)}{h}\right) = 0 \cdot f'(x) = 0.\end{aligned}$$

Therefore, $\lim_{h\to 0} f(x+h) = f(x)$, which means that f is continuous at x. **QED**

The converse of this theorem is terribly false. One counterexample appears in Example 36.3. Find two counterexamples of your own as Exercise 36.5. The following example demonstrates a subtle behavior by which the converse of Theorem 36.1 may again be falsified.

Example 36.4 We proved in Exercise 24.7 that $f(x) = x \sin \frac{1}{x}$ is continuous at $x = 0$ when we define $f(0) \equiv 0$. Its graph in Chapter 24 is magnified in Figure 36.4. Now we show that it is not differentiable at 0. The problem is that in *any* neighborhood of $x = 0$, secants from $\langle 0, 0\rangle$ to other points on the curve may have slopes varying anywhere from -1 to 1. But a derivative must be a unique number!

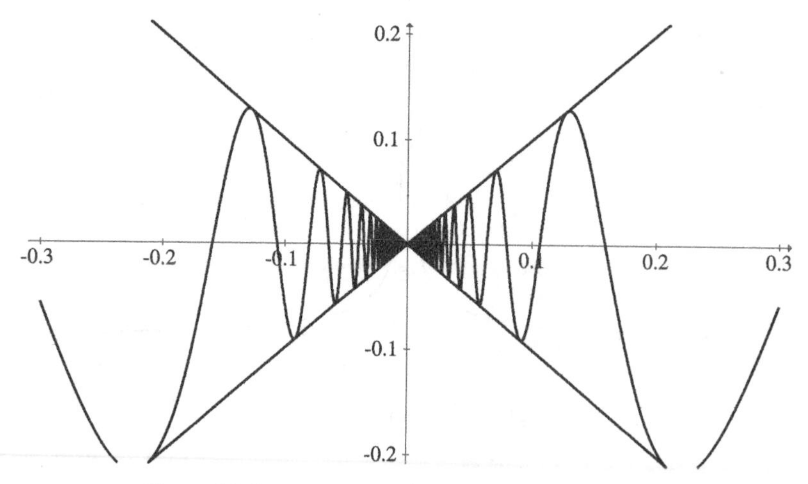

Figure 36.4. $f(x) = x \sin \frac{1}{x}$ with guide lines $y = \pm x$.

Thus, investigating

$$\lim_{h \to 0} Q_0(h) = \lim_{h \to 0} \frac{(0+h)\sin \frac{1}{0+h} - 0}{h} = \lim_{h \to 0} \frac{h \sin \frac{1}{h}}{h} = \lim_{h \to 0} \sin \frac{1}{h},$$

we see that $f'(0)$ does not exist. Compare this with a related function in Exercise 36.6. Also, see Note 3 in the next chapter. ■

Applying the definition of a limit found in Chapter 24 produces the equivalent definition of a derivative seen below. Although a bit more abstract, here we have at last freed the derivative from the troublesome idea of a quantity approaching anything.

> The **derivative** $f'(x)$ exists at a point $x \in \mathcal{D}_f$ if and only if given any $\varepsilon > 0$, there exists a corresponding $\delta > 0$ such that whenever $0 < |h| < \delta$ and $x + h \in \mathcal{D}_f$, then
>
> $$\left| \frac{f(x+h) - f(x)}{h} - f'(x) \right| < \varepsilon$$
>
> (geometrically, all secant slopes are in a neighborhood of $f'(x)$).

We end this chapter with a list of causes of non-differentiability at a point—the origin in all of the examples below.

CAUSES of NON-DIFFERENTIABILITY	*EXAMPLES*	
Non-removable Discontinuity	$y = \frac{1}{x}$	$y = \operatorname{sgn} x$
Removable Discontinuity	$y = \frac{\sin x}{x}$	$y = \frac{x^2}{x}$
Cusp	$y = \lvert x \rvert$	$y = x^{2/3}$
Vertical Tangent	$y = \sqrt{2x - x^2}$	$y = \sqrt[3]{x}$
Oscillation (but see Exercise 36.6)	$y = \sin \frac{1}{x}$	$y = x \sin \frac{1}{x}$
Isolated Point (see Figure 36.5)	$y = \sqrt{x^4 - x^2}$	

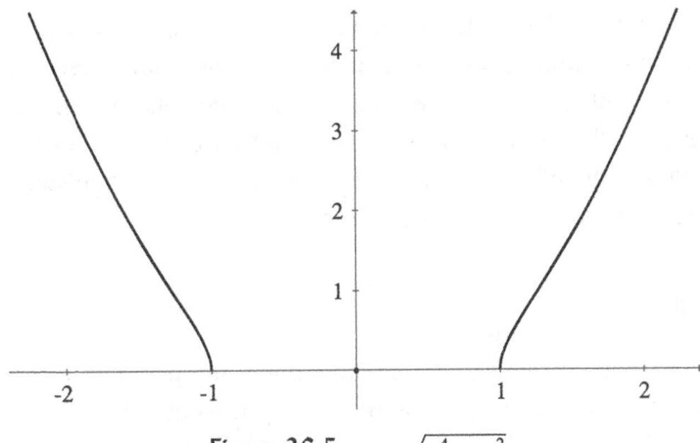

Figure 36.5. $y = \sqrt{x^4 - x^2}$

Note 3: The sometimes antagonistic fields of philosophy and mathematics came to a common ground in the intellect of René Descartes. He greatly advanced geometry by applying algebraic techniques to its problems, thus founding what we now call analytic geometry. This was disclosed in an appendix to his treatise with the grandiose title *Discours de la méthode pour bien conduire sa raison, et chercher la vérité dans les sciences* "A Discourse on the Method for Guiding One's Reason Well, and Finding Truth in the Sciences" (1637), in which he establishes rationalism (we find there his famous statement, "I think, therefore I am.")

On the other hand, intellectual swords crossed between John Wallis, the most accomplished mathematician in England prior to Newton, and his countryman, the materialist philosopher Thomas Hobbes. Wallis became the third Savilian Professor of Geometry at Oxford in 1649, remaining there for over 50 years. He was one of the first to study the "Imaginary Roots of Impossible Equations..." (that is, polynomial equations with complex roots), in *A Treatise of Algebra* (1685). His very influential *Arithmetica infinitorum* "Arithmetic of Infinities" (1656), revealed the equation $\frac{\pi}{2} = \frac{2\cdot 2}{1\cdot 3}\cdot\frac{4\cdot 4}{3\cdot 5}\cdot\frac{6\cdot 6}{5\cdot 7}\cdot\cdots$. This was a breakthrough in the millennia-long study of π since it was the first expression to yield π using only simple arithmetic operations. We know it today as "Wallis's product."

Hobbes became enamored with geometry, became fairly accomplished in it, and fell upon the celebrated problem of squaring the circle. When Hobbes (with much pride) published a solution, Wallis quickly detected a mistake. But Hobbes refused to admit any error, and instead attacked Wallis for his impropriety! He wrote a pamphlet entitled *Six lessons to the Professors of Mathematics at the Institute of Sir Henry Savile*. Wallis replied with his own pamphlet, *Due Correction for Mr. Hobbes, or School Discipline for not saying his Lessons Aright*, to which Hobbes responded with a ferocious broadside, *Marks of the Absurd Geometry, Rural Language, Scottish Church Politics, and Barbarisms of John Wallis, Professor of Geometry and Doctor of Divinity* (1657). Perhaps it's true, as someone has said, that academic squabbles are so vicious because the stakes are so small. (The titles of the pamphlets are in the biography of Wallis by J. J. O'Connor and E. F. Robertson at www-history.mcs.st-andrews.ac.uk/Biographies/Wallis.html, ©2002, which is an excellent resource for many mathematicians' lives.)

Hobbes had wrestled with the Cyclops of constructing a square of equal area to a circle by using only compass and straightedge. He could never have won the match, nor can anyone else. But many have refused to admit defeat, for the impossibility of the task is explained in the difficult proof that π is a transcendental number (see Chapter 8, Note 2). We should therefore commiserate with Hobbes for his mistake, but dismiss him for his pigheaded refusal to learn from a master.

36.1 Exercises

1. a. To find the equation of the tangent line to a point of a parabola $y - k = a(x - h)^2$, we first simplify the parabola's equation by shifting it so that its vertex coincides with the origin. Thus, the equation becomes $y = ax^2$, without loss of generality. Let a slope m_0 be given. Our goal is to find the unique line $y = m_0 x + b$ that intersects the parabola exactly once. Evidently, we must find b in terms of the given quantities a and m_0. Follow these steps.
   (1) Since the ordinates are equal at the point (or points) of intersection, solve $ax^2 = m_0 x + b$ for x.
   (2) Not surprisingly, there are usually two solutions to this equation. However, when a special condition is fulfilled, only one solution arises (call it x_0). What is that condition? What expression do you get for x_0?
   (3) Solve the condition that you found in (2) for b.
   (4) Write the equation of the tangent line to the parabola at x_0. You may also find the coordinates $\langle x_0, y_0 \rangle$ of the point of tangency. Aha! No calculus was needed!

   b. Use the procedure to find the equation of the tangent line to $y = 3(x - 5)^2 + 10$ with a slope $m_0 = 6$. Does your answer check with standard calculus?

2. Let $g(x) = \frac{1}{2}x^2 + 6x - 4$. Graph the divided difference function $Q_x(0.2)$ and compare it to $g'(x)$. Prove that $\lim_{h \to 0} Q_x(h) = g'(x)$.

3. Let $f(x) = \operatorname{sgn} x$ be the signum function seen in Chapter 23. Prove that f is not differentiable at $x = 0$. Prove more generally that

$$\frac{d \operatorname{sgn} x}{dx} = \lim_{h \to 0} Q_x(h) = \begin{cases} 0 & \text{if } x \neq 0 \\ \text{undefined} & \text{if } x = 0 \end{cases}.$$

4. Show that

$$y = \begin{cases} \sqrt{x} & \text{if } x \geq 0 \\ \sqrt{-x} & \text{if } x \leq 0 \end{cases}$$

has derivative

$$\frac{dy}{dx} = \lim_{h \to 0} Q_x(h) = \begin{cases} \frac{1}{2}x^{-1/2} & \text{if } x > 0 \\ \text{undefined} & \text{if } x = 0 \\ -\frac{1}{2}(-x)^{-1/2} & \text{if } x < 0 \end{cases},$$

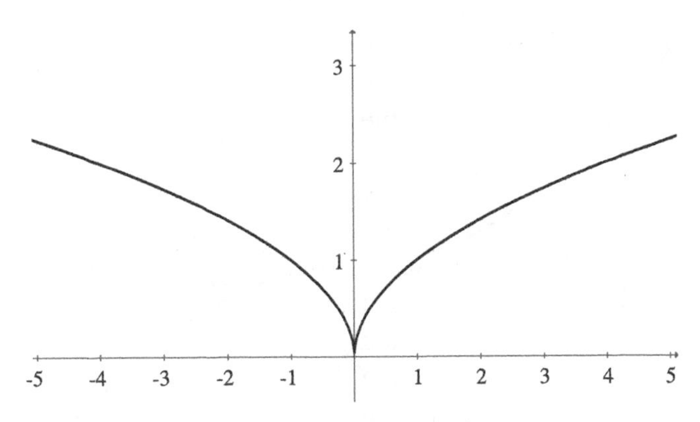

Figure 36.6. $y = \sqrt{|x|}$

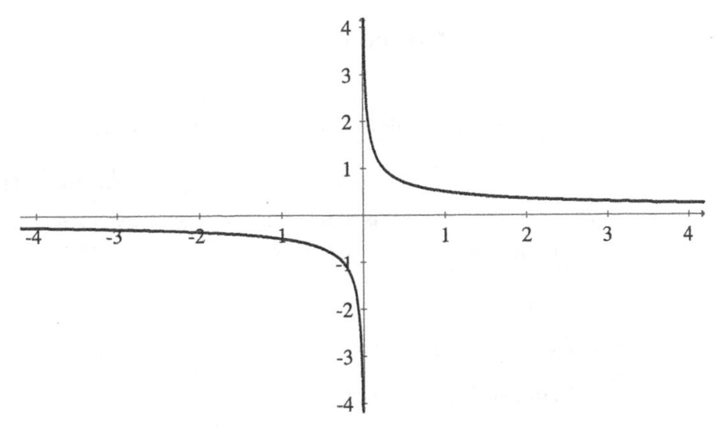

Figure 36.7. $\frac{dy}{dx} = \frac{\operatorname{sgn} x}{2\sqrt{|x|}}$

without using the power rule from calculus. The given function, concisely written as $y = \sqrt{|x|}$ and graphed in Figure 36.6, is continuous at $\langle 0, 0 \rangle$. But its derivative (Figure 36.7), concisely written as $\frac{dy}{dx} = \frac{\operatorname{sgn} x}{2\sqrt{|x|}}$, fails to exist at $\langle 0, 0 \rangle$.

5. Find two more counterexamples, besides Exercise 36.4 and Example 36.3, of the fatal converse (continuity implies differentiability) of Theorem 36.1.

6. In contrast to Example 36.4, prove that

$$f(x) = \begin{cases} x^2 \sin \frac{1}{x} & \text{if } x \neq 0 \\ 0 & \text{if } x = 0 \end{cases}$$

is differentiable at $x = 0$, and in fact, $f'(0) = 0$. The function is therefore continuous at $\langle 0, 0 \rangle$. Its graph appears in Figure 36.8 along with graphs of the envelope functions $y = x^2$ and $y = -x^2$. Compare it with the graph of $y = x \sin \frac{1}{x}$ in Figure 36.4.

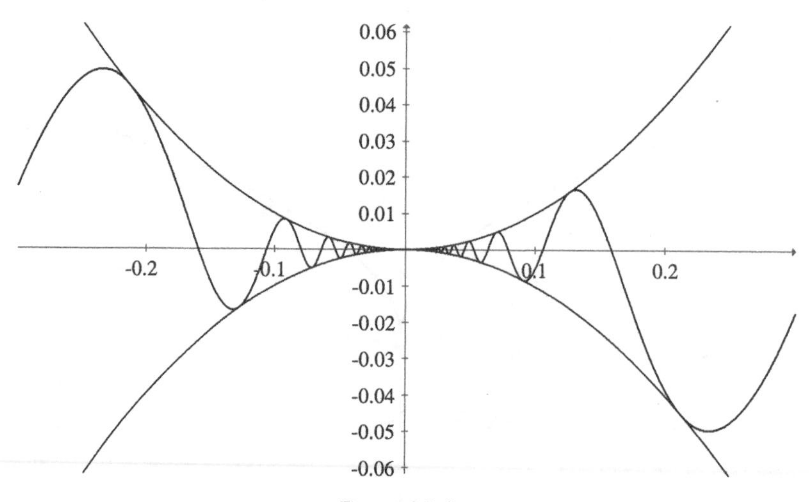

Figure 36.8.

7. The four functions below produce an interesting progression of cases:

Function	Found in...	Function exists at $\langle 0, 0\rangle$?	Function is continuous at $\langle 0, 0\rangle$?	Derivative exists at $\langle 0, 0\rangle$?	Derivative is continuous at $\langle 0, 0\rangle$?
$\sin\frac{1}{x}$, $f(0) = 0$	Exercise 25.1	yes	no	no	no
$x\sin\frac{1}{x}$, $f(0) = 0$	Example 36.4	yes	yes	no	no
$x^2\sin\frac{1}{x}$, $f(0) = 0$	Exercise 36.6	yes	yes	yes	?
$x^3\sin\frac{1}{x}$, $f(0) = 0$	—	yes	?	?	?

Show that the third function does not have a continuous derivative at $\langle 0, 0\rangle$. Next, show that the fourth function has a continuous derivative at $x = 0$, thus fulfilling the suspected pattern with a "yes" in three cells.

8. Let $y = |x|$. Its derivative is

$$y'(x) = \begin{cases} 1 & \text{if } x > 0 \\ \text{undefined} & \text{if } x = 0 \\ -1 & \text{if } x < 0 \end{cases}.$$

This function is identical to $y = \frac{1}{\operatorname{sgn} x}$. Put in other symbols, $\frac{d|x|}{dx} = \frac{1}{\operatorname{sgn} x}$.

Compare $y'(x)$ with $Q_x(0.1)$ by graphing them as a functions of x and explaining the discrepancy in the interval $[-0.1, 0]$. Is it true that $\lim_{h\to 0} Q_x(h) = y'(x)$? Hint: You will have to consider three intervals, $(-\infty, -0.1]$, $[-0.1, 0]$, and $[0, \infty)$.

9. Let $f(x) = \operatorname{sgn} x$. Its derivative was found in Exercise 36.3. Compare $f'(x)$ with $Q_x(0.1)$ by graphing them as a functions of x and explaining the discrepancy in the interval $[-0.1, 0]$. Is it true that $\lim_{h\to 0} Q_x(h) = f'(x)$? Hint: You will have to consider three intervals, $(-\infty, -0.1)$, $(-0.1, 0)$, and $(0, \infty)$, as well as $x = -0.1$ and $x = 0$.

10. Is the following argument an acceptable proof of Theorem 36.1? If not, why not?

    Proof. Theorem 24.1 applied to this situation states the three requirements for continuity at x this way: (1) $f(x)$ must exist, (2) $\lim_{h\to 0} f(x+h)$ must exist, and (3) $\lim_{h\to 0} f(x+h) = f(x)$. (1) is given as one of the requirements of differentiability. For (2) and (3), since $\lim_{h\to 0} \frac{f(x+h)-f(x)}{h}$ exists, then $\lim_{h\to 0}(f(x+h) - f(x)) = 0$. Hence, $\lim_{h\to 0} f(x+h) = f(x)$. Therefore, f is continuous at x. **QED?**

11. Which one of these is correct:

$$f'(x) = \lim_{h\to 0} \frac{f(x-h) - f(x)}{h}$$

    or

$$f'(x) = \lim_{h\to 0} \frac{f(x) - f(x-h)}{h}?$$

    Test your answer on a well-known function such as $f(x) = x^2 - x$. Is there any relationship between the incorrect one and $f'(x)$?

37

Theorems for Derivatives

. . . it was only after a considerable struggle that Leibniz arrived at the correct rules for the differentiation of products and quotients.

"Rigor and Proof in Mathematics: A Historical Perspective" by Israel Kleiner, in *Mathematics Magazine*, vol. 64, no. 5, pg. 295.

The definition of the derivative in the previous chapter becomes increasingly nasty to use as the functions of interest become more complicated. Thus, we now prove the standard theorems for what may be called the arithmetic of derivatives, along with some other derivative formulas.

Theorem 37.1 Let f and g be functions differentiable at x, and let $a, b \in \mathbb{R}$. Then any linear combination $af + bg$ of the functions is differentiable at x. In fact,

$$\frac{d(af+bg)(x)}{dx} = a\frac{df(x)}{dx} + b\frac{dg(x)}{dx}.$$

(In particular, $(f \pm g)'(x) = f'(x) \pm g'(x)$, so tangent slopes are additive.)

Proof. Given that $f'(x)$ and $g'(x)$ exist, we have

$$\begin{aligned}\frac{d(af+bg)(x)}{dx} &= \lim_{h\to 0}\frac{(af+bg)(x+h)-(af+bg)(x)}{h}\\ &= \lim_{h\to 0}\frac{af(x+h)+bg(x+h)-(af(x)+bg(x))}{h}\\ &= a\lim_{h\to 0}\frac{f(x+h)-f(x)}{h} + b\lim_{h\to 0}\frac{g(x+h)-g(x)}{h}\\ &= a\frac{df(x)}{dx} + b\frac{dg(x)}{dx}. \quad \textbf{QED}\end{aligned}$$

Theorem 37.2 The Product Rule If f and g are differentiable at x, then fg is differentiable at x. In fact,

$$\frac{d(fg)(x)}{dx} = f(x)\frac{dg(x)}{dx} + g(x)\frac{df(x)}{dx}, \quad \text{or} \quad (fg)'(x) = f(x)g'(x) + g(x)f'(x).$$

(Thus, tangent slopes are not multiplicative.)

Proof. Given that $f'(x)$ and $g'(x)$ exist, we have

$$\begin{aligned}
\frac{d(fg)(x)}{dx} &= \lim_{h\to 0}\frac{(fg)(x+h)-(fg)(x)}{h} = \lim_{h\to 0}\frac{f(x+h)g(x+h)-f(x)g(x)}{h} \\
&= \lim_{h\to 0}\frac{f(x+h)g(x+h)-f(x+h)g(x)+f(x+h)g(x)-f(x)g(x)}{h} \\
&= \lim_{h\to 0}(f(x+h)\frac{g(x+h)-g(x)}{h}) + g(x)\lim_{h\to 0}\frac{f(x+h)-f(x)}{h} \\
&= f(x)\frac{dg(x)}{dx} + g(x)\frac{df(x)}{dx}.
\end{aligned}$$

In the last equation, we justify that $\lim_{h\to 0} f(x+h) = f(x)$ by recalling Theorem 36.1. **QED**

The next, very powerful, theorem gives us the way to differentiate compositions of functions.

Theorem 37.3 The Chain Rule Let g be continuous on $(a,b) \subseteq \mathcal{D}_g$ and differentiable at $x \in (a,b)$. Furthermore, let the image $g((a,b)) = (c,d) \subseteq \mathcal{D}_f$, and let f be differentiable at $g(x)$. Then the composition $f \circ g$ is differentiable at x, and $(f \circ g)'(x) = f'(g(x)) \cdot g'(x)$. This is conveniently remembered in Leibniz notation as $\frac{df(y)}{dx} = \frac{df(y)}{dy} \cdot \frac{dy}{dx}$.

Proof. (We interpret the elegant proof in the textbook by Walter Rudin (1921–2010) [37], pg. 105.)

Let $y = g(x)$. Let V be a function on (c,d) that expresses the difference between the tangent slope $f'(y)$ and the secant slope $\frac{f(s)-f(y)}{s-y}$. Define $V(s)$ through the equation

$$f(s) - f(y) = (s-y)(f'(y) - V(s)), \tag{1}$$

where $s \in (c,d)$. Since f is differentiable at y, $\lim_{s\to y} V(s) = 0$.

Similarly, we define a function U by means of

$$g(t) - g(x) = (t-x)(g'(x) - U(t)), \tag{2}$$

where $t \in (a,b)$, so that $\lim_{t\to x} U(t) = 0$ because g is differentiable at x.

Next, complete the bridge between g and f by letting $s = g(t)$. We may now substitute from (1) and (2) to get $f(s) - f(y) = f(g(t)) - f(g(x)) = (g(t) - g(x))(f'(y) - V(s)) = (t-x)(g'(x) - U(t))(f'(y) - V(s))$.

Whenever $t \neq x$, we now have

$$\frac{f(g(t)) - f(g(x))}{t-x} = (g'(x) - U(t))(f'(y) - V(s)).$$

As $t \to x$ in the domain of g, we see that $s \to y$ in the range of g, by its continuity. It follows that

$$\lim_{t\to x}\frac{f(g(t)) - f(g(x))}{t-x} = (g'(x) - 0)(f'(y) - 0).$$

The left-hand side equals $(f \circ g)'(x)$ by the definition of the derivative, and the right-hand side equals $g'(x) \cdot f'(g(x))$. And so, the theorem is proved. **QED**

Note 1: An erroneous, but nevertheless instructive, argument for the chain rule uses the following definitions:

$$\frac{\Delta g}{\Delta x} = \frac{g(t) - g(x)}{t - x}, \frac{\Delta f}{\Delta y} = \frac{f(s) - f(y)}{s - y}, s \equiv g(t), \quad \text{and} \quad y \equiv g(x).$$

It follows that $\frac{\Delta f}{\Delta x} = \frac{\Delta f}{\Delta y}\frac{\Delta g}{\Delta x}$. On taking limits as $\Delta x \to 0$, we seem to obtain $\frac{d(f \circ g)(x)}{dx} = \frac{df}{dy}\frac{dg}{dx}$ right away! The subtle error lies in that it is possible for $\Delta y = g(t) - g(x)$ to equal zero within any deleted neighborhood of x (recall $y = x \sin \frac{1}{x}$ near the origin). This will make $\frac{\Delta f}{\Delta y}\frac{\Delta g}{\Delta x}$ a non-existent quantity in such neighborhoods, thus invalidating the limit, and the proof.

The proof of Theorem 37.3 purposely does not divide by $t - x$ or $s - y$ in the definitions of U and V, thus neatly avoiding the division by zero error.

Example 37.1 Find the derivative of $f(x) = \sin|x|$. Of course, $\frac{d \sin y}{dy} = \cos y$, where y represents any function of x whose range intersects with the domain of the sine function, in this case, $y = |x|$. In Exercise 36.8, we found that $\frac{d|x|}{dx} = \frac{1}{\operatorname{sgn} x}$. Hence, the chain rule gives

$$\frac{d \sin|x|}{dx} = \frac{d \sin y}{dy}\frac{dy}{dx} = (\cos y)\left(\frac{1}{\operatorname{sgn} x}\right) = \frac{\cos|x|}{\operatorname{sgn} x}.$$

This makes it clear that $f'(0)$ does not exist. The graphs appear in Figures 37.1 and 37.2. In all of this, we shouldn't abandon the convenience from calculus I of calling $f(y) = \sin y$ the "outside function," and $y = |x|$ the "inside function." Compare with Exercise 37.7. ■

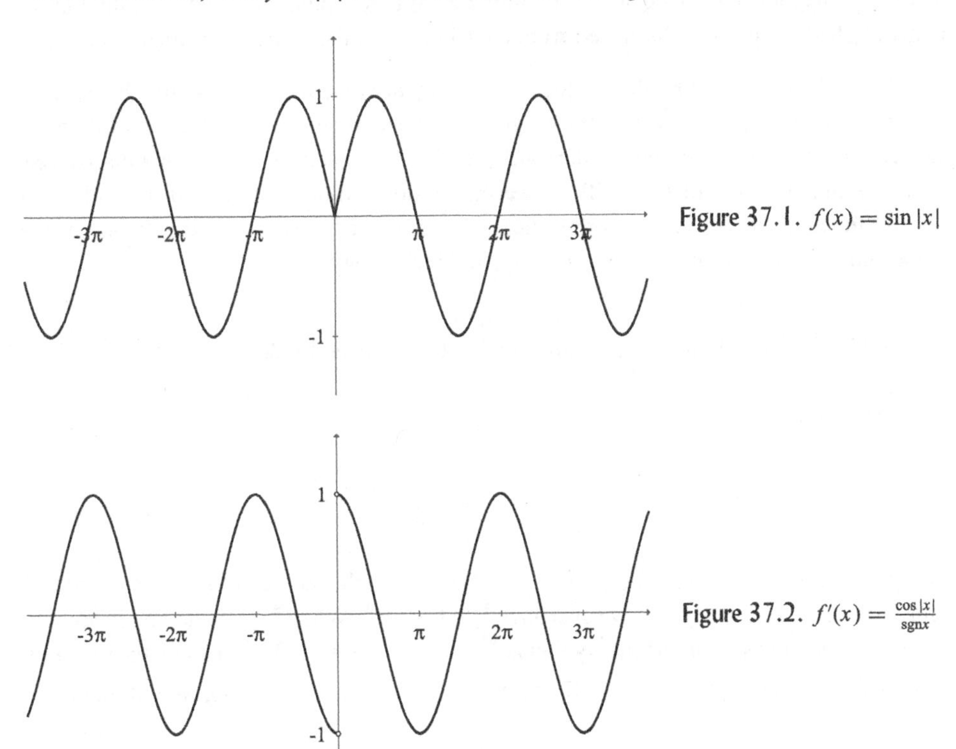

Figure 37.1. $f(x) = \sin|x|$

Figure 37.2. $f'(x) = \frac{\cos|x|}{\operatorname{sgn} x}$

Theorem 37.4 The Quotient Rule If f and g are differentiable at x and $g(x) \neq 0$, then $\frac{f}{g}$ is differentiable at x. In fact,

$$\frac{d(\frac{f}{g})(x)}{dx} = \frac{g(x)\frac{df(x)}{dx} - f(x)\frac{dg(x)}{dx}}{g^2(x)}, \quad \text{or} \quad \left(\frac{f}{g}\right)'(x) = \frac{g(x)f'(x) - f(x)g'(x)}{g^2(x)}.$$

Proof. Maria Agnesi (1718–1799) proved this in an efficient manner; see Exercise 37.3. **QED**

Note 2: Maria Gaëtana Agnesi was a child prodigy in languages, a mathematician, a philosopher, and a humanitarian. Her textbook, *Instituzioni analitiche ad uso della gioventù italiana* (Foundations of Analysis for the Use of Italian Youth), is a skilled, unified compilation of much mathematics done up to 1748, its date of publication. The first volume is a pedagogically excellent study of algebra and analytic geometry. The second, a study of infinitesimals, was considered to be the best introduction to analysis available. She was appointed a professor of mathematics at the University of Bologna in 1750, but there is debate as to whether she actually lectured there (remarkably, women were on the faculty at Bologna at the time).

S. I. B. Gray and Tagui Malakyan have written a fine article about Agnesi's book, with photos of it, "The Witch of Agnesi, a Lasting Contribution from the First Surviving Mathematical Work Written By a Woman," in *The College Mathematics Journal*, vol. 30, no. 4 (Sept. 1999), pp. 258–268. The "Witch" refers to a curve made famous by the mistranslation of one word in her text. The article's title is sobering. Before 1748, only mathematics written by men exists (and some of that as if by miracle). Where would we be now if fully half of humanity had not been so dissuaded from contributing to the edifice of mathematics?

In calculus I, it is observed that at $\langle 0, 1\rangle$, the tangent slope of $y = 2^x$ is much less than 1, while the tangent slope of $y = 3^x$ is a bit greater than 1. It seems reasonable that there be an exponential function $y = a^x$ whose derivative at $\langle 0, 1\rangle$ is exactly 1. As Leibniz discovered, $y = e^x$ is the function that fills the bill. However, we don't know how to find derivatives of exponential functions yet. It is therefore critical to prove that $\frac{de^x}{dx} = e^x$, for this is one of the most essential equations in science, engineering, and technology.

Theorem 37.5 The function $y = e^x$ is differentiable over its entire domain, and $\frac{de^x}{dx} = e^x$.

Proof. By definition,

$$\frac{de^x}{dx} = \lim_{h\to 0} \frac{e^{x+h} - e^x}{h} = e^x \lim_{h\to 0} \frac{e^h - 1}{h}.$$

Accordingly, we must prove that $\lim_{h\to 0} \frac{e^h-1}{h} = 1$. Begin with the inequality $1 + h \leq e^h$ for real h (Theorem 35.2). Substituting $-h$, we also have $1 - h \leq e^{-h}$. When $h > 0$, the first inequality yields $1 \leq \frac{e^h-1}{h}$ and the second inequality yields $\frac{e^h-1}{h} \leq e^h$. Let $h \to 0^+$ and apply the pinching theorem for functions to $1 \leq \frac{e^h-1}{h} \leq e^h$. Since $\lim_{h\to 0} e^h = 1$ (Lemma 35.1), we find that

$$\lim_{h\to 0+} \frac{e^h - 1}{h} = 1.$$

This implies that

$$\lim_{h\to 0-} \frac{e^{-h}-1}{-h} = 1.$$

Now,

$$1 = \left(\lim_{h\to 0-} e^h\right)\left(\lim_{h\to 0-} \frac{e^{-h}-1}{-h}\right) = \lim_{h\to 0-}\left(\frac{e^h}{1}\cdot\frac{e^{-h}-1}{-h}\right) = \lim_{h\to 0-}\frac{e^h-1}{h}.$$

Consequently, the full limit is $\lim_{h\to 0}\frac{e^h-1}{h} = 1$. The theorem's conclusion now follows. **QED**

Theorem 37.6 If $y = f(x)$ is a bijection, is differentiable at $x = a$, and $\left.\frac{df(x)}{dx}\right|_a \neq 0$, then f^{-1} is differentiable at $f(a) = b$, and

$$\left.\frac{df^{-1}(y)}{dy}\right|_b = \frac{1}{\left.\frac{df(x)}{dx}\right|_a}.$$

This is often contracted to $(f^{-1}(y))' = \frac{1}{f'(f^{-1}(y))}$.

Proof. Since f is a bijection, $x = f^{-1}(f(x))$ for all $x \in \mathcal{D}_f$. Taking derivatives of both sides, with the chain rule on the right-hand side,

$$1 = \frac{dx}{dx} = \frac{df^{-1}(f(x))}{dx} = \frac{df^{-1}(y)}{dy}\cdot\frac{df(x)}{dx}.$$

Because $\left.\frac{df(x)}{dx}\right|_a \neq 0$ as a premise, we infer that $\left.\frac{df^{-1}(y)}{dy}\right|_{f(a)}$ must exist, and the conclusion follows. **QED**

Example 37.2 The hyperbolic cosine $y = \cosh x$ is not a bijection until we restrict it to positive x. As shown in calculus, $\frac{d\cosh x}{dx} = \sinh x$, which is nonzero for positive x. Thus,

$$\left.\frac{d\cosh^{-1} y}{dy}\right|_b = \frac{1}{\sinh a}$$

at any value b given by $b = \cosh a$. For instance, at $a = \ln 3$, $\cosh(\ln 3) = \frac{5}{3}$, and its derivative there is $\sinh(\ln 3) = \frac{4}{3}$ (see Figure 37.3). Even if we didn't know the derivative formula for $x = \cosh^{-1} y$, we would still be able to evaluate

$$\left.\frac{d\cosh^{-1} y}{dy}\right|_{5/3} = \frac{1}{\left.\frac{d\cosh x}{dx}\right|_{\ln 3}} = \frac{1}{\sinh(\ln 3)} = \frac{3}{4}.$$

In Exercise 37.10, review your calculus skills by finding an expression for $\cosh^{-1} y$, and finding its derivative formula. ■

Corollary 37.1 For all $y \in \mathbb{R}$ except zero, $\frac{d\ln|y|}{dy} = \frac{1}{y}$.

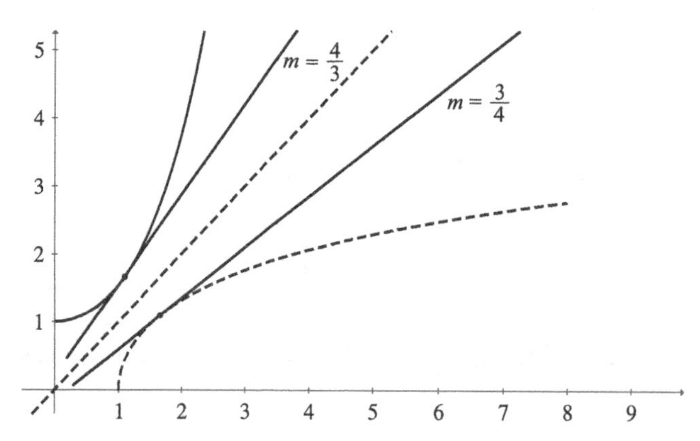

Figure 37.3. The tangent slope of $y = \cosh x$ (solid curve) at $\langle \ln 3, \frac{5}{3} \rangle$ is seen to be the reciprocal of the tangent slope of $y = \cosh^{-1} x$ (dashed curve) at the corresponding point $\langle \frac{5}{3}, \ln 3 \rangle$. The line of symmetry $y = x$ is also shown. The line passing through $\langle \ln 3, \frac{5}{3} \rangle$ and $\langle \frac{5}{3}, \ln 3 \rangle$ is, of course, perpendicular to $y = x$.

Proof. Write a proof as Exercise 37.5. **QED**

Note 3: Here is a classical proof of the corollary, found in Sec. 1755 of *Introduction a la Science de l'Ingénieur, Partie Théorique* (Introduction to the Science of the Engineer, Theoretical Part), sixth ed., by J. Joseph Claudel in 1875. He begins with $\Delta y = \ln(x + \Delta x) - \ln x = \ln(1 + \frac{\Delta x}{x})$, with $x > 0$ (and of course, $x + \Delta x > 0$). Then, in $\frac{\Delta y}{\Delta x} = \frac{1}{\Delta x} \ln(1 + \frac{\Delta x}{x})$, he substitutes $\Delta x = \frac{x}{m}$, where $m \in \mathbb{N}$. The expression becomes

$$\frac{\Delta y}{\Delta x} = \frac{m}{x} \ln\left(1 + \frac{1}{m}\right) = \frac{1}{x} \ln\left(1 + \frac{1}{m}\right)^m.$$

As $m \to \infty$, he writes that $\Delta x \to dx$, and arrives at

$$\frac{dy}{dx} = \lim_{\Delta x \to 0} \frac{\Delta y}{\Delta x} = \lim_{m \to \infty} \frac{1}{x} \ln\left(1 + \frac{1}{m}\right)^m = \frac{1}{x} \ln e = \frac{1}{x}.$$

Some flags of caution should be raised with this elegant proof. First, the continuity of $y = \ln x$ should be stated, since it was appealed to (where?). Second, although one commonly writes "$\Delta x \to dx$," this is not analytically meaningful (how big is dx?). What Claudel actually has is that $\Delta x = \frac{x}{m} \to 0^+$ as $m \to \infty$. Hence, he has found $\lim_{\Delta x \to 0^+} \frac{\Delta y}{\Delta x}$, rather than the full limit. This objection is easily answered by considering what happens as $m \to -\infty$.

However, yet more analytical care is needed. If we knew beforehand that $y = \ln x$ is differentiable, then the limit as $m \to \infty$ will yield $\frac{dy}{dx}$. Lacking that fact, a loophole opens. Setting $\Delta x = \frac{x}{m}$ actually defines a sequence $\Delta x_m = \frac{x}{m}$ that gives the sequence of slopes $\frac{\Delta y_m}{\Delta x_m} = \frac{1}{x} \ln(1 + \frac{1}{m})^m$. While it is true that $\lim_{m \to \infty} \frac{\Delta y_m}{\Delta x_m} = \frac{1}{x}$, we must guarantee that all sequences of points $\langle x + \Delta x_n, \ln(x + \Delta x_n) \rangle \to \langle x, \ln x \rangle$ will yield $\lim_{n \to \infty} \frac{\Delta y_n}{\Delta x_n} = \frac{1}{x}$.

As a counterexample to see what can happen, remember $f(x) = x \sin \frac{1}{x}$ in Example 36.4. Selecting $\langle \Delta x_m \rangle_1^\infty$ as the specific sequence of solutions to $x \sin \frac{1}{x} = x$ will give $\lim_{m \to \infty} \frac{\Delta y_m}{\Delta x_m} = 1$ as we approach $\langle 0, 0 \rangle$ from the right. But Example 36.4 shows that $f'(0)$ does not exist. The problem vanishes when we note that other sequences $\langle \Delta x_n \rangle_1^\infty$ approaching zero will yield different values for $\lim_{n \to \infty} \frac{\Delta y_n}{\Delta x_n}$ (or none at all). We would need to apply what might be called a sequential criterion for differentiability corresponding to the sequential criterion for continuity in Chapter 25.

A strategy that bypasses the problem with sequences is to redefine $\Delta x = \frac{x}{u}$, with u a nonzero real number. Using Claudel's steps, we now have

$$\lim_{\Delta x \to 0+} \frac{\Delta y}{\Delta x} = \lim_{u \to \infty} \frac{1}{x} \ln\left(1 + \frac{1}{u}\right)^u = \frac{1}{x} \ln e = \frac{1}{x},$$

where Exercise 35.5b and the continuity of $y = \ln x$ are used. Likewise,

$$\lim_{\Delta x \to 0-} \frac{\Delta y}{\Delta x} = \lim_{u \to -\infty} \frac{1}{x} \ln\left(1 + \frac{1}{u}\right)^u = \frac{1}{x}.$$

Therefore,

$$\frac{dy}{dx} = \lim_{\Delta x \to 0} \frac{\Delta y}{\Delta x} = \frac{1}{x}.$$

Of course, if $y \in \mathbb{R}^+$, Corollary 37.1 simplifies to $\frac{d \ln y}{dy} = \frac{1}{y}$. We are now able to handle exponential functions with any base $b > 0$. Appealing to the identity $b^x = \exp(\ln b^x)$, we derive that

$$\frac{db^x}{dx} = \exp(\ln b^x) \cdot \frac{d \ln b^x}{dx} = b^x \frac{d(x \ln b)}{dx} = b^x \ln b,$$

a result that should look familiar.

It is fairly easy to prove the power rule $\frac{dx^n}{dx} = nx^{n-1}$ for $n \in \mathbb{N}$ (do it by two methods, in Exercise 37.4). Proving it to be true for any rational number exponent takes more, and more careful, work. We now divulge what mathematician Peter Hilton (1923–2010) winkingly called "the principle of rational greed." This tenet holds that if a broader mathematical goal seems reasonably attainable, then one should go for it, since achieving it often takes no more effort than the lesser goal. Accordingly, we will prove the power rule for *all* real exponents α.

Theorem 37.7 The Power Rule If $f(x) = x^\alpha$ with domain $\mathcal{D}_f = \mathbb{R}^+$ and $\alpha \in \mathbb{R}$, then $\frac{dx^\alpha}{dx} = \alpha x^{\alpha - 1}$.

Proof. Using $x^\alpha = \exp \ln x^\alpha = \exp(\alpha \ln x)$, we find

$$\frac{dx^\alpha}{dx} = \exp(\alpha \ln x) \cdot \frac{d(\alpha \ln x)}{dx} = \exp(\alpha \ln x) \cdot \frac{\alpha}{x} = x^\alpha \frac{\alpha}{x} = \alpha x^{\alpha - 1}$$

by Theorem 37.5, the chain rule, and Corollary 37.1. **QED**

Whenever it is possible for x^α to be real over a negative portion of the X axis, then we may extend the domain of f accordingly (graphing calculators sometimes need to be reminded of this). For instance, $f(x) = x^{1/3}$ has $\mathcal{D}_f = \mathbb{R}$, and its derivative therefore exists for negative abscissas (see Example 36.3).

Example 37.3 Find the derivative of the **coupled exponential** class of functions $f(x) = u(x)^{v(x)}$, where $u(x) > 0$. We apply Theorem 37.7 and Corollary 37.1. Thus, since $u(x)^{v(x)} > 0$, we have

$\ln f(x) = v(x)\ln u(x)$, and we find the derivative using the chain rule:

$$\frac{f'(x)}{f(x)} = v(x)\frac{u'(x)}{u(x)} + v'(x)\ln u(x).$$

Therefore, we have a general derivative formula

$$\frac{du(x)^{v(x)}}{dx} = u(x)^{v(x)}\left(v(x)\frac{u'(x)}{u(x)} + v'(x)\ln u(x)\right).$$

Notice that this formula encompasses both the power rule (when v is constant) and the derivative of an exponential function (when u is constant). Regarding this, do Exercise 37.9. ■

37.1 Exercises

1. Prove that the derivative of an even function is an odd function (as seen in the graphs of Example 37.1), and that the derivative of an odd function is an even function.

2. Use the identity $|x| = x \operatorname{sgn} x$ and the product rule to once again prove that $\frac{d|x|}{dx} = \frac{1}{\operatorname{sgn} x}$. Is it true that $\frac{d|x|}{dx} = \frac{|x|}{x}$?

3. Prove the quotient rule using theorems prior to it in this chapter, as Maria Agnesi did.

4. Prove that $\frac{dx^n}{dx} = nx^{n-1}$, where $n \in \mathbb{N}$, by the binomial theorem (Exercise 4.7). Prove it again, by using induction.

5. Prove Corollary 37.1. Hint: Use Theorems 37.5 and 37.6, and the result of Exercise 37.2.

6. Prove the special case of Theorem 37.7 when $\alpha = \frac{m}{n} \in \mathbb{Q}$ reduced to lowest terms. Be careful of domains: if n is even, follow the theorem and use $\mathbb{R}^+$ as the domain. If n is odd, how can we justify using Theorem 37.7 over all $\mathbb{R}\backslash\{0\}$ as the domain?

7. The function $f(x) = |\sin x|$ is graphed in Figure 37.4. Find and graph $f'(x) = \frac{d|\sin x|}{dx}$.

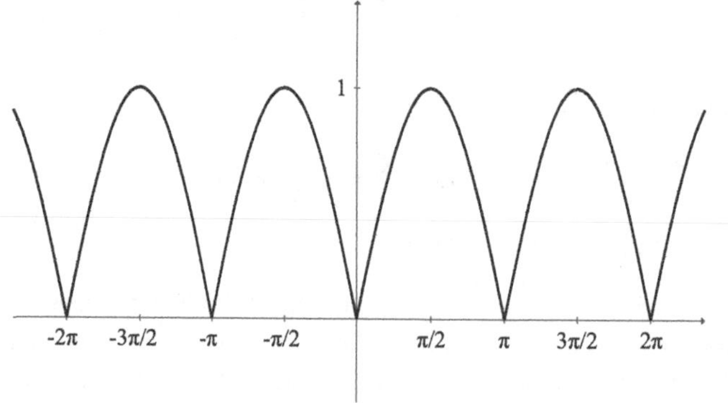

Figure 37.4.

8. Let $Q(x) = (x-3)^{3/4}$, which is a bijection over its domain $\{x \geq 3\}$. Evaluate $.\frac{dQ^{-1}}{dy}|_8$ in two ways: by Theorem 37.6, and by finding the inverse algebraically.

9. Use the formula

$$\frac{du(x)^{v(x)}}{dx} = u(x)^{v(x)}\left(v(x)\frac{u'(x)}{u(x)} + v'(x)\ln u(x)\right)$$

to find the derivatives of the following functions:

a. $y = x^a (x > 0)$ b. $y = b^x (b > 0)$

c. $y = x^x (x > 0)$ d. $y = |\sin x|^{\exp x}$

The results of **a** and **b** should be familiar! Graph the interesting function in **c** and find its absolute minimum point, then give your calculator a workout by having it graph the strange function in **d**.

10. Recall that $y = \cosh x = \frac{1}{2}(e^x + e^{-x})$ with $\mathcal{D}_{\cosh} = \mathbb{R}$, and that it is not one-to-one. We therefore restrict it to the domain $\{x \geq 0\}$, which creates a closely related function that we shall name $y = \text{Cosh}\, x$.

a. Solve $y = \frac{1}{2}(e^x + e^{-x})$ for x. Notice that you get two values of x for each y, but only one value is valid, since Cosh is a bijection. Call this function $x = \text{Cosh}^{-1} y$. Hint: $2ye^x = e^{2x} + 1$ is quadratic in form.

b. Use the expression from **a** to find $\frac{d\,\text{Cosh}^{-1} y}{dy}$, and show that it is, in fact, equal to $\frac{1}{\sinh x}$, where you must substitute $x = \text{Cosh}^{-1} y$ and simplify a lot. As a check, graph your expression for $\frac{d\text{Cosh}^{-1} y}{dy}$ and the function $y = \frac{1}{\sinh(\cosh^{-1} y)}$ over $y > 1$. Are the curves the same? Hint: You will need $\sinh u = \frac{1}{2}(e^u - e^{-u})$.

11. a. Prove an extension of the product rule, $[f_1(x)f_2(x)f_3(x)]' = f_1'(x)f_2(x)f_3(x) + f_1(x)f_2'(x)f_3(x) + f_1(x)f_2(x)f_3'(x)$.

b. Now extend **a** to any number of factors, by establishing this derivative formula:

$$\frac{d\prod_{i=1}^{n} f_i(x)}{dx} = \sum_{i=1}^{n}(f_i'(x)\prod_{j\neq i} f_j(x)).$$

38

Other Derivatives

Leibniz had by the early 1690s discovered most of the ideas present in current calculus texts, but had never written out a complete, coherent treatment of the material.

Victor J. Katz [5], pg. 481.

We have frequently investigated the limit of a function from the right or from the left, $\lim_{x\to a+} f(x)$ or $\lim_{x\to a-} f(x)$, beginning in Chapter 23. When these concepts are applied to derivatives, several new types emerge. We will discuss two pairs of types, followed by a fifth (quite distinct) type, and we will take a look at derivatives of higher order.

38.1 The Derivative From the Right and From the Left

If, in the definition of the derivative of f at a point $x \in \mathcal{D}_f$, we consider the limit from the right, namely

$$\lim_{h\to 0+} \frac{f(x+h)-f(x)}{h},$$

then we are restricted to secants drawn toward points to the right of $\langle x, f(x)\rangle$, that is, we are now considering only $[x, b) \subseteq \mathcal{D}_f$ (for some $b > x$). If the indicated limit of secant slopes exists, it is called the **derivative from the right** of f. As there is no standard symbol for this type of derivative, let's denote it by $f'^R(x)$. If we consider the limit from the left, namely

$$\lim_{h\to 0-} \frac{f(x+h)-f(x)}{h},$$

then we restrict ourselves to secants drawn toward points to the left of $\langle x, f(x)\rangle$, that is, we consider only $(a, x] \subseteq \mathcal{D}_f$ (for some $a < x$). If the indicated limit of secant slopes exists, it is called the **derivative from the left** of f. Denote it by $f'^L(x)$.

Figures 38.1–38.3 illustrate three situations involving derivatives from the right and from the left at $x = 2$, and continuity there.

Figure 38.1 demonstrates that a function can have wildly different derivatives from the right and left and still be continuous at a point. The figure also shows that the derivative at $x = 2$ fails to exist. Figure 38.2 hints at something better: equality of the derivatives from the right and

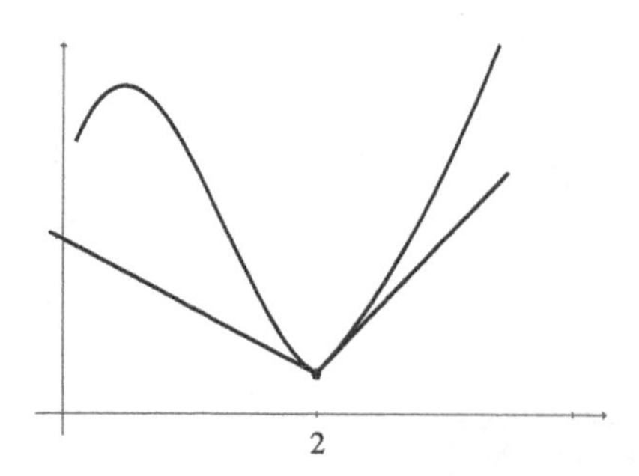

Figure 38.1.

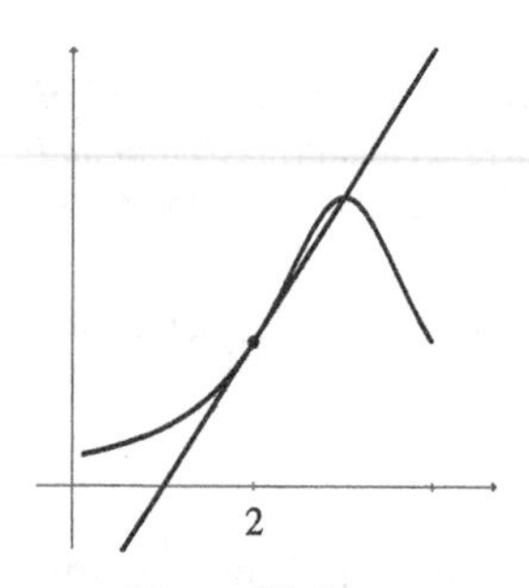

Figure 38.2.

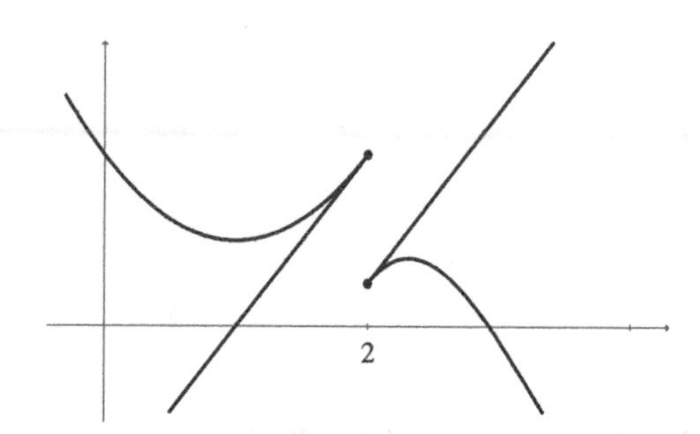

Figure 38.3.

left implies differentiability at a point. Notice carefully that if we decide to curtail the domain of the function in Figure 38.2 to $x \leq 2$, then $f'^L(2)$ would be unaffected, but the derivative would cease to exist there. This is because the derivative requires *neighborhoods* of $x = 2$ to be contained in the domain.

Figure 38.3 seems to show that having equal derivatives from the right and left doesn't guarantee continuity, but this figure is an example of how deceptive a graph can be! If both the derivative from the right and from the left were to exist at $x = 2$ as shown, then $f(2)$ would be double-valued. You are asked to correct the figure as Exercise 38.1.

Example 38.1 The function $v(x) = 3|x - 2|$ has no derivative at $x = 2$ (check its graph). But

$$v'^R(2) = \lim_{h \to 0+} \frac{3|2 + h - 2| - 0}{h} = \lim_{h \to 0+} \frac{3|h|}{h} = 3,$$

which is supported by visual evidence. Likewise,

$$v'^L(2) = \lim_{h \to 0-} \frac{3|2 + h - 2| - 0}{h} = \lim_{h \to 0-} \frac{3|h|}{h} = -3. \quad ■$$

Example 38.2 Return to Example 37.1, where we found that for

$$f(x) = \sin|x|, \qquad \left.\frac{d \sin|x|}{dx}\right|_0$$

didn't exist. Nevertheless,

$$f'^R(0) = \lim_{h \to 0+} \frac{\sin|0+h| - \sin|0|}{h} = 1, \quad \text{and} \quad f'^L(0) = \lim_{h \to 0-} \frac{\sin|0+h| - \sin|0|}{h} = -1.$$

■

The special case shown in Figure 38.2 leads us to the main theorem connecting the derivatives from the right and left, and the full derivative.

Theorem 38.1 A function f is differentiable at x if and only if $f'^R(x) = f'^L(x)$ (which carries the implication that both exist). In such cases, $f'(x) = f'^R(x) = f'^L(x)$.

Proof of Sufficiency. Given that

$$\lim_{h \to 0+} \frac{f(x+h) - f(x)}{h} \quad \text{and} \quad \lim_{h \to 0-} \frac{f(x+h) - f(x)}{h}$$

exist and are equal, the full limit $\lim_{h \to 0} \frac{f(x+h)-f(x)}{h}$ exists and equals the others, by Exercise 23.9. Hence, $f'(x) = f'^R(x) = f'^L(x)$.

Proof of Necessity. Let $f'(x)$ exist, meaning that

$$\lim_{h \to 0} \frac{f(x+h) - f(x)}{h}$$

exists. Then Exercise 23.9 implies that both $f'^R(x)$ and $f'^L(x)$ exist, and $f'^R(x) = f'^L(x) = f'(x)$. **QED**

Example 38.3 Reconsider $g(x) = e^{1/x}$ in Example 23.5, but extend the domain as follows:

$$g(x) = \begin{cases} e^{1/x} & \text{if } x \neq 0 \\ 0 & \text{if } x = 0 \end{cases}.$$

Using the analysis there, $g'(0)$ remains non-existent. But the graph in the example gives visual support that $g'^L(0) = 0$. Using the definition,

$$g'^L(0) = \lim_{h \to 0-} \frac{g(0+h) - g(0)}{h} = \lim_{h \to 0-} \frac{e^{1/h} - 0}{h} = 0,$$

evaluated in Exercise 23.15 using the pinching theorem for functions. And since $\frac{e^{1/h}}{h} > \frac{1}{h}$ for $h > 0$, it follows that

$$g'^R(0) = \lim_{h \to 0+} \frac{g(0+h) - g(0)}{h} = \lim_{h \to 0+} \frac{e^{1/h} - 0}{h}$$

doesn't exist. ■

Now, prove the lemma in Exercise 38.4, which is illustrated in Figure 38.1.

38.2 The Right-hand and Left-hand Derivative's Limit

The graph of $f(x) = \operatorname{sgn} x$ shows us that none of $f'(0)$, $f'^R(0)$, and $f'^L(0)$ exist for the signum function. Yet, its graph is completely flat in every deleted neighborhood of zero. It seems that another definition is needed to describe this behavior. The **right-hand derivative's limit** of f at a point x such that an open interval $(x, b) \subset \mathcal{D}_f$, is defined as $\lim_{\ell \to 0+} f'(x + \ell)$ (if it exists), and we shall denote it by $f'^+(x)$. Likewise, the **left-hand derivative's limit** of f at a point x such that $(a, x) \subset \mathcal{D}_f$ is defined as $\lim_{\ell \to 0-} f'(x + \ell)$ (if it exists), and we shall denote it by $f'^-(x)$. Accordingly, these two quantities are limits of *derivatives* of f as we approach the point x from one or the other direction. Note the order of operations required to find $f'^+(x)$: first, find the derivative at points to the right of x, and second, find the limit of those derivatives as we approach x from the right. The same order of operations is required for $f'^-(x)$, with "left" replacing "right." As a consequence, neither $f(x)$ nor $f'(x)$ need to exist at x.

Note 1: Expanded names for these derivatives are, respectively, the *right-hand limit of the derivative*, and the *left-hand limit of the derivative*.

Example 38.4 It is satisfying that for our showcase function $f(x) = \operatorname{sgn} x$, the right-hand derivative's limit exists and equals zero, as does the left-hand derivative's limit. To be precise,

$$f'^+(0) = \lim_{\ell \to 0+} f'(0 + \ell) = \lim_{\ell \to 0+} \frac{d \operatorname{sgn} \ell}{dx} = \lim_{\ell \to 0+} \frac{d1}{dx} = 0.$$

Similarly,

$$f'^-(0) = \lim_{\ell \to 0-} f'(0 + \ell) = \lim_{\ell \to 0-} \frac{d \operatorname{sgn} \ell}{dx} = \lim_{\ell \to 0-} \frac{d(-1)}{dx} = 0. \quad \blacksquare$$

Example 38.5 Recall

$$f(x) = \begin{cases} \frac{x^4 - 2x^3 - 3x + 6}{x - 2} & \text{if } x \neq 2 \\ 10 & \text{if } x = 2 \end{cases},$$

with domain $\mathbb{R}$, graphed in Example 23.2. Due to the (removable) discontinuity at $x = 2$, $f'(2)$ does not exist, nor do $f'^R(2)$ and $f'^L(2)$. However,

$$f'^+(2) = \lim_{\ell \to 0+} \left. \frac{df(x)}{dx} \right|_{2+\ell} = \lim_{\ell \to 0+} 3(2 + \ell)^2 = 12,$$

just as if the removable discontinuity had been filled in with the point $\langle 2, 5 \rangle$. Similarly,

$$f'^-(2) = \lim_{\ell \to 0-} 3(2 + \ell)^2 = 12.$$

Once again, the order of operations required to find these derivatives is first to find the derivative at points on one side or the other of $x = 2$, and second to find the derivative's limit as $x \to 2$ from one or the other direction. $\blacksquare$

Example 38.6 An interesting situation arises with function

$$f(x) = \begin{cases} x^2 \sin \frac{1}{x} & \text{if } x \neq 0 \\ 0 & \text{if } x = 0 \end{cases},$$

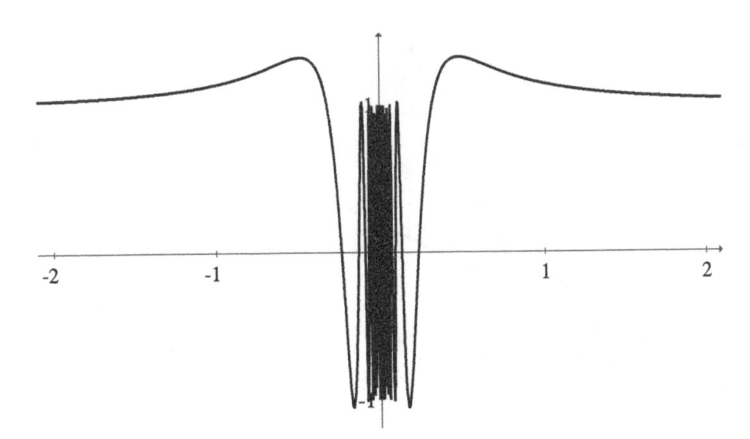

Figure 38.4. The graph of f'. Here, both $f'^{+}(0)$ and $f'^{-}(0)$ fail to exist, and f' is discontinuous at $x = 0$. But $f'(0) = 0$ exists.

graphed in Exercise 36.6. We saw there that f is differentiable at zero (and everywhere else):

$$f'(x) = \begin{cases} 2x \sin \frac{1}{x} - \cos \frac{1}{x} & \text{if } x \neq 0 \\ 0 & \text{if } x = 0 \end{cases}.$$

But in Exercise 36.7, we found that this derivative was not continuous at zero. To better understand this, consider that

$$f'^{+}(0) = \lim_{\ell \to 0+} \left(2\ell \sin \frac{1}{\ell} - \cos \frac{1}{\ell}\right)$$

fails to exist by oscillation. For the same reason, $f'^{-}(0)$ doesn't exist. The graph of f' in Figure 38.4 shows this behavior well. Now do Exercises 38.5 and 38.6. ■

38.3 The Symmetric Derivative

Figure 38.5 shows the tangent line of a function f at a point $\langle x, f(x)\rangle$, along with a typical secant drawn from it. The secant slope $Q_x(h) = \frac{f(x+h)-f(x)}{h}$ is very different from the tangent slope. This subsection is inspired by Figure 38.6, which shows that the tangent slope at x (that is, $f'(x)$), may be closely approximated by a secant's slope when drawn between two points *symmetrically* straddling x, at least for functions that are smooth in some sense.

Applying analytic geometry to this observation, the function $S_x(h) = \frac{f(x+h)-f(x-h)}{2h}$ gives us the secant slope of f at x in the symmetric or balanced Figure 38.6. This leads us directly to the definition of the **symmetric derivative** of f at a point x such that in some deleted neighborhood $\mathbf{D}_x^{\delta} \subset \mathcal{D}_f$:

$$f'^{S}(x) \equiv \lim_{h \to 0} \frac{f(x+h) - f(x-h)}{2h},$$

if this limit exists. As our first example will show, there is a significant difference between the symmetric and the standard derivative.

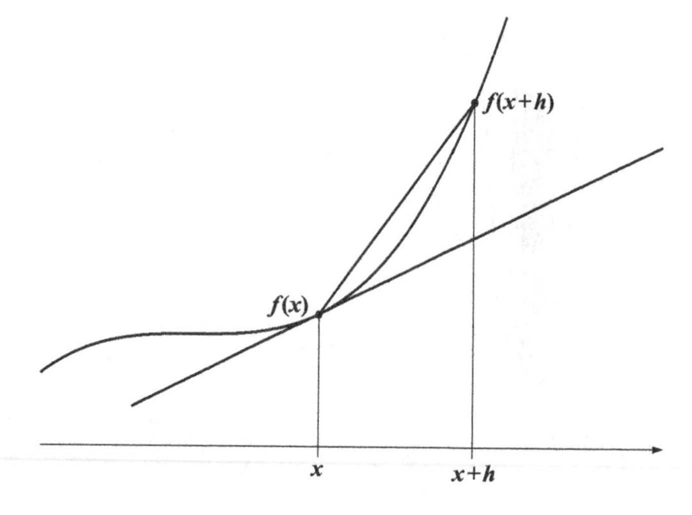

Figure 38.5.

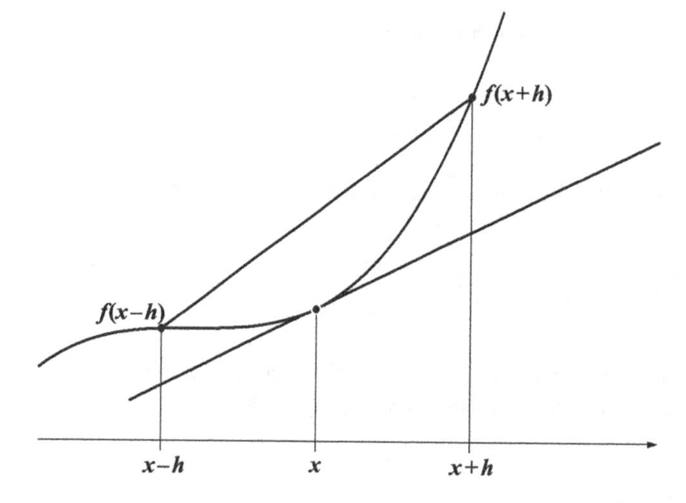

Figure 38.6.

Example 38.7 Consider $y = |x - 5|$, which isn't differentiable at $x = 5$. Contrast this with its symmetric derivative evaluated at $x = 5$:

$$y'^S(5) = \lim_{h \to 0} \frac{|(5+h)-5| - |(5-h)-5|}{2h} = 0.$$

This result becomes evident when you sketch the function and any secant between $x_1 = 5 - h$ and $x_2 = 5 + h$, for nonzero h. ■

As in the previous subsection, it isn't necessary for $f(x)$ or $f'(x)$ to exist when $f'^S(x)$ exists. On the other hand, in Exercise 38.10, you are asked to prove that if $f'(x)$ does exist, then so does $f'^S(x)$, and the two derivatives are equal.

Example 38.8 Compare the values of

$$Q_x(h), S_x(h), \quad \text{and} \quad \frac{dg}{dx} \quad \text{for} \quad g(x) = \ln x - x$$

at $x = 0.5$, using $h = 0.2$. We have

$$Q_{0.5}(0.2) = \frac{(\ln 0.7 - 0.7) - (\ln 0.5 - 0.5)}{0.2} \approx 0.6824$$

$$S_{0.5}(0.2) = \frac{(\ln 0.7 - 0.7) - (\ln 0.3 - 0.3)}{0.4} \approx 1.1182,$$

and $g'(0.5) = 1$. How good are the approximations of the true slope? The relative errors are

$$\frac{Q_{0.5}(0.2) - g'(0.5)}{g'(0.5)} \approx -0.32 \quad \text{and} \quad \frac{S_{0.5}(0.2) - g'(0.5)}{g'(0.5)} \approx 0.12.$$

It isn't surprising that when appropriate, $S_x(h)$ is the preferred estimate for the instantaneous rate of change of a function that is not known analytically. See Exercise 38.9. ■

Note 2: Some other types of derivatives are also recognized. For more information, see [18], pg. 89, and [33], pp. 318–319.

38.4 Derivatives of Higher Order

Derivatives of higher order are studied in calculus I. For later reference, we define them here as well. Considering $f'(x)$ as a function in its own right, the **second derivative** of f at a point x such that some neighborhood $\mathbf{N}_x^\delta \subseteq \mathcal{D}_{f'}$, is

$$f''(x) = \lim_{h \to 0} \frac{f'(x+h) - f'(x)}{h},$$

if the limit exists. Thus, the second derivative is to be found at a point in a neighborhood where f' exists.

We define successively higher order derivatives provided all previous derivatives exist by recursion. A commonly used notation is $f^{(n)}$, $n \in \mathbb{N} \cup \{0\}$, so that $f^{(0)} \equiv f$, $f^{(1)} \equiv f'$, and so on. Accordingly, for $n \in \mathbb{N}$ and $x \in \mathbf{N}_x^\delta \subseteq \mathcal{D}_{f^{(n-1)}}$, the n **th derivative** of f at x is

$$f^{(n)}(x) = \lim_{h \to 0} \frac{f^{(n-1)}(x+h) - f^{(n-1)}(x)}{h},$$

whenever the limit exists. Leibniz's notation $\frac{d^n f(x)}{dx^n}$ for $f^{(n)}(x)$ is also very common.

Example 38.9 Two curved sections of train track are to meet at a point. In the plan of the tracks shown in Figure 38.7, the junction is at $\langle 0, 1 \rangle$, where a circular section given by $u(x) = \sqrt{1 - x^2}$ ($\mathcal{D}_u = [-1, 0]$) meets a section given by $w(x) = \left(1 + x^2\right)^{-1}$ ($\mathcal{D}_w = [0, 1]$).

To prevent a catastrophe, the sections must join smoothly. As we can see, the first derivatives of the sections are equal at the junction: $u'(0) = w'(0) = 0$. But as model railroad aficionados know, this prevents derailment only at slow speeds. It turns out that the second derivatives must also agree, but $u''(0) = -1$, and $w''(0) = -2$ (verify both). Thus, the junction isn't safe at higher speeds.

Suppose that the circular section must be kept; we then require that $w''(0) = -1$. Given no further information about w'', the simplest function to use is $w''(x) = -1$. Of course, $w'(0) = 0$, and the plan suggests that we take $w(1) = \frac{1}{2}$. By calculus I methods we soon determine that $w(x) = -\frac{1}{2}x^2 + 1$ is a better curve for the track. Now, consider Exercise 38.11. ■

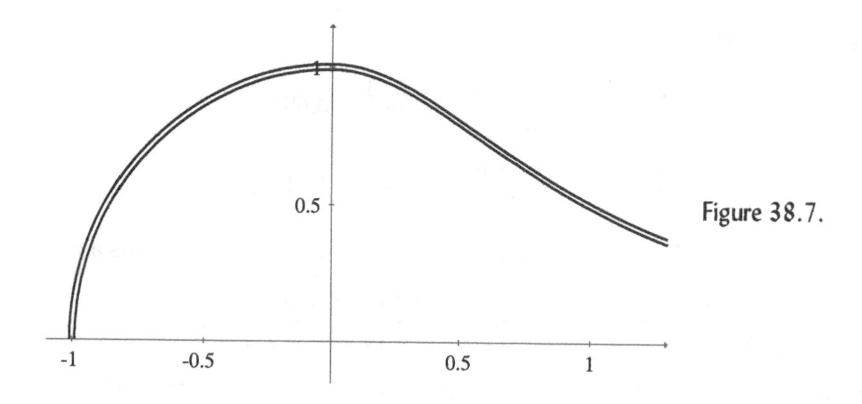

Figure 38.7.

Note 3: Once again, good notation allows leaps of imagination. We have $f^{(0)} = f$, and $f^{(1)} = f'$, and so on for integers $n = 0, 1, 2, \ldots$. And if $n = -1$, isn't it reasonable that $f^{(-1)}(x) = \int f(x)\,dx$? But what would happen if, say, $n = \frac{1}{2}$? What could $\frac{d^{1/2} f(x)}{dx^{1/2}}$ possibly mean? It may surprise you that this line of thought goes back perhaps two centuries. An excellent introduction to this subject is "A Child's Garden of Fractional Derivatives" by Marcia Kleinz and Thomas J. Osler in *The College Mathematics Journal*, vol. 31, no. 2 (March 2000), pp. 82–88.

38.5 Exercises

1. Correct the deceptive Figure 38.3 in any way you can think of.
2. Recall that $f(x) = \operatorname{sgn} x$ has no derivative at $x = 0$. Prove that neither $f'^R(0)$ nor $f'^L(0)$ exist. What would happen if we redefined $\operatorname{sgn} x$ so that $\operatorname{sgn} 0 = 1$ (admittedly, an odd idea)?
3. The graph in Exercise 37.7 gives evidence that $f(x) = |\sin x|$ is not differentiable at $x = k\pi, k \in \mathbb{Z}$, and the solution proves it. However, prove that $f'^R(k\pi) = 1$ and $f'^L(k\pi) = -1$.
4. Figure 38.1 hints at the lemma, if $f'^R(x)$ and $f'^L(x)$ both exist, then f is continuous at x. Prove it. Is the converse true? If yes, prove it; if no, find a counterexample.
5. For the function

$$f(x) = \begin{cases} x^2 \sin \frac{1}{x} & \text{if } x \neq 0 \\ 0 & \text{if } x = 0 \end{cases}$$

in Example 38.6, evaluate $f'^R(0)$ and $f'^L(0)$, if they exist.

6. Show that for

$$w(x) = \begin{cases} x^3 \sin \frac{1}{x} & \text{if } x \neq 0 \\ 0 & \text{if } x = 0 \end{cases}, \; w'^L(0) = w'^R(0) = w'^-(0) = w'^+(0) = 0.$$

Hint: Shorten your work by thinking about how this and f in the previous exercise are related.

7. Answer True or False, and support your decisions:
   a. If $f'^{+}(x)$ exists, then $f'^{R}(x)$ exists.
   b. If $f'^{R}(x)$ exists, then $f'^{+}(x)$ exists.
   c. If $f'(x)$ exists, then $f'^{+}(x)$ and $f'^{-}(x)$ exist and are equal.
   d. If $f'^{+}(x)$ and $f'^{-}(x)$ exist and are equal, then $f'(x)$ exists. If true, c and d together would form an analogue of Theorem 38.1.

8. Prove that $\lim_{h\to 0+} \frac{f(x)-f(x-h)}{h} = f'^{L}(x)$.

9. a. Suppose that the numbers of bacteria in a culture are charted as follows:

hours	2.45	5.89	9.33
number	177	234	312

   A researcher determines the average rate of growth as $\frac{312-177}{9.33-2.45} = 19.6$ bacteria/hr. Convince yourself that this is the value of the symmetric secant slope $S_{5.89}(h)$ at $x = 5.89$ hr by determining the value of h. Also evaluate the divided difference function $Q_{5.89}(h)$. What does $Q_{2.45}(h)$ represent? Which of the three ratios gives the best estimate of the instantaneous rate of growth at $x = 5.89$ hr ? Hint: Draw pictures, naturally.

   b. Evaluate $\frac{Q_{5.89}(h)+Q_{2.45}(h)}{2}$, and comment.

10. Prove that if $f'(x)$ exists, then $f'^{S}(x)$ exists and $f'^{S}(x) = f'(x)$. Hint:

$$\frac{f(x+h) - f(x) + f(x) - f(x-h)}{2h} = \frac{f(x+h) - f(x-h)}{2h}$$

and the left side has a limit as $h \to 0$. Also, see Exercise 36.11.

11. In Example 38.9, good engineering requires that $u^{(3)}(0) = w^{(3)}(0)$. Do the functions

$$u(x) = \sqrt{1-x^2} \quad \text{and} \quad w(x) = -\frac{1}{2}x^2 + 1$$

satisfy this requirement?

12. a. Prove that in terms of the original function f, the second derivative can be found as

$$f''(x) = \lim_{h\to 0} \frac{f(x+2h) - 2f(x+h) + f(x)}{h^2},$$

   if this limit exists and if f is differentiable over an interval containing x. Thus, the existence of this limit is only a necessary condition for the existence of the second derivative. Hint: In the definition of $f''(x)$, express the numerator's derivatives as limits themselves.

   b. Apply the limit in **a** to functions (i) $p(x) = 5x^2$, (ii) $y = e^x$, (iii) $q(x) = |x|$ at $x = 0$, and (iv) $g(x) = x\,|x|$ (be careful!).

39

The Mean Value Theorem

What's the best part of being a mathematician? I'm not a religious man, but it's almost like being in touch with God when you're thinking about mathematics.

Paul Halmos (1919–2006), in *In Touch with God: An Interview with Paul Halmos*, by Don Albers, in The College Mathematics Journal, vol. 35, no. 1 (Jan. 2004), pp. 2–3.

The extreme value theorem (Theorem 26.4) has far-reaching consequences in real analysis. For quick reference, the theorem states that if $f : \mathbb{R} \to \mathbb{R}$ is continuous over $[a, b] \subseteq \mathcal{D}_f$, then f attains both a minimum m and a maximum M over the interval. It's helpful (but hardly rigorous) to recall Euler's rough sense of continuity as the ability to graph f without lifting pencil from paper. With that image in mind, the extreme value theorem is pretty obvious. Keeping the image in mind, the following lemma blends the extreme value theorem with the derivative.

Lemma 39.1 Let $f : \mathbb{R} \to \mathbb{R}$ be continuous over $[a, b] \subseteq \mathcal{D}_f$ and differentiable over (a, b), and let f attain either its maximum or its minimum at $\xi \in (a, b)$. Then $f'(\xi) = 0$.

Proof. Considering first the case of the maximum, the extreme value theorem states that $f(\xi) = M$, the maximum of f over $[a, b]$. The divided difference function

$$Q_\xi(h) = \frac{f(\xi + h) - f(\xi)}{h} = \frac{f(\xi + h) - M}{h}$$

may be evaluated for any h such that $\xi + h \in (a, b)$. When h is negative, $Q_\xi(h) \geq 0$, and when h is positive, $Q_\xi(h) \leq 0$. Therefore, $f'^L(\xi) \geq 0$ and $f'^R(\xi) \leq 0$. But since f is differentiable at ξ, Theorem 38.1 requires that $f'^R(\xi) = f'^L(\xi)$. This implies that $f'(\xi) = 0$.

You should provide a parallel proof for the case of the minimum, $f(\xi) = m$, as Exercise 39.2. **QED**

The next statement is sometimes called Darboux's intermediate value theorem for derivatives, after the analyst Gaston Darboux. The conclusion may be surprising in light of the fact that a derivative need not be continuous (recall Example 36.3 and Exercise 36.7).

Theorem 39.1 Let $f : \mathbb{R} \to \mathbb{R}$ be a differentiable function over $[a, b] \subseteq \mathcal{D}_f$. Then f' takes on every value between $f'^R(a)$ and $f'^L(b)$, inclusive. In other words, the derivative f' has the intermediate value property over $[a, b]$.

Proof. There is nothing to prove if $f'^R(a) = f'^L(b)$, so we may begin with the premise that $f'^R(a) < f'^L(b)$. We will prove first that for every y_0 such that $f'^R(a) < y_0 < f'^L(b)$, there exists a $\xi \in (a, b)$ such that $f'(\xi) = y_0$.

Define a function $g(x) = f(x) - y_0 x$ over $\mathcal{D}_g = [a, b]$. The equation $g'(x) = f'(x) - y_0$ shows that g is differentiable over $[a, b]$, so by Theorem 36.1, g is continuous there. The extreme value theorem now applies, indicating that g attains a minimum $g(x_m)$ at some point $x_m \in [a, b]$. Since $g'(a) = f'^R(a) - y_0 < 0$, g cannot reach its minimum at $x = a$. Furthermore, g cannot reach its minimum at $x = b$ either, since $g'(b) = f'^L(b) - y_0 > 0$. (Prove these two statements as a lemma in Exercise 39.9.) Consequently, $x_m \in (a, b)$. Then by Lemma 39.1, $g'(x_m) = f'(x_m) - y_0 = 0$. Since $y_0 = f'(x_m)$, x_m qualifies as ξ and proves that f' has the intermediate value property over (a, b).

This property easily extends to $[a, b]$, for if $y_0 = f'^R(a)$ or $y_0 = f'^L(b)$, then $\xi = a$ or $\xi = b$, respectively.

A parallel proof using the maximum value of g applies to the alternative premise that $f'^R(a) > f'^L(b)$. **QED**

The contrapositive of Theorem 39.1 reveals an important fact about derivative functions. If a function f does not have the intermediate value property over $[a, b]$, then f cannot be the derivative of a function F over the entire interval $[a, b]$. A simple example of this begins with the function

$$f(x) = \begin{cases} 1 & \text{if } x \geq 0 \\ -1 & \text{if } x < 0 \end{cases},$$

which doesn't have the intermediate value property over, say, $[-3, 3]$. A candidate for F that comes to mind is $F(x) = |x|$, but then we realize that $F'(0)$ does not exist, whereas $f(0)$ does. We may redefine f as

$$f(x) = \begin{cases} 1 & \text{if } x > 0 \\ -1 & \text{if } x \leq 0 \end{cases},$$

or even as $f(x) = \operatorname{sgn} x$, but an antiderivative F will never be found over $[-3, 3]$, its existence eradicated by the contrapositive of Theorem 39.1. To put this in the context of a later chapter, $y = \operatorname{sgn} x$ can be integrated over $[-3, 3]$, but it cannot be antidifferentiated there!

Our next theorem, well-known to calculus students and actually only a lemma today, was first published by Michel Rolle (1652–1719) in 1691, a time of youthful development for calculus.

Theorem 39.2 Rolle's Theorem If $f : \mathbb{R} \to \mathbb{R}$ is continuous over $[a, b] \subseteq \mathcal{D}_f$, differentiable over (a, b), and if $f(a) = f(b)$, then there exists at least one point $\xi \in (a, b)$ at which $f'(\xi) = 0$.

Proof. If f is constant throughout $[a, b]$, then $f'(\xi) = 0$ for all $\xi \in (a, b)$. If f is not constant, then it attains a maximum $f(\xi_1) = M$ and a minimum $f(\xi_2) = m$ in the interval (a, b) by the extreme value theorem. By Lemma 39.1, we have $f'(\xi_1) = 0$ and also $f'(\xi_2) = 0$. **QED**

Geometrically, Rolle's theorem is speaking about a horizontal secant through f, between points $\langle a, f(a)\rangle$ and $\langle b, f(a)\rangle$. The next—and very important—theorem states that Rolle's theorem can be generalized to secant lines other than horizontal ones, as in Figure 39.1, where the secant has slope $\frac{f(b)-f(a)}{b-a}$. This theorem has roots as far back as 15th-century India. It was brought to the front-and-center of analysis by the great French mathematician Joseph-Louis Lagrange (1736–1813) in 1797.

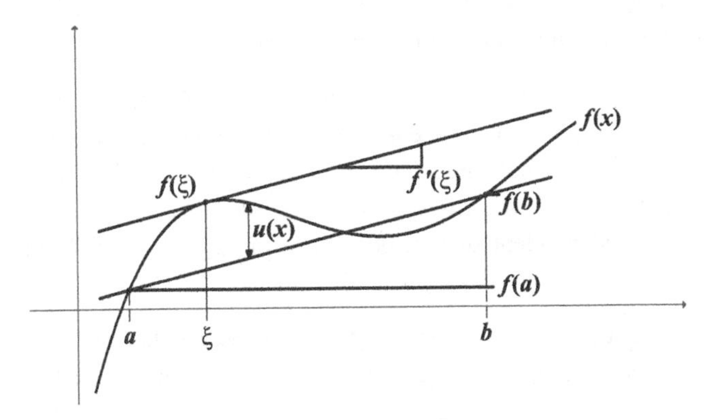

Figure 39.1. The derivative at ξ equals the slope of the secant over $[a, b]$.

Theorem 39.3 The Mean Value Theorem If $f : \mathbb{R} \to \mathbb{R}$ is continuous over $[a, b] \subseteq \mathcal{D}_f$ and differentiable over (a, b), then there exists at least one point $\xi \in (a, b)$ at which

$$f'(\xi) = \frac{f(b) - f(a)}{b - a}.$$

Proof. In Figure 39.1, a secant line with equation

$$y_{\text{sec}}(x) = \frac{f(b) - f(a)}{b - a}(x - a) + f(a)$$

has been drawn. For the purposes of this proof, we define a function $u : [a, b] \to \mathbb{R}$ that evaluates the vertical distance between f and the secant line:

$$u(x) \equiv f(x) - y_{\text{sec}}(x) = f(x) - \frac{f(b) - f(a)}{b - a}(x - a) - f(a).$$

By its definition, u is continuous over $[a, b]$, differentiable over (a, b), and $u(a) = u(b) = 0$ (all of which you should verify). Consequently, Rolle's theorem applies to u, telling us that there exists some point $\xi \in (a, b)$ at which $u'(\xi) = 0$. That is,

$$0 = u'(\xi) = f'(\xi) - y'_{\text{sec}}(\xi) = f'(\xi) - \frac{f(b) - f(a)}{b - a}.$$

The conclusion of the mean value theorem follows. **QED**

Two useful alternate forms of the mean value theorem now appear. For the function f as given in the theorem's premises, for every $x \in (a, b)$ there exists at least one $\xi \in (a, x)$ such that $f'(\xi) = \frac{f(x)-f(a)}{x-a}$. Hence,

$$f(x) = f(a) + f'(\xi)(x - a), \text{ for every } x \in (a, b) \subseteq \mathcal{D}_f \tag{1}$$

and some $\xi \in (a, x)$. Next, observe that by fixing endpoint b, the mean value theorem gives

$$f(x) = f(b) + f'(\xi)(x - b), \text{ for every } x \in (a, b) \subseteq \mathcal{D}_f \text{ and some } \xi \in (x, b). \quad (2)$$

In these equations, not much is known in general about the value of ξ besides the interval in which it lies, and that ξ usually varies with x.

A measure of how useful the preceding ideas are can be seen in the following examples.

Example 39.1 Since $f(x) = \sin x$ is both continuous and differentiable over $[0, \frac{\pi}{2}]$, (1) leads us to $\sin x = \sin 0 + (\cos \xi)(x - 0) = x \cos \xi$, where ξ varies with x subject to $0 < \xi < x$. However, $0 < \cos \xi < 1$, so we conclude that $\sin x < x$ over $(0, \frac{\pi}{2})$. This analytic demonstration reaffirms the geometric proof of $\sin \theta < \theta$ that was used in Lemma 23.1. (We must note that this result is not independent of that geometric proof, since $\frac{d \sin x}{dx} = \cos x$ depends on Lemma 23.1.) ■

Example 39.2 Prove that $\ln x \leq x - 1$ for all $x > 0$. We observe that $f(x) = \ln x$ is continuous and differentiable over $[1, \infty)$. By (1), for every $x \in (1, \infty)$, there exists some $\xi > 1$ such that

$$\ln x = \ln 1 + \frac{1}{\xi}(x - 1).$$

Hence, $\ln x < x - 1$ for $x > 1$.

Furthermore, $f(x) = \ln x$ is continuous and differentiable over $[x, 1]$, where $x > 0$. Thus, using (2) with $b = 1$, we have

$$\ln x = \ln 1 + \frac{1}{\xi}(x - 1),$$

where $\xi \in (x, 1)$. Since $\frac{1}{\xi} > 1$, we have $\ln x < x - 1$ again (verify). We conclude that $\ln x \leq x - 1$ for all $x > 0$, with equality only at $x = 1$. This was proved in a more roundabout way in Corollary 35.1. ■

Our last theorem may be thought of as an application of the mean value theorem to parametrically defined curves $\langle f(t), g(t) \rangle$. We will use it in Appendix E to demonstrate l'Hospital's rule. Notice that it simplifies to the mean value theorem when $g(x) = x$.

Theorem 39.4 Cauchy's Generalized Mean Value Theorem If $f, g : \mathbb{R} \to \mathbb{R}$ are continuous over $[a, b] \subseteq \mathcal{D}_f \cap \mathcal{D}_g$, differentiable over (a, b), and if $g'(x) \neq 0$ anywhere in (a, b), then there exists at least one point $\xi \in (a, b)$ at which

$$\frac{f'(\xi)}{g'(\xi)} = \frac{f(b) - f(a)}{g(b) - g(a)}.$$

Proof. To begin, it's given that the denominator $g'(\xi) \neq 0$ on the left side, but can $g(b) - g(a) = 0$ on the right side? No, because the mean value theorem applies to g over $[a, b]$, telling us that $g(b) - g(a) = g'(\xi)(b - a) \neq 0$. The rest of the proof makes a nice exercise (39.10) which is necessary for you to do to prepare for the truly ingenious proof of the upcoming Theorem 40.1. **QED**

39.1 Exercises

1. Use the mean value theorem to prove:
   a. There exists a number $\ell \in (1, b)$ for each $b > 1$ such that
   $$\frac{1}{3\ell^{2/3}} = \frac{\sqrt[3]{b} - 1}{b - 1}.$$
   When $b = 6$, find the value of ℓ to four decimal precision. Hint: Think about $y = x^{1/3}$.
   b. $|\sin x - \sin y| \leq |x - y|$ for all $x, y \in \mathbb{R}$.
   c. $|\cos x - \cos y| \leq |x - y|$ for all $x, y \in \mathbb{R}$.
   d. $1 + x \leq e^x$ for all $x \in \mathbb{R}$. Note how efficient this proof is as compared with that of Theorem 35.2, now that we know the derivative of $y = e^x$! Hint: Use $a = 0$ and $b = x$ in (1). Then use $a = x$ and $b = 0$ in (2).
   e. $\frac{x-1}{x} \leq \ln x$ for all $x > 0$. Together with Example 39.2, you will have shown that $\frac{x-1}{x} \leq \ln x \leq x - 1$ for all $x > 0$ (graph the functions forming the triple inequality!).

2. Prove the second case of Lemma 39.1, that is, where $f(\xi) = m$, the minimum of the function.

3. Use the mean value theorem's forms (1) and (2) to prove that
$$\sqrt{1 + x} \leq 1 + \frac{1}{2}x \quad \text{for all} \quad x \geq -1.$$

4. Suppose that two differentiable functions always differ by a constant, that is,
$$f_1(x) - f_2(x) = c \quad \text{for all} \quad x \in (a, b).$$
Prove that this occurs if and only if $f_1'(x) = f_2'(x)$.

5. Find a function for which the mean value theorem's conclusion cannot be satisfied.

6. a. Prove that if f is differentiable in a neighborhood of a, and for suitable h, there exists a number $\eta \in (0, 1)$ such that $f(a + h) = f(a) + hf'(a + \eta h)$. Hint: Let the final point in (1) be $a + h$ rather than x.
   b. Apply this form of the mean value theorem to $C(x) = \sqrt{4 - x^2}$ with domain $[-2, 2]$. Next, determine η to four decimal precision when $a = 1$ and $h = 0.1$. Check your value(s) of η in the formula in a.

7. Prove that even though the function $f(x) = \sqrt[3]{x}$ is not differentiable over $(-1, 1)$, the conclusion of the mean value theorem will be fulfilled. Is something logically wrong here? (Recall Exercise 1.10) Hint: The conclusion states that
$$\frac{b^{1/3} - a^{1/3}}{b - a} = \frac{1}{3\xi^{2/3}}.$$
Prove that this always has two solutions $\pm\xi_0$ when $b > 0$ and $a < 0$, which is the interesting case, since the mean value theorem applies for other cases of a and b. Now, can you prove that at least one of $-\xi_0, \xi_0$ belongs to (a, b)?

8. a. Recall Lemma 34.1, where after considerable work based on the arithmetic mean–geometric mean inequality, we proved the equivalent of

$$x^{\ell/m} \leq \frac{\ell}{m}x + \left(1 - \frac{\ell}{m}\right),$$

with $x \in \mathbb{R}^+$ (in that context), and where $\frac{\ell}{m}$ is a rational number in $(0, 1]$. Apply the mean value theorem's forms (1) and (2) to the function $f(x) = x^\alpha(\mathcal{D}_f = \{x \geq 0\})$ to arrive at $x^\alpha \leq \alpha x + (1 - \alpha)$ for *any* real $\alpha \in (0, 1]$. Draw a graphic example with $y = x^{0.7}$ and $y = 0.7x + 0.3$. Thus, the ability to differentiate x^α has greatly simplified a proof, and extended it to include irrational numbers! Hint: For (1), use $a = 1$, so that $\xi \in (1, x)$, meaning that $\xi > 1$. For (2), use $b = 1$, so that $\xi \in (x, 1)$, meaning that $0 < x < \xi < 1$.

b. Now prove that the inequality between x^α and $\alpha x + (1 - \alpha)$ is reversed when $\alpha > 1$, that is, $x^\alpha \geq \alpha x + (1 - \alpha)$. Draw a graphic example with $y = x^{1.7}$ and $y = 1.7x + 0.7$.

9. a. Prove the following lemma used in the proof of Theorem 39.1. Let u be a differentiable function over $[a, b]$. If $u'^R(a) < 0$, then $u(a)$ cannot be the minimum of u (guaranteed to exist by the extreme value theorem) over $[a, b]$, nor can u reach its minimum at $x = b$ either, if $u'^L(b) > 0$. Hint: For the first part,

$$u'^R(a) = \lim_{x \to a+} \frac{u(x) - u(a)}{x - a} < 0.$$

b. A contrapositive of Lemma 39.1 reads as follows: Given a function u differentiable over (a, b) (a minor premise), and if $u'(\xi) \neq 0$ for some $\xi \in (a, b)$, then u attains neither its maximum nor its minimum at ξ. Prove a more detailed lemma regarding $\xi \in (a, b)$: Given a function u differentiable over (a, b), if $u'(\xi) < 0$, then $u(\xi)$ cannot be the minimum of u, and if $u'(\xi) > 0$, then $u(\xi)$ cannot be the maximum of u. Hint:

$$u'(\xi) = \lim_{x \to \xi} \frac{u(x) - u(\xi)}{x - \xi},$$

and for the first part, show that there must be a neighborhood of ξ where $u(x) < u(\xi)$.

10. Finish the proof of Theorem 39.4 by following these steps:

(i) Let

$$K = \frac{f(b) - f(a)}{g(b) - g(a)},$$

which is a constant.. Define the ingenious function $\Delta(x) = f(x) - f(a) - K[g(x) - g(a)]$ over the domain $[a, b]$. Verify that Δ satisfies the three premises for Rolle's theorem (Theorem 39.2) over $[a, b]$.

(ii) Roll out Rolle's theorem, and conclude that K may also be evaluated as

$$K = \frac{f'(\xi)}{g'(\xi)},$$

with a condition on ξ.

(iii) Rewrite Δ with the new expression for K, and see what $\Delta(b)$ says.

11. The **cycloid** is an important curve in the early development of calculus (see [7], for instance). You can see it as the curve drawn by a reflector on a bicycle wheel at night as the bicycle

crosses a car's headlights. Its parametric equations are $x = g(t) = a(t - \sin t)$ and $y = f(t) = a(1 - \cos t)$, where a is the radius of the circle drawing the curve as it rolls (the wheel's radius). Graph one arch of the cycloid with $a = 2$. Show that there is a tangent slope to the cycloid,

$$\frac{dy}{dx} = \frac{f'(t)}{g'(t)},$$

equal to the secant slope between the points $\langle 0, 0 \rangle$ and $\langle 2\pi, 2 \rangle$. Hint: The points correspond to $t = 0$ and $t = \pi$, respectively.

40

Taylor's Theorem

Theorems are not often discovered historically in the order that logic would demand. This is the case with the extension of the mean value theorem named after the musician, artist, and mathematician Brook Taylor (1685–1731). The motivating idea was that the tabulation of difficult functions such as $f(x) = \cos x$ would be greatly simplified if the function could be approximated closely by a polynomial. Taking the first step in this direction, the mean value theorem yields a linear (first degree) polynomial that approximates f. Thus, under the conditions of that theorem, $f(x) \approx f(a) + f'(a)(x - a)$, where, in replacing the unknown number ξ by the initial value a, we have traded the theorem's statement of equality for an approximation.

40.1 Taylor Polynomials

Example 40.1 Let $f(x) = \cos x$, and choose $a = \frac{\pi}{3}$, for instance. The equation of the tangent line to

$$y = \cos x \quad \text{at} \quad \left\langle \frac{\pi}{3}, \frac{1}{2} \right\rangle$$

is

$$y = \cos \frac{\pi}{3} - \sin \frac{\pi}{3} \left(x - \frac{\pi}{3} \right) = \frac{1}{2} - \frac{\sqrt{3}}{2} \left(x - \frac{\pi}{3} \right).$$

Since the function is smooth within a small neighborhood of

$$\left\langle \frac{\pi}{3}, \frac{1}{2} \right\rangle,$$

we may say that

$$\cos x \approx \frac{1}{2} - \frac{\sqrt{3}}{2} \left(x - \frac{\pi}{3} \right)$$

within that neighborhood. Graphically, we see in Figure 40.1 that the approximation of the curve by the line is fairly good within a small neighborhood of $\frac{\pi}{3}$, say,

$$\left[\frac{\pi}{3} - 0.2, \frac{\pi}{3} + 0.2 \right] \approx [0.847, 1.247]. \quad \blacksquare$$

Taylor and his contemporaries would now assume that a polynomial of degree n, namely

$$P(x) = a_0 + a_1 (x - c) + a_2 (x - c)^2 + \cdots + a_n (x - c)^n,$$

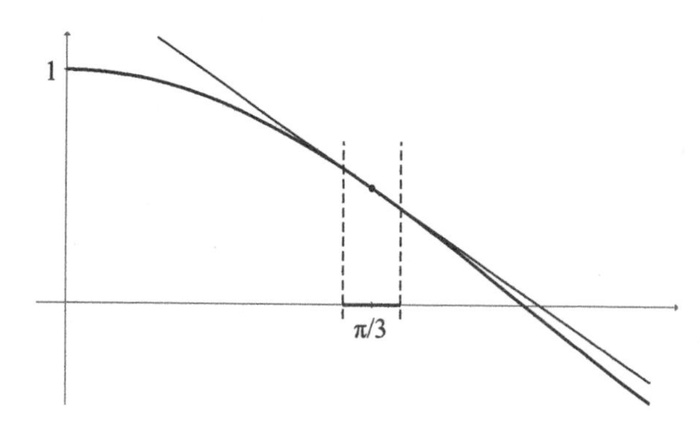

Figure 40.1. $y = \cos x$ is approximated well by the tangent line at $\langle \pi/3, 1/2 \rangle$, over a small neighborhood about $\pi/3 \approx 1.047$.

could be made to approximate $f(x)$ at the point $\langle c, f(c) \rangle$ with better and better precision by having the corresponding derivatives of f and P each agree at that point. Under this condition, a unique polynomial emerges as each coefficient a_k is determined. Accordingly, assuming that f is differentiable n times in a neighborhood of c, we develop the following sequence:

k	$f^{(k)}(c) = P^{(k)}(c)$		a_k
0	$f(c) = P(c) = a_0$	$\rightarrow$	$a_0 = f(c)$
1	$f'(c) = P'(c) = a_1$	$\rightarrow$	$a_1 = f'(c)$
2	$f''(c) = P''(c) = 2a_2$	$\rightarrow$	$a_2 = \frac{1}{2} f''(c)$
3	$f^{(3)}(c) = P^{(3)}(c) = 6a_3$	$\rightarrow$	$a_3 = \frac{1}{6} f^{(3)}(c)$
4	$f^{(4)}(c) = P^{(4)}(c) = 4!a_4$	$\rightarrow$	$a_4 = \frac{1}{4!} f^{(4)}(c)$
...	...		...
n	$f^{(n)}(c) = P^{(n)}(c) = n!a_n$	$\rightarrow$	$a_n = \frac{1}{n!} f^{(n)}(c)$

The result is the unique **Taylor polynomial** of degree n for f about $x = c$:

$$P_n(x) = f(c) + f'(c)(x-c) + f''(c)2!(x-c)^2 + \cdots + \frac{f^{(n)}(c)}{n!}(x-c)^n = \sum_{k=0}^{n} \frac{f^{(k)}(c)}{k!}(x-c)^k.$$

Notice that $P_1(x) = f(c) + f'(c)(x-c)$ is the equation of the tangent line to f at $x = c$, as illustrated in Figure 40.1.

If f is differentiable n times at $c = 0$, then the Taylor polynomial takes the simple form

$$P_n(x) = f(0) + f'(0)x + \frac{1}{2!}f''(0)x^2 + \cdots + \frac{1}{n!}f^{(n)}(0)x^n$$

known as the **Maclaurin polynomial** for f.

Example 40.2 Once again, let $f(x) = \cos x$ and $c = \frac{\pi}{3}$. Three Taylor polynomials for the cosine function about $\left\langle \frac{\pi}{3}, \frac{1}{2} \right\rangle$ appear below, along with their graphs in Figure 40.2. The graphs give evidence that as the degree n increases, the neighborhood within which a good approximation

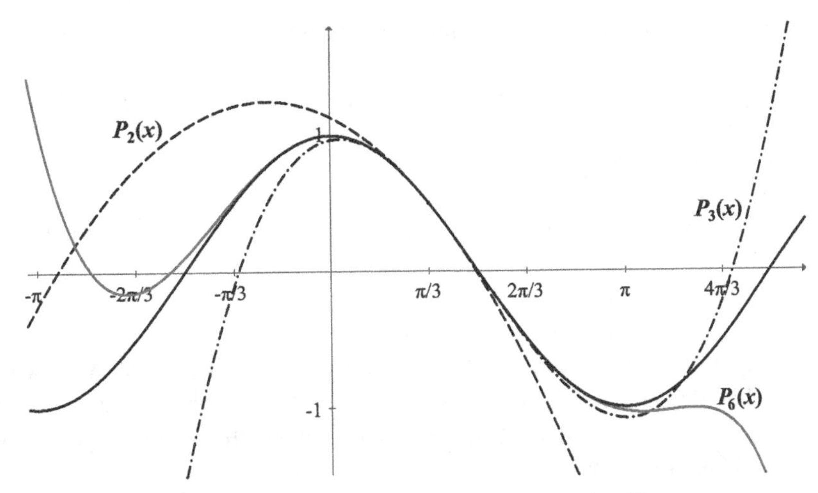

Figure 40.2. The three Taylor polynomials above for $y = \cos x$ about $\left(\frac{\pi}{3}, \frac{1}{2}\right)$. P_6 is the thin line, P_3 is the dot-dash, and P_2 (the parabola) is the dashed line. $P_6(x)$ is quite close to the cosine curve over a wide neighborhood of $\frac{\pi}{3}$, roughly $[-0.7, 2.7]$.

n	$P_n(x)$ at $c = \frac{\pi}{3}$
2	$P_2(x) = \frac{1}{2} - \frac{\sqrt{3}}{2}\left(x - \frac{\pi}{3}\right) - \frac{1}{2!}\frac{1}{2}\left(x - \frac{\pi}{3}\right)^2$
3	$P_3(x) = \frac{1}{2} - \frac{\sqrt{3}}{2}\left(x - \frac{\pi}{3}\right) - \frac{1}{2!}\frac{1}{2}\left(x - \frac{\pi}{3}\right)^2 + \frac{1}{3!}\frac{\sqrt{3}}{2}\left(x - \frac{\pi}{3}\right)^3$
6	$P_6(x) = \frac{1}{2} - \frac{\sqrt{3}}{2}\left(x - \frac{\pi}{3}\right) - \frac{1}{2!}\frac{1}{2}\left(x - \frac{\pi}{3}\right)^2 + \frac{1}{3!}\frac{\sqrt{3}}{2}\left(x - \frac{\pi}{3}\right)^3 + \frac{1}{4!}\frac{1}{2}\left(x - \frac{\pi}{3}\right)^4$ $- \frac{1}{5!}\frac{\sqrt{3}}{2}\left(x - \frac{\pi}{3}\right)^5 - \frac{1}{6!}\frac{1}{2}\left(x - \frac{\pi}{3}\right)^6$

is obtained will increase in radius. Also, in a fixed neighborhood of $c = \frac{\pi}{3}$, the difference between $f(x)$ and $P_n(x)$ diminishes as n increases. ■

40.2 Taylor's Theorem

The crucial details about the difference—or remainder—between $f(x)$ and $P_n(x)$ would have to wait until Joseph-Louis Lagrange's studies at the beginning of the 19th century. An expression for the remainder will emerge from Lagrange's refinement of the theorem named for Brook Taylor.

Theorem 40.1 Taylor's Theorem Let the function $f : \mathbb{R} \to \mathbb{R}$ and each of its derivatives $f', f'', \ldots, f^{(n)}$ be continuous over a common domain $\mathcal{D}$, and let $[a, b] \subseteq \mathcal{D}$. Let $f^{(n+1)}$ exist over (a, b). Then

$$f(b) = f(a) + f'(a)(b-a) + \frac{f''(a)}{2!}(b-a)^2 + \cdots + \frac{f^{(n)}(a)}{n!}(b-a)^n + \frac{f^{(n+1)}(\xi)}{(n+1)!}(b-a)^{n+1}, \quad (1)$$

where ξ is some (at least one) number between a and b.

Proof. (Have you done Exercise 39.10?) Define the constant K by

$$K = \frac{(n+1)!}{(b-a)^{n+1}}\left[f(b) - f(a) - f'(a)(b-a) - \frac{1}{2!}f''(a)(b-a)^2 - \cdots - \frac{1}{n!}f^{(n)}(a)(b-a)^n\right]. \quad (2)$$

Use it to define a function Δ with $\mathcal{D}_\Delta = [a, b]$ by

$$\Delta(x) = f(b) - f(x) - f'(x)(b-x) - \frac{1}{2!}f''(x)(b-x)^2 - \cdots - \frac{1}{n!}f^{(n)}(x)(b-x)^n - \frac{K}{(n+1)!}(b-x)^{n+1}.$$

Interestingly, this function satisfies the premises of Rolle's Theorem: Δ is continuous over $[a, b]$, Δ is differentiable over (a, b), and $\Delta(a) = \Delta(b)$ (both equal zero). Therefore, there exists some $\xi \in (a, b)$ at which $\Delta'(\xi) = 0$. What's more, the derivative of $\Delta(x)$ is quite remarkable (see Exercise 40.5):

$$\begin{aligned}\Delta'(x) &= -f'(x) + f'(x) - f''(x)(b-x) + f''(x)(b-x) \\ &\quad - \frac{1}{2!}f'''(x)(b-x)^2 + \cdots - \frac{1}{n!}f^{(n+1)}(x)(b-x)^n + \frac{K}{n!}(b-x)^n \\ &= \frac{K}{n!}(b-x)^n - \frac{1}{n!}f^{(n+1)}(x)(b-x)^n,\end{aligned}$$

due to the telescoping of all the other terms. It follows that

$$\Delta'(\xi) = \frac{K}{n!}(b-\xi)^n - \frac{1}{n!}f^{(n+1)}(\xi)(b-\xi)^n = 0.$$

Hence $K = f^{(n+1)}(\xi)$, and in view of (2), the theorem's conclusion follows. **QED**

The first thing to note is that if we replace b by $x \in (a, b) \subseteq \mathcal{D}_f$ and a by any fixed $c \in (a, b)$ in (1), we obtain a most useful form of Taylor's theorem,

$$f(x) = f(c) + f'(c)(x-c) + \frac{f''(c)}{2!}(x-c)^2 + \cdots + \frac{f^{(n)}(c)}{n!}(x-c)^n + \frac{f^{(n+1)}(\xi)}{(n+1)!}(x-c)^{n+1}, \quad (3)$$

where again ξ is between x and c, and is generally undetermined. This is the analogue of the mean value theorem form (1) in Chapter 39. (3) delivers what we have been looking for: an analytic expression for the difference, the remainder, between $f(x)$ and $P_n(x)$. The last term in (3) is called the **Lagrange form** of the remainder:

$$R_n(x) \equiv \frac{f^{(n+1)}(\xi)}{(n+1)!}(x-c)^{n+1} = f(x) - P_n(x) \text{ (where } \xi \text{ lies between } x \text{ and } c).$$

$R_n(x)$ conveniently has the form of the next term of the nth degree Taylor polynomial

$$P_n(x) = \sum_{k=0}^{n} \frac{1}{k!} f^{(k)}(c)(x-c)^k.$$

Example 40.3 In Example 40.2, put a bound on the error in using Taylor polynomial $P_3(x)$ in place of $f(x) = \cos x$ over the interval

$$\left[\frac{\pi}{3} - 1, \frac{\pi}{3} + 1\right] \approx [0.0472, 2.0472].$$

We are considering

$$|R_3(x)| = \left|\frac{1}{4!} f^{(4)}(\xi)\left(x - \frac{\pi}{3}\right)^4\right| = \frac{1}{24}|\cos\xi|\left(x - \frac{\pi}{3}\right)^4.$$

First, take

$$x \in \left[\frac{\pi}{3}, \frac{\pi}{3} + 1\right],$$

so that

$$\xi \in \left(\frac{\pi}{3}, x\right).$$

Simplifying,

$$|R_3(x)| \leq \frac{1}{24}\left(x - \frac{\pi}{3}\right)^4,$$

which is a maximum at $x = \frac{\pi}{3} + 1$. Thus,

$$|R_3(x)| \leq \frac{1}{24}\left(\frac{\pi}{3} + 1 - \frac{\pi}{3}\right)^4 = \frac{1}{24} \approx 0.0417.$$

Moreover, the same expression is obtained for $|R_3(x)|$ when

$$x \in \left[\frac{\pi}{3} - 1, \frac{\pi}{3}\right],$$

with

$$\xi \in \left(x, \frac{\pi}{3}\right).$$

In this case, the same maximum is obtained. We conclude that

$$|R_3(x)| = |\cos x - P_3(x)| < 0.0417$$

over the interval, as seen in Figure 40.3. Now, do Exercise 40.3. ■

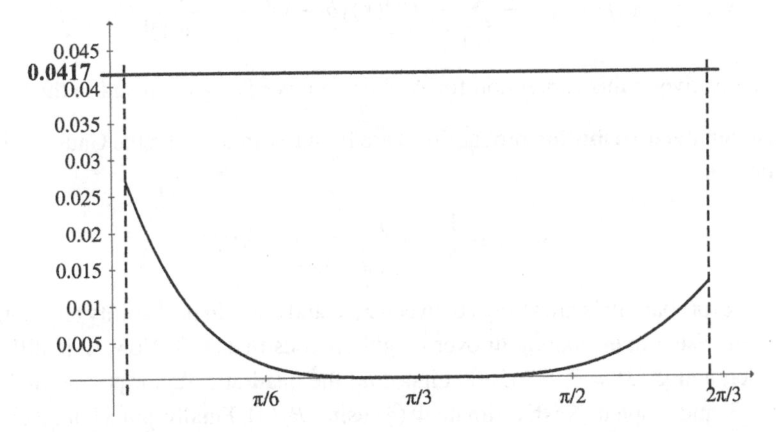

Figure 40.3. The graph of the absolute remainder $|R_3(x)| = |\cos x - P_3(x)|$. P_3 was found in Example 40.2. We see that $|R_3(x)| < 0.0417$ over $\left[\frac{\pi}{3} - 1, \frac{\pi}{3} + 1\right]$. Of course, $\left|R_3\left(\frac{\pi}{3}\right)\right| = 0$ exactly.

40.3 Exercises

1. In Example 40.2, determine the Taylor polynomial $P_4(x)$ and graph it along with the others. What is the maximum absolute error that $P_4(x)$ yields over the neighborhood $\left(\frac{\pi}{3}-1, \frac{\pi}{3}+1\right)$? Answer the same question for $P_6(x)$.

2. Continuing Exercise 38.11, recall that $u(x)=\sqrt{1-x^2}$ and $w(x)=-\frac{1}{2}x^2+1$ were two functions such that $u^{(k)}(0)=w^{(k)}(0)$ for $k=0,1,2,3$. Apply Taylor's theorem to u to find a fourth degree Taylor polynomial about $\langle 0, 1\rangle$. Hint: Since $\langle 0, 1\rangle$ is an endpoint of the domain of u, extend its domain to the natural one, $[-1,1]$.

3. a. Put a bound on the error in using the sixth degree Taylor polynomial (see Example 40.2) in place of $f(x)=\cos x$ over the neighborhood

$$\left[\frac{\pi}{3}-1, \frac{\pi}{3}+1\right] \approx [0.0472, 2.0472].$$

   b. It was mentioned in Example 40.2 that $P_6(x)$ about $c=\frac{\pi}{3}$ was "quite close to the cosine curve over a wide neighborhood, roughly $[-0.7, 2.7]$." Put a bound on how greatly the curves will differ.

4. a. Prove that over the interval $\left[-\frac{\pi}{2}, \frac{\pi}{2}\right]$ about zero,

$$1-\frac{1}{2}x^2 \leq \cos x \leq 1-\frac{1}{2}x^2+\frac{1}{24}x^4.$$

   b. **(i)** Prove that

$$x-\frac{1}{6}x^3 \leq \sin x \leq x-\frac{1}{6}x^3+\frac{1}{120}x^5$$

   over $\left[0, \frac{\pi}{2}\right]$, so that this is an improved version of $\sin\theta \leq \theta$ from Theorem 23.2. **(ii)** Prove that the inequalities in **(i)** are reversed over $\left[-\frac{\pi}{2}, 0\right]$.

5. In the proof of Taylor's theorem we see

$$\Delta(x)=f(b)-f(x)-\sum_{k=1}^{n}\frac{1}{k!}f^{(k)}(x)(b-x)^k-\frac{K}{(n+1)!}(b-x)^{n+1}.$$

   Find the derivative of this expression for Δ, therefore verifying the result in the proof.

6. **Standard normal distribution** probabilities are found as areas under the Gaussian bell curve by the function

$$\Phi(x)=\frac{1}{2}+\frac{1}{\sqrt{2\pi}}\int_0^x e^{-t^2/2}dt,$$

   where $x \in \mathbb{R}$ (probabilities are always between zero and one). In probability and statistics, we are often interested in evaluating Φ over neighborhoods of $x=3$. Now, Φ is differentiable to any order, and $\Phi(3)=0.99865\ldots$. First, find the quadratic Taylor polynomial for $\Phi(x)$ about $c=3$, and graph it. Next, estimate $\Phi(4)$ using $P_2(4)$. Finally, put a bound on the error when using the polynomial in the interval $[2,4]$. Hint: $\Phi'(x)=\frac{1}{\sqrt{2\pi}}e^{-x^2/2}$ by the fundamental theorem of integral calculus, still to come.

7. Nineteenth century physics assumed that an electron with a given energy would circulate at a fixed distance around a proton, but it was wrong! Quantum physicists discovered that the square of a "wave function" describes the motion of the electron correctly by giving the probability that it is within a thin spherical shell of radius r centered at the proton. The average probability is $\overline{P(r)} = \frac{2r^2}{a^3} e^{-2r/a}$, where a is a physical constant. (Source: *Physics Part II* by David Halliday and Robert Resnick, Wiley, 1962, pg. 1209.) We may take $a = 1$. First, graph $\overline{P(r)}$ over $[0, 4]$. Next, find the cubic Taylor polynomial for $\overline{P(r)}$ about $c = 1$, and graph it with $\overline{P(r)}$. Finally, find an error bound when using the polynomial in the interval $[0.5, 2.0]$.

41

Infinite Sequences of Functions

All our work in previous chapters on infinite sequences of constants $\langle a_n \rangle_1^\infty$ can be brought to bear on sequences whose terms are subscripted functions $f_n(x)$ of an independent variable x. We have then a **sequence of functions** $\langle f_n(x) \rangle_{n=1}^\infty = \langle f_1(x), f_2(x), f_3(x), \cdots \rangle$, where the values of x must belong to a common domain. If no ambiguities arise, we may denote the sequence as $\langle f_n(x) \rangle_1^\infty$, or as $\langle f_n \rangle_1^\infty$. As an example, we have $\langle \cos \frac{x}{n} \rangle_1^\infty = \langle \cos x, \cos \frac{x}{2}, \cos \frac{x}{3}, \cdots \rangle$.

41.1 Pointwise Convergence

It is important to determine which values of x allow $\langle f_n(x) \rangle_1^\infty$ to converge. The order of operations implied by this statement is crucial: first, select x in the common domain, and second, investigate $\lim_{n \to \infty} f_n(x)$. For instance, if $f_n(x) = nx$, then $\langle nx \rangle_1^\infty$ converges only for $x = 0$. In contrast, $\langle \cos \frac{x}{n} \rangle_1^\infty$ converges to 1 for all real x. That is, first select $x = x_0$, and second, $\lim_{n \to \infty} \cos \frac{x_0}{n} = 1$. In Figure 41.1, $x_0 = 4.2$, and we see the sequence of ordinates $\langle \cos \frac{4.2}{n} \rangle_1^\infty$ converging upwards towards the point $\langle 4.2, 1 \rangle$. Graphically, the overall effect is that the sequence of curves $y = \cos \frac{x}{n}$, $n \in \mathbb{N}$, consistently approaches the horizontal line $F(x) = 1$ over any interval $[a, b]$.

These observations, especially the "consistently approaches" concept, need a clear definition before we can continue the analysis. It is the definition of pointwise convergence.

> Let $\langle f_n(x) \rangle_1^\infty$ be a sequence of functions with a common domain $\mathcal{D}$. Then $\langle f_n(x) \rangle_1^\infty$ **converges pointwise** over a set $A \subseteq \mathcal{D}$ to a function $F(x)$ if and only if for any $\varepsilon > 0$, there exists an $N > 0$ that may depend upon both ε and x, such that whenever $x \in A$ and $n > N$, then $|f_n(x) - F(x)| < \varepsilon$.

The function $F(x)$ is called the **limiting function** of the sequence, and we write equivalently $\lim_{n \to \infty} f_n(x) = F(x)$, $\langle f_n(x) \rangle_1^\infty \to F(x)$, $f_n(x) \to F(x)$, or just $f_n \to F$. We say that $\langle f_n(x) \rangle_1^\infty$ converges pointwise (or is pointwise convergent) over A. It is possible for A to be the entire domain $\mathcal{D}$

Figure 41.2 illustrates the definition of pointwise convergence in a more general manner. We see an ε-channel from $F(x) - \varepsilon$ to $F(x) + \varepsilon$. Three functions, f_n, f_{n+1}, and f_{n+2} lie within it over most of set A (here, an interval). The ordinates $f_n(x_0)$, $f_{n+1}(x_0)$, and $f_{n+2}(x_0)$ lie within the ε-channel. Thus, the inequality $|f_n(x_0) - F(x_0)| < \varepsilon$ is satisfied at x_0 for n, $n+1$, and $n+2$.

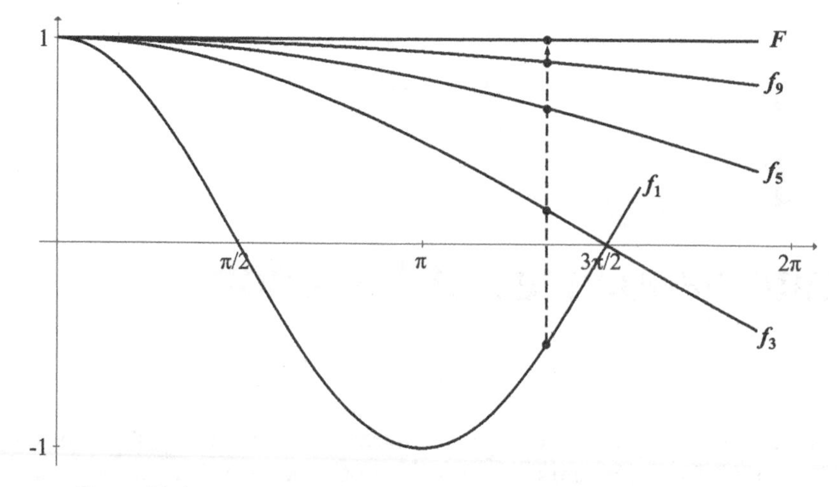

Figure 41.1. $f_9(x) = \cos\frac{x}{9}$, $f_5(x) = \cos\frac{x}{5}$, $f_3(x) = \cos\frac{x}{3}$, $f_1(x) = \cos x$.

Pointwise convergence tells us that $\lim_{n\to\infty} f_n(x_0) = F(x_0)$, as hinted by the unlabeled fourth curve in the figure. But at x_1, $|f_n(x_1) - F(x_1)| > \varepsilon$, so that the inequality $|f_n(x) - F(x)| < \varepsilon$ depends upon x as well as ε.

Example 41.1 Considering $f_n(x) = \cos\frac{x}{n}$ once again, we saw that $\cos\frac{x}{n} \to 1$ for each $x \in \mathbb{R}$, so that $\langle\cos\frac{x}{n}\rangle_1^\infty$ converges pointwise to $F(x) = 1$ over the entire real line.

Following the details required by the definition, given any $\varepsilon > 0$, to assert that $|\cos\frac{x}{n} - 1| < \varepsilon$, a trigonometric identity gives us $|\cos\frac{x}{n} - 1| = 2\sin^2\frac{x}{2n}$. Furthermore,

$$2\sin^2\frac{x}{2n} \le 2\left(\frac{x}{2n}\right)^2 = \frac{x^2}{2n^2}.$$

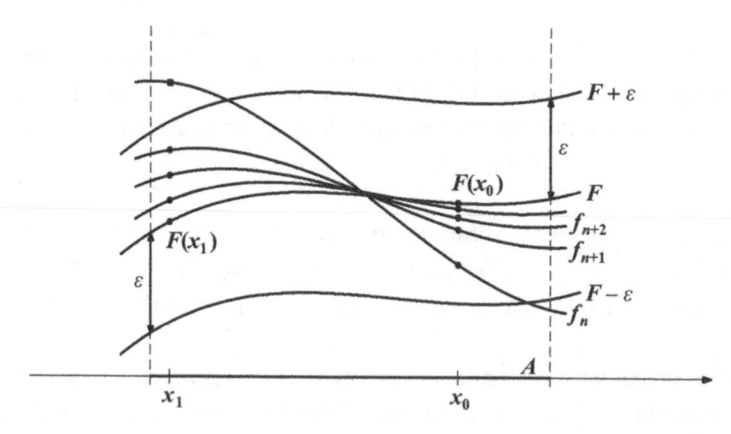

Figure 41.2. The sequence of functions $\langle f_n\rangle_1^\infty$ converges pointwise to F over A, but which functions f_n fall within the ε-channel depends on x and ε.

Thus, it's sufficient to take $\frac{x^2}{2n^2} < \varepsilon$. So, for any fixed x, when $n > \frac{|x|}{\sqrt{2\varepsilon}}$, then $|\cos\frac{x}{n} - 1| \le \frac{x^2}{2n^2} < \varepsilon$. This identifies the number N in the definition as $N = \frac{|x|}{\sqrt{2\varepsilon}}$. Hence, $\langle\cos\frac{x}{n}\rangle_1^\infty \to 1$. Study this further in Exercise 41.5. ■

Example 41.2 Let $f_n(x) = \frac{ne^x+(-1)^n}{n}$ over $\mathbb{R}$. For every real x, $\lim_{n\to\infty}(\frac{ne^x+(-1)^n}{n}) = e^x$. It follows that $f_n(x) \to e^x$ pointwise over the entire real line, that is, $F(x) = e^x$ over $\mathbb{R}$. To be precise, given any $\varepsilon > 0$, we can find an $N > 0$ such that whenever $n > N$, then $|\frac{ne^x+(-1)^n}{n} - e^x| = |\frac{(-1)^n}{n}| = \frac{1}{n} < \varepsilon$. This tells us to use $N = \frac{1}{\varepsilon}$ in the definition of pointwise convergence. Here, N depends upon ε only, as opposed to Example 41.1, where N depended upon both ε and x.

Selecting $x_0 = -1.5$, the sequence of ordinates $\langle\frac{ne^{-1.5}+(-1)^n}{n}\rangle_1^\infty$ in Figure 41.3 can be seen to converge to $F(-1.5) = e^{-1.5}$, alternating above and below the point $\langle -1.5, e^{-1.5}\rangle$. ■

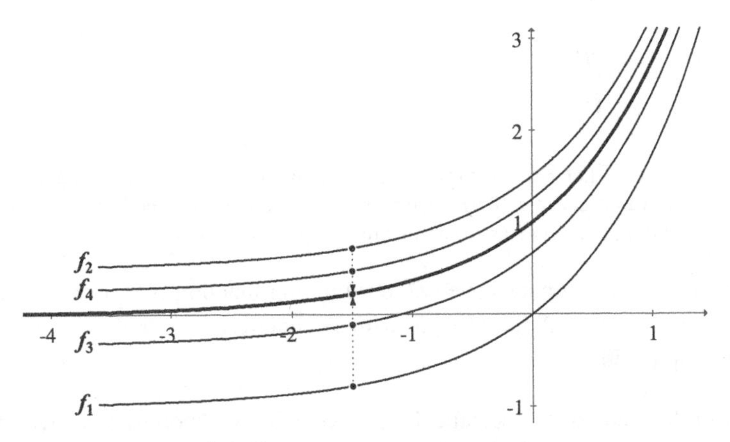

Figure 41.3. Graphs of $f_n(x) = \frac{ne^x+(-1)^n}{n}$ for $n = 1, 2, 3$, and 4. The limiting function $F(x) = e^x$ is the thick line. The ordinates $\frac{ne^{-1.5}+(-1)^n}{n}$ form a sequence converging to $e^{-1.5} \approx 0.223$ from above and below.

In the two examples above, all the functions in the sequences $\langle f_n(x)\rangle_1^\infty$ are continuous over their common domain, and the limiting functions were also continuous. It seems reasonable that pointwise convergence of continuous functions can assure us of a limiting function that is continuous. But consider the next example.

Example 41.3 Let $f_n(x) = x^{2n-1}$ over $\mathcal{D} = [-1, 1]$. Figure 41.4 shows graphs of $f_1(x) = x$, $f_2(x) = x^3$, $f_6(x) = x^{11}$, and $f_{15}(x) = x^{29}$ over $\mathcal{D}$. The graphs suggest a limiting function

$$F(x) = \begin{cases} -1 & \text{if } x = -1 \\ 0 & \text{if } -1 < x < 1 \\ 1 & \text{if } x = 1 \end{cases}$$

From Theorem 20.6, we know that $\lim_{n\to\infty} x^{2n-1} = 0$ for $|x| < 1$. Also, all $f_n(-1) = -1$, implying that $\lim_{n\to\infty} f_n(-1) = -1$, and for similar reasons, $\lim_{n\to\infty} f_n(1) = 1$. Hence, $\langle x^{2n-1}\rangle_1^\infty \to F(x)$ over $\mathcal{D}$, as expected. In the figure, at $x_0 = 0.9$ the vertical sequence of points $\langle 0.9, 0.9^{2n-1}\rangle_1^\infty$ converges downwards to $\langle 0.9, 0\rangle$, and indeed, $F(0.9) = 0$.

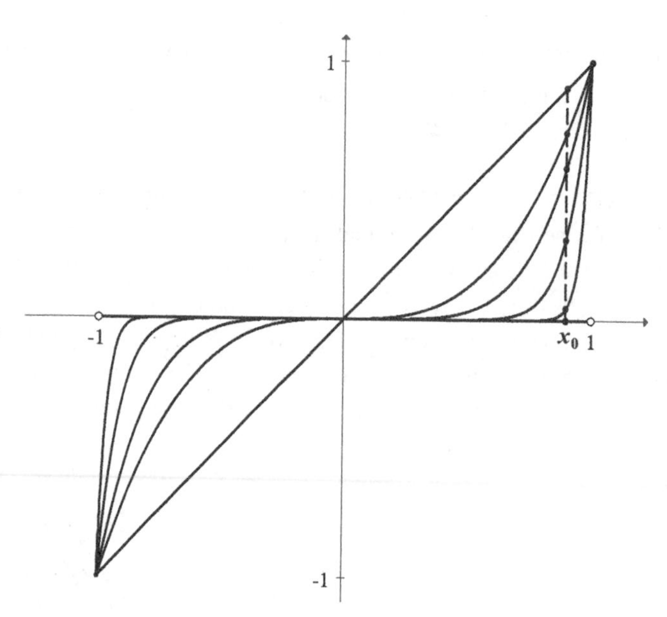

Figure 41.4. Graphs of $f_n(x) = x^{2n-1}$ for $n = 1, 2, 6,$ and 15. Selecting for example $x_0 = 0.9$, we see the vertical sequence of ordinates $f_n(0.9)$ converging downward towards the X axis. The convergence of $\langle x^{2n-1}\rangle_{n=1}^{\infty}$ to the limiting function $F(x)$ (the thick line) is pointwise over $[-1, 1]$, even at the endpoints.

Though each f_n is continuous over $\mathcal{D}$, our limiting function F is not continuous over $\mathcal{D}$. Hence, the sequence $\langle x^{2n-1}\rangle_1^{\infty}$ furnishes us with a counterexample to the conjecture posed just before this example. ■

Note 1: In the study of mathematical statistics, one finds pointwise convergence applied to cumulative distribution functions (CDFs). If a sequence of CDFs $\langle F_n(\omega)\rangle_1^{\infty} \to F(\omega)$ at each $\omega \in \mathbb{R}$ to a continuous CDF $F(\omega)$, then an associated sequence of random variables $\langle X_n\rangle_1^{\infty}$ is said to *converge in distribution* to a random variable X as a limit. This definition is essential to what is called the central limit theorem (the adjective "central" doesn't have mathematical meaning, but conveys how important this theorem is for statistics).

41.2 Uniform Convergence

Example 41.2 displays a distinctly different behavior of convergence than Examples 41.1 and 41.3, even though in all three examples we have pointwise convergence to limiting functions. In Example 41.2, given any $\varepsilon > 0$, the corresponding $N = \frac{1}{\varepsilon}$ is free of x. This means that this value of N is sufficient to guarantee that the *entire* function $f_n(x) = \frac{ne^x+(-1)^n}{n}$ will fall within the ε-channel from $e^x - \varepsilon$ to $e^x + \varepsilon$, regardless of the value of x, so long as $n > N$. This special type of convergence is called uniform convergence.

Let $\langle f_n(x)\rangle_1^{\infty}$ be a sequence of functions with a common domain $\mathcal{D}$. Then $\langle f_n(x)\rangle_1^{\infty}$ **converges uniformly** over a set $A \subseteq \mathcal{D}$ to a function $F(x)$ if and only if for any $\varepsilon > 0$, there exists an $N_\varepsilon > 0$ that depends *only* on ε, such that whenever $x \in A$ and $n > N_\varepsilon$, then $|f_n(x) - F(x)| < \varepsilon$.

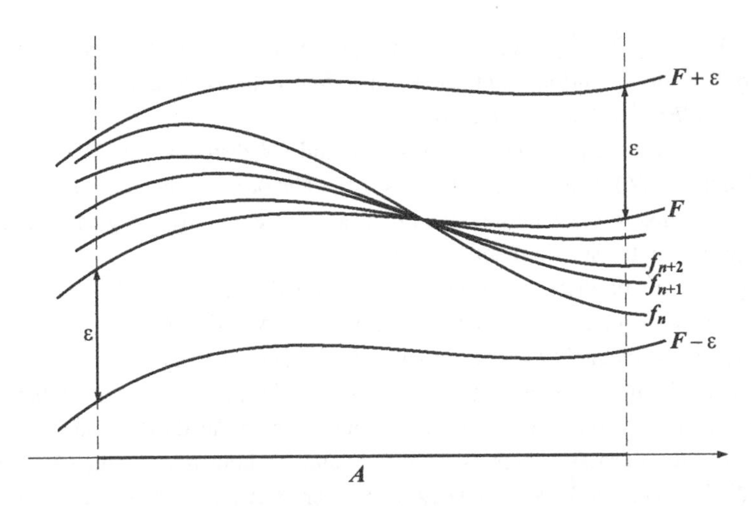

Figure 41.5. The sequence of functions $\langle f_n(x)\rangle_1^\infty$ converges uniformly to F over A.

The function $F(x)$ is still called the limiting function of the sequence. We introduce a new notation to distinguish this new form of convergence: $\lim\limits_{n\to\infty} f_n(x) \overset{u}{=} F(x)$. Equivalently,

$$\lim_{n\to\infty} f_n \overset{u}{=} F,\ \langle f_n(x)\rangle_1^\infty \overset{u}{\to} F(x),\ f_n(x) \overset{u}{\to} F(x), \text{ or just } f_n \overset{u}{\to} F.$$

Because of the definition, first, it's pointless to worry about uniform convergence at a single point x_0, and second, uniform convergence over A implies pointwise convergence over A, but not conversely. The order of the quantifiers in the definition is important. First, $\varepsilon > 0$ is premised; next, the assertion is made that N_ε exists dependent only on ε; and last, the inequality is asserted about all $x \in A$ and all $n > N_\varepsilon$.

Looking at Figure 41.5, we see that claiming that $|f_n(x) - F(x)| < \varepsilon$for every value of $x \in A$ implies that the entire graph of f_n will lie within the ε-channel over A, in contrast to Figure 41.2. Now, we may be able to guarantee that for *any* ε-channel, regardless of how thin, there will be found a function f_N of the sequence that lies entirely within it, and likewise for every function f_n after f_N in the sequence. If this is possible, then we will have proved that $\langle f_n(x)\rangle_1^\infty$ converges uniformly to F over the entire interval.

Example 41.4 In Example 41.2, $N_\varepsilon = \frac{1}{\varepsilon}$ is independent of x. This means that starting at the subscript $M = \lceil N_\varepsilon \rceil$, the entire graph of $f_M(x) = \frac{Me^x+(-1)^M}{M}$ lies within ε units of $F(x) = e^x$, regardless of the value of x, and this is also true for $f_{M+1}(x)$, $f_{M+2}(x)$, and so on. Therefore, $\frac{ne^x+(-1)^n}{n} \overset{u}{\to} e^x$ for all real x. ■

Example 41.5 Returning to Example 41.3, we are looking for $N > 0$ such that $|x^{2n-1} - F(x)| < \varepsilon$ for all subscripts $n > N$, and over $\mathcal{D} = [-1, 1]$. For $x \in \{-1, 0, 1\}$, any N will do (why?), so consider only $x \in A = (-1, 1)\backslash\{0\}$. We must now investigate $|x^{2n-1} - 0| = |x|^{2n-1} < \varepsilon$. We may simplify this by using $|x|^{2n-1} \le |x|^n < \varepsilon$. Solving for n, $n > \frac{\ln\varepsilon}{\ln|x|}$. Thus, take $N = \max\{\frac{\ln\varepsilon}{\ln|x|}, 1\}$, to avoid negative numbers and zero. The relationship cannot be made independent of x, and we conclude that $\langle |x^{2n-1}|\rangle_1^\infty$ does not converge uniformly to its limiting function $F(x)$ over A. Continue this investigation in Exercise 41.8. ■

Note 2: We have just discovered a convergence that is not uniform. It is interesting to symbolize the definition of uniform convergence logically and then negate it. Thus, $\langle f_n(x)\rangle_1^\infty \xrightarrow{u} F(x)$ over a set A means:

$\forall\varepsilon$, if $\varepsilon > 0$ then [$\exists N$ such that [$N > 0$ and (N depends only on ε) and ($\forall x, n$, if ($x \in A$ and $n > N$) then $|f_n(x) - F(x)| < \varepsilon$)]. (The order of the quantifiers shows that N depends on ε, not on x, so the statement that says this explicitly is logically redundant, but very important in the standard definition.)

Negating this produces: $\exists\varepsilon$ such that $\varepsilon > 0$ and $\{\forall N$, either $N \le 0$ or (N does not depend only on ε) or [$\exists x, n$ such that ($x \in A$ and $n > N$ and $|f_n(x) - F(x)| \ge \varepsilon$)]$\}$. (More precisely, the last inequality would actually read "$|f_n(x) - F(x)| < \varepsilon$ is false.)

We may discard the statement that $N \le 0$ as irrelevant. The statement that N does not depend only on ε becomes instantly true because it is prefaced by the universal quantifier. Hence, only the third statement is operative, and fleshing things out, we arrive at this: $\langle f_n(x)\rangle_1^\infty$ *does not converge uniformly* to F over A if and only if there exists an $\varepsilon^* > 0$ such that for any $N > 0$, we can find an $x_0 \in A$ and a subscript $n_0 > N$ such that $|f_{n_0}(x_0) - F(x_0)| \ge \varepsilon^*$.

In the previous example, we found that N did not depend only on ε. Reconsidering Figure 41.4, select, say, $\varepsilon^* = \frac{1}{4}$, and take, say, $x_0 = 1 - \frac{0.1}{N+1} \in (-1, 1)$ and $n_0 = \lceil N \rceil + 1 > N$. Then, for any $N > 0$ (since $F(1 - \frac{0.1}{N+1}) = 0$), we find that $|(1 - \frac{0.1}{N+1})^{2n_0-1} - 0| = (1 - \frac{0.1}{N+1})^{2\lceil N \rceil+1} > 0.5 > \varepsilon^*$. Thus, we have contrived to select x_0 so close to 1 (but less than 1) that $|f_{n_0}(x_0) - F(x_0)| > 0.5$, regardless of how large N is, and this difference is outside of the tolerance allowed by ε^*.

We come now to an important reason that uniform convergence was identified. This theorem, as well as the concept of uniform continuity, is by Eduard Heine, a brilliant student of Dirichlet and of analysis.

Theorem 41.1 If a sequence of functions $\langle f_n(x)\rangle_1^\infty$, each continuous over a set A, converges uniformly over A to a limiting function $F(x)$, then F is continuous over A.

Proof. Our goal is to show that for every $x_0 \in A$, given any $\varepsilon > 0$, there exists a corresponding $\delta > 0$ such that whenever $|x - x_0| < \delta$, then $|F(x) - F(x_0)| < \varepsilon$. The ingenious argument to that end uses the triangle inequality as follows: for any $a, b, c, d \in \mathbb{R}$, $|a - d| \le |a - b| + |b - c| + |c - d|$.

First of all, since $f_n(x) \xrightarrow{u} F(x)$ over A, we are sure that so large an N_ε exists that $|F(x) - f_n(x)| < \frac{\varepsilon}{3}$ for all $n > N_\varepsilon$. This statement applies in particular to $x_0 \in A$: $|f_n(x_0) - F(x_0)| < \frac{\varepsilon}{3}$. Now, select any subscript $M > N_\varepsilon$. Since f_M is continuous at x_0, we are able to find a $\delta > 0$ small enough that if $|x - x_0| < \delta$, then $|f_M(x) - f_M(x_0)| < \frac{\varepsilon}{3}$.

At this point, we are sure that given any $\varepsilon > 0$, if $|x - x_0| < \delta$, then

$$|F(x)-F(x_0)| \le |F(x)-f_M(x)| + |f_M(x) - f_M(x_0)| + |f_M(x_0) - F(x_0)| < \frac{\varepsilon}{3}+\frac{\varepsilon}{3}+\frac{\varepsilon}{3} = \varepsilon.$$

This means that F is continuous at each and every $x_0 \in A$. **QED**

The following corollary overcomes an unfortunate shortcoming of pointwise convergence $\langle f_n(x)\rangle_1^\infty \to F(x)$. Suppose that we integrate the limiting function, $\int_a^b F(x)dx = A$. This should equal the limit of the integrations done one by one along the sequence, $\langle \int_a^b f_n(x)dx\rangle_1^\infty \to A$, shouldn't it? Alas, no. We have interchanged the order of two operations, integration and limit, and this may lead to disaster, as you will experience in Exercise 41.10. The corollary's demonstration depends on three basic properties that will be proved in Chapter 48: if a function w is continuous over $[a, b]$, then it is integrable there; if w is integrable over $[a, b]$, then $|\int_a^b w(x)dx| \leq \int_a^b |w(x)|dx$; and if w is integrable over $[a, b]$ and $|w(x)| < M$ over the interval, then $\int_a^b w(x)dx < \int_a^b Mdx = M(b-a)$.

Corollary 41.1 Let $\langle f_n(x)\rangle_1^\infty$ be a sequence of continuous functions that converges uniformly to a limiting function F over an interval $[a, b]$. Then $\lim\limits_{n\to\infty} \int_a^b f_n(x)dx = \int_a^b \lim\limits_{n\to\infty} f_n(x)dx = \int_a^b F(x)dx$. In words, given the premises, the limit of the areas under a sequence of functions equals the area under the limiting function.

Proof. By Theorem 41.1, $F(x)$ is continuous over $[a, b]$. Hence, $\langle f_n(x) - F(x)\rangle_1^\infty$ is a sequence of integrable functions over $[a, b]$. By uniform convergence, we can find an N_ε large enough so that for all $n > N_\varepsilon$ and all $x \in [a, b]$, $|f_n(x) - F(x)| < \frac{\varepsilon}{b-a}$. Hence, whenever $n > N_\varepsilon$, we have

$|\int_a^b f_n(x)dx - \int_a^b F(x)dx| = |\int_a^b (f_n(x) - F(x))dx| \leq \int_a^b |f_n(x) - F(x)|dx < \int_a^b \frac{\varepsilon}{b-a}dx = \varepsilon.$

Therefore, $\int_a^b f_n(x)dx \to \int_a^b F(x)dx$, and the interchange of operations is allowed. **QED**

Up to now, we have had the benefit of knowing in advance the limiting function $F(x)$ of $\langle f_n(x)\rangle_1^\infty$. This is analogous to proposing that the limit of a sequence $\langle a_n\rangle_1^\infty$ is L, and then going on to prove it. But $F(x)$ is not often explicitly available. Thus, the definition can't be applied in such cases. The solution to this problem was the Cauchy convergence criterion, Theorem 21.1. That theorem can be adapted to sequences of functions.

Theorem 41.2 The Cauchy Criterion for Uniform Convergence Let $\langle f_n(x)\rangle_1^\infty$ be a sequence of real-valued functions with a common domain $\mathcal{D}$. Then $\langle f_n(x)\rangle_1^\infty$ converges uniformly over a set $A \subseteq \mathcal{D}$ if and only if $\langle f_n(x)\rangle_1^\infty$ is a Cauchy sequence at each x in A (this means that for any $\varepsilon > 0$, and for all $x \in A$, there exists an $N_\varepsilon > 0$ such that $|f_n(x) - f_m(x)| < \varepsilon$ whenever we choose any two subscripts $m, n > N_\varepsilon$).

Proof of necessity. The "only if" condition is easy, and mimics the proof in Theorem 21.1. Assume that $\langle f_n(x)\rangle_1^\infty \xrightarrow{u} F(x)$ over A. Then for any $\varepsilon > 0$, we are guaranteed an N_ε such that when $m, n > N_\varepsilon$, then simultaneously $|f_n(x) - F(x)| < \frac{\varepsilon}{2}$ and $|f_m(x) - F(x)| < \frac{\varepsilon}{2}$, for all $x \in A$. It follows that $|f_n(x) - f_m(x)| \leq |f_n(x) - F(x)| + |F(x) - f_m(x)| < \varepsilon$, for all $x \in A$.

Proof of sufficiency. Let $\langle f_n(x)\rangle_1^\infty$ be a Cauchy sequence for each $x \in A$. By Theorem 21.1, $\lim\limits_{n\to\infty} f_n(x)$ exists at each x, which means that $\langle f_n(x)\rangle_1^\infty \to F(x)$ pointwise to a limiting function $F(x)$. Since $\langle f_n(x)\rangle_1^\infty$ is a Cauchy sequence, for any $\varepsilon > 0$ and for each $x \in A$ there exists a number $N_{\varepsilon/2}$ large enough that $|f_n(x) - f_m(x)| < \frac{\varepsilon}{2}$ whenever $m, n > N_{\varepsilon/2}$. Now, select $n > N_{\varepsilon/2}$, and consider that

$$|f_n(x) - F(x)| = \lim_{m\to\infty} |f_n(x) - f_m(x)| \leq \frac{\varepsilon}{2} < \varepsilon.$$

Since this is true for all $n > N_{\varepsilon/2}$, independently of $x \in A$, it follows that $\langle f_n(x)\rangle_1^\infty \xrightarrow{u} F(x)$ over A. The equivalence claimed in the theorem is now proved. **QED**

Example 41.6 Determine whether the sequence of functions $\langle \frac{2nx}{nx+3}\rangle_1^\infty$ converges uniformly over $\mathcal{D} = [1, \infty)$. We try Theorem 41.2, and consider

$$|f_n(x) - f_m(x)| = \left|\frac{2nx}{nx+3} - \frac{2mx}{mx+3}\right| = \frac{6x|n-m|}{(nx+3)(mx+3)} = \frac{6x(n-m)}{(nx+3)(mx+3)}$$
$$< \frac{6x(n-m)}{nmx^2} = \frac{1}{x}\left(\frac{6}{m} - \frac{6}{n}\right) < \frac{6}{m},$$

where we have allowed $n > m$. Hence, if we set $\varepsilon = \frac{6}{N}$, then we establish that $N_\varepsilon = \frac{6}{\varepsilon}$. Now, when $m > N_\varepsilon = \frac{6}{\varepsilon}$, then $n, m > N_\varepsilon$, and it follows that $|f_n(x) - f_m(x)| < \frac{6}{m} < \varepsilon$. Observe that this is true for all $x \in \mathcal{D}$, since N_ε depends only on ε. Therefore, $\langle \frac{2nx}{nx+3}\rangle_1^\infty$ is a Cauchy sequence at each $x \in \mathcal{D}$, and by Theorem 41.2, $\langle \frac{2nx}{nx+3}\rangle_1^\infty$ is uniformly convergent over $\mathcal{D}$.

Notice that we did not need to specify the limiting function F. In this example, it's easy to determine F because $\lim\limits_{n\to\infty} \frac{2nx}{nx+3} = 2$ regardless of the value of x in $\mathcal{D}$, implying that $F(x) = 2$. It's always a good idea to graph several $f_n(x)$ in the sequence to see the convergence behavior. When you do this, you will notice that our sequence seems to converge uniformly to F over points $0 < x < 1$. Show that this is true in Exercise 41.7, and also show there that uniform convergence is lost if $\mathcal{D} = [0, \infty)$. ■

41.3 Exercises

1. Graph several functions in the sequence $\langle \frac{1}{1+|x|^n}\rangle_{n=1}^\infty$, over the domain $A = [-2, 2]$. Does the sequence converge pointwise to a limiting function $F(x)$? If so, state the limiting function, and prove your conjecture by taking limits. Hint:Fix x and consider what happens as $n$increases.

2. Graph several functions in the sequence $\langle x^{2n/(n+1)}\rangle_{n=1}^\infty$, over $\mathbb{R}$. Does the sequence converge pointwise to a limiting function $F(x)$? If so, state the limiting function, and prove your conjecture by taking limits. Hint: Fix x and consider what happens as $n$increases.

3. Graph several functions in the sequence $\langle |\frac{x-n}{x+2n}|\rangle_{n=1}^\infty$, over $\mathbb{R}^+$. Does the sequence converge pointwise to a limiting function $F(x)$? Explain in the case that $\varepsilon = 0.1$. Hint: Fix x and consider what happens as n increases. Then, think in terms of ε-channels.

4. Show that $\langle n \sin \frac{x}{n}\rangle_1^\infty \to x$, for all real x.

5. We have seen in Example 41.1 that when $n > \frac{|x|}{\sqrt{2\varepsilon}}$, then $|\cos \frac{x}{n} - 1| \le \frac{x^2}{2n^2} < \varepsilon$.
   a. For a particular x, say $x = 3.7$, we can find the number N corresponding to any given ε. Do this for $\varepsilon = 0.002$, and verify that $|\cos \frac{3.7}{n} - 1| \le \frac{3.7^2}{2n^2} < 0.002$ for any $n \ge N$.
   b. Keep the same ε and N as in **a**, but check the inequality $|\cos \frac{x}{n} - 1| < \varepsilon$ at $x = 8.9$. Explain what happened, and find the new value of N that is needed to reassert the inequality.

6. a. Prove that $g_n(x) = \frac{\sin nx}{n}$ converges uniformly over the entire real line. Graph a few terms of the sequence to see what the limiting function seems to be (g_1, g_3, and g_9 together make a delightful graph).
   b. Evaluate $\lim\limits_{n\to\infty} \int_1^5 (\frac{\sin nx}{n} + 1)^2 dx$.

7. a. Extend the result of Example 41.6 to prove that the sequence $\langle \frac{2nx}{nx+3} \rangle_1^\infty$ converges uniformly over $\mathcal{D} = [a, \infty)$, where a is a positive endpoint. Hint: Eventually, you will need the fact that $\frac{1}{x} < \frac{1}{a}$ over $[a, \infty)$.
   b. Show that uniform convergence is lost if $\mathcal{D} = [0, \infty)$.
   c. Evaluate $\lim\limits_{n\to\infty} \int_{1/2}^2 e^x \sqrt{\frac{2nx}{nx+3}} dx$.

8. In Example 41.5, let $\varepsilon^* = 0.003$, and find the corresponding N for (a) $x_0 = \frac{1}{4}$, and (b) $x_0 = 0.99$. This will emphasize that $\langle x^{2n-1} \rangle_1^\infty$ does not converge uniformly to F over $(-1, 1)$. Does $\langle x^{2n-1} \rangle_1^\infty \xrightarrow{u} F(x)$ over $[-1, 1]$? Why or why not?

9. Prove that the sequence in Exercise 41.1 does not converge uniformly to its limiting function. Hints: Think about what you did in the previous exercise. All that is necessary is to choose $\varepsilon^* > 0$ and x_0, find the corresponding N, and then show that at some other x_1, $f_n(x_1)$is outside the ε-channel even when $n > N$. Also, points $\langle \pm 1, \frac{1}{2} \rangle$ of $F$are type 2 isolated points.

10. Define the following sequence of functions with common domain $[0, 2]$:

$$w_n(x) = \begin{cases} n - n^2|x - \frac{1}{n}| & \text{if } 0 \le x \le \frac{2}{n} \\ 0 & \text{if } \frac{2}{n} < x \le 2. \end{cases}$$

   a. Draw a graph $w_1(x)$ by thinking about a translation (shift) of $y = -|x|$. Then make clear graphs of $w_2(x)$ through $w_5(x)$ with similar shifts.
   b. Prove that $\langle w_n(x) \rangle_1^\infty \to 0$ (the limiting function is $F(x) = 0$), the convergence being pointwise but not uniform.
   c. Use your graphs to show that $\int_0^2 w_n(x)dx = 1$ for $n = 1, 2, 3, 4, 5$. Then prove by integration or by geometry that $\int_0^2 w_n(x)dx = 1$ for every $n \in \mathbb{N}$.
   d. According to Corollary 41.1, $\lim\limits_{n\to\infty} \int_0^2 w_n(x)dx = \int_0^2 \lim\limits_{n\to\infty} w_n(x)dx = \int_0^2 F(x)dx$, but by c, $\lim\limits_{n\to\infty} \int_0^2 w_n(x)dx = 1$, while by b, $\int_0^2 F(x)dx = 0$. Hence $1 = 0$, all the theory of integration disintegrates, and Axiom RQZ7 is canceled! Or did we slip up somewhere? Explain.

42

Infinite Series of Functions

In mathematics beyond the calculus we use functions that are known to us only as infinite series.

Morris Kline, *Calculus*, 2nd ed.

42.1 Pointwise Convergence

All of our work in previous chapters on infinite series of constants $\sum_{n=1}^{\infty} a_n$ can be brought to bear on infinite series whose terms are functions $f_n(x)$ of an independent variable x, provided that we first fix the value of x. This will be our mode of operation from now on. At each particular x, $f_n(x)$ acts as term a_n. We then have the indicated sum $\sum_{n=1}^{\infty} f_n(x)$, which for some values of x may actually converge. For instance, let $f_n(x) = nx$. Then

$$\sum_{n=1}^{\infty} nx = x + 2x + 3x + \cdots$$

converges for no x except $x_0 = 0$. To be precise about this, we apply Theorem 28.4 in a form applicable to the type of series in this chapter: if the sequence of terms $\langle nx \rangle_{n=1}^{\infty} \not\to 0$, then the series $\sum_{n=1}^{\infty} nx$ will diverge. Therefore, the series diverges for all $x \neq 0$.

As always, a clear definition of the new concept must be made.

> Let $\langle f_n(x) \rangle_1^{\infty}$ be a sequence of functions with a common domain $\mathcal{D}$. Then $\sum_{n=1}^{\infty} f_n(x)$ **converges pointwise** over a set $A \subseteq \mathcal{D}$ to a function $F(x)$ if and only if, at each $x \in A$, the sequence of partial sums
>
> $$\langle \sum_{k=1}^{n} f_k(x) \rangle_{n=1}^{\infty} = \langle f_1(x),\ f_1(x) + f_2(x),\ f_1(x) + f_2(x) + f_3(x), \ldots \rangle$$
>
> converges pointwise to $F(x)$.

Function F is defined by the series and is called the **sum function** (or just the sum) of the series. It may appear as a closed form expression for $\sum_{n=1}^{\infty} f_n(x)$. We will write equivalently

$$\lim_{n \to \infty} \sum_{k=1}^{n} f_k(x) = F(x), \sum_{n=1}^{\infty} f_n(x) = F(x), \sum f_n(x) \to F(x),$$

or briefly, $\sum f_n = F$.

The sequence of partial sums of our series is written as

$$\begin{aligned} s_1(x) &= f_1(x), \\ s_2(x) &= f_1(x) + f_2(x), \\ s_3(x) &= f_1(x) + f_2(x) + f_3(x), \\ &\dots \\ s_n(x) &= \sum_{k=1}^{n} f_k(x) = s_{n-1}(x) + f_n(x) \qquad (n \geq 2), \\ &\dots \end{aligned}$$

so that $\langle \sum_{k=1}^{n} f_k(x) \rangle_{n=1}^{\infty} = \langle s_1(x), s_2(x), s_3(x), \dots \rangle$. Note that $\langle s_1(x), s_2(x), s_3(x), \dots \rangle$ is a sequence of *functions* in the sense of Chapter 41 that converges pointwise over the set A to the sum function $F(x)$.

All the theorems pertaining to infinite series may be applied to the study of convergence of $\sum_{n=1}^{\infty} f_n(x)$. We have already used Theorem 28.4. As another example, the ratio test (Theorem 29.4) reads as follows: let $\sum_{n=1}^{\infty} f_n(x)$ be a series of positive functions over a set $A \subseteq \mathcal{D}$. Form the ratio $\frac{f_{n+1}(x)}{f_n(x)}$, and consider $\rho = \lim_{n \to \infty} \frac{f_{n+1}(x)}{f_n(x)}$. When $0 \leq \rho < 1$, then $\sum f_n(x) \to F(x)$ on A. When $1 < \rho < +\infty$, $\sum_{n=1}^{\infty} f_n(x)$ diverges; and when $\rho = 1$ or doesn't exist, the test concludes nothing.

Example 42.1 Let $f_n(x) = x^n$ over common domain $\mathcal{D} = \mathbb{R}$. We are interested in the convergence of $\sum_{n=1}^{\infty} x^n$ over $\mathcal{D}$. For any fixed $x_0 \in (-1, 1)$, $\sum_{n=1}^{\infty} x_0^n$ is a geometric series that converges to $\frac{x_0}{1-x_0}$. Hence, we have the answer to convergence: $\sum x^n \to \frac{x}{1-x}$ only for $x \in (-1, 1)$, so that $A = (-1, 1)$. And $F(x) = \frac{x}{1-x}$ with domain A.

It's interesting to see in Figure 42.1 the sequence of partial sums $s_1(x) = x$, $s_2(x) = x + x^2$, $s_3(x) = x + x^2 + x^3, \dots$ converging pointwise to the sum function $F(x)$.

Interesting behavior occurs at the endpoints $x = -1$ and $x = 1$. At the left endpoint, the series diverges by oscillation, but $F(-1) = -\frac{1}{2}$. At the right endpoint, the series diverges to $+\infty$, while $F(1)$ doesn't exist. ■

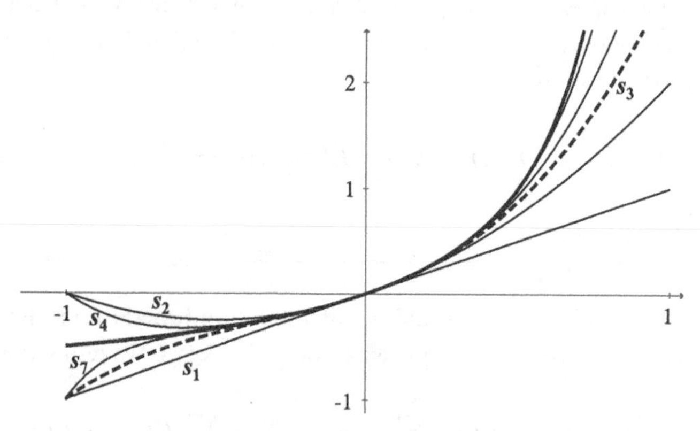

Figure 42.1. The partial sums $s_1(x) = x$, $s_2(x) = x + x^2$, $s_3(x) = x + x^2 + x^3$, $s_4(x) = x + x^2 + x^3 + x^4$, and $s_7(x) = \sum_{n=1}^{7} x^n$ (all thin lines) are representative of the sequence of partial sums of $\sum x^n$. They graphically show pointwise convergence towards the sum $F(x) = \frac{x}{1-x}$ (the thick line) over $(-1, 1)$.

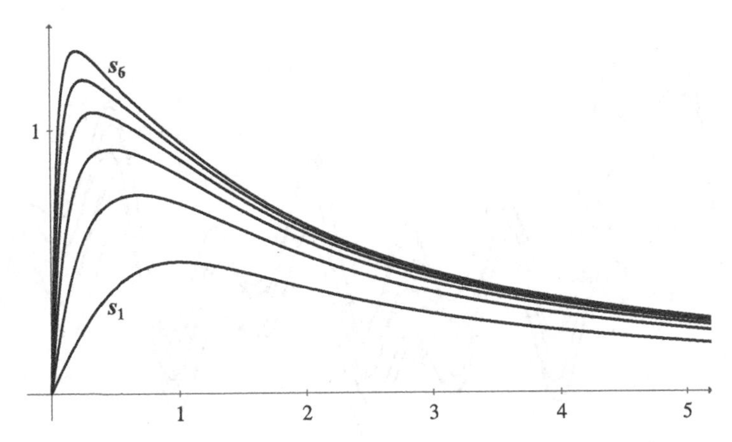

Figure 42.2. The partial sums $s_1(x)$ through $s_6(x)$ seem to be converging pointwise to a sum $F(x)$ (not shown).

Example 42.2 We investigate $\sum_{n=1}^{\infty} \frac{nx}{1+n^3x^2}$ for $x \in [0, \infty)$. In this example, $f_n(x) = \frac{nx}{1+n^3x^2}$. Does the series converge pointwise? As is often the case, graphing the first few partial sums (Figure 42.2) tells much.

Successive $s_n(x)$ seem to be crowding together, although the maximum of each $s_n(x)$may be increasing without bound as $n \to \infty$. Thus, we have some evidence that $\sum_{n=1}^{\infty} \frac{nx}{1+n^3x^2}$ converges. Let's apply the limit comparison test, Theorem 30.2. Choose $\sum_{n=1}^{\infty} \frac{1}{n^2x}$, a convergent series for any nonzero x (see Euler's series in Exercise 42.4). Then

$$\rho = \lim_{n\to\infty} \left(\frac{nx}{1+n^3x^2} \cdot \frac{n^2x}{1} \right) = 1,$$

so $\sum_{n=1}^{\infty} \frac{nx}{1+n^3x^2}$ does converge at each $x > 0$. And at $x = 0$ every term $f_n(0) = 0$, so we conclude that the series converges pointwise over $[0, \infty)$. (Finding the sum function F is beyond our scope.) ■

Example 42.3 Let $f_n(x) = \sin \frac{x}{n}$ over $\mathbb{R}^+ \cup \{0\}$. Does $\sum_{n=1}^{\infty} \sin \frac{x}{n}$ converge? The first few partial sums, $s_1(x) = \sin x$ through $s_6(x) = \sum_{n=1}^{6} \sin \frac{x}{n}$, show no evidence of convergence in Figure 42.3. But graphs can be deceiving—we must rely on our theorems! The comparison test (Theorem 28.6) is appealing since we have proved that $\sin \frac{x}{n} \le \frac{x}{n}$ for $\frac{x}{n} \in [0, \frac{\pi}{2}]$. However, if we are going to prove the divergence of $\sum_{n=1}^{\infty} \sin \frac{x}{n}$, comparison with the divergent series $\sum_{n=1}^{\infty} \frac{x}{n}$ is fruitless. Digging deeper, from Exercise 40.4b, $0 < \frac{x}{n} - \frac{1}{6}(\frac{x}{n})^3 \le \sin \frac{x}{n}$ for $\frac{x}{n} \in [0, \frac{\pi}{2}]$. Still, the comparison with $\sum_{n=1}^{\infty} (\frac{x}{n} - \frac{1}{6}(\frac{x}{n})^3)$ seems tricky. But fortunately, $\frac{x}{2n} \le \frac{x}{n} - \frac{1}{6}(\frac{x}{n})^3$ for $\frac{x}{n} \in [0, \frac{\pi}{2}]$ (verify), so we will try the comparison test with $\sum_{n=1}^{\infty} \frac{x}{2n}$.

A critical requirement in the comparison test is that the terms of the series involved must be nonnegative; thus, $f_n(x) = \sin \frac{x}{n}$ is of concern. As long as $\frac{x}{n} \in [0, \pi]$, the requirement is fulfilled. For any fixed, positive x, we can say that $\frac{x}{n} \in [0, \frac{\pi}{2}]$ for all subscripts $n \ge \frac{2x}{\pi}$. Thus, let $N = \lceil \frac{2x}{\pi} \rceil$. We will therefore compare $\sum_{n=N}^{\infty} \frac{x}{2n}$ and $\sum_{n=N}^{\infty} \sin \frac{x}{n}$, both of which now contain only nonnegative terms. Now, $\sum_{n=N}^{\infty} \frac{x}{2n} = \frac{x}{2} \sum_{n=N}^{\infty} \frac{1}{n}$ involves the tail end of the harmonic series, implying by the comparison test that $\sum_{n=N}^{\infty} \sin \frac{x}{n}$ diverges at any $x > 0$. The finite front end of a series never affects convergence, consequently, $\sum_{n=1}^{\infty} \sin \frac{x}{n}$ diverges at any $x > 0$. As a last detail, the given series does converge to zero at $x = 0$. ■

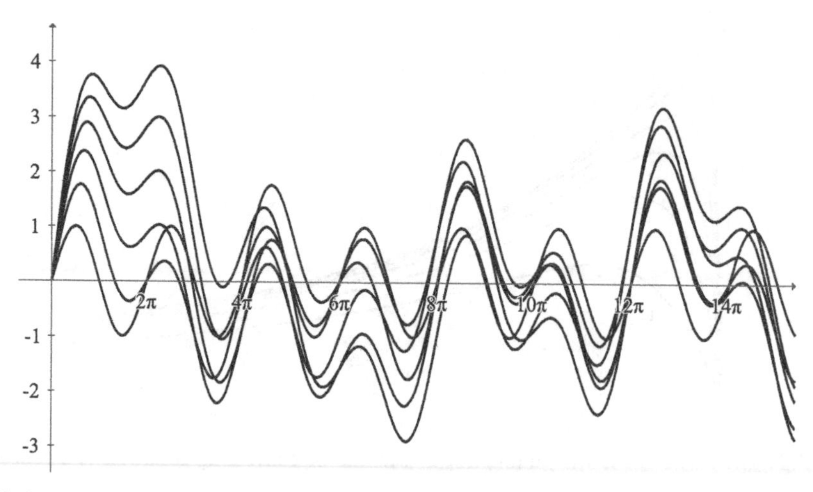

Figure 42.3. The partial sums $s_1(x)$ through $s_6(x)$ of $\sum_1^\infty \sin \frac{x}{n}$ do not seem to be converging towards a sum function.

42.2 Uniform Convergence

Since the convergence of an infinite series of functions completely depends on the convergence of its sequence of partial sums, and since we have defined uniform convergence of a sequence of functions in the previous chapter, we now apply that definition to series of functions.

> Let $\langle f_n(x)\rangle_1^\infty$ be a sequence of functions with a common domain $\mathcal{D}$. Then $\sum_{n=1}^\infty f_n(x)$ **converges uniformly** to a function $F(x)$ over a set $A \subseteq \mathcal{D}$ if and only if the sequence of partial sums
>
> $$\langle \sum_{k=1}^{n} f_k(x)\rangle_{n=1}^\infty = \langle f_1(x),\ \ f_1(x)+f_2(x),\ \ f_1(x)+f_2(x)+f_3(x), \ldots\rangle$$
>
> converges uniformly to $F(x)$ over A.

The function F is still called the sum of the series. We write equivalently

$$\lim_{n\to\infty} \sum_{k=1}^{n} f_k(x) \overset{u}{=} F(x), \sum_{n=1}^{\infty} f_n(x) \overset{u}{=} F(x), \sum f_n(x) \overset{u}{\to} F(x),$$

or just $\sum f_n \overset{u}{\to} F$.

The condition of uniform convergence in this context becomes this: given any $\varepsilon > 0$, we are able to find a positive N_ε depending solely upon ε and such that for all subscripts $n > N_\varepsilon$ and for all $x \in A$, it is true that $|\sum_{k=1}^n f_k(x) - F(x)| < \varepsilon$. To illustrate this, Figure 42.5 from the previous chapter only needs updating, as seen here.

The functions f_n, f_{n+1}, and f_{n+2} in the former figure have been replaced by partial sums s_n, s_{n+1}, and s_{n+2} in Figure 42.4. The limiting function F is now the sum of the series, that is, $F(x) = \sum_{n=1}^\infty f_n(x) = \lim_{n\to\infty} s_n(x)$.

Note 1: The phrase "$\sum_{n=1}^\infty f_n(x)$ converges" is nothing more than a contraction of "$\sum_{n=1}^\infty f_n(x)$ converges pointwise," and cannot imply uniform convergence.

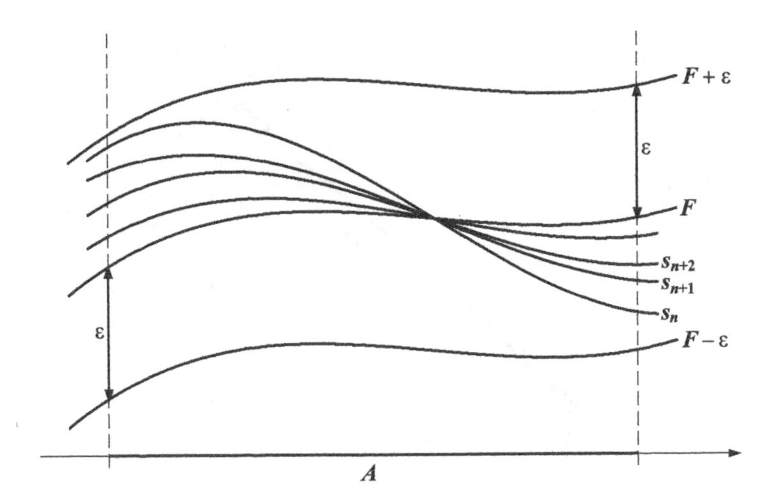

Figure 42.4. Partial sum functions $s_n(x), s_{n+1}(x), \ldots$ converging uniformly towards their sum $F(x)$.

Note 2: It turns out that uniform convergence will be a sufficient condition for distributing the operation of integration over the terms of an infinite series (in short, $\int(\Sigma f_n(x))dx = \Sigma \int f_n(x)dx$) as we will see in a later chapter. It is a necessary condition for distributing the operation of differentiation over the terms of an infinite series.

Example 42.4 Let $f_n(x) = x^n(1-x)$ over $A = [-\frac{1}{2}, \frac{1}{2}]$. Thinking about the geometric series in Example 42.1,

$$\sum_{n=1}^{\infty} x^n(1-x) = (1-x)\sum_{n=1}^{\infty} x^n = (1-x)\frac{x}{1-x} = x$$

where the convergence is pointwise. Can we show that the sequence of partial sums $s_1(x) = x(1-x)$,

$$s_2(x) = x(1-x) + x^2(1-x), \ldots, s_n(x) = \sum_{k=1}^{n} x^k(1-x), \ldots$$

converges uniformly to $F(x) = x$ over A? In other words, can we find an N_ε, depending only upon ε, such that for all $n > N_\varepsilon$, $|\sum_{k=1}^{n} x^k(1-x) - x| < \varepsilon$ throughout A? We have

$$\left|(1-x)\sum_{k=1}^{n} x^k - x\right| = \left|(1-x)\frac{x - x^{n+1}}{1-x} - x\right| = \left|x^{n+1}\right| \leq \left|x^n\right| \leq \left(\frac{1}{2}\right)^n.$$

Thus, in order that $|\sum_{k=1}^{n} x^k(1-x) - x| < \varepsilon$, it is sufficient to require $(\frac{1}{2})^n < \varepsilon$. This inequality becomes true when $n > \frac{\ln \varepsilon}{-\ln 2}$. Thus, set $N_\varepsilon = \max\{\frac{\ln \varepsilon}{-\ln 2}, 1\}$. Very important here is that N_ε depends only on ε, and not on values of x in A. Consequently, $\sum_{n=1}^{\infty} x^n(1-x) \overset{u}{=} x$ over A, and $F(x) = x$.

Figure 42.5 shows the first four partial sums and the sum of the series $F(x) = x$ over A. An ε-channel with $\varepsilon = 0.1$ also appears. Notice that s_1 and s_2 fail to fall within the channel throughout A, but s_3 and s_4 do comply. To apply the analysis in the previous paragraph,

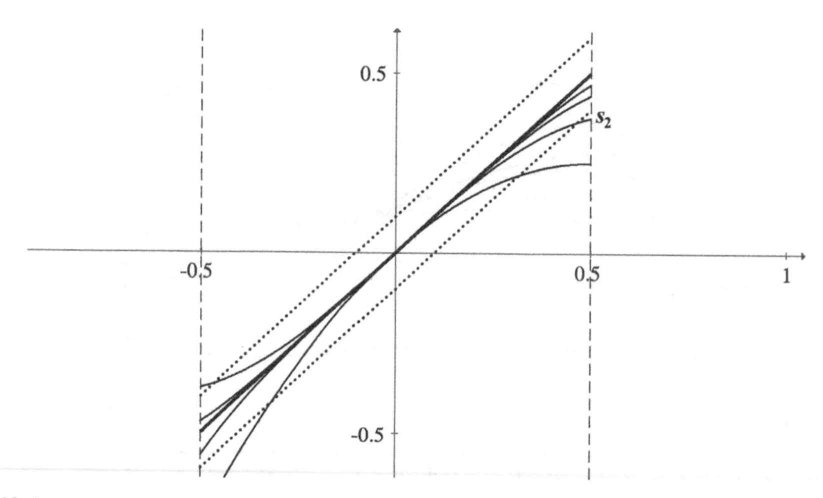

Figure 42.5. Series $\sum_{n=1}^{\infty} x^n(1-x)$ over $[-0.5, 0.5]$. The ε-channel from $x - 0.1$ to $x + 0.1$ (dashed parallel lines) contains the partial sums s_3 and s_4. The cubic polynomial s_2, symmetric across the origin, just barely escapes out of the channel. The sum of the series, $F(x) = x$, also appears (thick line).

$N_{0.1} = \max\{\frac{\ln 0.1}{-\ln 2}, 1\} \approx 3.32$, implying $n \geq 4$. That is, for all $n \geq 4$, and all $x \in A$, $|\sum_{k=1}^{n} x^k(1-x) - x| < 0.1$. (It's irrelevant, but nice, that s_3 also satisfies this inequality.)

We will investigate the series $\sum_{n=1}^{\infty} x^n(1-x)$ quite a bit more as this chapter develops. ■

In the example just finished, recognizing that the partial sums $\sum_{k=1}^{n} x^k$ have a closed form expression $\frac{x-x^{n+1}}{1-x}$ quickly revealed the sum of the series, and allowed a short proof of uniform convergence. Of course, most series $\sum_{n=1}^{\infty} f_n(x)$ will not have such a convenience. We need some theorems for uniform convergence of series of functions.

Lemma 42.1 The Comparison Test for Series of Functions Let $\langle f_n(x)\rangle_1^{\infty}$ and $\langle g_n(x)\rangle_1^{\infty}$ have a common domain $\mathcal{D}$, and suppose that over $\mathcal{D}$ and for every subscript $n \in \mathbb{N}$, $|f_n(x)| \leq g_n(x)$. If

$$\sum_{n=1}^{\infty} g_n(x) \xrightarrow{u} G(x)$$

over $\mathcal{D}$, then there exists a function F over $\mathcal{D}$ such that $\sum f_n(x) \xrightarrow{u} F(x)$.

Proof. The comparison test (Theorem 28.6) shows us that $\sum_{n=1}^{\infty} f_n(x_0)$ converges pointwise at every $x_0 \in \mathcal{D}$. Thus, let $F(x) = \sum_{n=1}^{\infty} f_n(x)$ over $\mathcal{D}$. Then

$$\left|\sum_{k=1}^{n} f_k(x) - F(x)\right| = \left|\sum_{k=n+1}^{\infty} f_k(x)\right| \leq \sum_{k=n+1}^{\infty} g_k(x) = \left|\sum_{k=1}^{n} g_k(x) - G(x)\right|.$$

Now, because $\sum_{n=1}^{\infty} g_n(x)$ converges uniformly, for any $\varepsilon > 0$ there certainly exists an N_ε such that for every $x \in \mathcal{D}$ and all $n > N_\varepsilon$, $|\sum_{k=1}^{n} g_k(x) - G(x)| < \varepsilon$. Then perforce $|\sum_{k=1}^{n} f_k(x) - F(x)| < \varepsilon$ (for every $x \in \mathcal{D}$ and all $n > N_\varepsilon$), which means that $\sum_{n=1}^{\infty} f_n(x)$ converges uniformly to $F(x)$ over $\mathcal{D}$. **QED**

The next test for uniform convergence, devised by Karl Weierstrass, follows as a corollary, but it is important and useful enough to be elevated to theorem status.

Theorem 42.1 The Weierstrass M-Test Let $\sum_{n=1}^{\infty} f_n(x)$ be defined on a set A. Let $\sum_{n=1}^{\infty} M_n$ be a convergent series of nonnegative *constants* (that is, free of x), such that $|f_n(x)| \le M_n$ for every $x \in A$ and all $n \in \mathbb{N}$. Then there exists a function F such that $\sum f_n(x) \xrightarrow{u} F(x)$ over A.

Proof. Every convergent series of constants converges uniformly over any set A whatsoever (convince yourself!). Thus, in Lemma 42.1, let $g_n(x) = M_n$. The theorem follows. **QED**

Example 42.5 Does $\sum_{n=1}^{\infty} n^2x^2e^{-nx}$ converge uniformly over $[1, \infty)$? This series converges if and only if $\sum_{n=1}^{\infty} \frac{n^2}{(e^x)^n}$ converges, and we see that

$$|f_n(x)| = \frac{n^2}{(e^x)^n} \le \frac{n^2}{e^n} < \frac{n^2}{n^4} = \frac{1}{n^2}$$

beginning at $n = 9$, and for any $x \in [1, \infty)$. Therefore, we choose $M_n = \frac{1}{n^2}$. Then $\frac{n^2}{(e^x)^n} < \frac{1}{n^2}$, for all $x \in [1, \infty)$ and $n \ge 9$. Since $\sum_{n=9}^{\infty} \frac{1}{n^2}$ converges, we conclude that $\sum_{n=9}^{\infty} n^2e^{-nx}$ converges uniformly over $[1, \infty)$ by the Weierstrass M-test. Consequently, so do $\sum_{n=9}^{\infty} n^2x^2e^{-nx}$ and $\sum_{n=1}^{\infty} n^2x^2e^{-nx}$.

One of the premises of the M-test is "for all $n \in \mathbb{N}$," not $n \ge 9$ as we have just affirmed. This alteration isn't critical at all, as you should show in Exercise 42.6. ■

Example 42.6 In Example 42.4, we discovered the uniform convergence of $\sum_{n=1}^{\infty} x^n(1-x)$ over $A = [-\frac{1}{2}, \frac{1}{2}]$. Since $|f_n(x)| = |x^n||1-x| \le (\frac{1}{2})^n(\frac{3}{2})$ over A, we select $M_n = (\frac{3}{2})(\frac{1}{2})^n$. Noting that $\frac{3}{2}\sum_{n=1}^{\infty}(\frac{1}{2})^n$ is convergent, we quickly conclude by the Weierstrass M-test that the given series converges uniformly over A. How efficient, compared to Example 42.4!

If we expand A to $B = [-\beta, \beta]$, where β is a fixed, positive number *less* than one, we can still find a convergent series $\sum_{n=1}^{\infty} M_n$ such that $|x^n||1-x| \le M_n$ for all $x \in B$ and all $n \in \mathbb{N}$, namely, $\sum_{n=1}^{\infty} \beta^n(1+\beta)$. We conclude that $\sum x^n(1-x) \xrightarrow{u} x$ over B.

In Figure 42.6, $B = [-0.9, 0.9]$ and the partial sums $s_n(x) = \sum_{k=1}^{n} x^k(1-x)$ are graphed for $n = 1, 2, 3, 8,$and 59. The sum of the series, the line $F(x) = x$, is very closely approximated by s_{59} over B. You should verify that, for instance, $s_{59}(0.9) \approx 0.898$. ■

In the next example, we will find it convenient to use the version of Theorem 41.1 that applies to series of functions.

Corollary 42.1 If in a series $\sum_{n=1}^{\infty} f_n(x)$, each partial sum $s_n(x)$ is continuous over a set C, and the sequence of partial sums $\langle s_n(x)\rangle_1^{\infty}$ converges uniformly over C to a sum $F(x)$, then F will be continuous over C.

Proof. In Theorem 41.1, function s_n takes the role of function f_n, and $\langle s_n(x)\rangle_1^{\infty} \xrightarrow{u} F(x)$ over C. Thus, F is continuous over C. **QED**

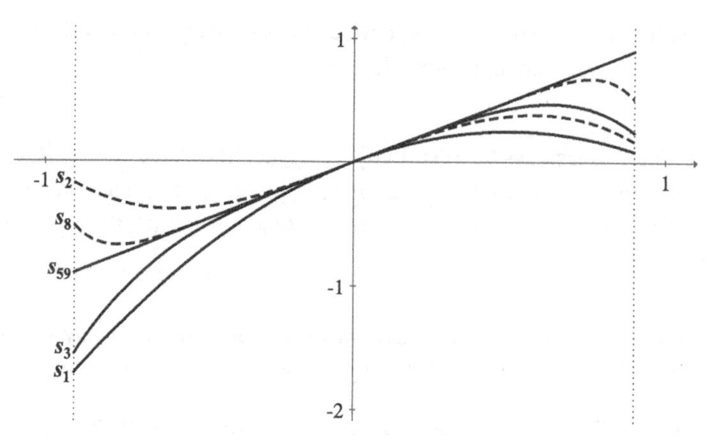

Figure 42.6. Partial sums s_1, s_2, s_3, s_8, and s_{59} of $\sum x^n(1-x)$ appear over $B = [-0.9, 0.9]$. Odd-numbered s_n are the solid lines; even-numbered s_n are the dashed lines. In this scale, s_{59} is indistinguishable from $F(x) = x$, so F isn't graphed at all.

Example 42.7 Continuing the study of series $\sum_{n=1}^{\infty} x^n(1-x)$ in the previous example, if we expand the interval further to $C = [-1, 1]$, then things change drastically. A short sequence of partial sums, graphed in Figure 42.7, shows unusual behavior at the endpoints.

We apply Corollary 42.1 in its contrapositive form. In this case, every partial sum is a (continuous) polynomial over C. But

$$F(x) = \begin{cases} \text{undefined} & \text{if } x = -1 \\ x & \text{if } -1 < x < 1 \\ 0 & \text{if } x = 1 \end{cases}$$

is not continuous over C. Therefore, uniform convergence is not possible. Indeed, this must be so because when $x = -1$, the series diverges by oscillation between points $\langle -1, 0 \rangle$ and $\langle -1, -2 \rangle$. The partial sums show this in Figure 42.7. In Exercise 42.8, investigate the convergence of the series at $x = 1$. ■

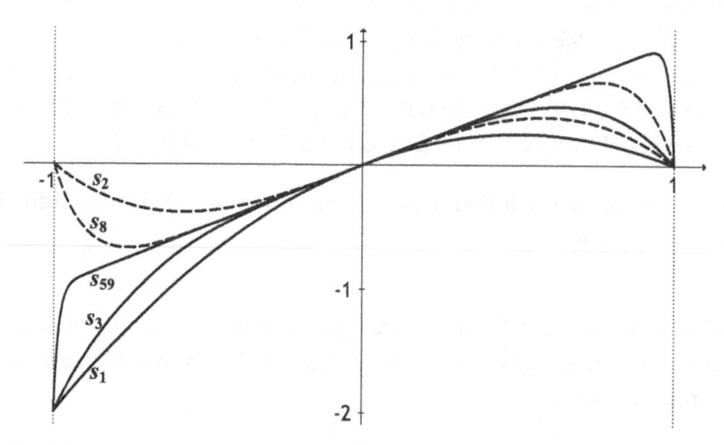

Figure 42.7. Partial sums s_1, s_2, s_3, s_8, and s_{59} of $\sum x^n(1-x)$ appear over $C = [-1, 1]$. Odd-numbered s_n are solid lines; even-numbered s_n are dashed lines. The series diverges at the left endpoint, and does not converge uniformly over any interval containing $x = 1$.

If we now remove the offending endpoints of interval C in the example just finished, do we reestablish uniform convergence? Disappointingly, our strategy in Example 42.6 for finding series $\sum_{n=1}^{\infty} M_n$ fails because it yields $M_n = 2$. Of course, this only shows that our strategy fails; can a series somehow be found that lets us apply the Weierstrass M-test? No, for the hope of uniform convergence vanishes as we see in Figure 42.7 some ordinates of s_n that are stubbornly distant from $F(x) = x$, even as n increases. We need to state this observation analytically.

Reread the ε-N_ε definition of uniform convergence in the paragraph just before Figure 42.4, for we are going to negate it. The result is that $\sum_{k=1}^{\infty} f_k(x)$ **does not converge uniformly** to a function $F(x)$ over $A \subseteq \mathcal{D}$ if and only if there exists an $\varepsilon^* > 0$ such that for any positive N, we can find some subscript $n_0 > N$ and some $x_0 \in A$ such that

$$\left| \sum_{k=1}^{n_0} f_k(x_0) - F(x_0) \right| \geq \varepsilon^*$$

(compare this with Note 2 in the previous chapter).

Example 42.8 Reconsider $\sum_{n=1}^{\infty} x^n(1-x)$, this time over $D = (-1, 1)$. To prove the lack of uniform convergence, let $\varepsilon^* = \frac{1}{2}$, guiding ourselves by the graphs in Figure 42.7. For any $N > 0$, we must confirm that *(1)* $|\sum_{k=1}^{n_0} x_0^k(1-x_0) - x_0| \geq \frac{1}{2}$ for some $n_0 > N$ and some $x_0 \in D$. As before, we have

$$\left| (1-x_0)\frac{x_0 - x_0^{n_0+1}}{1-x_0} - x_0 \right| = |x_0^{n_0+1}| = |x_0|^{n_0+1}.$$

This expresses the absolute deviation between $s_{n_0}(x_0)$ and x_0, the sum of the series. For some $n_0 > N$ and some $x_0 \in D$, can we force $|x_0|^{n_0+1} \geq \frac{1}{2}$? Yes, when $|x_0| \geq (\frac{1}{2})^{1/(n_0+1)}$. Thus, for any $N > 0$, create an expression for n_0 that is consistently greater than N, such as $n_0 = 3\lceil N \rceil - 1$. This yields $|x_0| \geq (\frac{1}{2})^{1/(3\lceil N \rceil)}$, indicating that we may select some $x_0 \in (-1, -(\frac{1}{2})^{1/(3\lceil N \rceil)}]$ or else some $x_0 \in [(\frac{1}{2})^{1/(3\lceil N \rceil)}, 1)$. Do Exercise 42.7, where values will be assigned to the quantities.

Thus, we have found an $n_0 > N$ and an $x_0 \in D$ (actually, short intervals of x_0s) with which the deviation *(1)* is certainly possible. This means that $\sum_{n=1}^{\infty} x^n(1-x)$ is not uniformly convergent over $(-1, 1)$. ■

Last but not least, we apply Theorem 41.2, the Cauchy criterion for uniform convergence, to series of functions.

Theorem 42.2 Let $\langle f_n(x) \rangle_1^{\infty}$ be a sequence of functions with a common domain $\mathcal{D}$. Then $\sum_{n=1}^{\infty} f_n(x)$ converges uniformly over $A \subseteq \mathcal{D}$ if and only if the sequence of partial sums $\langle s_n(x) \rangle_1^{\infty}$of $\sum_{n=1}^{\infty} f_n(x)$ is a Cauchy sequence at each x in A (in detail, for any $\varepsilon > 0$, and for all $x \in A$, there exists an $N_\varepsilon > 0$ such that $|s_n(x) - s_m(x)| < \varepsilon$ whenever we choose two subscripts $m, n \geq N_\varepsilon$).

Proof. As Exercise 42.12, prove this theorem by following the proof of Theorem 41.2. **QED**

42.3 Exercises

1. Determine whether $\sum_{n=1}^{\infty} \frac{nx}{1+nx^2}$ converges over $[0, \infty)$. Graph a few partial sums $s_n(x)$, and notice how similar it is to the series in Example 42.2.

2. Theorem 29.3, the integral test, reads this way for infinite series of functions: let $\langle f(x, n)\rangle_{n=1}^{\infty} \equiv \langle f_n(x)\rangle_{n=1}^{\infty}$ be a sequence of functions that is positive and decreasing with respect to $n \in \mathbb{N}$. Then $\sum_{k=1}^{\infty} f_k(x)$ (that is, $\sum_{k=1}^{\infty} f(x, k)$) converges at each x if and only if the improper integral $\int_{n=1}^{\infty} f(x, n)dn$ converges.
   a. Does $\sum_{n=1}^{\infty} \frac{n}{1+n^3x}$ converge over $[1, \infty)$? For positive fixed x,

   $$\langle f_n(x)\rangle_1^{\infty} = \left\langle \frac{n}{1+n^3x} \right\rangle_1^{\infty}$$

   is a decreasing sequence of positive terms, so use the integral test. Hint:

   $$\int_1^{\infty} \frac{n}{1+n^3x} dn < \int_1^{\infty} \frac{1}{n^2} dn.$$

   b. What can be said about this series over $[\alpha, \infty)$, where α is a *positive* endpoint? And finally, over $[0, \infty)$?

3. What does the Weierstrass M-test say about the convergence of $\sum_{n=1}^{\infty} \frac{n}{1+n^3x}$ of the previous exercise, over the intervals $[1, \infty)$, $[\alpha, \infty)$, and $(0, \infty)$? (We exclude $x = 0$ since this causes divergence, and once again, α is a positive endpoint.) Hint: For $(0, \infty)$, look at $0 < x < 1$ first. Show that there is no convergent series $\sum_{n=1}^{\infty} M_n$ such that $\frac{n}{1+n^3x} \leq M_n$ for all positive x and natural numbers n, by assuming that there is.

4. A famous series of Euler is $\sum_{n=1}^{\infty} \frac{1}{n^2} = \frac{\pi^2}{6}$. Hence,

   $$\sum_{n=1}^{\infty} \frac{x}{n^2(1+x^2)} = \frac{\pi^2 x}{6(1+x^2)}$$

   for all $x \geq 0$. Three partial sums (thin lines) and the sum of the series $F(x) = \frac{\pi^2 x}{6(1+x^2)}$ are shown in Figure 42.8. Determine if $\sum_{n=1}^{\infty} \frac{x}{n^2(1+x^2)}$ converges uniformly over $[0, \infty)$.

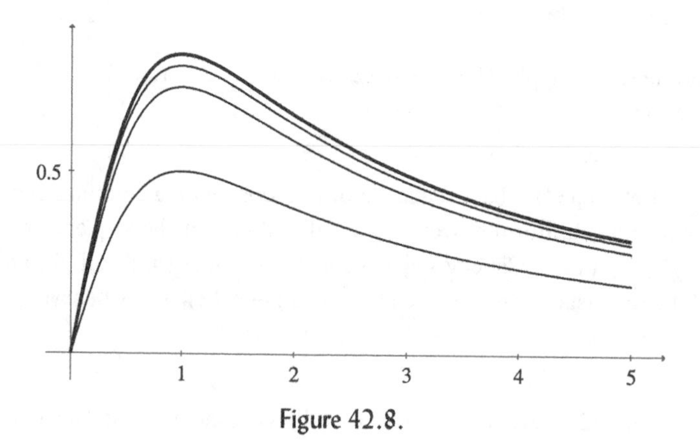

Figure 42.8.

5. Determine the type of convergence of $\sum_{n=1}^{\infty} n^2xe^{-nx}$ over $[1,\infty)$. What about over (α,∞), where α is a positive endpoint? And what about over $[0,\infty)$? Hint: For $[0,\infty)$, Corollary 29.2 applied to series of functions, allows us to conclude that $\int_{m+1}^{\infty} f(x,n)dn + s_m(x) \leq F(x)$. What does the integral yield?

6. In Example 42.5, the Weierstrass M-test seems to be shoddily applied. Give an argument to prove that beginning at $n = 9$ is a harmless adjustment.

7. In Example 42.8, for any $N > 0$, we showed that *(1)* $|\sum_{k=1}^{n_0} x_0^k(1-x_0) - x_0| \geq \frac{1}{2} = \varepsilon^*$ is true when $|x_0| \geq (\frac{1}{2})^{1/(3\lceil N\rceil)}$, where $n_0 = 3\lceil N\rceil - 1$. As a partial check, choose, say, $N = 123.4$, and verify that *(1)* is correct for the corresponding n_0 and a properly chosen value of x_0. What two expressions now deviate from each other by at least $\frac{1}{2}$?

8. Show that $\sum_{k=1}^{\infty} x^k(1-x)$ does not converge uniformly over $[0,1]$, but does converge pointwise over that interval.

9. Use the Cauchy criterion to prove the uniform convergence of $\sum_{n=1}^{\infty} x^n$ over $A = (-\frac{1}{2},\frac{1}{2})$. Notice that we do not need to know the sum $F(x)$ of the series.

10. Show that uniform convergence and absolute convergence are distinct ideas for series of functions, as follows.
   a. Prove that $\sum_{n=1}^{\infty} \frac{(-1)^n}{x^2+n}$ converges uniformly over $\mathbb{R}$, but not absolutely. Hint:

$$\sum_{n=1}^{\infty} \frac{(-1)^n}{x^2+n} = \left(\frac{-1}{x^2+1} + \frac{1}{x^2+2}\right) + \left(\frac{-1}{x^2+3} + \frac{1}{x^2+4}\right) + \left(\frac{-1}{x^2+5} + \frac{1}{x^2+6}\right) + \cdots.$$

   b. Prove that for

$$f_n(x) = \begin{cases} 0 & \text{if } 0 \leq x \leq 1 \\ \frac{1}{n^2(1+x)} & \text{if } x > 1 \end{cases},$$

   $\sum_{n=1}^{\infty} f_n(x)$ converges absolutely over $[0,\infty)$, but not uniformly. Hint: Look at several partial sums.
   c. Explain why the Weierstrass M-test proves *both* absolute and uniform convergence of a series of functions at the same time.

11. Find the set S of points of convergence of $\sum_{n=1}^{\infty} \frac{\sin^n x}{n}$, and comment on absolute and uniform convergence. Graph several partial sums to visualize your analysis. Hint: Let $\sin x = y$, and use the comparison test.

12. Prove Theorem 42.2 by following the proof of Theorem 41.2.

13. Reconsider Exercise 33.10b, and show why $\sum_{n=1}^{\infty} \frac{\sin(n\theta)}{n}$ converges uniformly everywhere except at $\theta = 2k\pi$, $k \in \mathbb{Z}$. Hint: Look at typical graphs of the situation, such as those of

$$\sum_{1}^{5} \frac{\sin(n\theta)}{n}, \quad \sum_{1}^{20} \frac{\sin(n\theta)}{n} \quad \text{and} \quad \sum_{1}^{40} \frac{\sin(n\theta)}{n},$$

and see what's going on near $\theta = 0$. To locate absolute maxima, use one of Lagrange's trigonometric identities:

$$\sum_{n=1}^{N} \cos(n\theta) = \frac{\sin((2N+1)\frac{\theta}{2})}{2\sin\frac{\theta}{2}} - \frac{1}{2}.$$

Are the maxima increasing or decreasing as N increases? Therefore, what type of convergence occurs in neighborhoods of $\theta = 2k\pi$?

The height of the maxima as $N \to \infty$ is known as the Gibbs phenomenon, after the brilliant American mathematician and thermodynamicist Josiah Willard Gibbs (1839–1903).

43

Power Series

We now focus on a special subset of the collection of series of functions, the **power series in** x, $\sum_{n=0}^{\infty} a_n x^n$, where the a_n are constant coefficients. For example, the power series $\sum_{n=0}^{\infty} x^n$ has all $a_n = 1$. A more general form is $\sum_{n=0}^{\infty} a_n (x-c)^n$, called the **power series in** $x - c$. Whenever encountered, 0^0 shall equal 1.

As in the previous chapter, $\sum_{n=0}^{\infty} a_n (x-c)^n$ may converge or diverge at individual values $x = x_0$. It is our job to discover what values of x cause the power series to converge. It must have been a comfort to early researchers to discover that for power series, the collection of points of convergence is always an interval of some kind. This, then, is where our analysis of power series begins.

Theorem 43.1 A power series $\sum_{n=0}^{\infty} a_n x^n$ converges over a set of points S, called the **interval of convergence**, which may be one of these intervals: $(-\beta, \beta)$, $[-\beta, \beta]$, $(-\beta, \beta]$, or $[-\beta, \beta)$, or else either $S = \{0\}$ or $S = \mathbb{R}$. Furthermore, if S is $(-\beta, \beta)$, $\{0\}$, or $\mathbb{R}$, then the power series converges absolutely there.

Proof. First, $x = 0$ (the case $S = \{0\}$) is always a point of convergence. Second, if any theorem about series shows that convergence occurs at every real x, then $S = \mathbb{R}$.

The third, and final, possibility is that $\sum_{n=0}^{\infty} a_n x^n$ converges at some $x_0 \neq 0$ and diverges at some other value y, (we aren't assuming that S is an interval). For ease of writing the proof, let x_0 be positive (otherwise, use $|x_0|$). Then $a_n x_0^n \to 0$ as $n \to \infty$. This implies that $\langle a_n x_0^n \rangle_{n=1}^{\infty}$ is a bounded sequence, that is, there exists a B such that $|a_n x_0^n| \leq B$ for all $n \in \mathbb{N}$. Now, take $x_1 \in (-x_0, x_0)$. Then

$$|a_n x_1^n| = \left|a_n x_0^n\right| \left|\frac{x_1}{x_0}\right|^n \leq B \left|\frac{x_1}{x_0}\right|^n.$$

Since $|\frac{x_1}{x_0}| < 1$, the series of constants $\sum_{n=0}^{\infty} B|\frac{x_1}{x_0}|^n$ is a convergent geometric series. Therefore, by the Weierstrass M-test, $\sum_{n=0}^{\infty} |a_n x_1^n|$ converges at each $x_1 \in (-x_0, x_0)$. We have proved absolute convergence over $[-x_0, x_0]$ and uniform convergence over the open interval $(-x_0, x_0)$. We have also shown that S is an interval.

But the interval just found may not be the largest one, since x_0 wasn't selected for that purpose. Instead, we know that $[-x_0, x_0] \subseteq S$, and that the power series diverges at some $y \notin [-x_0, x_0]$. Thus, S is nonempty, and has an upper bound $|y|$. By the completeness axiom R11, S has a supremum β, with $\beta \geq x_0$. It follows now that $(-\beta, \beta)$ is the interval with maximum length contained in S. At either endpoint, convergence or divergence may occur, as will be seen in the examples and exercises. Hence, the third possibility is that S is one of the four intervals stated in the theorem. **QED**

We have just found that endpoint convergence is not predictable in general for power series. Consequently, uniform convergence can't be guaranteed over $(-\beta, \beta)$. Nevertheless, uniform convergence does occur on any interval *interior* to $(-\beta, \beta)$, that is, in any interval contained in $(-\beta, \beta)$. (It's reasonable that endpoints $-\beta$ and β do not belong to the interior of $(-\beta, \beta)$, but a rigorous definition of the interior of a set must wait for a later chapter.) The following corollary states the case for this delicate situation.

Corollary 43.1 A power series $\sum_{n=0}^{\infty} a_n x^n$ converges uniformly over any closed interval $[\alpha_1, \alpha_2] \subset (-\beta, \beta)$, where $(-\beta, \beta)$ is the interval found in the proof of the theorem above.

Proof. Devise a proof as Exercise 43.8. **QED**

The various intervals of convergence in Theorem 43.1 have a unique length 2β, thought of as the interval's diameter. Because of this, β is called the **radius of convergence** of the power series. Even the last two possibilities for S in that theorem are just special intervals, in this way: $S = \{0\} = [0, 0]$ using $\beta = 0$, and $S = (-\infty, \infty) = \mathbb{R}$ using $\beta = +\infty$ (meaning that S has no supremum). With these ideas in mind, an explicit formula for β may be derived from the root test (Theorem 30.1).

Theorem 43.2 Given a power series $\sum_{n=0}^{\infty} a_n x^n$, its radius of convergence β is

$$\beta = \begin{cases} +\infty & \text{if } \lim_{n\to\infty} \sqrt[n]{|a_n|} = 0 \\ \dfrac{1}{\lim_{n\to\infty} \sqrt[n]{|a_n|}} & \text{if } \lim_{n\to\infty} \sqrt[n]{|a_n|} > 0 \\ 0 & \text{if } \lim_{n\to\infty} \sqrt[n]{|a_n|} = +\infty \end{cases}.$$

Proof. The nth term of the power series is $a_n x^n$, so we consider $\sigma = \lim_{n\to\infty} \sqrt[n]{|a_n x^n|} = |x| \lim_{n\to\infty} \sqrt[n]{|a_n|}$. According to the root test, if $0 \le \sigma < 1$, then the power series converges absolutely. If it is the case that $\lim_{n\to\infty} \sqrt[n]{|a_n|} > 0$, then $\sigma = \frac{|x|}{\beta}$, and the sufficient condition for convergence is that $0 \le \frac{|x|}{\beta} < 1$, or $0 \le |x| < \beta$, or $x \in (-\beta, \beta)$. This shows us that

$$\beta = \frac{1}{\lim\limits_{n\to\infty} \sqrt[n]{|a_n|}}$$

is in fact the radius of convergence of the power series. In case $\lim_{n\to\infty} \sqrt[n]{|a_n|} = 0$ (corresponding to $\beta = +\infty$) then $\sigma = |x| \lim_{n\to\infty} \sqrt[n]{|a_n|} = 0$ for all real x, and the power series converges for all $\mathbb{R}$. And finally, in case $\lim_{n\to\infty} \sqrt[n]{|a_n|} = +\infty$ (corresponding to $\beta = 0$), then the only possible route to convergence is to have $x = 0$, since $\sigma = \lim_{n\to\infty} \sqrt[n]{|a_n \cdot 0|} = 0$. **QED**

A less refined form of Theorem 43.2 uses

$$\beta = \frac{1}{\lim_{n\to\infty}|\frac{a_{n+1}}{a_n}|} = \lim_{n\to\infty}\left|\frac{a_n}{a_{n+1}}\right|,$$

which is based on the ratio test. On balance, this form is popular due to its ease of use.

A more refined form of Theorem 43.2 uses $\beta = \frac{1}{\limsup \sqrt[n]{|a_n|}}$. This form is useful when the limit of $\langle a_n x^n\rangle_1^\infty$ doesn't exist due to the presence of more than one subsequential limit. On the other hand, this form is trickier to use.

With the two theorems in hand, we can find the intervals of convergence for a variety of power series.

Example 43.1 To find the radius of convergence of $\sum_{n=1}^\infty x^n$, we note that $a_n = 1$ for all $n \geq 1$. Therefore, $\beta = (\lim_{n\to\infty}\sqrt[n]{|a_n|})^{-1} = 1$. Using the less refined form above, we also find $\beta = \lim_{n\to\infty}|\frac{a_n}{a_{n+1}}| = 1$. Thus, the series converges absolutely over $(-1, 1)$ by Theorem 43.1. At the endpoints, the series diverges. Hence, the interval of convergence is $S = (-1, 1)$. Outside of the interval, the power series diverges. See Exercise 43.1. ■

A simple extension of Theorem 43.1 is: suppose that the power series $\sum_{n=0}^\infty a_n x^n$ has an interval of convergence with endpoints $-\beta$ and β. Then the more general power series $\sum_{n=0}^\infty a_n(x-c)^n$ will have the corresponding interval of convergence with endpoints $c-\beta$ and $c+\beta$. This is the effect of a translation (or shift) of the series by c units. For instance, if we change the power series in the previous example to $\sum_{n=1}^\infty (x+10)^n$, then the interval of convergence becomes $(-1-10, 1-10)$, or $(-11, -9)$. Of course, the radius of convergence is not affected by translations, nor is the convergence or divergence at the endpoints.

Example 43.2 We return to our old friend $\sum_{n=1}^\infty x^n(1-x)$ from Example 42.8. In this form it isn't a power series, but we can study the equivalent $(1-x)\sum_{n=1}^\infty x^n$. From the previous example, $\sum_{n=1}^\infty x^n$ converges absolutely over $(-1, 1)$, therefore the interval of (absolute) convergence of $\sum_{n=1}^\infty x^n(1-x)$ is $S = (-1, 1]$.

The series $\sum_{n=1}^\infty (x-3)^n(4-x)$ may be rewritten as $\sum_{n=1}^\infty (x-3)^n(1-(x-3))$, which shows it to be a translation of the original series. Hence, the interval of convergence shifts to $(2, 4]$. ■

Example 43.3 What is the interval of convergence of $\sum_{n=1}^\infty \frac{x^n}{n}$? We have

$$\beta = \lim_{n\to\infty}\left|\frac{a_n}{a_{n+1}}\right| = \lim_{n\to\infty}\frac{n+1}{n} = 1$$

(or with the help of Exercise 30.10,

$$\beta = \left(\lim_{n\to\infty}\sqrt[n]{\frac{1}{n}}\right)^{-1} = \lim_{n\to\infty}\sqrt[n]{n} = 1).$$

Therefore, the power series converges absolutely over $(-1, 1)$. At $x = -1$, we obtain the opposite of the alternating harmonic series, $\sum_{n=1}^\infty \frac{(-1)^n}{n} = -\ln 2$. At $x = 1$, we find the harmonic series. Therefore, the interval of convergence of this power series is $S = [-1, 1)$, with conditional convergence at the left endpoint. For x outside this interval, divergence occurs. ■

Example 43.4 What interval of convergence does $\sum_{n=1}^{\infty}(\frac{3nx}{n+1})^n$ have? First, identify it as a legitimate power series,

$$\sum_{n=1}^{\infty}\left(\frac{3n}{n+1}\right)^n x^n \quad \text{with} \quad a_n = \left(\frac{3n}{n+1}\right)^n.$$

Then

$$\beta = \left(\lim_{n\to\infty}\sqrt[n]{|a_n|}\right)^{-1} = \left(\lim_{n\to\infty}\frac{3n}{n+1}\right)^{-1} = \frac{1}{3}.$$

Hence, the power series converges absolutely over $(-\frac{1}{3}, \frac{1}{3})$. (It is possible to find β as $\lim_{n\to\infty}|\frac{a_n}{a_{n+1}}|$, but only if you have time for some weight-lifting with limit theorems.)

At the right endpoint $x = \frac{1}{3}$, we find the series

$$\sum_{n=1}^{\infty}\left(\frac{n}{n+1}\right)^n.$$

Here,

$$a_n = \frac{1}{\left(1+\frac{1}{n}\right)^n} \to \frac{1}{e},$$

from which we conclude that the power series diverges there. Finally, at the endpoint $x = -\frac{1}{3}$, we have

$$a_n = \frac{(-1)^n}{\left(1+\frac{1}{n}\right)^n},$$

which has the two subsequential limits $\pm\frac{1}{e}$ (verify!). Hence, the series

$$\sum_{n=1}^{\infty}\frac{(-1)^n}{\left(1+\frac{1}{n}\right)^n}$$

diverges, meaning that the power series diverges at $x = -\frac{1}{3}$. The interval of convergence of the given power series is therefore $S = (-\frac{1}{3}, \frac{1}{3})$.

Don't confuse the discovery of divergence at the left endpoint with the statement about the refinement of Theorem 43.2. That idea pertains to evaluating the radius of convergence β, not to determining the limit of a series of constants. ■

Example 43.5 What is the interval of convergence of $\sum_{n=0}^{\infty}\frac{x^n}{n!} = 1 + x + \frac{x^2}{2!} + \frac{x^3}{3!} + \cdots$? We have $\beta = \lim_{n\to\infty}|\frac{(n+1)!}{n!}| = +\infty$. (Also,

$$\beta = \left(\lim_{n\to\infty}\sqrt[n]{\frac{1}{n!}}\right)^{-1} = \lim_{n\to\infty}\sqrt[n]{n!} = +\infty,$$

using Exercise 18.8.) Therefore, this famous power series converges absolutely over all $\mathbb{R}$. Many calculus textbooks show graphs of several partial sums of this power series, giving strong visual evidence of the convergence (to $F(x) = e^x$, as will be proved in Chapter 45). ■

Example 43.6 Find the interval of convergence of the power series

$$\frac{1}{3}+\frac{1}{5}x+\frac{1}{3^2}x^2+\frac{1}{5^2}x^3+\frac{1}{3^3}x^4+\frac{1}{5^3}x^5+\frac{1}{3^4}x^6+\frac{1}{5^4}x^7+\cdots.$$

In this case, $\lim_{n\to\infty}|\frac{a_n}{a_{n+1}}|$ fails to exist (verify this as Exercise 43.2a). Things get worse, for if n is even, then

$$\lim_{n\to\infty}\sqrt[n]{|a_n|}=\lim_{n\to\infty}\sqrt[n]{\frac{1}{5^{n/2}}}=\frac{1}{\sqrt{5}},$$

whereas if n is odd,

$$\lim_{n\to\infty}\sqrt[n]{|a_n|}=\lim_{n\to\infty}\sqrt[n]{\frac{1}{3^{(n+1)/2}}}=\frac{1}{\sqrt{3}}.$$

In other words, $\lim_{n\to\infty}\sqrt[n]{|a_n|}$ doesn't exist. This is the time to use the refined form of Theorem 43.2, where we find that $\limsup\sqrt[n]{|a_n|}=\frac{1}{\sqrt{3}}$, the maximum subsequential limit of $\langle\sqrt[n]{|a_n|}\rangle_1^\infty$. Therefore, $\beta=\sqrt{3}$, and the given series converges absolutely over $(-\sqrt{3},\sqrt{3})$.

Convergence at the endpoints is left for you to investigate in Exercise 43.2b. ■

Example 43.7 Find the interval of convergence of the power series $\sum_{n=0}^{\infty}\frac{x^{4n}}{2^n}$. This is a power series in x^4, not in x. Thus, we make the change of variable $4n=m$, and arrive at $\sum_{m=0}^{\infty}\frac{x^m}{2^{m/4}}$. The effect of this is to form a new power series with the same radius of convergence (verify). Now we obtain $\beta=(\lim_{m\to\infty}\sqrt[m]{|a_m|})^{-1}=(\lim_{m\to\infty}\sqrt[m]{2^{-m/4}})^{-1}=\sqrt[4]{2}$. Consequently, the power series converges absolutely over $(-\sqrt[4]{2},\sqrt[4]{2})$.

At $x=-\sqrt[4]{2}$, we uncover the series

$$\sum_{n=0}^{\infty}\frac{1}{2^n}(-\sqrt[4]{2})^{4n}=\sum_{n=0}^{\infty}1$$

(or use $\sum_{m=0}^{\infty}\frac{1}{2^{m/4}}(-\sqrt[4]{2})^m=\sum_{m=0}^{\infty}(-1)^m$). At $x=\sqrt[4]{2}$, a similar result happens. Therefore, the interval of convergence is precisely $S=(-\sqrt[4]{2},\sqrt[4]{2})$.

In Figure 43.1, the partial sum polynomials s_2, s_3, s_5, and s_{11} of the series appear above its interval of convergence. As a warning, these graphs don't hint at the interval of convergence. Investigate the behavior of the partial sums near an endpoint in Exercise 43.5.

From another point of view, the power series is, at each x, a geometric series. That is, $\sum_{n=0}^{\infty}\frac{x^{4n}}{2^n}$ has common ratio $r=\frac{x^4}{2}$, and therefore converges to a sum $F(x)=\frac{2}{2-x^4}$ if and only if $|\frac{x^4}{2}|<1$. These results reinforce two facts: convergence occurs over $(-\sqrt[4]{2},\sqrt[4]{2})$, and divergence occurs at the endpoints. ■

So far, our theorems have not said anything general about convergence at endpoints of the interval of convergence. Thus, it is a welcome discovery by Abel that if a power series converges at one of its endpoints, then it converges uniformly out to that endpoint. The proof serves as a good application of a number of past ideas!

Theorem 43.3 Abel's Theorem If a power series $\sum_{n=0}^{\infty}a_nx^n$ converges over $(-\beta,\beta]$ (implying $\beta>0$), then it will converge *uniformly* over $[0,\beta]$.

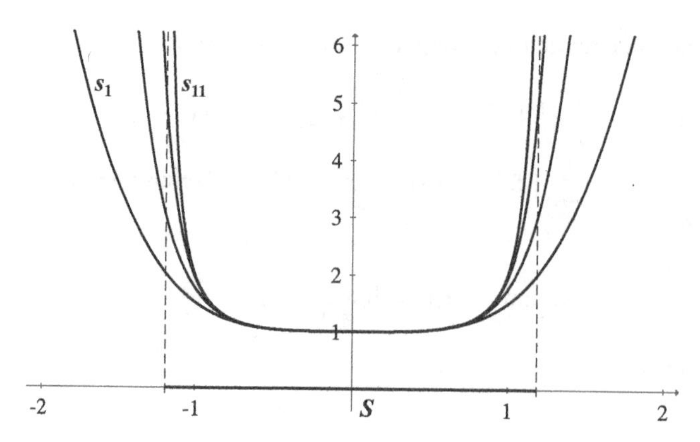

Figure 43.1. The partial sums s_2, s_3, s_5, and s_{11} show little evidence of convergence over their interval of convergence $(-\sqrt[4]{2}, \sqrt[4]{2})$

Proof. Begin with the change of variable $y = \frac{x}{\beta}$, forming a power series $\sum_{n=0}^{\infty} a_n y^n$ that converges over $y \in (-1, 1]$. In particular, at $y = 1$, $\sum_{n=0}^{\infty} a_n$ converges.

We will prove that Theorem 42.2, the Cauchy criterion for uniform convergence, is satisfied. The criterion is that for any $\varepsilon > 0$ and all $y \in [0, 1]$, we must be able to find a positive M_ε such that if $m \geq n > M_\varepsilon$, then $|\sum_{k=0}^{m} a_k y^k - \sum_{k=0}^{n-1} a_k y^k| < \varepsilon$, that is, $|\sum_{k=n}^{m} a_k y^k| < \varepsilon$.

For the moment, use only $y \in [0, 1)$. Bring in Abel's lemma (Lemma 33.1) in the following useful form. Consider the sequences $\langle a_n \rangle_0^\infty$ and $\langle y^n \rangle_0^\infty$ that appear in $\sum_{n=0}^{\infty} a_n y^n$, and let the partial sums of $\langle a_n \rangle_0^\infty$ be $A_k = \sum_{j=0}^{k} a_j$, then for all $m \geq n \geq 0$, we have

$$\left|\sum_{k=n}^{m} a_k y^k\right| \leq y^{m+1}|A_m| + y^n|A_{n-1}| + \sum_{k=n}^{m}(y^k - y^{k+1})|A_k| \text{ (set } A_{-1} \equiv 0). \tag{1}$$

Now, $\langle A_k \rangle_0^\infty$ is a bounded sequence, so let $|A_k| \leq B$ for all $k \in \mathbb{N} \cup \{0\}$. Therefore,

$$\begin{aligned} &y^{m+1}|A_m| + y^n|A_{n-1}| + \sum_{k=n}^{m}(y^k - y^{k+1})|A_k| \\ &\leq B\left[y^{m+1} + y^n + \sum_{k=n}^{m}(y^k - y^{k+1})\right] = 2By^n. \end{aligned} \tag{2}$$

(The inequality is true because all the differences $y^k - y^{k+1}$ are nonnegative.)

Next, given any $\varepsilon > 0$, and since $y \in [0, 1)$, there does exist a large enough positive N_ε such that for all $n > N_\varepsilon$, we have $y^n < \frac{\varepsilon}{2B}$.

Putting inequalities *(1)* and *(2)* together, we arrive at $|\sum_{k=n}^{m} a_k y^k| \leq 2By^n < \varepsilon$ for any $y \in [0, 1)$. For the special case $y = 1$, we have $|\sum_{k=n}^{m} a_k| < \varepsilon$ when $m \geq n > N'_\varepsilon$, because the partial sums of $\sum_{n=0}^{\infty} a_n$ form a Cauchy sequence.

Therefore, for any $\varepsilon > 0$ and for any $y \in [0, 1]$, take $M_\varepsilon = \max\{N_\varepsilon, N'_\varepsilon\}$, and we are guaranteed that when $m \geq n > M_\varepsilon$, then $|\sum_{k=n}^{m} a_k y^k| < \varepsilon$. Consequently, $\sum_{n=0}^{\infty} a_n y^n$ converges uniformly over $[0, 1]$.

We transport this conclusion back to $\sum_{n=0}^{\infty} a_n x^n$, which now converges uniformly for any $x \in [0, \beta]$. **QED**

Now that Abel's famous theorem is part of our arsenal, we can easily extend the interval of uniform convergence of a power series from $[0, \beta]$ to $[\alpha, \beta]$, where α is any left endpoint such that $-\beta < \alpha$, by using Corollary 43.1. Finally, we can affirm the following statement, which extends Corollary 43.1.

Corollary 43.2 If a power series $\sum_{n=0}^{\infty} a_n x^n$ has a closed interval of convergence $[-\beta, \beta]$, then it converges uniformly over the interval.

Proof. Create a short proof as Exercise 43.9. **QED**

Abel's theorem used $y = \frac{x}{\beta}$. This transformation, together with all translations, fall into the more general **linear transformation** $x \mapsto mx + b$. Investigate these in Exercise 43.10.

43.1 Exercises

1. What differences and likenesses do you find in the analyses in Example 43.1 and Example 42.1?
2. a. Verify that $\lim_{n\to\infty} |\frac{a_n}{a_{n+1}}|$ fails to exist for the power series in Example 43.6.
   b. In the same example, we found that the series converged over $(-\sqrt{3}, \sqrt{3})$. Determine whether or not the series converges at the endpoints, thus establishing the interval of convergence of this unusual power series.
3. Find the intervals of convergence S of the following power series.
   a. $\sum_{n=1}^{\infty} n(x+5)^n$
   b. $\sum_{n=1}^{\infty} \dfrac{x^n}{n(n+1)}$
   c. $\sum_{n=1}^{\infty} \dfrac{(-x)^n}{n(n+1)}$
   d. $x - \dfrac{x^3}{3} + \dfrac{x^5}{5} - \dfrac{x^7}{7} + \cdots$
   e. $\dfrac{x^2}{2} - \dfrac{x^4}{4} + \dfrac{x^6}{6} - \dfrac{x^8}{8} + \cdots$
   f. $\sum_{n=0}^{\infty} \dfrac{x^{3n}}{5^n}$
   g. $\sum_{n=0}^{\infty} \dfrac{(3x)^n}{5^n}$
   h. $\sum_{n=1}^{\infty} \dfrac{(x-1)^n}{n^2}$
   i. $\sum_{n=1}^{\infty} \dfrac{n^n x^n}{n!}$
   j. $\sum_{n=0}^{\infty} \dfrac{(2x+6)^{2n+1}}{2n+1}$
   k. $\sum_{n=1}^{\infty} \dfrac{(2x)^n}{\sqrt{n}}$
   l. $\sum_{n=0}^{\infty} \dfrac{(2n)!}{2^n n!} x^n = 1 + (1)x + (1\cdot3)x^2 + (1\cdot3\cdot5)x^3 + (1\cdot3\cdot5\cdot7)x^4 + \cdots$.
4. **Bessel functions** are solutions to an important differential equation in the history of astronomy, discovered in 1817 by the astronomer and mathematician Friedrich Wilhelm Bessel (1784–1846), a close friend of Carl F. Gauss. A particular Bessel function is

$$J_1(x) = \frac{x}{2} - \frac{x^3}{2^3(1!2!)} + \frac{x^5}{2^5(2!3!)} - \frac{x^7}{2^7(3!4!)} + \cdots,$$

with $\mathcal{D}_{J_1} = \{x \geq 0\}$, see Figure 43.2. Find its interval of convergence.

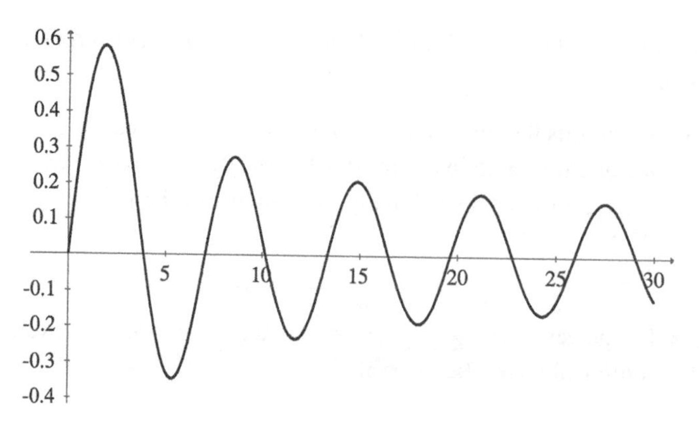

Figure 43.2.

5. The partial sum polynomials graphed in the figure for Example 43.7 don't show any unusual behavior near the endpoints of the interval of convergence $(-\sqrt[4]{2}, \sqrt[4]{2})$. Nevertheless, use a calculator to investigate the power series in a neighborhood of $x = \sqrt[4]{2} \approx 1.18921$ by evaluating the following partial sums (use five or six significant digits) at the indicated values of x. Comment on the rates of convergence or divergence.

	Evaluate at:		
Partial sums	$x = 1.12$	$x = 1.18$	$x = 1.19$
$\sum_{n=0}^{5} \frac{x^{4n}}{2^n}$			
$\sum_{n=0}^{50} \frac{x^{4n}}{2^n}$			
$\sum_{n=0}^{100} \frac{x^{4n}}{2^n}$			
$\sum_{n=0}^{200} \frac{x^{4n}}{2^n}$			
Series sum $\sum_{n=0}^{\infty} \frac{x^{4n}}{2^n}$			

6. Prove that if a power series converges absolutely at one of its endpoints, then it converges absolutely at its other endpoint as well.

7. The power series $\sum_{n=1}^{\infty} \frac{x^n}{n(n+1)}$ in Exercise 43.3b has the interval of convergence $[-1, 1]$. According to Corollary 43.2, it converges uniformly over the closed interval. Graph partial sums s_5, s_{10}, and s_{29} over a wider interval such as $[-1.3, 1.2]$, and look closely at the behavior near $x = \pm 1$ to observe uniform convergence. As will be seen in Exercise 45.7, the unlikely sum function is

$$F(x) = \begin{cases} -\ln(1-x) + \frac{\ln(1-x)}{x} + 1 & \text{if } x \in [-1, 0) \cup (0, 1) \\ 0 & \text{if } x = 0 \\ 1 & \text{if } x = 1 \end{cases}.$$

Show that F is a continuous function at $x = 0$ and $x = 1$, as Theorem 41.1 demands.

8. Prove Corollary 43.1. Hint: Look at the use of the Weierstrass M-test in the proof of Theorem 43.1.

9. Prove Corollary 43.2.

10. Let $\sum_{n=0}^{\infty} a_n y^n$ have an interval of convergence $(-\beta, \beta)$ (as usual, $\beta \geq 0$, and $\beta = 0$ implies that the interval is $[0, 0]$). Then the power series $\sum_{n=0}^{\infty} a_n(mx + b)^n$ found by substituting $y = mx + b$ will have an interval of convergence that is both translated and dilated.
    a. Starting with the interval of convergence $-\beta < y < \beta$ find the interval of convergence of $\sum_{n=0}^{\infty} a_n(mx + b)^n$, assuming first that $m > 0$. What do you get when $m < 0$? Prove your results again, by starting from the definition of β in Theorem 43.2. Hint: $\sum_{n=0}^{\infty} a_n(mx + b)^n = \sum_{n=0}^{\infty} a_n m^n (x - \frac{-b}{m})^n$, with cases when m is positive and negative.
    b. Find the interval of convergence of $\sum_{n=1}^{\infty} \frac{(-5x+2.3)^n}{n}$, knowing that $\sum_{n=1}^{\infty} \frac{x^n}{n}$ has interval of convergence $[-1, 1)$.

44

Operations with Power Series

Now that power series have been well established, we need to consider some operations involving them. For instance, how do we add, $\sum_{n=0}^{\infty} a_n x^n + \sum_{n=0}^{\infty} b_n x^n$? The addition is meaningless for any x at which either series diverges. Thus, we limit ourselves to x in the intersection of their intervals of convergence. Going further, if $\sum_{n=0}^{\infty} a_n x^n$ has a sum $F_A(x)$ over an interval of convergence S_A, and $\sum_{n=0}^{\infty} b_n x^n$ has a sum $F_B(x)$ over an interval of convergence S_B, then $\sum_{n=0}^{\infty} a_n x^n + \sum_{n=0}^{\infty} b_n x^n = \sum_{n=0}^{\infty}(a_n + b_n)x^n$ and has the sum function $F_A(x) + F_B(x)$ over $S_A \cap S_B$. How are we sure of this? It is a direct application of Theorem 28.2. $S_A \cap S_B$ is centered at zero, as is necessary for intervals of convergence of $\sum_{n=0}^{\infty}(a_n + b_n)x^n$ by Theorem 43.1. We have proved our first theorem, about the **sum of power series**.

Theorem 44.1 Let $\sum_{n=0}^{\infty} a_n x^n$ have the sum function $F_A(x)$ over an interval of convergence S_A, and let $\sum_{n=0}^{\infty} b_n x^n$ have the sum function $F_B(x)$ over an interval of convergence S_B. Then $\sum_{n=0}^{\infty} a_n x^n + \sum_{n=0}^{\infty} b_n x^n = \sum_{n=0}^{\infty}(a_n + b_n)x^n$ and this power series has the sum function $F_A(x) + F_B(x)$ over the interval of convergence $S_A \cap S_B$.

Example 44.1 From the previous chapter, we know that $\sum_{n=1}^{\infty} x^n = \frac{x}{1-x} = F_A(x)$ over $S_A = (-1, 1)$. Consequently, $\sum_{n=1}^{\infty}(-1)^n x^n$, or $\sum_{n=1}^{\infty}(-x)^n$, has the sum $F_B(x) = \frac{-x}{1+x}$ over $S_B = S_A$. By Theorem 44.1, we conclude first that $\sum_{n=1}^{\infty} x^n + \sum_{n=1}^{\infty}(-1)^n x^n = \sum_{n=1}^{\infty}(1 + (-1)^n)x^n = 0 + 2x^2 + 0 + 2x^4 + 0 + 2x^6 + \cdots$ over $(-1, 1)$, and second, that the new series has sum function $F_A(x) + F_B(x) = \frac{2x^2}{(1-x)(1+x)}$ over $(-1, 1)$.

Changing our viewpoint, consider $2x^2 + 2x^4 + \cdots = 2\sum_{n=1}^{\infty} x^{2n}$. This is a geometric power series with common ratio $r = x^2$, which therefore has the sum $F(x) = 2(\frac{x^2}{1-x^2})$ over $S = (-1, 1)$. This agrees with $F_A(x) + F_B(x)$. ■

Not surprisingly, the next theorem is about **multiplication of power series**. We deal with Cauchy products of series (see Chapter 33).

Theorem 44.2 Let $\sum_{n=0}^{\infty} a_n x^n$ have the sum $F_A(x)$ over an interval of convergence S_A, and $\sum_{n=0}^{\infty} b_n x^n$ have the sum $F_B(x)$ over an interval of convergence S_B. Then the Cauchy product $\sum_{n=0}^{\infty} c_n x^n$ (where $c_n = \sum_{k=0}^{n} a_k b_{n-k}$) has the sum function $F_A(x)F_B(x)$ over $S_A \cap S_B$.

Proof. We have proved by Theorem 43.1 that both $\sum_{n=0}^{\infty} a_n x^n$ and $\sum_{n=0}^{\infty} b_n x^n$ converge absolutely at each x in the interiors of their respective intervals of convergence. Thus, at each $x \in S_A \cap S_B$, Mertens' theorem (Theorem 33.1) indicates that the Cauchy product $\sum_{n=0}^{\infty} c_n x^n$ converges, with sum function $F_A(x)F_B(x)$. **QED**

Example 44.2 From the previous example, $F_A(x) = \frac{x}{1-x}$ is the sum function of $\sum_{n=1}^{\infty} x^n$ over $S_A = (-1, 1)$. As we shall see in Exercise 45.8c, $F_B(x) = -\ln(1-x)$ is the sum function of $\sum_{n=1}^{\infty} \frac{x^n}{n}$ over $[-1, 1)$. Therefore, by our latest theorem,

$$F_A(x)F_B(x) = \frac{x\ln(1-x)}{x-1}$$

is the sum of the Cauchy product $(\sum_{n=1}^{\infty} x^n)(\sum_{n=1}^{\infty} \frac{x^n}{n})$ over $S_A \cap S_B = (-1, 1)$.

Let's expand this product. We write $(0 + x + x^2 + x^3 + \cdots)(0 + x + \frac{x^2}{2} + \frac{x^3}{3} + \cdots)$ and proceed with the evaluations

$$c_0 = a_0 b_0 = 0,\ c_1 = a_0 b_1 + a_1 b_0 = 0,\ c_2 = a_0 b_2 + a_1 b_1 + a_2 b_0 = 1,$$
$$c_3 = \sum_{k=0}^{3} a_k b_{n-k} = \frac{1}{2} + 1,\ c_4 = \sum_{k=0}^{4} a_k b_{n-k} = \frac{1}{3} + \frac{1}{2} + 1,$$

and so on. The curious power series that emerges is

$$x^2 + \left(1 + \frac{1}{2}\right)x^3 + \left(1 + \frac{1}{2} + \frac{1}{3}\right)x^4 + \cdots = \sum_{n=2}^{\infty}\left(\sum_{k=1}^{n-1} \frac{1}{k}\right)x^n.$$

To summarize,

$$\left(\sum_{n=1}^{\infty} x^n\right)\left(\sum_{n=1}^{\infty} \frac{x^n}{n}\right) = \sum_{n=2}^{\infty}\left(\sum_{k=1}^{n-1} \frac{1}{k}\right)x^n = \frac{x\ln(1-x)}{x-1},$$

and the sum holds for $x \in (-1, 1)$. ■

Example 44.3 In Example 43.7, we found that

$$\sum_{n=0}^{\infty} \frac{x^{4n}}{2^n} = F(x) = \frac{2}{2-x^4}$$

over $(-\sqrt[4]{2}, \sqrt[4]{2})$. Hence, translating to $c = -\frac{1}{2}$ yields the series $\sum_{n=0}^{\infty} \frac{1}{2^n}\left(x + \frac{1}{2}\right)^{4n}$, which has the sum

$$F_A(x) = \frac{2}{2 - \left(x + \frac{1}{2}\right)^4}$$

over $(-\sqrt[4]{2} - \frac{1}{2}, \sqrt[4]{2} - \frac{1}{2})$. From Example 43.2, when we translate $\sum_{n=1}^{\infty} x^n(1-x)$ to $c = 1$, we arrive at the series $\sum_{n=1}^{\infty}(x-1)^n(2-x)$ with an interval of convergence $(0, 2]$, and which has the sum $F_B(x) = x - 1$ (verify these two claims). By Theorem 44.2, the Cauchy product

$$\left(\sum_{n=0}^{\infty} \frac{1}{2^n}\left(x + \frac{1}{2}\right)^{4n}\right)\left(\sum_{n=1}^{\infty}(x-1)^n(2-x)\right)$$

will have the sum

$$F_A(x)F_B(x) = \frac{2(x-1)}{2-\left(x+\frac{1}{2}\right)^4}$$

over the interval of convergence $(-\sqrt[4]{2}-\frac{1}{2}, \sqrt[4]{2}-\frac{1}{2}) \cap (0, 2] = (0, \sqrt[4]{2}-\frac{1}{2})$.

The Cauchy product's expansion is considered in Exercise 44.2. ■

Division of power series is not a theoretically difficult idea, just lengthy to do by hand. Given $\sum_{n=0}^{\infty} a_n x^n$ and $\sum_{n=0}^{\infty} b_n x^n$, we define the **quotient of two power series** as the series $\sum_{n=0}^{\infty} q_n x^n$ such that $(\sum_{n=0}^{\infty} b_n x^n)(\sum_{n=0}^{\infty} q_n x^n) = \sum_{n=0}^{\infty} a_n x^n$, and write

$$\sum_{n=0}^{\infty} q_n x^n = \frac{\sum_{n=0}^{\infty} a_n x^n}{\sum_{n=0}^{\infty} b_n x^n}.$$

The following theorem describes the conditions for such a division.

Theorem 44.3 Let $\sum_{n=0}^{\infty} a_n x^n$ and $\sum_{n=0}^{\infty} b_n x^n$ have intervals of convergence S_A and S_B, and sum functions $F_A(x)$ and $F_B(x)$, respectively. Let x_B equal the minimum of the absolute values of the roots of $\sum_{n=0}^{\infty} b_n x^n$ (if no roots exist, take $x_B = +\infty$). If $b_0 \neq 0$, then the quotient power series

$$\sum_{n=0}^{\infty} q_n x^n = \frac{\sum_{n=0}^{\infty} a_n x^n}{\sum_{n=0}^{\infty} b_n x^n} = \frac{F_A(x)}{F_B(x)}$$

has the interval of convergence $(-x_B, x_B) \cap S_A \cap S_B$. Furthermore, $\sum_{n=0}^{\infty} q_n x^n$ has the following coefficients:

$$\begin{aligned}
q_0 &= \frac{a_0}{b_0}, \\
q_1 &= \frac{a_1}{b_0} - \frac{a_0}{b_0^2} b_1, \\
q_2 &= \frac{a_2}{b_0} - \frac{a_1}{b_0^2} b_1 + \frac{a_0}{b_0^3}(b_1^2 - b_2 b_0), \\
q_3 &= \frac{a_3}{b_0} - \frac{a_2}{b_0^2} b_1 + \frac{a_1}{b_0^3}(b_1^2 - b_2 b_0) - \frac{a_0}{b_0^4}(b_1^3 - 2b_2 b_1 b_0 + b_3 b_0^2), \\
q_4 &= \frac{a_4}{b_0} - \frac{a_3}{b_0^2} b_1 + \frac{a_2}{b_0^3}(b_1^2 - b_2 b_0) - \frac{a_1}{b_0^4}(b_1^3 - 2b_2 b_1 b_0 + b_3 b_0^2) \\
&\quad + \frac{a_0}{b_0^5}(b_1^4 - 3b_2 b_1^2 b_0 + 2b_3 b_1 b_0^2 + b_2^2 b_0^2 - b_4 b_0^3), \\
q_5 &= \frac{a_5}{b_0} - \frac{a_4}{b_0^2} b_1 + \frac{a_3}{b_0^3}(b_1^2 - b_2 b_0) - \frac{a_2}{b_0^4}(b_1^3 - 2b_2 b_1 b_0 + b_3 b_0^2) \\
&\quad + \frac{a_1}{b_0^5}(b_1^4 - 3b_2 b_1^2 b_0 + 2b_3 b_1 b_0^2 + b_2^2 b_0^2 - b_4 b_0^3) \\
&\quad - \frac{a_0}{b_0^6}(b_1^5 - 4b_1^3 b_2 b_0 + 3b_1^2 b_3 b_0^2 + 3b_1 b_2^2 b_0^2 - 2b_1 b_4 b_0^3 - 2b_2 b_3 b_0^3 + b_5 b_0^4), \ldots .
\end{aligned}$$

Proof. At each $x \in S_A \cap S_B$, both of the given series converge, and wherever the denominator series $\sum_{n=0}^{\infty} b_n x^n \neq 0$, the ratio $\frac{\sum_{n=0}^{\infty} a_n x^n}{\sum_{n=0}^{\infty} b_n x^n}$ exists, implying that $\sum_{n=0}^{\infty} q_n x^n$ converges at those x. Successively equating the coefficients of the Cauchy product $(\sum_{n=0}^{\infty} b_n x^n)(\sum_{n=0}^{\infty} q_n x^n)$ with those of $\sum_{n=0}^{\infty} a_n x^n$ gives

$$b_0 q_0 = a_0, b_0 q_1 + b_1 q_0 = a_1, \ldots, \sum_{k=0}^{n} b_k q_{n-k} = a_n,$$

and so on. Solving the equations for the q_n gives

$$q_0 = \frac{a_0}{b_0}, \quad q_1 = \frac{a_1 - b_1 q_0}{b_0}, \quad q_2 = \frac{a_2 - b_1 q_1 - b_2 q_0}{b_0}, \ldots, \quad q_n = \frac{a_n - \sum_{k=1}^{n} b_k q_{n-k}}{b_0}, \ldots.$$

(These equations may be used in an algorithm to evaluate the q_n recursively.) Substitutions then give the explicit formulas in the theorem.

We have seen that whenever $x \in S_A \cap S_B$ and is not a root of $\sum_{n=0}^{\infty} b_n x^n$, the quotient series converges. Since $b_0 \neq 0$, it follows that $x_B \neq 0$ and $(-x_B, x_B) \neq \varnothing$. The interval of convergence of the quotient series cannot contain any $x \notin (-x_B, x_B)$, or else it would contain x_B also, by Theorem 43.1. Hence, the interval of convergence for the quotient series is $(-x_B, x_B) \cap S_A \cap S_B$, and for any x in this interval, $\sum_{n=0}^{\infty} q_n x^n$ converges to $\frac{F_A(x)}{F_B(x)}$. **QED**

Example 44.4 Rational functions make easy power series via long division or otherwise. Thus, $\frac{1}{-x+3}$, performed as the long division $1 \div (3 - x)$, gives

$$1 \div (3 - x) = \frac{1}{3} + \frac{1}{9}x + \frac{1}{27}x^2 + \frac{1}{81}x^3 + \cdots,$$

and

$$\frac{5x+2}{2x-3} = (2 + 5x) \div (-3 + 2x) = -\frac{2}{3} - \frac{19}{9}x - \frac{38}{27}x^2 - \frac{76}{81}x^3 - \cdots.$$

Let's see what the division $\frac{1}{-x+3} \div \frac{5x+2}{2x-3}$ is like with power series. We will use

$$a_0 = \frac{1}{3} \quad a_1 = \frac{1}{9} \quad a_2 = \frac{1}{27}$$
$$b_0 = -\frac{2}{3} \quad b_1 = -\frac{19}{9} \quad b_2 = -\frac{38}{27}.$$

Thus, $q_0 = \frac{a_0}{b_0} = -\frac{1}{2}$, and evaluating two more coefficients from Theorem 44.3, we get

$$\begin{aligned} q_0 + q_1 x + q_2 x^2 &= -\frac{1}{2} + \left[\frac{1}{9} \cdot \frac{-3}{2} - \frac{1}{3} \cdot \left(\frac{-3}{2}\right)^2 \cdot \frac{-19}{9}\right] x \\ &\quad + \left[\frac{1}{27} \cdot \frac{-3}{2} - \frac{1}{9} \cdot \left(\frac{-3}{2}\right)^2 \cdot \frac{-19}{9} + \frac{1}{3} \cdot \left(\frac{-3}{2}\right)^3 \left(\left(\frac{-19}{9}\right)^2 - \frac{-38}{27} \cdot \frac{-2}{3}\right)\right] x^2 \\ &= -\frac{1}{2} + \frac{17}{12}x - \frac{251}{72}x^2. \end{aligned}$$

Compare this with the power series resulting from the long division $\frac{2x-3}{(-x+3)(5x+2)}$, performed as $(-3 + 2x) \div (6 + 13x - 5x^2)$, and we get perfect agreement.

The numerator $F_A(x) = \frac{1}{-x+3}$ has an asymptote at $x = 3$, so that its power series converges over $(-3, 3)$. For the same reason, the denominator $F_B(x) = \frac{5x+2}{2x-3}$ has a power series converging over $(-\frac{3}{2}, \frac{3}{2})$. $F_B(x)$ also has a root at $x = -\frac{2}{5}$, so $x_B = \frac{2}{5}$. Therefore, the interval of convergence of the quotient power series

$$-\frac{1}{2} + \frac{17}{12}x - \frac{251}{72}x^2 + \cdots \text{ is } \left(-\frac{2}{5}, \frac{2}{5}\right) \cap (-3, 3) \cap \left(-\frac{3}{2}, \frac{3}{2}\right) = \left(-\frac{2}{5}, \frac{2}{5}\right). \quad \blacksquare$$

Note 1: If the premise that $b_0 \neq 0$ is denied in Theorem 44.3, then logic tells us that the conclusions of the theorem may not hold. For example, $x - \frac{1}{6}x^3 + \frac{1}{120}x^5 - \cdots$ has the sum function $F_A(x) = \sin x$, and $0 + x + 0x^2 + 0x^3 + \cdots$ is the series with the (obvious) sum function $F_B(x) = x$. The quotient series for $\frac{\sin x}{x}$ exists, and is given by

$$\left(x - \frac{1}{6}x^3 + \frac{1}{120}x^5 - \cdots\right) \Big/ x = 1 - \frac{1}{6}x^2 + \frac{1}{120}x^4 - \cdots,$$

despite the fact that $b_0 = 0$.

Without the help of Theorem 44.3, however, we must now worry about whether the quotient series converges to $\frac{\sin x}{x}$ at each x-value within an interval. This is not so at $x = 0$, for instance. We are facing the problem of *representation* of a function by a series, which will be studied in the next chapter.

Compare with Exercise 44.5.

Composition of power series is a useful operation. As an example of what we will do, recall that stating that $e^u = \sum_{m=0}^{\infty} \frac{u^m}{m!}$ means that $F(u) = e^u$ is the series sum, with $\mathcal{D}_F = \mathbb{R}$. Now, suppose that we know that

$$\sum_{n=1}^{\infty} \frac{(-1)^{n+1}x^n}{n}$$

has sum $u(x) = \ln(x + 1)$ over $(-1, 1]$. That is,

$$u(x) = \sum_{n=1}^{\infty} \frac{(-1)^{n+1}x^n}{n} = \ln(x + 1)$$

with $\mathcal{D}_u = (-1, 1]$. The question arises, "Does $(F \circ u)(x) = e^{\ln(x+1)} = x + 1$ equal the composition of the two power series at each $x \in (-1, 1]$?"

The composition is formally written as

$$\sum_{m=0}^{\infty}\left(\frac{1}{m!}\left(\sum_{n=1}^{\infty}\frac{(-1)^{n+1}x^n}{n}\right)^m\right) = 1 + \left(x - \frac{x^2}{2} + \cdots\right) + \frac{1}{2!}\left(x - \frac{x^2}{2} + \cdots\right)^2 + \frac{1}{3!}\left(x - \frac{x^2}{2} + \cdots\right)^3 + \cdots. \quad (1)$$

But the composition should equal $\sum_{m=0}^{\infty} \frac{1}{m!}(\ln(x + 1))^m = 1 + \ln(x + 1) + \frac{1}{2!}(\ln(x + 1))^2 + \cdots = e^{\ln(x+1)} = x + 1$ for all $x \in (-1, 1]$, since the outer series converges for all real values of $u(x) = \ln(x + 1)$.

However, what happens to convergence if the terms in (1) are rearranged so as to form a standard power series in x? To be specific, the Cauchy products in (1) are expanded and rearranged to yield $1 + x + (-\frac{1}{2} + \frac{1}{2})x^2 + (\frac{1}{3} - \frac{1}{2} + \frac{1}{6})x^3 + (-\frac{1}{4} + \frac{11}{24} - \frac{1}{4} + \frac{1}{24})x^4 + \cdots$. For the five terms visible here, the result is $1 + x$. Can we expect *all* the coefficients in parentheses to be zero, so that the whole power series collapses to $1 + x$? The next theorem confirms that the answer will be "Yes." (What follows is essentially from [18], pp. 256–257.)

Theorem 44.4 Let $\sum_{m=0}^{\infty} a_m u^m = F(u)$ have a radius of convergence α, and let $\sum_{n=1}^{\infty} b_n x^n = G(x) = u$ have a radius of convergence β, where either radius may be infinite. Then the composition $\sum_{m=0}^{\infty}(a_m(\sum_{n=1}^{\infty} b_n x^n)^m)$, explicitly written

$$a_0 + a_1 \sum_{n=1}^{\infty} b_n x^n + a_2 \left(\sum_{n=1}^{\infty} b_n x^n\right)^2 + a_3 \left(\sum_{n=1}^{\infty} b_n x^n\right)^3 + \cdots, \tag{2}$$

will expand into the power series

$$\begin{aligned} &a_0 + (a_1 b_1)x + (a_1 b_2 + a_2 b_1^2)x^2 + (a_1 b_3 + 2a_2 b_1 b_2 + a_3 b_1^3)x^3 + (a_1 b_4 + a_2 b_2^2 + 2a_2 b_1 b_3 \\ &\quad + 3a_3 b_1^2 b_2 + a_4 b_1^4)x^4 + (a_1 b_5 + 2a_2 b_2 b_3 + 2a_2 b_1 b_4 + 3a_3 b_1^2 b_3 + 3a_3 b_1 b_2^2 \\ &\quad + 4a_4 b_1^3 b_2 + a_5 b_1^5)x^5 + \cdots, \end{aligned} \tag{3}$$

with sum function $(F \circ G)(x)$ over an interval of convergence with radius $\gamma = \alpha\beta/(P + \alpha)$, where P is defined in the proof, $\gamma = \alpha$ when $\beta = \infty$, and $\gamma = \beta$ when $\alpha = \infty$. (the endpoints must be tested for convergence independently).

Proof. Suppose that both α and β are positive and finite. **(2)** converges if $|\sum_{n=1}^{\infty} b_n x^n| < \alpha$. Because of the absolute convergence of $\sum_{n=1}^{\infty} b_n x^n$ over $(-\beta, \beta)$ (Theorem 43.1), $|\sum_{n=1}^{\infty} b_n x^n| \le \sum_{n=1}^{\infty} |b_n x^n|$ (Exercise 31.3). For any $0 < r < \beta$, $\{|b_n r^n|\}_1^{\infty}$ is bounded, and hence has a supremum, P. Thus, if $\sum_{n=1}^{\infty} \frac{P}{|r^n|}|x^n| = P\sum_{n=1}^{\infty} \left|\frac{x^n}{r^n}\right| < \alpha$, then **(2)** converges. Now, for $|x| < r$, $\sum_{n=1}^{\infty} \left|\frac{x^n}{r^n}\right| = \frac{|x|}{r-|x|}$ (verify), so if $\frac{P|x|}{r-|x|} < \alpha$ with $|x| < r$, then **(2)** converges. Solving this for $|x|$ gives $|x| < \frac{\alpha r}{P+\alpha}$ (which is less than r). Therefore, the radius of convergence of the series $F \circ G$ in **(2)** is $\gamma = \frac{\alpha\beta}{P+\alpha}$.

As extreme cases, due to the composition, if $\beta = \infty$ then $\gamma = \alpha$, if $\alpha = \infty$ then $\gamma = \beta$, and if either α or β is zero, then $\gamma = 0$. Notice that $(-\gamma, \gamma) \subseteq (-\beta, \beta)$ in every case.

We now form an infinite array by expanding every term in (2) and situating each expansion in a row of its own:

$$\begin{array}{ccccccc}
a_0 & 0 & 0 & 0 & 0 & 0 & \cdots \\
0 & a_1 b_1 x & a_1 b_2 x^2 & a_1 b_3 x^3 & a_1 b_4 x^4 & a_1 b_5 x^5 & \cdots \\
0 & 0 & a_2 b_1^2 x^2 & 2a_2 b_1 b_2 x^3 & a_2(b_2^2 + 2b_1 b_3)x^4 & a_2(2b_2 b_3 + 2b_1 b_4)x^5 & \cdots \\
0 & 0 & 0 & a_3 b_1^3 x^3 & 3a_3 b_1^2 b_2 x^4 & a_3(3b_1^2 b_3 + 3b_1 b_2^2)x^5 & \cdots \\
0 & 0 & 0 & 0 & a_4 b_1^4 x^4 & 4a_4 b_1^3 b_2 x^5 & \cdots \\
0 & 0 & 0 & 0 & 0 & a_5 b_1^5 x^5 & \cdots \\
0 & 0 & 0 & 0 & 0 & 0 & \cdots \\
\cdots & \cdots & \cdots & \cdots & \cdots & \cdots & \ddots
\end{array}$$

For instance, the row-two series is the expansion

$$a_2(G(x))^2 = a_2\left(\sum_{n=1}^{\infty} b_n x^n\right)^2 = a_2 b_1^2 x^2 + 2a_2 b_1 b_2 x^3 + a_2(b_2^2 + 2b_1 b_3)x^4 + \cdots$$

done as a Cauchy product, and it is absolutely convergent since $x \in (-\gamma, \gamma)$.

Indeed, every row series converges absolutely, since $x \in (-\gamma, \gamma)$. Furthermore, any finite number of terms of the array, when added, are terms of $\sum_{m=0}^{N} a_m(G(x))^m$ for some large enough N, and $\sum_{m=0}^{N} a_m(G(x))^m$ is itself a partial sum of the convergent series $\sum_{m=0}^{\infty} a_m(G(x))^m$ in (2) (remembering that $x \in (-\gamma, \gamma)$). Thus, the full array has sum $F(G(x))$ when $x \in (-\gamma, \gamma)$. The lemma in Note 2 below may now be applied to the array, implying that we will obtain the same sum $F(G(x))$ when we add column by column. These columns contain precisely the terms in power series (3), and from what we have seen, its radius of convergence is γ, and $\gamma \le \beta$. **QED**

Note 2: The following lemma is needed in the proof above. Let the terms of the infinite array below be such that every row series $t_{i1} + t_{i2} + t_{i3} + \cdots$ converges absolutely: $\sum_{j=1}^{\infty} |t_{ij}| = T_i$, for $i \in \mathbb{N}$. Suppose also that $\sum_{i=1}^{\infty} T_i$ converges. Then $\sum_{i=1}^{\infty}\sum_{j=1}^{\infty} t_{ij} = \sum_{j=1}^{\infty}\sum_{i=1}^{\infty} t_{ij}$, meaning that the entire array can be summed across by rows or down by columns. The lemma and its proof are found as Theorem 8.3 in [37], pp. 175–176.

$$\begin{array}{cccc} t_{11} & t_{12} & t_{13} & \cdots \\ t_{21} & t_{22} & t_{23} & \cdots \\ t_{31} & t_{32} & t_{33} & \cdots \\ \cdots & \cdots & \cdots & \ddots \end{array}$$

Example 44.5 We determine the first five terms of the composition of

$$F(u) = \sum_{m=0}^{\infty} \frac{1}{3^m} u^m = \frac{3}{3-u} \quad \text{with} \quad G(x) = \sum_{n=1}^{\infty} x^n = \frac{x}{1-x},$$

and the resulting interval of convergence. First, the respective intervals of convergence are $S_F = (-3, 3)$ and $S_G = (-1, 1)$, implying that $\alpha = 3$ and $\beta = 1$. The composition is

$$\sum_{m=0}^{\infty} \frac{1}{3^m}\left(\frac{x}{1-x}\right)^m, \quad \text{or} \quad \sum_{m=0}^{\infty} \frac{1}{3^m}\left(\sum_{n=1}^{\infty} x^n\right)^m.$$

Often, Theorem 44.4 is the easiest way to find the resulting power series. To that end, we write

$$\begin{array}{ccccc} a_0 = 1 & a_1 = \dfrac{1}{3} & a_2 = \dfrac{1}{9} & a_3 = \dfrac{1}{27} & a_4 = \dfrac{1}{81}, \\ & b_1 = 1 & b_2 = 1 & b_3 = 1 & b_4 = 1 \end{array}$$

with which we evaluate the expansion (3) as

$$\begin{aligned} &1 + \frac{1}{3}x + \left(\frac{1}{3} + \frac{1}{9}\right)x^2 + \left(\frac{1}{3} + 2\cdot\frac{1}{9} + \frac{1}{27}\right)x^3 + \left(\frac{1}{3} + 2\cdot\frac{1}{9} + \frac{1}{9} + 3\cdot\frac{1}{27} + \frac{1}{81}\right)x^4 + \cdots \\ &= 1 + \frac{1}{3}x + \frac{4}{9}x^2 + \frac{16}{27}x^3 + \frac{64}{81}x^4 + \cdots. \end{aligned}$$

By virtue of the same theorem, this series converges to

$$F(G(x)) = \frac{3}{3 - \frac{x}{1-x}} = \frac{3x-3}{4x-3}.$$

Furthermore, the interval of convergence is in the interior of $S_G = (-1, 1)$. To find it, recall that $\mathcal{D}_{F\circ G}$ is the set of $x \in \mathcal{D}_G$ such that $G(x) \in \mathcal{D}_F$. Since $\mathcal{D}_G = (-1, 1)$ and $\mathcal{D}_F = (-3, 3)$, we solve $-3 < \frac{x}{1-x} < 3$, removing all solutions not belonging to $(-1, 1)$. A bit of algebra yields $x < \frac{3}{4}$ and $x < \frac{3}{2}$, and still, $x \in (-1, 1)$. Thus, so far, $x \in (-1, \frac{3}{4})$. Now we must remember that power series about zero have intervals of convergence with center at zero. Hence, the composition $1 + \frac{1}{3}x + \frac{4}{9}x^2 + \frac{16}{27}x^3 + \frac{64}{81}x^4 + \cdots$ has the interval of convergence $(-\frac{3}{4}, \frac{3}{4})$. Verify that the series diverges at the endpoints. To use the theorem's formula for γ, $|\beta_n r^n| = |r^n|$ for series G, so that for $0 < r < 1$, $\{|r^n|\}|r^n|_1^\infty$ has $P = 1$. Therefore, $\gamma = \frac{3}{4}$. Now, consider Exercise 44.6. ■

Something interesting has evolved in the example just finished. The five terms of the composed series equal $1 + \frac{1}{4}\sum_{n=1}^{4} \frac{4^n}{3^n}x^n$. Is it then possible that $\sum_{n=0}^{\infty} \frac{1}{3^n}(\frac{x}{1-x})^n = 1 + \frac{1}{4}\sum_{n=1}^{\infty} \frac{4^n}{3^n}x^n$, despite the dissimilarities in form? The right side, involving a geometric series, converges to

$$1 + \frac{1}{4}\left(\frac{\frac{4}{3}x}{1 - \frac{4}{3}x}\right) = \frac{3x - 3}{4x - 3}$$

(aha!), with interval of convergence $-1 < \frac{4}{3}x < 1$, which yields $(-\frac{3}{4}, \frac{3}{4})$ (aha again!). Hence, we have two equal expressions, $1 + \frac{1}{4}\sum_{n=1}^{\infty} \frac{4^n}{3^n}x^n$, and $\sum_{n=0}^{\infty} \frac{1}{3^n}(\frac{x}{1-x})^n$ from the exercise above, with identical domains. This means that they are the same function. As will be studied in the next chapter, the function $f(x) = \frac{3x-3}{4x-3}$ is *represented* by these two series over $(-\frac{3}{4}, \frac{3}{4})$.

Example 44.6 Find the first three nonzero terms of the composition of

$$F(x) = e^x = \sum_{m=0}^{\infty} \frac{x^m}{m!}$$

with $G(x) = \sum_{n=0}^{\infty} \frac{(-1)^n x^{2n}}{(2n)!} = 1 + 0x - \frac{1}{2!}x^2 + 0x^3 + \frac{1}{4!}x^4 + 0x^5 - \frac{1}{6!}x^6 + \cdots$. We use

$$F(u) = e^u = 1 + u + \frac{u^2}{2!} + \frac{u^3}{3!} + \cdots,$$

but observe that the series for $G(x)$ in Theorem 44.4 begins at $n = 1$, not zero. Thus, we rewrite

$$G(x) = 1 + \sum_{n=1}^{\infty} \frac{(-1)^n x^{2n}}{(2n)!},$$

and the composition becomes

$$F(G(x)) = \exp\left(1 + \sum_{n=1}^{\infty} \frac{(-1)^n x^{2n}}{(2n)!}\right) = e\sum_{m=0}^{\infty}\left(\frac{1}{m!}\left(\sum_{n=1}^{\infty} \frac{(-1)^n x^{2n}}{(2n)!}\right)^m\right).$$

Next, identify

$$a_0 = 1 \quad a_1 = 1 \quad a_2 = \frac{1}{2!} \quad a_3 = \frac{1}{3!} \quad a_4 = \frac{1}{4!}$$
$$b_1 = 0 \quad b_2 = -\frac{1}{2!} \quad b_3 = 0 \quad b_4 = \frac{1}{4!}.$$

Expansion (3) then yields

$$F(G(x)) = e\left[1 + 0x + \left(\frac{-1}{2} + 0\right)x^2 + (0+0+0)x^3 + \left(\frac{1}{4!} + 0 + \frac{1}{2!}\left(\frac{-1}{2}\right)^2 + 0 + 0\right)x^4 + \cdots\right] = e - \frac{e}{2}x^2 + \frac{e}{6}x^4 - \cdots.$$

Our work agrees with a computer algebra system that also supplies the next two nonzero terms, $\frac{-31e}{720}x^6$ and $\frac{379e}{40320}x^8$. The radius of convergence is $\gamma = \infty$ since both α and β are ∞.

If, in addition to all this, we happen to recognize that $u = G(x) = \cos x$ for all real x, then we have found a power series that converges to $F(G(x)) = e^{\cos x}$ over all $\mathbb{R}$. ■

We now consider the **integration of power series**. Formally, this is done by term by term integration, which is one reason why power series are so useful. However, wantonly exchanging the integration and limiting processes may lead to embarrassing conclusions, as was seen in Exercise 41.10. The key ingredient is uniform convergence, and the following theorem is a direct application of Corollary 41.1. (The fundamental theorem of integral calculus will be used justifiably in $\int_a^b x^k dx = \frac{1}{k+1}(b^{k+1} - a^{k+1})$.)

Theorem 44.5 Let the power series $F(x) = \sum_{n=0}^{\infty} a_n x^n$ have interval of convergence S. If the closed interval $[c, d] \subseteq S$, then $\int_c^d (\sum_{n=0}^{\infty} a_n x^n) dx = \sum_{n=0}^{\infty} \int_c^d a_n x^n dx = \sum_{n=0}^{\infty} \frac{a_n}{n+1}(d^{n+1} - c^{n+1})$, so that the integration and summation operations may be exchanged. In other words, a power series may be integrated term by term to yield the integral of its sum function, within its interval of convergence.

Proof. Let $s_n(x) = \sum_{k=0}^{n} a_k x^k$ be the partial sums of the series. Then $\langle s_n(x)\rangle_0^\infty$ is a sequence of (continuous) polynomials that converges to $F(x)$. By Corollary 43.1, $\sum a_n x^n \xrightarrow{u} F(x)$ over $[c, d]$. Then Corollary 41.1 affirms that

$$\lim_{n\to\infty} \int_c^d s_n(x)dx = \int_c^d \lim_{n\to\infty} s_n(x)dx = \int_c^d F(x)dx.$$

But here,

$$\int_c^d \lim_{n\to\infty} s_n(x)dx = \int_c^d \left(\sum_{n=0}^{\infty} a_n x^n\right)dx,$$

while

$$\lim_{n\to\infty} \int_c^d s_n(x)dx = \lim_{n\to\infty} \int_c^d \left(\sum_{k=0}^{n} a_k x^k\right)dx = \lim_{n\to\infty} \sum_{k=0}^{n} \int_c^d a_k x^k dx$$
$$= \lim_{n\to\infty} \sum_{k=0}^{n} \frac{a_k}{k+1}(d^{k+1} - c^{k+1}).$$

Prove that this limit of a sequence of partial sums exists, in Exercise 44.10.

Once that is done, we will have shown that

$$\lim_{n\to\infty} \int_c^d s_n(x)dx = \sum_{n=0}^{\infty} \frac{a_n}{n+1}(d^{n+1} - c^{n+1}).$$

But this lets us conclude that

$$\int_c^d \left(\sum_{n=0}^{\infty} a_n x^n\right) dx = \sum_{n=0}^{\infty} \frac{a_n}{n+1}(d^{n+1} - c^{n+1}) = \sum_{n=0}^{\infty} \int_c^d a_n x^n dx,$$

implying that the term by term integration of the power series is justified over $[c, d]$. **QED**

Example 44.7 The theorem just proved allows us to do the following interesting work. Since $\int \frac{1}{1-x} dx = -\ln(1-x) + c$, first obtain the power series for $y = \frac{1}{1-x}$, then integrate term by term to obtain a power series for $-\ln(1-x)$. One way to obtain the power series (can you name another?) is by long division: $1 \div (1-x) = 1 + x + x^2 + x^3 + x^4 + \cdots$. By the theorem just proved,

$$\int \frac{1}{1-x} dx = \int (1 + x + x^2 + x^3 + x^4 + \cdots) dx = x + \frac{1}{2}x^2 + \frac{1}{3}x^3 + \frac{1}{4}x^4 + \frac{1}{5}x^5 + \cdots$$
$$= \sum_{n=1}^{\infty} \frac{x^n}{n}$$

over $[-1, 1)$ (see Example 43.4), and the constant of integration is zero.

For instance, choosing $[0, \frac{1}{3}] \subset [-1, 1)$, we have

$$\ln \frac{3}{2} = \int_0^{1/3} \frac{1}{1-x} dx = \sum_{n=0}^{\infty} \frac{1}{n+1}((\tfrac{1}{3})^{n+1} - 0^{n+1}) = \sum_{n=1}^{\infty} \frac{1}{n3^n}.$$

This is a noteworthy result because $\ln \frac{3}{2}$ is the sum of this non-geometric series. Exercises 44.8 and 44.9 reinforce the procedure in this example. ■

In Exercise 45.8c, we will use Maclaurin series to show again that $F(x) = -\ln(1-x)$ is the sum function for the power series $\sum_{n=1}^{\infty} \frac{x^n}{n}$, with interval of convergence $[-1, 1)$. This will provide another avenue to show that $\sum_{n=1}^{\infty} \frac{1}{n3^n} = \ln \frac{3}{2}$.

Our final operation is the **differentiation of power series**. As might be suspected, the derivative of a power series may be done term by term, or loosely speaking, the derivative of a power series equals the power series of the derivatives. (The proof will once again need the fundamental theorem of integral calculus.)

Theorem 44.6 Let power series $F(x) = \sum_{n=0}^{\infty} a_n x^n$ have interval of convergence S. If x is an interior point of S (x is not an endpoint), then $F'(x) = \sum_{n=1}^{\infty} n a_n x^{n-1}$, that is,

$$\frac{d}{dx} \sum_{n=0}^{\infty} a_n x^n = \sum_{n=0}^{\infty} \frac{d a_n x^n}{dx},$$

so that the differentiation and summation operations may be exchanged. In other words, a power series may be differentiated term by term in the *interior* of its interval of convergence. Convergence at endpoints must be checked separately.

Proof. The radius of convergence of $\sum_{n=1}^{\infty} n a_n x^{n-1}$ is

$$\beta = \frac{1}{\lim_{n\to\infty} \sqrt[n]{n a_n}} = \frac{1}{\lim_{n\to\infty} \sqrt[n]{a_n}},$$

since $\lim_{n\to\infty} \sqrt[n]{n} = 1$. Thus, $\sum_{n=1}^{\infty} na_n x^{n-1}$ has the same radius of convergence as the original power series, implying that the series exists at any point in the interior of S. Furthermore, letting $\beta = \sup S$, $\sum_{n=1}^{\infty} na_n x^{n-1}$ converges uniformly to a sum function f over any $[a, b] \subset (-\beta, \beta)$ by Corollary 43.1. That is, $\sum_{n=1}^{\infty} na_n x_0^{n-1} \xrightarrow{u} f(x_0)$ at any $x_0 \in (-\beta, \beta)$. Since the partial sums of the differentiated series are continuous, f is continuous over $[a, b]$ (Theorem 41.1). Thus, with $x_0 \in (-\beta, \beta)$, we may integrate according to Theorem 44.5:

$$\int_0^{x_0} f(x)dx = \int_0^{x_0} \left(\sum_{n=1}^{\infty} na_n x^{n-1}\right) dx = \sum_{n=1}^{\infty} \int_0^{x_0} na_n x^{n-1} dx$$

$$= \sum_{n=1}^{\infty} \left(a_n.x^n \Big|_0^{x_0}\right) = F(x_0) - a_0 = F(x_0) - F(0).$$

This demonstrates that F is an antiderivative of f, and by the fundamental theorem of integral calculus (Theorem 49.2), $F'(x_0) = f(x_0)$. Since the conclusion is the same for any $x_0 \in (-\beta, \beta)$, the theorem is proved. **QED**

Note 3: The exchange of the order of operations in the theorem just proved, namely, the derivative of a power series for the power series of the derivatives, may be false for series that are not power series. The following instance of this is made easier because the individual functions $f_n(x)$ are not explicitly identified, but only their partial sums.

Let the nth partial sum of a series of functions be

$$s_n(x) = \frac{x}{1 + nx^2}$$

with common domain $[-1, 1]$. Then $s_n(x) \xrightarrow{u} 0$ over $[-1, 1]$. Therefore, the derivative of the sum function must be the function $y = 0$ over $[-1, 1]$. Unfortunately, the derivative sequence has

$$s_n'(x) = \frac{1 - nx^2}{1 + 2nx^2 + n^2x^4},$$

which yields

$$\lim_{n\to\infty} s_n'(x) = \begin{cases} 1 & \text{if } x = 0 \\ 0 & \text{if } x \neq 0 \end{cases}.$$

Thus, the sum function of the derivatives is not even continuous over $[-1, 1]$. We have discovered that

$$0 = \frac{d}{dx} \sum_{n=0}^{\infty} f_n(x) \neq \sum_{n=1}^{\infty} \frac{d}{dx} f_n(x)$$

over $[-1, 1]$. The culprit here is the non-uniform convergence of the derivatives $s_n'(x)$over the domain. You should graph $s_n'(x)$ for $n = 1, 3, 10, 60$, and 500 for a strong visual confirmation of non-uniform convergence. This example is found in [18], pg. 281.

Example 44.8 Recycle the Example 44.7, where we found that $-\ln(1-x) = \sum_{n=1}^{\infty} \frac{1}{n} x^n$ over $[-1, 1)$. Taking the derivative of both sides,

$$\frac{1}{1-x} = \frac{d}{dx} \sum_{n=0}^{\infty} \frac{1}{n} x^n = \sum_{n=0}^{\infty} \frac{d}{dx}\left(\frac{1}{n} x^n\right) = \sum_{n=1}^{\infty} x^{n-1},$$

by Theorem 44.6. Note that the last summation begins at one, not zero (otherwise, something bad happens at $x = 0$ if $n = 0$). But $\sum_{n=1}^{\infty} x^{n-1} = \sum_{n=0}^{\infty} x^n$, and we return to the geometric series $\frac{1}{1-x} = 1 + x + x^2 + x^3 + \cdots$ of that example, as expected. We notice that this series diverges at $x_0 = -1$, so we must heed the premise of Theorem 44.6 that $x_0 \in S \backslash \{-1, 1\} = (-1, 1)$. On the other hand, at $x_0 = \frac{1}{2} \in (-1, 1)$, the slope of the tangent to $F(x) = -\ln(1-x)$ is correctly evaluated as $\sum_{n=0}^{\infty} (\frac{1}{2})^n = 2$. ■

Example 44.9 In Note 1, it was determined by formal manipulation that the power series $\sum_{n=0}^{\infty} \frac{(-1)^n}{(2n+1)!} x^{2n} = 1 - \frac{1}{6}x^2 + \frac{1}{120}x^4 - \cdots$ converges to $y = \frac{\sin x}{x}$. We extend this to the continuous sum function

$$F(x) = \begin{cases} \dfrac{\sin x}{x} & \text{if } x \neq 0 \\ 1 & \text{if } x = 0 \end{cases},$$

and we assume that the interval of convergence is all $\mathbb{R}$. Then Theorem 44.6 guarantees that

$$F'(x) = \left\{ \begin{array}{ll} \dfrac{\cos x}{x} - \dfrac{\sin x}{x^2} & \text{if } x \neq 0 \\ 0 & \text{if } x = 0 \end{array} \right\} = \frac{d}{dx} \sum_{n=0}^{\infty} \frac{(-1)^n}{(2n+1)!} x^{2n} = \sum_{n=1}^{\infty} \frac{(-1)^n}{(2n+1)!} \frac{d}{dx} x^{2n}$$

$$= \sum_{n=1}^{\infty} \frac{2n(-1)^n}{(2n+1)!} x^{2n-1} = -\frac{1}{3}x + \frac{1}{30}x^3 - \frac{1}{840}x^5 + \cdots.$$

Check by graphing that $y = -\frac{1}{3}x + \frac{1}{30}x^3 - \frac{1}{840}x^5$ is a good approximation to $F'(x)$ near the origin. ■

44.1 Exercises

1. Return to our dog-eared example $\sum_{n=1}^{\infty} x^n(1-x)$ from Example 43.2 and earlier, with interval of convergence $(-1, 1]$ and sum

$$F(x) = \begin{cases} x & \text{if } -1 < x < 1 \\ 0 & \text{if } x = 1 \end{cases}.$$

   Find the interval of convergence of $\sum_{n=1}^{\infty} (x^n - x^{n+1})$ and its sum by using Theorem 44.1.

2. Find the first four terms of the Cauchy product in Example 44.3 by expanding

$$\left(\sum_{n=0}^{1} \frac{1}{2^n} \left(x + \frac{1}{2}\right)^{4n}\right)\left(\sum_{n=1}^{4} (x-1)^n\right)$$

in powers of x, and then multiplying by $2-x$. Graph your resulting polynomial along with the sum function

$$F(x)=\frac{2(x-1)}{2-(x+\frac{1}{2})^4}$$

over the domain $(0, \sqrt[4]{2}-\frac{1}{2})$ in order to judge how accurate the partial sum is.

3. a. Find a power series for $f(x)=\ln\frac{1+x}{1-x}$, done as $\ln(1+x)-\ln(1-x)$. Be sure to attach its interval of convergence. Hint: See Example 43.8.
   b. Evaluate $\ln 10$ to four decimal place precision using the series in **a** (you will need a calculator with summation capability). Although the power series converges quickly for $|x|<0.5$, convergence slows down drastically as $|x|$ approaches 1.

4. a. Use Theorem 44.3 to find the reciprocal

$$\frac{1}{\sum_{n=0}^{\infty} b_n x^n}$$

   of a power series. Hint: $1=1+0x+0x^2+\cdots$.
   b. Use the results of **a** to find the first few terms of the power series for $f(x)=\sec x$ given that $\cos x = 1-\frac{1}{2}x^2+\frac{1}{24}x^4-\frac{1}{720}x^6+\cdots$.

5. Theorem 44.3 alerts us that $b_0\neq 0$ for division of power series. But alterations to the denominator power series may be made so that we *can* work with the proper type of series. To find a series for $f(x)=\cot x$ about zero, we may consider the ratio $\frac{\cos x}{\sin x}$. However, the power (Maclaurin) series for $\sin x$ has $b_0=0$. After casting about a little, we alter our ratio to

$$\frac{\cos x}{x(\frac{\sin x}{x})}=\frac{1}{x}\left(\frac{1-\frac{1}{2}x^2+\frac{1}{24}x^4-\cdots}{1-\frac{1}{6}x^2+\frac{1}{120}x^4-\cdots}\right),$$

   with convergence for $x\neq 0$. Apply Theorem 44.3 to the division in parentheses, and obtain the quotient power series. Then multiply it by $\frac{1}{x}$ as indicated, to arrive at a well-known series for the cotangent function about $x=0$. As you will see, Theorem 44.3 hasn't given us a free ride, for our final series will not be a *power* series. (Compare with Note 1.)

6. In truth, the result in Example 44.5 is formally fairly easy to determine, since the substitution

$$\sum_{m=0}^{\infty}\left(\frac{1}{3^m}\left(\sum_{n=1}^{\infty}x^n\right)^m\right)=\sum_{m=0}^{\infty}\frac{1}{3^m}\left(\frac{x}{1-x}\right)^m$$

   yields a geometric series. Arrive at the same conclusions in the example by using this point of view.

7. The 16 terms shown in the array in the proof of Theorem 44.4 may look a bit obscure. Use software or a calculator with algebraic capabilities to expand the following partial

sum of (3):

$$a_0 + a_1\left(\sum_{n=1}^{5} b_n x^n\right)^1 + a_2\left(\sum_{n=1}^{4} b_n x^n\right)^2 + a_3\left(\sum_{n=1}^{3} b_n x^n\right)^3$$
$$+ a_4\left(\sum_{n=1}^{2} b_n x^n\right)^4 + a_5\left(\sum_{n=1}^{1} b_n x^n\right)^5.$$

Among the 32 resulting terms, find the 16 terms. Give a reasonable explanation why, if higher powers or more terms are summed in this expression, the initial 16 terms will never change.

8. Perform a long division on $\frac{1}{1+x^2}$ to arrive at the power series

$$1 - x^2 + x^4 - x^6 + \cdots = \sum_{n=0}^{\infty}(-1)^n x^{2n}.$$

Confirm that its interval of convergence is $(-1, 1)$. Use Theorem 44.5 to find a power series for $f(x) = \arctan x$. Use the series to estimate the value of $\arctan(\frac{1}{\sqrt{3}})$ to four decimal places by putting a bound on the error.

9. Prove that

$$\sum_{n=1}^{\infty} \frac{1}{nM^n} = \ln \frac{M}{M-1},$$

where $M \in (-\infty, -1] \cup (1, \infty)$. Note that this extends the scope of the results of Exercise 29.9, since $q = -1$ is covered by this exercise.

10. In Theorem 44.5, prove that

$$\lim_{n\to\infty} \sum_{k=0}^{n} \frac{a_k}{k+1}\left(d^{k+1} - c^{k+1}\right)$$

exists. Hint: Rewrite the partial sums as

$$d\sum_{k=0}^{n} \frac{1}{k+1} a_k d^k - c\sum_{k=0}^{n} \frac{1}{k+1} a_k c^k.$$

11. Demonstrate Theorem 44.6 by first evaluating the derivatives of the following functions by standard calculus techniques, and then the derivatives of their corresponding power series:

a. $E(-x) = e^{-x}$ at $x = 1$ b. $F(x) = \ln(\frac{1+x}{1-x})$ at $x = \frac{1}{2}$ (see Exercise 44.3a)
c. $F(x) = \sin x$ at $x = \pi$ d. $F(x) = \sinh x = x + \frac{1}{3!}x^3 + \frac{1}{5!}x^5 + \frac{1}{7!}x^7 + \cdots$ at $x = 0$

12. Exercise 29.9 introduced series of constants of the form $T_q = \sum_{n=1}^{\infty} \frac{n^q}{M^n}$, where q was a nonnegative integer constant and M was a real number constant, $|M| > 1$. A formula for evaluating the sum T_q was found. We saw that $T_0 = \sum_{n=1}^{\infty} \frac{1}{M^n} = \frac{1}{M-1}$. To make this into a

recognizable power series, let $M = \frac{1}{x}$, implying that $x \in (-1, 1)\backslash\{0\}$. The result is that

$$T_0 = \sum_{n=1}^{\infty} x^n = \frac{1}{\frac{1}{x} - 1} = \frac{x}{1-x},$$

where, of course, $x \neq 0$. Now, we extend this to the familiar power series $T_0(x) = \sum_{n=1}^{\infty} x^n = \frac{x}{1-x}$ with the interval of convergence $(-1, 1)$. Generalizing this to follow Exercise 29.9, we define the power series $T_q(x) \equiv \sum_{n=1}^{\infty} n^q x^n$, where $q \in \mathbb{N} \cup \{0\}$ and $x \in (-1, 1)$.

Exercise 29.9 also determined that

$$T_1 = \sum_{n=1}^{\infty} \frac{n}{M^n} = \frac{M}{(M-1)^2}.$$

Replacing M by $\frac{1}{x}$ again, we have

$$T_1(x) = \sum_{n=1}^{\infty} n x^n = \frac{x}{(1-x)^2}$$

with a domain again extended to $(-1, 1)$.

a. Take the derivative of $\sum_{n=1}^{\infty} x^n$ and show that $T_1(x) = x\frac{dT_0(x)}{dx}$. What theorem justifies what you did?

b. It was also found that

$$T_2 = \sum_{n=1}^{\infty} \frac{n^2}{M^n} = \frac{M^2 + M}{(M-1)^3}.$$

Find the corresponding power series $T_2(x)$ over the interval of convergence $(-1, 1)$. Show and justify that $T_2(x) = x\frac{dT_1(x)}{dx}$ by using their power series.

In general, prove that $T_{q+1}(x) = x\frac{dT_q(x)}{dx}$, over $(-1, 1)$. (This gives us a formula for T_q in terms of T_{q-1}, an alternative to what appears in Exercise 29.9.)

45

Taylor Series

It is now time to blend the theory of power series with that of Taylor polynomials from Chapter 40. An interesting observation is that $\sum_{n=0}^{\infty} a_n x^n$ has the form of an endless polynomial, or a polynomial of "infinite degree." But we are no longer deceived by such ideas, because infinity isn't an integer, and consequently, polynomials of infinite degree are just chimeras.

The proper analysis is the study of Taylor series. Being a special subset of the set of power series, all the theorems developed in Chapters 42 to 44 apply to Taylor series. As usual, we set $0^0 \equiv 1$.

We begin with a function f that has derivatives $f^{(n)}$ of all orders, and for consistency we put $f^{(0)}(x) \equiv f(x)$.

Let f be a function with derivatives $f^{(n)}$ of all orders over a common domain $\mathcal{D}$. The power series

$$\sum_{n=0}^{\infty} \frac{f^{(n)}(c)}{n!}(x-c)^n = f(c) + f'(c)(x-c) + \frac{f''(c)}{2!}(x-c)^2 + \frac{f'''(c)}{3!}(x-c)^3 + \cdots$$

is called the **Taylor series** of f about, or at, $x = c$. When $c = 0$, the resulting Taylor series

$$\sum_{n=0}^{\infty} \frac{f^{(n)}(0)}{n!}x^n = f(0) + f'(0)x + \frac{f''(0)}{2!}x^2 + \cdots$$

is called the **Maclaurin series** of f.

Note 1: A Taylor series looks much like an endless version of the Taylor polynomial defined in Chapter 40. Brook Taylor (1685–1731) and his contemporaries actually did think of power series in this way, and applied them successfully. Taylor's *Methodus incrementorum directa et inversa*, "Methods of Direct and Inverse Growth" of 1715 has his development of the series named for him, including the special case of the Maclaurin series. Weird as it may seem, the *Treatise of Fluxions* by Colin Maclaurin (1698–1746), published 27 years after Taylor's book, and even acknowledging Taylor's work, contained the special case that would after all carry Maclaurin's name.

Going farther back in time, by 1668 both Isaac Newton and the Scottish mathematician James Gregory had discovered a few Maclaurin series. But according to Victor Katz [5], pg. 451, some of these power series had been derived centuries earlier by Indian mathematicians Madhava (1340–1425), Nīlakantha (1445–1545) and Jyesthadeva (1500–1610).

Taylor series would increase in importance during the 18th century, reaching a pinnacle of sorts with Joseph-Louis Lagrange (1736–1813). He attempted to use Taylor series as the foundation for all of calculus in his books *Théorie des fonctions analytiques*, "Theory of Analytical Functions" (1797) and *Leçons sur le calcul des fonctions*, "Lessons on the Calculus of Functions" (1801). Although Lagrange's approach wasn't completely successful, it was the best attempt up to that time to once and for all get rid of the "vanishingly small quantities" that had bedeviled calculus since its discovery.

We have seen a number of examples of Taylor series in past chapters, the classic being the Maclaurin series $\sum_{n=0}^{\infty} \frac{1}{n!}x^n$ generated by $E(x) = e^x$. We know from Example 43.5 (and previous experience) that the Maclaurin series $\sum_{k=0}^{n} \frac{x^k}{k!}$ converges for all real x, but we have simply assumed that at each x, $\sum \frac{1}{n!}x^n \to e^x$. Although correct for this series (we will prove it), the assumption is false in general, as will be demonstrated with a counterexample devised by Cauchy. Thus, we come to the definition of representation of functions by series.

45.1 Representation of Functions

A Taylor series $\sum_{n=0}^{\infty} \frac{f^{(n)}(c)}{n!}(x-c)^n$ with interval of convergence S **represents** the function f in an interval containing $x = c$ if and only if at every x in the interval, the series converges pointwise to $f(x)$. Equivalently, f **is represented by** the Taylor series over the interval.

There are two characters in the scene that opens here. First, there is $F(x)$, the sum function defined by $\sum_{n=0}^{\infty} \frac{1}{n!} f^{(n)}(c)(x-c)^n$, and to which it converges (see Theorem 43.1 and Corollary 43.2). These two expressions are equal functions, that is to say, $\sum_{n=0}^{\infty} \frac{1}{n!} f^{(n)}(c)(x-c)^n = F(x)$ with the *same* domain $S = \mathcal{D}_F$, as we saw in Chapter 42. Second, there is the function f with its full complement of derivatives, from which we generate the Taylor series, sometimes laboriously! Here, f need not be defined by a series at all, and may very well have a different domain than that of its Taylor series, that is, sometimes $S \neq \mathcal{D}_f$. Furthermore, as was mentioned, it is possible for the resulting series to converge to values different than f, in symbols: $F(x) \neq f(x)$ for some $x \in S \cap \mathcal{D}_f$. However, it is implicit in the definition that whenever $\sum_{n=0}^{\infty} \frac{1}{n!} f^{(n)}(c)(x-c)^n$ *represents* f, then $F(x) = f(x)$ over an interval containing c.

Some examples are certainly in order.

Example 45.1 The function $f(x) = \frac{x}{1-x}$, with domain $\mathcal{D}_f = \mathbb{R}\backslash\{1\}$, generates the Maclaurin series $\sum_{n=1}^{\infty} x^n$. Verify this as Exercise 45.1. In Example 42.1, we found the interval of convergence to be $S = (-1, 1)$. But does $\sum_{n=1}^{\infty} x^n$ represent f over some neighborhood of $x = 0$? Fortunately, in that example we showed that $\sum_{n=1}^{\infty} x^n \to \frac{x}{1-x}$ over S. In other words, pointwise convergence of the series to the function was proved. It follows that $F(x) = \sum_{n=1}^{\infty} x^n$ represents $f(x) = \frac{x}{1-x}$ over S (and nowhere else). ■

The question of representation of $E(x) = e^x$ by its Maclaurin series is more difficult to answer than in the example above, since the series is not geometric. Fortunately, we have already studied Lagrange's work on Taylor polynomials.

Reviewing Chapter 40, we used Taylor's theorem (Theorem 40.1) to arrive at the useful form

$$f(x) = \underbrace{f(c) + f'(c)(x-c) + \frac{f''(c)}{2!}(x-c)^2 + \cdots + \frac{f^{(n)}(c)}{n!}(x-c)^n}_{P_n(x)}$$

$$+ \underbrace{\frac{f^{(n+1)}(\xi_n)}{(n+1)!}(x-c)^{n+1}}_{R_n(x)}, \tag{1}$$

or concisely, $f(x) = P_n(x) + R_n(x)$. Here, $P_n(x)$ is both the Taylor polynomial for f about c as well as the nth partial sum of the Taylor series for f. Last but not least (figuratively speaking), $R_n(x)$ is Lagrange's form of the remainder, where ξ_n falls between x and c, and is usually undetermined (the subscript reminds us that the value of ξ may change with n). The remainder term is the key to the problem of representation.

Theorem 45.1 Let the function f and all its derivatives exist over an interval containing point $x = c$. Let the Taylor series for f about $x = c$ have an interval of convergence S (always containing c). Then the Taylor series represents f at a point $x_0 \in S$ if and only if $\lim_{n\to\infty} R_n(x_0) = 0$.

Proof of sufficiency. The Taylor polynomials $P_n(x)$are the partial sums of the Taylor series for $f(x)$, and $R_n(x_0) = f(x_0) - P_n(x_0)$. Therefore, if $\lim_{n\to\infty} R_n(x_0) = 0$, then $\lim_{n\to\infty} P_n(x_0) = f(x_0)$ and the definition of representation is fulfilled at x_0.

Proof of necessity. If the Taylor series represents f at x_0, then the partial sums converge to $f(x_0)$, meaning that $\lim_{n\to\infty} P_n(x_0) = f(x_0)$, and $\lim_{n\to\infty} R_n(x_0) = 0$. **QED**

Example 45.2 Use Lagrange's form of the remainder to prove that $E(x) = e^x$ is represented by the Maclaurin series $\sum_{k=0}^{n} \frac{x^k}{k!}$ over $\mathbb{R}$. We find that

$$R_n(x_0) = \frac{f^{(n+1)}(\xi_n)}{(n+1)!}(x_0 - c)^{n+1} = \frac{e^{\xi_n}}{(n+1)!}x_0^{n+1}$$

where ξ_n, whose value depends upon n, nevertheless falls between 0 and x_0, meaning that $|\xi_n| < |x_0|$. Hence, e^{ξ_n} is bounded above by $e^{|x_0|}$. Thus, it is enough to verify that

$$e^{|x_0|} \lim_{n\to\infty} \left| \frac{x_0^{n+1}}{(n+1)!} \right| = 0$$

for all real x_0. The limit was proved to be zero in Exercise 17.8, so we may conclude that the classic series $\sum_{k=0}^{n} \frac{x^k}{k!}$ actually represents the equally classic function $E(x) = e^x$ over $\mathbb{R}$. ■

Lagrange's remainder term also establishes the representation of the sine and cosine functions by their respective Maclaurin series over $\mathbb{R}$, see Exercises 45.2 and 45.3.

Example 45.3 Generate the Taylor series for $f(x) = \ln x$ about $c = 3$. We sequentially find

$f^{(n)}(x)$	$\dfrac{f^{(n)}(3)}{n!}$
$f(x) = \ln x$	$\dfrac{f(3)}{0!} = \ln 3$
$f'(x) = \dfrac{1}{x}$	$\dfrac{f'(3)}{1!} = \dfrac{1}{3}$
$f''(x) = -\dfrac{1}{x^2}$	$\dfrac{f''(3)}{2!} = -\dfrac{1}{3^2 2!} = -\dfrac{1}{18}$
$f'''(x) = \dfrac{2}{x^3}$	$\dfrac{f'''(3)}{3!} = \dfrac{2!}{3^3 3!} = \dfrac{1}{81}$
$f^{(4)}(x) = -\dfrac{3!}{x^4}$	$\dfrac{f^{(4)}(3)}{4!} = -\dfrac{3!}{3^4 4!} = -\dfrac{1}{324}$
$\cdots$	$\cdots$
$f^{(n)}(x) = \dfrac{(-1)^{n-1}(n-1)!}{x^n}$	$\dfrac{f^{(n)}(3)}{n!} = \dfrac{(-1)^{n-1}(n-1)!}{n!3^n} = \dfrac{(-1)^{n-1}}{n3^n}$ $(n \geq 1)$.

From this, we obtain the Taylor series

$$\ln 3 + \sum_{n=1}^{\infty} \frac{(-1)^{n-1}}{n3^n}(x-3)^n = \ln 3 + \frac{1}{3}(x-3) - \frac{1}{18}(x-3)^2 + \frac{1}{81}(x-3)^3 - \frac{1}{324}(x-3)^4 + \cdots. \tag{2}$$

The partial sums (Taylor polynomials) $s_1(x) = \ln 3 + \frac{1}{3}(x-3)$, $s_2(x) = \ln 3 + \frac{1}{3}(x-3) - \frac{1}{18}(x-3)^2$, $s_5(x)$, and $s_{13}(x)$, are graphed in Figure 45.1. Recall that s_1 is nothing more than the tangent line, or linear approximation, to $f(x) = \ln x$ at $c = 3$. These graphs display the typical convergence behavior seen since Chapter 40.

The ratio test is a good choice for testing convergence of the series we have just created. Apply it in Exercise 45.4a to find that the Taylor series converges for $0 < x < 6$. Convergence

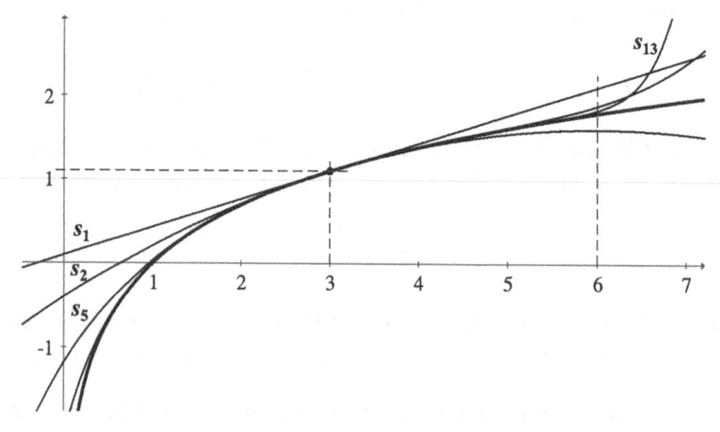

Figure 45.1. Taylor polynomials s_1, s_2, s_5, and s_{13} (the thin lines), taken from the Taylor series for $f(x) = \ln x$ about $c = 3$. The function itself is the heavy line.

is easily established at the right endpoint, too. The result is that

$$F(x) = \ln 3 + \sum_{n=1}^{\infty} \frac{(-1)^{n-1}}{n3^n}(x-3)^n$$

has domain $(0, 6]$. The question that immediately arises is, "Does $F(x) = \ln x$ over that interval, that is, does the series represent the logarithmic function over that interval?" The next example will investigate this question. But first, compare this example with Exercise 45.5. ■

Example 45.4 Find an interval over which the Taylor series (2) represents $f(x) = \ln x$. The absolute value of the Lagrange remainder is

$$|R_n(x_0)| = \frac{|f^{(n+1)}(\xi_n)|}{(n+1)!}|x_0 - 3|^{n+1} = \frac{n!}{(n+1)!(\xi_n)^{n+1}}|x_0 - 3|^{n+1} = \frac{1}{n+1}\left|\frac{x_0 - 3}{\xi_n}\right|^{n+1},$$

where ξ_n lies between x_0 and 3 and $x_0 \in (0, 6]$ from the previous example. Now, $\lim_{n\to\infty} |R_n(x_0)| = 0$ if and only if $|\frac{x_0-3}{\xi_n}| \leq 1$, which implies $\xi_n \geq |x_0 - 3|$. We must look into two subintervals, (a) $x_0 \in (0, 3)$, and (b) $x_0 \in (3, 6]$. In (a), $x_0 < \xi_n < 3$, and if furthermore, we take $\frac{3}{2} \leq x_0 < 3$, then $\xi_n > 3 - x_0$ (draw a number line to verify!), and it follows that $\lim_{n\to\infty} |R_n(x_0)| = 0$. In (b), $3 < \xi_n < x_0$, and since $x_0 \leq 6$, then $\xi_n \geq x_0 - 3$ (also verify), so that again $\lim_{n\to\infty} |R_n(x_0)| = 0$. But unfortunately, under (a), if $0 < x_0 < \frac{3}{2}$, it is possible for $\xi_n < 3 - x_0$, and we can't conclude that $R_n(x_0) \to 0$.

Thus, the Lagrange remainder is capable of guaranteeing representation of $f(x) = \ln x$ by its Taylor series $\ln 3 + \sum_{n=1}^{\infty} \frac{(-1)^{n-1}}{n3^n}(x-3)^n$ over the interval $[\frac{3}{2}, 6]$. The question of representation over $(0, \frac{3}{2})$ is still unresolved, although the graphs in Figure 45.1 give evidence of it. ■

There are other forms for the remainder term $R_n(x)$ for a Taylor polynomial, some tailor-made to the series being investigated. The following one—by Cauchy—is as useful as Lagrange's form. Its proof is patterned after that of Theorem 40.1.

Theorem 45.2 Cauchy's Form of the Remainder Let f and all of its derivatives f', $f'', \ldots, f^{(n)}$ be continuous over a common domain $\mathcal{D}$, and let $f^{(n+1)}$ exist over $\mathcal{D}$. Let $x_0 \in \mathcal{D}$, and let c be a point interior to $\mathcal{D}$. Then the remainder $R_n(x_0)$ in Taylor's theorem about $x = c$ has the form

$$R_n(x_0) = \frac{f^{(n+1)}(\xi_n)}{n!}(x_0 - \xi_n)^n(x_0 - c),$$

where ξ_n falls between c and x_0. This is called the **Cauchy form of the remainder**.

Proof. If $x_0 = c$, then $\xi_n = x_0$ and $R_n(x_0) = 0$, and there's nothing more to prove. Whenever $x_0 \neq c$, define the function Δ, with either $\mathcal{D}_\Delta = [c, x_0]$ or $\mathcal{D}_\Delta = [x_0, c]$, by

$$\Delta(x) = f(x_0) - f(x) - f'(x)(x_0 - x) - \frac{f''(x)}{2!}(x_0 - x)^2 - \cdots$$
$$- \frac{f^{(n)}(x)}{n!}(x_0 - x)^n - (x_0 - x)K,$$

where K is a constant that we will now determine. We have $\Delta(x_0) = 0$, and we wish to have $\Delta(c) = 0$ also. For this to happen, the value of K must be

$$K = \frac{f(x_0) - f(c) - f'(c)(x_0 - c) - \frac{1}{2!}f''(c)(x_0 - c)^2 - \cdots - \frac{1}{n!}f^{(n)}(c)(x_0 - c)^n}{x_0 - c}. \qquad (3)$$

Rolle's theorem (Theorem 39.2) applies to $\Delta(x)$ over the closed interval between c and x_0. Thus, there exists a ξ_n between c and x_0 such that $\Delta'(\xi_n) = 0$. Carefully obtaining the derivative of $\Delta(x)$, we find $\Delta'(x) = -\frac{1}{n!}(x_0 - x)^n f^{(n+1)}(x) + K$. Therefore, $\Delta'(\xi_n) = -\frac{1}{n!}(x_0 - \xi_n)^n f^{(n+1)}(\xi_n) + K = 0$, so that in a new version, $K = \frac{1}{n!}(x_0 - \xi_n)^n f^{(n+1)}(\xi_n)$. Substituting this into (3) and isolating $f(x_0)$ gives us

$$f(x_0) = \underbrace{f(c) + f'(c)(x_0 - c) + \frac{f''(c)}{2!}(x_0 - c)^2 + \cdots + \frac{f^{(n)}(c)}{n!}(x_0 - c)^n}_{P_n(x_0)}$$

$$+ \underbrace{\frac{f^{(n+1)}(\xi_n)}{n!}(x_0 - \xi_n)^n (x_0 - c)}_{R_n(x_0)}.$$

The first terms form the Taylor polynomial $P_n(x_0)$ of f about $x = c$, which means that the last term expresses the remainder at x_0 as $R_n(x_0) = \frac{f^{(n+1)}(\xi_n)}{n!}(x_0 - \xi_n)^n (x_0 - c)$. **QED**

Example 45.5 Use the Cauchy form of the remainder to find an interval over which the Taylor series (2) represents $f(x) = \ln x$. In the table of Example 45.3, we found that

$$f^{(n)}(x) = \frac{(-1)^{n-1}(n-1)!}{x^n}.$$

Therefore, the Cauchy remainder, in absolute value, is

$$|R_n(x_0)| = \frac{|f^{(n+1)}(\xi_n)|}{n!}|x_0 - \xi_n|^n |x_0 - c| = \frac{|x_0 - \xi_n|^n |x_0 - 3|}{(\xi_n)^{n+1}}$$

where x_0 is positive and ξ_n falls between x_0 and 3. It's useful to rewrite this as

$$|R_n(x_0)| = \frac{|x_0 - 3|}{\xi_n} \left| \frac{x_0 - \xi_n}{\xi_n} \right|^n$$

in the case that $0 < x_0 < \xi_n < 3$, because $|\frac{x_0 - \xi_n}{\xi_n}| = |1 - \frac{x_0}{\xi_n}|$, where $0 < \frac{x_0}{\xi_n} < 1$. Furthermore, $\frac{|x_0 - 3|}{\xi_n} < \frac{|x_0 - 3|}{x_0}$ in this case. And now, since

$$\lim_{n\to\infty} \frac{|x_0 - 3|}{x_0} \left|1 - \frac{x_0}{\xi_n}\right|^n = \frac{|x_0 - 3|}{x_0} \lim_{n\to\infty} \left|1 - \frac{x_0}{\xi_n}\right|^n = 0,$$

it follows that $\lim_{n\to\infty} |R_n(x_0)| = 0$ for any $x_0 \in (0, 3)$.

We have gone beyond the conclusion obtained by the Lagrange remainder form in Example 45.4, and since that example showed representation over $[\frac{3}{2}, 6]$, we now may conclude that the Taylor series (2) represents $f(x) = \ln x$ over $(0, 6]$. Things can't get better than this, for we recall that $(0, 6]$ is the entire domain of the Taylor series! Now consider Exercise 45.8. ■

Needless to say, establishing representation is not often easy. Similar to integration, there is no single technique that always works, but the forms of the remainder by Lagrange and Cauchy are often fruitful.

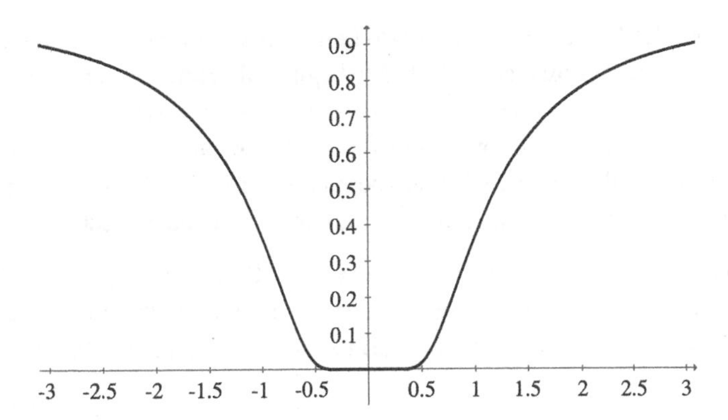

Figure 45.2. The function $f(x) = e^{-1/x^2}$ (if $x \neq 0$), $f(0) = 0$. Notice the extreme flatness of the curve near the origin. All Maclaurin polynomials for f are identically zero, as is the Maclaurin series. Caution: although $f(x) > 0$ for all nonzero x, many calculators will give for instance, $f(0.01) = 0$, because of loss of precision.

Example 45.6 The function

$$f(x) = \begin{cases} e^{-1/x^2} & \text{if } x \neq 0 \\ 0 & \text{if } x = 0 \end{cases}$$

has a disturbing property discovered by Cauchy: its Maclaurin series does not represent the function itself except at the origin! To see this, we first determine that successive derivatives are continuous at $\langle 0, 0\rangle$ if we set $f^{(n)}(0) \equiv 0$ (show this in Exercise 45.6).

Thus, we have $f^{(n)}(0) = 0$ for all $n \in \mathbb{N} \cup \{0\}$. It follows that the Maclaurin series is $0 + 0x + 0x^2 + \cdots$, which of course is identically the function $F(x) = 0$, with interval of convergence $\mathbb{R}$. Therefore, the series represents the function only at $x = 0$. ■

Note 2: The problem of representation arises any time we form an infinite series of functions $\sum_{n=1}^{\infty} f_n(x) \equiv F(x)$ and ask if F equals some other function f over a common set. Recall Example 42.6, where we showed that $\sum_{n=1}^{\infty} x^n(1-x)$ represented $f(x) = x$ over $(-1, 1)$. However, representation becomes critically important with Taylor series because of their extensive use.

45.2 Exercises

1. Verify by induction that for $f(x) = \frac{x}{1-x}$, $f^{(n)}(0) = n!$ for every $n \in \mathbb{N}$.

2. Prove that the Maclaurin series $\sum_{n=0}^{\infty} \frac{(-1)^n}{(2n+1)!} x^{2n+1} = x - \frac{1}{3!}x^3 + \frac{1}{5!}x^5 - \cdots$ represents $f(x) = \sin x$ over all real numbers.

3. The direct expansion of $f(x) = \cos x$ is the easiest way to obtain its Maclaurin series. However, use the trigonometric identity $\cos x = 1 - 2\sin^2 \frac{x}{2}$ and the series in the previous exercise to rediscover the Maclaurin series for the cosine function. Does it represent $f(x) = \cos x$ for all real x? Hint: Cauchy product.

4. a. Use the ratio test to find the interval of convergence of $\sum_{n=1}^{\infty} \frac{(-1)^{n-1}}{n3^n}(x-3)^n$ found in (2) in Example 45.3.

b. It might seem futile to use the series (2) to find approximations to natural logarithms, since we would need an approximation for ln 3 to begin with! However, since representation was proved for $\ln x = \ln 3 + \frac{1}{3}(x-3) - \frac{1}{18}(x-3)^2 + \frac{1}{81}(x-3)^3 - \frac{1}{324}(x-3)^4 + \cdots$ over $(0, 6]$, we shall estimate ln 3 by letting $x = e$. (In Chapter 34, we found $e \approx 2.718282$.) Thus, use the fourth degree Taylor polynomial here to find a five-digit approximation for ln 3, and put an estimate on the error. (Next, how would you estimate ln 12, for instance?)

5. Generate the Maclaurin series for $f(x) = \ln(x+3)$. Compare the graphs of the Maclaurin polynomials s_1, s_2, and s_5 with those in Figure 45.1. What is the only difference that you see? Substitute $x = y - 3$ into the Maclaurin series and compare with the results of (2) in that example.

6. With regard to Example 45.6, show that $f(x) = e^{-1/x^2}$ and all its derivatives will be continuous at the origin if we set $f^{(n)}(0) = 0$ for $n = 0, 1, 2, \ldots$. Hint: Show by induction that the derivatives have the form $f^{(n)}(x) = e^{-1/x^2}\frac{Q_n(x)}{x^{3n}}(x \neq 0)$, where the $Q_n(x)$are polynomials (it's crucial that they are polynomials).

7. In Exercise 43.7, it was stated that $\sum_{n=1}^{\infty}\frac{1}{n(n+1)}x^n = -\ln(1-x) + \frac{\ln(1-x)}{x} + 1$ over $[-1, 0) \cup (0, 1)$. Show that this representation is correct, as follows.
   a. Show that the series converges over $[-1, 1]$.
   b. Use a partial fraction decomposition to separate the series into $\sum_{n=1}^{\infty}\frac{1}{n}x^n - \sum_{n=1}^{\infty}\frac{1}{n+1}x^n$.
   c. Show that $F(x) = \sum_{n=1}^{\infty}\frac{1}{n}x^n$ represents $f(x) = -\ln(1-x)$ over $[-1, 1)$. (Convergence was proved in Example 43.4.) What interesting sum can you evaluate as $-F(-1)$?
   d. Verify that when convergence occurs, $G(x) = -1 + \frac{1}{x}\sum_{n=1}^{\infty}\frac{1}{n}x^n$ is equal to $\sum_{n=1}^{\infty}\frac{1}{n+1}x^n$ with $x \neq 0$, and use this to save labor in finding that $G(x)$ represents $g(x) = \frac{-\ln(1-x)}{x} - 1$ over $[-1, 1)\setminus\{0\}$.
   e. Use a theorem from Chapter 44 to prove that $\sum_{n=1}^{\infty}\frac{1}{n(n+1)}x^n$ represents $f(x) - g(x) = -\ln(1-x) + \frac{\ln(1-x)}{x} + 1$ over $[-1, 0) \cup (0, 1)$.
   f. As a quick partial check, evaluate $\sum_{n=1}^{10}\frac{1}{n(n+1)}x^n$ at $x = -0.7$, and compare with $f(-0.7) - g(-0.7)$. Graphs of partial sums would also help to check your work.

8. In Example 45.5, we proved that the Taylor series (2) represented of $f(x) = \ln x$ over $(0, 3)$ by using the Cauchy form of the remainder. Prove representation over $(3, 6)$ using this form as well. Hint: Use the other case $3 < \xi_n < x_0 \leq 6$, and this time, $\frac{|x_0-3|}{\xi_n} < \frac{|x_0-3|}{3}$.

46

Taylor Series, Part II

James Gregory in a letter to [John] Collins ... of November 23, 1670 ... got $(1+d)^x = 1 + dx + \frac{x(x-1)}{1\cdot 2}d^2 + \frac{x(x-1)(x-2)}{1\cdot 2\cdot 3}d^3 + \cdots$. *Thus Gregory obtained the binomial expansion for general* x.

Morris Kline [6], pp. 440–441.

Since all Taylor series are power series, much of our analysis of Taylor series has already been done. We will consider some applications, a few more topics, and then study the binomial series discovered by Isaac Newton. For ease of presentation, the theorems below will use Maclaurin series, but they can be easily generalized to Taylor series (see Exercise 46.1). As usual, we stipulate that $0^0 \equiv 1$.

Theorem 46.1 Let $F(x) = \sum_{n=0}^{\infty} a_n x^n$ be any power series with interval of convergence S that is not just $[0, 0]$. Then $a_n = \frac{F^{(n)}(0)}{n!}$ for all $n \in \mathbb{N} \cup \{0\}$. In other words, the power series will be the Maclaurin series for F.

Proof. We will prove by induction that the derivative of a power series is always a power series. In fact, we will prove specifically that at any x interior to S,

$$F^{(n)}(x) = \sum_{k=0}^{\infty} \frac{(n+k)!a_{n+k}}{k!}x^k = n!a_n + (n+1)!a_{n+1}x + \frac{(n+2)!a_{n+2}}{2!}x^2 + \cdots.$$

The anchor is the case $n = 0$, $F^{(0)}(x) = \sum_{k=0}^{\infty} a_k x^k = F(x)$, given as a premise. Now, assume that the series for $F^{(n)}(x)$ is true. Then, by Theorem 44.6, we find

$$F^{(n+1)}(x) = \frac{d}{dx}\sum_{k=0}^{\infty} \frac{(n+k)!a_{n+k}}{k!}x^k = \sum_{k=1}^{\infty} \frac{(n+k)!a_{n+k}}{(k-1)!}x^{k-1}$$

$$= (n+1)!a_{n+1} + (n+2)!a_{n+2}x + \frac{(n+3)!a_{n+3}}{2!}x^2 + \cdots,$$

which completes the induction. Therefore, in particular, $F^{(n)}(0) = n!a_n$, and the conclusion follows. **QED**

Corollary 46.1 Let $\sum_{n=0}^{\infty} a_n x^n$ and $\sum_{n=0}^{\infty} b_n x^n$ represent function f over some interval S. Then $a_n = b_n$ for all $n \in \mathbb{N}\cup\{0\}$. In other words, representation of functions is unique.

Proof. Create a proof as Exercise 46.2. **QED**

Example 46.1 Reviewing Example 44.5, we found that $\sum_{n=0}^{\infty} \frac{1}{3^n}(\frac{x}{1-x})^n$ was the result of the composition of $F(u) = \frac{3}{3-u}$ with $G(x) = \frac{x}{1-x}$, over $S = (-\frac{3}{4}, \frac{3}{4})$. The analysis there yielded a power series, and immediately thereafter we realized that the series could be written as $1 + \frac{1}{4}\sum_{n=1}^{\infty} \frac{4^n}{3^n} x^n$. Thus, we were gratified when the composition and the latter series both represented the *same* function, $f(x) = \frac{3x-3}{4x-3}$. Now we realize that this is necessarily so, by Corollary 46.1. ■

Theorem 46.2 Let f be represented by its Maclaurin series $\sum_{n=0}^{\infty} \frac{f^{(n)}(0)}{n!}x^n$. Then the series can be integrated term by term over any closed interval within its interval of convergence S. That is,

(1) $\int f(x)dx = \int(\sum_{n=0}^{\infty} \frac{f^{(n)}(0)}{n!}x^n)dx = \sum_{n=0}^{\infty}(\frac{f^{(n)}(0)}{n!}\int x^n dx) = \sum_{n=0}^{\infty} \frac{f^{(n)}(0)}{(n+1)!}x^{n+1}$ over an interval of convergence interior to S, and

(2) $\int_a^b f(x)dx = \int_a^b(\sum_{n=0}^{\infty} \frac{f^{(n)}(0)}{n!}x^n)dx = \sum_{n=0}^{\infty}(\frac{f^{(n)}(0)}{n!}\int_a^b x^n dx) = \sum_{n=0}^{\infty} \frac{f^{(n)}(0)}{(n+1)!}(b^{n+1} - a^{n+1})$, where $[a, b] \subset S$.

Proof. A direct application of Theorem 44.5. **QED**

Theorem 46.3 Let f be represented by its Maclaurin series $\sum_{n=0}^{\infty} \frac{f^{(n)}(0)}{n!}x^n$. Then the series can be differentiated term by term at any point interior to its interval of convergence S. That is,

$$f'(x) = \frac{d}{dx}\sum_{n=0}^{\infty} \frac{f^{(n)}(0)}{n!}x^n = \sum_{n=1}^{\infty} \frac{f^{(n)}(0)}{(n-1)!}x^{n-1}$$

at any $x \in S$ that is not an endpoint.

Proof. A direct application of Theorem 44.6. **QED**

Example 46.2 We have worked with the standard normal distribution's probability, $\Phi(x) = \frac{1}{2} + \frac{1}{\sqrt{2\pi}}\int_0^x e^{-t^2/2}dt$ over $(-\infty, \infty)$, in Exercise 40.6. This integral doesn't have an elementary function as an antiderivative, so how were evaluations done before computers came on the scene? Simpson's rule and Taylor series were typical tools for this. Taylor series have the advantage that the integrand does not have to be evaluated many times. Let's look at the typical case of $\Phi(0.7) = \frac{1}{2} + \frac{1}{\sqrt{2\pi}}\int_0^{0.7} e^{-t^2/2}dt$. The Maclaurin series for $f(t) = e^{-t^2/2}$ is tiresome to find if done by direct application of the definition. Instead, substitute $x = -\frac{t^2}{2}$ into both sides of

$e^x = \sum_{n=0}^{\infty} \frac{1}{n!}x^n$ and we efficiently get

$$e^{-t^2/2} = \sum_{n=0}^{\infty} \frac{(-1)^n}{n!2^n} t^{2n} = 1 - \frac{1}{2}t^2 + \frac{1}{8}t^4 - \frac{1}{48}t^6 + \frac{1}{384}t^8 - \cdots,$$

(guaranteed by Corollary 46.1). Integrating each term, we easily arrive at

$$\frac{1}{2} + \frac{1}{\sqrt{2\pi}} \int_0^x e^{-t^2/2} dt = \frac{1}{2} + \frac{1}{\sqrt{2\pi}} \left[t - \frac{t^3}{2(3)} + \frac{t^5}{8(5)} - \frac{t^7}{48(7)} + \frac{t^9}{384(9)} - \cdots \right]_0^x.$$

Convergence is fast when $0 \le x \le 1$, so we may expect that the seventh (and eighth) degree Maclaurin polynomial will give a good approximation of the probability we seek. Thus,

$$\begin{aligned}\Phi(0.7) &= \frac{1}{2} + \frac{1}{\sqrt{2\pi}} \int_0^{0.7} e^{-t^2/2} dt \\ &\approx 0.5 + \frac{1}{\sqrt{2\pi}} \left[t - \frac{t^3}{2(3)} + \frac{t^5}{8(5)} - \frac{t^7}{48(7)} \right]_0^{0.7} = 0.758031\ldots.\end{aligned}$$

And we know that the error cannot be more than the next term's absolute value:

$$\frac{1}{\sqrt{2\pi}} \frac{0.7^9}{384(9)} \approx 5 \times 10^{-6}.$$

This means that the sixth decimal place is not known, but we do know that $0.75803 < \Phi(0.7) < 0.75804$. Use Simpson's rule to obtain similar precision as Exercise 46.3, for comparison's sake. ■

46.1 The Binomial Series of Newton

The binomial theorem (see Exercise 4.7), the formula for expanding a binomial, probably has been known since ancient times, but without algebraic symbolism, of course. Applying it to the purposes of this section, if $m \in \mathbb{N}$, then

$$\begin{aligned}(1+x)^m = 1 + mx &+ \frac{m(m-1)}{2!}x^2 + \frac{m(m-1)(m-2)}{3!}x^3 + \cdots \\ &+ \frac{m(m-1)(m-2)\cdots(3)}{(m-2)!}x^{m-2} + mx^{m-1} + x^m.\end{aligned}$$

The coefficients of this polynomial are the binomial coefficients, written

$$\binom{m}{n} = \frac{m(m-1)(m-2)\cdots(m-n+1)}{n!},$$

where n is an integer between zero and m, inclusive, and $\binom{m}{0} \equiv 1$ for consistency. With this notation, the binomial theorem appears as

$$(1+x)^m = \sum_{n=0}^{m} \binom{m}{n} x^n \quad (m \in \mathbb{N}).$$

If instead of a natural number m we use any real number $p \notin \mathbb{N} \cup \{0\}$, then the theorem yields an infinite series since $\binom{p}{n}$ never zeros out. Hence, we have Newton's (or perhaps Gregory's)

binomial series

$$(3)\quad (1+x)^p = 1 + px + \frac{p(p-1)}{2!}x^2 + \frac{p(p-1)(p-2)}{3!}x^3 + \cdots$$
$$+ \frac{p(p-1)(p-2)\cdots(p-n+1)}{n!}x^n + \cdots = \sum_{n=0}^{\infty}\binom{p}{n}x^n.$$

When $p \notin \mathbb{N}\cup\{0\}$, questions of convergence and representation immediately arise for the binomial series. They are answered in the following important theorem.

Theorem 46.4 The binomial series (3), where $p \in \mathbb{R}$, represents the function $f(x) = (1+x)^p$ over an interval of convergence S. If $p \in \mathbb{N}\cup\{0\}$, then the series terminates, and the representation occurs over $S = \mathbb{R}$ (put $0^0 \equiv 1$). If $p \notin \mathbb{N}\cup\{0\}$, then the series represents the function over $S = (-1, 1)$. Convergence at endpoints $x = \pm 1$ depends on the value of p, as follows.

	Endpoint $x = -1$	Endpoint $x = 1$
$p > 0$	Absolute convergence: $\sum_{n=0}^{\infty}(-1)^n\binom{p}{n} = f(-1) = 0^p = 0$	Absolute convergence: $\sum_{n=0}^{\infty}\binom{p}{n} = f(1) = 2^p$
$-1 < p < 0$	Divergence	Conditional convergence: $\sum_{n=0}^{\infty}\binom{p}{n} = f(1) = 2^p$
$p \le -1$	Divergence	Divergence

Proof. We have already seen that if $p \in \mathbb{N}$, then the binomial series becomes the binomial theorem, which converges instantly for all $x \in \mathbb{R}$. If $p = 0$, the series converges, and represents $f(x) = (1+x)^0 = 1$ for all real x (set $0^0 \equiv 1$). Thus, proceed with $p \notin \mathbb{N}\cup\{0\}$.

Beginning with $f(x) = (1+x)^p$, we have $f'(x) = p(1+x)^{p-1}$, $f''(x) = p(p-1)(1+x)^{p-2}$, and in general, $f^{(n)}(x) = p(p-1)(p-2)\cdots(p-n+1)(1+x)^{p-n}$. The sequence of derivatives then yields the binomial series as the Maclaurin series $\sum_{n=0}^{\infty}\binom{p}{n}x^n$ corresponding to $f(x) = (1+x)^p$, where

$$\frac{f^{(n)}(0)}{n!} = \binom{p}{n} = \frac{p(p-1)(p-2)\cdots(p-n+1)}{n!}.$$

Since $p \notin \mathbb{N}\cup\{0\}$, we may apply the ratio test to our series:

$$\rho = \lim_{n\to\infty}\left|\frac{\binom{p}{n+1}x^{n+1}}{\binom{p}{n}x^n}\right|$$
$$= |x| \lim_{n\to\infty}\frac{|p(p-1)(p-2)\cdots(p-(n+1)+1)|}{(n+1)!}\cdot\frac{n!}{|p(p-1)(p-2)\cdots(p-n+1)|}$$
$$= |x| \lim_{n\to\infty}\frac{|p-n|}{n+1} = |x|.$$

This means that (3) converges absolutely when $x \in (-1, 1)$, for any $p \notin \mathbb{N} \cup \{0\}$. Together with the first paragraph of the proof, we now know that when $x \in (-1, 1)$, the binomial series (3) converges for any real exponent p.

The proof that the binomial series actually represents $f(x) = (1 + x)^p$ over $(-1, 1)$ is appropriate for a more advanced course. Reading a proof in, for instance, [13] § 8.6B, you will find it to be an application of a special form of Cauchy's form of the remainder of the series (3).

As for convergence at endpoints $x = \pm 1$, proving the results in the table above makes for good review, in Exercises 46.7 and 46.8. **QED**

Note 1: The extension of the binomial theorem to a series is one of Isaac Newton's major contributions to mathematics. He discovered it around the age of 22, and it served to hone his enormous mathematical talents. From pp. 394–395 of [3], Newton's work on this series "made clear to him that one could operate with infinite series in much the same way as with finite polynomial expressions.... Newton had discovered something far more important than the binomial theorem [our binomial series]; he had found that the analysis of infinite series had the same inner consistency and was subject to the same general laws as the algebra of finite quantities."

In other words, Newton found that algebraic and calculus operations with $y = (1 + x)^p$ would yield results in accordance with those operations on the binomial series. Thus, he was perhaps the first to realize that infinite series could be handled as functions, and not just as a method for approximating functions.

Historical details surrounding the discovery of the binomial series, with an excerpt from its first appearance in publication in 1685, can be found on pp. 379–382 of "Isaac Newton: Man, Myth, and Mathematics" by V. Frederick Rickey, in *The College Mathematics Journal*, vol. 18, no. 5 (Nov. 1987), pp. 362–389 (the article is an excellent review of Newton's various studies).

Example 46.3 Let's look at an early application of the binomial series that convinced Newton of its efficacy (see Note 1). Performing a long division on $\frac{1}{1-x^2}$, he gets $1 + x^2 + x^4 + x^6 + x^8 + \cdots$. Applying his brand-new binomial series to $(1 - x^2)^{-1}$, he finds $1 + (-1)(-x^2) + \frac{1}{2!}(-1)(-2)(-x^2)^2 + \frac{1}{3!}(-1)(-2)(-3)(-x^2)^3 + \cdots = 1 + x^2 + x^4 + x^6 + \cdots$. There must have been an "Eureka!" somewhere around that moment. Although Newton did not recognize representation, he surely knew that the series could be used only for $x \in (-1, 1)$. ■

Example 46.4 On page 45 of Albert Einstein's *Relativity, the Special and General Theory*, we find a revolutionary formula for the energy of an object that has rest mass m_0 and is moving with velocity

$$v : E_v = \frac{m_0 c^2}{\sqrt{1 - v^2/c^2}}.$$

For a very good reason, he says, "This expression approaches infinity as the velocity v approaches the velocity of light c. The velocity must therefore always remain less than c..." (c is almost 300000 km / s see Figure 46.1). Using the binomial series with $p = -\frac{1}{2}$ and $x = -\frac{v^2}{c^2}$ to expand

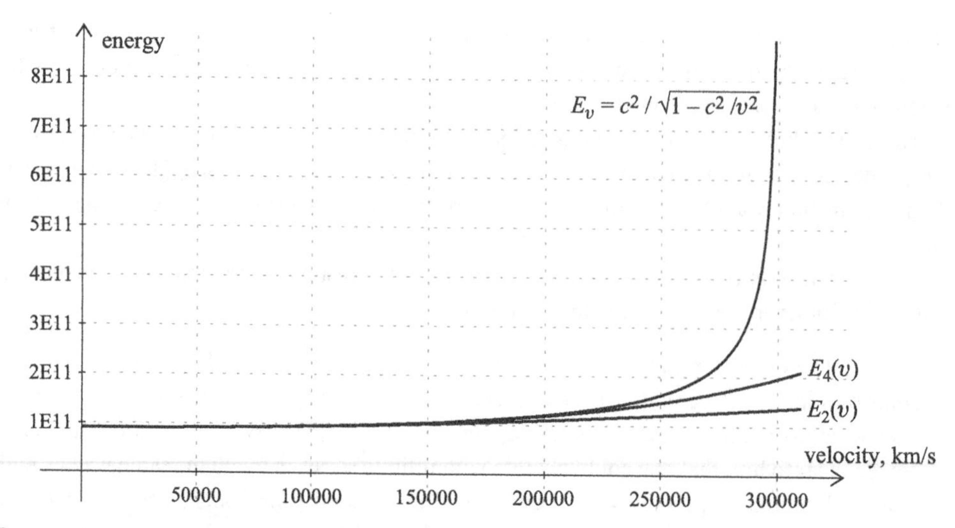

Figure 46.1. The relativistic energy formula Even the two-term binomial approximation agrees well with E_v up to about $v = 150000$ km/s, or 335 million mph, about half the speed of light!

$\frac{1}{\sqrt{1-v^2/c^2}}$, we find that

$$\left(1-\frac{v^2}{c^2}\right)^{-1/2} = 1+\left(\frac{-1}{2}\right)\left(\frac{-v^2}{c^2}\right)+\frac{1}{2!}\left(\frac{-1}{2}\right)\left(\frac{-3}{2}\right)\left(\frac{-v^2}{c^2}\right)^2 + \frac{1}{3!}\left(\frac{-1}{2}\right)\left(\frac{-3}{2}\right)\left(\frac{-5}{2}\right)\left(\frac{-v^2}{c^2}\right)^3+\cdots$$
$$= 1+\frac{1}{2}\frac{v^2}{c^2}+\frac{3}{8}\frac{v^4}{c^4}+\frac{5}{16}\frac{v^6}{c^6}+\cdots.$$

This series converges whenever $|-\frac{v^2}{c^2}| < 1$, that is, whenever $|v| < c$, or $v \in [0, c)$ since negative velocities make no difference in his study. Hence, the series represents $f(v) = \frac{1}{\sqrt{1-v^2/c^2}}$ over $[0, c)$. We therefore confirm Einstein's application of the binomial series,

$$E(v) = m_0c^2 + \frac{1}{2}m_0v^2 + \frac{3}{8}m_0\frac{v^4}{c^2} + \frac{5}{16}m_0\frac{v^6}{c^4} + \cdots,$$

and notice the famous first term. He continues, "When $\frac{v^2}{c^2}$ is small compared with unity, the third of these terms is always small in comparison to the second, which last is alone considered in classical mechanics." That is, $E = \frac{1}{2}m_0v^2$ was the one and only term that Newtonian dynamics disclosed for kinetic energy.

We also see in Figure 46.1 evidence that the series $E(v)$ converges uniformly towards E_v over $0 \le v < c$. ■

Example 46.5 The length of an arc of a circle is simple to find, so one might imagine that finding the arclength of an ellipse would be fairly straightforward, but it's not! An ellipse with vertical semimajor axis length a and semiminor axis length b is given by parametric equations $x = x(\theta) = a\sin\theta$ and $y = y(\theta) = b\cos\theta$. Its arclength (parametric form) in Quadrant I is therefore found as

$$s = \int_0^{\pi/2} \sqrt{a^2\cos^2\theta + b^2\sin^2\theta}\,d\theta = a\int_0^{\pi/2} \sqrt{1-\epsilon^2\sin^2\theta}\,d\theta,$$

where $\epsilon = \frac{\sqrt{a^2-b^2}}{a}$ is the eccentricity of the ellipse (verify the several steps, and remember that when $\epsilon = 0$, the ellipse becomes a circle).

As it happens, the integral does not have an antiderivative that can be expressed as an elementary function. (In fact, $E(\epsilon, t) = \int_0^t \sqrt{1-\epsilon^2\sin^2\theta}\,d\theta$ is an entirely new function called the **incomplete elliptic integral of the second kind**.) Fortunately, the binomial theorem rescues us with $x = -\epsilon^2\sin^2\theta$ and $p = \frac{1}{2}$ in (3):

$$\begin{aligned}(1-\epsilon^2\sin^2\theta)^{1/2} &= 1 + \frac{1}{2}(-\epsilon^2\sin^2\theta) + \frac{1}{2!}\left(\frac{1}{2}\right)\left(\frac{-1}{2}\right)(-\epsilon^2\sin^2\theta)^2 \\ &\quad + \frac{1}{3!}\left(\frac{1}{2}\right)\left(\frac{-1}{2}\right)\left(\frac{-3}{2}\right)(-\epsilon^2\sin^2\theta)^3 \\ &\quad + \frac{1}{4!}\left(\frac{1}{2}\right)\left(\frac{-1}{2}\right)\left(\frac{-3}{2}\right)\left(\frac{-5}{2}\right)(-\epsilon^2\sin^2\theta)^4 + \cdots \\ &= 1 - \frac{1}{2}\epsilon^2\sin^2\theta - \frac{1}{8}\epsilon^4\sin^4\theta - \frac{1}{16}\epsilon^6\sin^6\theta - \frac{5}{128}\epsilon^8\sin^8\theta - \cdots.\end{aligned}$$

With regard to convergence and representation, verify that indeed $|-\epsilon^2\sin^2\theta| < 1$. Hence, by Theorem 46.2,

$$s = a\left(\int_0^{\pi/2} d\theta - \frac{1}{2}\epsilon^2\int_0^{\pi/2}\sin^2\theta\,d\theta - \frac{1}{8}\epsilon^4\int_0^{\pi/2}\sin^4\theta\,d\theta - \frac{1}{16}\epsilon^6\int_0^{\pi/2}\sin^6\theta\,d\theta - \cdots\right).$$

Next, use your favorite calculus text to find the last two antiderivatives seen here, and finally arrive at

$$s = a\left(\frac{\pi}{2} - \frac{1}{2}\epsilon^2\cdot\frac{\pi}{4} - \frac{1}{8}\epsilon^4\cdot\frac{3\pi}{16} - \frac{1}{16}\epsilon^6\cdot\frac{5\pi}{32} - \cdots\right).$$

When $\epsilon = 0$, this checks out perfectly as the arclength of a quarter circle. If we now select an eccentricity $\epsilon = 0.5$, we obtain a rapidly converging series whose limit is the exact value of $a\int_0^{\pi/2}\sqrt{1-0.5^2\sin^2\theta}\,d\theta$. Using the four-term expansion just derived, we calculate that $s \approx 1.4675a$ as the arclength of the ellipse in Quadrant I. It's natural to evaluate the integral by calculator and see how accurate our approximation is. ■

46.2 Exercises

1. Write Theorems 46.1 through 46.4 in terms of Taylor series about $x = c$.

2. Prove Corollary 46.1.

3. Use Simpson's rule with $n = 2$ and $n = 4$ to evaluate the integral in Example 46.2. Compare the number of operations done with that in the Maclaurin approximation in that example.

4. The graph of $s(\theta) = \frac{\sin\theta}{\theta}$ is found just before Theorem 23.2. Find a series approximation for the area under the curve and in Quadrant I (the integral does not have an elementary antiderivative). What is the maximum error if four terms of the series are used?

5. a. Expand $\sqrt[3]{1+\tan\theta}$ as a binomial series. What restrictions are necessary for θ? Approximate $\sqrt[3]{1+\tan\frac{\pi}{8}}$ with terms up to $\tan^4\theta$, and determine how many decimal places are correct.
   b. Graph $g(x) = \sqrt[3]{1+\tan\theta}$ and the series with terms up to $\tan^4\theta$ that you found in **a**, over the interval $-1 \le \theta \le 1$. What detail about g definitively sets the interval of convergence for this binomial (power) series, in accordance with that prescribed by Theorem 46.4?

6. Use the binomial series to expand $f(y) = (1+\frac{1}{y})^{ty}$, where t is a constant, and representation of f requires $\frac{1}{y} \in (-1, 1)$ by Theorem 46.4, so $y \in (-\infty, -1) \cup (1, \infty)$. In the expansion, find the limit as $y \to +\infty$. What do you see?

7. Prove the results for the left endpoint $x = -1$ stated in the table of Theorem 46.4. Follow these steps:
   (i) For $p > 0$ but not a natural number or zero, verify that the series of constants $\sum_{n=0}^{\infty}(-1)^n\binom{p}{n}$ eventually stops alternating.
   (ii) Given

$$|a_n| = \left|\frac{p(p-1)(p-2)\cdots(p-n+1)}{n!}\right|,$$

   use Raabe's test to prove that the binomial series converges absolutely for $p > 0$, that is, $\sum_{n=0}^{\infty}|(-1)^n\binom{p}{n}|$ converges. Does this imply that $\sum_{n=0}^{\infty}(-1)^n\binom{p}{n}$ converges?
   (iii) When $p = 0$, what happens to the binomial series?
   (iv) For $p < 0$, convince yourself that $\sum_{n=0}^{\infty}(-1)^n\binom{p}{n}$ is a series of strictly positive terms. Apply Raabe's test to prove divergence.

8. Prove the results for the right endpoint $x = 1$ stated in the table of Theorem 46.4. Follow these steps:
   (i) For p not a natural number or zero, verify that the series of constants $\sum_{n=0}^{\infty}\binom{p}{n}$ eventually begins to alternate.
   (ii) Show that for $p > -1$, the absolute values of the terms,

$$|a_n| = \left|\frac{p(p-1)(p-2)\cdots(p-n+1)}{n!}\right|,$$

   make an eventually monotone decreasing sequence, by showing that $|\frac{a_{n+1}}{a_n}| < 1$ after n becomes larger than p.
   (iii) Next, for $p > 0$, prove that $\lim_{n\to\infty}|a_n| = 0$ by induction, as follows. Prove that if $1 < p+1 < n$, then $|a_n| < \frac{1}{n}$. Show first the anchor at case $n = 2$: if $1 < p+1 < 2$, then $|a_2| < \frac{1}{2}$.
   (iv) With the facts in **(i)**, **(ii)**, and **(iii)** in hand, what powerful theorem allows us to conclude that $\sum_{n=0}^{\infty}\binom{p}{n}$ converges when $p > 0$?
   (v) Showing convergence when $-1 < p < 0$ is the most challenging part of this exercise. The induction in **(iii)** won't work: for instance, at $p = -0.8$, calculate that $|a_2| > \frac{1}{2}$. Fortunately, **(i)** and **(ii)** still hold. We shall prove that for the constant $M = \frac{p+2}{p+1}$, ($M > 2$ over this range of values of p), $\sum_{n=0}^{\infty}|a_n|^M$ will converge, which implies that $|a_n|^M \to 0$ (by which theorem?), and hence, $|a_n| \to 0$.

Raabe's test seeks

$$\mu = \lim_{n\to\infty} n\left(1 - \frac{|a_{n+1}|^M}{|a_n|^M}\right)$$

for the series in question. Show that

$$\mu = \lim_{n\to\infty} n\left(1 - \left(\frac{|p-n|}{n+1}\right)^M\right) = \lim_{n\to\infty} n\left(1 - \left(\frac{n-p}{n+1}\right)^M\right) = p+2.$$

What can we conclude?

(vi) Since **(i)**, **(ii)**, and **(v)** apply to $\sum_{n=0}^{\infty} \binom{p}{n}$ for $-1 < p < 0$, what powerful theorem once again allows us to conclude that $\sum_{n=0}^{\infty} \binom{p}{n}$ converges when $p > 0$? Is this absolute or conditional convergence?

(vii) Finally, show that if $p \leq -1$, $\sum_{n=0}^{\infty} \binom{p}{n}$ diverges, by considering $|\frac{a_{n+1}}{a_n}|$.

47
The Riemann Integral

. . . at the suggestion of Schmalfuss [the director of the Gymnasium at Lüneburg, young Bernhard Riemann] carried off Legendre's Théorie des Nombres (Theory of Numbers). *This is a mere trifle of 859 large quarto pages, many of them crabbed with very close reasoning indeed. Six days later Riemann returned the book. "How far did you read?" Schmalfuss asked. . . . "That is certainly a wonderful book. I have mastered it." And in fact he had.*

E. T. Bell [2], pg. 487.

We have in previous chapters stated theorems involving integration, and it is now time to consider this operation analytically. On November 21, 1675, Gottfried Wilhelm Leibniz for the first time placed the symbol "dx" after an elongated "s" to denote the sum of infinitely many infinitesimal areas under a curve simply denoted as "y," and thus created the eternally enduring notation $\int y dx$. For the mathematicians of the time, the curve defined by y had to be a continuous curve such as that followed by a fired cannonball or described by a point on a rolling wheel.

Note 1: The concept of integration has evolved tremendously since its discovery. In the 18th century, the integral was usually considered as an antiderivative. Later, in Cauchy's writings, the integral reverted to its role as the limit of a sum. He also extended integration to functions of a complex variable, pretty much founding the study of complex analysis. As time went on, the integral had to be successively generalized to stay abreast of the increasingly abstract definitions of a function. Along with it, the concept of area, which once seemed to be intuitively obvious, slowly revealed its very complex nature (how can we define the area inside the Mandelbrot set, for instance?). Early in the 20th century Henri Lebesgue (1875–1941) published two papers that completely redefined the integral. Perhaps the latest chapter in the study of integration is the discovery of the generalized Riemann integral by Ralph Henstock (1923–2007) and Jaroslav Kurzweil (1926–) in recent decades.

The originality and influence of the mathematics written by Georg Friedrich Bernhard Riemann (1826–1866) is breathtaking. As one example, the ideas pronounced for the first time in his doctoral dissertation of 1854 regarding the structure of higher-dimensional spaces (called manifolds) were employed six decades later by Albert Einstein to establish general relativity. In the same year, with a good sense of how complicated a function could be, Riemann generalized the definition of the definite integral $\int_a^b f(x)dx$. The following steps apply to a function f over a closed interval $[a, b]$, and their order is crucial:

(a) Partition (slice) the interval $[a, b]$ into n subintervals $[x_{i-1}, x_i]$, $i = 1, 2, \ldots, n$, where $x_{i-1} < x_i$, $x_0 = a$, and $x_n = b$. We shall call the collection of subintervals $\{[x_{i-1}, x_i]\}_1^n$ a **net**, and symbolize it by $\mathcal{N}$. The subintervals have positive lengths $\Delta x_i = x_i - x_{i-1}$. Let $\Delta\chi$ be the maximum subinterval length in $\mathcal{N}$, that is, $\Delta\chi \equiv \max\{\Delta x_i\}_1^n$. ($\Delta\chi$ is sometimes called the **norm** of the net.) We may also identify $\mathcal{N}$ by its ordered set of points: $\mathcal{N} = \langle x_0, x_1, \ldots, x_{n-1}, x_n\rangle$.
(b) Select a number $X_i \in [x_{i-1}, x_i]$ in each subinterval such that $X_i \in \mathcal{D}_f$. Let's democratically call X_i the *representative* of its subinterval.
(c) Add the finite number of terms $\sum_{i=1}^n f(X_i)\Delta x_i$, called a **Riemann sum**, which we recognize as a sum of n rectangular areas.
(d) Consider nets $\mathcal{N}$ such that $\Delta\chi \to 0$ (forcing $n \to \infty$), and evaluate $\lim_{\Delta\chi\to 0} \sum_{i=1}^n f(X_i)\Delta x_i$ if it exists.

> If, following the steps above, $\lim_{\Delta\chi\to 0} \sum_{i=1}^n f(X_i)\Delta x_i$ exists, we call this limit the **Riemann integral** of f over $[a, b]$, say that f is **Riemann integrable** over $[a, b]$, and write $\lim_{\Delta\chi\to 0} \sum_{i=1}^n f(X_i)\Delta x_i = \int_a^b f(x)dx$. (1)

In terms similar to the definition of a limit, we say that the Riemann integral (1) exists if and only if given any $\varepsilon > 0$, there exists a corresponding $\delta > 0$ such that for any net with $\Delta\chi < \delta$, and for any choice of representatives $X_i \in [x_{i-1}, x_i]$, $i = 1, 2, \ldots, n$, it is true that $|\sum_{i=1}^n f(X_i)\Delta x_i - \int_a^b f(x)dx| < \varepsilon$.

The definitions in **(a)** above require that $b > a$, and that $\Delta x_i > 0$. To make sense of $\int_b^a f(x)dx$, begin at $x_n = b$, form $\Delta x_i = x_{i-1} - x_i < 0$, and end at $x_0 = a$. The effect is that all Riemann sums will change signs, and we will have $\int_b^a f(x)dx = -\int_a^b f(x)dx$ for any b and a. Last of all, define $\int_a^a f(x)dx \equiv 0$. Because a net over $[a, b]$ is defined with complete generality, the requirement in **(b)** can be fulfilled only if the domain of f does not have type 1 isolated points (see Exercise 24.9). On the other hand, it isn't necessary for $[a, b] \subseteq \mathcal{D}_f$, as is hinted in Figure 47.1.

Note 2: The verb "to partition" as used in **(a)** doesn't quite yield the noun "partition" in the proper sense of Chapter 5, but no confusion should arise.

In **(d)**, just letting $n \to \infty$ does not force $\Delta\chi \to 0$, for we could mischievously fix $\Delta\chi = (b - a)/2$ for instance, and still have increasingly many subintervals in a net. Hence, $\lim_{n\to\infty} \sum_{i=1}^n f(X_i)\Delta x_i$ isn't sufficient for what we want.

The Riemann integral is intended to be the *definition* of the area between f and $[a, b]$. In Figure 47.1, a net of $n = 7$ subintervals partitions the interval $[0, 4.5]$, with seven randomly selected representatives. Here, $\Delta\chi = 1.5$ since $\mathcal{N} = \langle 0, 0.5, 0.8, 1.0, 2.0, 2.5, 3.0, 4.5\rangle$. This particular Riemann sum $\sum_{i=1}^7 f(X_i)\Delta x_i$ doesn't seem to be a good approximation of $\int_0^{4.5} f(x)dx$.

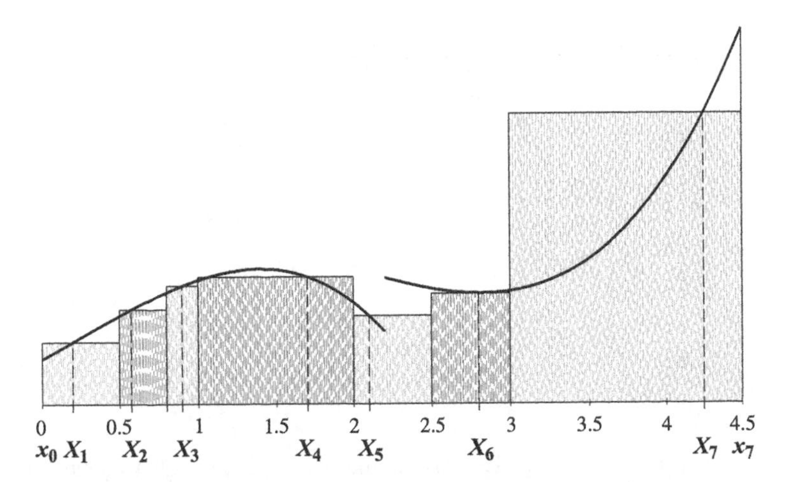

Figure 47.1. A typical Riemann sum with $n = 7$ subintervals over $[0, 4.5]$, for a discontinuous function. Isolated discontinuities don't affect the Riemann integral.

Admittedly, much is going on in Riemann's definition. The limit denoted $\int_a^b f(x)dx$ is to be found for *all* possible nets of $[a, b]$ as $\Delta\chi \to 0$, and for *any* representatives X_i in *each* subinterval $[x_{i-1}, x_i]$! An example of how all this variability can be taken into account follows.

Example 47.1 We apply Riemann's definition to

$$f(x) = \operatorname{sgn} x + 1 = \begin{cases} 0 & \text{if } x < 0 \\ 1 & \text{if } x = 0 \\ 2 & \text{if } x > 0 \end{cases}$$

over $[-1, 2]$, graphed in Figure 47.2. The discontinuity at $x = 0$ immediately concerns us. As the first of two cases, suppose subinterval P is made to contain $x = 0$ in its interior, that is, $0 \in (x_{P-1}, x_P)$. Then, regardless of how we partition the segment $[-1, x_{P-1}]$ to the left of $[x_{P-1}, x_P]$ into subintervals, we have $\sum_{i=1}^{P-1} f(X_i)\Delta x_i = 0$. Likewise, regardless of how the segment $[x_P, 2]$ to the right of $[x_{P-1}, x_P]$ is partitioned into subintervals, we have $\sum_{i=P+1}^{n} f(X_i)\Delta x_i = 2(2 - x_P)$.

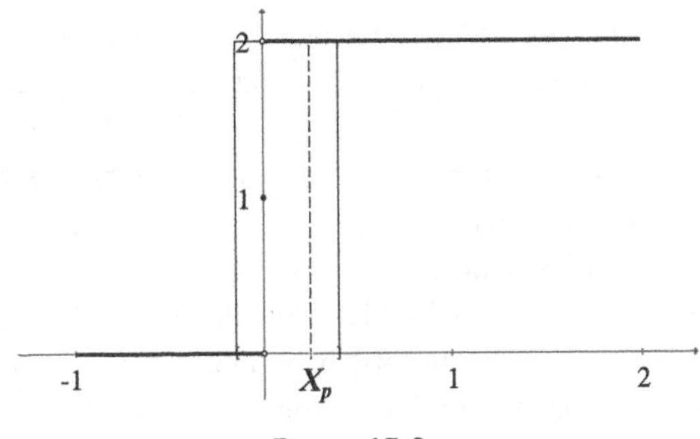

Figure 47.2.

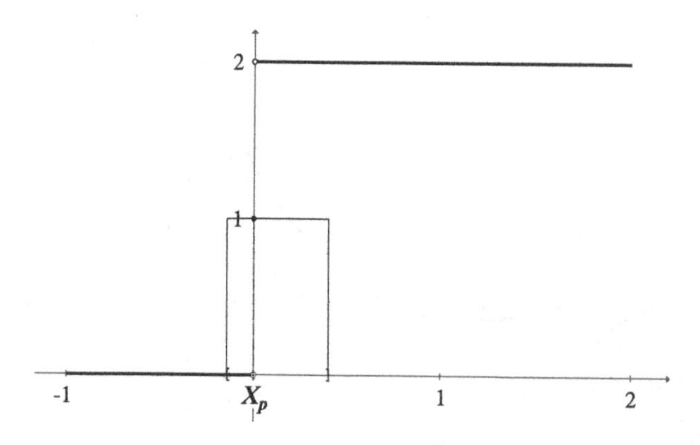

Figure 47.3.

$f(x) = \operatorname{sgn} x + 1$, seen over $[-1, 2]$. A subinterval $[x_{P-1}, x_P]$ that contains the origin is drawn in the two figures. A representative $X_P > 0$ is in Figure 47.2. In Figure 47.3, $X_P = 0$. The resulting rectangles over $[x_{P-1}, x_P]$ are drawn.

If within the subinterval $[x_{P-1}, x_P]$, we select a representative $X_P \in (0, x_P]$ as shown in Figure 47.2, then $f(X_P)\Delta x_P = 2\Delta x_P$. Consequently,

$$\sum_{i=1}^{n} f(X_i)\Delta x_i = \sum_{i=1}^{P-1} f(X_i)\Delta x_i + 2\Delta x_P + \sum_{i=P+1}^{n} f(X_i)\Delta x_i = 2\Delta x_P + 2(2 - x_P).$$

Or, if we select any representative $X_P \in [x_{P-1}, 0)$, then $f(X_P)\Delta x_P = 0\Delta x_P = 0$, and thus,

$$\sum_{i=1}^{n} f(X_i)\Delta x_i = \sum_{i=1}^{P-1} f(X_i)\Delta x_i + 0\Delta x_P + \sum_{i=P+1}^{n} f(X_i)\Delta x_i = 2(2 - x_P).$$

If we select the representative $X_P = 0$ as the remaining possibility (Figure 47.3), then $f(X_P)\Delta x_P = 1\Delta x_P$, and $\sum_{i=1}^{n} f(X_i)\Delta x_i = \Delta x_P + 2(2 - x_P)$.

Finally, the limit operation requires that $\Delta x_P \to 0$, which implies that $x_P \to 0$. Therefore,

$$\lim_{\Delta\chi\to 0} 2(2 - x_P) = \lim_{\Delta\chi\to 0} (2\Delta x_P + 2(2 - x_P)) = \lim_{\Delta\chi\to 0} (\Delta x_P + 2(2 - x_P)) = 4.$$

This ends the case that $\int_{-1}^{2} (\operatorname{sgn} x + 1)dx = 4$ when $0 \in (x_{P-1}, x_P)$.

In Exercise 47.1, show that the limit still equals 4 when one of the net's points is $x_i = 0$. This will complete the second case in the proof that $\int_{-1}^{2} (\operatorname{sgn} x + 1)dx = 4$. (Most of us never doubted that this was the area under the graph of f, but remember, we are evaluating a subtle limit!) ■

Analysts are always testing the implications of a mathematical definition by pushing the envelope. Thus, is it possible for a function $f : \mathbb{R} \to \mathbb{R}$ not to be Riemann integrable over any $[a, b] \subseteq \mathcal{D}_f$? Yes, but Riemann's definition is so comprehensive that such a function would have to be quite strange. In this sense, Dirichlet's function from Chapter 25 is strange indeed (redrawn in Figure 47.4). Show that it is not Riemann integrable in Exercise 47.3.

It is reasonable to argue that the problem with Dirichlet's function is that there are infinitely many discontinuities, one at every point of $[a, b]$, causing the Riemann integral to be overwhelmed. But a surprising example now appears.

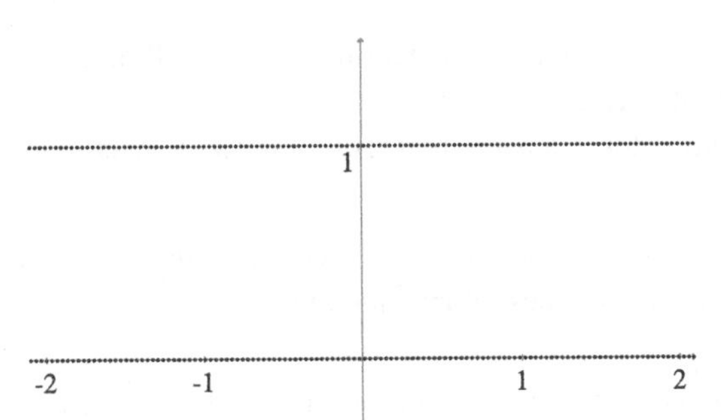

Figure 47.4. An unavoidably sketchy drawing of Dirichlet's function $y = \begin{cases} 1 & \text{if } x \text{ is rational} \\ 0 & \text{if } x \text{ is irrational} \end{cases}$

Review the ruler function from Chapter 25, where we found that it was discontinuous only at points with rational abscissas. Thus, this function does have infinitely many discontinuities, but the critical difference with Dirichlet's function is that this set of discontinuities is *countably* infinite.

Although tricky to prove directly from the definition, the ruler function is Riemann integrable, and in fact, $\int_a^b f(x)dx = 0$ (see Appendix C). We conjecture that Riemann integration is overwhelmed by uncountably many discontinuities on a closed interval, but apparently, a countable number doesn't cause a problem.

We will have to leave the search for precisely which functions are Riemann integrable to a more advanced course. However, an answer is at the end of Chapter 55. Also, read [4], which contains an excellent historical and mathematical saga of the search, beginning in Chapter 10.

Logically speaking, the negation of Riemann integrability is: $\int_a^b f(x)dx$ is not the Riemann integral of f over $[a, b]$ if and only if there is an $\varepsilon^* > 0$ such that for any $\delta > 0$, there exists some net with $\Delta\chi < \delta$ and some choice of representatives $X_i \in [x_{i-1}, x_i]$, $i = 1, 2, \ldots, n$, such that

$$\left| \sum_{i=1}^{n} f(X_i)\Delta x_i - \int_a^b f(x)dx \right| \geq \varepsilon^*.$$

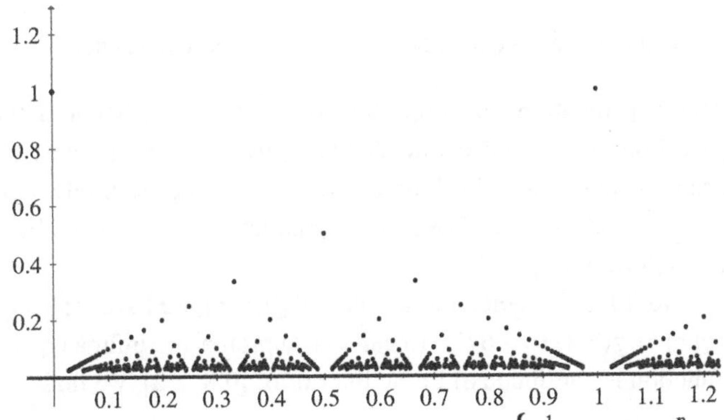

Figure 47.5. Some points of the ruler function $f(x) = \begin{cases} \frac{1}{|q|} & \text{if } x = \frac{p}{q} \text{ in lowest terms} \\ 0 & \text{if x is irrational} \end{cases}$.

Closely related to this is the blanket statement that f is not Riemann integrable at all over a given $[a, b]$. Formulate it as Exercise 47.2.

A necessary condition for Riemann integrability is that a function be bounded over $[a, b]$.

Lemma 47.1 If a function f is not bounded over $[a, b]$, then it is not Riemann integrable there. (This includes unboundedness at an endpoint.)

Proof. Let f be unbounded in any neighborhood of $c \in [a, b]$ (c may or may not belong to $\mathcal{D}_f$), consider a net over $[a, b]$, and the subinterval $[x_{i-1}, x_i]$ which contains c. Then given any $N > 0$, there exists an $x^* \in [x_{i-1}, x_i] \cap \mathcal{D}_f$ such that $|f(x^*)| > N$. We may always choose x^* as the representative of $[x_{i-1}, x_i]$ (we have complete freedom in choosing representatives), giving us $|f(x^*)|\Delta x_i > N\Delta x_i$. It follows that $|f(x^*)|\Delta x_i$ can be made larger than any number we wish, and it doesn't matter how small $\Delta\chi$ is. Therefore, $\lim_{\Delta\chi\to 0}\sum_{i=1}^{n} f(X_i)\Delta x_i$ will not exist, meaning that f is not Riemann integrable over $[a, b]$. **QED**

47.1 Upper and Lower Riemann Sums

The definition of Riemann integrability (1) is often difficult to work with. Another approach begins with some simple functions, those that are piecewise constant; they are called **step functions**. A typical one is $y = \lfloor x \rfloor$, seen in Figure 47.6. The function in Figure 47.7 has four steps, if we stretch the idea of a step a little.

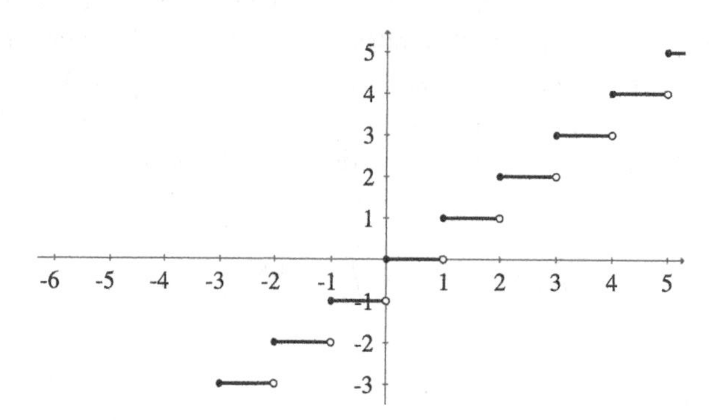

Figure 47.6. The floor function $y = \lfloor x \rfloor$ is a step function.

The reason that step functions are important is that they can bound a function at every subinterval of a net, both above and below. Actually, we want the tightest bounds possible, which in general require suprema and infima of the function. For example, consider the polynomial $g(x) = (x-2)^3 - 2(x-2) + 3$ over $[0, 4]$, and let a net $\mathcal{G}$ be given by the set of points $\langle 0, 0.8, 2.2, 4.0\rangle$, as shown in Figure 47.8.

Then, in the interior $(0, 0.8)$ of the first subinterval $[0, 0.8]$, we have $\inf_{(0,0.8)} g(x) = g(0) = -1$, and $\sup_{(0,0.8)} g(x) = g(0.8) = 3.672$. (Recall that suprema and infima equal actual function values when the function is continuous.) In the interior of $[0.8, 2.2]$, we find

$$\inf_{(0.8,2.2)} g(x) = g(2.2) = 2.608, \text{ and } \sup_{(0.8,2.2)} g(x) = g\left(2 - \sqrt{\frac{2}{3}}\right) \approx 4.089,$$

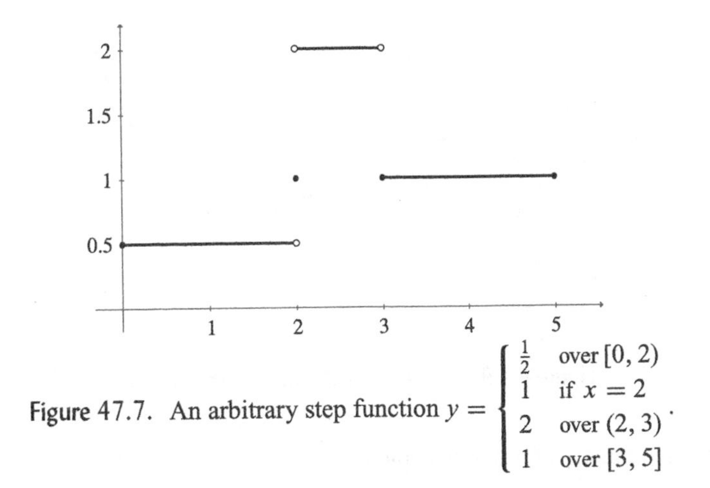

Figure 47.7. An arbitrary step function $y = \begin{cases} \frac{1}{2} & \text{over } [0,2) \\ 1 & \text{if } x = 2 \\ 2 & \text{over } (2,3) \\ 1 & \text{over } [3,5] \end{cases}$.

which should be verified with a bit of calculus. In the interior of [2.2, 4], we find

$$\inf_{(2.2,4)} g(x) = g\left(2 + \sqrt{\frac{2}{3}}\right) \approx 1.911, \text{ and } \sup_{(2.2,4)} g(x) = g(4) = 7.$$

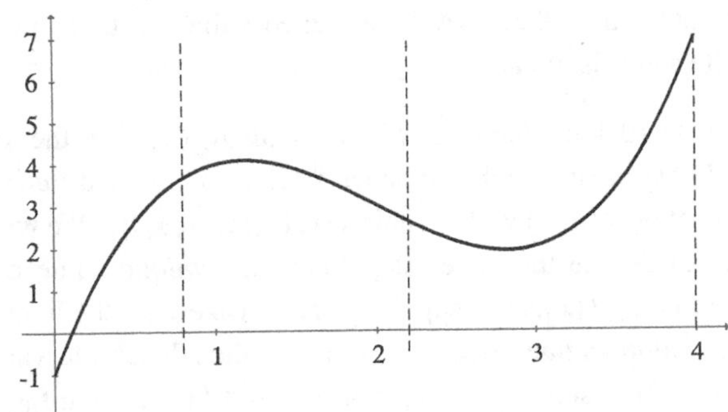

Figure 47.8. Polynomial $g(x) = (x-2)^3 - 2(x-2) + 3$ with $\mathcal{G}$, a rather coarse net of three subintervals.

We now define the **lower step function** for a function f on a net as the step function of the infima of f in the interior of each subinterval of the net, and denote it $\ell(f(x))$ or just $\ell(f)$. Similarly, define the **upper step function** for a function f on a net as the step function of the suprema of f in the interior of each subinterval of the net, and denote it $u(f(x))$ or just $u(f)$. If f is unbounded on a net, then either $u(f) = \infty$ or $\ell(f) = -\infty$ (or both) over at least one subinterval of the net. Thus, in accord with Lemma 47.1, we will need f to be bounded over $[a,b]$.

The lower step function for our example is

$$\ell(g(x)) = \begin{cases} -1 & \text{if } x \in (0, 0.8) \\ 2.608 & \text{if } x \in (0.8, 2.2) \\ \approx 1.911 & \text{if } x \in (2.2, 4) \end{cases},$$

seen in Figure 47.9.

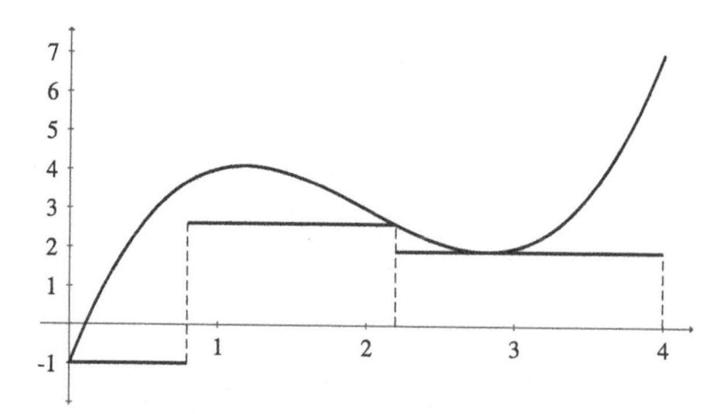

Figure 47.9. $\ell(g)$ is the thick line, g is the thin line.

The upper step function for our example is

$$u(g(x)) = \begin{cases} 3.672 & \text{if } x \in (0, 0.8) \\ \approx 4.089 & \text{if } x \in (0.8, 2.2) \\ 7 & \text{if } x \in (2.2, 4) \end{cases},$$

seen in Figure 47.10.

Note 3: You may have noticed that the upper and lower step functions are not defined at the points of the net. Thus, they may be assigned values at those points purely for the convenience of the analysis at hand.

Figures 47.9 and 47.10 look familiar: $\ell(g(x))$ and $u(g(x))$ are the graphs for specific Riemann sums for g. Let's investigate this in general. Given a bounded function f and a closed interval $[a, b]$, form a net over $[a, b]$ with n subintervals $\{[x_{i-1}, x_i]\}_1^n$. We will use the notations $U(f(x), i)$ or $U(f, i)$ to denote the upper step function's *unique* value over the interior of the ith subinterval, that is, $U(f(x), i) = \sup_{(x_{i-1}, x_i)} f(x)$. Likewise, the notations $L(f(x), i)$ or $L(f, i)$will denote the *unique* value $\inf_{(x_{i-1}, x_i)} f(x)$ over the ith subinterval. For example, our polynomial g yields $U(g, 2) = \sup_{(0.8, 2.2)} g(x) \approx 4.089$, and $L(g, 2) = \inf_{(0.8, 2.2)} g(x) = 2.608$. We now proceed to add all the rectangular areas bounded by the upper step function's values for a function f over a net: $\sum_{i=1}^n U(f(x), i)\Delta x_i$. Similarly, add all the rectangular areas bounded

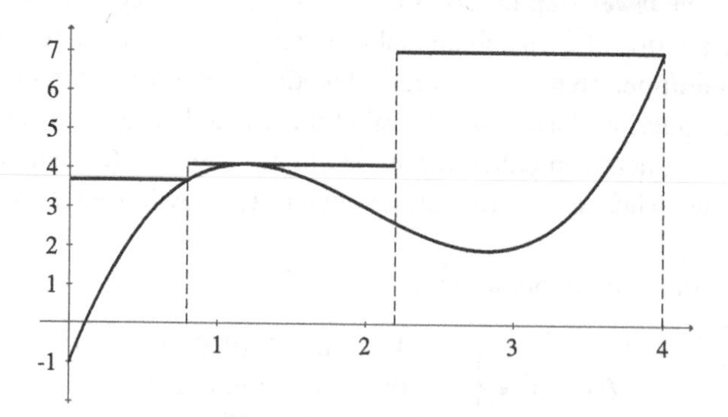

Figure 47.10. $u(g)$ is the thick line, g is the thin line.

by the lower step function's values for f over the net:

$$\sum_{i=1}^{n} L(f(x), i)\Delta x_i.$$

These two sums depend upon the particular net that we use (and upon f, of course), so let us concisely write $\sum_{i=1}^{n} U(f, i)\Delta x_i$ as $U(f, \mathcal{N})$, and $\sum_{i=1}^{n} L(f, i)\Delta x_i$ as $L(f, \mathcal{N})$. Let's introduce names for these important concepts.

The **upper Riemann sum** of a function f over a net $\mathcal{N}$ is $\sum_{i=1}^{n} U(f(x), i)\Delta x_i \equiv U(f, \mathcal{N})$, and the **lower Riemann sum** of f over the net $\mathcal{N}$ is $\sum_{i=1}^{n} L(f(x), i)\Delta x_i \equiv L(f, \mathcal{N})$.

Applying the new definitions to our example (see Figures 47.9 and 47.10), we evaluate the area

$$U(g, \mathcal{G}) = \sum_{i=1}^{3} U(g(x), i)\Delta x_i = \sup_{(0,0.8)} g(x)(0.8) + \sup_{(0.8,2.2)} g(x)(1.4) + \sup_{(2.2,4)} g(x)(1.8)$$

$$\approx (3.672)(0.8) + (4.089)(1.4) + (7)(1.8) = 21.262,$$

and the area

$$L(g, \mathcal{G}) = \sum_{i=1}^{3} L(g(x), i)\Delta x_i = \inf_{(0,0.8)} g(x)(0.8) + \inf_{(0.8,2.2)} g(x)(1.4) + \inf_{(2.2,4)} g(x)(1.8)$$

$$\approx (-1)(0.8) + (2.608)(1.4) + (1.911)(1.8) = 6.291.$$

Just to keep things in focus, we note that $\int_0^4((x-2)^3 - 2(x-2) + 3)dx = 12$, which falls neatly between $L(g, \mathcal{G})$ and $U(g, \mathcal{G})$. Incidentally, why did this example generate such a large difference between the two Riemann sums?

Because of what suprema and infima are, we must agree that for a given net, no Riemann sum can be less than the lower Riemann sum, nor can it be more than the upper Riemann sum. Using our new notation,

$$L(f, \mathcal{N}) \leq \sum_{i=1}^{n} f(X_i)\Delta x_i \leq U(f, \mathcal{N}).$$

In Exercise 47.4, you will evaluate another Riemann sum for g over $\mathcal{G}$, and see how it falls between the areas $L(g, \mathcal{G})$ and $U(g, \mathcal{G})$. Geometrically, then, the lower and upper Riemann sums are, respectively, the least and greatest areas associated with f and a given net.

We need one more ingredient to compose the Riemann integral anew. A **refinement** of a net is obtained by partitioning some (which allows all, per logic) of its subintervals into thinner subintervals. In other words, if the net is defined by the points $\langle x_0, x_1, \ldots, x_n\rangle$, then a refinement will contain at least one more point y such that $x_0 < y < x_n$ and $y \notin \{x_0, x_1, \ldots, x_n\}$. Hence, the number of subintervals increases by as many as the number of points introduced. For notation, use $\overline{\mathcal{N}}$ to denote any refinement of $\mathcal{N}$ whatsoever.

Lemma 47.2 A refinement of a net cannot decrease the lower Riemann sum nor increase the upper Riemann sum of a bounded function f over $[a, b]$. That is, $L(f, \mathcal{N}) \leq L(f, \overline{\mathcal{N}})$ and $U(f, \mathcal{N}) \geq U(f, \overline{\mathcal{N}})$.

Proof. We are given a net $\mathcal{N} = \langle x_0, x_1, \ldots, x_n \rangle$. Consider what happens when just one point w is inserted to make the refinement $\overline{\mathcal{N}}$. To be specific, let $w \in [x_{h-1}, x_h]$, splitting subinterval h into two subintervals, $[x_{h-1}, w]$ and $[w, x_h]$, whose respective lengths we will denote by Δw_0 and Δw_1 in order to distinguish them from the other lengths Δx_i $(i \neq h)$ inherited from $\mathcal{N}$. So now, $\overline{\mathcal{N}}$ has $n + 1$ subintervals.

Since $\Delta x_h = \Delta w_0 + \Delta w_1$, we know that

$$\inf_{(x_{h-1}, x_h)} f(x)\Delta x_h = \inf_{(x_{h-1}, x_h)} f(x)\Delta w_0 + \inf_{(x_{h-1}, x_h)} f(x)\Delta w_1.$$

Convince yourself that $\inf_{(x_{h-1}, x_h)} f(x) \leq \inf_{(x_{h-1}, w)} f(x)$ and $\inf_{(x_{h-1}, x_h)} f(x) \leq \inf_{(w, x_h)} f(x)$. It follows that

$$L(f, h)\Delta x_h = \inf_{(x_{h-1}, x_h)} f(x)\Delta x_h \leq \inf_{(x_{h-1}, w)} f(x)\Delta w_0 + \inf_{(w, x_h)} f(x)\Delta w_1.$$

Thinking in terms of areas, the inequality says that splitting the subinterval at the base of a given rectangle results in two rectangles that have at least as much area in total as the area of the original rectangle. Consequently,

$$\begin{aligned} L(f, \mathcal{N}) &= \sum_{i=1}^{h-1} L(f, i)\Delta x_i + L(f, h)\Delta x_h + \sum_{i=h+1}^{n} L(f, i)\Delta x_i \\ &\leq \sum_{i=1}^{h-1} L(f, i)\Delta x_i + \inf_{(x_{h-1}, w)} f(x)\Delta w_0 + \inf_{(w, x_h)} f(x)\Delta w_1 \\ &\quad + \sum_{i=h+1}^{n} L(f, i)\Delta x_i = L(f, \overline{\mathcal{N}}). \end{aligned}$$

For suprema, the inequalities go the other way:

$$\sup_{(x_{h-1}, x_h)} f(x) \geq \sup_{(x_{h-1}, w)} f(x)$$

and

$$\sup_{(x_{h-1}, x_h)} f(x) \geq \sup_{(w, x_h)} f(x).$$

By parallel reasoning, we conclude that

$$\begin{aligned} U(f, \mathcal{N}) &= \sum_{i=1}^{h-1} U(f,i)\Delta x_i + U(f,h)\Delta x_h + \sum_{i=h+1}^{n} U(f,i)\Delta x_i \\ &\geq \sum_{i=1}^{h-1} U(f,i)\Delta x_i + \sup_{(x_{h-1},w)} f(x)\Delta w_0 + \sup_{(w,x_h)} f(x)\Delta w_1 \\ &\quad + \sum_{i=h+1}^{n} U(f,i)\Delta x_i = U(f,\overline{\mathcal{N}}). \end{aligned}$$

Since any refinement of $\mathcal{N}$ may be constructed one point at a time, the lemma is proved. **QED**

The lemma supports a completely general statement comparing any lower Riemann sum with any upper Riemann sum for a function f.

Lemma 47.3 For a bounded function f over $[a, b]$, any lower Riemann sum cannot be greater than any upper Riemann sum, regardless of the nets used. That is, $L(f, \mathcal{N}) \leq U(f, \mathcal{M})$, whether or not one net is a refinement of the other.

Proof. Let $\mathcal{N}$ be defined by $\langle a, x_1, \ldots, x_{n-1}, b\rangle$ and $\mathcal{M}$ by $\langle a, y_1, \ldots, y_{m-1}, b\rangle$, and for anything interesting to happen, let $\mathcal{N} \neq \mathcal{M}$. A clever idea is to form the new net $\mathcal{N} \cup \mathcal{M} = \langle a, x_1, \ldots, x_{n-1}, b\rangle \cup \langle a, y_1, \ldots, y_{m-1}, b\rangle$. It follows that $\mathcal{N} \cup \mathcal{M}$ is a refinement of at least one of $\mathcal{N}$ and $\mathcal{M}$. Knowing that $L(f, \mathcal{N}) \leq U(f, \mathcal{N})$ for the *same* net, finish the proof as Exercise 47.5. **QED**

The two lemmas just proved set up the scenario that lower Riemann sums must monotonically increase as $\Delta\chi$ decreases, but every $L(f, \mathcal{N})$ is bounded above by any single $U(f, \mathcal{N})$, while upper Riemann sums must monotonically decrease as $\Delta\chi$ decreases, but every $U(f, \mathcal{N})$ is bounded below by any single $L(f, \mathcal{N})$. Therefore, by the monotone convergence theorem (Theorem 19.1), the limits $\lim_{\Delta\chi\to 0} L(f, \mathcal{N})$ and $\lim_{\Delta\chi\to 0} U(f, \mathcal{N})$ actually exist. Moreover, since limits are unique, these limits are independent of previous nets, and depend only upon f and interval $[a, b]$. This justifies writing $\lim_{\Delta\chi\to 0} L(f, \mathcal{N})$ with a new symbol,

$$\lim_{\Delta\chi\to 0} L(f, \mathcal{N}) \equiv \underline{\int_a^b} f(x)dx,$$

called the **lower Riemann integral** of f over $[a, b]$. Similarly,

$$\lim_{\Delta\chi\to 0} U(f, \mathcal{N}) \equiv \overline{\int_a^b} f(x)dx$$

is called the **upper Riemann integral** of f over $[a, b]$. After all is said and done, it's good to remember that these symbols are real numbers, as seen in Figure 47.11.

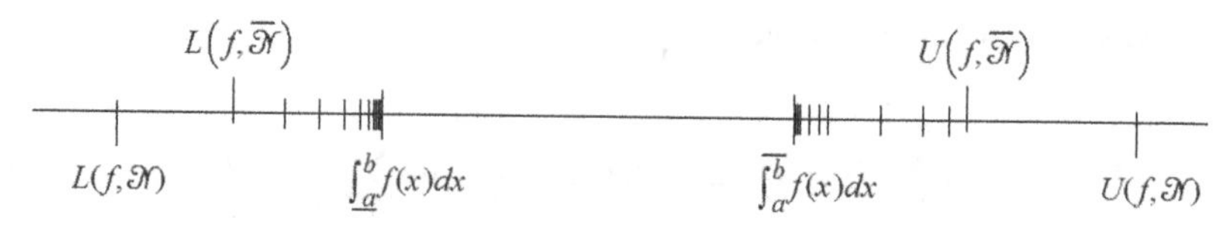

Figure 47.11. Upper Riemann sums decrease towards the upper Riemann integral, and lower Riemann sums increase towards the lower Riemann integral.

Figure 47.11 exposes the important possibility that the lower and upper Riemann integrals may be equal. Under this critical condition, function f becomes Riemann integrable. Thus,

A bounded function f is Riemann integrable over $[a, b]$ if and only if $\underline{\int_a^b} f(x)dx = \overline{\int_a^b} f(x)dx$.

(2)

The symbols suit their purpose perfectly:

$$\underline{\int_a^b} f(x)dx = \overline{\int_a^b} f(x)dx = \int_a^b f(x)dx,$$

the Riemann integral. This condition is equivalent to the definition of Riemann integrability stated in (1), and it is sometimes taken as the definition itself (see Note 4). It was the geometer and analyst Gaston Darboux who first proposed this definition of the Riemann integral, in 1875.

Darboux's definition of Riemann integrability has a clear geometric interpretation. The graph of our exemplary polynomial $g(x) = (x-2)^3 - 2(x-2) + 3$ is in Figure 47.12, with the step functions $\ell(g)$ and $u(g)$ from Figures 47.9 and 47.10.

From our previous calculations, the shaded rectangles have total area $U(g, \mathcal{N}) - L(g, \mathcal{N}) \approx 21.262 - 6.291 = 14.971$ square units.

In the general setting, we see that $U(f, \mathcal{N}) - L(f, \mathcal{N})$ indicates how different the upper and lower Riemann sums are as a function of the net $\mathcal{N}$ being used. Thus, if $U(f, \mathcal{N}) - L(f, \mathcal{N}) \to 0$ as nets become more and more refined, then f will be Riemann integrable, and otherwise,

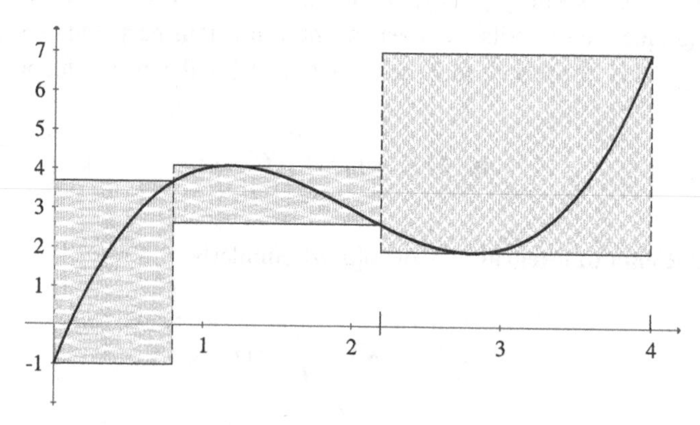

Figure 47.12. Step functions $u(g)$ and $\ell(g)$ shown with polynomial g over $[0, 4]$. When added together, the shaded areas illustrate $U(g, \mathcal{N}) - L(g, \mathcal{N})$

not. Bringing us back to definition (2), if

$$\lim_{\Delta\chi\to 0} U(f, \mathcal{N}) = \lim_{\Delta\chi\to 0} L(f, \mathcal{N}),$$

then the common value of the two limits is the Riemann integral $\int_a^b f(x)dx$.

Once a function is established as being Riemann integrable by (2), then *any* succession of increasingly refined nets must approach the unique value of $\int_a^b f(x)dx$. For the sake of programming efficiency, let alone our patience, we naturally opt for the simplest type of nets! They are the ones with equal subintervals, $\Delta x_i = \Delta\chi = \frac{b-a}{n}$, found in all calculus texts. Even more economy of effort is obtained by selecting as representatives X_i the midpoints of the subintervals, the left endpoints, or the right endpoints. Do Exercises 47.6 and 47.7 regarding this.

As shown in Figure 47.11, it is quite possible for $L(f, \mathcal{N})$ to be strictly less than $U(f, \mathcal{N})$ for a bounded function f over $[a, b]$, regardless of how small $\Delta\chi$ becomes (even if we use different nets $\mathcal{M}$ as $\mathcal{N}$ as in Lemma 47.4). As (2) requires, a function f is not Riemann integrable over $[a, b]$ if and only if

$$\underline{\int_a^b} f(x)dx < \overline{\int_a^b} f(x)dx.$$

This occurs, for instance, with Dirichlet's function (Figure 47.4), in which $L(f, \mathcal{N}) = 0$ and $U(f, \mathcal{N}) = 1$, regardless of the size of $\Delta\chi$.

Note 4: Riemann's and Darboux's definitions of integrability are equivalent. We shall prove this corollary by showing that the statements of non-integrability are equivalent. Starting with (1) and using the second statement of non-integrability in Exercise 47.2, suppose that for two different choices of representatives, denoted by $\{X_i\}_1^n$ and $\{Y_i\}_1^n$, we have

$$\lim_{\Delta\chi\to 0} \sum_{i=1}^{n} f(X_i)\Delta x_i < \lim_{\Delta\chi\to 0} \sum_{i=1}^{n} f(Y_i)\Delta x_i.$$

Since

$$\begin{aligned} L(f, \mathcal{N}) = \sum_{i=1}^{n} L(f, i)\Delta x_i &\le \sum_{i=1}^{n} f(X_i)\Delta x_i < \sum_{i=1}^{n} f(Y_i)\Delta x_i \\ &\le \sum_{i=1}^{n} U(f, i)\Delta x_i = U(f, \mathcal{N}), \end{aligned}$$

the strict inequality survives as we take limits:

$$\lim_{\Delta\chi\to 0} L(f, \mathcal{N}) < \lim_{\Delta\chi\to 0} U(f, \mathcal{N}).$$

This implies that f is not integrable by (2).

Conversely, start with

$$\underline{\int_a^b} f(x)dx < \overline{\int_a^b} f(x)dx$$

in (2), meaning that $\lim_{\Delta\chi\to 0} L(f, \mathcal{N}) < \lim_{\Delta\chi\to 0} U(f, \mathcal{N})$. By Lemma 47.3, $L(f, \mathcal{N}) \le \lim_{\Delta\chi\to 0} L(f, \mathcal{N})$. This means that we can always find representatives $\{X_i\}_1^n$ such that

$$L(f, \mathcal{N}) \le \sum_{i=1}^{n} f(X_i)\Delta x_i \le \lim_{\Delta\chi\to 0} L(f, \mathcal{N}).$$

Likewise, we can find representatives $\{Y_i\}_1^n$ such that

$$\lim_{\Delta\chi\to 0} U(f, \mathcal{N}) \le \sum_{i=1}^{n} f(Y_i)\Delta x_i \le U(f, \mathcal{N}).$$

Hence, we can always find two different choices of representatives such that

$$\sum_{i=1}^{n} f(X_i)\Delta x_i \le \lim_{\Delta\chi\to 0} L(f, \mathcal{N}) < \lim_{\Delta\chi\to 0} U(f, \mathcal{N}) \le \sum_{i=1}^{n} f(Y_i)\Delta x_i.$$

And from this,

$$\lim_{\Delta\chi\to 0} \sum_{i=1}^{n} f(X_i)\Delta x_i < \lim_{\Delta\chi\to 0} \sum_{i=1}^{n} f(Y_i)\Delta x_i.$$

Therefore, f is not integrable by (1). **QED**

47.2 Exercises

1. Finish the second case in Example 47.1, finally proving that $\int_{-1}^{2} (\operatorname{sgn} x + 1)dx = 4$.

2. Find a function that fails to be Riemann integrable for each of these reasons:
   (1) $\sum_{i=1}^{n} f(X_i)\Delta x_i \to +\infty$ over some $[a, b]$ as $\Delta\chi \to 0$.
   (2) For two different choices of representatives $\{X_i\}_1^n$ and $\{Y_i\}_1^n$ in a sequence of nets,

   $$\lim_{\Delta\chi\to 0} \sum_{i=1}^{n} f(X_i)\Delta x_i \ne \lim_{\Delta\chi\to 0} \sum_{i=1}^{n} f(Y_i)\Delta x_i.$$

   Hint: See the next exercise.

3. Show that Dirichlet's function

   $$y = \begin{cases} 1 & \text{if } x \text{ is rational} \\ 0 & \text{if } x \text{ is irrational} \end{cases}$$

   is not Riemann integrable over any interval $[a, b]$. Hint: Can $\lim_{\Delta\chi\to 0} \sum_{i=1}^{n} f(X_i)\Delta x_i$ be made to evaluate to two different answers?

4. Using $g(x) = (x-2)^3 - 2(x-2) + 3$ and the net $\mathcal{G} = \langle 0, 0.8, 2.2, 4.0 \rangle$, choose the midpoints as representatives: $X_1 = 0.4$, $X_2 = 1.5$, $X_3 = 3.1$, and evaluate the Riemann sum. This becomes an example of the well-known midpoint rule of numerical integration found in calculus texts. Compare your answer with $L(g, \mathcal{G}) \approx 6.291$, $U(g, \mathcal{G}) \approx 21.262$, and $\int_0^4 g(x)dx$.

5. Finish the proof of Lemma 47.4.

6. Seen in Figure 47.13 is $g(x) = (x-2)^3 - 2(x-2) + 3$, by now a celebrity polynomial, with upper and lower step functions over a net $\mathcal{M}$ with 16 equal subintervals, so that $\Delta\chi = \frac{1}{4}$. Shade the area corresponding to $U(g, \mathcal{M}) - L(g, \mathcal{M})$, and evaluate this difference. Hint: Rather than finding the 32 rectangular areas, note that over subintervals where g is increasing or decreasing, the summations telescope. For instance,

$$\sum_{i=1}^{4} \frac{1}{4} U(g(x), i) - \sum_{i=1}^{4} \frac{1}{4} L(g(x), i) = \frac{1}{4} \sum_{i=1}^{4} \left(g\left(\tfrac{i}{4}\right) - g\left(\frac{i-1}{4}\right) \right) = \tfrac{1}{4}(g(1) - g(0)).$$

However, two subintervals are not like the others.

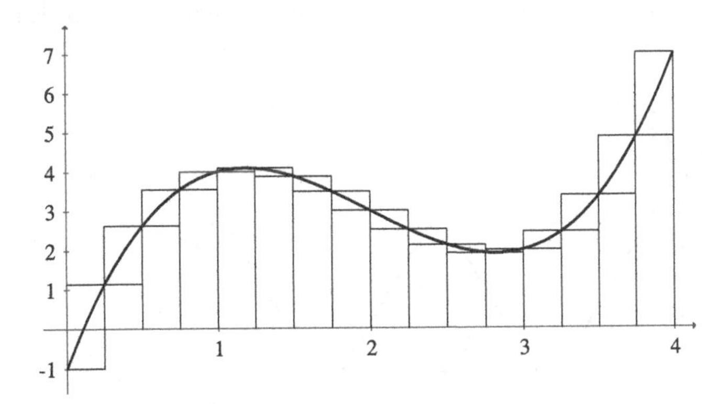

Figure 47.13.

7. Assume that $g(x) = (x-2)^3 - 2(x-2) + 3$ is Riemann integrable over $[0, 4]$ (we won't need to assume this after the next chapter). We have mentioned that any sequence of nets whatsoever will yield Riemann sums that converge to $\int_0^4 g(x)dx$ as $\Delta\chi \to 0$. Thus, choose nets with $\Delta x_i = \frac{4-0}{n} = \frac{4}{n}$ for all subintervals, select representatives X_i conveniently at left endpoints of subintervals, and let $n \to \infty$. You are verifying that your calculator is correct when it says that $\int_0^4 g(x)dx = 12$. Hint: This exercise could be found in a typical calculus text.

8. Prove that every step function with a finite number of steps over $[a, b]$ is Riemann integrable there. Hint: Every step function with a finite number of steps sets up a net on the X axis. There may be some degenerate subintervals where $x_{i-1} = x_i$, but this won't matter.

9. Prove that if $f(x) \geq 0$ and is Riemann integrable over $[a, b]$, then $\int_a^b f(x)dx \geq 0$. Show that the converse, if $\int_a^b f(x)dx \geq 0$ then $f(x) \geq 0$ over $[a, b]$, is false (the minor premise being that f is integrable over $[a, b]$).

10. Prove that if f is Riemann integrable over $[a, b]$, then modifying f by changing its ordinates at a finite number of points will not change the value of $\int_a^b f(x)dx$ (perforce, it remains integrable). (See also Exercise 48.4.)

11. Answer true or false:

    a. $u(f(x)) = \bigcup_1^n U(f(x), i)$.
    b. $U(f(x), i) = f(X_i)$ for some $X_i \in [x_{i-1}, x_i]$.
    c. All bounded functions are Riemann integrable.
    d. All continuous functions are Riemann integrable.
    e. The lower Riemann integral equals $\sup_{\mathcal{N}} L(f, \mathcal{N})$.
    f. The upper Riemann integral equals $\inf_{\mathcal{N}} U(f, \mathcal{N})$.

12. (This exercise is from Problem 481 by Leonard Gillman in *The College Mathematics Journal*, vol. 24, no. 4 (Sept. 1993), with a solution by H. K. Krishnapriyan.) Seen in Figure 47.14 is the continuous sawtooth function $Q(x) = 2|\frac{5}{2}x - \frac{1}{2} - \lfloor\frac{5}{2}x\rfloor|$ over $[0, 1]$. Show that $L(Q, \mathcal{N}) > L(Q, \mathcal{M})$ and $U(Q, \mathcal{N}) < U(Q, \mathcal{M})$ by evaluating the Riemann sums when $\mathcal{N}$ is the net of four equal subintervals ($\Delta x = \frac{1}{4}$), and $\mathcal{M}$ is the net of five equal subintervals ($\Delta x = \frac{1}{5}$). This seems to violate Lemma 47.3. Explain the apparent contradiction.

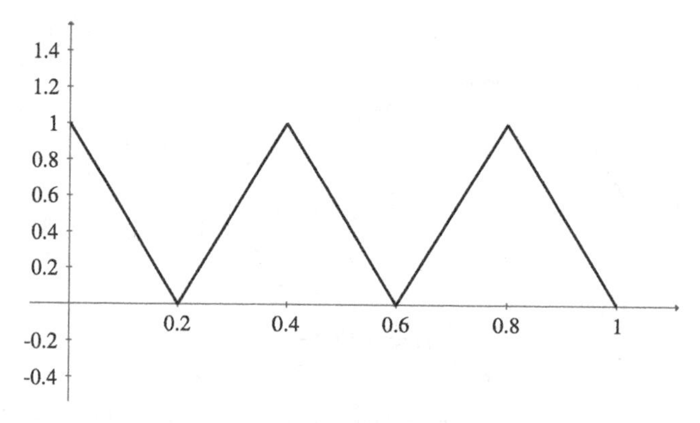

Figure 47.14.

48

The Riemann Integral, Part II

Now that Riemann integration has been well established, we continue with several theorems describing its familiar properties. As you study the proofs, note how versatile is the concept of upper and lower Riemann sums. The first theorem lets us identify many Riemann integrable functions, including all polynomials.

Theorem 48.1 If function f is continuous over $[a, b]$, then it is Riemann integrable there.

Proof. Given the premises, f is bounded, and Theorem 27.1 assures us that f is uniformly continuous over $[a, b]$. Thus, given any $\varepsilon > 0$, there exists a $\delta_\varepsilon > 0$ such that for any $u, v \in [a, b]$, if $|u - v| < \delta_\varepsilon$, then $|f(u) - f(v)| < \frac{\varepsilon}{b-a}$. We will now apply this to a net $\mathcal{N}$ of $[a, b]$ such that $\Delta\chi < \delta_\varepsilon$.

We know that

$$0 \le U(f, \mathcal{N}) - L(f, \mathcal{N}) = \sum_{i=1}^{n} U(f, i)\Delta x_i - \sum_{i=1}^{n} L(f, i)\Delta x_i$$
$$= \sum_{i=1}^{n} [U(f, i) - L(f, i)]\Delta x_i. \quad (1)$$

Now, $U(f, i)$ and $L(f, i)$ are the supremum and infimum of f over (x_{i-1}, x_i), which implies that they are also the supremum and infimum of f over the closed subinterval $[x_{i-1}, x_i]$, since f is continuous over $[a, b]$. Therefore, by the extreme value theorem (Theorem 26.5), there actually exist points $x_m, x_M \in [x_{i-1}, x_i]$ such that $f(x_m) = L(f, i)$ and $f(x_M) = U(f, i)$. Because of the net that we have chosen, $|x_m - x_M| < \delta_\varepsilon$. By the uniform continuity of f, this implies that

$$f(x_M) - f(x_m) < \frac{\varepsilon}{b-a},$$

regardless of the subinterval we are in (note that $|f(x_M) - f(x_m)| = f(x_M) - f(x_m)$).

Continuing the line of reasoning in (1),

$$\sum_{i=1}^{n} [U(f, i) - L(f, i)]\, \Delta x_i = \sum_{i=1}^{n} (f(x_M) - f(x_m))\, \Delta x_i < \frac{\varepsilon}{b-a} \sum_{i=1}^{n} \Delta x_i$$
$$= \frac{\varepsilon}{b-a}(b-a) = \varepsilon.$$

In summary, $0 \leq U(f, \mathcal{N}) - L(f, \mathcal{N}) < \varepsilon$ whenever $\Delta\chi < \delta_\varepsilon$, which means that $\lim_{\Delta\chi \to 0}(U(f, \mathcal{N}) - L(f, \mathcal{N})) = 0$. By the second definition in the previous chapter, f is Riemann integrable over $[a, b]$. **QED**

Theorems 48.2 through 48.4 and the corollary in between are old friends from calculus.

Theorem 48.2 If f is Riemann integrable over $[a, b]$, then $\int_a^b kf(x)dx = k\int_a^b f(x)dx$, where k is any real number constant.

Proof. If $k = 0$, then $\int_a^b 0dx = 0 = 0\int_a^b f(x)dx$ and we are done. If $k \neq 0$, since f is Riemann integrable, given any $\varepsilon > 0$, there exists a $\delta > 0$ such that for any net with $\Delta\chi < \delta$, we can certify that

$$\left|\sum_{i=1}^{n} f(X_i)\,\Delta x_i - \int_a^b f(x)dx\right| < \frac{\varepsilon}{|k|}.$$

Multiplying both sides by $|k|$,

$$\left|\sum_{i=1}^{n} kf(X_i)\,\Delta x_i - k\int_a^b f(x)dx\right| < \varepsilon.$$

This tells us two things: the function $kf(x)$ is integrable, that is, $\int_a^b kf(x)dx$ exists, and its value is $k\int_a^b f(x)dx$. **QED**

Theorem 48.3 If both f and g are Riemann integrable over $[a, b]$, then $f + g$ and $f - g$ are Riemann integrable functions, and

$$\int_a^b [f(x) \pm g(x)]\,dx = \int_a^b f(x)dx \pm \int_a^b g(x)dx.$$

Proof. For convenience, let $I_f = \int_a^b f(x)\,dx$ and $I_g = \int_a^b g(x)dx$. We prove the case for added functions. For any $\varepsilon > 0$, there exists a $\delta > 0$ such that for any net with $\Delta\chi < \delta$, we have

$$\left|\sum_{i=1}^{n} f(X_i)\,\Delta x_i - I_f\right| < \frac{\varepsilon}{2}$$

and

$$\left|\sum_{i=1}^{n} g(X_i)\,\Delta x_i - I_g\right| < \frac{\varepsilon}{2}.$$

Now,

$$\left|\sum_{i=1}^{n}[f(X_i) + g(X_i)]\Delta x_i - (I_f + I_g)\right| = \left|\sum_{i=1}^{n} f(X_i)\Delta x_i - I_f + \sum_{i=1}^{n} g(X_i)\Delta x_i - I_g\right|.$$

Hence, by the triangle inequality,

$$\left|\sum_{i=1}^{n}[f(X_i)+g(X_i)]\,\Delta x_i-(I_f+I_g)\right|\leq\left|\sum_{i=1}^{n}f(X_i)\,\Delta x_i-I_f\right|$$
$$+\left|\sum_{i=1}^{n}g(X_i)\,\Delta x_i-I_g\right|<\frac{\varepsilon}{2}+\frac{\varepsilon}{2}=\varepsilon.$$

Consequently,

$$\lim_{\Delta\chi\to 0}\sum_{i=1}^{n}[f(X_i)+g(X_i)]\,\Delta x_i=I_f+I_g,$$

which means that

$$\int_a^b[f(x)+g(x)]\,dx=\int_a^b f(x)dx+\int_a^b g(x)dx.$$

Prove the case for subtracted functions as Exercise 48.1. **QED**

Corollary 48.1 If f and g are Riemann integrable over $[a, b]$, and $f(x)\leq g(x)$ over that interval, then

$$\int_a^b f(x)dx\leq\int_a^b g(x)\,dx.$$

Proof. Prove this as Exercise 48.2. **QED**

Theorem 48.4 If f is Riemann integrable over $[a, c]$ and $[c, b]$, then f is Riemann integrable over $[a, b]$, and

$$\int_a^b f(x)dx=\int_a^c f(x)\,dx+\int_c^b f(x)dx.$$

Proof. First, it is necessary for you to show in Exercise 48.3 that f is bounded over $[a, b]$. With that done, consider any net $\mathcal{L}$ for f over $[a, b]$. If one of the net's points is c, we may write $\mathcal{L}=\langle a, x_1,\ldots,c,x_k,\ldots b\rangle$. In this case, define $\mathcal{N}$ for f over $[a, c]$ by $\mathcal{N}=\langle a, x_1,\ldots,c\rangle$, and define $\mathcal{M}$ for f over $[c, b]$ by $\mathcal{M}=\langle c, x_k,\ldots,b\rangle$, so $\mathcal{L}=\mathcal{N}\cup\mathcal{M}$. Relative to $\mathcal{N}$ and $\mathcal{M}$, the upper Riemann sums $U(f,\mathcal{N})$ and $U(f,\mathcal{M})$ are well-defined real numbers, so that (a) $U(f,\mathcal{N})+U(f,\mathcal{M})=U(f,\mathcal{L})$. Since f is Riemann integrable over $[a, c]$ and $[c, b]$, the upper Riemann integrals exist for each interval. This, together with (a), implies that

$$\overline{\int_a^c}f(x)dx+\overline{\int_c^b}f(x)dx=\overline{\int_a^b}f(x)dx.$$

By similar reasoning, $L(f,\mathcal{N})+L(f,\mathcal{M})=L(f,\mathcal{L})$ implies

$$\underline{\int_a^c}f(x)\,dx+\underline{\int_c^b}f(x)dx=\underline{\int_a^b}f(x)dx.$$

By the premises,

$$\overline{\int_a^c} f(x)dx + \overline{\int_c^b} f(x)dx = \underline{\int_a^c} f(x)\,dx + \underline{\int_c^b} f(x)dx,$$

which tells us that

$$\overline{\int_a^b} f(x)dx = \underline{\int_a^b} f(x)dx.$$

Therefore, $\int_a^b f(x)\,dx$ exists and equals

$$\int_a^c f(x)dx + \int_c^b f(x)dx.$$

The only other possibility is that $c \notin \mathcal{L}$. This means that $c \in (x_{k-1}, x_k)$ for some subinterval of the net. We are then able to make the refinement $\overline{\mathcal{L}} = \mathcal{L} \cup \{c\}$. Now, we are able to recycle the argument in the first paragraph of the proof (this is one of mathematicians' favorite techniques), replacing $\mathcal{L}$ with $\overline{\mathcal{L}}$. The conclusion once again is that $\int_a^b f(x)dx$ exists and equals

$$\int_a^c f(x)dx + \int_c^b f(x)dx. \quad \textbf{QED}$$

The following theorem seems intuitively obvious, except that f may be discontinuous at up to a countable infinity of points! A slightly different upper and lower step function that is suitable for this situation is introduced.

Theorem 48.5 If f is monotonic over a closed interval $[a, b] \subseteq \mathcal{D}_f$, then it is Riemann integrable there.

Proof. As required, f is bounded over $[a, b]$. As the first of two cases, suppose that f is monotone increasing over $[a, b]$. Let $\mathcal{N} = \langle a, x_1, \ldots, x_{n-1}, b\rangle$ be any net over $[a, b]$, where we are given that all $f(x_i)$, $i = 0, 1, \ldots, n$, exist. Using $\mathcal{N}$, and since f is monotone increasing, we have

$$0 \le \sum\nolimits_{i=1}^{n} [f(x_i) - f(x_{i-1})]\,\Delta x_i \le \Delta\chi \sum\nolimits_{i=1}^{n} [f(x_i) - f(x_{i-1})].$$

The last sum telescopes to $f(x_n) - f(x_0) = f(b) - f(a)$. Thus, we arrive at

$$0 \le \sum\nolimits_{i=1}^{n} f(x_i)\,\Delta x_i - \sum\nolimits_{i=1}^{n} f(x_{i-1})\,\Delta x_i \le \Delta\chi\,(f(b) - f(a)). \tag{2}$$

Next, taking further advantage of the monotonicity, we redefine the upper and lower step functions of the previous chapter so that $\ell(f)$ and $u(f)$ are evaluated respectively at the left and right endpoints of each subinterval $[x_{i-1}, x_i]$. Specifically, $L(f, 1) = f(a)$, $U(f, 1) = f(x_1)$, $L(f, 2) = f(x_1)$, $U(f, 2) = f(x_2)$, and $L(f, i) = f(x_{i-1})$, $U(f, i) = f(x_i)$ in general, until $L(f, n) = f(x_{n-1})$, $U(f, n) = f(b)$, as seen in Figure 48.1.

We now have

$$\sum_{i=1}^{n} U(f, i)\,\Delta x_i = \sum_{i=1}^{n} f(x_i)\,\Delta x_i$$

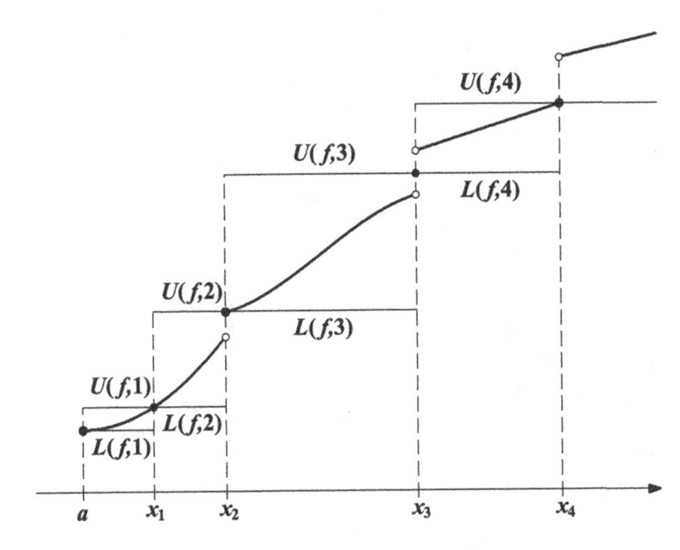

Figure 48.1. A monotone increasing function f over the first four subintervals of a net, and its upper and lower functions. At x_1, f is continuous. The three allowable types of discontinuities are seen at x_2, x_3, and x_4 (see Lemma 24.1).

as the upper Riemann sum for our $\mathcal{N}$. Hence, as $\Delta\chi \to 0$, the sums decrease but are bounded below by $f(a)(b-a)$ since $f(a)$ is the absolute minimum of f over $[a, b]$. It follows that the limit $\overline{\int_a^b} f(x)dx$ exists. Likewise,

$$\sum\nolimits_{i=1}^{n} L(f, i)\,\Delta x_i = \sum\nolimits_{i=1}^{n} f(x_{i-1})\,\Delta x_i$$

is the lower Riemann sum for $\mathcal{N}$, so as $\Delta\chi \to 0$, these sums increase but are bounded above by $f(b)(b-a)$. Hence, the limit $\underline{\int_a^b} f(x)\,dx$ also exists.

Now, letting $\Delta\chi \to 0$ in the inequalities (b),

$$0 \leq \lim_{\Delta\chi\to 0} \sum\nolimits_{i=1}^{n} f(x_i)\,\Delta x_i - \lim_{\Delta\chi\to 0} \sum\nolimits_{i=1}^{n} f(x_{i-1})\,\Delta x_i \leq \lim_{\Delta\chi\to 0} \Delta\chi\,(f(b) - f(a)) = 0,$$

and we recognize the resulting upper and lower Riemann integrals from our analysis in the previous paragraph. Therefore,

$$0 = \overline{\int_a^b} f(x)dx - \underline{\int_a^b} f(x)dx,$$

which means finally that $\int_a^b f(x)dx$ exists.

As Exercise 48.5, make a proof for the case of a monotone decreasing function f, and then we are done. **QED**

Theorem 48.6 If f is bounded over $[a, b]$, and Riemann integrable over every closed interval $[c, d] \subset [a, b]$, then f is Riemann integrable over the full interval $[a, b]$.

Proof. Since f is bounded over $[a, b]$, there is a positive constant B such that $|f(x)| < B$ over $[a, b]$. Because of integrability over every $[c, d]$, given any $\varepsilon > 0$, there are $\delta \in \left(0, \frac{1}{2}(b-a)\right)$ for which we are able to find nets $\mathcal{N}$ over $[c, d]$ with $\Delta\chi < \delta$, and such that $U(f, \mathcal{N}) - L(f, \mathcal{N}) < \varepsilon$. For what follows, we will need to select an ε small enough so that $\frac{\varepsilon}{B} < b - a$.

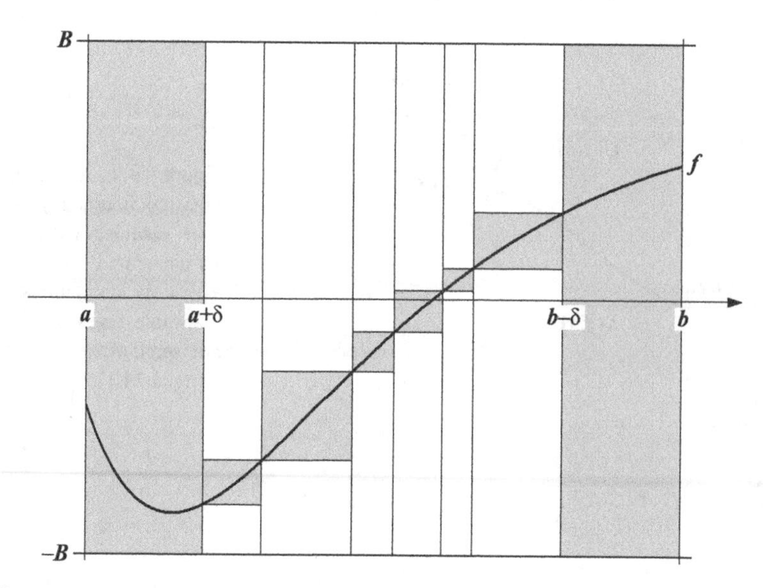

Figure 48.2. A function bounded over $[a, b]$ and integrable over $[c, d]$, where $c = a + \delta$ and $d = b - \delta$. For the net $\mathcal{N}$ over $[c, d]$, $\Delta\chi < \delta$.

Now, in particular, take $c = a + \delta$ and $d = b - \delta$, implying that $c < d$, and δ is the common width of the subintervals $[a, c]$ and $[d, b]$ (see Figure 48.2). Define the one-step step functions $\ell([a, c]) = \ell([d, b]) = -B$ and $u([a, c]) = u([d, b]) = B$. Form the net $\mathcal{M} = [a, c] \cup \mathcal{N} \cup [d, b]$, which covers the interval $[a, b]$. Then $U(f, \mathcal{M}) = B\delta + U(f, \mathcal{N}) + B\delta$, and $L(f, \mathcal{M}) = -B\delta + L(f, \mathcal{N}) - B\delta$. Therefore, $0 \leq U(f, \mathcal{M}) - L(f, \mathcal{M}) = 4B\delta + U(f, \mathcal{N}) - L(f, \mathcal{N}) < 4B\delta + \varepsilon$. We now choose $\delta < \frac{\varepsilon}{4B}$ (which keeps $c < d$), so that we have $0 \leq U(f, \mathcal{M}) - L(f, \mathcal{M}) < 2\varepsilon$. Thus, $U(f, \mathcal{M}) - L(f, \mathcal{M})$ can be made as small as we wish, which is another way to say that

$$\overline{\int_a^b} f(x)\,dx = \underline{\int_a^b} f(x)dx.$$

We conclude that f is Riemann integrable over $[a, b]$. **QED**

Example 48.1 It is surprising that

$$f(x) = \begin{cases} \sin\frac{1}{x} & \text{if } x \neq 0 \\ 0 & \text{if } x = 0 \end{cases}$$

graphed in Figure 48.3 is Riemann integrable over $[0, 1]$. Since this function is bounded and continuous over any $[c, 1]$, where $0 < c \leq 1$, Theorem 48.1 and the theorem just proved allow us to state that $\int_0^1 \sin\frac{1}{x}dx$ exists, despite the infinity of oscillations. ■

Theorem 48.7 The Mean Value Theorem for Integrals If f is continuous over $[a, b]$, then there exists a point ξ between a and b such that $\int_a^b f(x)dx = f(\xi)(b - a)$.

Proof. By the extreme value theorem (Theorem 26.5), there exist points $m, M \in [a, b]$ such that $f(m)$ is the minimum and $f(M)$ is the maximum, respectively, of f over the interval. Also, f

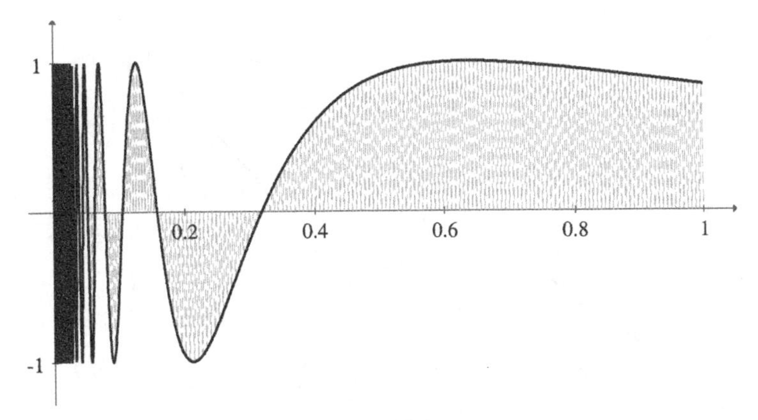

Figure 48.3.

is Riemann integrable over $[a, b]$ (Theorem 48.1). Thus,

$$f(m)(b-a) \leq \int_a^b f(x)dx \leq f(M)(b-a).$$

But then,

$$f(m) \leq \frac{\int_a^b f(x)dx}{b-a} \leq f(M).$$

Assume for the moment that f is not a constant function over $[a, b]$. Then the inequalities become strict, and by the intermediate value theorem (Theorem 26.4), there exists a number ξbetween m and M such that

$$f(\xi) = \frac{\int_a^b f(x)\,dx}{b-a}.$$

Since m and M belong to $[a, b]$, the theorem's conclusion follows. And, if f is constant throughout $[a, b]$, then

$$\int_a^b f(x)dx = f(\xi)(b-a)$$

for any $\xi \in (a, b)$, and the conclusion follows again. **QED**

The quantity

$$f_{avg} \equiv \frac{\int_a^b f(x)dx}{b-a}$$

is called the **average value** of the function over $[a, b]$. The mean value theorem for integrals therefore gives the condition that guarantees that at least once over $[a, b]$, $f(x) = f_{avg}$. Geometrically, this means that f is able to output the height of an equivalent rectangle: one equal in area to $\int_a^b f(x)dx$ and with the same base $[a, b]$, seen in Figure 48.4. Seventeenth century mathematicians would have called this the "quadrature" of the curve.

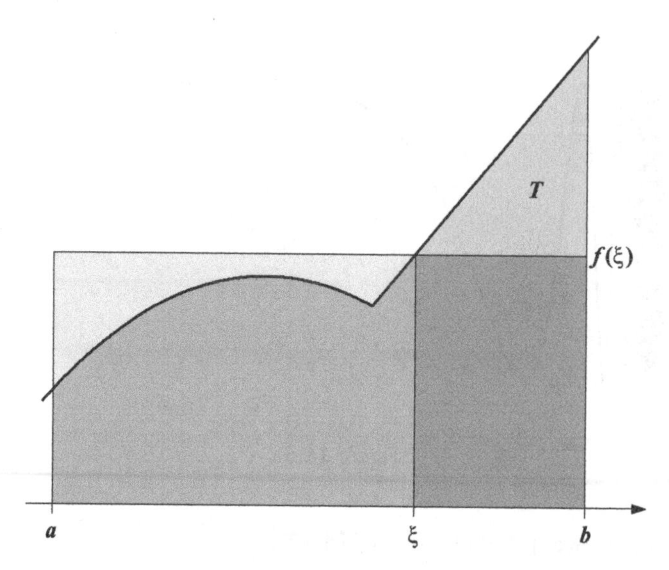

Figure 48.4. The equivalent rectangle with base $[a, b]$ and with an area equal to the shaded area $\int_a^b f(x)dx$. The lightly shaded triangle T has area equal to the lightly shaded area above $[a, \xi]$.

Example 48.2 Find the average value of $E(x) = e^x$ over $[-1, 1]$, and evaluate the number $\xi \in (-1, 1)$ such that $E(\xi) = E_{avg}$. We have

$$E_{avg} = \frac{1}{2} \int_{-1}^{1} e^x dx = \frac{1}{2} \left(e - e^{-1}\right).$$

Since $E(x)$ is continuous, the mean value theorem for integrals guarantees that a solution to $e^\xi = \frac{1}{2}\left(e - e^{-1}\right)$ will exist within $(-1, 1)$. In fact,

$$\xi = \ln\left(\frac{e}{2} - \frac{1}{2e}\right) \approx 0.161. \quad \blacksquare$$

Our next theorem is about integrating the composition of two functions. As may be expected, the proof is subtle, but the theorem will loose a flood of corollaries about integration (perhaps rational greed is at work again). This is an adaptation of the proof in [37], pg. 127.

Theorem 48.8 Let f be Riemann integrable over $[a, b]$ and let g be continuous over $[c, d]$, where $f([a, b]) \subseteq [c, d]$. Let $h(x) = g(f(x))$ be the composition of g with f. Then h is Riemann integrable over $[a, b]$.

Proof. We begin with two properties of g. First, g is bounded over $[c, d]$, meaning that for some positive constant B, $|g(y)| \leq B$ for all $y \in [c, d]$. Second, $g(y)$ is uniformly continuous over $[c, d]$, so that for any $\varepsilon > 0$, there will be a δ_ε with $0 < \delta_\varepsilon < \varepsilon$ such that if $u, v \in [c, d]$ and $|u - v| < \delta_\varepsilon$, then $|g(u) - g(v)| < \varepsilon$.

Since f is integrable, set up a net $\mathcal{L}$ over $[a, b]$ such that (c) $U(f, \mathcal{L}) - L(f, \mathcal{L}) < \delta_\varepsilon^2$. It will be necessary to distinguish two mutually exclusive collections of subintervals of $\mathcal{L}$. Let $\mathcal{N}$ contain the subintervals where $U(f, i) - L(f, i) < \delta_\varepsilon$, and let $\mathcal{M}$ contain the subintervals where $U(f, k) - L(f, k) \geq \delta_\varepsilon$.

For any subinterval $[x_{i-1}, x_i]$ in $\mathcal{N}$, if $x', x'' \in [x_{i-1}, x_i]$, then we have $\left|f(x') - f(x'')\right| \leq U(f, i) - L(f, i) < \delta_\varepsilon$. But now, let $u = f(x')$ and $v = f(x'')$. Since $|u - v| < \delta_\varepsilon$, then

$|g(u) - g(v)| = |g(f(x')) - g(f(x''))| = |h(x') - h(x'')| < \varepsilon$. This implies that $U(h,i) - L(h,i) \leq \varepsilon$ for any of the subintervals in $\mathcal{N}$ (explain the possible equality with ε). It is therefore sufficient to require that $\Delta\chi < \delta_\varepsilon$ for $\mathcal{N}$ in order that $U(h,i) - L(h,i) \leq \varepsilon$.

Turning now to collection $\mathcal{M}$, the boundedness of g over $[c,d]$ implies that $|h(x)| = |g(f(x))| \leq B$ over $[a,b]$. Thus, $U(h,k) \leq B$ and $-B \leq L(h,k)$ for any $[x_{k-1}, x_k] \in \mathcal{M}$. Therefore, $U(h,k) - L(h,k) \leq 2B$.

Next,

$$\delta_\varepsilon \sum_{k\in\mathcal{M}} \Delta x_k = \sum_{k\in\mathcal{M}} \delta_\varepsilon \Delta x_k \leq \sum_{k\in\mathcal{M}} [U(f,k) - L(f,k)]\,\Delta x_k < \delta_\varepsilon^2,$$

the last inequality because of the overarching condition (c). Thus, $\sum_{k\in\mathcal{M}} \Delta x_k < \delta_\varepsilon$, a most interesting statement saying that $\mathcal{M}$ has as small a total length as we wish.

Reuniting $\mathcal{N}$ and $\mathcal{M}$ together ($\mathcal{L} = \mathcal{N} \cup \mathcal{M}$), we consider the upper and lower Riemann sums of h:

$$\begin{aligned} U(h,\mathcal{L}) - L(h,\mathcal{L}) &= \sum_{i\in\mathcal{N}} [U(h,i) - L(h,i)]\,\Delta x_i + \sum_{k\in\mathcal{M}} [U(h,k) - L(h,k)]\,\Delta x_k \\ &\leq \varepsilon \sum_{i\in\mathcal{N}} \Delta x_i + 2B \sum_{k\in\mathcal{M}} \Delta x_k < \varepsilon(b-a) + 2B\delta_\varepsilon < \varepsilon(b-a) + 2B\varepsilon. \end{aligned}$$

(The last inequality follows from $\delta_\varepsilon < \varepsilon$.) Thus, we have

$$0 \leq U(h,\mathcal{L}) - L(h,\mathcal{L}) < \varepsilon(b-a) + 2B\varepsilon.$$

We now guarantee the existence of three limits when ε is allowed to diminish towards zero: (1) $\Delta\chi \to 0$ for the net $\mathcal{N}$ because $\Delta\chi < \delta_\varepsilon < \varepsilon$, (2) all $\Delta x_k \to 0$ for $\mathcal{M}$ because

$$\sum_{k\in\mathcal{M}} \Delta x_k < \delta_\varepsilon < \varepsilon,$$

and (3) $U(h,\mathcal{L}) - L(h,\mathcal{L}) \to 0$. We therefore arrive at

$$\overline{\int_a^b} h(x)dx = \underline{\int_a^b} h(x)dx,$$

meaning that $\int_a^b h(x)\,dx$ exists. **QED**

Corollary 48.2 If f is Riemann integrable over $[a,b]$, then f^2 is Riemann integrable there.

Proof. The polynomial $g(y) = y^2$ is continuous over any closed interval. The composition $g(f(x)) = f^2(x)$ must therefore be integrable over $[a,b]$ by the previous theorem. **QED**

Corollary 48.3 If f and g are Riemann integrable over $[a,b]$, then fg is Riemann integrable there.

Proof. Create a proof as Exercise 48.8. **QED**

Corollary 48.4 If f is Riemann integrable over $[a, b]$, then $|f|$ is Riemann integrable there, and

$$\left|\int_a^b f(x)dx\right| \leq \int_a^b |f(x)|\, dx \leq B\,(b-a),$$

where

$$B = \sup_{[a,b]} |f(x)|\,.$$

Proof. Let $g(y) = |y|$, which is continuous over any $[a, b]$. Then $h\,(x) = g\,(f(x)) = |f\,(x)|$ is Riemann integrable by Theorem 48.8. Compare with Exercise 48.9.

Since $-|f(x)| \leq f\,(x) \leq |f(x)|$, Corollary 48.1 and Theorem 48.2 (used to extract the -1 factor) imply that $-\int_a^b |f(x)|\, dx \leq \int_a^b f(x)dx \leq \int_a^b |f\,(x)|\, dx$. This means that

$$\left|\int_a^b f(x)dx\right| \leq \int_a^b |f(x)|\, dx.$$

By Lemma 47.1, the supremum B exists. Then $B\,(b-a) = U\,(|f|\,, \mathcal{N})$ for the coarsest net possible: the one with $[a, b]$ as its single subinterval. Therefore,

$$\int_a^b |f\,(x)|\, dx \leq B\,(b-a). \quad \textbf{QED}$$

Corollary 48.5 If f is Riemann integrable over $[a, b]$ and there is some positive α such that $f(x) \geq \alpha$ over $[a, b]$, then $\frac{1}{f}$ is Riemann integrable over $[a, b]$. The same conclusion holds if $f(x) \leq -\alpha$ over $[a, b]$.

Proof. The function $g(y) = \frac{1}{y}$ is continuous over any closed interval not containing $y = 0$. Since f is bounded away from zero over $[a, b]$, then $f\,([a, b]) \cap [0, \alpha) = \varnothing$. But f is also bounded over $[a, b]$, so that $f\,([a, b]) \subseteq [\alpha, B]$ for some positive constant B. By Theorem 48.8, the composition $g\,(f(x)) = \frac{1}{f(x)}$ is integrable over $[a, b]$.

In case $f(x) \leq -\alpha$ over $[a, b]$, then $f\,([a, b]) \subseteq [-B, -\alpha]$, and Theorem 48.8 applies once again. **QED**

Corollary 48.6 If f and g are Riemann integrable over $[a, b]$, and there is some positive α such that $g\,(x) \geq \alpha$ over $[a, b]$, then $\frac{f}{g}$ is Riemann integrable over $[a, b]$. The same conclusion holds if $g(x) \leq -\alpha$ over $[a, b]$.

Proof. Create a proof as Exercise 48.10. **QED**

48.1 Exercises

1. Finish the proof of Theorem 48.3 by proving that

$$\int_a^b [f(x) - g(x)]\, dx = \int_a^b f(x)dx - \int_a^b g(x)dx.$$

2. Prove Corollary 48.1. Hint: Use Exercise 47.9.

3. Fill in the missing step in Theorem 48.4: show that f is a bounded function over $[a, b]$.

4. Prove that if f is integrable over $[a, b]$, then modifying f to

$$f^*(x) = \begin{cases} f(x) & \text{if } a \leq x \leq c \\ f(x) + k & \text{if } c < x \leq b \end{cases},$$

where k is a constant, keeps f^* integrable, but

$$\int_a^b f^*(x)dx = k(b - c) + \int_a^b f(x)dx.$$

Hint: Draw a sketch of the relationship between f and f^*. Use Exercise 47.8.

5. Create a short proof to cover the case of a monotone decreasing function f in Theorem 48.5.

6. Prove that the Riemann integral is a linear operator, meaning that for any real numbers m and n,

$$\int_a^b [mf(x) + ng(x)]\,dx = m\int_a^b f(x)dx + n\int_a^b g(x)dx$$

(of course, the integrals on the right are assumed to exist).

7. In electric circuit theory, alternating current has a voltage V that varies with time as $V(t) = V_p \sin(2\pi \omega t)$, where V_p is the peak voltage, ω is the frequency of oscillation, and t is in seconds. If you were to accidentally touch a live wire in your house, you fortunately would not feel the peak voltage, but rather the RMS, or root-mean-square, voltage. To define the RMS voltage, set $\omega = \frac{1}{2\pi}$, and then

$$V_{RMS} \equiv \sqrt{\frac{\int_0^\pi (V_p \sin t)^2\,dt}{\pi}}.$$

The "mean-square" terminology indicates the average value of $V^2(t)$ over $[0, \pi]$, which is

$$\frac{\int_0^\pi (V_p \sin t)^2\,dt}{\pi - 0}.$$

Find the RMS voltage in terms of V_p. Given that $V_p = 115$ volts is a common house voltage, evaluate V_{RMS}. Is there a time $t \in [0, \pi]$ at which $V^2(t) = V_{RMS}^2$?

8. Prove Corollary 48.3. Hint: $(f + g)^2 = f^2 + 2fg + g^2$.

9. Convince yourself that $|f(x)| = f(x)\operatorname{sgn} f(x)$. Use this to prove the first assertion of Corollary 48.4 once again.

10. Prove Corollary 48.6.

11. Answer true or false, and give a short reason.
    a. If f is Riemann integrable over $[a, b]$, then $h(x) = f^n(x)$ is Riemann integrable there.
    b. If f is Riemann integrable over $[a, b]$, then $h(x) = \sqrt[n]{f(x)}$ is Riemann integrable there, provided that when n is even, $f(x) \geq 0$ over $[a, b]$.

c. If f is positive, bounded away from zero, and Riemann integrable over $[a, b]$, then $h(x) = [f(x)]^\alpha$ is Riemann integrable there, for any real α.

12. In Example 46.5, the integral $\int_0^{\pi/2} \sqrt{1 - \epsilon^2 \sin^2 \theta d\theta}$ arose in the process of finding the arclength of a quarter-ellipse. We complained there that although the function

$$F(x) = \int_0^x \sqrt{1 - \epsilon^2 \sin^2 \theta d\theta}$$

exists, there is no closed-form expression for it, so the evaluation $F\left(\frac{\pi}{2}\right) - F(0)$ gets short-circuited. How do we know that F exists?

49

The Fundamental Theorem of Integral Calculus

Leibniz realized rather early, probably from studying the work of [Isaac] Barrow, that differentiation and integration as a summation must be inverse processes; so area, when differentiated, must give a length.

Morris Kline [6], pp. 373–374.

We now go on to investigate some relationships between differentiation and integration. Gottfried Leibniz and Isaac Newton understood as a general principle that the operations were capable of reversing each other, as in $\frac{d}{dx}\int y dx = y$, and $\int \frac{dy}{dx} dx = y$. Newton and Leibniz are called the discoverers of calculus because they understood the inverse role of differentiation and integration in addition to knowing the use of each operation separately. They also saw that if $F(x) = \int f(x)dx$, then $\int_a^b f(x)dx$ could be evaluated as $F(b) - F(a)$. We appreciate the enormous saving of effort in this evaluation after having studied the previous two chapters! However, some ideas have to be made clear or problems will arise, as we will see in the two examples below.

Example 49.1 A difficulty arises with the bare statement $\int \frac{dy}{dx} dx = y$ when we apply it to the continuous function

$$y(x) = \begin{cases} x^2 \sin \frac{1}{x^2} & \text{if } x \neq 0 \\ 0 & \text{if } x = 0 \end{cases}$$

over $[0, 1]$, seen over a more revealing interval in Figure 49.1. Its derivative, for $x \neq 0$, is routine:

$$\frac{d\,y(x)}{dx} = y'(x) = 2x \sin \frac{1}{x^2} - \frac{2}{x} \cos \frac{1}{x^2},$$

seen in Figure 49.2. Although the figure can't show it well, in any neighborhood of zero, y' is unbounded, and its frequency of oscillation increases unboundedly. Amazingly,

$$y'(0) = \left.\frac{d\,y(x)}{dx}\right|_{x=0} = \lim_{h \to 0} \left(\frac{y(h) - y(0)}{h} \right) = 0.$$

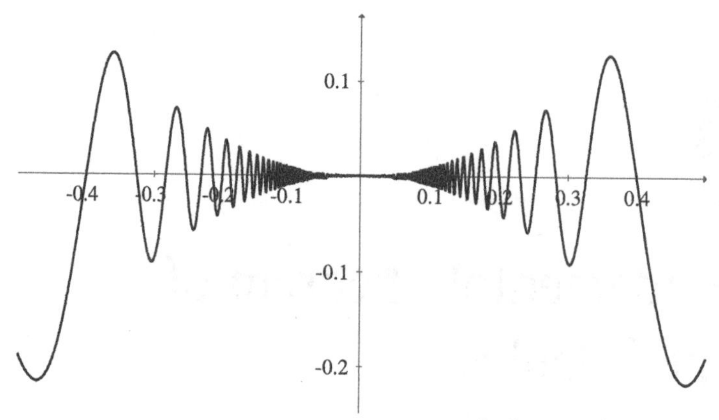

Figure 49.1. The continuous function $y(x) = \begin{cases} x^2 \sin(1/x^2) & \text{if } x \neq 0 \\ 0 & \text{if } x = 0 \end{cases}$

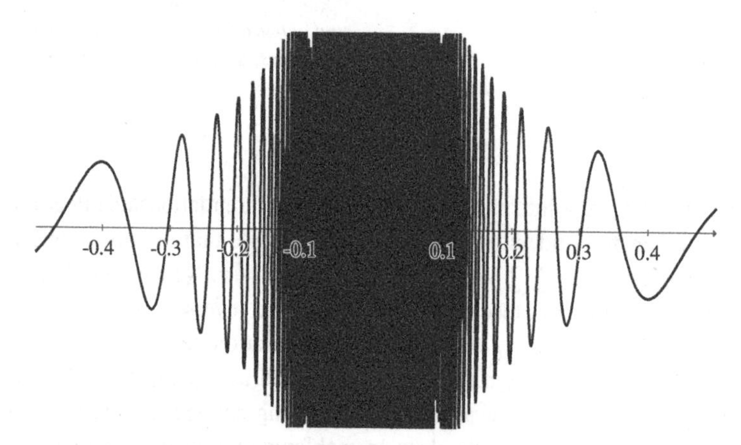

Figure 49.2. The nightmarish, unbounded derivative of y.

Consequently, we have a function y' which is defined over $[0, 1]$, and which has a continuous antiderivative y over $[0, 1]$. But since y' is not bounded over $[0, 1]$, it cannot be Riemann integrable there. Symbolically, even though $\frac{d\,y(x)}{dx}$ exists and $\int \frac{d\,y(x)}{dx}dx = y(x)$ over $[0, 1]$, we find that $\int_0^1 \frac{d\,y(x)}{dx}dx$ fails to exist! ■

Example 49.2 Is there a difficulty with the statement $\frac{d}{dx}\int y dx = y$? Let

$$y = f(x) = \begin{cases} 0 & \text{if } -1 \leq x < 0 \\ 1 & \text{if } x = 0 \\ 2 & \text{if } 0 < x \leq 1 \end{cases},$$

which is the function $1 + \operatorname{sgn} x$ with domain restricted to $[-1, 1]$, seen in Figure 49.3. Consider the area under f as a function F with $\mathcal{D}_F = [-1, 1]$. Function f is Riemann integrable (review

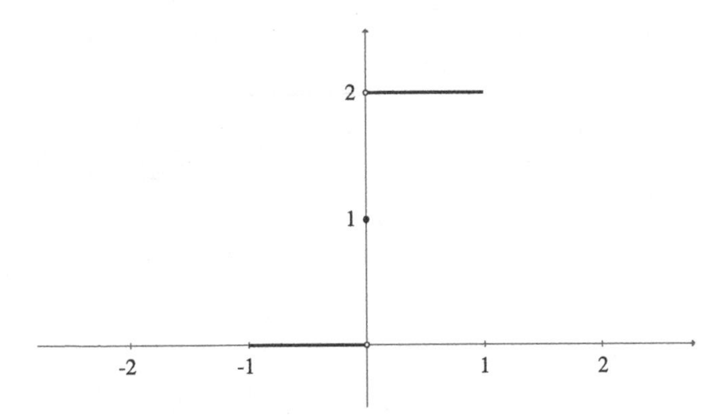

Figure 49.3. $f(x) = 1 + \operatorname{sgn} x$, $\mathcal{D}_f = [-1, 1]$.

Example 47.1), therefore

$$F(x) = \int_{-1}^{x} f(t)dt = \begin{cases} \int_{-1}^{x} 0dt = 0 & \text{if } -1 \le x \le 0 \\ \int_{0}^{x} 2dt = 2x & \text{if } 0 < x \le 1 \end{cases},$$

a continuous function (see Figure 49.4). Next, differentiate:

$$\frac{dF(x)}{dx} = \frac{d}{dx}\int_{-1}^{x} f(t)dt = \begin{cases} 0 & \text{if } -1 \le x < 0 \\ \text{undefined} & \text{if } x = 0 \\ 2 & \text{if } 0 < x \le 1 \end{cases}.$$

This yields a function (Figure 49.5) which is nearly, but *not exactly*, the function f that we started with.

It follows that $\frac{d}{dx}\int y dx = y$ cannot be a universally valid "fundamental theorem of calculus."

In fairness to Leibniz and Newton, they would have recognized $y = 1 + \operatorname{sgn} x$ not as a function, but as two disjoint curves (and an isolated point). The masters would be correct

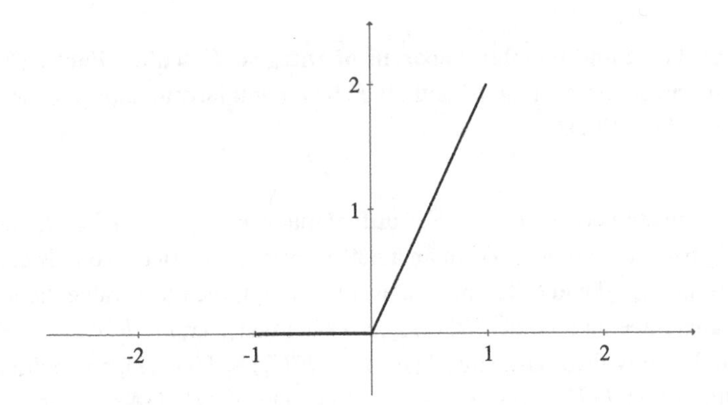

Figure 49.4. $F(x) = \int_{-1}^{x} f(t)dt$, $\mathcal{D}_F = [-1, 1]$.

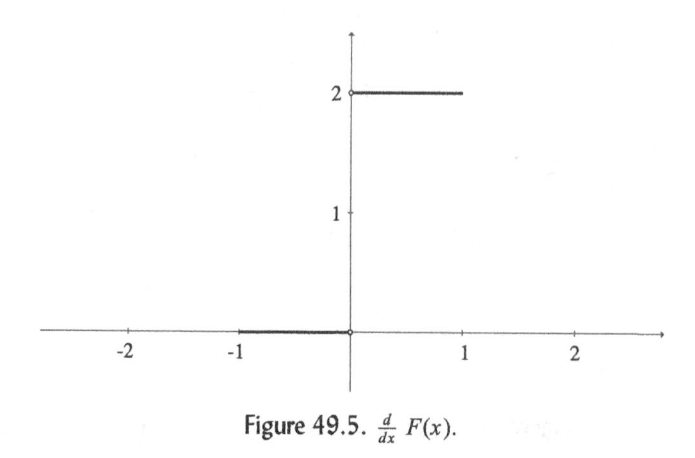

Figure 49.5. $\frac{d}{dx}F(x)$.

in noting that the problem is that f is discontinuous at $x = 0$, although the rigorous definition of continuity was still in the distant future for them. Interestingly, it seems that integration has a way of smoothing over discontinuous functions, or at least, functions with isolated discontinuities.... ■

We have seen in our first example that a function may have an antiderivative and yet not be integrable. The second example is equally disturbing: a function may be integrable and yet have no antiderivative. As always, confusion reigns when careful definitions are ignored. Thus, it is time to state and prove the key theorems that prevent us from being entangled by such functions. Suppose that a function f is Riemann integrable over $[a, b]$. We have been considering that a Riemann integral may define a function, $F(x) = \int_a^x f(t)dt$, by treating one of the limits of integration as an independent variable. The number $F(b) = \int_a^b f(x)\,dx$ is named the **definite integral** of f over $[a, b]$. We realize that the domain of F is precisely the closed interval $[a, b]$ over which f is Riemann integrable, that is, $\mathcal{D}_F = [a, b]$. Also, $F(a) = \int_a^a f(t)dt = 0$, and $\int_x^a f(t)dt = -F(x)$, recalling the preliminary definitions of Chapter 47.

On the other hand, given a function $f : [a, b] \to \mathbb{R}$, if there exists a function F with the same domain, $F : [a, b] \to \mathbb{R}$, such that $\frac{dF(x)}{dx} = f(x)$ over $[a, b]$, then we call F an **antiderivative** of f, and write $F(x) = \int f(x)dx$. Two other synonyms for F are the **indefinite integral** or **primitive** of f.

Theorem 49.1 The Fundamental Theorem of Integral Calculus, Part I (FTIC-1) Let f be Riemann integrable over $[a, b]$ and let F be an antiderivative of f over $[a, b]$. Then $\int_a^b f(x)dx = F(b) - F(a)$.

Proof. Since F is an antiderivative of f, F is differentiable over $[a, b]$, $F' = f$, and F is continuous over $[a, b]$ by Theorem 36.1. Form an a net $\mathcal{N}$ over $[a, b]$. Because F is continuous over each subinterval $[x_{i-1}, x_i]$ and differentiable over (x_{i-1}, x_i), the mean value theorem (Theorem 39.3) applies, guaranteeing a point $\xi_i \in (x_{i-1}, x_i)$ such that $F(x_i) - F(x_{i-1}) = F'(\xi_i)\Delta x_i$, for $i = 1, 2, \ldots, n$. But it is always true that $L(F', i) \leq F'(\xi_i) \leq U(F', i)$, so, multiplying through by Δx_i, it follows that $L(F', i)\Delta x_i \leq F(x_i) - F(x_{i-1}) \leq U(F', i)\Delta x_i$. This is equivalent to $L(f, i)\Delta x_i \leq F(x_i) - F(x_{i-1}) \leq U(f, i)\Delta x_i$.

Next, form Riemann sums:

$$\sum_{i=1}^{n} L(f,i)\Delta x_i \leq \sum_{i=1}^{n}[F(x_i) - F(x_{i-1})] \leq \sum_{i=1}^{n} U(f,i)\Delta x_i$$

becomes (a) $L(f,\mathcal{N}) \leq F(b) - F(a) \leq U(f,\mathcal{N})$, since $\sum_{i=1}^{n}[F(x_i) - F(x_{i-1})]$ telescopes to $F(x_n) - F(x_0)$.

Since f is Riemann integrable over $[a,b]$, we know that

$$\lim_{\Delta\chi\to 0} L(f,\mathcal{N}) = \lim_{\Delta\chi\to 0} U(f,\mathcal{N}) = \int_a^b f(x)dx.$$

And this, together with (a) and the pinching theorem, implies that $\int_a^b f(x)dx = F(b) - F(a)$. **QED**

The equation $\int \frac{dy}{dx}dx = y$ was doomed to fail in Example 49.1 because it violated one of the premises in the FTIC-1, namely, integrability. Put in the standard logical manner, the example shows that the premise that f has an antiderivative, even a continuous one, is *not sufficient* for the famous evaluation $\int_a^b f(x)dx = F(b) - F(a)$.

Theorem 49.2 The Fundamental Theorem of Integral Calculus, Part II (FTIC-2) Let f be Riemann integrable over $[a,b]$ and define $F(x) \equiv \int_a^x f(x)dx$, where $x \in [a,b]$. Then

(1) F is continuous over $[a,b]$, and
(2) if f is continuous at a point $c \in [a,b]$, then F will be differentiable at $x = c$, and $F'(c) = f(c)$.

Proof. To prove **(1)**, since f is bounded over $[a,b]$, let B be a positive constant such that $|f(x)| \leq B$ over $[a,b]$. For any u, v such that $a \leq u < v \leq b$, Theorem 48.4 implies that (b)

$$F(v) - F(u) = \int_a^v f(x)dx - \int_a^u f(x)dx = \int_u^v f(x)dx.$$

Now,

$$\left|\int_v^u f(x)dx\right| \leq \int_v^u |f(x)|dx \leq B(v-u)$$

by Corollary 48.4. Thus, $|F(v) - F(u)| \leq B|v-u|$, and this is a sufficient condition for the continuity of F throughout $[a,b]$(verify this).

To prove **(2)**, first suppose that c is in the interior of $[a,b]$. We know that given any $\varepsilon > 0$ there exists a $\delta > 0$ such that if $|x-c| < \delta$, then $|f(x) - f(c)| < \varepsilon$. In fact, we will use an ε small enough that the neighborhood $(c-\varepsilon, c+\varepsilon) \subset [a,b]$, and let $\delta_1 = \min\{\delta, \varepsilon\}$. Then we have $f(c) - \varepsilon < f(x) < f(c) + \varepsilon$ for any x such that $|x-c| < \delta_1$. By Corollary 48.1,

$$\int_c^{c+\varepsilon}(f(c)-\varepsilon)dx < \int_c^{c+\varepsilon} f(x)dx < \int_c^{c+\varepsilon}(f(c)+\varepsilon)dx,$$

$$\text{or}\qquad (f(c)-\varepsilon)\varepsilon < \int_c^{c+\varepsilon} f(x)dx < (f(c)+\varepsilon)\varepsilon,$$

since $f(c) \pm \varepsilon$ are constants. Using (b) and dividing through by ε, we arrive at

$$f(c) - \varepsilon < \frac{F(c+\varepsilon) - F(c)}{\varepsilon} < f(c) + \varepsilon.$$

Hence,

$$\lim_{\varepsilon \to 0+} \left(\frac{F(c+\varepsilon) - F(c)}{\varepsilon} \right) = f(c).$$

This tells us that the derivative of F from the right exists at $x = c$.

With some changes of signs, show that

$$\lim_{\varepsilon \to 0+} \left(\frac{F(c) - F(c-\varepsilon)}{\varepsilon} \right) = f(c)$$

as Exercise 49.1. This means that the derivative of F from the left exists at $x = c$. By Theorem 38.1, the full derivative $F'(c)$ exists, and equals $f(c)$.

As the last detail, if c is at either endpoint of $[a, b]$, then only the corresponding derivative from the right or left would be used to prove that $F'^R(a) = f(a)$ or $F'^L(b) = f(b)$, respectively. **QED**

Now we see why the equation $\frac{d}{dx} \int y dx = y$ in Example 49.2 was also doomed to fail. In order for $\frac{dF(x)}{dx}|_c = f(c)$, it isn't sufficient to define $F(x)$ as $\int_a^x f(x)dx$, and then take its derivative. It is necessary that f be continuous at $x = c$. The example would have worked out by the FTIC-2 over any interval not containing $c = 0$. The proof of the FTIC-2 also shows us why F smooths over a discontinuity in f, as Figures 49.3 and 49.4 show.

The next example demonstrates a link between Riemann integration and certain infinite sequences.

Example 49.3 Consider the unusual sequence

$$a_1 = \frac{\ln 2}{2}, \quad a_2 = \frac{\ln(3/2)}{3} + \frac{\ln 2}{4}, \quad a_3 = \frac{\ln(4/3)}{4} + \frac{\ln(5/2)}{5} + \frac{\ln 2}{6}, \ldots, a_n = \sum_{k=1}^{n} \left(\frac{\ln \frac{n+k}{n}}{n+k} \right).$$

Thus, a_n is itself the sum of n terms. The problem is to evaluate the limit

$$\lim_{n \to \infty} \left(\frac{\ln \frac{n+1}{n}}{n+1} + \frac{\ln \frac{n+2}{n}}{n+2} + \frac{\ln \frac{n+3}{n}}{n+3} + \cdots + \frac{\ln 2}{2n} \right) = \lim_{n \to \infty} \sum_{k=1}^{n} \left(\frac{\ln \frac{n+k}{n}}{n+k} \right).$$

The novel idea is to reexpress term a_n as

$$\sum_{k=1}^{n} \left(\frac{\ln \frac{n+k}{n}}{n+k} \right) = \sum_{k=1}^{n} \left(\frac{\ln\left(1 + \frac{k}{n}\right)}{n+k} \right) = \sum_{k=1}^{n} \frac{1}{n} \left(\frac{\ln\left(1 + \frac{k}{n}\right)}{1 + \frac{k}{n}} \right),$$

so as to recognize a Riemann sum $\sum_{k=1}^{n} f(X_i)\Delta x_i$. This will be the case if we set $\Delta x_i = \frac{1}{n}$ in a net with n equal subintervals, and let $f(x) = \frac{\ln x}{x}$, implying that the representative of subinterval $[x_{k-1}, x_k]$ is $X_k = 1 + \frac{k}{n}$, $k = 1, 2, \ldots, n$. This in turn is possible if we set $x_0 = a = 1$ and $x_n = b = 1 + \frac{n}{n} = 2$. Thus, we have associated the given sequence with a sequence of Riemann sums for the expression $\int_1^2 \frac{\ln x}{x} dx$.

Relying now on the FTIC-1, we note that $f(x) = \frac{\ln x}{x}$ is Riemann integrable over $[1, 2]$ (by Theorem 48.1), and f has an antiderivative $\int \frac{\ln x}{x} dx = \frac{1}{2} \ln^2 x$ that is continuous over $[1, 2]$ (any antiderivative will do for the FTIC-1). Therefore, $\int_1^2 \frac{\ln x}{x} dx = \frac{1}{2} \ln^2 x|_1^2 = \frac{1}{2} \ln^2 2$.

We conclude that

$$\lim_{n \to \infty} \sum_{k=1}^{n} \left(\frac{\ln \frac{n+k}{n}}{n+k} \right) = \lim_{\Delta\chi \to 0} \sum_{k=1}^{n} f(X_i)\Delta x_i = \int_1^2 \frac{\ln x}{x} dx = \frac{1}{2} \ln^2 2,$$

which is the exact limit of this sequence of sums. As a quick partial check, $\frac{1}{2} \ln^2 2 = 0.240226\ldots$, in comparison to

$$a_{10} = \sum_{k=1}^{10} \left(\frac{\ln \frac{10+k}{10}}{10+k} \right) \approx 0.2568, \; a_{50} = \sum_{k=1}^{50} \left(\frac{\ln \frac{50+k}{50}}{50+k} \right) \approx 0.2437,$$

and

$$a_{200} = \sum_{k=1}^{200} \left(\frac{\ln \frac{200+k}{200}}{200+k} \right) \approx 0.2411.$$

Exercise 49.3 asks an important question about the numerical results. ■

49.1 Two Essential Methods of Integration

In this subsection we will show that the methods of substitution (or change of variable) and integration by parts are analytically sound under certain conditions. It will be good to remember two theorems as you read the proofs: a function that is differentiable over an interval is continuous there (Theorem 36.1), and a continuous function over a closed interval is Riemann integrable there (Theorem 48.1).

Theorem 49.3 Integration by Substitution, Part I Let $y = v(x)$ be differentiable over $[a, b]$, with a continuous derivative $v'(x)$ there. Let the image $v([a, b])$ be either $[c, d]$ or $[d, c]$ (although $v(a) = c$ and $v(b) = d$, it isn't implied that $c < d$). Select a continuous function $f(v)$ over $[c, d]$ (or $[d, c]$). Then $\int_{v(a)}^{v(b)} f(v)dv = \int_a^b f(v(x)) \cdot v'(x)dx$.

Proof. The continuity of $v(x)$ over $[a, b]$ implies two facts: first, that $[c, d]$ (or $[d, c]$) exists by Exercise 26.11, and second, that the composition $f(v(x)) = f(y)$ is continuous over $[a, b]$ (Theorem 26.2), and hence, integrable there. Selecting a constant $\mu \in [c, d]$ (or $[d, c]$), define $F(t) = \int_\mu^t f(v)dv$, where $t \in [c, d]$ (or $[d, c]$) as well. Then, by the FTIC-2, $F'(t) = f(t)$. Define $u(x) = F(v(x))$, and use the chain rule (Theorem 37.3) to get $u'(x) = F'(v(x)) \cdot v'(x) = f(v(x)) \cdot v'(x)$. This is the product of two integrable functions over $[a, b]$, which implies that $u'(x)$ is integrable there. Moreover, its antiderivative, namely $F(v(x))$, is continuous over $[a, b]$, since $u'(x)$ exists there. With these two facts, the FTIC-1 applies to u', giving us

$$\int_a^b u'(x)dx = u(b) - u(a) = F(v(b)) - F(v(a))$$

$$= \int_\mu^{v(b)} f(v)dv - \int_\mu^{v(a)} f(v)dv = \int_{v(a)}^{v(b)} f(v)dv.$$

But also, $\int_a^b u'(x)dx = \int_a^b f(v(x)) \cdot v'(x)dx$, which gives the theorem's conclusion. **QED**

Example 49.4 Find $\int_a^b 3^x \tan(3^{1+x})dx$, noting any restrictions on a and b.

The composition of functions is quite clear here, so let $y = v(x) = 3^{1+x}$ be the inside function, and $v'(x) = 3^{1+x} \ln 3$. In accordance with Theorem 49.3, v certainly has a continuous derivative over any $[a, b]$. The outside function must be $f(v) = \tan v$, which is continuous over intervals not containing $(2k+1)\frac{\pi}{2}$, $k \in \mathbb{Z}$. Thus, $[c, d]$ cannot contain odd multiples of $\frac{\pi}{2}$. With this restriction in mind, the change of variable transforms the integral into

$$\frac{1}{3\ln 3}\int_{3^{1+a}}^{3^{1+b}} \tan v\,dv = \frac{-1}{3\ln 3}\ln(\cos v)\Big|_{3^{1+a}}^{3^{1+b}} = \frac{1}{3\ln 3}(\ln(\cos 3^{1+a}) - \ln(\cos 3^{1+b})).$$

If we choose the basic interval $(-\frac{\pi}{2}, \frac{\pi}{2})$ for $f(v) = \tan v$, then we need to solve $-\frac{\pi}{2} < 3^{1+x} < \frac{\pi}{2}$, which implies that $3^{1+x} < \frac{\pi}{2}$. Thus, $x < \frac{\ln(\pi/2)}{\ln 3} - 1 = -0.58895\ldots$. Hence,

$$a, b \in \left(-\infty, \frac{\ln(\pi/2)}{\ln 3} - 1\right)$$

in order for the integral to exist.

For instance,

$$\int_{-3}^{-1} 3^x \tan(3^{1+x})dx = \frac{1}{3\ln 3}\int_{1/9}^{1} \tan v\,dv = 0.184912\ldots,$$

but $\int_{-3}^{0} 3^x \tan(3^{1+x})dx$ is meaningless. Continuing with the interval $[-3, -1]$, it's interesting to see the transformation of areas effected by the substitution $v(x) = 3^{1+x}$. The left-hand area in Figure 49.6 matches the original integral $\int_{-3}^{-1} 3^x \tan(3^{1+x})dx$; to the right of it is the transformed area found by $\frac{1}{3\ln 3}\int_{1/9}^{1} \tan v\,dv$. Theorem 49.3 guarantees that the areas are equal. ■

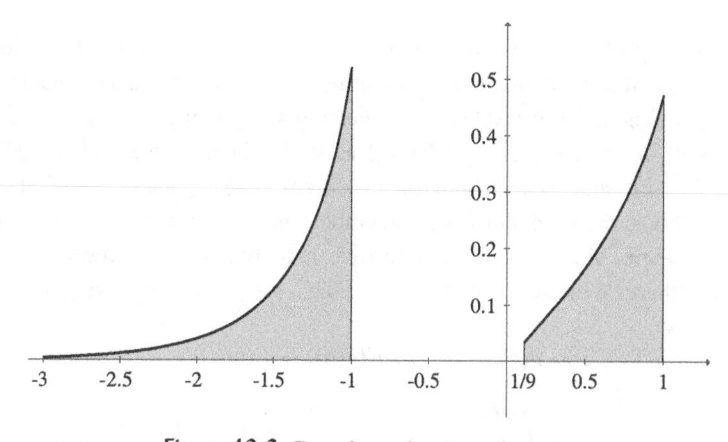

Figure 49.6. Transformed, yet equal, areas.

The next theorem justifies the substitutions that involve a function and its inverse.

Theorem 49.4 Integration by Substitution, Part II Let $y = v(x)$ be differentiable over $[a, b]$, with a continuous derivative such that $v'(x) \neq 0$ there. Let the image $v([a, b])$ be either $[c, d]$ or $[d, c]$ (although $v(a) = c$ and $v(b) = d$, it isn't implied that $c \leq d$), so that the inverse function $w(y) = v^{-1}(y)$ exists over $[c, d]$ or $[d, c]$, respectively. Select a continuous function $f(v)$ over $[c, d]$ (or $[d, c]$). Then $\int_a^b f(v(x))dx = \int_{v(a)}^{v(b)} f(y){\cdot}w'(y)dy$.

Proof. Since $v'(x) \neq 0$ over $[a, b]$, v is a monotone, one-to-one function there, so that $x = w(y)$ exists over $[c, d]$ (or $[d, c]$). Moreover, w has a continuous derivative there (Theorem 37.6). Let $G(x)$ be any antiderivative of the continuous function $f(v(x))$ over $[a, b]$. Then $G'(w(y)) = G'(x) = f(v(x)) = f(y)$ over $[c, d]$ (or $[d, c]$). So, by the chain rule, $(G \circ w)'(y) = G'(w){\cdot}w'(y) = G'(w(y)){\cdot}w'(y) = f(y){\cdot}w'(y)$ over $[c, d]$ (or $[d, c]$).

We now know two facts about the function $f(y)w'(y)$: first, it is Riemann integrable over $[c, d]$ (or $[d, c]$), and second, it possesses a continuous antiderivative $(G \circ w)(y)$ there. Therefore, by the FTIC-1,

$$\int_{v(a)}^{v(b)} f(y){\cdot}w'(y)dy = (G \circ w)(v(b)) - (G \circ w)(v(a)) = G(b) - G(a)$$

(remembering that w and v are inverses).

But also, $f(v(x))$ is Riemann integrable over $[a, b]$, and $G'(x) = f(v(x))$ there, so again by the FTIC-1,

$$\int_a^b f(v(x))dx = G(b) - G(a).$$

It follows that

$$\int_a^b f(v(x))dx = \int_{v(a)}^{v(b)} f(y){\cdot}w'(y)dy. \quad \textbf{QED}$$

Example 49.5 Find $\int_1^3 xe^{x^2}dx$. For convenience in applying Theorem 49.4, let's use $\int_1^3 ye^{y^2}dy$. Observing the composition, let $y = v(x) = x^2$ be an inside function with continuous derivative over $[1, 3]$, and $v'(x) \neq 0$ there. Also, $v([1, 3]) = [1, 9] = [c, d]$. Then $w(y) = \sqrt{y}$ exists and is continuous over $[1, 9]$ (Theorem 26.6). The necessary outside function must be $f(y) = \sqrt{y}e^y$ for the composition $f(v(x))$ to produce the original integrand, and $f(y)$ is continuous over $[1, 9]$. Now Theorem 49.4 tells us that

$$\int_a^b f(v(x))dx = \int_1^3 xe^{x^2}dx$$

is equal to

$$\int_{v(a)}^{v(b)} f(y){\cdot}w'(y)dy = \int_1^9 \sqrt{y}e^y \frac{1}{2\sqrt{y}}dy = \frac{1}{2}\int_1^9 e^y dy = \frac{1}{2}(e^9 - e).$$

(Exercise 49.9 adds a slight twist to this example.) ■

Example 49.6 Find

$$\int_0^2 \frac{1}{\sqrt{9-y^2}}dy.$$

The indication that Theorem 49.4 applies here is that a trigonometric substitution is needed, along with the inverse of the chosen function.

Let $y = v(x) = \sin^{-1}\frac{x}{3}$, as the admittedly obscure inside function with a continuous derivative over $[0, 2]$ (verify this), and $v'(x) \neq 0$ there. Also,

$$v([0,2]) = \left[0, \sin^{-1}\frac{2}{3}\right] = [c, d].$$

Then $x = w(y) = 3\sin y$ is the inverse, over $[0, \sin^{-1}\frac{2}{3}]$. We may proceed with Theorem 49.4.

Select the outside function as $f(y) = \frac{1}{3\cos y}$, continuous over $[0, \sin^{-1}\frac{2}{3}]$, so that the composition

$$f(v(x)) = \frac{1}{3\cos(\sin^{-1}\frac{x}{3})} = \frac{1}{\sqrt{9-x^2}}$$

is precisely the original integrand. Then

$$\int_a^b f(v(x))dx = \int_0^2 \frac{1}{\sqrt{9-x^2}}dx$$

is equal to

$$\int_{v(a)}^{v(b)} f(y)\cdot w'(y)dy = \int_0^{\sin^{-1}(2/3)} \frac{1}{3\cos y}\cdot 3\cos y dy = \sin^{-1}\frac{2}{3}. \quad \blacksquare$$

Example 49.7 Find $\int_0^3 \sqrt{9-y^2}dy$. With the experience from the last example, let $y = v(x) = \sin^{-1}\frac{x}{3}$. But now we see that $v(x)$ is continuous and differentiable only over $[0, 3)$ (and $v'(x) \neq 0$ there). Thus, to apply Theorem 49.4, we use the temporary interval $[0, 3-\varepsilon]$, where $0 < \varepsilon < 3$. With this in place,

$$v([0, 3-\varepsilon]) = \left[0, \sin^{-1}\frac{3-\varepsilon}{3}\right],$$

and let $\mu = \frac{3-\varepsilon}{3}$ just for convenience. Consequently, $w(y) = 3\sin y$ exists over $[0, \sin^{-1}\mu]$. We select $f(y) = 3\cos y$, continuous over $[0, \sin^{-1}\mu]$, so that the composition $f(v(x)) = 3\cos(\sin^{-1}\frac{x}{3}) = \sqrt{9-x^2}$ is precisely the original integrand.

Then, by Theorem 49.4,

$$\int_a^b f(v(x))dx = \int_0^3 \sqrt{9-x^2}dx$$

is equal to

$$\int_{v(a)}^{v(b)} f(y)\cdot w'(y)dy = \int_0^{\sin^{-1}\mu} 3\cos y(3\cos y)dy = \frac{9}{2}\int_0^{\sin^{-1}\mu}(1+\cos 2y)dy$$
$$= \frac{9}{2}\sin^{-1}\mu + \frac{9}{4}\sin(2\sin^{-1}\mu).$$

Our last step is to let $\varepsilon \to 0^+$ so that we may cover the interval $[0,3]$. Equivalently, let $\mu \to 1^-$. Thus,

$$\int_0^3 \sqrt{9-y^2}dy = \lim_{\mu \to 1-} \frac{9}{2}\sin^{-1}\mu + \lim_{\mu \to 1-} \frac{9}{4}(\sin(2\sin^{-1}\mu)) = \frac{9\pi}{4}.$$

It's been a long haul, but we can't forget that $\int_0^3 \sqrt{9-y^2}dy$ equals the area in Quadrant I of the circle with radius $r = 3$ centered at the origin – which is $\frac{9\pi}{4}$, of course! ■

The analyses in the last three examples are not meant to supersede the streamlined substitution techniques from calculus for evaluating such integrals. Rather, we are showing the rigorous application of the powerful Theorem 49.4.

Note 1: We did not evaluate an improper integral in the example above. It demonstrated a procedure if an interval was not closed when using the method of substitution.

Theorem 49.5 Integration by Parts Let $[a,b]$ be contained in the domains of u and v, and let them have continuous derivatives u' and v' over $[a,b]$. Then the products uv' and vu' are Riemann integrable, and $\int_a^b u(x)v'(x)dx = u(b)v(b) - u(a)v(a) - \int_a^b v(x)u'(x)dx$.

Proof. Since u' exists at every $x \in [a,b]$, then u is continuous over $[a,b]$, and the same for v. Hence the product uv is continuous over $[a,b]$. Because u, v, u', and v' are Riemann integrable over $[a,b]$ (Theorem 48.1), the products uv' and vu' are Riemann integrable there as well (Corollary 48.3). By the product rule, we have $(uv)' = uv' + vu'$ over $[a,b]$. This now implies that $(uv)'$ is integrable over $[a,b]$. Therefore, applying the FTIC-1 to $(uv)'$, $\int_a^b (uv)'(x)dx = u(b)v(b) - u(a)v(a)$. But simultaneously, $\int_a^b (uv)'(x)dx = \int_a^b u(x)v'(x)dx + \int_a^b v(x)u'(x)dx$. The theorem's conclusion follows. **QED**

Example 49.8 Evaluate $\int_0^1 \sin^{-1} x dx$, if it exists. The two methods below emphasize the interplay between the FTIC-1 and Riemann integration.

Method 1. To use the FTIC-1 directly, we must first determine if $f(x) = \sin^{-1} x$ is Riemann integrable over $[0,1]$. Since the function is continuous over $[-1,1]$, Theorem 48.1 tells us that $\int_0^1 \sin^{-1} x dx$ does exist. We next need to find an antiderivative F of f that is continuous over $[0,1]$. Here, integration by parts is effective as a technique for finding antiderivatives. Use

$$u(x) = \sin^{-1} x, v'(x) = 1, \quad \text{and} \quad u'(x) = \frac{1}{\sqrt{1-x^2}}, v(x) = x.$$

This yields

$$\int \sin^{-1} x dx = x\sin^{-1} x - \int \frac{x}{\sqrt{1-x^2}}dx = x\sin^{-1} x + \sqrt{1-x^2},$$

a continuous antiderivative over $[0,1]$. By the FTIC-1, it follows that

$$\int_0^1 \sin^{-1} x dx = .x\sin^{-1} x + \sqrt{1-x^2}|_0^1 = \frac{\pi}{2} - 1.$$

Method 2. Again, $f(x) = \sin^{-1} x$ is Riemann integrable over [0, 1]. Our first attempt at integration by parts naturally uses $u(x) = \sin^{-1} x$ and $v'(x) = 1$. Unfortunately, $u'(x) = \frac{1}{\sqrt{1-x^2}}$, which is not continuous over [0, 1] as Theorem 49.5 requires.

We try something else, such as the substitution $x = \sin\theta$. This yields $\int_0^{\pi/2} \theta \cos\theta d\theta$ (verify). Begin integration by parts, with $u(\theta) = \theta$ and $v'(\theta) = \cos\theta$. Hence, $u'(\theta) = 1$ and $v(\theta) = \sin\theta$. In accordance with Theorem 49.5, $[0, \frac{\pi}{2}]$ is within the domains of both $u(\theta)$ and $v(\theta)$, and this time, their derivatives are continuous over $[0, \frac{\pi}{2}]$. We are now justified to continue the integration by parts:

$$\int_0^{\pi/2} \theta \cos\theta d\theta = \theta \sin\theta\Big|_0^{\pi/2} - \int_0^{\pi/2} \sin\theta d\theta = \frac{\pi}{2} - 1.$$

Hence, $\int_0^1 \sin^{-1} x dx = \frac{\pi}{2} - 1$.

A third method to evaluate the given integral will be seen in Exercise 50.3. ■

Note 2: An investigation of two functions that purposely derail integration by parts is found in "A Counterexample to Integration by Parts" by A. Kheifets and J. Propp, in *Mathematics Magazine*, vol. 83, no. 3 (June 2010), pp. 222–225. The authors' goal is not to deconstruct, of course. They carefully dissect Theorem 49.5, and find less restrictive premises for it. For instance, they show that the continuity of even one of f' or g' is sufficient for integration by parts. The graphs there will look familiar!

49.2 Exercises

1. Show that

$$\lim_{\varepsilon \to 0+} \left(\frac{F(c) - F(c-\varepsilon)}{\varepsilon} \right) = f(c)$$

under the conditions of Theorem 49.2. Hint: Start with $-f(c) - \varepsilon < -f(x) < -f(c) + \varepsilon$, and note that $c - \varepsilon < c$ when using *(b)*. See Exercise 38.7.

2. Prove the corollary: Let $f(x)$ be continuous over $[a, b]$ and define $F(x) \equiv \int_a^x f(x)dx$, where $x \in [a, b]$. Then F is differentiable over $[a, b]$, and $F'(x) = f(x)$ there. What is the significance of this corollary? Hint: Which theorem is this a corollary of?

3. In the numerical check of Example 49.3, why are terms a_{10}, a_{50}, and a_{200}steadily *decreasing* towards the integral's value of 0.240226...?

4. a. Let $\langle a_n \rangle_1^\infty$ be given by

$$a_1 = \frac{1}{2}, a_2 = \frac{1}{3} + \frac{1}{4}, a_3 = \frac{1}{4} + \frac{1}{5} + \frac{1}{6}, a_n = \sum_{k=1}^{n} \frac{1}{n+k}.$$

Evaluate

$$\lim_{n \to \infty} \sum_{k=1}^{n} \frac{1}{n+k}$$

exactly. Hint:

$$a_n = \frac{1}{n}\left(\frac{1}{1+1/n} + \frac{1}{1+2/n} + \frac{1}{1+3/n} + \cdots + \frac{1}{1+n/n}\right).$$

b. For the sequence in **a**, what happens if we take as representatives $X_k = \frac{k}{n}$ instead?

c. Evaluate

$$\lim_{n\to\infty} \sum_{k=1}^{n} \sqrt{\frac{2}{n^2} + \frac{k}{n^3}}$$

exactly. Check your answer with a calculator, as in Example 49.3.

5. Show that in general, for a continuous function f over $[a, b]$, $\lim_{n\to\infty}(\frac{b-a}{n}\sum_{k=1}^{n} f(a + \frac{k(b-a)}{n}))$ is equal to the Riemann integral $\int_a^b f(x)dx$.

6. The integral $\int_1^3 \frac{\ln x}{x}dx$ is to be evaluated by substitution. Apply Theorem 49.3 rigorously, as follows.
   a. What substitution $y = v(x)$ should be used? Does it fulfill the premises? What is $[c, d]$? What function $f(v)$ will be used? Is it continuous over $[c, d]$?
   b. Explicitly write $\int_a^b f(v(x))\cdot v'(x)dx$ and $\int_{v(a)}^{v(b)} f(v)dv$. Evaluate the transformed integral.

7. Evaluate $\int_1^4 \frac{1}{3+\sqrt{y}}dy$, applying Theorem 49.4 rigorously as follows.
   a. What substitution $y = v(x)$ should be used? Does it fulfill the premises? What is $[c, d]$? What is the inverse function? What function $f(v)$ will be used? Is it continuous over $[c, d]$?
   b. Explicitly write $\int_a^b f(v(x))dx$ and $\int_{v(a)}^{v(b)} f(y)\cdot w'(y)dy$. Evaluate the transformed integral.

8. Evaluate $\int_0^{\pi/2} \frac{1}{3+2\cos x}dx$ by using the substitution $\tan\frac{x}{2} = v(x)$, applying Theorem 49.4 rigorously as in the previous exercises.

9. Evaluate $\int_{-3}^{-1} xe^{x^2}dx$ following Example 49.5. Hint: The variable is negative in the integrand. Since the integrand is an odd function, you can check your answer with that of Example 49.5.

10. Evaluate $\int_0^1 x\sin^{-1} x dx$ by two methods, as in Example 49.6.

11. a. Tables of integrals always have a selection of **reduction formulas**, with which an integral may be reduced to a simpler one. Prove this one:

$$\int \sin^m x\cos^n x dx = \frac{\sin^{m+1} x\cos^{n-1} x}{m+n} + \frac{n-1}{m+n}\int \sin^m x\cos^{n-2} x dx,$$

   where $m + n \neq 0$.

   b. Evaluate $\int_{\pi/4}^{\pi/2} \frac{\cos^3 x}{\sin^2 x}dx$ (use a calculator only for checking!).

12. Let

$$g(x) = \begin{cases} \frac{1}{\sqrt{5-x}} & \text{if } 1 \leq x < 5 \\ \frac{1}{2\sqrt{x}} & \text{if } 0 < x < 1 \end{cases}.$$

Graph g over its domain. To evaluate $\int_{1/4}^{3} g(x)dx$, an antiderivative of g is

$$G(x) = \begin{cases} -2\sqrt{5-x} & \text{if } 1 \leq x < 5 \\ \sqrt{x} & \text{if } 0 < x < 1 \end{cases},$$

but if we state that

$$\int_{1/4}^{3} g(x)dx = G(3) - G\left(\frac{1}{4}\right) = -2\sqrt{2} - \frac{1}{2},$$

we will be quite embarrassed. Explain the problem, and evaluate the integral correctly. Hint: Graph G over its domain.

50

Improper Integrals

In most sciences one generation tears down what another has built, and what one has established, another undoes. In mathematics alone each generation adds a new storey to the old structure.

Hermann Hankel (1839–1873)

There are certain instances when a Riemann integral does not exist by definition. One is when f is unbounded over $[a, b]$. A second is when the interval over which function f is studied is unbounded, either $[a, \infty)$, or $(-\infty, b]$, or $(-\infty, \infty)$. Other cases involve combinations of them. Nevertheless, the Riemann integral can be extended meaningfully in some situations.

50.1 Unbounded Functions Over a Closed Interval

We look first at the case of unbounded functions over $[a, b]$. To be clear, for any $\varepsilon \in (0, b-a)$, let the function f have no bound in $[a, a+\varepsilon]$, but let f be Riemann integrable over $[a+\varepsilon, b]$. (a need not belong to $\mathcal{D}_f$.) If $\lim_{\varepsilon \to 0+} \int_{a+\varepsilon}^{b} f(x)dx$ exists, we call this limit the **improper integral** of f over $[a, b]$.

Likewise, for any $\varepsilon \in (0, b-a)$, let g have no bound in $[b-\varepsilon, b]$, but let g be Riemann integrable over $[a, b-\varepsilon]$. (b need not belong to $\mathcal{D}_f$.) If $\lim_{\varepsilon \to 0+} \int_{a}^{b-\varepsilon} g(x)dx$ exists, we again call this limit the **improper integral** of g over $[a, b]$.

We also say that the integrals in the two previous paragraphs are **convergent integrals** over $[a, b]$, and write the respective limits concisely as $\int_a^b f(x)dx$ and $\int_a^b g(x)dx$. It is also correct to say that f and g are **improperly integrable** over $[a, b]$. If the improper integrals do not exist, we say that they are **divergent integrals**.

Suppose that a function h is unbounded at both endpoints of $[a, b]$, but is Riemann integrable over $[a+\varepsilon, b-\varepsilon]$, for any

$$\varepsilon \in \left(0, \frac{b-a}{2}\right).$$

Let $c \in (a, b)$. If both

$$\lim_{\varepsilon \to 0+} \int_{a+\varepsilon}^{c} h(x)dx$$

and

$$\lim_{\varepsilon \to 0+} \int_c^{b-\varepsilon} h(x)dx$$

exist, then

$$\lim_{\varepsilon \to 0+} \int_{a+\varepsilon}^{c} h(x)dx + \lim_{\varepsilon \to 0+} \int_c^{b-\varepsilon} h(x)dx = \lim_{\varepsilon \to 0+} \int_{a+\varepsilon}^{b-\varepsilon} h(x)dx$$

exists. We write this limit as $\int_a^b h(x)dx$, and it's called the value of the improper integral over $[a, b]$. As above, $\int_a^b h(x)dx$ is a convergent integral over $[a, b]$, and h is improperly integrable there. The negation of this occurs when either $\int_{a+\varepsilon}^{c} h(x)dx$ or

$$\lim_{\varepsilon \to 0+} \int_c^{b-\varepsilon} h\,(x)\,dx$$

fails to exist, resulting in a divergent integral $\int_a^b h(x)dx$ over $[a, b]$.

These types of improper integrals are discussed in Examples 50.1 to 50.3.

Example 50.1 Find

$$\int_1^5 \frac{1}{\sqrt{x-1}} dx.$$

This is improper due to the unbounded integrand at $a = 1$. Because

$$\lim_{\varepsilon \to 0+} \int_{1+\varepsilon}^{5} \frac{1}{\sqrt{x-1}} dx = \lim_{\varepsilon \to 0+} \left(2\sqrt{x-1}\Big|_{1+\varepsilon}^{5} \right) = \lim_{\varepsilon \to 0+} \left(4 - 2\sqrt{1+\varepsilon-1} \right) = 4,$$

the integral is convergent, and

$$\int_1^5 \frac{1}{\sqrt{x-1}} dx = 4.$$

Notice that the limit operation is performed after the expansion of

$$2\sqrt{x-1}\Big|_{1+\varepsilon}^{5}. \quad \blacksquare$$

Example 50.2 Find

$$\int_0^3 \frac{1}{\sqrt{9-y^2}} dy.$$

The integrand

$$f(x) = \frac{1}{\sqrt{9-y^2}}$$

is unbounded at $b = 3$. Thus, consider

$$\int_0^{3-\varepsilon} \frac{1}{\sqrt{9-y^2}} dy,$$

where $\varepsilon \in (0, 1)$. As in Example 49.6, use $y = v(x) = \sin^{-1} \frac{x}{3}$, which is differentiable over $[0, 3 - \varepsilon]$. With this interval, all the steps in Example 49.6 may be taken, and we arrive at

$$\int_0^{3-\varepsilon} \frac{1}{\sqrt{9 - x^2}} dx = \int_0^{\sin^{-1}(3-\varepsilon)/3} \frac{1}{3 \cos y} \cdot 3 \cos y dy = \sin^{-1} \frac{3 - \varepsilon}{3}.$$

Therefore, the improper integral

$$\int_0^3 \frac{1}{\sqrt{9 - y^2}} dy = \lim_{\varepsilon \to 0+} \sin^{-1} \frac{3 - \varepsilon}{3} = \frac{\pi}{2}. \quad \blacksquare$$

Example 50.3 Find

$$\int_{-1}^1 \frac{1}{\sqrt{1 - x^2}} dx.$$

This integrand is unbounded at both endpoints of $[-1, 1]$, but is integrable over any closed interval within $[-1, 1]$. Thus, choose any interior point, such as $c = 0$, and evaluate

$$\lim_{\varepsilon \to 0+} \int_{-1+\varepsilon}^0 \frac{1}{\sqrt{1 - x^2}} dx = \lim_{\varepsilon \to 0+} \left(\sin^{-1} x \Big|_{-1+\varepsilon}^0 \right) = \frac{\pi}{2},$$

and

$$\lim_{\varepsilon \to 0+} \int_0^{1-\varepsilon} \frac{x}{\sqrt{1 - x^2}} dx = \lim_{\varepsilon \to 0+} \left(\sin^{-1} x \Big|_0^{1-\varepsilon} \right) = \frac{\pi}{2}.$$

Hence, the value of the integral is

$$\int_{-1}^1 \frac{1}{\sqrt{1 - x^2}} dx = \frac{\pi}{2} + \frac{\pi}{2} = \pi. \quad \blacksquare$$

Other cases under this subsection occur when the integrand is unbounded, or behaves in unfamiliar ways, within $[a, b]$, as the next two examples show.

Example 50.4 Find $\int_{-1}^2 \frac{1}{x^{2/3}} dx$. Making the quick evaluation

$$3x^{1/3} \Big|_{-1}^2 = 3\sqrt[3]{2} - 3$$

is a fatal error, since $f(x) = \frac{1}{x^{2/3}}$ is unbounded at $0 \in [-1, 2]$, thereby violating the premises of the FTIC-1. On the other hand, f is Riemann integrable over $[-1, 0 - \varepsilon]$ and $[0 + \varepsilon, 2]$ for any $\varepsilon > 0$. Thus, we first determine

$$\int_{-1}^{0-\varepsilon} \frac{1}{x^{2/3}} dx + \int_{0+\varepsilon}^2 \frac{1}{x^{2/3}} dx = 3x^{1/3} \Big|_{-1}^{0-\varepsilon} + 3x^{1/3} \Big|_{0+\varepsilon}^2 = 3 \left(-\sqrt[3]{\varepsilon} + 1 \right) + 3 \left(\sqrt[3]{2} - \sqrt[3]{\varepsilon} \right),$$

and after this, evaluate

$$\lim_{\varepsilon \to 0+} \left(3 \left(-\sqrt[3]{\varepsilon} + 1 \right) + 3 \left(\sqrt[3]{2} - \sqrt[3]{\varepsilon} \right) \right) = 3 + 3\sqrt[3]{2}.$$

Consequently,

$$\int_{-1}^2 \frac{1}{x^{2/3}} dx$$

is improperly integrable, with value $3 + 3\sqrt[3]{2}$.

The convergence that we have been investigating may be observed by graphing

$$I(t) = \int_{-1}^{-t} \frac{1}{x^{2/3}} dx + \int_{t}^{2} \frac{1}{x^{2/3}} dx$$

over a domain such as $\mathcal{D}_I = (0, 2]$. Does $I(t)$ approach $3 + 3\sqrt[3]{2}$ as $t \to 0^+$? (Some graphing utilities may have difficulty with the limit.) ■

Example 50.5 Let

$$g(x) = \begin{cases} \dfrac{1}{\sqrt{5-x}} & \text{if } 1 \leq x < 5 \\ \dfrac{1}{2\sqrt{x}} & \text{if } 0 < x < 1 \end{cases}.$$

Find

$$\int_0^5 g(x)dx,$$

if it exists. An antiderivative of g is

$$G(x) = \begin{cases} -2\sqrt{5-x} & \text{if } 1 \leq x < 5 \\ \sqrt{x} & \text{if } 0 < x < 1 \end{cases},$$

and we may be tempted to write

$$\int_0^5 g(x)dx = \lim_{x \to 5-} G(x) - \lim_{x \to 0+} G(x) = 0.$$

This is an incorrect application of the FTIC-1 (review Exercise 49.12), and furthermore, the integral is improper over $[0, 5]$.

For the correct evaluation of $\int_0^5 g(x)dx$, we must first discover that G has a jump discontinuity at $x = 1$, so that two improper integrals are necessary. Hence, we consider

$$\lim_{\varepsilon \to 0+} \int_{\varepsilon}^{1} \frac{1}{2\sqrt{x}} dx + \lim_{\varepsilon \to 0+} \int_{1}^{5-\varepsilon} \frac{1}{\sqrt{5-x}} dx = \lim_{\varepsilon \to 0+} \left(\sqrt{x}\Big|_{\varepsilon}^{1} \right) - 2 \lim_{\varepsilon \to 0+} \left(\sqrt{5-x}\Big|_{1}^{5-\varepsilon} \right) = 5. \quad ■$$

50.2 Bounded Functions Over Unbounded Intervals

In this subsection, we study the possibility of integrating a bounded function over an unbounded interval such as $[a, \infty)$, $(-\infty, b]$, or $(-\infty, \infty)$. Let the function f be bounded over $[a, \infty)$. Suppose $\int_a^c f(x)dx$ exists over every closed interval $[a, c]$, then consider

$$\lim_{c \to \infty} \int_a^c f(x)dx.$$

If the limit exists, call it the **improper integral** of f over $[a, \infty)$, and denote it as $\int_a^\infty f(x)dx$. We say that this is a **convergent integral**, and if the improper integral does not exist, we may say that it **diverges**.

An analogous definition applies to a function g bounded over $(-\infty, b]$. If $\int_c^b g(x)dx$ exists over every $[c, b]$, and if

$$\lim_{c \to -\infty} \int_c^b g(x)dx$$

exists, then denote this limit as $\int_{-\infty}^b g(x)dx$. The same terminology is used as above.

The third case involves a function h bounded over the entire real line, $(-\infty, \infty)$. Define the improper integral

$$\int_{-\infty}^{\infty} h(x)dx$$

as

$$\int_{-\infty}^{c} h(x)dx + \int_{c}^{\infty} h(x)dx,$$

which requires that each of these integrals exist *independently*, the first as

$$\lim_{a \to -\infty} \int_a^c h(x)\, dx$$

and the second as

$$\lim_{b \to \infty} \int_c^b h(x)dx.$$

Logically, if either diverges, then $\int_{-\infty}^{\infty} h(x)dx$ is divergent. (Often, $c = 0$ is chosen for ease of evaluation.)

There is another possibility. If it exists, the limit

$$\lim_{a \to \infty} \int_{-a}^{a} h(x)dx$$

is called the **Cauchy principal value** of the integral, denoted $\wp \int_{-\infty}^{\infty} h(x)dx$. Note that $\wp \int_{-\infty}^{\infty} h(x)dx$ is not necessarily equal to $\int_{-\infty}^{\infty} h(x)dx$.

For instance,

$$\wp \int_{-\infty}^{\infty} \sin x dx = \lim_{a \to \infty} \int_{-a}^{a} \sin x dx = 0,$$

because as an odd function, $\int_{-a}^{a} \sin x dx = 0$ for any a. On the other hand, $\int_{-\infty}^{\infty} \sin x dx$ diverges, since the component $\lim_{b \to \infty} \int_0^b \sin x dx$ doesn't exist (nor does the other one).

See Exercise 50.9 and Note 1 below for more about the relationship between $\int_{-\infty}^{\infty} h(x)dx$ and $\wp \int_{-\infty}^{\infty} h(x)dx$.

Example 50.6 Determine if $\int_a^{\infty} e^{-3x} dx$ exists for some, or any, constants a. We investigate

$$\lim_{c \to \infty} \int_a^c e^{-3x} dx,$$

recognizing that $f(x) = e^{-3x}$ is Riemann integrable over any $[a, c]$. Thus,

$$\lim_{c \to \infty} \int_a^c e^{-3x} dx = \lim_{c \to \infty} \left(-\frac{1}{3} e^{-3c} + \frac{1}{3} e^{-3a} \right) = \frac{1}{3} e^{-3a},$$

which does indeed exist for any constant a. We summarize this as

$$\int_{a}^{\infty} e^{-3x} dx = \frac{1}{3} e^{-3a}. \quad \blacksquare$$

Example 50.7 Determine if

$$\int_{-\infty}^{\infty} \frac{1}{1+x^2} dx$$

converges. We consider

$$\int_{-\infty}^{0} \frac{1}{1+x^2} dx$$

and

$$\int_{0}^{\infty} \frac{1}{1+x^2} dx$$

independently. For the first improper integral,

$$\lim_{a \to -\infty} \int_{a}^{0} \frac{1}{1+x^2} dx = \lim_{a \to -\infty} \left(\tan^{-1} 0 - \tan^{-1} a\right) = \frac{\pi}{2},$$

and for the second,

$$\lim_{b \to \infty} \int_{0}^{b} \frac{1}{1+x^2} dx = \frac{\pi}{2}.$$

Therefore,

$$\int_{-\infty}^{\infty} \frac{1}{1+x^2} dx$$

converges to $\frac{\pi}{2} + \frac{\pi}{2} = \pi$. Note that in this case,

$$\int_{-\infty}^{\infty} \frac{1}{1+x^2} dx = \wp \int_{-\infty}^{\infty} \frac{1}{1+x^2} dx.$$

(The related function

$$P(c) = \frac{1}{\pi} \int_{-\infty}^{c} \frac{1}{1+x^2} dx,$$

which we now know exists, is found in probability theory.) $\blacksquare$

There are improper integrals that combine this subsection's definitions with those of the previous subsection.

Example 50.8 Determine if

$$\int_{0}^{\infty} \frac{1}{x^p} dx$$

converges for some, or any, constants p. If $p = 0$ or $p = 1$, the integral diverges.

The integrand is unbounded over $[0, a]$ when $p > 0$, but it is Riemann integrable over any $[\varepsilon, a]$, where $\varepsilon > 0$. Thus, we consider

$$\lim_{\varepsilon \to 0+} \int_{\varepsilon}^{a} \frac{1}{x^p} dx = \frac{1}{1-p} \lim_{\varepsilon \to 0+} \left(a^{1-p} - \varepsilon^{1-p} \right).$$

The limit exists if and only if we further constrain $p < 1$, so that the improper integral

$$\int_0^a \frac{1}{x^p} dx = \frac{a^{1-p}}{1-p}$$

for $0 < p < 1$. Our next step is to consider

$$\lim_{a \to \infty} \int_0^a \frac{1}{x^p} dx = \lim_{a \to \infty} \frac{a^{1-p}}{1-p}.$$

Since the limit doesn't exist,

$$\int_0^\infty \frac{1}{x^p} dx$$

diverges for all $p \geq 0$.

For $p < 0$,

$$\int_0^\infty \frac{1}{x^p} dx = \int_0^\infty x^{|p|} dx$$

is a divergent improper integral. In conclusion,

$$\int_0^\infty \frac{1}{x^p} dx$$

hopelessly diverges for all constants p. ■

Note 1. Be careful not to abuse the definition of

$$\int_{-\infty}^{\infty} f(x) dx,$$

or else ambiguity results. For example,

$$I_1 = \int_{-\infty}^{\infty} \frac{x}{1+x^2} dx$$

diverges, since

$$\lim_{a \to -\infty} \int_a^0 \frac{x}{1+x^2} dx = -\infty$$

and

$$\lim_{b\to\infty}\int_0^b \frac{x}{1+x^2}dx = \infty.$$

But if I_1 is evaluated as, let's say,

$$\lim_{a\to\infty}\int_{-a}^{3a} \frac{x}{1+x^2}dx,$$

we discover $\ln 3$ to be the limit! The problem is that $3a$ is not independent of a, allowing for simplification of

$$\int_{-a}^{3a} \frac{x}{1+x^2}dx.$$

What was probably intended in the first place was

$$I_2 = \wp\int_{-\infty}^{\infty} \frac{x}{1+x^2}dx = 0,$$

which is visually in accord with the graph of $y = \frac{x}{1+x^2}$.

50.3 Exercises

1. Determine if $\int_0^1 \ln x dx$ converges.

2. Determine if

$$\int_1^3 \frac{\ln(2-x)}{2-x}dx$$

converges.

3. Example 49.8 method 2 attempted to evaluate $\int_0^1 \sin^{-1} x dx$ by integration by parts with $u(x) = \sin^{-1} x$ and $v'(x) = 1$. But we saw that

$$u'(x) = \frac{1}{\sqrt{1-x^2}}$$

isn't continuous over $[0, 1]$ as Theorem 49.5 requires. However, it is continuous over $[0, 1-\varepsilon]$, where $0 < \varepsilon < 1$. Use this interval to perform the integration by parts, evaluating an improper integral.

4. a. For what values of the constant p does $\int_1^\infty \frac{1}{x^p}dx$ converge? When it converges, what is the limit's value?

   b. For what values of the constant p does $\int_0^1 \frac{1}{x^p}dx$ converge? When it converges, what is the limit's value?

5. Determine if

$$\int_{-\infty}^{\infty} \frac{1}{(x-1)^2}dx$$

converges.

6. Determine if

$$\int_1^\infty \frac{1}{\sqrt{x-1}}dx$$

converges.

7. Determine if

$$\int_1^\infty \frac{1}{x\sqrt{x-1}}dx$$

converges.

8. Graph $h(x) = e^{-|x|}$ and show that

$$\int_{-\infty}^{\infty} e^{-|x|}dx = \wp \int_{-\infty}^{\infty} e^{-|x|}dx.$$

(The related function $P(c) = \frac{1}{2}\int_{-\infty}^{c} e^{-|x|}dx$, which we now know exists, occurs in probability theory.)

9. Prove the lemma that if $\int_{-\infty}^{\infty} f(x)dx$ converges, then $\wp \int_{-\infty}^{\infty} f(x)dx$ exists, and the two values are equal. Show that the converse is false.

51

The Cauchy-Schwarz and Minkowski Inequalities

We introduce two famous inequalities of analysis named after Cauchy and the brilliant Hermann Minkowski (1864–1909). Historically, Cauchy was the discoverer of Theorem 51.1, in 1821. An extension to integrals was discovered by Victor Bunyakovskiĭ (1804–1889) in 1859, and rediscovered by analyst Hermann Schwarz (1843–1921) much later. Bunyakovskiĭ was a professor at the St. Petersburg Academy of Sciences for many years, and published in number theory, mechanics, hydrostatics, and probability as well as analysis. Our first theorem is sometimes named the Cauchy-Bunyakovskiĭ-Schwarz inequality.

Theorem 51.1 The Cauchy-Schwarz Inequality Let $\{a_1, a_2, \ldots, a_n\}$ and $\{b_1, b_2, \ldots, b_n\}$ be any collections of real numbers. Then $(\sum_{i=1}^n a_i b_i)^2 \le (\sum_{i=1}^n a_i^2)(\sum_{i=1}^n b_i^2)$.

Proof. First, we need to prove **Lagrange's identity**, named for Joseph-Louis Lagrange:

$$\left(\sum_{i=1}^n a_i b_i\right)^2 = \left(\sum_{i=1}^n a_i^2\right)\left(\sum_{j=1}^n b_j^2\right) - \frac{1}{2}\sum_{i=1}^n\sum_{j=1}^n (a_i b_j - a_j b_i)^2.$$

Expanding the double summation on the right side,

$$\begin{aligned}
\sum_{i=1}^n\sum_{j=1}^n (a_i b_j - a_j b_i)^2 &= \sum_{i=1}^n\sum_{j=1}^n \left(a_i^2 b_j^2 - 2a_i b_j a_j b_i + a_j^2 b_i^2\right) \\
&= \sum_{i=1}^n\sum_{j=1}^n a_i^2 b_j^2 - 2\sum_{i=1}^n\sum_{j=1}^n a_i b_i a_j b_j + \sum_{i=1}^n\sum_{j=1}^n a_j^2 b_i^2 \\
&= \sum_{i=1}^n a_i^2 \sum_{j=1}^n b_j^2 - 2\sum_{i=1}^n a_i b_i \sum_{j=1}^n b_j a_j + \sum_{i=1}^n b_i^2 \sum_{j=1}^n a_j^2 \\
&= 2\sum_{i=1}^n a_i^2 \sum_{j=1}^n b_j^2 - 2\left(\sum_{i=1}^n a_i b_i\right)^2.
\end{aligned}$$

Thus, the full right side equals

$$\sum_{i=1}^{n} a_i^2 \sum_{j=1}^{n} b_j^2 - \frac{1}{2}\left(2\sum_{i=1}^{n} a_i^2 \sum_{j=1}^{n} b_j^2 - 2\left(\sum_{i=1}^{n} a_i b_i\right)^2\right),$$

which simplifies to exactly the left side of the identity.

The Cauchy-Schwarz inequality follows immediately from Lagrange's identity. **QED**

An extension of the Cauchy-Schwarz inequality to infinite series appears in Exercise 51.8.

Note 1: A clever geometric demonstration of Theorem 51.1 in the specific case

$$|x_1y_1 + x_2y_2| \le \sqrt{x_1^2 + x_2^2}\sqrt{y_1^2 + y_2^2}$$

appears in "Proof Without Words: Cauchy-Schwarz Inequality" by Claudi Alsina, in *Mathematics Magazine*, vol. 77, no. 1 (Feb. 2004), pg. 30.

Theorem 51.2 The Minkowski Inequality Let $\{a_1, a_2, \ldots, a_n\}$ and $\{b_1, b_2, \ldots, b_n\}$ be any collections of real numbers. Then

$$\sqrt{\sum_{i=1}^{n}(a_i + b_i)^2} \le \sqrt{\sum_{i=1}^{n} a_i^2} + \sqrt{\sum_{i=1}^{n} b_i^2}.$$

Proof. Start with the Cauchy-Schwarz inequality in the form

$$\sum_{i=1}^{n} a_i b_i \le \sqrt{\sum_{i=1}^{n} a_i^2}\sqrt{\sum_{i=1}^{n} b_i^2}.$$

Now, expand $\sum_{i=1}^{n}(a_i + b_i)^2$ and we arrive at

$$\sum_{i=1}^{n} a_i^2 + 2\sum_{i=1}^{n} a_i b_i + \sum_{i=1}^{n} b_i^2 \le \sum_{i=1}^{n} a_i^2 + 2\sqrt{\sum_{i=1}^{n} a_i^2}\sqrt{\sum_{i=1}^{n} b_i^2} + \sum_{i=1}^{n} b_i^2.$$

The rest of the steps are left for you to do in Exercise 51.3. **QED**

An extension of the Minkowski inequality to infinite series appears in Exercise 51.9.

Note 2: Theorem 51.2 can be generalized beyond the squares and square roots that appear here. Given any real number $p \ge 1$, and *nonnegative* sets of real numbers $\{a_1, a_2, \ldots, a_n\}$ and $\{b_1, b_2, \ldots, b_n\}$, the generalized Minkowski inequality reads as

$$\left[\sum_{i=1}^{n}(a_i + b_i)^p\right]^{1/p} \le \left(\sum_{i=1}^{n} a_i^p\right)^{1/p} + \left(\sum_{i=1}^{n} b_i^p\right)^{1/p}.$$

The proof of this and other famous inequalities can be found in [10].

51.1 Exercises

1. Prove that for any real numbers, $(w^2+x^2)(y^2+z^2) \geq (wy+xz)^2$ in two ways, by direct algebra, and by the Cauchy-Schwarz inequality.

2. Prove the Cauchy-Schwarz inequality by induction. Hints: For ease of work, let $A \equiv \sum_{i=1}^n a_i^2$, $B \equiv \sum_{i=1}^n b_i^2$, and $C \equiv \sum_{i=1}^n a_i b_i$. In this way, for instance, the induction hypothesis appears as $C^2 \leq AB$. Now, write case $n+1$ in terms of A, B, and C, and prove it.

3. Finish the proof of Theorem 51.2, the Minkowski inequality.

4. Prove that for any real numbers, $(\sqrt{w^2+y^2}+\sqrt{x^2+z^2})^2 \geq (w+x)^2+(y+z)^2$ in two ways, by direct algebra, and by the Minkowski inequality.

5. In statistics, the standard deviation of a sample is given by

$$s = \sqrt{\frac{n\sum_{i=1}^n x_i^2 - (\sum_{i=1}^n x_i)^2}{n(n-1)}},$$

where $n \geq 2$ is the size of the sample, and $\{x_1, x_2, \ldots, x_n\}$ are the data values (thus, arbitrary real numbers). Prove that the numerator is never negative (hence, s is never imaginary), by using the Cauchy-Schwarz inequality.

6. In statistics, the well-known Pearson correlation coefficient r is defined as

$$r = \frac{\sum_{i=1}^n (x_i - a)(y_i - b)}{\sqrt{\sum_{i=1}^n (x_i - a)^2}\sqrt{\sum_{i=1}^n (y_i - b)^2}},$$

where a and b are the means of the data sets $\{x_i\}_1^n$ and $\{y_i\}_1^n$ respectively (we view them as collections of real numbers), and not all $x_i = a$, and not all $y_i = b$ (this implies that $n \geq 2$). Prove that $0 \leq r^2 \leq 1$. The famous property $-1 \leq r \leq 1$ follows.

7. a. What does $\sum_{i=1}^n \sum_{j \neq i} a_i a_j$ mean? For example, evaluate it when $a_1 = 2$, $a_2 = 3$, and $a_3 = -4$. How does $\sum_{i=1}^n \sum_{j \neq i} a_i a_j$ compare with $\sum_{i=1}^n \sum_{j>i} a_i a_j$?
   b. Prove the identity $\sum_{i=1}^n a_i^2 + \sum_{i=1}^n \sum_{j \neq i} a_i a_j = \sum_{i=1}^n \sum_{j=1}^n a_i a_j$. To see what this identity describes, try it with $n = 2$. Hint: $\sum_{i=1}^n \sum_{j=1}^n a_i a_j = \sum_{i=1}^n a_i \sum_{j=1}^n a_j = (\sum_{i=1}^n a_i)^2$.

8. The **extended Cauchy-Schwarz inequality** is

$$\left|\sum_{i=1}^{\infty} a_i b_i\right| \leq \sqrt{\sum_{i=1}^{\infty} a_i^2}\sqrt{\sum_{i=1}^{\infty} b_i^2},$$

for any sequences of real numbers $\langle a_1, a_2, \ldots\rangle$ and $\langle b_1, b_2, \ldots\rangle$, provided that the infinite series on the right converge. The Cauchy-Schwarz inequality can be extended to series by following these steps for a proof.

   (i) We may assume that not all a_i are zero. As a finite sum, $\sum_{i=1}^n (xa_i + b_i)^2 \geq 0$, where x is a parameter used only for the purposes of this classic, but ingenious, proof. Expand the binomial and rearrange the summations to arrive at $Ax^2 + Bx + C \geq 0$. What summations do A, B, and C represent?

(ii) Since $A > 0$, let $x = -\frac{B}{2A}$, the axis of symmetry of the parabola $y = Ax^2 + Bx + C$. Derive that $B^2 \leq 4AC$.

(iii) Use the inequality just derived to show that

$$\sum_{i=1}^{n} |a_i b_i| \leq \sqrt{\sum_{i=1}^{\infty} a_i^2} \sqrt{\sum_{i=1}^{\infty} b_i^2}.$$

Hint: The premises allow use of any real a_i and b_i. Select your substitutions wisely.

(iv) Since the infinite series on the right converge, state something about the sequence of partial sums $\sum_{i=1}^{n} |a_i b_i|$.

(v) Finish the proof of the extended Cauchy-Schwarz inequality by first proving absolute convergence

$$\sum_{i=1}^{\infty} |a_i b_i| \leq \sqrt{\sum_{i=1}^{\infty} a_i^2} \sqrt{\sum_{i=1}^{\infty} b_i^2},$$

and then using Exercise 31.3.

9. The **extended Minkowski inequality** is

$$\sqrt{\sum_{i=1}^{\infty} (a_i + b_i)^2} \leq \sqrt{\sum_{i=1}^{\infty} a_i^2} + \sqrt{\sum_{i=1}^{\infty} b_i^2},$$

for any sequences of real numbers $\langle a_1, a_2, \ldots \rangle$ and $\langle b_1, b_2, \ldots \rangle$, provided that the infinite series on the right converge. Prove this extension to series by following the ideas in the proof of Theorem 51.2. You will need the extended Cauchy-Schwarz inequality above.

52

Metric Spaces

There are things which seem incredible to most men who have not studied mathematics.

Archimedes

Three-dimensional space and the Euclidean geometry used to study it are characteristics of the space of common experience. This geometry was named after Euclid of Alexandria (active in 300 B.C.), author of the *Elements*, one of civilization's most influential written works (it is comprised of thirteen books on geometry and algebra). In the early 1800s, mathematicians became aware that Euclidean geometry was far from unique (see Note 3, Chapter 21). Analysts began clarifying and extending the definition of distance, and that of dimension. The abstract definition of distance will bring us to a special function called a "metric." The term **space** is generalized to mean a universal set of objects under consideration, but it carries the connotation of some distinguishing analytic property of that set. For example, one may study probability spaces or vector spaces. The property that we will focus on is that of distance, so that we will analyze several metric spaces.

Example 52.1 We consider ordered collections of real numbers. **Two-dimensional space** is the set of ordered pairs $\langle x_1, x_2 \rangle \equiv \mathbf{x}_2$; these quantities are commonly known as **vectors** in 2-space. Two-space is visualized as the Cartesian plane, and analytically described as the Cartesian product $\mathbb{R}\times\mathbb{R} = \mathbb{R}^2$ (pronounced "r-two"). **Three-dimensional space** is the Cartesian product $\mathbb{R}\times\mathbb{R}\times\mathbb{R} = \mathbb{R}^3$ ("r-three") that consists of all ordered triples $\langle x_1, x_2, x_3 \rangle \equiv \mathbf{x}_3$, called vectors in 3-space. Generalizing beyond the sensory world, $\mathbb{R}^n$ is ***n*-dimensional space** (n-space for short), the set of ordered n-tuples $\langle x_1, x_2, \ldots, x_n \rangle \equiv \mathbf{x}_n$. (From now on, for vectors, we will delete the subscripts but keep the bold font.) We identify the real number line as the **one-dimensional space** $\mathbb{R}^1$ (or just $\mathbb{R}$), whose vectors are simply real numbers x. ■

Example 52.2 As often happens, good notation suggests new concepts. The astonishing possibility of **infinite-dimensional space**, $\mathbb{R}^\infty$, arises. It consists of vectors with an infinite number of coordinates, $\langle x_1, x_2, x_3 \cdots \rangle \equiv \langle x_n \rangle_1^\infty$, whence we recognize $\mathbb{R}^\infty$as none other than the space of all infinite sequences. ■

Example 52.3 The space of all open intervals (a, b) in $\mathbb{R}$ is a more abstract example of a universal set of objects. ■

Given a universal set of objects $\mathbb{M}$, a **metric** on $\mathbb{M}$ is a function $\rho : \mathbb{M}\times\mathbb{M} \to \mathbb{R}$ that has these properties for any members $x, y, z \in \mathbb{M}$:

(a) nonnegativity: $\rho(x, y) \geq 0$,
(b) identity: $\rho(x, y) = 0$ if and only if $x = y$,
(c) symmetry: $\rho(x, y) = \rho(y, x)$,
(d) the triangle inequality: $\rho(x, y) \leq \rho(x, z) + \rho(z, y)$.

We shall refer to the pair $\langle \mathbb{M}, \rho \rangle$ as a **metric space**. It is possible to have different metrics on the same space, say $\langle \mathbb{M}, \rho \rangle$ and $\langle \mathbb{M}, \tau \rangle$. The guiding concept is that $\mathbb{M}$, along with a metric ρ, will follow the four properties in analogy to classical two-dimensional space and its distance formula. To that end, we prove first that $\mathbb{R}^2$, together with the standard distance formula d_2 ("d-two") between points in the plane, forms the metric space $\langle \mathbb{R}^2, d_2 \rangle$ (pronounced exactly as the name of the little robot in *Star Wars*).

Example 52.4 We will prove that $\langle \mathbb{R}^2, d_2 \rangle$, with the classical distance formula

$$d_2(\mathbf{x}, \mathbf{y}) = d_2(\langle x_1, x_2 \rangle, \langle y_1, y_2 \rangle) = \sqrt{(x_1 - y_1)^2 + (x_2 - y_2)^2} = \sqrt{\sum_{i=1}^{2} (x_i - y_i)^2},$$

is a metric space. We must verify properties **(a)** through **(d)** for the distance formula d_2, applied to points $\mathbf{x}$, $\mathbf{y}$ and $\mathbf{z}$ of $\mathbb{R}^2$.

a. is true: since $(d_2(\mathbf{x}, \mathbf{y}))^2$ is the sum of squared quantities, $d_2(\mathbf{x}, \mathbf{y}) \geq 0$.
b. always requires two proofs. If $d_2(\mathbf{x}, \mathbf{y}) = 0$, then $\sum_{i=1}^{2} (x_i - y_i)^2 = 0$, which implies that $(x_1 - y_1)^2 = 0$ and $(x_2 - y_2)^2 = 0$ separately. Hence, $x_1 = y_1$ and $x_2 = y_2$, which means that $\mathbf{x} = \mathbf{y}$. Conversely, if $\mathbf{x} = \mathbf{y}$, then

$$d_2(\mathbf{x}, \mathbf{y}) = \sqrt{\sum_{i=1}^{2} 0^2} = 0.$$

c. is true because

$$\sqrt{\sum_{i=1}^{2} (x_i - y_i)^2} = \sqrt{\sum_{i=1}^{2} (y_i - x_i)^2}.$$

d. is often difficult to prove. In this case, we must prove that $d_2(\mathbf{x}, \mathbf{y}) \leq d_2(\mathbf{x}, \mathbf{z}) + d_2(\mathbf{z}, \mathbf{y})$ for any three points in $\mathbb{R}^2$. That is, show that

$$\sqrt{(x_1 - y_1)^2 + (x_2 - y_2)^2} \leq \sqrt{(x_1 - z_1)^2 + (x_2 - z_2)^2} + \sqrt{(z_1 - y_1)^2 + (z_2 - y_2)^2}.$$

A direct assault leads to terribly tedious algebra, but we can simplify things without loss of generality. Given three points $\mathbf{x}$, $\mathbf{y}$, and $\mathbf{z}$, draw the X axis through $\mathbf{x}$ and $\mathbf{z}$, and the Y axis so as to pass through $\mathbf{y}$ (it won't matter if the three points are collinear). As shown in Figure 52.1, this allows us to assign coordinates $\langle x_1, 0 \rangle$ to $\mathbf{x}$, $\langle z_1, 0 \rangle$ to $\mathbf{z}$, and $\langle 0, y_2 \rangle$ to $\mathbf{y}$. We will only stipulate that $z_1 \geq x_1$; nevertheless, the triangle formed is completely general. Then $d_2(\mathbf{x}, \mathbf{y}) \leq d_2(\mathbf{x}, \mathbf{z}) + d_2(\mathbf{z}, \mathbf{y})$ becomes

$$\sqrt{x_1^2 + y_2^2} \leq (z_1 - x_1) + \sqrt{z_1^2 + y_2^2}.$$

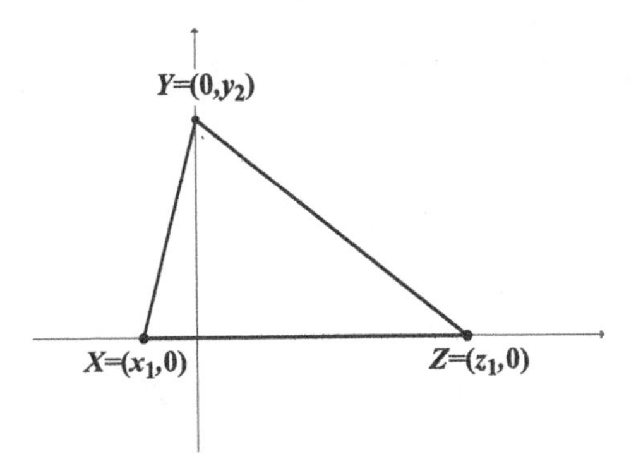

Figure 52.1.

Squaring both sides and simplifying, we get

$$0 \le z_1^2 - x_1 z_1 + (z_1 - x_1)\sqrt{z_1^2 + y_2^2},$$

or

$$0 \le (z_1 - x_1)\left(z_1 + \sqrt{z_1^2 + y_2^2}\right). \tag{1}$$

This inequality is always true, since, regardless of the sign of z_1, the quantity $z_1 + \sqrt{z_1^2 + y_2^2}$ cannot be negative. As is often done, we now begin with (1) and reverse our steps to conclude that $d_2(\mathbf{x}, \mathbf{y}) \le d_2(\mathbf{x}, \mathbf{z}) + d_2(\mathbf{z}, \mathbf{y})$ is true for any three points in $\mathbb{R}^2$. Having proved **(a)** through **(d)**, we are sure that $\langle \mathbb{R}^2, d_2 \rangle$ is a metric space. **QED**

Similar work could be done to prove that $\langle \mathbb{R}^3, d_2 \rangle$ is also a metric space, using the distance formula

$$d_2(\mathbf{x}, \mathbf{y}) = \sqrt{(x_1 - y_1)^2 + (x_2 - y_2)^2 + (x_3 - y_3)^2} = \sqrt{\sum_{i=1}^{3} (x_i - y_i)^2}.$$

Proving $\langle \mathbb{R}^4, d_2 \rangle$ to be a metric space by this method is a bit strained, since this is now four-dimensional space. In fact, all $\langle \mathbb{R}^n, d_2 \rangle$ are metric spaces, as Theorem 52.1 will show. But first, note that the expression

$$d_2(\mathbf{x}, \mathbf{y}) = \sqrt{\sum_{i=1}^{n} (x_i - y_i)^2}, \mathbf{x} = \langle x_1, x_2, \ldots, x_n \rangle, \mathbf{y} = \langle y_1, y_2, \ldots, y_n \rangle, n \in \mathbb{N} \tag{2}$$

is called the **Euclidean metric** for $\mathbb{R}^n$ since it is precisely the Pythagorean theorem in n dimensions, and this ancient theorem defined the only metric that Euclid recognized. Consequently, all $\langle \mathbb{R}^n, d_2 \rangle$ are called **Euclidean metric spaces**. To put it succinctly, Euclidean metric spaces are spaces where the Pythagorean theorem defines distance.

We are going to use the law of cosines in n dimensions. That it actually exists in higher dimensions is proved in the following lemma.

Lemma 52.1 Given any three points $\mathbf{x}$, $\mathbf{y}$, $\mathbf{z} \in \mathbb{R}^n$, the law of cosines is $\|\mathbf{x} - \mathbf{y}\|^2 = \|\mathbf{z} - \mathbf{x}\|^2 + \|\mathbf{y} - \mathbf{z}\|^2 - 2\|\mathbf{z} - \mathbf{x}\|\,\|\mathbf{y} - \mathbf{z}\| \cos\alpha$.

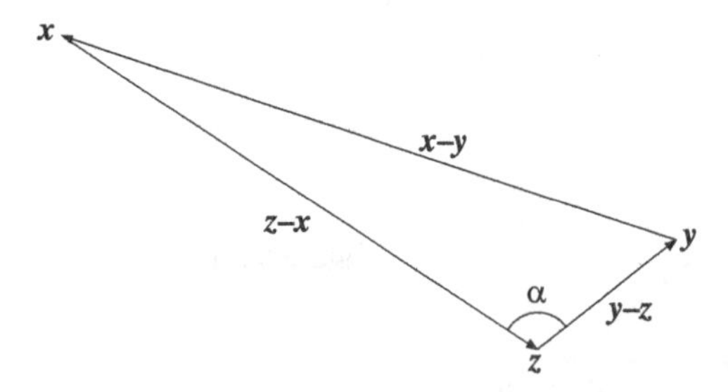

Figure 52.2. Vectors $\mathbf{x}-\mathbf{y}$, $\mathbf{z}-\mathbf{x}$, and $\mathbf{y}-\mathbf{z}$ form a triangle in $\mathbb{R}^n$. Collinearity is possible.

Proof. $\mathbf{x}, \mathbf{y}$, and $\mathbf{z}$ form the vector triangle shown in Figure 52.2. Borrowing a notation from vector analysis, $d_2(\mathbf{x}, \mathbf{y}) \equiv \|\mathbf{x}-\mathbf{y}\|$, and $\cos\alpha$ is unambiguously *defined* as $\cos\alpha \equiv \frac{\mathbf{x}-\mathbf{z}}{\|\mathbf{x}-\mathbf{z}\|} \cdot \frac{\mathbf{y}-\mathbf{z}}{\|\mathbf{y}-\mathbf{z}\|}$, the dot (or inner) product of two unit vectors. Since $\|\mathbf{x}-\mathbf{z}\|\,\|\mathbf{y}-\mathbf{z}\|\cos\alpha = (\mathbf{x}-\mathbf{z})\cdot(\mathbf{y}-\mathbf{z})$, we therefore ask if

$$\|\mathbf{x}-\mathbf{y}\|^2 = \|\mathbf{z}-\mathbf{x}\|^2 + \|\mathbf{y}-\mathbf{z}\|^2 - 2(\mathbf{x}-\mathbf{z})\cdot(\mathbf{y}-\mathbf{z})$$

is correct. Writing out the squared magnitudes and the dot product, we get

$$\sum_{i=1}^{n}(x_i-y_i)^2 = \sum_{i=1}^{n}(z_i-x_i)^2 + \sum_{i=1}^{n}(y_i-z_i)^2 - 2\sum_{i=1}^{n}(x_i-z_i)(y_i-z_i).$$

After carefully expanding the products and summations, we find this to be an identity, and the lemma follows. **QED**

Theorem 52.1 $\langle\mathbb{R}^n, d_2\rangle$ is an Euclidean metric space for $n \in \mathbb{N}$.

First Proof. Since (2) involves the sum of squared quantities, $d_2(\mathbf{x}, \mathbf{y}) \geq 0$, so **(a)** is true. Analogous reasoning to Example 52.4 shows that **(b)** is true. **(c)** is clearly true. To prove the triangle inequality **(d)** in n dimensions, consider three points $\mathbf{x}$, $\mathbf{y}$, and $\mathbf{z}$ in n-space as in Figure 52.2. Using the notation $d_2(\mathbf{x}, \mathbf{y}) = \|\mathbf{x}-\mathbf{y}\|$, we must prove that $\|\mathbf{x}-\mathbf{y}\| \leq \|\mathbf{x}-\mathbf{z}\| + \|\mathbf{z}-\mathbf{y}\|$ (not by coincidence, this looks just like the triangle inequality for real numbers).

The law of cosines applies by Lemma 52.1: $\|\mathbf{x}-\mathbf{y}\|^2 = \|\mathbf{z}-\mathbf{x}\|^2 + \|\mathbf{y}-\mathbf{z}\|^2 - 2\|\mathbf{z}-\mathbf{x}\|\,\|\mathbf{y}-\mathbf{z}\|\cos\alpha$. Therefore,

$$\|\mathbf{x}-\mathbf{y}\|^2 \leq \|\mathbf{z}-\mathbf{x}\|^2 + \|\mathbf{y}-\mathbf{z}\|^2 + 2\|\mathbf{z}-\mathbf{x}\|\,\|\mathbf{y}-\mathbf{z}\| = (\|\mathbf{z}-\mathbf{x}\| + \|\mathbf{y}-\mathbf{z}\|)^2,$$

from which we get $\|\mathbf{x}-\mathbf{y}\| \leq \|\mathbf{z}-\mathbf{x}\| + \|\mathbf{y}-\mathbf{z}\|$. We conclude that $\langle\mathbb{R}^n, d_2\rangle$ is an Euclidean metric space. **QED**

Second Proof. The first three properties of a metric have been verified in the first proof of the theorem. For **(d)**, take the Minkowski inequality (Theorem 51.2) and select $a_i = x_i - z_i$ and $b_i = z_i - y_i$. Immediately, we find that

$$d_2(\mathbf{x}, \mathbf{y}) = \sqrt{\sum_{i=1}^{n}(x_i - y_i)^2} \leq \sqrt{\sum_{i=1}^{n}(x_i - z_i)^2} + \sqrt{\sum_{i=1}^{n}(z_i - y_i)^2} = d_2(\mathbf{x}, \mathbf{z}) + d_2(\mathbf{z}, \mathbf{y}).$$ **QED**

Example 52.5 Let's call $\mathbb{O}$ the space of all open intervals (a, b) in $\mathbb{R}$, and define $D[(a, b), (c, d)] = |c - a| + |d - b|$. Then D acts as a metric for open intervals as follows. Property **(a)**, $D[(a, b), (c, d)] \geq 0$ is clear. For **(b)**, if $D[(a, b), (c, d)] = |c - a| + |d - b| = 0$, then $c = a$ and $d = b$, so $(a, b) = (c, d)$, and conversely, $D[(a, b), (a, b)] = 0$. Property **(c)** is true: $D[(a, b), (c, d)] = D[(c, d), (a, b)]$. To prove **(d)**, verify for yourself that $D[(a, b), (c, d)] \leq D[(a, b), (p, q)] + D[(p, q), (c, d)]$. Put another way, is it true that $|c - a| + |d - b| \leq (|p - a| + |q - b|) + (|c - p| + |d - q|)$? ■

Example 52.6 In each space $\mathbb{R}^n$, define the function

$$d_\infty(\mathbf{x}, \mathbf{y}) = \max\{|x_k - y_k|\}_1^n.$$

We shall prove that $\langle \mathbb{R}^n, d_\infty \rangle$ are metric spaces. To begin, properties **(a)** through **(c)** are fairly straightforward. Property **(d)** appears as $\max\{|x_k - y_k|\}_1^n \leq \max\{|x_k - z_k|\}_1^n + \max\{|z_k - y_k|\}_1^n$ and we must prove it. On the left side, suppose that $\max\{|x_k - y_k|\}_1^n = |x_a - y_a|$ for some pair $x_a \in \{x_k\}_1^n$ and $y_a \in \{y_k\}_1^n$. We know that $|x_a - y_a| \leq |x_a - z_a| + |z_a - y_a|$. But also, $|x_a - z_a| + |z_a - y_a| \leq \max\{|x_k - z_k|\}_1^n + \max\{|z_k - y_k|\}_1^n$. We have proved **(d)**, and it follows that $\langle \mathbb{R}^n, d_\infty \rangle$ is a metric space. **QED**

Taking $n = 2$ in Example 52.6, two geometric considerations demonstrate how strange the metric space $\langle \mathbb{R}^2, d_\infty \rangle$ is. First, we shall graph the unit circle, which is now defined as $\{\langle x, y \rangle \in \mathbb{R}^2 : d_\infty(\langle x, y \rangle, \langle 0, 0 \rangle) = 1\}$. This indicates the relation $\max\{|x|, |y|\} = 1$, seen in Figure 52.3, and incredibly, all points of the graph are at unit distance from the origin!

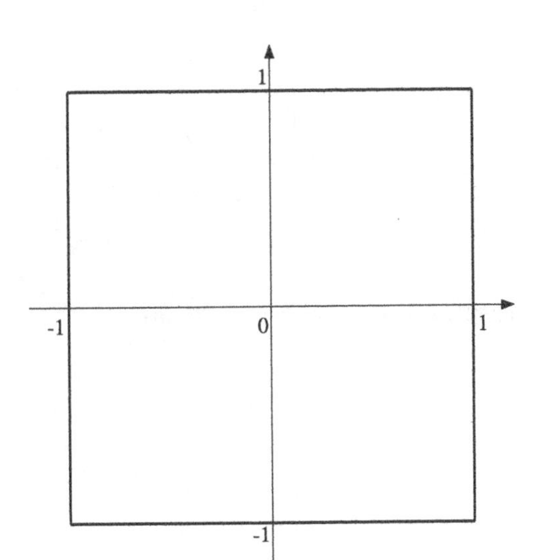

Figure 52.3. The unit circle in $\langle \mathbb{R}^2, d_\infty \rangle$.

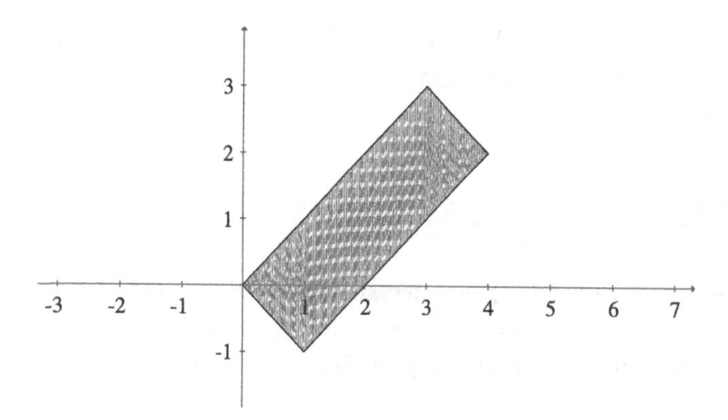

Figure 52.4. A line segment in $\langle \mathbb{R}^2, d_\infty \rangle$.

Secondly, consider collinearity in $\langle \mathbb{R}^2, d_\infty \rangle$. Three points $\mathbf{x}, \mathbf{z}, \mathbf{y}$, in that order, are collinear if and only if $d_\infty(\mathbf{x}, \mathbf{y}) = d_\infty(\mathbf{x}, \mathbf{z}) + d_\infty(\mathbf{z}, \mathbf{y})$. What points are collinear with $\mathbf{x} = \langle 0, 0 \rangle$ and $\mathbf{y} = \langle 4, 2 \rangle$? We are looking for the solution region to $4 = \max\{|z_1|, |z_2|\} + \max\{|z_1 - 4|, |z_2 - 2|\}$. After some work, the collinear points are found to fill the rectangle shown in Figure 52.4. We would therefore be justified in calling this graph the "line segment" between $\langle 0, 0 \rangle$ and $\langle 4, 2 \rangle$. All in all, $\langle \mathbb{R}^2, d_\infty \rangle$ would be a distressing metric space to live in....

Example 52.7 In each $\mathbb{R}^n$, define $d_1(\mathbf{x}, \mathbf{y}) = \sum_{k=1}^{n} |x_k - y_k|$. You will find properties **(a)** through **(d)** easy to prove, and that is the task in Exercise 52.1. With that done, every $\langle \mathbb{R}^n, d_1 \rangle$ becomes a metric space. Interestingly, metric $d_1(x_1, y_1) = |x_1 - y_1|$ for $\langle \mathbb{R}, d_1 \rangle$ is the same as the metric for $\langle \mathbb{R}, d_2 \rangle$ in (2). **QED**

Taking $n = 2$ in Example 52.7, we consider the unit circle in $\langle \mathbb{R}^2, d_1 \rangle$. Since $d_1(\mathbf{x}, \mathbf{y}) = |x_1 - y_1| + |x_2 - y_2|$, the unit circle $\{\langle x, y \rangle \in \mathbb{R}^2 : d_1(\langle x, y \rangle, \langle 0, 0 \rangle) = 1\}$ yields the relation $|x| + |y| = 1$, graphed in Figure 52.5.

The unit circle in $\langle \mathbb{R}^2, d_1 \rangle$ is even stranger than it looks. Its circumference is $C = 8$, and its diameter is $d = 2$ (verify both). Thus, the well-known ratio $\frac{C}{d} = 4$ yields a new, whole

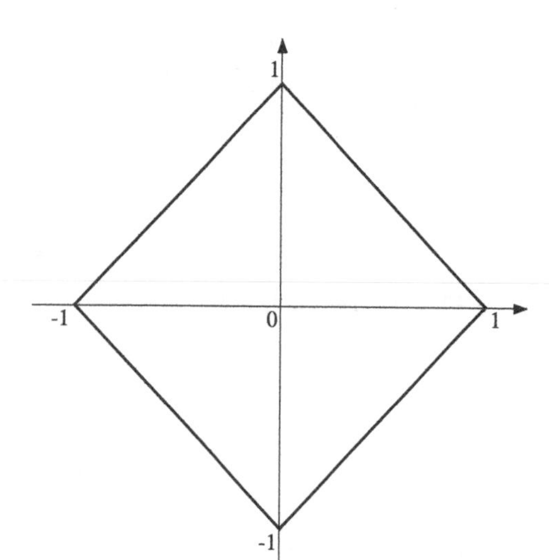

Figure 52.5. The unit circle in $\langle \mathbb{R}^2, d_1 \rangle$.

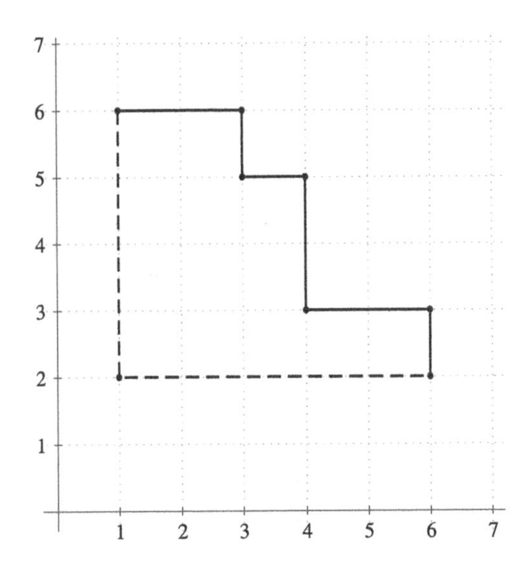

Figure 52.6. Two shortest paths between $\langle 1, 6\rangle$ and $\langle 6, 2\rangle$ in $\langle \mathbb{Z}^2, d_1\rangle$.

number, value for π! Once again, changing metrics produces really uncanny results. See also Exercise 52.2.

Example 52.8 The space $\{\langle m, n\rangle : m, n \in \mathbb{Z}\} \equiv \mathbb{Z} \times \mathbb{Z}$, or $\mathbb{Z}^2$, is the set of lattice points on the plane (a sort of discrete or "digitized" version of the plane). Applying Exercise 52.5 to Example 52.7, we find that $\langle \mathbb{Z}^2, d_1\rangle$ is a metric space. Geometry looks very strange in $\langle \mathbb{Z}^2, d_1\rangle$. Distances can only be measured along zigzag paths from point to point, as in Figure 52.6. The illustrated paths have $d_1(\langle 1, 6\rangle, \langle 6, 2\rangle) = 9$, and we see that the shortest distance between two points is not unique or straight! The circle of radius 2, centered at the origin, consists of exactly eight points, graphed in Figure 52.7 (verify). There is an interesting visualization of $\langle \mathbb{Z}^2, d_1\rangle$ called the "taxicab geometry." If one imagines driving a cab in the streets of a city whose city blocks are all square, then the distance between two street corners $\mathbf{x}$ and $\mathbf{y}$ would be $d_1(\mathbf{x}, \mathbf{y})$. Do Exercise 52.6, and visit taxicab geometry web sites. ■

Example 52.9 In $\mathbb{R}^2$, define the function $d_3(\mathbf{x}, \mathbf{y}) = \sqrt[3]{|x_1 - y_1|^3 + |x_2 - y_2|^3}$. Properties **(a)** through **(c)** of a metric are again easy to show for $d_3(\mathbf{x}, \mathbf{y})$. Property **(d)** now appears as

$$\sqrt[3]{|x_1 - y_1|^3 + |x_2 - y_2|^3} \le \sqrt[3]{|x_1 - z_1|^3 + |x_2 - z_2|^3} + \sqrt[3]{|z_1 - y_1|^3 + |z_2 - y_2|^3}.$$

Gratefully, we won't need to slog our way through a proof of this, because of the following general result.

In any $\mathbb{R}^n$, the function

$$d_r(\mathbf{x}, \mathbf{y}) = \sqrt[r]{|x_1 - y_1|^r + \cdots + |x_n - y_n|^r} = \left(\sum_{k=1}^{n} |x_k - y_k|^r\right)^{1/r}$$

is a metric for any $r \ge 1$. As usual, properties **(a)** through **(c)** are not difficult to prove. A direct proof of **(d)** rests on the Minkowski inequality; see Exercise 52.9. Once this is done, we will know that every such $\langle \mathbb{R}^n, d_r\rangle$is a metric space. **QED**

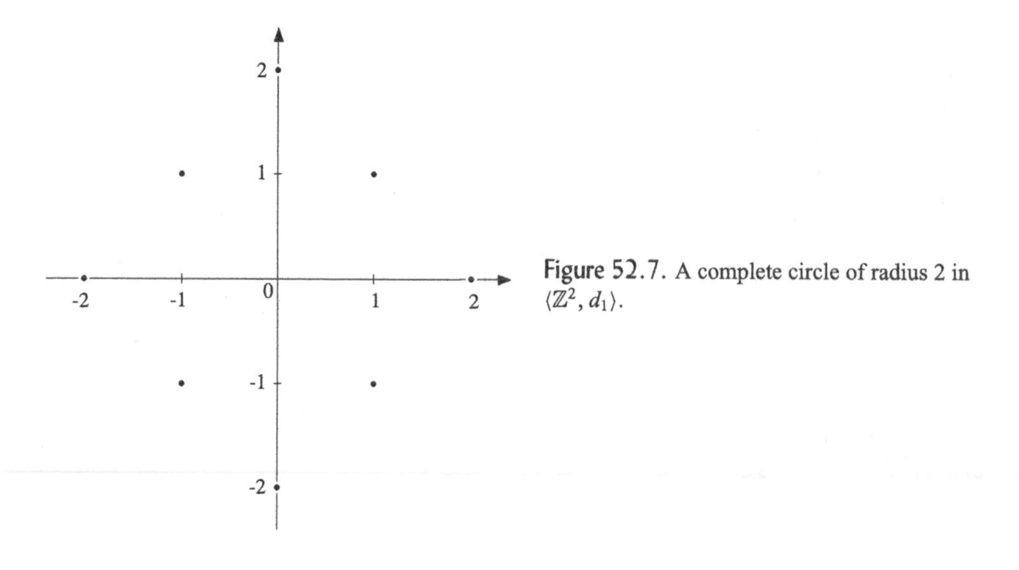

Figure 52.7. A complete circle of radius 2 in $\langle \mathbb{Z}^2, d_1 \rangle$.

Note 1: We found that $\pi = 4$ in the metric space of Figure 52.5. An insightful article about the varying values of π as a function of r in metric spaces $\langle \mathbb{R}^2, d_r \rangle$, $r \geq 1$, is "π is the Minimum Value for Pi," by Charles Adler and James Tanton, in *The College Mathematics Journal*, vol. 31, no. 2 (March 2000), pp. 102–106.

Example 52.10 As a more abstract example of a metric space, define the set ℓ^2 ("little-ell two") as the set containing all sequences $\langle a_k \rangle_1^\infty$ such that $\sum_{k=1}^\infty a_k^2$ converges (often and conveniently called "square summable" sequences). For example, $\langle k^{-1} \rangle_1^\infty \in \ell^2$ since $\sum_{k=1}^\infty a_k^2 = \sum_{k=1}^\infty \frac{1}{k^2}$ converges. Given two square summable sequences $\langle a_k \rangle_1^\infty$ and $\langle b_k \rangle_1^\infty$, define the function

$$\rho\left(\langle a_k \rangle_1^\infty, \langle b_k \rangle_1^\infty\right) = \sqrt{\sum_{k=1}^\infty (a_k - b_k)^2}.$$

Knowing very little about this function, we ask, "Does this quantity even exist?" In other words, is $\langle a_k - b_k \rangle_1^\infty \in \ell^2$ whenever $\langle a_k \rangle_1^\infty, \langle b_k \rangle_1^\infty \in \ell^2$? In

$$\sum_{k=1}^{\infty} a_k^2 - 2\sum_{k=1}^\infty a_k b_k + \sum_{k=1}^{\infty} b_k^2,$$

the first and last series converge (they are in ℓ^2), and by the extended Cauchy-Schwarz inequality (Exercise 51.8), $\sum_{k=1}^\infty a_k b_k$ converges too. Thus, $\sum_{k=1}^\infty (a_k - b_k)^2$ converges, meaning that $\langle a_k - b_k \rangle_1^\infty \in \ell^2$.

It is interesting that ρ is a metric for ℓ^2. To prove it, property **(a)** is true. For property **(b)**, if

$$\rho\left(\langle a_k \rangle_1^\infty, \langle b_k \rangle_1^\infty\right) = \sqrt{\sum_{k=1}^\infty (a_k - b_k)^2} = 0,$$

then all terms $(a_k - b_k)^2 = 0$, whence $a_k = b_k$ for all $k \in \mathbb{N}$, and $\langle a_k \rangle_1^\infty = \langle b_k \rangle_1^\infty$. The converse is also true. Property **(c)** is clearly true. Property **(d)** states that $\rho(\langle a_k \rangle_1^\infty, \langle b_k \rangle_1^\infty) \leq$

$\rho(\langle a_k\rangle_1^\infty, \langle c_k\rangle_1^\infty) + \rho(\langle c_k\rangle_1^\infty, \langle b_k\rangle_1^\infty)$, or

$$\sqrt{\sum_{k=1}^{\infty}(a_k - b_k)^2} \le \sqrt{\sum_{k=1}^{\infty}(a_k - c_k)^2} + \sqrt{\sum_{k=1}^{\infty}(c_k - b_k)^2}.$$

This may seem impossible to prove, but look no farther than the extended Minkowski inequality in Exercise 51.9! Show that **(d)** is satisfied by ρ as Exercise 52.12. Therefore, $\langle \ell^2, \rho\rangle$ is a metric space. **QED**

If we think of $\langle \ell^2, \rho\rangle$ in terms of Example 52.2, we have just established an Euclidean distance formula for the infinite-dimensional space ℓ^2, where the vectors $\langle a_k\rangle_1^\infty$ have infinitely many components ($\ell^2 \subset \mathbb{R}^\infty$). Metric space $\langle \ell^2, \rho\rangle$ is a type of **Hilbert space**, named for David Hilbert.

Note 2: David Hilbert (1862–1943) was one of the last universal mathematicians, meaning that he contributed deeply to all the main branches of mathematics. At the International Congress of Mathematicians (a meeting in Paris in 1900), and shortly thereafter, Hilbert proposed twenty-three problems that would greatly influence all of 20th century mathematics. A few of them are still unresolved. He also advanced the physics of the day, closely supporting Einstein in his research on general relativity. Hilbert was a fitting successor to Gauss at the University of Göttingen, where he maintained its status of worldwide preeminence in mathematics research until around 1930, when he retired. But then, the rising tyranny of Nazism slowly destroyed that marvellous castle of intellect.

52.1 Exercises

1. Carry out the proof for Example 52.7. Hint: Induction!
2. In $\mathbb{R}^2$, determine which points are between and collinear with $\langle 0, 0\rangle$ and $\langle 4, 2\rangle$ when the d_1 metric is used.
3. a. Prove that $\langle \mathbb{R}^+, \rho\rangle$ is a metric space with $\rho(x, y) = |\ln(\frac{x}{y})|$.
   b. Under this metric, what can be said about a set of equivalent rational numbers in $\mathbb{R}^+$ such as $\{2, \frac{6}{3}, \frac{1}{1/2}, \ldots\}$? If S_α is the set of rational numbers equal to α, and likewise for S_β, can the metric be applied to the collection of subsets of $\mathbb{R}^+$, namely $\mathbb{S} = \{S_\alpha : \alpha \in \mathbb{Q}^+\}$, by defining $\rho(S_\alpha, S_\beta) = |\ln(\frac{\alpha}{\beta})|$? If so, we have a new metric space, $\langle \mathbb{S}, \rho\rangle$.
4. Determine whether the function $\rho(\mathbf{x}, \mathbf{y}) = \min\{|x_1 - y_1|, |x_2 - y_2|\}$ defines a metric for $\mathbb{R}^2$.
5. Prove that if $\langle \mathbb{M}, \rho\rangle$ is a metric space and $\mathbb{P} \subset \mathbb{M}$, then $\langle \mathbb{P}, \rho\rangle$ is also a metric space, which we may call a **metric subspace**.
6. As we have seen, metric spaces really challenge our notions of geometry. As another example of this, we commonly state that "the shortest path between two points is a straight line." True, if the metric space is Euclidean! However, in the taxicab geometry $\langle \mathbb{Z}^2, d_1\rangle$, there are many shortest paths between, say, $\langle 0, 0\rangle$ and $\langle 4, 2\rangle$. How many different paths

between those two points have distance equal to 6 (the minimal distance)? Hint: Think of city streets, and use only lattice points in Exercise 52.2.

7. Given two intervals of real numbers S and T of any type, define $d(S, T) = \inf\{|s - t| : s \in S \text{ and } t \in T\} \equiv \inf(|S - T|)$. Is this a metric for the space $\mathbb{I}$ of all intervals in $\mathbb{R}$? Note that we are trying to generalize the idea of the distance between a point and a set, which was defined in Chapter 14. Does it also generalize the function D in Example 52.5?

8. The oddest of all metrics may be the **discrete metric**

$$\rho_{01}(\xi, \zeta) = \begin{cases} 0 & \text{if } \xi = \zeta \\ 1 & \text{if } \xi \neq \zeta \end{cases},$$

where ξ and ζ are elements of any space $\mathbb{M}$ one desires. The function ρ_{01} acts as a binary flag to indicate distinct versus identical elements of $\mathbb{M}$. Prove that $\langle \mathbb{M}, \rho_{01} \rangle$ is a metric space.

9. In Example 52.9, we started to show that every $\langle \mathbb{R}^n, d_r \rangle$ where $r \geq 1$, is a metric space. Theorem 52.1 proved the particular case $n = 2, r = 2$. The proof for all $r \geq 1$ requires the generalized Minkowski inequality,

$$\left(\sum_{k=1}^{n} |a_k + b_k|^r \right)^{1/r} \leq \left(\sum_{k=1}^{n} |a_k|^r \right)^{1/r} + \left(\sum_{k=1}^{n} |b_k|^r \right)^{1/r},$$

seen in Note 2 of the previous chapter. Use this generalized form to prove Property **(d)** for $\langle \mathbb{R}^n, d_r \rangle$. Together with the verified properties **(a)** to **(c)**, this shows that $\langle \mathbb{R}^n, d_r \rangle$ is a metric space as claimed.

10. a. We may reasonably imagine that for instance, for $r = \frac{2}{3}$, the function $d_{2/3}(\mathbf{x}, \mathbf{y}) = (|x_1 - y_1|^{2/3} + |x_2 - y_2|^{2/3})^{3/2}$ defines a metric in $\mathbb{R}^2$. Surprisingly, show that the triangle inequality **(d)** fails for the three points $\langle 0, 0 \rangle$, $\langle 1, 0 \rangle$, and $\langle 1, 1 \rangle$.
    b. Does the problem with property **(d)** persist for all $d_r(\mathbf{x}, \mathbf{y})$ with $0 < r < 1$ in $\mathbb{R}^2$? Does it persist for

$$d_r(\mathbf{x}, \mathbf{y}) = \sqrt[r]{|x_1 - y_1|^r + \cdots + |x_n - y_n|^r} = \left(\sum_{k=1}^{n} |x_k - y_k|^r \right)^{1/r}$$

with $0 < r < 1$ in higher dimensions $\mathbb{R}^n$? Hint: Generalize the three points from **a** to higher dimensions.

11. Given positive real constants a and b, and a space $\mathbb{M}$ with two metrics ρ and τ on it, what does the equation $(a\rho + b\tau)(x, y) = a\rho(x, y) + b\tau(x, y)$ mean? (Of course, $x, y \in \mathbb{M}$.) Prove that $\langle \mathbb{M}, a\rho + b\tau \rangle$ is a new metric space.

12. Prove that **(d)** is satisfied by $\rho(\langle a_k \rangle_1^\infty, \langle b_k \rangle_1^\infty)$ in the metric space $\langle \ell^2, \rho \rangle$ of Example 52.10.

13. In the metric space $\langle \ell^2, \rho \rangle$, how would the length of a vector $\langle a_k \rangle_1^\infty$ be defined? What would be the equation for the unit ball in infinite dimensions?

14. We introduce a metric ℓ^∞ ("little-ell infinity") closely related to the one in Examples 52.6 and 52.10. The metric space $\langle \mathbb{B}, \ell^\infty \rangle$ contains all the bounded sequences $\mathbb{B} = \{\textit{bounded } \langle a_n \rangle_1^\infty : a_n \in \mathbb{R}\}$, with metric $\ell^\infty(\langle a_n \rangle_1^\infty, \langle b_n \rangle_1^\infty) = \sup_n(|a_n - b_n|)$. Evaluate $\ell^\infty(\langle 2^{-n} \rangle_1^\infty, \langle 3^{-n} \rangle_1^\infty)$.
    Show that ℓ^∞ is indeed a metric for $\mathbb{B}$. Next, define $\mathbb{E} = \{\textit{bounded } \langle r_n \rangle_1^\infty : r_n \in \mathbb{Q}\}$. Show that $\langle \mathbb{E}, \ell^\infty \rangle$ is a metric space.

15. a. It is interesting that the unit circles in $\langle \mathbb{R}^2, d_r \rangle$, $r > 2$, have caught the eye of architects through the work of mathematician, designer, inventor, and poet Piet Hein (1905–1996) of Denmark. Use a graphing calculator to draw the unit circle in $\langle \mathbb{R}^2, d_3 \rangle$ and $\langle \mathbb{R}^2, d_6 \rangle$. Hein called these "supercircles," and they seemed to him better suited for human use than ellipses. Hint: The easiest way to graph these curves is parametrically.
    b. Supercircles are a subset of a larger class of **Lamé curves** given by $|\frac{x}{a}|^r + |\frac{y}{b}|^r = 1$, and studied by the French mathematician, thermodynamicist, and engineer Gabriel Lamé (1795–1870). Here, r is any real number, $a > 0$ and $b > 0$. Graph for instance the footballish curve $|\frac{x}{4}|^{1.3} + |\frac{y}{2}|^{1.3} = 1$. Hein called such curves "superellipses." A fine article about Piet Hein and Lamé curves is in Martin Gardner's *Mathematical Carnival*, published by Alfred A. Knopf, pp. 240–254.

16. Set to rest analytically Paradox 0.5, the refuted Pythagorean theorem. Hint: What metrics are involved?

53

Functions and Limits in Metric Spaces

After reconciling ourselves with the ideas in the previous chapter about the endless variety of metrics, we come to the unsettling realization that much of our work with functions and all of it with limits is based on the d_1 metric on $\mathbb{R}$, $d_1(x, y) = |x - y|$. (Actually, on the real line $\mathbb{R}$, all the d_r metrics are equivalent, so d_1 is chosen for simplicity.) A general statement for a function is therefore $f : \langle \mathbb{M}, \rho_1 \rangle \to \langle \mathbb{P}, \rho_2 \rangle$, where $\mathbb{M}$ and $\mathbb{P}$ are arbitrary sets of objects, and the domain is measured with metric ρ_1 while the range (or codomain) uses metric ρ_2.

53.1 Functions and Metric Spaces

This being real analysis, we focus on functions $f : \langle \mathbb{R}, \rho_1 \rangle \to \langle \mathbb{R}, \rho_2 \rangle$. We investigate the common slope formula for a secant between points $\langle x_1, y_1 \rangle$ and $\langle x_2, y_2 \rangle$ on the function; it is of course

$$m_{\text{sec}} = \frac{y_2 - y_1}{x_2 - x_1}.$$

But writing the secant slope as

$$m_{\text{sec}} = \frac{\operatorname{sgn}(y_2 - y_1)}{\operatorname{sgn}(x_2 - x_1)} \frac{|y_2 - y_1|}{|x_2 - x_1|}$$

uncovers the use of the d_1 metric in both the domain and range. We shall now allow the metrics of the domain and range to be arbitrary. Thus, the general definition of secant slope for functions

$$f : \langle \mathbb{R}, \rho_1 \rangle \to \langle \mathbb{R}, \rho_2 \rangle$$

becomes

$$m_{\text{sec}} = \frac{\operatorname{sgn}(y_2 - y_1)}{\operatorname{sgn}(x_2 - x_1)} \frac{\rho_2(y_2, y_1)}{\rho_1(x_2, x_1)}.$$

Suppose that we keep the domain space of f as $\langle \mathbb{R}, d_1 \rangle$, but change the range space to $\langle \mathbb{R}, \rho_{01} \rangle$, where

$$\rho_{01}(u, v) = \begin{cases} 0 & \text{if } u = v \\ 1 & \text{if } u \neq v \end{cases}$$

is the discrete metric first seen in Exercise 52.8. Thus, while $f : \mathbb{R} \to \mathbb{R}$, the way we measure distances in the range is radically different. Consider the linear function $f(x) = 3x + 6$. The slope of the line now becomes

$$m_{\sec} = \frac{\operatorname{sgn}(y_2 - y_1)\,\rho_{01}(y_2, y_1)}{\operatorname{sgn}(x_2 - x_1)\,d_1(x_2, x_1)} = \frac{\operatorname{sgn}((3x_2 + 6) - (3x_1 + 6))}{\operatorname{sgn}(x_2 - x_1)} \frac{\rho_{01}(3x_2 + 6,\ 3x_1 + 6)}{|x_2 - x_1|}$$
$$= \frac{\operatorname{sgn}(x_2 - x_1)}{\operatorname{sgn}(x_2 - x_1)} \frac{1}{|x_2 - x_1|} = \frac{1}{|x_2 - x_1|}.$$

Suddenly, a line does not have a constant slope! Reflecting a bit more, we see that

$$m_{\sec} = \begin{cases} \dfrac{1}{|x_2 - x_1|} & \text{if } y_1 \neq y_2 \\ 0 & \text{if } y_1 = y_2 \end{cases}$$

regardless of the function being inspected. Furthermore, the geometry has been so disrupted that the derivative of any function, except the constant functions, does not exist, since

$$\lim_{x_2 \to x_1} \frac{1}{|x_2 - x_1|}$$

doesn't exist.

The choice of the metric in the investigation above may have been motivated by its shock value, but are there metric spaces that are used in a natural way? Yes, for example, when we study functions $f : \mathbb{R} \to \mathbb{R}^2$, commonly called vector-valued functions. For instance, an ellipse

$$\frac{x^2}{a^2} + \frac{y^2}{b^2} = 1$$

can be graphed as the set of vectors $\langle a\cos\theta, b\sin\theta\rangle$, often expressed by the pair of parametric equations $x(\theta) = a\cos\theta$ and $y(\theta) = b\sin\theta$. Thus, $\mathcal{D}_f = \langle \mathbb{R}, d_1\rangle$, but the natural range space is $\mathcal{R}_f = \langle \mathbb{R}^2, d_2\rangle$. In more detail, the distance between two points (vectors) $\mathbf{x} = \langle x(\theta_1), y(\theta_1)\rangle$ and $\mathbf{y} = \langle x(\theta_2), y(\theta_2)\rangle$ on the ellipse is

$$d_2(\mathbf{x}, \mathbf{y}) = \sqrt{[x(\theta_1) - x(\theta_2)]^2 + [y(\theta_1) - y(\theta_2)]^2}.$$

Similar thinking enters into the study of vector-valued functions $f : \mathbb{R} \to \mathbb{R}^3$, which have graphs that look like curves in space.

In multivariable calculus, graphs called surfaces are studied. A typical surface in space is given by the function $z = 3x^2 - y^3 + 1$. This is a function whose domain is $\mathbb{R}^2$, the set of vectors $\langle x, y\rangle$, and whose range is real numbers z. The natural metric for the domain is d_2, and that of the range is d_1, so that we describe this function as $f : \langle \mathbb{R}^2, d_2\rangle \to \langle \mathbb{R}, d_1\rangle$.

53.2 Limits of Sequences in Metric Spaces

In Chapter 17, Definition 2 states that a sequence $\langle a_n\rangle_1^\infty$ converges to a number L if and only if for any $\varepsilon > 0$, there exists a corresponding number N_ε such that for all $n > N_\varepsilon$, we have $|a_n - L| < \varepsilon$. Hence, we are measuring the distance between term a_n and the limit L in metric space $\langle \mathbb{R}, d_1\rangle$. We can therefore expect drastic—but logical—changes in the study of convergence by changing the metric.

For instance, what happens to the convergence of sequences if we switch to the discrete metric ρ_{01}? We would say that $\langle a_n \rangle_1^\infty$ converges to L if and only if

$$\rho_{01}(a_n, L) = \begin{cases} 0 & \text{if } a_n = L \\ 1 & \text{if } a_n \neq L \end{cases} < \varepsilon,$$

for all $n > N_\varepsilon$. This brings us to the strange conclusion that the only sequences that converge to L are those that eventually become identical to L, that is, those of the form $\langle a_1, a_2, \cdots, a_n, L, L, L, \cdots \rangle$ (fill in the reasoning behind this as Exercise 53.1). Thus, for example, $\left\langle \frac{1}{n} \right\rangle_1^\infty$ is a *divergent* sequence in $\langle \mathbb{R}, \rho_{01} \rangle$.

As always, we need a clear definition of the convergence of a sequence under the new circumstances:

> A sequence $\langle \mu_n \rangle_1^\infty$ of objects $\mu_n \in \mathbb{M}$ converges to a limit $L \in \mathbb{M}$ in a metric space $\langle \mathbb{M}, \rho \rangle$ if and only if for any $\varepsilon > 0$, there exists a corresponding positive N_ε such that for all $n \geq N_\varepsilon$, we have $\rho(\mu_n, L) < \varepsilon$.

This is a generalization of the definition of a limit of a sequence in Chapter 17. We summarize the definition by stating that $\lim_{n \to \infty} \rho(\mu_n, L) = 0$. It is important to note that convergence requires that $L \in \mathbb{M}$ by this definition.

The analogue of the definition of a Cauchy sequence in Chapter 21 is:

> A sequence $\langle \mu_n \rangle_1^\infty$ is a Cauchy sequence in a metric space $\langle \mathbb{M}, \rho \rangle$ if and only if for any $\varepsilon > 0$, there is a corresponding positive N_ε such that for any $m, n \geq N_\varepsilon$, we have $\rho(\mu_n, \mu_m) < \varepsilon$.

We summarize this by writing $\lim_{m,n \to \infty} \rho(\mu_n, \mu_m) = 0$.

Theorem 21.1, the Cauchy convergence criterion, would transfer to this chapter in this way: (1) a sequence $\langle \mu_n \rangle_1^\infty$ is convergent in metric space $\langle \mathbb{M}, \rho \rangle$ if and only if it is a Cauchy sequence there. However, a fundamental difference arises in that now the statement of equivalence is lost. Only necessity remains: if $\langle \mu_n \rangle_1^\infty$ is convergent in the metric space $\langle \mathbb{M}, \rho \rangle$, then it is a Cauchy sequence. The problem with sufficiency is that even if a sequence is a Cauchy sequence, its limit may exist *outside* of the metric space the sequence belongs to. Consider now some examples where the limit is and is not an element of the given space.

Example 53.1 Example 52.5 introduced the metric space $\langle \mathbb{O}, D \rangle$ of open intervals in $\mathbb{R}$, where

$$D[(a, b), (c, d)] = |a - c| + |b - d|.$$

The sequence

$$s_n = \bigcup_{k=1}^{n} \left(1 + \frac{1}{k}, 4 - \frac{1}{k}\right) = \left(1 + \frac{1}{n}, 4 - \frac{1}{n}\right) \in \mathbb{O}$$

has the limit $L = (1, 4)$ because of what union is, and also because

$$\lim_{n \to \infty} D\left[\left(1 + \frac{1}{n}, 4 - \frac{1}{n}\right), (1, 4)\right] = \lim_{n \to \infty} \left(\left|1 + \frac{1}{n} - 1\right| + \left|4 - \frac{1}{n} - 4\right|\right) = 0.$$

In this case, the limit $(1, 4) \in \mathbb{O}$; therefore, $\langle s_n \rangle_1^\infty$ converges to a limit in $\langle \mathbb{O}, D \rangle$. ■

Example 53.2 Contrast Example 53.1 with Exercise 21.8, in which the related sequence of open intervals

$$s_n = \bigcap_{k=1}^{n} \left(1 - \frac{1}{k}, 4 + \frac{1}{k}\right) = \left(1 - \frac{1}{n}, 4 + \frac{1}{n}\right) \in \mathbb{O}$$

was found to have limit $L = [1, 4]$, which does not belong to $\langle \mathbb{O}, D \rangle$. The conclusion there (and here) is that the sequence does *not* converge in $\langle \mathbb{O}, D \rangle$. In part **c** of that exercise, we proved moreover that

$$\left\langle \left(1 - \frac{1}{n}, 4 + \frac{1}{n}\right) \right\rangle_1^\infty$$

was a Cauchy sequence within its space $\mathbb{O}$. Hence, $\langle s_n \rangle_1^\infty$ does not follow the "if" (sufficiency) part of the Cauchy convergence criterion. ■

The next example is important because it discusses sequences in the most common metric space, $\langle \mathbb{R}, d_1 \rangle$.

Example 53.3 In Corollary 34.1, $e = 2.71828\ldots$ was found to be irrational, and the sequence

$$\langle e_n \rangle_1^\infty = \left\langle \left(1 + \frac{1}{n}\right)^n \right\rangle_1^\infty = \left\langle 2, \frac{9}{4}, \frac{64}{27}, \cdots \right\rangle,$$

which converges to e by Theorem 34.1, is a sequence of rational numbers. It is easy to verify that $\langle \mathbb{Q}, d_1 \rangle$ is a metric space. Hence, we have a sequence $\langle e_n \rangle_1^\infty \subset \mathbb{Q}$ which converges in $\langle \mathbb{R}, d_1 \rangle$, but diverges in $\langle \mathbb{Q}, d_1 \rangle$ because its limit is outside of $\mathbb{Q}$.

The same thing can be said about $\langle a_n \rangle_1^\infty = \langle 3.1, 3.14, 3.141, \ldots \rangle$, where a_n is the decimal expansion of π truncated to n places. Although we don't have an explicit formula for a_n, $\langle a_n \rangle_1^\infty \subset \mathbb{Q}$, $\langle a_n \rangle_1^\infty$ is a Cauchy sequence, and yet $\langle a_n \rangle_1^\infty \to \pi \notin \mathbb{Q}$. ■

In the example, we were able to demonstrate that $\langle e_n \rangle_1^\infty$ converges because the metric d_1 is able to measure distances between any two real numbers (for instance, $d_1(e_3, e) = e - e_3$), not just between rational numbers. The question arises whether a sequence like $\langle e_n \rangle_1^\infty$ could be detected to have a limit (or a cluster point) from within the metric space $\langle \mathbb{Q}, d_1 \rangle$ itself, that is, without reference to the limit e outside of $\mathbb{Q}$. One reason that Cauchy sequences are important is that they give us an affirmative answer to this question because in the definition of a Cauchy sequence, only members $\mu_n, \mu_m \in \mathbb{M}$ are used. Thus, in Example 53.3, we need only show that $\lim_{m,n\to\infty} |e_m - e_n| = 0$ to prove that $\langle e_n \rangle_1^\infty$ has a limit.

There is a unifying principle underlying the convergence and divergence results in the examples above. It is the completeness of a metric space. A metric space $\langle \mathbb{M}, \rho \rangle$ is **complete** if and only if every Cauchy sequence $\langle \mu_n \rangle_1^\infty \subset \mathbb{M}$ has a limit $L \in \mathbb{M}$. Hence, for instance, we have just found that $\langle \mathbb{Q}, d_1 \rangle$ is not a complete metric space. Another way to see this is that the Cauchy convergence criterion *(1)* holds only for complete metric spaces.

That our familiar metric space $\langle \mathbb{R}, d_1 \rangle$ *is* complete ultimately depends upon Axiom R11 in Chapter 12 (aptly called the completeness axiom). Recall that every Cauchy sequence $\langle a_n \rangle_1^\infty \subset \mathbb{R}$

is a convergent sequence in $\mathbb{R}$ by Theorem 21.1. Such a sequence converges to a limit $L \in \mathbb{R}$, but to arrive at the conclusion, Theorem 21.1 uses Axiom R11, as Note 2 explains there.

53.3 Continuity in Metric Spaces

Corresponding to the definition of continuity for functions $f : \mathbb{R} \to \mathbb{R}$ in Chapter 24, we have the following one, whose only novelty is to incorporate the metrics of the domain and range of f.

> A function $f : \langle \mathbb{M}, \rho_1 \rangle \to \langle \mathbb{P}, \rho_2 \rangle$ is continuous at $a \in \mathcal{D}_f \subseteq \mathbb{M}$ if and only if given any $\varepsilon > 0$, we can find a corresponding $\delta > 0$ such that if $x \in \mathcal{D}_f$ and $\rho_1(x, a) < \delta$, then $\rho_2(f(x), f(a)) < \varepsilon$.

We may summarize this by writing $\lim_{\rho_1(x,a)\to 0} f(x) = f(a)$. If f is continuous at every element of some subset $A \subset \mathbb{M}$, then we say that f is continuous over A. (If a isn't a cluster point of the domain, then $\rho_1(x, a) < \delta$ is possible only for $x = a$, and continuity is automatic at a.)

Example 53.4 Show that $f : \mathbb{R}^2 \to \mathbb{R}$ given by $f(\mathbf{x}) = x_1 + x_2$ is a continuous function over all $\mathbb{R}^2$. In concise notation, we need to show that $\lim_{\mathbf{x}\to\mathbf{a}} f(\mathbf{x}) = f(\mathbf{a})$. We use the natural metrics of the domain and range spaces, and prove that if $d_2(\mathbf{x}, \mathbf{a}) < \delta$, then $d_1(f(\mathbf{x}), f(\mathbf{a})) < \varepsilon$, for δ depending on ε. That is, find the δ corresponding to ε such that if

$$\sqrt{(x_1 - a_1)^2 + (x_2 - a_2)^2} < \delta,$$

then $|(x_1 + x_2) - (a_1 + a_2)| < \varepsilon$.

Begin by noting that in the domain,

$$\sqrt{(x_1 - a_1)^2 + (x_2 - a_2)^2} \leq \sqrt{2} \max \{|x_1 - a_1|, |x_2 - a_2|\}.$$

Now, $\sqrt{2} \max \{|x_1 - a_1|, |x_2 - a_2|\} < \delta$ if and only if $\sqrt{2}\,|x_1 - a_1| < \delta$ and $\sqrt{2}\,|x_2 - a_2| < \delta$. Also, in the range, we have

$$|(x_1 + x_2) - (a_1 + a_2)| \leq |x_1 - a_1| + |x_2 - a_2|,$$

so it is sufficient to show that

$$|x_1 - a_1| + |x_2 - a_2| < \varepsilon.$$

Thus, we set $\delta = \frac{\varepsilon}{\sqrt{2}}$, and we can state that if, in the domain,

$$\sqrt{2} \max \{|x_1 - a_1|, |x_2 - a_2|\} < \delta = \frac{\varepsilon}{\sqrt{2}},$$

then

$$|x_1 - a_1| < \frac{\varepsilon}{2} \text{and} \, |x_2 - a_2| < \frac{\varepsilon}{2},$$

implying that

$$|(x_1 + x_2) - (a_1 + a_2)| \le |x_1 - a_1| + |x_2 - a_2| < \varepsilon$$

in the range. Consequently, $\lim_{\mathbf{x}\to\mathbf{a}} f(\mathbf{x}) = f(\mathbf{a})$, and $f(\mathbf{x})$ is continuous over all $\mathbb{R}^2$. ■

We can go on to show for instance that if functions $f : \langle \mathbb{M}, \rho\rangle \to \langle \mathbb{R}, d_1\rangle$ and $g : \langle \mathbb{M}, \rho\rangle \to \langle \mathbb{R}, d_1\rangle$ are continuous at $a \in \mathbb{M}$, then $f \pm g$, fg, and $\frac{f}{g}$ will be continuous there (provided that $g(a) \neq 0$ in the quotient function). This extension of Theorem 26.1 won't be proved here. You can read about this and other extensions in [13], pp. 132–0134. Example 53.4, and Exercises 53.6 to 53.8, give a flavor of the analysis of continuity in metric spaces.

53.4 Exercises

1. Use the ε-δ definition of a limit to explain why the only convergent sequences in the metric space $\langle \mathbb{R}, \rho_{01}\rangle$ have the form $\langle a_1, a_2, \ldots, a_n, L, L, L, \ldots\rangle$, as described above.

2. A useful function $A : \mathbb{R}^3 \to \mathbb{R}$ is the arithmetic mean of three numbers:

$$m = A(\mathbf{x}) = A(\langle x_1, x_2, x_3\rangle) = \frac{x_1 + x_2 + x_3}{3}.$$

   For any two vectors $\mathbf{x}, \mathbf{y} \in \mathbb{R}^3$, how is the distance between $\mathbf{x}$ and $\mathbf{y}$ typically measured? How is the distance between $A(\mathbf{x})$ and $A(\mathbf{y})$ measured? Fill in the blanks: $A : (\mathbb{R}^3, \quad) \to (\mathbb{R}, \quad)$.

3. Functions over the field of complex numbers $\mathbb{C}$ are functions $f : \mathbb{C} \to \mathbb{C}$, where the complex number $z = x + yi = \langle x, y\rangle$ is visualized as a vector in $\mathbb{R}^2$. Thus, in terms of vectors, $f : \mathbb{R}^2 \to \mathbb{R}^2$, given by $f(\langle x, y\rangle) = \langle \gamma, \delta\rangle = \langle f_1(\langle x, y\rangle), f_2(\langle x, y\rangle)\rangle$, but this presents a great problem for graphing. What is it? To partially graph such a function, it is sometimes restricted to $f : \mathbb{R}^1 \to \mathbb{R}^2$. How is such a partial graph related to the full graph of f?

   The metric used to determine the distance $\|z_2 - z_1\|$ between complex numbers is d_2, as follows:

$$d_2(z_1, z_2) = \|z_1 - z_2\| = \sqrt{(x_1 - x_2)^2 + (y_1 - y_2)^2}$$

   (how convenient!). Can this metric be simplified for the partial graph suggested?

4. Prove the theorem that in any metric space $\langle \mathbb{M}, \rho\rangle$, if a sequence $\langle \mu_n\rangle_1^\infty$ of objects $\mu_n \in \mathbb{M}$ converges to a limit $L \in \mathbb{M}$, then L is unique. This is the analogue to Theorem 17.1.

   In Example 53.1, we saw that

$$\left\langle \left(1 + \frac{1}{n}, 4 - \frac{1}{n}\right)\right\rangle_1^\infty$$

   converges to $(1, 4)$. But as sequences, $1 + \frac{1}{n} \to 1$ and $4 - \frac{1}{n} \to 4$, so we can say that the limit is $[1, 4]$. How does this not contradict the uniqueness theorem just proved?

5. Prove that the arithmetic mean function in Exercise 53.2 is continuous over all $\mathbb{R}^3$. Hint: Similar to Example 53.4, note that this time, in the domain,

$$\sqrt{\sum_{k=1}^{3}(x_k - y_k)^2} \leq \sqrt{3}\max\{|x_k - y_k|\}_1^3.$$

6. Determine which functions $g : \langle\mathbb{R}, d_1\rangle \rightarrow \langle\mathbb{R}, \rho_{01}\rangle$ are continuous. Going in reverse, which functions $h : \langle\mathbb{R}, \rho_{01}\rangle \rightarrow \langle\mathbb{R}, d_1\rangle$ are continuous? Hint: Consider the contrapositive of "if $\rho_{01}(x, a) < \delta$ then $|h(x) - h(a)| < \varepsilon$".

7. Consider the parametric function

$$u : \left[0, \frac{\pi}{2}\right] \rightarrow \mathbb{R}^2$$

given by

$$u(\theta) = \left\langle\sqrt[3]{\cos\theta}, \sqrt[3]{\sin\theta}\right\rangle,$$

which draws the part of the supercircle $x^6 + y^6 = 1$ in Quadrant I, as in Exercise 52.15. Show that u is continuous over its domain, using the natural metrics d_1 in the domain and d_2 in the range. Hint: Similar to Example 53.4, but in the range this time,

$$\sqrt{(x_1 - x_2)^2 + (y_1 - y_2)^2} \leq \sqrt{2}\max\{|x_1 - x_2|, |y_1 - y_2|\},$$

where $\langle x_1, y_1\rangle = \langle\sqrt[3]{\cos\theta_1}, \sqrt[3]{\sin\theta_1}\rangle$, and likewise for $\langle x_2, y_2\rangle$. Verify that $|x_1 - x_2| < 2|\theta_1 - \theta_2|$ and $|y_1 - y_2| < 2|\theta_1 - \theta_2|$ in Quadrant I.

8. In Example 52.6, the metric $d_\infty(\mathbf{x}, \mathbf{y}) = \max\{|x_k - y_k|\}_{k=1}^n$ was introduced for $\mathbb{R}^n$. The infinity subscript is appropriate for the following reason. We know from Example 52.9 that in $\mathbb{R}^n$, the function

$$d_r(\mathbf{x}, \mathbf{y}) = \sqrt[r]{|x_1 - y_1|^r + \cdots + |x_n - y_n|^r}$$

is a metric for any $r \geq 1$. As $r \rightarrow \infty$, it is interesting to prove that $d_r(\mathbf{x}, \mathbf{y}) \rightarrow d_\infty(\mathbf{x}, \mathbf{y})$. To simplify things, let $d_r = \left(a_1^r + a_2^r + \cdots + a_n^r\right)^{1/r}$, where the a_i are nonnegative constants. Let $a_M = \max\{a_i\}_1^n$.

   Prove that $\lim_{r\rightarrow\infty} d_r = a_M$. Hint: Show that

$$a_M \leq \left(a_1^r + a_2^r + \cdots + a_n^r\right)^{1/r} \leq n^{1/r}a_M.$$

9. Consider the space $\langle\mathbb{D}, d_1\rangle$, where $\mathbb{D} = \{\langle x, y\rangle : x^2 + y^2 < 1\}$. How do we know it is a metric space? Does every convergent sequence of points in $\mathbb{D}$ have a limit belonging to $\mathbb{D}$? That is, is $\langle\mathbb{D}, d_1\rangle$ a complete metric space? If it isn't, how can $\mathbb{D}$ be modified to make the space complete? Hint: Graph the set $\mathbb{D}$.

10. a. As we know, the set of irrational numbers $\mathbb{I}$ is uncountable. How do we know that $\langle \mathbb{I}, d_1 \rangle$ is a metric space? Is it a complete metric space?
    b. We met the quadratic extension $\mathbb{Q}(\sqrt{2})$ in Chapter 8, where $p + q\sqrt{2} \in \mathbb{Q}(\sqrt{2})$ if and only if p and q are rational numbers. How do we know that $\langle \mathbb{Q}(\sqrt{2}), d_1 \rangle$ is a metric space? Is it complete?
    c. Are there any, perhaps simple, complete metric subspaces of $\langle \mathbb{R}, d_1 \rangle$ at all?

54

Some Topology of the Real Number Line

In ordinary parlance, when applied to doors, windows, and minds, the words "open" and "closed" are antonyms. However, when applied to subsets of R^p, these words are not antonyms.

Robert G. Bartle [9], pg. 72.

Now, near the end of this book, we return to the real number line. We have come across many concepts relating to sets of points, or real numbers, while studying functions, sequences, and series. For instance, Taylor series converge absolutely inside of a closed interval $[a, b]$, and the limit of a function can be defined at a cluster point of the function's domain. In this chapter we will look at some basic theory of open and closed sets, from a branch of analysis often called **point set topology**. The theory is readily extended to $\mathbb{R}^n$, or to any universal set that we like in the most general setting, but soon we will return to the set of points of the real number line, $\mathbb{R}$.

Given a universal set $\mathbb{M}$, a **topology** is any family of sets $\mathcal{T}$ in $\mathbb{M}$, distinguished by being called the open sets of the topology, that

(a) contains the empty set and $\mathbb{M}$ itself,
(b) contains the intersection of any *finite* number of sets from $\mathcal{T}$, and
(c) contains the union of *any* collection of sets from $\mathcal{T}$.

Symbolically, **(a)** affirms that $\varnothing, \mathbb{M} \in \mathcal{T}$, and **(b)** affirms that if $\{A_1, A_2, \ldots, A_n\} \subset \mathcal{T}$, then $\bigcap_1^n A_i \in \mathcal{T}$. Returning to the notation in Chapter 5, **(c)** affirms that given any subscripting set S, not restricted to finite or even countable size, if $A_\alpha \in \mathcal{T}$ for every $\alpha \in S$, then $\bigcup_{\alpha \in S} A_\alpha \in \mathcal{T}$.

The pair $\langle \mathbb{M}, \mathcal{T} \rangle$ is called a **topological space**, keeping with the theme from Chapter 52 about naming spaces. It's interesting that a topological space is a more primitive construction than a metric space, in the sense that every metric space is a topological space, but not conversely. This is true because once we establish the open sets in a metric space, they will satisfy the three properties above. Our first theorem will demonstrate this for $\mathbb{R}$. (There are topological spaces that are not metric spaces, that is, that have no metric for defining the open sets of the topology. See [29], pp. 166–167.)

Note 1: From here on, any mathematical term written in italics has been defined previously, in Chapters 5, 13, 14, or 23.

It is possible to define neighborhoods in a general topological space, but we will return to the set of reals, with its natural metric $d_1 = |x - y|$. Recall from Chapter 14 that a *neighborhood* of a real number x is the open interval $\{t : |t - x| < \varepsilon,\ \varepsilon \in \mathbb{R}^+\}$ (d_1 was used!), which is written more simply as $(x - \varepsilon, x + \varepsilon)$, and very concisely as $\mathbf{N}_x^\varepsilon$. An **open set** in $\mathbb{R}$ shall be a set A such that for every $x \in A$, there is a neighborhood $\mathbf{N}_x^\varepsilon \subset A$. For instance, every open interval $(a, b) = \{x : a < x < b\}$ is an open set, as you must prove in Exercise 54.2. However, think simply and you will find it easy to create an open set that isn't an open interval.

Theorem 54.1 The collection $\mathcal{T}$ of all open sets in $\langle \mathbb{R}, d_1 \rangle$ is a topology for $\mathbb{R}$.

Proof. Regarding **(a)**, both $\varnothing$ and $\mathbb{R}$ belong to $\mathcal{T}$ (see Exercise 54.1). As for **(b)**, let A and B be open in $\mathbb{R}$. If $A \cap B = \varnothing$, then nothing more needs proving. Suppose that $x \in A \cap B$. Since $x \in A$ and $x \in B$, there exists a $\mathbf{N}_x^\varepsilon \subset A$ and a possibly different $\mathbf{N}_x^\gamma \subset B$. Looking more closely at their structures, $\mathbf{N}_x^\varepsilon = (x - \varepsilon, x + \varepsilon)$ and $\mathbf{N}_x^\varepsilon = (x - \gamma, x + \gamma)$. We choose $\beta = \min\{\varepsilon, \gamma\}$, and conclude that the possibly shorter interval $(x - \beta, x + \beta) \subset A \cap B$. Hence, x has a neighborhood $\mathbf{N}_x^\beta$ that is contained in $A \cap B$, for any choice of $x \in A \cap B$. This means that $A \cap B \in \mathcal{T}$. The intersection operation may be done repeatedly with n open sets $\{A_i\}_1^n$, and the final result is that $\bigcap_1^n A_i \in \mathcal{T}$, as **(b)** requires.

Property **(c)** is always subtle. Let $\{A_\alpha\}$ be any collection of open sets in $\mathbb{R}$, where $\alpha \in S$ as above. We may assume that $\bigcup_{\alpha \in S} A_\alpha$ is not $\varnothing$. Suppose that $x \in \bigcup_{\alpha \in S} A_\alpha$. Then for some open set A_o in $\{A_\alpha\}$, $x \in A_o$. (Interesting notation: $x \in A_o \in \{A_\alpha\} \subseteq \mathcal{T}$.) This implies that $\mathbf{N}_x^\varepsilon$ exists such that $\mathbf{N}_x^\varepsilon \subset A_o$. Then $\mathbf{N}_x^\varepsilon \subset \bigcup_{\alpha \in S} A_\alpha$. We come to the same conclusion for any x in $\bigcup_{\alpha \in S} A_\alpha$. This tells us that $\bigcup_{\alpha \in S} A_\alpha$ is an open set, and **(c)** is fulfilled. Hence, the collection of all open sets in $\langle \mathbb{R}, d_1 \rangle$ is a topology for $\mathbb{R}$. **QED**

It might seem that the intersection of an infinite collection of nonempty open sets in $\mathbb{R}$ should be open, but this is false. Recall Exercise 21.8, where a sequence of open intervals $s_n = \bigcap_1^n (1 - \frac{1}{k}, 4 + \frac{1}{k})$ yielded $\lim_{n \to \infty} s_n = [1, 4]$. We recognize $[1, 4]$ as closed, but this term has not been topologically defined yet. More logically, $[1, 4]$ is not open, since for any $\varepsilon > 0$, $\mathbf{N}_1^\varepsilon \not\subset [1, 4]$ and $\mathbf{N}_4^\varepsilon \not\subset [1, 4]$.

One way to topologically define a closed set in $\mathbb{R}$ is that a set C is **closed** if and only if its complement $\mathbb{R} \backslash C$ is open. It follows that $[1, 4]$ is closed, since $\overline{[1, 4]} = (-\infty, 1) \cup (4, \infty)$ is definitely open. The set $\mathbb{Z}$ is closed since $\overline{\mathbb{Z}}$ is the union of countably many open sets (notice that $\mathbb{Z}$ is composed of isolated points, so that it contains no neighborhoods at all). Interestingly, **(a)** makes $\varnothing$ closed as well as open, and the same for $\mathbb{R}$ (the odd name "clopen" has been suggested for these sets). No other sets in the topology have this dual property, as you should show in Exercise 54.11. In contrast, the set $[1, 2)$ is neither open nor closed (might we say it's "disclopen"?).

The set of rational numbers $\mathbb{Q}$ is not open, since if $r \in \mathbb{Q}$, every neighborhood of r contains numbers not in $\mathbb{Q}$, namely, irrational numbers. This so is because the irrational numbers are *dense* in $\mathbb{Q}$ (see Exercise 13.7). Is $\mathbb{Q}$ closed, or equivalently, is $\mathbb{R} \backslash \mathbb{Q}$ open? Since $\mathbb{R} \backslash \mathbb{Q} = \mathbb{I}$, the set of irrational numbers, and $\mathbb{I}$ is not open (verify), we conclude that $\mathbb{Q}$ is one of those sets that is neither open nor closed.

We have needed to talk about the interior of an interval in previous chapters. A point x is an **interior point** of a set A in $\mathbb{R}$ if and only if some neighborhood of x is a subset of A: $\mathbf{N}_x^\varepsilon \subset A$. As a special case, every point in an open interval is an interior point, since if $x \in (a, b)$, then there always exists an $\mathbf{N}_x^\varepsilon$ such that $\mathbf{N}_x^\varepsilon \subset (a, b)$, in fact, infinitely many such neighborhoods exist (did you do Exercise 54.2?). The **interior** of a set A is the set $i(A)$ of all points that are interior to A. It must follow that $i(A)$ is open. Thus, for any open set A, $i(A) = A$, and conversely. Convince yourself that for an arbitrary set A, $i(A) \subseteq A$, a sensible fact!

Example 54.1 If $A = [a, b]$, then $i(A) = (a, b)$. If $A = \{a_1, a_2, \ldots, a_n\}$, then $i(A) = \varnothing$; and if $A = \{1, \frac{1}{2}, \frac{1}{3}, \ldots, \frac{1}{n}, \ldots\}$, then $i(A) = \varnothing$ again. Interestingly, $\mathbb{Q}$ has no interior (that is, $i(\mathbb{Q}) = \varnothing$), and the reason is once again that irrational numbers always intrude. ■

A point x is a *boundary point* (or frontier point) of a set A in $\mathbb{R}$ if and only if every neighborhood of x contains one point in A and one point not in A. The **boundary** (or **frontier**) of a set A is the set $b(A)$ of all boundary points of A. Thus, $x \in b(A)$ if and only if $\mathbf{N}_x^\varepsilon \cap A \neq \varnothing$ and $\mathbf{N}_x^\varepsilon \cap \overline{A} \neq \varnothing$ for any neighborhood of x.

Example 54.2 If $A = (a, b)$, then $b(A) = \{a, b\}$; if $A = [a, b]$, then $b(A) = \{a, b\}$ again. If $A = \{a_1, a_2, \ldots, a_n\}$, then $b(A) = A$, because at each point a_k in A, $a_k \in \mathbf{N}_{a_k}^\varepsilon \cap A$ for all $\varepsilon > 0$, and $\mathbf{N}_{a_k}^\varepsilon \cap \overline{A}$ contains infinitely many points. Thus, a finite set of points is identical to its boundary. If $A = \{1, \frac{1}{2}, \frac{1}{3}, \ldots, \frac{1}{n}, \ldots\}$, then $b(A) = A \cup \{0\}$. Two instances here demonstrate that a boundary point need not belong to its associated set. ■

A point x is a *cluster point* of a set A in $\mathbb{R}$ if and only if every neighborhood of x contains a point in A that is distinct from x. This is put symbolically as $\mathbf{D}_x^\varepsilon \cap A \neq \varnothing$ for every $\varepsilon > 0$, where $\mathbf{D}_x^\varepsilon$ represents a *deleted neighborhood* of x, that is, $\mathbf{D}_x^\varepsilon = \mathbf{N}_x^\varepsilon \backslash \{x\}$ (Chapter 23 has other formulations). An equivalent, intuitive, statement is that a cluster point of A has infinitely many points in A clustered nearer and nearer to it (see Chapter 14). The analytic definition of this appears in Exercise 54.4. The **set of cluster points** associated with set A will be denoted $c(A)$.

Example 54.3 If $A = (a, b)$, then $c(A) = [a, b]$; but if $A = [a, b]$, then $c(A) = A$, which says that a closed interval already contains all of its cluster points. If $A = \{a_1, a_2, \ldots, a_n\}$, then $c(A) = \varnothing$ since at each a_k in A, $\mathbf{D}_{a_k}^\varepsilon \cap A = \varnothing$ for a small enough ε. If $A = \{1, \frac{1}{2}, \frac{1}{3}, \ldots, \frac{1}{n}, \ldots\}$, then the only cluster point is $x = 0$, so that $c(A) = \{0\}$. Two of the instances here show that a cluster point need not be a member of its associated set.

What set is $c(\mathbb{Q})$? Due to the density of $\mathbb{Q}$ in $\mathbb{R}$, we have that for each real x, every $\mathbf{D}_x^\varepsilon$ contains lots of points of $\mathbb{Q}$, implying that $c(\mathbb{Q}) = \mathbb{R}$. ■

Consider now closed sets. We have found that $\mathbb{Z}$ is closed, and we know of two classes of closed sets, intervals $[a, b]$ and finite collections of points $\{a_1, a_2, \ldots, a_n\}$ (see Exercise 54.3). $\mathbb{Z}$ and $\{a_1, a_2, \ldots, a_n\}$ have no cluster points, while $[a, b]$ is made of nothing but cluster points. Thus, all these closed sets contain all of their cluster points. This turns out to be a theorem in topology, and a useful alternative way to describe closed sets in $\mathbb{R}$.

Theorem 54.2 A set C is closed in $\mathbb{R}$ if and only if it contains all its cluster points: $c(C) \subseteq C$.

Proof of necessity. The special case of the closed set $C = \varnothing$ is easy, since $c(\varnothing) = \varnothing$. Given a nonempty closed set C in $\mathbb{R}$, let $x \in c(C)$. This means that every $\mathbf{D}_x^\varepsilon$ contains points in C. To cause a contradiction, suppose that $x \notin C$. Then $x \in \mathbb{R}\backslash C$, where $\mathbb{R}\backslash C$ is an open set. Thus, there is a $\mathbf{N}_x^\varepsilon \subset \mathbb{R}\backslash C$, which is impossible, since any $\mathbf{N}_x^\varepsilon$ contains at least one point aside from x that belongs to C. Hence $x \in C$, that is, C contains all of its cluster points.

Proof of sufficiency. Now we are given that $c(C) \subseteq C$. We would like to prove that $\mathbb{R}\backslash C$ is an open set. Let $y \in \mathbb{R}\backslash C$, which tells us that y is not a cluster point of C. Hence, there must be a $\mathbf{N}_y^\varepsilon$ that doesn't contain any points of C. Thus, $\mathbf{N}_y^\varepsilon \subset \mathbb{R}\backslash C$. Since this can be affirmed for *any* $y \in \mathbb{R}\backslash C$, it follows that $\mathbb{R}\backslash C$ is an open set. We conclude that C is closed. **QED**

Continuing with closed sets, the next theorem shows that closed sets behave in a photo-negative sense to open sets in Theorem 54.1.

Theorem 54.3 (1) The union of any finite collection of closed sets $\bigcup_1^n C_k$ in $\mathbb{R}$ is closed. (2) The intersection $\bigcap_{\alpha \in S} C_\alpha$ of any collection (countable or uncountable) of closed sets in $\mathbb{R}$ is closed.

Proof. Create a proof based on Theorem 5.3, as Exercise 54.5. **QED**

The **closure** of a set A is the union of A with its cluster points $c(A)$. Denote the closure by $\tilde{A}$, and the definition states that $\tilde{A} = A \cup c(A)$ (the closure of A is sometimes written $\overline{A}$, but we have already used $\overline{A}$ to express the complement of A, as is common).

Example 54.4 If $A = (a, b)$, then $\tilde{A} = [a, b]$; but if $A = [a, b]$, then $\tilde{A} = A$. If $A = \{a_1, a_2, \ldots, a_n\}$, then $\tilde{A} = A$, since we saw in Example 54.3 that such sets have no cluster points at all. If $A = \{1, \frac{1}{2}, \frac{1}{3}, \ldots, \frac{1}{n}, \ldots\}$, then $\tilde{A} = \{0, 1, \frac{1}{2}, \frac{1}{3}, \ldots, \frac{1}{n}, \ldots\}$. Also, $\tilde{\mathbb{Q}} = \mathbb{R}$. ■

In view of the definition of closure, we must have $A \subseteq \tilde{A}$ in general. However, any closed set is identical to its closure (verify). And the word "closure" is completely appropriate, for the closure of a set always produces a closed set (see Exercise 54.8).

Our next—and rather pretty—theorem gives a relationship among a set A, its boundary $b(A)$, its interior $i(A)$, and its cluster points $c(A)$.

Theorem 54.4 For any set A, $i(A) \cup b(A) = A \cup c(A)$. In other symbols, $i(A) \cup b(A) = \tilde{A}$.

Proof. The special case $i(A) \cup b(A) = \varnothing$ is equivalent to $A = \varnothing$, and we know that $c(\varnothing) = \varnothing$. So, the theorem is true in this case.

Now, for nonempty $i(A) \cup b(A)$, we prove that $i(A) \cup b(A) \subseteq A \cup c(A)$. Suppose $x \in i(A) \cup b(A)$. If on the one hand, $x \in A$, then $x \in A \cup c(A)$, and we are done with the first inclusion. If on the other hand, $x \notin A$, then $x \notin i(A)$, so we must have $x \in b(A)$. For x to be a boundary point, for every $\mathbf{N}_x^\varepsilon$, $\mathbf{N}_x^\varepsilon \cap A \neq \varnothing$ (and $\mathbf{N}_x^\varepsilon \cap \overline{A} \neq \varnothing$, which is automatically true here).

But since $x \notin A$, then it is true that for every $\mathbf{D}_x^\varepsilon$, $\mathbf{D}_x^\varepsilon \cap A \neq \varnothing$. This means that x is a cluster point of A, so that $x \in c(A)$. Once again, $x \in A \cup c(A)$. This proves that $i(A) \cup b(A) \subseteq A \cup c(A)$.

Next, we prove that $A \cup c(A) \subseteq i(A) \cup b(A)$. Suppose $x \in A \cup c(A)$. If $x \in b(A)$, then $x \in i(A) \cup b(A)$ and we are done with the second inclusion. If on the other hand, $x \notin b(A)$, then there exists a $\mathbf{N}_x^\varepsilon$ such that either $\mathbf{N}_x^\varepsilon \cap A = \varnothing$ or $\mathbf{N}_x^\varepsilon \cap \overline{A} = \varnothing$. Looking at the first possibility, $\mathbf{N}_x^\varepsilon \cap A = \varnothing$ implies that $x \notin A$ and $x \notin c(A)$, which contradicts the premise $x \in A \cup c(A)$. We must therefore turn to the second possibility, which is that $\mathbf{N}_x^\varepsilon \cap \overline{A} = \varnothing$, meaning that $\mathbf{N}_x^\varepsilon$ and $\overline{A}$ are disjoint. Hence, $\mathbf{N}_x^\varepsilon \subseteq A$, where if $\mathbf{N}_x^\varepsilon = A$, then certainly $\mathbf{N}_x^{\varepsilon/2} \subset A$. Either way, $x \in i(A)$, and once again, $x \in i(A) \cup b(A)$. We have proved that $A \cup c(A) \subseteq i(A) \cup b(A)$.

The two set inclusions have been proved, so the theorem is proved. **QED**

The various relationships discussed in the proof above and others, are summarized in Figure 54.1, a Venn diagram about a set of points A. The set A is made up of three subsets: the interior of A (shaded darkly), and any boundary points and cluster points that belong to it. We see, for example, that $i(A) \subseteq c(A)$. Also, Theorem 54.4 is directly illustrated. The three regions $b(A) \cap \overline{A} \cap \overline{c(A)}$, $c(A) \cap \overline{A} \cap \overline{b(A)}$, and $A \cap \overline{c(A)} \cap \overline{b(A)}$ are always empty. You are asked to prove this in Exercise 54.14.

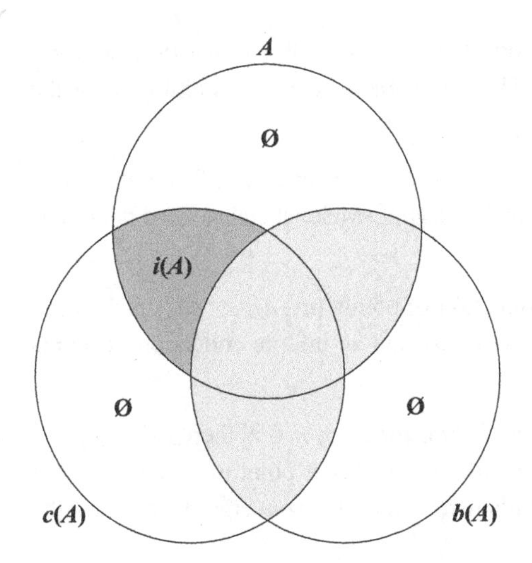

Figure 54.1. Venn diagram showing the relationships among point set A, its interior $i(A)$ (the shaded subset of A), its set of cluster points $c(A)$, and its boundary $b(A)$. For instance, the closure of A is represented as the central trefoil, $\tilde{A} = i(A) \cup b(A)$ per Theorem 54.4.

To finish things off, we define and denote the **exterior** of a set A by $e(A) \equiv \overline{b(A)} \cap \overline{A} \cap \overline{c(A)}$ (the region outside the circles).

Corollary 54.1 The exterior of a set is the complement of the set's closure. Symbolically, for any set A, $e(A) = \overline{\tilde{A}}$.

Proof. By De Morgan's laws, $e(A) = \overline{b(A) \cup A \cup c(A)}$. By Theorem 54.4, we have $(i(A) \cup b(A)) \cup A \cup c(A) = (A \cup c(A)) \cup A \cup c(A)$, which simplifies to $b(A) \cup A \cup c(A) = A \cup c(A)$. And since $A \cup c(A) = \tilde{A}$ by Theorem 54.4, we come to $e(A) = \overline{\tilde{A}}$. **QED**

Example 54.5 Draw a fairly accurate picture of the set $E = \{\frac{1}{2}, \frac{1}{3}, \ldots, \frac{1}{n}, \ldots\} \cup [1, 2) \cup \{3\}$. We see that $i(E) = (1, 2)$, so $i(E) \subset E$, as symbolized in Figure 54.1.

Next, $c(E) = \{0\} \cup [1, 2]$, so that $\tilde{E} = \{0\} \cup \{\frac{1}{2}, \frac{1}{3}, \ldots, \frac{1}{n}, \ldots\} \cup [1, 2] \cup \{3\}$. Hence, $E \subset \tilde{E}$ as the Venn diagram indicates. We also infer from the Venn diagram that some boundary points may not belong to a set, and this is true here because $b(E) = \{0, 1, 2, 3\} \cup \{\frac{1}{2}, \frac{1}{3}, \ldots, \frac{1}{n}, \ldots\}$. Also, Theorem 54.4 is demonstrated by $i(E) \cup b(E) = \tilde{E}$. Last but not least, $e(E) = (-\infty, 0) \cup \bigcup_1^\infty (\frac{1}{n+1}, \frac{1}{n}) \cup (2, 3) \cup (3, \infty)$ by the corollary just proved. ■

Note 2: A detailed account of the struggles encountered in sifting out the concepts in this chapter (as well as many other topological ideas) is in "The Emergence of Open Sets, Closed Sets, and Limit Points in Analysis and Topology," by Gregory H. Moore, in *Historia Mathematica*, vol. 35, no. 3 (August 2008), pp. 220–241. The long road to clarity begins with Bernhard Bolzano in the early 1800s, continues past Weierstrass, Cantor, Lebesgue and many others, and arrives at its present form by the 1950s.

54.1 Exercises

1. By **(a)**, the sets $\varnothing$ and $\mathbb{R}$ belong to the family of open sets $\mathcal{T}$. Prove that they conform to the definition of an open set in $\mathbb{R}$. Hint: The proof for $\varnothing$ requires thinking about the contrapositive of the definition of "open."

2. Show that every open interval $(a, b) = \{x : a < x < b\}$ is an open set. Then show that $(a, b) \cup (b, c)$ is open, without using Theorem 54.1! Hint: Make sketches with neighborhoods of x.

3. Prove in two different ways that any finite collection of points $\{a_1, a_2, \ldots, a_n\}$ is a closed set. Find an infinite collection of points that is closed, and an infinite collection of points that isn't closed.

4. Show that a point x is a cluster point of A if and only if for every $n \in \mathbb{N}$ there exists a point $x_n \in A$ such that $0 < |x_n - x| < \frac{1}{n}$. In other words, x is a cluster point of A if and only if there exists a sequence of points $\langle x_n \rangle_1^\infty \subset A$ distinct from x that converges to x (using the d_1 metric). Hint: Review Theorem 14.1.

5. Prove Theorem 54.3, using Theorem 5.3.

6. Answer as always, sometimes (but not always), or never, true. Give a convincing reason (or a proof) of the always true statements, and an example and a counterexample for the sometimes true statements.
   a. An open set is composed of nothing but interior points.
   b. If x is a cluster point of A, then there is a $\mathbf{D}_x^\varepsilon$ that contains no points of A.
   c. If x is a cluster point of A, then for all $\mathbf{D}_x^\varepsilon$, there is a point $a \in A$ such that $a \notin \mathbf{D}_x^\varepsilon$.
   d. If $a \in \mathbf{N}_b^\varepsilon$, then $b \in \mathbf{N}_a^\varepsilon$.
   e. Boundary points are cluster points of a set.
   f. The exterior of a set is the set's complement.
   g. The boundary of a set are the points neither in its interior nor in its exterior.
   h. The exterior of a set is always open. Hint: Use Exercise 54.8.

7. Fill in the following chart (for example, $\tilde{\mathbb{Q}} = \mathbb{R}$).

SET	$\varnothing$	$\mathbb{Q}$	$\mathbb{Z}$	$\mathbb{R}$	$[a,b)\cup(b,c]$	$[-1,0)\cup\{\frac{1}{2},\frac{2}{3},\frac{3}{4},\ldots,\frac{n}{n+1},\ldots\}$
Complement						
Interior						
Boundary						
Cluster Points						
Closure		$\mathbb{R}$				
Exterior						

8. Create a short proof that the closure of a set is a closed set.

9. Show that a set $C \subset \mathbb{R}$ is closed if and only if no point in $\overline{C}$ is a cluster point of C. Hint: Sufficiency is easy to prove. For necessity, let C be closed, and assume that $y \in \overline{C}$ and is a cluster point of C.

10. Can the topological definitions in this chapter be applied to physical objects? For instance, is it possible for a physical solid to behave like $\mathbb{Q}$ and be dense but have no interior? Can a physical object have an interior but no boundary?

11. a. Show that the interval $[1, 2)$ is neither open nor closed.
    b. Show that if A is both open and closed in $\mathbb{R}$, then A is either $\varnothing$ or $\mathbb{R}$. This will mean that $\varnothing$ and $\mathbb{R}$ are the only sets in our topology that are both open and closed. Hint: Think of $b(A)$. The only sets with empty boundary are $\varnothing$ and R.
    c. Show that there are only two sets such that $\overline{(\tilde{A})} = \widetilde{(\overline{A})}$. Hint: What types of sets are they?

12. We have seen that the intersection of an infinite collection of nonempty open sets in $\mathbb{R}$ may be closed. Likewise, the union of an infinite collection of nonempty closed sets in $\mathbb{R}$ may be open. Create an example of this.

13. a. Prove that $i(A)$ and $b(A)$ form a partition of any closed (nonempty) set A, as illustrated in Figure 54.1. Now, explain why if A is any set (not necessarily closed), then $A \subseteq i(A) \cup b(A)$.
    b. Prove that the boundary of a set is figuratively speaking what lies between its interior and its exterior; to be analytically precise, $b(A) = \overline{i(A)} \cap \overline{e(A)}$.
    c. Prove that the boundary of any set is closed. Hint: Use Exercise 54.6.

14. Three regions of interest in Figure 54.1 are always empty: (1) $b(A) \cap \overline{A} \cap \overline{c(A)}$, (2) $c(A) \cap \overline{A} \cap \overline{b(A)}$, and (3) $A \cap \overline{c(A)} \cap \overline{b(A)}$. (3) asserts an interesting situation: if a point is neither a cluster point nor a boundary point, then it isn't part of the set. Prove the statements in that order (the order of difficulty), using Theorem 54.4 whenever possible. Hint: For (1) and (2), if a set of points is claimed to be empty, its complement should be $\mathbb{R}$. Recall De Morgan's laws. For (3), Theorem 54.4 doesn't seem to help, so show that a contradiction is produced by assuming $x \in A \cap \overline{c(A)} \cap \overline{b(A)}$.

55

The Cantor Ternary Set

If something happens except on a set of measure zero, it is said to happen almost everywhere...

Ralph P. Boas, Jr. [26], pg. 70.

In 1883, Georg Cantor introduced a very peculiar set to mathematics in a note to the mathematical journal *Grundlagen*. (This means "Foundations", a shortened version of the full title. Great mathematicians often work high above us when they are strengthening foundations.) Cantor had in mind a construction that would demonstrate how complex and counterintuitive a set of real numbers could actually be.

55.1 Ternary Expansions

To do this, Cantor delved into the unfamiliar territory of the **ternary expansion** of numbers, the expression of numbers in base three. This is similar to binary (base two) expansion, but allows the digits 0, 1, and 2. For integers, the expansion is fairly easy. Here are some equivalents:

Base 10	0	1	2	3	4	5	6	7	8	9	10	11	12
Base 2	0	1	10	11	100	101	110	111	1000	1001	1010	1011	1100
Base 3	0	1	2	10	11	12	20	21	22	100	101	102	110

The translation of a base three integer to base ten is done with powers of 3: $2201_3 = 2(3^3) + 2(3^2) + 0(3^1) + 1(3^0) = 73$ (unsubscripted numbers will indicate base ten). The inverse procedure is shown in the algorithm below.

Example 55.1 Find the base three integer equivalent to 586. Divide 586 by 3 (the algorithm always divides by 3): $\frac{586}{3} = 195.333\ldots$. The fractional part is $\frac{1}{3}$, so the ternary expansion *ends* with a ternary 1. Next, divide the integer part 195 by 3: $\frac{195}{3} = 65.0$. The fractional part is 0, so the next-to-last digit of the expansion is 0. Divide the integer part 65by 3: $\frac{65}{3} = 21.666\ldots$. The fractional part is $\frac{2}{3}$, so the next digit to the left in the expansion is 2. Continuing, $\frac{21}{3} = 7.0$, so the next digit to the left is 0. Next, $\frac{7}{3} = 2.333\ldots$, so the next digit to the left is 1. Finally, $\frac{2}{3} = 0.666\ldots$, so the left-most digit is 2. Hence, $586 = 210201_3$. Check this result by expanding 210201_3 in powers of 3, and explain to yourself why the algorithm works by considering the expansion. ■

Ternary fractions correspond to decimal fractions. Thus, just as $0.73 = 7(10^{-1}) + 3(10^{-2}) = \frac{73}{100}$, so too are $0.21_3 = 2(3^{-1}) + 1(3^{-2}) = \frac{7}{9}$ and $0.0101_3 = 1(3^{-2}) + 1(3^{-4}) = \frac{10}{81}$. It's reasonable that all terminating ternary expansions $0.t_1t_2 \ldots t_n$, (where $t_n = 1$ or 2), are rational numbers of the form $\frac{a}{3^n}$, $a \in \mathbb{N}$ (see Exercise 55.1). Since 3 and 10 are relatively prime, any terminating decimal in base ten will be non-terminating in base three, and conversely. For example, $\frac{1}{10} = 0.1$ becomes $0.002200220022\ldots_3$, and $\frac{1}{8} = 0.125 = 0.010101\ldots_3$. Before we proceed to calculate these expansions via an algorithm, recall how to do the inverse procedure by geometric series, and otherwise.

Example 55.2 To convert $0.010101\ldots_3$ into its decimal equivalent, treat the expansion as a geometric series. Thus,

$$0.010101\ldots_3 = \sum_{n=1}^{\infty} \frac{1}{3^{2n}} = \frac{\frac{1}{9}}{1 - \frac{1}{9}} = \frac{1}{8} = 0.125.$$

Involving slightly more detail,

$$0.002200220022\ldots_3 = \sum_{n=1}^{\infty} \left(\frac{2}{3^{4n-1}} + \frac{2}{3^{4n}} \right) = \sum_{n=1}^{\infty} \left(\frac{8}{3^{4n}} \right) = \frac{\frac{8}{3^4}}{1 - \frac{1}{3^4}} = \frac{1}{10} = 0.1.$$

All very nice. Now, the xi algorithm, used with base ten decimals in Chapter 28, also works for base three. Given $x = 0.002200220022\ldots_3$, multiply both sides by 3^4: $3^4x = 22.00220022\ldots_3$. Subtract: $3^4x - x = 22.00220022\ldots_3 - 0.00220022\ldots_3$, and simplify: $80x = 22_3 = 8$, giving us $x = \frac{1}{10}$. Eureka! ■

The following algorithm inverts the process of the previous example. It's essentially the same algorithm as in Exercise 10.5.

Example 55.3 To convert $\frac{1}{10}$ to its ternary equivalent, divide by the lowest power of 3 that will that will give at least 1. Express the result as a mixed number. Thus, $\frac{1}{10} \div (3^{-3}) = \frac{27}{10} = 2\frac{7}{10}$. The integer 2 is the first nonzero ternary digit, located in the place value for 3^{-3}, so that $\frac{1}{10} = 0.002\ldots_3$. Repeat the steps with the remaining fraction $\frac{7}{10}$. These steps and some succeeding ones appear in the table.

3^{-1}	3^{-2}	3^{-3}	3^{-4}	3^{-5}	3^{-6}	3^{-7}
$\rightarrow$	$\rightarrow$	$\frac{1}{10} \div 3^{-3} = \underset{\downarrow}{2}\frac{7}{10} \rightarrow$	$\frac{7}{10} \div 3^{-1} = \underset{\downarrow}{2}\frac{1}{10} \rightarrow$	$\rightarrow$	$\rightarrow$	$\frac{1}{10} \div 3^{-3} = \underset{\downarrow}{2}\frac{7}{10} \rightarrow$
0	0	2	2	0	0	2

3^{-8}	3^{-9}	...
$\frac{7}{10} \div 3^{-1} = \underset{\downarrow}{2}\frac{1}{10}$	$\rightarrow$	...
2	0	...

.

We have found that $\frac{1}{10} = 0.002200220\ldots_3$, as expected. Now, do Exercise 55.2. ■

Try Exercise 55.4 to see another method for finding ternary expansions.

55.2 The Cantor Ternary Set

Cantor divided the ternary expansions of real numbers in $[0, 1]$ into two classes, those that contain the digit 1 and those that don't. A point of clarification: any expansion terminating with a 1 will have that digit replaced by the nonterminating sequence $022222\ldots$. For instance, $\frac{7}{9} = 0.21_3 = 0.202222\ldots_3$. The expansions are equal, as can be shown with a geometric series.

> The **Cantor Ternary Set** $\mathbb{K}$ is the set of all real numbers $x \in [0, 1]$ whose ternary expansion may be written without the digit 1.

Beginning to consider the effects of this definition, observe that $\frac{1}{3} = 0.1_3 = 0.0222\ldots_3 \in \mathbb{K}$ and $\frac{2}{3} = 0.2 \in \mathbb{K}$. But no $x \in (\frac{1}{3}, \frac{2}{3})$ belongs to $\mathbb{K}$, since all such numbers have ternary expansions $0.1t_2t_3t_4\ldots$ in which the initial 1 is inescapable. Thus, the open middle third is removed from $[0, 1]$, and our first step in the construction of $\mathbb{K}$ appears in Figure 55.1. Verify that the endpoints $x = 0$ and $x = 1$ also remain in $\mathbb{K}$.

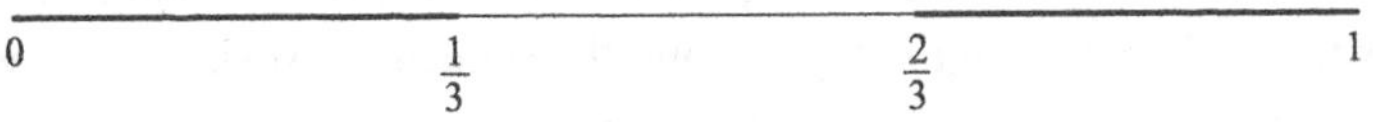

$0 \quad \frac{1}{3} \quad \frac{2}{3} \quad 1$

Figure 55.1.

For the second step, $\frac{1}{9} = 0.01_3 = 0.00222\ldots_3 \in \mathbb{K}$, and $\frac{2}{9} = 0.02 \in \mathbb{K}$. But no $x \in (\frac{1}{9}, \frac{2}{9})$ belongs to $\mathbb{K}$, since all those numbers have ternary expansions $0.01t_3t_4t_5\ldots$ in which the initial 1 cannot be removed. Also new is that no $x \in (\frac{7}{9}, \frac{8}{9})$ belongs to $\mathbb{K}$, since all such numbers have ternary expansions $0.21t_3t_4t_5\ldots$ in which the 1 is entrenched. The effect is to remove the open middle ninth from the remaining segments, leaving what we see in Figure 55.2. Verify that all eight endpoints belong to $\mathbb{K}$.

$0 \quad \frac{1}{9} \quad \frac{2}{9} \quad \frac{1}{3} \quad \frac{2}{3} \quad \frac{7}{9} \quad \frac{8}{9} \quad 1$

Figure 55.2.

Step three removes the four middle 27ths, corresponding to points with ternary expansions $0.001t_4t_5t_6\ldots$, $0.021t_4t_5t_6\ldots$, $0.201t_4t_5t_6\ldots$, and $0.221t_4t_5t_6\ldots$. The view, which includes all 16 endpoints, is in Figure 55.3.

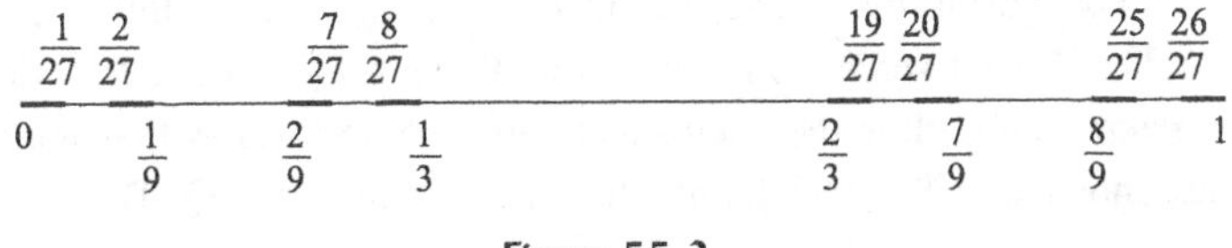

$\frac{1}{27} \; \frac{2}{27} \quad \frac{7}{27} \; \frac{8}{27} \quad \frac{19}{27} \; \frac{20}{27} \quad \frac{25}{27} \; \frac{26}{27}$

$0 \quad \frac{1}{9} \quad \frac{2}{9} \quad \frac{1}{3} \quad \frac{2}{3} \quad \frac{7}{9} \quad \frac{8}{9} \quad 1$

Figure 55.3.

All of Cantor's steps proceed in the same fashion. At step n, exactly 2^{n-1} open intervals containing points with ternary expansions $0.q_1q_2\ldots q_{n-1}1t_{n+1}t_{n+2}\ldots$ are extracted, where the 1 appears as the nth digit, and all digits q_i are either 0 or 2 (verify the 2^{n-1} count by an easy argument). Also at step n, each removed open interval has length $\frac{1}{3^n}$. The intervals that remain are closed, since no endpoints can ever be removed. The extractions continue *ad infinitum*.

The limit as $n \to \infty$ is the Cantor ternary set $\mathbb{K}$. The sum of the lengths of the open intervals removed is

$$\frac{1}{3} + \frac{2}{9} + \frac{4}{27} + \cdots = \sum_{n=1}^{\infty} \frac{2^{n-1}}{3^n} = 1.$$

This is a huge surprise! Apparently, *all* the length of $[0, 1]$ has been removed. But we mustn't hastily assume that nothing is left. What remains? A vast, disconnected, dusting of points, invisible at any magnification. First of all, the endpoints of the intervals from each successive step remain, since their expansions are 1-free. This is a countable collection of points (verify). But there are more points left behind, for example, $\frac{1}{4} = 0.020202\ldots_3 \in \mathbb{K}$, and also $\frac{1}{10}$, per Example 55.3.

Since the union of the removed open sets (intervals) is open, then its complement relative to $[0, 1]$ is closed; therefore, $\mathbb{K}$ is a closed set. And by Theorem 54.2, this means that $\mathbb{K}$ contains *all* its cluster points. Here is a way to show that $\frac{1}{4}$ is a cluster point of $\mathbb{K}$ (we have seen that it belongs to $\mathbb{K}$). Since $\mathbb{K}$ is closed, by Exercise 54.4, there is a sequence of Cantor points $\langle k_n \rangle_1^\infty \subset \mathbb{K}$ that converges to $\frac{1}{4}$. We use Cantor's steps to find the sequence. At step one (see Figure 55.1), select the closest endpoint to $\frac{1}{4}$, namely, $\frac{1}{3}$, and $k_1 = \frac{1}{3}$. At step two, select the closest ninth (in lowest terms) to $\frac{1}{4}$, which is $\frac{2}{9} = k_2$. At step three, we find $\frac{7}{27} = k_3$ is closest. Alas, we can't continue for long using graphs. But we observe that $k_2 = \frac{1}{3} - \frac{1}{9}$, and $k_3 = \frac{1}{3} - \frac{1}{9} + \frac{1}{27}$, and it can be shown by induction that $k_n = \sum_{i=1}^{n} \frac{(-1)^{i-1}}{3^i}$. Therefore,

$$\langle k_n \rangle_1^\infty \to \frac{\frac{1}{3}}{1 + \frac{1}{3}} = \frac{1}{4},$$

proving that $x = \frac{1}{4}$ is a cluster point of $\mathbb{K}$. An easier way to show this is by forming a sequence out of the ternary expansion of $\frac{1}{4}$: $\langle 0.02, 0.0202, 0.020202, \ldots \rangle$. Each term t_n of the sequence of course belongs to $\mathbb{K}$, and $t_n \to \frac{1}{4}$.

What we have just done generalizes nicely to a lemma that tells us more about $\mathbb{K}$.

Lemma 55.1 $\mathbb{K} = c(\mathbb{K})$, that is, the Cantor ternary set is identical to its set of cluster points.

Proof. We have just shown that $\mathbb{K}$ is closed, implying by Theorem 54.2 that $c(\mathbb{K}) \subseteq \mathbb{K}$. Thus, all that's left to show is that $\mathbb{K} \subseteq c(\mathbb{K})$.

Consider the ternary expansion of any $k \in \mathbb{K}$. If the expansion is non-terminating, then $k = 0.k_1k_2,\ldots$ (in base 3) is free of the digit 1, and yields the sequence $\langle 0.k_1, 0.k_1k_2, \ldots, 0.k_1k_2\ldots k_n, \ldots \rangle \to k$. Furthermore, every term $t_n = 0.k_1k_2\ldots k_n$ of this sequence is 1-free, meaning that $\langle t_n \rangle_1^\infty \subset \mathbb{K}$. By Exercise 54.4, it follows that $k \in c(\mathbb{K})$. If, on the other hand, $k \in \mathbb{K}$ has a terminating expansion, fill in the details as Exercise 55.5 to show that again, $k \in c(\mathbb{K})$.

We have now shown that $\mathbb{K} \subseteq c(\mathbb{K})$, and the lemma is proved. **QED**

Note 1: A precise, and very readable, description of the rational numbers belonging to $\mathbb{K}$ can be found in Ioana Mihaila's article, "The Rationals of the Cantor Set," in *The College Mathematics Journal*, vol. 35, no. 4 (Sept. 2004), pp. 251–255.

Are there perhaps a countable infinity of cluster points in $\mathbb{K}$? Cantor's next discovery is amazing.

Theorem 55.1 The Cantor ternary set is uncountably infinite.

Proof. We will form a one-to-one correspondence with an uncountable subset of [0, 1] and part of $\mathbb{K}$. Every number in $\mathbb{K}$ has a 1-free ternary expansion. If we then divide each 0 or 2 ternary digit in half, we get a binary expansion, which corresponds to some real $b \in [0, 1]$. This simple procedure *almost* sets up the 1-1 correspondence needed, but there is a difficulty to be addressed. Certain endpoints included in $\mathbb{K}$ yield the same binary expansion. For instance, $\frac{7}{9} = 0.21_3 = 0.20222\ldots_3 \in \mathbb{K}$, and the procedure yields $0.10111\ldots_2$, but since $0.10111\ldots_2 = 0.11_2$, doubling the digits gives $0.22_3 = \frac{8}{9} \in \mathbb{K}$. Hence, both $\frac{7}{9}, \frac{8}{9} \in \mathbb{K}$ are mapped to the single point $0.11_2 = \frac{3}{4} \in [0, 1]$.

Fortunately, the difficulty only occurs with the endpoints of $\mathbb{K}$ as identified in its construction. This subset is countable; let's name it E. Hence, the set of binary numbers in [0, 1] with two equal expansions is countable, call it B. It follows that $\mathbb{K}\backslash E$ is in 1-1 correspondence with $[0, 1]\backslash B$. But $[0, 1]\backslash B$ is uncountable, implying that $\mathbb{K}\backslash E$, and therefore $\mathbb{K}$, is uncountable. **QED**

We now know that $\mathbb{K}$ has the same cardinality as the entire real number line, is bounded, but is so pulverized that it is invisible at any magnification. It is totally counterintuitive that a set can exist this way. We will discuss three topological properties that clarify this sense that $\mathbb{K}$ is dusty or scattered.

First, the interior of Cantor's set is empty, $i(\mathbb{K}) = \varnothing$. To see this, let $x \in \mathbb{K}$, so that its ternary expansion is free of the digit 1. To show that no $\mathbf{N}_x^\varepsilon \subset \mathbb{K}$, we show that every $\mathbf{N}_x^\varepsilon$ contains real numbers y whose ternary expansions contain an embedded 1. Choose $n \in \mathbb{N}$ so large that $\frac{1}{3^n} < \varepsilon$ and $x \pm \frac{1}{3^n} \in (0, 1)$. Then $x \pm \frac{1}{3^n} \in \mathbf{N}_x^\varepsilon$, but $x \pm \frac{1}{3^n}$ will display the digit 1 somewhere. Therefore, $\mathbf{N}_x^\varepsilon \not\subseteq \mathbb{K}$ for any $\varepsilon > 0$. Thus, $\mathbb{K}$ has no interior. Consequently, every point of $\mathbb{K}$ is a boundary point or a cluster point (Theorem 54.4). In fact, for $\mathbb{K}$, every point is both a boundary point and cluster point.

Second, the Cantor ternary set is nowhere dense in $\mathbb{R}$. A set A is **nowhere dense** in $\mathbb{R}$ if and only if its closure $\widetilde{A}$ contains no open intervals; in other words, such a set's closure has empty interior. Roughly speaking, A is nowhere dense if it is scattered enough that we can stuff an open interval between any two of its elements. A finite collection of points $\{a_1, a_2, \ldots, a_n\}$, and the countable collection $\{1, \frac{1}{2}, \frac{1}{3}, \ldots, \frac{1}{n}, \ldots\}$ are nowhere dense, and so is $\varnothing$. The conclusion that $\mathbb{K}$ is nowhere dense follows from the facts that $\mathbb{K}$ is closed, and that $i(\mathbb{K}) = \varnothing$. See Exercise 55.6 for the comparison between a nowhere dense set and a dense set.

The third property develops in what follows.

55.3 The Measure of a Set

In order to study a third property of Cantor's set, one that shows that it is vanishingly thin, we investigate the idea of a set's measure. A set A has **measure zero**, or is of measure zero, if and only if it can be covered by a countable union of open intervals whose total length can be made arbitrarily small (open *sets* replace open intervals in a more general topology than this, and remember, a countable union of intervals does allow a finite union). To be precise, (1) if $A \subseteq \bigcup_1^\infty (a_i, b_i)$, where the intervals (a_i, b_i) need not be disjoint (this is called an "open covering" of A), and (2) if the total interval length $M = \sum_{i=1}^{\infty}(b_i - a_i)$ has a limit of zero because all $a_i \to b_i$, then A has measure zero. A set of measure zero has no length or footprint on the real number line, figuratively speaking. As examples, consider a single point $\{a\}$ and the countable collection $T = \{1, \frac{1}{2}, \frac{1}{3}, \ldots, \frac{1}{n}, \ldots\}$. If $A = \{a\}$, then we use one neighborhood $\mathbf{N}_a^{\varepsilon}$ to cover A. Hence $M = 2\varepsilon$, which can be made as small as we wish. Therefore, $\{a\}$ has measure zero. For T, we can surround each member $\frac{1}{n}$ by the neighborhood $\mathbf{N}_{1/n}^{\varepsilon/2^n}$ whose length is $\frac{2\varepsilon}{2^n}$, where $\varepsilon > 0$. Then $T \subset \bigcup_1^{\infty} \mathbf{N}_{1/n}^{\varepsilon/2^n}$, and $M = \sum_{n=1}^{\infty} \frac{2\varepsilon}{2^n} = 2\varepsilon$. We may diminish ε and therefore the length M of the open cover as much as we like. Thus, $\{1, \frac{1}{2}, \frac{1}{3}, \ldots, \frac{1}{n}, \ldots\}$ has measure zero. This makes sense, since the individual members of T have measure zero. Then again, T has only a countably infinite number of members. For a set that is uncountable, the measure may, or may not, be zero. Now, consider Exercise 55.7.

The measure of a set is a vast generalization of the length of an interval. A **measure** M is a function whose domain is practically all the subsets of $\mathbb{R}$, and whose range is $[0, \infty)$. Some of its properties, the ones we will need, are:

(a) Define $M(\varnothing) \equiv 0$, and $M(\mathbb{R}) \equiv \infty$.
(b) If B is an interval with endpoints a and b, then $M(B)$ equals its length, $b - a$. For instance, $M([2, 6)) = 4$.
(c) For measurable sets A and B of reals such that $A \subset B$, $M(A) \leq M(B)$. Note that even though $(3, 9) \subset [3, 9]$, we have $M((3, 9)) = M([3, 9])$.
(d) For measurable sets A and B of reals such that $A \subseteq B$, the measure of the relative complement is $M(B \backslash A) = M(B) - M(A)$. This implies that all singleton sets $\{a\}$ have measure zero, since $M([a, b) \backslash (a, b)) = M([a, b)) - M((a, b)) = 0$, and on the other hand, $M([a, b) \backslash (a, b)) = M(\{a\})$. This agrees completely with the analysis above. (With this property, we can derive (c), but (c) is worth stating anyway.)
(e) A measure M is additive, defined as follows. Let $\{B_i\}_1^{\infty}$ be a countable collection of measurable sets that are pairwise disjoint (so $B_i \cap B_j = \varnothing$ when $i \neq j$), then additivity means that $M(\bigcup_1^{\infty} B_i) = \sum_{i=1}^{\infty} M(B_i)$. (When only a finite number of B_i are nonempty, then we just have $\{B_i\}_1^n$, leading to $M(\bigcup_1^n B_i) = \sum_{i=1}^{n} M(B_i)$.)

In other words, the measure of a disjoint union of sets equals the sum of the individual measures. For intervals, this property is comfortably acceptable. For example,

$$M((-1, 0) \cup [2, 5]) = \sum_{i=1}^{2}(b_i - a_i) = 1 + 3 = 4.$$

Notice that the property dovetails nicely with our conclusion that

$$M\left(\left\{1, \frac{1}{2}, \frac{1}{3}, \ldots, \frac{1}{n}, \ldots\right\}\right) = M\left(\bigcup_1^{\infty}\left\{\frac{1}{n}\right\}\right) = \sum_{n=1}^{\infty} M\left(\left\{\frac{1}{n}\right\}\right) = 0 + 0 + \cdots = 0.$$

(Aha! With this property, we can derive (d), but (d) is also worth stating anyway.)

Now it is time to find the measure of a more complicated set.

Theorem 55.2 The Cantor ternary set has measure zero.

Proof. We found that $\mathbb{K}$ is the set that remains when a countable collection of open middle thirds are removed from [0, 1] according to the procedure illustrated above. For instance, $O_2 = (\frac{1}{9}, \frac{2}{9}) \cup (\frac{7}{9}, \frac{8}{9})$ is removed in Figure 55.2, and the collection is written as $\bigcup_1^\infty O_i$. Finish the proof as Exercise 55.9. **QED**

Thus, the third, and perhaps strangest, property of Cantor's set is that it has measure zero, even though it is uncountably infinite. As Ralph P. Boas (1912–1992) says, "If something happens except on a set of measure zero, then it is said to happen **almost everywhere**, or for almost all points." Using this idea, the Cantor ternary set does not exist almost everywhere!

The measure function that we have been using is the one that Henri Lebesgue created in 1904 in his *Leçons sur l'intégration* (Lessons on Integration), to reestablish integration on a much broader foundation. How he did this is informally discussed in William Dunham's excellent book [4], while a fully analytical introduction makes a fitting story for advanced analysis texts, such as [30] or [37]. With Lebesgue's theory, the question raised in Chapter 47 about which functions are Riemann integrable can now at least be answered in a proper context.

> A function is Riemann integrable over $[a, b]$ if and only if it is bounded over $[a, b]$ and continuous almost everywhere in $[a, b]$.

55.4 Exercises

1. Prove that any terminating ternary expansion $0.t_1t_2\ldots t_n$, where the last digit $t_n = 1$ or 2, is equal to a rational number of the form $\frac{a}{3^n}$, $a \in \mathbb{N}$. Hint: What happens when you multiply $0.t_1t_2\ldots t_n$ by 3^n?

2. Find the ternary expansion rounded to four digits for **a.** 0.875 and **b.** π. Hint: Use $\frac{355}{113} \approx \pi$ for easier work. The expansion for 0.875 will eventually repeat, whereas that for π will never repeat (the approximation will be correct to at least four digits!).

3. In Example 55.2, you found that $\frac{1}{10} = 0.002200220022\ldots_3$. Multiply this by 4 on the left and 11_3 on the right to quickly find the ternary expansion for $\frac{2}{5}$ (addition also works nicely).

4. If the ternary expansion of a rational number $\frac{1}{\lambda}$ ($\lambda \in \mathbb{N}$) is non-terminating, its expansion can be expressed as a geometric series whose sum is

$$\frac{\frac{a}{3^m}}{1 - \frac{1}{3^m}},$$

for some natural numbers a and m (see Example 55.2). Simplifying

$$\frac{1}{\lambda} = \frac{\frac{a}{3^m}}{1 - \frac{1}{3^m}}$$

gives us $a = \frac{3^m - 1}{\lambda}$, an equation to be solved by substituting $m = 1, 2, \ldots$ and seeing if the result is an integer a. For a usefully large collection of rationals $\frac{1}{\lambda}$, a can be found without overflowing a calculator. The natural number a is then expressed in ternary form by the algorithm in Example 55.1, and this will lead us to the coefficients of the geometric series yielding $\frac{1}{\lambda}$ in base three, in the form $\sum_{n=1}^{\infty} \frac{a}{(3^m)^n}$. From this, we can read off the ternary digits for $\frac{1}{\lambda}$.

Use this procedure to find $\frac{1}{7}$ in base three. Next, how would, say, $\frac{5}{7}$ be found? Hint: Eventually, you will reach $a = 104 = 1(3^4) + 2(3^2) + 1(3^1) + 2(3^0)$ in ternary form.

5. Finish the proof of Lemma 55.1 by proving that if $k \in \mathbb{K}$ has a terminating ternary expansion, then $k \in c(\mathbb{K})$.

6. Recall that a set A is *dense* in $\mathbb{R}$ (Chapter 13) if and only if between any real numbers $x < y$, there exists an $a \in A$ such that $x < a < y$. With more topological flavor, the definition says that every interval (x, y) contains a point of A. Hence, $(x, y) \cap A \neq \varnothing$ is always the case when A is dense in $\mathbb{R}$.

   Not being dense is *not* equivalent to being nowhere dense. Find an example of a non-dense set in $\mathbb{R}$ such that it is also not nowhere dense (logically, we might call this a "somewhere dense set").

7. The proof for establishing that set $T = \{1, \frac{1}{2}, \frac{1}{3}, \ldots, \frac{1}{n}, \ldots\}$ has measure zero is the standard, slick, technique. Consider the following attempted proof. We can surround each member $\frac{1}{n}$ by the neighborhood $\mathbf{N}_{1/n}^{1/2^n}$ whose length is $\frac{2}{2^n}$. Then

$$T \subset \bigcup_1^\infty \mathbf{N}_{1/n}^{1/2^n},$$

   and

$$M_1 = \sum_{n=1}^{\infty} \frac{2}{2^n} = 2.$$

   But we may also use neighborhoods $\mathbf{N}_{1/n}^{1/3^n}$, yielding

$$M_2 = \sum_{n=1}^{\infty} \frac{2}{3^n} = 1 < M_1.$$

   Continuing with $\mathbf{N}_{1/n}^{1/4^n}$ and so on, we may diminish the length M of the open cover as much as we like. Thus, T has measure zero.

   Give reasons why you think this is or isn't a valid proof.

8. a. Show that any finite collection of points $\{a_1, a_2, \ldots, a_n\}$ is of measure zero. Use careful notation.

   b. Create two proofs that any denumerable set, such as $\mathbb{Q}$, has measure zero.

9. Use the additivity of measure **(e)** and some key facts about $\mathbb{K}$ to finish proving that $\mathbb{K}$ is of measure zero. Use careful notation.

10. We have seen that $\mathbb{K}$ has the same cardinality as the entire real number line, and is bounded. But so is the rather mundane set $[0, 1]$. What makes $\mathbb{K}$ different than $[0, 1]$?

    We have seen that $\mathbb{K}$ is so scattered that it is invisible at any magnification. But the set $\{1, \frac{1}{2}, \frac{1}{3}, \ldots, \frac{1}{n}, \ldots\}$ is also scattered and invisible. What makes $\mathbb{K}$ different?

    We have seen that $\mathbb{K}$ has measure zero. But it can be shown that the set of rational numbers in $[0, 1]$, that is, $\mathbb{Q} \cap [0, 1]$, also has measure zero. What makes $\mathbb{K}$ different?

11. What does the Cartesian product $\mathbb{K} \times \mathbb{K}$ look like? Try to draw it on graph paper. What would a square dart board look like if only the points in $\mathbb{K} \times \mathbb{K}$ were colored red?

Note that just as measure zero extends the idea of zero length in $\mathbb{R}$, measure zero extends the concept of zero area in $\mathbb{R}^2$. It's probably not surprising that $\mathbb{K} \times \mathbb{K}$ is uncountably infinite but of measure zero. Suppose that you had a dart that could hit exactly one point of "Cantor's dart board." What do you suspect would be your chances of hitting a (red) point of $\mathbb{K} \times \mathbb{K}$? Would your chances increase if you could throw a million darts? Hint: Start with the set in Figure 55.1, which we may call K_1, and draw $K_1 \times K_1$.

Farey Sequences

The **Farey sequence** $\mathcal{F}_n$ is the sequence of all reduced rational numbers in $[0, 1]$ with denominators less than or equal to n, listed in ascending order ($n \in \mathbb{N}$). We shall call the members of $\mathcal{F}_n$ **Farey fractions**. We begin with $\mathcal{F}_1 = \langle \frac{0}{1}, \frac{1}{1} \rangle$ and $\mathcal{F}_2 = \langle \frac{0}{1}, \frac{1}{2}, \frac{1}{1} \rangle$. Soon, however, it becomes tedious to list Farey fractions in order. Fortunately, $\mathcal{F}_n$ can be constructed recursively by a peculiar algorithm. The **mediant** of two rational numbers $\frac{p}{q}$ and $\frac{r}{s}$ is $\frac{p+r}{q+s}$ (curiously, it's exactly what youngsters would like addition of fractions to be!). The construction of $\mathcal{F}_{n+1}$ requires us to form the mediant of *successive* pairs of members $\frac{p}{q}$ and $\frac{r}{s}$ of $\mathcal{F}_n$ if and only if $q + s = n + 1$, and to insert it in its proper position in $\mathcal{F}_n$. Once such a mediant is created, it is immortal; it joins the ranks of all future generations of mediants. Thus, $\mathcal{F}_3 = \langle \frac{0}{1}, \frac{1}{3}, \frac{1}{2}, \frac{2}{3}, \frac{1}{1} \rangle$, where two new mediants have been created from $\mathcal{F}_2$.

More Farey sequences appear below, with suggestive spacing. You are invited to fill in the 23 elements of $\mathcal{F}_8$. Many interesting patterns begin to emerge (some have called mathematics the investigation of patterns. . .).

$\mathcal{F}_1$	$\frac{0}{1}$																		$\frac{1}{1}$
$\mathcal{F}_2$	$\frac{0}{1}$									$\frac{1}{2}$									$\frac{1}{1}$
$\mathcal{F}_3$	$\frac{0}{1}$						$\frac{1}{3}$			$\frac{1}{2}$			$\frac{2}{3}$						$\frac{1}{1}$
$\mathcal{F}_4$	$\frac{0}{1}$				$\frac{1}{4}$		$\frac{1}{3}$			$\frac{1}{2}$			$\frac{2}{3}$		$\frac{3}{4}$				$\frac{1}{1}$
$\mathcal{F}_5$	$\frac{0}{1}$			$\frac{1}{5}$	$\frac{1}{4}$		$\frac{1}{3}$	$\frac{2}{5}$		$\frac{1}{2}$		$\frac{3}{5}$	$\frac{2}{3}$		$\frac{3}{4}$	$\frac{4}{5}$			$\frac{1}{1}$
$\mathcal{F}_6$	$\frac{0}{1}$		$\frac{1}{6}$	$\frac{1}{5}$	$\frac{1}{4}$		$\frac{1}{3}$	$\frac{2}{5}$		$\frac{1}{2}$		$\frac{3}{5}$	$\frac{2}{3}$		$\frac{3}{4}$	$\frac{4}{5}$	$\frac{5}{6}$		$\frac{1}{1}$
$\mathcal{F}_7$	$\frac{0}{1}$	$\frac{1}{7}$	$\frac{1}{6}$	$\frac{1}{5}$	$\frac{1}{4}$	$\frac{2}{7}$	$\frac{1}{3}$	$\frac{2}{5}$	$\frac{3}{7}$	$\frac{1}{2}$	$\frac{4}{7}$	$\frac{3}{5}$	$\frac{2}{3}$	$\frac{5}{7}$	$\frac{3}{4}$	$\frac{4}{5}$	$\frac{5}{6}$	$\frac{6}{7}$	$\frac{1}{1}$
$\mathcal{F}_8$																			

Lemma A.1 Let $\frac{a}{b} < \frac{c}{d}$ be any two rational numbers. Then their mediant is unique and behaves this way: $\frac{a}{b} < \frac{a+c}{b+d} < \frac{c}{d}$.

Proof. The proof is Exercise A.1 below. **QED**

Theorem A.1 A fundamental property of Farey sequences is that if $\frac{p}{q}$ and $\frac{r}{s}$ are *successive* fractions in $\mathcal{F}_n$, then $qr - ps = 1$.

Proof. For a proof by induction, our anchor is that the two members of $\mathcal{F}_1$ satisfy $1(1) - 0(1) = 1$. Now, assume that each pair of successive fractions $\frac{p}{q} < \frac{r}{s}$, both in $\mathcal{F}_n$, satisfy $qr - ps = 1$. Whenever their mediant $\frac{p+r}{q+s} \in \mathcal{F}_{n+1}$, Lemma A.1 tells us that $\frac{p}{q} < \frac{p+r}{q+s} < \frac{r}{s}$. Thus $\frac{p}{q}, \frac{p+r}{q+s}$, and $\frac{r}{s}$ are successive members of $\mathcal{F}_{n+1}$. We also find that $q(p+r) - p(q+s) = qr - ps + qp - pq = 1$ and $r(q+s) - s(p+r) = rq - sp + rs - sr = 1$, using the induction hypothesis. The induction is complete; hence, any pair of successive fractions $\frac{p}{q} < \frac{r}{s}$ in any $\mathcal{F}_n$ will obey $qr - ps = 1$. **QED**

Corollary A.1 Farey fractions are reduced fractions.

Proof. We call on a theorem from number theory: two integers p and q are relatively prime (their greatest common factor is 1) if and only if there exist integers α and β such that $\alpha q + \beta p = 1$. Consider any pair of successive fractions $\frac{p}{q} < \frac{r}{s}$ in $\mathcal{F}_n$. By Theorem A.1, $qr - ps = 1$, which implies that the pairs p, q and r, s are relatively prime. The corollary follows. **QED**

Theorem A.2 The algorithm described in the first paragraph creates all Farey fractions except $\frac{0}{1}$ and $\frac{1}{1}$.

Proof. We must prove that mediants will form all Farey fractions except $\frac{0}{1}$ and $\frac{1}{1}$. Beginning the induction, $\frac{0+1}{1+1} = \frac{1}{2} \in \mathcal{F}_2$ and it is the only mediant in $\mathcal{F}_2$. The induction hypothesis is that all Farey fractions in $\mathcal{F}_n$ $(n \geq 2)$ are formed by the algorithm. Consider now any reduced $\frac{a}{n+1}$ in $(0, 1)$, so by definition, $\frac{a}{n+1} \in \mathcal{F}_{n+1}$. Find the unique pair of Farey fractions $\frac{p}{q}, \frac{r}{s} \in \mathcal{F}_n$ such that $\frac{p}{q} < \frac{a}{n+1} < \frac{r}{s}$. This is always possible because $q, s \leq n$. From Lemma A.1 and Corollary A.1, the mediant $\frac{p+r}{q+s}$ is a unique, reduced fraction with $\frac{p}{q} < \frac{p+r}{q+s} < \frac{r}{s}$. And if $q + s = n + 1$, then $\frac{p+r}{q+s} \in \mathcal{F}_{n+1}$ and $\frac{p+r}{q+s} = \frac{a}{n+1}$. This implies that the algorithm has formed every Farey fraction in $\mathcal{F}_{n+1}$. The induction is complete, and the theorem follows. **QED**

Lemma A.2 Every rational number in $[0, 1]$ will eventually be created by the algorithm.

Proof. By definition, all fractions in $[0, 1]$, reduced and with denominators at most n, belong to $\mathcal{F}_n$. And by Theorem A.2, all such fractions are formed by the algorithm. **QED**

Theorem A.3 The set of rational numbers in $[0, 1]$ is denumerable.

Proof. Do this in Exercise A.4. **QED**

Note 1: More information about Farey fractions and continued fractions (Appendix D), with an application, is found in "Small Denominators: No Small Problem," by Scott J. Beslin, Douglas J. Baney, and Valerio De Angelis, in *Mathematics Magazine*, vol. 71, no. 2 (April 1998), pp. 132–138.

A.1 Exercises

1. Prove Lemma A.1. Hint: Would it be easier if b and d were positive?

2. How can the difference between any two successive Farey fractions be found, lightning fast?

3. A statistician extracts two independent samples of sizes n_1 and n_2 from a large population. Suppose the first sample has a proportion $\frac{x}{n_1}$ exhibiting an attribute, and the second sample has a proportion $\frac{y}{n_2}$ with the attribute. Statisticians often use the pooled proportion $\frac{x+y}{n_1+n_2}$ for the attribute in the combined samples. What does our analysis say about the value of the pooled proportion?

4. Prove Theorem A.3, which provides another way to prove that $\mathbb{Q}$ is countably infinite. Hint: Consider the cardinality of each $\mathcal{F}_n$ (in Exercise A8 you will find the actual number of fractions in $\mathcal{F}_n$).

5. How does the mediant in Lemma A.1 compare to the arithmetic mean of the two fractions?

6. Create the symbol $\frac{1}{0}$, and imbue it with the condition that all rational numbers $\frac{p}{q} < \frac{1}{0}$. Define $\mathfrak{S}_1 = \langle \frac{0}{1}, \frac{1}{0} \rangle$. This time, insert all possible mediants to create $\mathfrak{S}_2$, and then $\mathfrak{S}_3$, and so on. $\mathfrak{S}_1, \mathfrak{S}_2, \mathfrak{S}_3, \ldots$ are called **Stern-Brocot sequences**. Find $\mathfrak{S}_2$, $\mathfrak{S}_3$, and $\mathfrak{S}_4$. What fractions are you getting now? Recall the famous Fibonacci sequence $F = \langle 1, 1, 2, 3, 5, 8, \ldots \rangle$, where $F_1 = F_2 = 1$, and $F_{n+1} = F_n + F_{n-1}$. Show that the largest numerator and denominator appearing in $\mathfrak{S}_n$ are F_n.

7. Recall, in Chapter 25, the ruler function

$$R(x) = \begin{cases} \frac{1}{|q|} & \text{if } x = \frac{p}{q}, \text{ reduced} \\ 0 & \text{if } x \text{ isirrational} \end{cases}$$

and look at only $x \in [0, 1]$. Compare its graph with the display of Farey sequences. What connections can be made? Show that the largest difference between two successive mediants in any $\mathcal{F}_n$ is $\frac{1}{n}$.

8. Show that the number of Farey fractions in $\mathcal{F}_n$ is given by $1 + \sum_{k=1}^{n} \phi(k)$, where $\phi(k)$ is the count of natural numbers not exceeding k and relatively prime to k (it's called **Euler's ϕ Function**). For example, $\phi(5) = 4$ since $\{1, 2, 3, 4\}$ are less than and relatively prime to 5, and $\phi(12) = 4$ since $\{1, 5, 7, 11\}$ are less than and relatively prime to 12.

9. Let fractions

$$\frac{a_1}{b_1} < \frac{a_2}{b_2} \quad \text{and} \quad \frac{c_1}{d_1} < \frac{c_2}{d_2}.$$

While of course

$$\frac{a_1}{b_1} + \frac{c_1}{d_1} < \frac{a_2}{b_2} + \frac{c_2}{d_2}$$

by the real number axioms, the mediants of corresponding sides of these inequalities may yield the striking result that

$$\frac{a_1 + c_1}{b_1 + d_1} > \frac{a_2 + c_2}{b_2 + d_2}!$$

This goes sufficiently against common sense that it is called **Simpson's paradox**, although there is no paradox at all. Such an outcome is far from being an obscure mathematical exercise, as the following famous case in point shows.

Seen below are the voting proportions in favor of the Civil Rights Act of 1964 for Republicans and Democrats in the House of Representatives, distinguished between northern and southern states.

in favor	Repub.	Dem.
southern	$\frac{0}{10}$	$\frac{7}{94}$
northern	$\frac{138}{162}$	$\frac{145}{154}$

Source: www.answers.com/topic/simpson-s-paradox

The total proportion of votes in favor for each party is naturally calculated as a mediant, seen below. The Democrats had higher proportions in favor in both the north and the south. Voting as a block, however, the Republican party was far more in favor of the act than the Democratic party. Simpson's paradox occurred! (Political science students undoubtedly study the effects of this "paradox.")

	Repub. Dem.
southern	$\frac{a_1}{b_1} = \frac{0}{10} < \frac{7}{94} = \frac{a_2}{b_2}$
northern	$\frac{c_1}{d_1} = \frac{138}{162} < \frac{145}{154} = \frac{c_2}{d_2}$
Totals	$\frac{a_1 + c_1}{b_1 + d_1} = \frac{138}{172} > \frac{152}{248} = \frac{a_2 + c_2}{b_2 + d_2}$

We are given that $\frac{a_1}{b_1} < \frac{a_2}{b_2}$ and $\frac{c_1}{d_1} < \frac{c_2}{d_2}$. Prove that if furthermore,

$$\frac{a_1}{b_1} < \frac{c_1}{d_1} \quad \text{and} \quad \frac{a_2}{b_2} > \frac{c_2}{d_2},$$

then Simpson's paradox cannot occur. Again, given

$$\frac{a_1}{b_1} < \frac{a_2}{b_2} \quad \text{and} \quad \frac{c_1}{d_1} < \frac{c_2}{d_2},$$

but supposing that $\frac{a_2}{b_2} < \frac{c_2}{d_2}$, can the paradox be averted?

B

Proving that $\sum_{k=0}^{n} \frac{1}{k!} < (1 + \frac{1}{n})^{n+1}$

To complete the proof of Theorem 34.2, we need to verify that

$$\sum_{k=0}^{n} \frac{1}{k!} < \left(1 + \frac{1}{n}\right)^{n+1}$$

for all $n \in \mathbb{N}$. We align expansions (5) and (6) in Chapter 34, and look for the excesses between matching terms. The first few cases can be easily checked by hand. For instance, for any $n \geq 1$, the excess of term $\frac{1}{2!}$ over term number $m = 2$ of $\left(1 + \frac{1}{n}\right)^{n+1}$ is

$$\frac{1}{2!} - \frac{(n+1)(n)(n-1)}{2!n^3} = \frac{1}{2n^2}.$$

Some excesses can be seen below, from $m = 0$ to $m = n$. Only the first two are negative. The rest are positive, for in general,

$$0 < \frac{1}{m!} - \frac{(n+1)(n)(n-1)\cdots(n-(m-1))}{m!n^{m+1}}$$

if and only if

$$0 < 1 - \frac{n+1}{n}\frac{n-1}{n}\cdot\cdots\cdot\frac{n-(m-1)}{n},$$

and already at $m = 2$, it's clear that

$$0 < 1 - \frac{n+1}{n}\frac{n-1}{n}.$$

The excesses become messy rational expressions, as one might expect. However, they are bounded by the explicit terms named a_m, as will be proved.

mth term	of $\sum_{k=0}^{n} \frac{1}{k!}$,	of $(1+\frac{1}{n})^{n+1}$,	the mth excess, and its bound a_m
$m=0$	1	$\frac{n+1}{n}$	$1-\frac{n+1}{n} = -\frac{1}{n} = a_0$
$m=1$	1	$\frac{n+1}{n}$	$1-\frac{n+1}{n} = -\frac{1}{n} = a_1$
$m=2$	$\frac{1}{2!}$	$\frac{(n+1)(n)(n-1)}{2!n^3}$	$\frac{1}{2!} - \frac{(n+1)(n)(n-1)}{2!n^3} = \frac{1}{2n^2} \le \frac{1}{4n} = a_2$
$m=3$	$\frac{1}{3!}$	$\frac{(n+1)(n)(n-1)(n-2)}{3!n^4}$	$\frac{2n^2+n-2}{6n^3} < \frac{1}{2n} = a_3$
$m=4$	$\frac{1}{4!}$	$\frac{(n+1)(n)(n-1)(n-2)(n-3)}{4!n^5}$	$\frac{5n^3-5n^2-5n+6}{24n^4} < \frac{5}{24n} = a_4$
$m=5$	$\frac{1}{5!}$	$\frac{(n+1)(n)(n-1)(n-2)(n-3)(n-4)}{5!n^6}$	$\frac{9n^4-25n^3+15n^2+26n-24}{120n^5} < \frac{9}{120n} = a_5$
$m=6$	$\frac{1}{6!}$	$\frac{(n+1)!}{6!n^7(n-6)!}$	$\frac{1}{6!} - \frac{(n+1)!}{6!n^7(n-6)!} < \frac{14}{720n} = a_6$
...	...	...	...
m	$\frac{1}{m!}$	$\frac{(n+1)!}{m!n^{m+1}(n-m)!}$	$\frac{1}{m!} - \frac{(n+1)!}{m!n^{m+1}(n-m)!} < \frac{m^2-m-2}{2m!n} = a_m$
...	...	...	...
$m=n$	$\frac{1}{n!}$	$\frac{n+1}{n^{n+1}}$	$\frac{1}{n!} - \frac{n+1}{n^{n+1}} < \frac{n^2-n-2}{2n!n} = a_n$

Note 1: The numerator polynomials of the excesses in the table follow the recursion relation $P_{m+1}(n) = (n-m)P_m(n) + mn^m$, beginning with $P_2(n) = 1$.

From this vantage point, we would like to show that $\sum_{m=2}^{n} a_m < \frac{2}{n}$, for any $n \ge 2$, which is the missing piece in the proof of Theorem 34.2.

The terms a_m form a sequence of bounds

$$\left\langle \frac{5}{24n}, \frac{9}{120n}, \frac{14}{720n}, \ldots, \frac{n^2-n-2}{2n!n} \right\rangle$$

with a claimed formula

$$a_m = \frac{m^2-m-2}{2m!n}$$

for $4 \le m \le n$. For instance, as seen above,

$$a_5 = \frac{25-5-2}{2(5!)n} = \frac{9}{120n},$$

for any $n \ge 5$.

In general, for a fixed $n \ge 4$ and for $4 \le m \le n$, the excesses are expressed as

$$\frac{1}{m!} - \frac{(n+1)!}{m!n^{m+1}(n-m)!}.$$

We want to prove, then, that

$$\frac{1}{m!} - \frac{(n+1)!}{m!n^{m+1}(n-m)!} < \frac{m^2-m-2}{2m!n}.$$

Simplifying this a little, we will prove that

$$-m^2 + m + 2 + 2n < \frac{2(n+1)!}{n^m(n-m)!} \text{ for } 4 \le m \le n. \tag{1}$$

For any $n \ge 4$ and $m = 4$, (1) yields

$$-10 + 2n < \frac{2(n+1)!}{n^4(n-4)!},$$

or $0 < 5n^2 + 5n - 6$, which is true. This becomes the anchor case for a short induction on m. That is, assuming (1) is true, we must now prove that

$$-(m+1)^2 + (m+1) + 2 + 2n < \frac{2(n+1)!}{n^{m+1}(n-(m+1))!},$$

not for all m, but only for $4 \le m \le n-1$, where $n \ge 5$ is arbitrary but fixed.

We have

$$-(m+1)^2 + (m+1) + 2 + 2n = -m^2 - m + 2 + 2n < (-m^2 + m + 2 + 2n)\left(\frac{n-m}{n}\right)$$

over $4 \le m \le n$. With the help of the induction hypothesis (1),

$$\begin{aligned} -m^2 - m + 2 + 2n &< (-m^2 + m + 2 + 2n)\left(\frac{n-m}{n}\right) < \frac{2(n+1)!}{n^m(n-m)!}\left(\frac{n-m}{n}\right) \\ &= \frac{2(n+1)!}{n^{m+1}(n-(m+1))!}, \end{aligned}$$

and the induction is complete for $4 \le m \le n-1$. Consequently, (1) is true, and the bounds $a_m = \frac{m^2-m-2}{2m!n}$ are correct for $4 \le m \le n$, where n is the number of terms.

Reviewing the table above, the total *positive* excess from terms m, where $2 \le m \le n$, is therefore less than

$$\sum_{m=2}^{n} a_m = \frac{1}{4n} + \frac{1}{2n} + \frac{5}{24n} + \frac{9}{120n} + \frac{14}{720n} + \cdots + \frac{n^2-n-2}{2n!n}$$

(there are exactly $n-1$ terms here). Write the sum as

$$\frac{1}{4n} + \frac{\operatorname{sgn}(n-2)}{2n} + \frac{1}{n}\sum_{m=4}^{n} \frac{m^2-m-2}{2m!} \text{ for } n \ge 2, \tag{2}$$

where, by convention, the summation is zero when $n \le 3$. The partial sums of

$$\sum_{m=4}^{\infty} \frac{m^2-m-2}{2m!}$$

are always less than the corresponding partial sums of

$$\sum_{m=4}^{\infty} \frac{m^2}{3(2^m)} = \frac{9}{8}$$

(see Exercise 29.9).

Hence, using (2),

$$\sum_{m=2}^{n} a_m = \frac{1}{4n} + \frac{\operatorname{sgn}(n-2)}{2n} + \frac{1}{n}\sum_{m=4}^{n} \frac{m^2 - m - 2}{2m!} < \frac{1}{4n} + \frac{1}{2n} + \frac{9}{8n} < \frac{2}{n},$$

for any $n \geq 2$.

We have therefore demonstrated that $\sum_{m=2}^{n} a_m < \frac{2}{n}$, for any $n \geq 2$, which implies that

$$\sum_{k=0}^{n} \frac{1}{k!} < \left(1 + \frac{1}{n}\right)^{n+1}$$

for all $n \geq 1$. **QED**

C

The Ruler Function Is Riemann Integrable

The ruler function

$$f(x) = \begin{cases} \frac{1}{|q|} & \text{if } x = \frac{p}{q} \text{ in lowest terms} \\ 0 & \text{if } x \text{ is irrational} \end{cases}$$

was introduced in Chapter 25. Because of its periodic nature, we only need to prove its Riemann integrability over [0, 1]. Partition [0, 1] into a net, with norm $\Delta\chi$. Since every subinterval contains irrational abscissas from which we may select representatives X_i, it is always possible to obtain the Riemann sum

$$\sum_{i=1}^{n} f(X_i)\Delta x_i = 0$$

for any $n \in \mathbb{N}$. Hence for irrational X_i,

$$\lim_{\Delta\chi \to 0} \sum_{i=1}^{n} f(X_i)\Delta x_i = 0.$$

If the ruler function is to be Riemann integrable, all other such limits must therefore equal zero. Nets with a mixture of irrational and rational representatives will have Riemann sums less than nets where the irrational representatives are replaced by rational ones. We thus turn our attention to nets with none but rational representatives $X_i = \frac{p_i}{q_i}$. Applying the ε-δ definition that appears after (1) in Chapter 47, given any $\varepsilon > 0$, we must find a corresponding $\delta > 0$ such that for all nets with $\Delta\chi < \delta$ and for any representatives $\{\frac{p_i}{q_i}\}_{i=1}^{n}$, we can certify that

$$\sum_{i=1}^{n} f\left(\frac{p_i}{q_i}\right)\Delta x_i = \sum_{i=1}^{n} \frac{1}{q_i}\Delta x_i < \varepsilon.$$

To begin with, if $\varepsilon \geq 1$, then take $\delta = \frac{1}{2}$ because all resulting nets will produce Riemann sums $\sum_{i=1}^{n} \frac{1}{q_i}\Delta x_i < 1$ (Figure C.1 will help you to see this).

In case $0 < \varepsilon < 1$, there are at least two, but only a finite number m_ε, of points $\langle \frac{p_i}{q_i}, \frac{1}{q_i} \rangle$ of the ruler function in [0, 1] with $\frac{1}{q_i} \geq \varepsilon, i = 1, 2, \ldots, m_\varepsilon$ (see Figure A.1). Given any net, each such fraction will be found in a subinterval, that is, $\frac{p_i}{q_i} \in [x_{j-1}^*, x_j^*]$, where $j \in \{1, 2, \ldots, n\}$, and let's denote the lengths of the starred subintervals by $\Delta x_j^* = x_j^* - x_{j-1}^*$ (they need not be contiguous in [0, 1]). Now, select one (whichever you like best) of the $\frac{p_i}{q_i} \in [x_{j-1}^*, x_j^*]$ as the representative

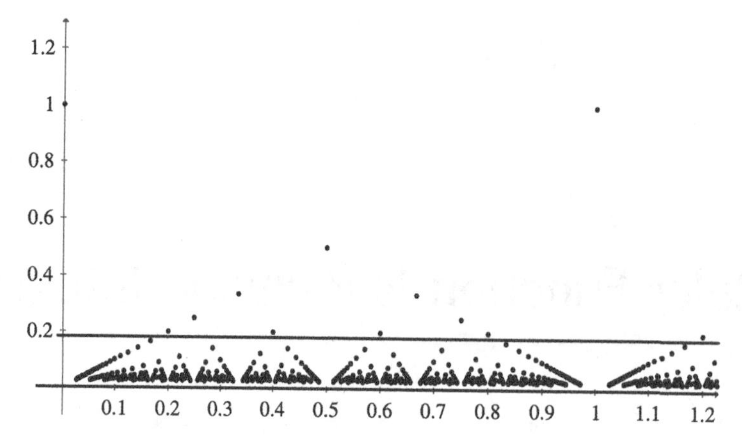

Figure C.1. The horizontal line at $\varepsilon = 0.18$ is shown. Exactly $m_{0.18} = 11$ points of the ruler function have ordinates at least as high as ε, over [0, 1].

of its subinterval, renaming it $\frac{p_j^*}{q_j^*}$. Let L be the collection of the starred subintervals, at most m_ε of them. Then

$$\sum_L \frac{1}{q_j^*} \Delta x_j^* \leq \Delta\chi \sum_L \frac{1}{q_j^*} \leq m_\varepsilon \Delta\chi,$$

because all $\frac{1}{q_j^*} \leq 1$. This gives us a hint that we should use $\Delta\chi < \frac{\varepsilon}{m_\varepsilon}$, since doing so will then yield

$$\sum_L \frac{1}{q_j^*} \Delta x_j^* \leq m_\varepsilon \Delta\chi < \varepsilon.$$

Alternatively, if we *don't* consistently choose among the $\frac{p_i}{q_i}$ as representatives, opting for some other representatives

$$\frac{r_j}{s_j} \in [x_{j-1}^*, x_j^*]$$

instead, then certainly

$$\sum_L \frac{1}{s_j} \Delta x_j^* < \sum_L \frac{1}{q_j^*} \Delta x_j^* < \varepsilon.$$

Roughly speaking, we have found that the contribution to the Riemann sum made by the rectangles with the tallest heights can be made arbitrarily small.

If there are any subintervals not belonging to L – the non-L subintervals – they necessarily have total length less than the length of the entire interval [0, 1], that is,

$$\sum_{non\text{-}L} \Delta x_i < 1.$$

If any, they all have representatives $\frac{r_i}{s_i}$ with ordinates $\frac{1}{s_i} < \varepsilon$. Therefore,

$$\sum_{non\text{-}L} \frac{1}{s_i} \Delta x_i \leq \sum_{non\text{-}L} \varepsilon \Delta x_i < \varepsilon \cdot 1 = \varepsilon.$$

Hence, the contribution to the Riemann sum made by the rectangles over the non-L intervals is arbitrarily small.

We now realize that by setting

$$\delta = \begin{cases} \frac{\varepsilon}{m_\varepsilon} & \text{if } \varepsilon < 1 \\ \frac{1}{2} & \text{if } \varepsilon \geq 1 \end{cases}$$

and taking $\Delta\chi < \delta$, the Riemann sum

$$\sum_{L} \frac{1}{q_j^*} \Delta x_j^* + \sum_{non\text{-}L} \frac{1}{s_i} \Delta x_i < \varepsilon + \varepsilon = 2\varepsilon.$$

Since 2ε may be made arbitrarily small,

$$\lim_{\Delta\chi \to 0} \sum_{i=1}^{n} f(X_i)\Delta x_i = 0,$$

where X_i is a rational number in [0, 1]. The conclusion, per our definition, is that regardless of the nets and representatives (rational or irrational) that we use,

$$\lim_{\Delta\chi \to 0} \sum_{i=1}^{n} f(X_i)\Delta x_i = \int_0^1 f(x)dx = 0,$$

the Riemann integral of the ruler function over [0, 1]. **QED**

Intervals are the simplest subsets of $\mathbb{R}$, and their lengths are simple to calculate. For this function, we segregated each net into subintervals of two kinds, the L type and the non-L type. Since the net was composed of intervals, we could easily imagine the lengths Δx_j^* and Δx_i.

But as subsets become more difficult to describe, so do their lengths. For instance, how does one assign a length to the subset of irrational numbers in $[a, b]$? In Chapter 55, you begin to learn about the *measure* of a set, which is a real number that extends the definition of length to many non-interval sets (as can be expected, the measure of an interval is still its regular length). With the measure of a subset of $[a, b]$ in hand, extremely general nets for $[a, b]$ can be imagined, no longer being restricted to subintervals. Subsets may be segregated into those of measure zero, and those of positive measure. Then, a more general definition of area can be defined for a vast class of functions. The tools for this are the Lebesgue integral and the generalized Riemann integral, excellent topics for advanced real analysis!

D

Continued Fractions

A continued fraction is an amazing discovery. Here is an example of one,

$$5 + \cfrac{1}{1 + \cfrac{1}{6+\frac{1}{3}}},$$

which equals $\frac{129}{22}$ when simplified. As a general form, a **finite continued fraction** is the expression

$$a_0 + \cfrac{1}{a_1 + \cfrac{1}{a_2 + \ddots + \cfrac{1}{a_{n-1} + \frac{1}{a_n}}}},$$

where all $a_i \in \mathbb{N}$ (more general settings allow the numerators to be any natural numbers and $a_0 \in \mathbb{Z}$). To save space, we collapse the expression to $[a_0; a_1, \ldots, a_n]$. The process of going from, say, $[5; 1, 6, 3]$ to $\frac{129}{22}$, is straightforward. The reverse process goes as follows,

$$\frac{129}{22} = 5 + \frac{19}{22} = 5 + \frac{1}{\frac{22}{19}} = 5 + \frac{1}{1 + \frac{3}{19}} = 5 + \frac{1}{1 + \frac{1}{\frac{19}{3}}} = 5 + \cfrac{1}{1 + \cfrac{1}{6+\frac{1}{3}}},$$

and illustrates the general algorithm for converting any rational number into its continued fraction. At every step, the algorithm insists that the largest integer be extracted from an improper fraction, and that all fractions be in lowest terms. In our example, it is required that we extract $\lfloor \frac{129}{22} \rfloor = 5$, $\lfloor \frac{22}{19} \rfloor = 1$, and $\lfloor \frac{19}{3} \rfloor = 6$. Thus, the algorithm turns any fraction $\frac{p}{q}$ into $[a_0; a_1, \ldots, a_n]$, but not quite uniquely. For instance, $\frac{129}{22}$ may be converted into two continued fractions: $[5; 1, 6, 3]$ and

$$[5; 1, 6, 2, 1] = 5 + \cfrac{1}{1 + \cfrac{1}{6+\cfrac{1}{2+\frac{1}{1}}}}.$$

However, no other continued fractions can equal $\frac{129}{22}$. In general, any $\frac{p}{q}$ will generate the pair $[a_0; a_1, \ldots, a_n]$ and $[a_0; a_1, \ldots, a_n - 1, 1]$. Aside from this harmless ambiguity, the continued fraction expansion for any $\frac{p}{q}$ is unique (this won't be proved here). It follows that $\mathbb{Q}$ can be put into one-to-one correspondence with the set of finite continued fractions.

The algorithm looks a bit strange when applied to decimal expressions. For instance, with a calculator, we have

$$23.675 = 23 + 0.675 \approx 23 + \frac{1}{1 + 0.481481} \approx 23 + \frac{1}{1 + \frac{1}{2+0.0769252}} \approx 23 + \frac{1}{1 + \frac{1}{2+\frac{1}{12.999641}}}.$$

We suspect that the last denominator is actually 13. In fact, it is 13, as you can easily verify since $23.675 = \frac{947}{40}$. And we know that unavoidable rounding is the reason the algorithm didn't terminate with exactly 13.

Tangling with non-terminating decimals puts us in a vast landscape, for consider

$$\pi = 3 + 0.14159265\ldots = 3 + \frac{1}{7 + 0.0625135\ldots} = 3 + \frac{1}{7 + \frac{1}{15+0.996545\ldots}},$$

and so on, yielding $\pi = [3; 7, 15, 1, 292, 1, 1, 1, 2, \ldots]$. It becomes clear that π cannot have a terminating continued fraction. We have seen that things get tricky: if a decimal that the algorithm is applied to (including the original one, naturally) is rounded off to too few significant digits, then terrible roundoff errors will soon accrue, resulting in incorrect integers a_n. Try the algorithm on 3.1416, an otherwise excellent approximation of π, to see how short the list of correct terms a_n is (and do Exercise D.1a).

It is now time to ask the converse question, "What number, if any, does the **infinite continued fraction**

$$a_0 + \frac{1}{a_1 + \frac{1}{a_2 + \ddots}} \equiv [a_0; a_1, a_2, \ldots]$$

represent?" First of all, the expression is bounded above by $a_0 + \frac{1}{a_1}$ and below by 0. To investigate further, we consider the nth **convergent** $\frac{p_n}{q_n}$ of $[a_0; a_1, a_2 \cdots]$, defined as the front end fraction $\frac{p_n}{q_n} = [a_0; a_1, \ldots, a_n]$. Specifically,

$$\frac{p_0}{q_0} = [a_0] = \frac{a_0}{1}, \frac{p_1}{q_1} = [a_0; a_1] = a_0 + \frac{1}{a_1} = \frac{a_0 a_1 + 1}{a_1},$$

and so forth. (It can be proved that all convergents are reduced fractions, so that for instance, $p_1 = a_0 a_1 + 1$ and $q_1 = a_1$.) If $\langle \frac{p_n}{q_n} \rangle_0^\infty$ converges to a limit L, then the infinite continued fraction $[a_0; a_1, a_2 \cdots]$ shall equal L by definition. If L exists, it must be irrational, by what was found about finite continued fractions.

One of the most important properties about convergents is stated in the following theorem, which won't be proved here. It implies that successive convergents differ by smaller and smaller amounts.

Theorem D.1 Successive convergents of a continued fraction follow

$$\frac{p_n}{q_n} - \frac{p_{n-1}}{q_{n-1}} = \frac{(-1)^{n-1}}{q_n q_{n-1}}$$

for $n \in \mathbb{N}$, where $1 = q_0 \leq q_1 < q_2 < q_3 < \cdots$.

For example, for [1; 2, 3, 4, . . .] we have

$$\frac{p_2}{q_2} = [1; 2, 3] = \frac{10}{7}$$

and

$$\frac{p_3}{q_3} = [1; 2, 3, 4] = \frac{43}{30},$$

so that

$$\frac{p_3}{q_3} - \frac{p_2}{q_2} = \frac{43}{30} - \frac{10}{7} = \frac{(-1)^2}{30 \cdot 7}.$$

Notice that $q_0 = 1 < q_1 = 2 < q_2 = 7 < q_3 = 30 < \cdots$. Very nice!

Lemma D.1 Successive even-numbered convergents form an increasing subsequence, and successive odd-numbered convergents form a decreasing subsequence of $\langle \frac{p_n}{q_n} \rangle_0^\infty$.

Proof. Devise a proof as Exercise D.3b, using Figure D.1 as a guide. **QED**

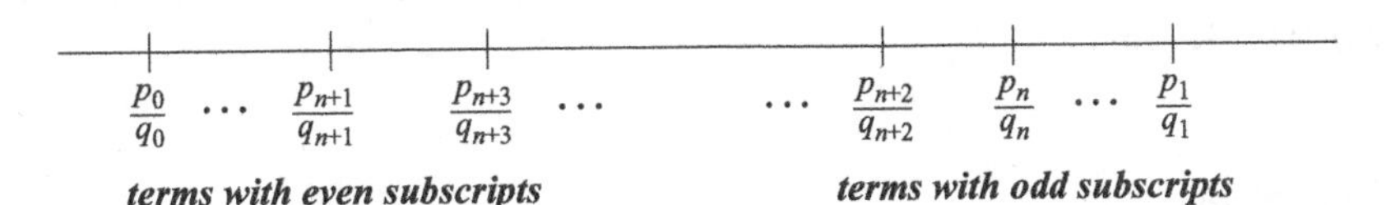

Figure D.1. The two key subsequences in $\langle \frac{p_i}{q_i} \rangle_0^\infty$.

We will now prove that $\langle \frac{p_n}{q_n} \rangle_1^\infty$ converges.

Theorem D.2 The convergents $\langle \frac{p_n}{q_n} \rangle_1^\infty$ of any continued fraction form a Cauchy sequence, and therefore converge.

Proof. To show that $\langle \frac{p_n}{q_n} \rangle_1^\infty$ is a Cauchy sequence, we must prove that for any $\varepsilon > 0$, there exists an N_ε such that for any $m, n \in \mathbb{N}$ with $m > n \geq N_\varepsilon$, $|\frac{p_m}{q_m} - \frac{p_n}{q_n}| < \varepsilon$. Let n be a fixed natural number, and let $m > n$, where if n is odd then m is even, and vice versa (we say they are of opposite parity). As Lemma D.1 and Exercise D.3b prove, and Figure D.1 shows, even convergents $\frac{p_m}{q_m}$ increase and don't surpass any odd convergent, and odd convergents $\frac{p_m}{q_m}$ decrease and don't fall behind any even convergent. Therefore, with help from Theorem D.1,

$$\left| \frac{p_m}{q_m} - \frac{p_n}{q_n} \right| \leq \left| \frac{p_{n+1}}{q_{n+1}} - \frac{p_n}{q_n} \right| = \frac{1}{q_{n+1} q_n}$$

in both cases.

In case n and m are of the same parity (keeping $m > n$), it is still true that

$$\left| \frac{p_m}{q_m} - \frac{p_n}{q_n} \right| \leq \left| \frac{p_{n+1}}{q_{n+1}} - \frac{p_n}{q_n} \right| = \frac{1}{q_{n+1} q_n}$$

because now $n+1$ is of opposite parity to both n and m, and Exercise D.3b applies again. We now see that the convergents are relentlessly packing closer and closer together, because $\langle q_i \rangle_0^\infty$ is a monotone increasing sequence.

Continuing, in general we have $q_i \geq i$ (in fact, q_i is often much larger than i), so that

$$\frac{1}{q_{n+1}q_n} \leq \frac{1}{(n+1)n} < \frac{1}{n},$$

excluding $n = 0$. Thus, for any $\varepsilon > 0$, establish $N_\varepsilon = \frac{1}{\varepsilon}$. We now retrace our logic, and state that for any $\varepsilon > 0$, let $N_\varepsilon = \frac{1}{\varepsilon}$, and take $m > n \geq N_\varepsilon$. Then

$$\left| \frac{p_m}{q_m} - \frac{p_n}{q_n} \right| \leq \frac{1}{q_{n+1}q_n} < \frac{1}{n} \leq \frac{1}{N_\varepsilon} = \varepsilon.$$

It follows that $\langle \frac{p_n}{q_n} \rangle_1^\infty$ is a Cauchy sequence.

We conclude that all infinite continued fractions $[a_0; a_1, a_2 \cdots]$ converge in $\mathbb{R}$ by the Cauchy convergence criterion (Theorem 21.1). **QED**

Note 1: More information about continued fractions and Farey fractions (Appendix A), with an application, is found in "Small Denominators: No Small Problem," by Scott J. Beslin, Douglas J. Baney, and Valerio De Angelis, in *Mathematics Magazine*, vol. 71, no. 2 (April 1998), pp. 132–138.

D.1 Exercises

1. a. Find the continued fractions for: (i) $\frac{355}{113}$ (ii) $\sqrt{2}$ (iii) the golden ratio $\frac{1+\sqrt{5}}{2}$.
   b. Start with a calculator value of e and see how many terms of your continued fraction agree with $e = [2; 1, 2, 1, 1, 4, 1, 1, 6, 1, 1, 8, 1, 1, 10, \ldots]$. Isn't it amazing that e, a transcendental number, possesses a pattern as a continued fraction? To find a formula for a_n in this sequence, follow these steps. (i) Beginning with $a_2 = 2$, the sequence of subscripts $\langle 2, 5, 8, 11, \ldots \rangle$ yields the subsequence $\langle 2, 4, 6, 8, \ldots \rangle$. Use this to find a linear function $\ell(n)$ that gives $\ell(2) = 2$, $\ell(5) = 4, \ldots$. (ii) To obtain the intervening 1s, it seems best to use an exponent, as in $(\ell(n))^{f(n)}$, where you must design $f(n)$ so that

$$f(n) = \begin{cases} 1 & \text{for } n \in \{2, 5, 8, 11, \ldots\} \\ 0 & \text{for } n \in \mathbb{N} \backslash \{2, 5, 8, 11, \ldots\} \end{cases}.$$

   A trigonometric function $f(n) = A\cos(B(n-2)) + C$ can actually do this. To make things easier, set $f(2) = 1$.

2. Find some convergents for the following, and check the claim of Lemma D.1:
   a. $\sqrt{2}$
   b. the golden ratio $\frac{1+\sqrt{5}}{2}$ Hint: Fibonacci numbers!
   c. e.

3. a. Explain how it is that successive convergents differ by smaller and smaller amounts. How small do these amounts become?

b. (i) Prove Lemma D.1, that odd convergents form a decreasing sequence, while even convergents form an increasing sequence, as in Figure D1. Hint:

$$\frac{p_{n+1}}{q_{n+1}} - \frac{p_n}{q_n} = \frac{(-1)^n}{q_{n+1}q_n}$$

and

$$\frac{p_{n+2}}{q_{n+2}} - \frac{p_{n+1}}{q_{n+1}} = \frac{(-1)^{n+1}}{q_{n+2}q_{n+1}}.$$

You need to consider

$$\frac{p_{n+2}}{q_{n+2}} - \frac{p_n}{q_n}.$$

(ii) Prove that the decreasing sequence of odd convergents is bounded below by any even convergent $\frac{p_{2n}}{q_{2n}}$, and that the increasing sequence of even convergents is bounded above by any odd convergent $\frac{p_{2n+1}}{q_{2n+1}}$. Hint: Suppose, on the contrary, that there is a first, odd convergent such that

$$\frac{p_{2m+1}}{q_{2m+1}} < \frac{p_{2n}}{q_{2n}}.$$

c. With the observations from **a** and **b**, may we conclude that all sequences of convergents converge? Hint: Does a figure in Chapter 19 look something like Figure D1 here?

E

L'Hospital's Rule

Sometimes one finds in calculation singular expressions that cannot be... considered as converging towards fixed limits. Such are, for example, the expressions 0×0, $\frac{0}{0}$, $\infty \times \infty$, $\frac{\infty}{\infty}$, $0 \times \infty$, 0^0, 1^∞, *&c....*

Augustin-Louis Cauchy, *Cours d'analyse de l'École royale polytechnique*, 1821, pg. 68.

Historically, this helpful theorem was discovered by Johann Bernoulli around 1694, during a period when discoveries in the new calculus came very fast. At that time, the brilliant Bernoulli earned a steady salary as the tutor of Guillaume, Marquis de l'Hospital (1661–1704), a member of the French nobility and himself a good mathematician. But Bernoulli also had contracted to supply all his mathematical discoveries exclusively to l'Hospital! In 1696, the marquis published the first—and extremely influential—textbook on calculus, his *Analyse des infiniment petits pour l'intelligence des lignes courbes* (Analysis of the Infinitesimally Small for the Understanding of Curved Lines). As it turned out, the book was basically a finely written compilation of Bernoulli's lessons, all under l'Hospital's name. In this way l'Hospital's rule, and most of Bernoulli's discoveries, became widely known. Nevertheless, the two men were cordial to each other, and because Bernoulli long outlived l'Hospital, it became clear who the master was, and who the expert writer.

E.1 Indeterminate Forms

In every calculus book we encounter the **indeterminate forms** $\frac{0}{0}$, $\frac{\infty}{\infty}$, $0 \cdot \infty$, 1^∞, 0^0, ∞^0, and $\infty - \infty$. None are real numbers, so for instance, $1^\infty = 1$ is completely unwarranted. Begin with two functions f and g appearing in the corresponding operations, that is, $\frac{f}{g}$, $f \cdot g$, f^g, and $f - g$. The indeterminate forms are *shorthands* for the respective operations done to the limits of f and g. In each case, the (attempted) operation on the limits doesn't yield a real number. Instead, we should be investigating the limit of the operations. For example, given $f(x) = 3 + \frac{1}{x}$ and $g(x) = \frac{1}{\pi x}$, we know that $\lim_{x \to 0+} f(x) = \infty$ and $\lim_{x \to 0+} g(x) = \infty$. Dividing these gives $\frac{\infty}{\infty}$, a shorthand for the division of the limits

$$\frac{\lim_{x \to 0+} f(x)}{\lim_{x \to 0+} g(x)}.$$

Saying that $\frac{\infty}{\infty} = 1$ is baseless; what is needed is the limit of the division: $\lim_{x \to 0+} \frac{f(x)}{g(x)} = \pi$. Replace π in $g(x)$ by any other positive constant a, and we get "$\frac{\infty}{\infty} = a$," as indeterminate as can be, and a misuse of the shorthand. Just as bad is writing "$\frac{\infty}{\infty} = \infty$," a shorthand that may be strained ($\lim_{x \to \infty} \frac{x^2}{3x} = \infty$), or false ($\lim_{x \to \infty} \frac{3x+1}{2x+5} = \frac{3}{2}$).

Some of the confusion surrounding indeterminate forms comes from how similar they look to expressions like $\frac{1}{\infty}$, $\frac{\infty}{0}$, $\infty \cdot \infty$, 0^∞, and $\infty + \infty$. These shorthands also are not real numbers, but are called *determinate* since they actually yield results: $\frac{1}{\infty}$ yields 0, $\infty \cdot \infty$ yields ∞, $\frac{\infty}{0}$ yields $\pm\infty$, 0^∞ yields 0, and $\infty + \infty$ yields ∞. (Note that "yields" doesn't mean "equals.") Determinacy implies that the form describes a unique limit (possibly $+\infty$ or $-\infty$). For example, if we think of

$$\lim_{x \to \infty} \left(\frac{1}{x}\right)^x \quad \text{as} \quad \left(\lim_{x \to \infty} \frac{1}{x}\right)^{\lim_{x \to \infty} x},$$

we obtain the determinate form 0^∞. With analytical care, we have

$$\lim_{x \to \infty} \left(\frac{1}{x}\right)^x = \frac{1}{\lim_{x \to \infty} x^x} = 0.$$

It's in this sense that we interpret "$\frac{1}{\infty}$ yields 0" and "0^∞ yields 0."

As is hinted with the examples at the end of this appendix, all indeterminate forms can be reduced to either $\frac{0}{0}$ or $\pm\frac{\infty}{\infty}$. For instance, if $f - g$ yields the form $\infty - \infty$, we can at least in principle reexpress

$$f - g = \frac{\frac{1}{g} - \frac{1}{f}}{\frac{1}{fg}},$$

which then indicates $\frac{0}{0}$. Or, let's say $f > 0$ over $(0, b)$, and suppose f^g yields the form 0^0 when $x = 0$. We let $y = f^g$ and then $\ln y = g \ln f$. In this case, $g \ln f$ will yield the form $0 \cdot \infty$ or $-(0 \cdot \infty)$, so that $\frac{\ln f}{1/g}$ yields $\frac{\infty}{\infty}$ or $-\frac{\infty}{\infty}$.

E.2 L'Hospital's Rule

L'Hospital's rule will be proved in four parts. Parts I and II cover the indeterminate form $\frac{0}{0}$, and parts III and IV investigate the indeterminate form $\frac{\infty}{\infty}$. Within each case there will be several subcases, 30 in total. All of this makes l'Hospital's rule very flexible, but be careful, it shouldn't become our first tool for determining limits (as is said, now that we have a hammer, every problem begins to look like a nail).

The proof of the theorem relies heavily on Cauchy's (generalized) mean value theorem (Theorem 39.4). As noted there, $g(x) \neq 0$ throughout the open intervals in question here by applying to g the (regular) mean value theorem (Theorem 39.3).

Theorem E.1 L'Hospital's Rule, Part I Let $f, g : \mathbb{R} \to \mathbb{R}$ be continuous over $[a, b] \subseteq \mathcal{D}_f \cap \mathcal{D}_g$, differentiable over (a, b), and let $g'(x) \neq 0$ anywhere in (a, b). If $\lim_P f(x) = \lim_P g(x) = 0$, and if $\lim_P \frac{f'(x)}{g'(x)} = L$, where P is one of $x \to a^+$, $x \to b^-$, and $x \to c \in (a, b)$, then the conclusion is that $\lim_P \frac{f(x)}{g(x)} = L$ also. Furthermore, L may be $+\infty$ or $-\infty$.

Proof if P is $x \to a^+$ and L is finite. Since f and g are continuous at a and

$$\lim_{x \to a+} f(x) = \lim_{x \to a+} g(x) = 0,$$

we have $f(a) = g(a) = 0$. Because f and g are continuous over any closed interval $[a, d] \subset [a, b]$ and differentiable over (a, d), Cauchy's generalized mean value theorem (Theorem 39.4) applies to any $x \in (a, d)$, and we get that there exists a $\xi \in (a, x)$ (see Figure E.1) at which

$$\frac{f'(\xi)}{g'(\xi)} = \frac{f(x) - f(a)}{g(x) - g(a)} = \frac{f(x)}{g(x)}.$$

As $x \to a^+$, so too does $\xi \to a^+$. We are given that

$$\lim_{x \to a+} \frac{f'(x)}{g'(x)} = L,$$

meaning that

$$\lim_{\xi \to a+} \frac{f'(\xi)}{g'(\xi)} = L.$$

Therefore,

$$\lim_{x \to a+} \frac{f(x)}{g(x)} = L$$

also.

Figure E.1. The order of points.

Proof if P is $x \to b^-$ and L is finite. This is proved parallel to the first case, using intervals $[d, b]$ and (d, b), and $\xi \in (x, b) \subset (d, b)$. The points ξ, x, and d will be in reverse order in Figure E.1.

Proof if P is $x \to c \in (a, b)$ and L is finite. By the two previous cases,

$$\lim_{x \to c+} \frac{f(x)}{g(x)} = \lim_{x \to c-} \frac{f(x)}{g(x)} = L,$$

and Exercise 23.9 gives us $\lim_{x \to c} \frac{f(x)}{g(x)} = L$.

Proofs if P is $x \to a^+$ and L is $+\infty$ or $-\infty$. If we find that $\lim_{x \to a+} \frac{f'(x)}{g'(x)} = L = +\infty$, this means that $\frac{f'(x)}{g'(x)}$ is unbounded in every interval $(a, d) \subset [a, b]$. But for each $x \in (a, d)$, Theorem 39.4 *applies, and we still have* $\frac{f'(\xi)}{g'(\xi)} = \frac{f(x)}{g(x)}$ for some $\xi \in (a, x)$ (see Figure E.1). Therefore,

$$\lim_{x \to a+} \frac{f(x)}{g(x)} = \lim_{\xi \to a+} \frac{f'(\xi)}{g'(\xi)} = +\infty.$$

Parallel thinking proves the case when $L = -\infty$.

The remaining cases are proved similarly. **QED**

Theorem E.2 L'Hospital's Rule, Part II For some $a > 0$, let $f, g : \mathbb{R} \to \mathbb{R}$ be continuous over $[a, \infty) \subseteq \mathcal{D}_f \cap \mathcal{D}_g$, differentiable over (a, ∞), and let $g'(x) \neq 0$ over (a, ∞). If $\lim_{x\to\infty} f(x) = \lim_{x\to\infty} g(x) = 0$, and if $\lim_{x\to\infty} \frac{f'(x)}{g'(x)} = L$, then $\lim_{x\to\infty} \frac{f(x)}{g(x)} = L$ also. Furthermore, L may be $+\infty$ or $-\infty$.

Similarly, for some $a < 0$, let $f, g : \mathbb{R} \to \mathbb{R}$ be continuous over $(-\infty, a] \subseteq \mathcal{D}_f \cap \mathcal{D}_g$, differentiable over $(-\infty, a)$, and let $g'(x) \neq 0$ over $(-\infty, a)$. If $\lim_{x\to-\infty} f(x) = \lim_{x\to-\infty} g(x) = 0$, and if $\lim_{x\to-\infty} \frac{f'(x)}{g'(x)} = L$, then $\lim_{x\to-\infty} \frac{f(x)}{g(x)} = L$ also. Furthermore, L may be $+\infty$ or $-\infty$.

Proof if $x \to \infty$ and L is finite. We use a change of variables $x = \frac{1}{u}$. Define

$$F(u) = \begin{cases} f\left(\frac{1}{u}\right) & \text{if } u \in \left(0, \frac{1}{a}\right) \\ 0 & \text{if } u = 0 \end{cases} \quad \text{and} \quad G(u) = \begin{cases} g\left(\frac{1}{u}\right) & \text{if } u \in \left(0, \frac{1}{a}\right) \\ 0 & \text{if } u = 0 \end{cases}.$$

Since $\lim_{u\to0+} F(u) = \lim_{x\to\infty} f(x) = 0$, F is continuous at $u = 0$, and likewise for G. Thus, F and G are continuous over $\left[0, \frac{1}{a}\right]$ and differentiable over $\left(0, \frac{1}{a}\right)$. Also,

$$G'(u) = g'\left(\frac{1}{u}\right)\left(\frac{-1}{u^2}\right) \neq 0$$

over $u \in \left(0, \frac{1}{a}\right)$ (since $g'(x) \neq 0$ over (a, ∞)). We can therefore apply Theorem 39.4 to F and G, and get that for each $u \in \left(0, \frac{1}{a}\right)$, there exists some $\xi \in (0, u)$ such that

$$\frac{F'(\xi)}{G'(\xi)} = \frac{F(u) - F(0)}{G(u) - G(0)} = \frac{F(u)}{G(u)}.$$

This means that

$$\frac{f'\left(\frac{1}{\xi}\right)\left(\frac{-1}{\xi^2}\right)}{g'\left(\frac{1}{\xi}\right)\left(\frac{-1}{\xi^2}\right)} = \frac{f'\left(\frac{1}{\xi}\right)}{g'\left(\frac{1}{\xi}\right)} = \frac{f\left(\frac{1}{u}\right)}{g\left(\frac{1}{u}\right)},$$

and hence,

$$\lim_{\xi\to0+} \frac{f'\left(\frac{1}{\xi}\right)}{g'\left(\frac{1}{\xi}\right)} = \lim_{u\to0+} \frac{f\left(\frac{1}{u}\right)}{g\left(\frac{1}{u}\right)}.$$

Returning to x, we have that

$$\lim_{\xi\to\infty} \frac{f'(\xi)}{g'(\xi)} = \lim_{x\to\infty} \frac{f(x)}{g(x)},$$

and therefore,

$$\lim_{x\to\infty} \frac{f'(x)}{g'(x)} = \lim_{x\to\infty} \frac{f(x)}{g(x)} = L.$$

Proof if $x \to -\infty$ and L is finite. This case is proved in a similar manner, with $a < 0$ and F and G defined over $\left[\frac{1}{a}, 0\right]$ and differentiable over $\left(\frac{1}{a}, 0\right)$.

Proofs if $x \to \infty$ and L *is* $+\infty$ *and* $-\infty$. If

$$\lim_{x\to\infty} \frac{f'(x)}{g'(x)} = L = +\infty$$

or

$$\lim_{x\to\infty} \frac{f'(x)}{g'(x)} = L = -\infty,$$

we know that

$$Q(u) = \frac{f'\left(\frac{1}{u}\right)}{g'\left(\frac{1}{u}\right)}$$

(see the case above) is unbounded in every interval $\left(0, \frac{1}{a}\right)$. Thus, for any $N > 0$, there is a $u \in \left(0, \frac{1}{a}\right)$ such that

$$\left|\frac{f'\left(\frac{1}{u}\right)}{g'\left(\frac{1}{u}\right)}\right| > N.$$

Hence, there is a $\xi \in (0, u)$ such that

$$\left|\frac{f'\left(\frac{1}{\xi}\right)}{g'\left(\frac{1}{\xi}\right)}\right| = \left|\frac{f\left(\frac{1}{u}\right)}{g\left(\frac{1}{u}\right)}\right| > N,$$

or equivalently, $\left|\frac{f(x)}{g(x)}\right| > N$ for some $x \in (a, \infty)$. Therefore,

$$\lim_{x\to\infty} \frac{f(x)}{g(x)} = +\infty \qquad \text{or} \qquad \lim_{x\to\infty} \frac{f(x)}{g(x)} = -\infty,$$

respectively.

Proofs if $x \to -\infty$ and L *is* $+\infty$ *and* $-\infty$. This is similar to the last case, but with $a < 0$, and $Q(u)$ unbounded in every interval $\left(\frac{1}{a}, 0\right)$. The conclusions that

$$\lim_{x\to-\infty} \frac{f'(x)}{g'(x)} = +\infty \qquad \text{or} \qquad \lim_{x\to-\infty} \frac{f'(x)}{g'(x)} = -\infty$$

follow by similar reasoning. **QED**

We will need the following lemma in parts III and IV of l'Hospital's rule. The proof of part III is guided by [18], pp. 118–119.

Lemma E.1 For any $\varepsilon > 0$ and any real constant L, if the expressions A and B are such that $|A - L| < \varepsilon$ and $|B - 1| < \varepsilon$, then $|AB - L| < (|L| + \varepsilon + 1)\,\varepsilon$, a number as small as we please.

Proof. $|AB - L| = |AB - A + A - L| \le |A|\,|B - 1| + |A - L|$ by the triangle inequality. Also, $|A - L| < \varepsilon$ implies $|A| < |L| + \varepsilon$ by Corollary 15.1. Thus, $|AB - L| < (|L| + \varepsilon)\,\varepsilon + \varepsilon = (|L| + \varepsilon + 1)\,\varepsilon$. **QED**

Theorem E.3 L'Hospital's Rule, Part III Let $f, g : \mathbb{R} \to \mathbb{R}$ be differentiable over $[a, \infty) \subseteq \mathcal{D}_f \cap \mathcal{D}_g$, and let $g'(x) \neq 0$ over $[a, \infty)$. If $\lim_{x\to\infty} f(x) = \lim_{x\to\infty} g(x) = \infty$, and if $\lim_{x\to\infty} \frac{f'(x)}{g'(x)} = L$, then $\lim_{x\to\infty} \frac{f(x)}{g(x)} = L$ also. Furthermore, L may be $+\infty$ or $-\infty$.

Similarly, let $f, g : \mathbb{R} \to \mathbb{R}$ be continuous over $(-\infty, a] \subseteq \mathcal{D}_f \cap \mathcal{D}_g$, differentiable over $(-\infty, a)$, and let $g'(x) \neq 0$ over $(-\infty, a)$. If $\lim_{x\to-\infty} f(x) = \lim_{x\to-\infty} g(x) = \infty$, and if $\lim_{x\to-\infty} \frac{f'(x)}{g'(x)} = L$, then $\lim_{x\to-\infty} \frac{f(x)}{g(x)} = L$ also. Furthermore, L may be $+\infty$ or $-\infty$.

Proof if $x \to \infty$ and L is finite. Since f and g increase without bound, we can choose $M > a$ large enough that $f(x) \neq 0$ and $g(x) \neq 0$ over $[M, \infty)$. Next, take $m > M$ large enough that for all $x > m$, $f(x) \neq f(M)$. By Cauchy's generalized mean value theorem, for each $x > m$ there is a $\xi \in (M, x)$ (see Figure E.2) such that

$$\frac{f(x) - f(M)}{g(x) - g(M)} = \frac{f'(\xi)}{g'(\xi)}$$

(verify that neither $g(x) - g(M)$ nor $f(x) - f(M)$ will be zero). Reexpressing, this is

$$\frac{f(x)}{g(x)} \cdot \frac{1 - f(M)/f(x)}{1 - g(M)/g(x)} = \frac{f'(\xi)}{g'(\xi)},$$

or

$$\frac{f(x)}{g(x)} = \frac{f'(\xi)}{g'(\xi)} \left[\frac{1 - g(M)/g(x)}{1 - f(M)/f(x)} \right] \equiv \frac{f'(\xi)}{g'(\xi)} U(x), \tag{1}$$

where the function in brackets defines $U(x)$. Note that $\lim_{x\to\infty} U(x) = 1$. (1) tells us that since f and g increase without bound and $U(x) \to 1$ as $x \to \infty$,

$$\frac{f(x)}{g(x)} \to \frac{f'(\xi)}{g'(\xi)} \to L.$$

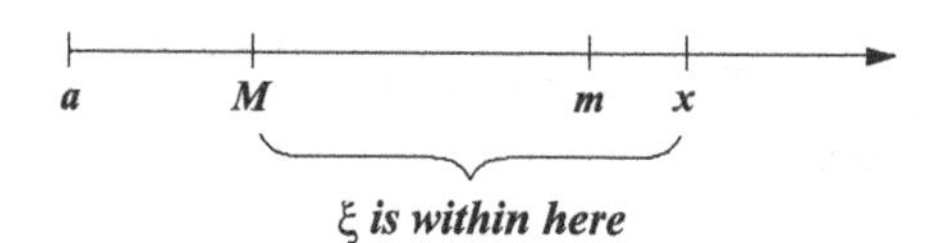

Figure E.2. The order of points.

To be analytically precise about all this, given any $\varepsilon > 0$, let $M > a$ and let $m > M$ be so large that for any $x > m > M$ and the matching $\xi \in (M, x)$ (see Figure E.2) guaranteed by Theorem 39.4, both

$$|U(x) - 1| < \varepsilon \quad \text{and} \quad \left| \frac{f'(\xi)}{g'(\xi)} - L \right| < \varepsilon.$$

We need to show that $|\frac{f(x)}{g(x)} - L|$ can become as small as we wish. Apply Lemma E1 with $A = \frac{f'(\xi)}{g'(\xi)}$ and $B = U(x)$. Then given any $\varepsilon > 0$, we have found an M (and an $m > M$) such that for our x and ξ, we have

$$\left| \frac{f(x)}{g(x)} - L \right| = \left| \frac{f'(\xi)}{g'(\xi)} U(x) - L \right| < (|L| + \varepsilon + 1)\,\varepsilon,$$

a number as small as we please. Hence, $\left|\frac{f(x)}{g(x)} - L\right| \to 0$ as M, and thus x, increase unboundedly. Concisely put, $\lim_{x\to\infty} \frac{f(x)}{g(x)} = L$.

Proof if $x \to \infty$ and L is $+\infty$ and $-\infty$. Using the order of points in Figure E.2, we still have

$$\frac{f(x)}{g(x)} = \frac{f'(\xi)}{g'(\xi)} \left[\frac{1 - g(M)/g(x)}{1 - f(M)/f(x)} \right] = \frac{f'(\xi)}{g'(\xi)} U(x),$$

and $U(x) \to 1$ as $x \to \infty$. But now, $\frac{f'(\xi)}{g'(\xi)}$ is unbounded as $M, m, x \to \infty$. So, for any $N > 0$, $m > M$ can be found so that

$$\left| \frac{f'(\xi)}{g'(\xi)} \right| U(x) > N.$$

It follows that $\left| \frac{f(x)}{g(x)} \right| > N$, meaning that

$$\lim_{x\to\infty} \frac{f(x)}{g(x)} = +\infty \qquad \text{or} \qquad \lim_{x\to\infty} \frac{f(x)}{g(x)} = -\infty,$$

respectively.

Proof if $x \to -\infty$ and L is finite. The proof follows the first case under this part, but with a, M, m, and x in reverse order in Figure E.2, and with the arrow pointing to the left. Now $M, m, x \to -\infty$.

Proofs if $x \to -\infty$ and L is $+\infty$ and $-\infty$. The proofs follow from the case above with a, M, m, and x in reverse order in Figure E.2, where as $M, m, x \to -\infty$, $\left| \frac{f'(\xi)}{g'(\xi)} \right| U(x) > N$. Therefore, $\lim_{x\to\infty} \frac{f(x)}{g(x)} = +\infty$ or $\lim_{x\to\infty} \frac{f(x)}{g(x)} = -\infty$, respectively. **QED**

Theorem E.4 L'Hospital's Rule, Part IV Let $f, g : \mathbb{R} \to \mathbb{R}$ be continuous over $[a, b] \subseteq \mathcal{D}_f \cap \mathcal{D}_g$, differentiable over (a, b), and let $g'(x) \neq 0$ over (a, b). If $\lim_P f(x) = \lim_P g(x) = \infty$, and if $\lim_P \frac{f'(x)}{g'(x)} = L$, where P is one of $x \to a^+, x \to b^-$, and $x \to c \in (a, b)$, then the conclusion is that $\lim_P \frac{f(x)}{g(x)} = L$ also. Furthermore, L may be $+\infty$ or $-\infty$.

Proof if $x \to a^+$ and L is finite. Since f and g increase without bound near a, we can choose $M \in (a, b)$ close enough to a that $f(x) \neq 0$ and $g(x) \neq 0$ over $(a, M]$. Take $m \in (a, M]$ such that for all $x \in (a, m]$, $f(x) \neq f(M)$. The proof now follows that of part III when L is finite, but using Figure E.3, with $M, m, x \to a^+$.

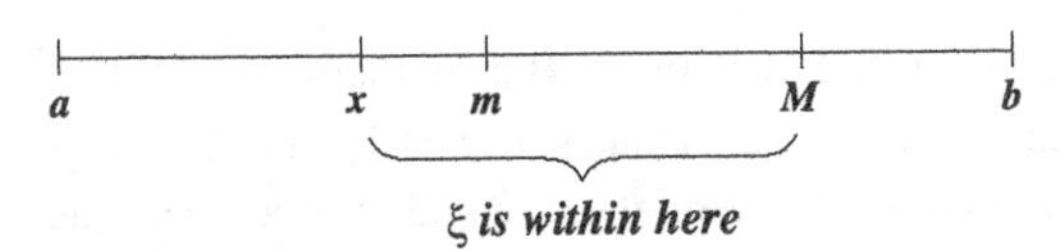

Figure E.3. The order of points.

Proof if $x \to a^+$ and L is $+\infty$ and $-\infty$. Using the order of points in Figure E.3, we still have

$$\frac{f(x)}{g(x)} = \frac{f'(\xi)}{g'(\xi)} \left[\frac{1 - g(M)/g(x)}{1 - f(M)/f(x)} \right] = \frac{f'(\xi)}{g'(\xi)} U(x),$$

and $U(x) \to 1$ as $M, m, x \to a^+$. As in the corresponding case in part III, x can be found such that $\left| \frac{f'(\xi)}{g'(\xi)} \right| U(x) > N$ for any $N > 0$. It follows that $\left| \frac{f(x)}{g(x)} \right| > N$, meaning that

$$\lim_{x\to a+} \frac{f(x)}{g(x)} = +\infty \qquad \text{or} \qquad \lim_{x\to a+} \frac{f(x)}{g(x)} = -\infty,$$

respectively.

Proof if $x \to b^-$ and L is finite, $+\infty$ and $-\infty$. For finite L, the proof follows the first case of this part, but with x, m, and M in reverse order in Figure E.3. To prove this case when L is $+\infty$ or $-\infty$, follow the steps in the case above, but with $M, m, x \to b^-$.

Proof if $x \to c \in (a, b)$ and L is finite, $+\infty$ and $-\infty$. These cases follow from the cases above applied to $x \to c^+$ and $x \to c^-$. **QED**

Example E.1 Find $\lim_{x\to 0}(1+x)^{3/x}$. Because the indeterminate form is 1^∞, let $y = (1+x)^{3/x}$. Then $\ln y = \frac{3}{x}\ln(1+x)$ and we are interested in $\lim_{x\to 0}\ln y = \lim_{x\to 0}\frac{3\ln(1+x)}{x}$. By the continuity of the logarithmic function at any $x > 0$, we have $\lim_{x\to 0}\ln y = \ln(\lim_{x\to 0} y)$ if the limit exits. Now, $\lim_{x\to 0}\frac{3\ln(1+x)}{x}$ displays the indeterminate form $\frac{0}{0}$, the functions $f(x) = 3\ln(1+x)$ and $g(x) = x$ are continuous and differentiable over $[0, 1]$, and $g'(x) \neq 0$ over $(0, 1)$. Thus, l'Hospital's rule part I applies, and we find that

$$\lim_{x\to 0}\frac{3\ln(1+x)}{x} = \lim_{x\to 0}\frac{3}{1+x} = 3.$$

Then $\ln(\lim_{x\to 0} y) = 3$, so $\lim_{x\to 0}(1+x)^{3/x} = e^3$.

For comparison, a l'Hospital-free method has

$$\lim_{x\to 0+}(1+x)^{3/x} = \left(\lim_{x\to 0+}(1+x)^{1/x}\right)^3 = \left(\lim_{y\to\infty}\left(1+\frac{1}{y}\right)^y\right)^3 = e^3$$

(and similarly for $x \to 0^-$) by Chapter 35. ■

Example E.2 Find $\lim_{x\to\infty}\frac{\sqrt{x^2+1}}{x}$. The indeterminate form is $\frac{\infty}{\infty}$, and checking that over $[0, \infty)$ the premises of l'Hospital's rule part III are in order, we begin:

$$\lim_{x\to\infty}\frac{\sqrt{x^2+1}}{x} = \lim_{x\to\infty}\frac{x}{\sqrt{x^2+1}}.$$

The right hand side has the form $\frac{\infty}{\infty}$, and l'Hospital's rule part III applies again, giving

$$\lim_{x\to\infty}\frac{x}{\sqrt{x^2+1}} = \lim_{x\to\infty}\frac{\sqrt{x^2+1}}{x}.$$

This is nasty, and l'Hospital's rule may have wasted our time. But perhaps not, for if $\lim_{x\to\infty}\frac{\sqrt{x^2+1}}{x} = L$ exists, then we have found that $L = \frac{1}{L}$. Hence, $L = \pm 1$, and the negative value is rejected. But still, we should have avoided l'Hospital's rule and used a technique from Chapter 20 applied to continuous functions:

$$\lim_{x\to\infty}\frac{\sqrt{x^2+1}}{x} = \lim_{x\to\infty}\sqrt{\frac{x^2+1}{x^2}} = 1. \quad ■$$

Example E.3 Find $\lim_{\theta\to 0}(\csc^2\theta - \theta\csc^3\theta)$. This involves the indeterminate forms $\infty - 0\cdot\infty$. We have

$$\frac{1}{\sin^2\theta} - \frac{\theta}{\sin^3\theta} = \frac{\sin\theta - \theta}{\sin^3\theta},$$

which leads to $\frac{0}{0}$. By l'Hospital's rule part I,

$$\lim_{\theta\to 0}\frac{\sin\theta-\theta}{\sin^3\theta}=\lim_{\theta\to 0}\frac{\cos\theta-1}{3\sin^2\theta\cos\theta}=\lim_{\theta\to 0}\frac{1-\sec\theta}{3\sin^2\theta}.$$

We are able to use the theorem again, and

$$\lim_{\theta\to 0}\frac{1-\sec\theta}{3\sin^2\theta}=\lim_{\theta\to 0}\frac{-\frac{\sin\theta}{\cos^2\theta}}{6\sin\theta\cos\theta}=\lim_{\theta\to 0}\frac{-1}{6\cos^3\theta}=-\frac{1}{6}.\quad\blacksquare$$

Example E.4 L'Hospital's rule is very helpful in Exercises 35.9 and 46.8e. ■

F

Symbols, and the Greek Alphabet

F.1 Special Symbols in Real Analysis

SYMBOL	*NAME*	*CHAPTER*
$\equiv$	equivalence	1
$\mathbb{N}$	set of natural numbers	2
$\mathbb{Z}$	set of integers	3
$\mathbb{Q}$	set of rational numbers	3
$\langle a, b\rangle$	point	5
$\varnothing$	empty set	5
$\overline{A}$	complement	5
$A\backslash B$	relative complement	5
$\bigcap$	generalized intersection	5
$\bigcup$	generalized union	5
$D\times R$	Cartesian product	6
$\overset{f}{\to}$	function	6
$\mathcal{D}$	domain	6
$\mathcal{R}$	range	6
$\rightarrow$	bijection	7
$\mathbb{T}$	set of terminating decimals	8
$\mathbb{Q}(\sqrt{p})$	quadratic extension	8
$\mathbb{A}$	set of algebraic numbers	8
$\aleph$	transfinite cardinal number	11
c	cardinality of $\mathbb{R}$	11
$\mathcal{P}(M)$	power set of M	11
$\mathbb{I}$	set of irrational numbers	12
$\mathbb{R}$	set of real numbers	12
sup	supremum	12
inf	infimum	12
$\langle A\vert B\rangle$	Dedekind cut	12
$\lfloor\ \rfloor$	floor function	13
$\lceil\ \rceil$	ceiling function	13
$\mathbf{N}_x^{\varepsilon}$	neighborhood	14
$\langle a_n\rangle_1^{\infty}$	infinite sequence	16
$\rightarrow$	convergence	17

SYMBOL	*NAME*	*CHAPTER*
a_{n_k}	sub-subscript	19
lim sup	limit superior	22
lim inf	limit inferior	22
$\mathbf{D}_x^{\delta}$	deleted neighborhood	23
$f'^{L}(x)$	derivative from the left	38
$f'^{R}(x)$	derivative from the right	38
$f'^{-}(x)$	left-hand deriv's. limit	38
$f'^{+}(x)$	right-hand deriv's. limit	38
$f'^{S}(x)$	symmetric derivative	38
$\overset{u}{=}, \overset{u}{\to}$	uniform convergence	42
$\mathcal{N}$	net	47
$\overline{\mathcal{N}}$	refinement of a net	47
$\Delta\chi$	norm of a net	47
$U(f, \mathcal{N})$	upper Riemann sum	47
$L(f, \mathcal{N})$	lower Riemann sum	47
$\overline{\int_a^b} f(x)dx$	upper Riemann integral	47
$\underline{\int_a^b} f(x)dx$	lower Riemann integral	47
$\wp$	Cauchy principal value	50
$\langle\mathbb{M}, \rho\rangle$	metric space	52
d_1, d_r, d_∞	Euclidean metrics	52
ℓ^2, ℓ^∞	sequence metrics	52
ρ_{01}	discrete metric	52
$\mathcal{T}$	topology	54
$i(A)$	set of interior points	54
$c(A)$	set of cluster points	54
$b(A)$	set of boundary points	54
$\widetilde{A}$	closure	54
$\mathbb{K}$	Cantor ternary set	55
$\mathcal{F}_n$	Farey sequence	Appx. A
$[a_0; a_1, \cdots]$	continued fraction	Appx. D

F.2 The Greek Alphabet

LETTER	*NAME*	*ENGLISH EQUIVALENT*	*LETTER*	*NAME*	*ENGLISH EQUIVALENT*
$\mathrm{A}\ \alpha$	alpha	a	$\mathrm{N}\ \nu$	nu	n
$\mathrm{B}\ \beta$	beta	b	$\Xi\ \xi$	xi	x
$\Gamma\ \gamma$	gamma	g	$\mathrm{O}\ \mathrm{o}$	omicron	o
$\Delta\ \delta$	delta	d	$\Pi\ \pi$	pi	p
$\mathrm{E}\ \varepsilon\ \epsilon$	epsilon	e	$\mathrm{P}\ \rho\ \varrho$	rho	r
$\mathrm{Z}\ \zeta$	zeta	z	$\Sigma\ \sigma\ \varsigma$	sigma	s
$\mathrm{H}\ \eta$	eta	ee	$\mathrm{T}\ \tau$	tau	t
$\Theta\ \theta\ \vartheta$	theta	th	$\Upsilon\ \upsilon$	upsilon	u
$\mathrm{I}\ \iota$	iota	i	$\Phi\ \phi\ \varphi$	phi	ph
$\mathrm{K}\ \kappa$	kappa	k	$\mathrm{X}\ \chi$	chi	ch
$\Lambda\ \lambda$	lambda	l	$\Psi\ \psi$	psi	ps
$\mathrm{M}\ \mu$	mu	m	$\Omega\ \omega$	omega	oo

Thus: ή γεωμετρία = geometry, and ό 'Αρχιμήδης = Archimedes.

Annotated Bibliography

History of Mathematics

(These books range from pure biography to detailed exposition of mathematics from the historical viewpoint. None are by journalists.)

1. Beckmann, Petr. *A History of π (Pi)*, St. Martin's Press, New York, 1971. Pi is so pervasive that this is actually a history of mathematics and civilization; written with personal viewpoints throughout.
2. Bell, Eric Temple. *Men of Mathematics*, Simon and Schuster, New York, 1937. Selected biographies written in an engaging, sometimes sensationalist, style. The essay on Galois has been updated in "Genius and Biographers: The Fictionalization of Evariste Galois," by Tony Rothman, in *The American Mathematical Monthly*, vol. 89 (1982), pp. 84–106.
3. Boyer, Carl B. *A History of Mathematics*, 2nd ed., revised by Uta C. Merzbach, John Wiley & Sons, New York, 1991. A standard in the field.
4. Dunham, William. *The Calculus Gallery–Masterpieces from Newton to Lebesgue*, Princeton University Press, Princeton, 2005. Mathematics and its history combined superbly by an expert hand.
5. Katz, Victor J. *A History of Mathematics–An Introduction*, New York, HarperCollins, 1993. A textbook with exercises as well as a history.
6. Kline, Morris. *Mathematical Thought from Ancient to Modern Times*, 3 vols., Oxford University Press, New York, 1972. A tremendously detailed history, with a bibliography at the end of each of its 51 chapters.
7. Kramer, Edna E. *The Nature and Growth of Modern Mathematics*, Princeton University Press, Princeton, 1981. A panoramic and erudite history, written in a personal and warm style.
8. Struik, Dirk J. *A Concise History of Mathematics*, 3rd revised ed., Dover Publications, New York, 1967. As concise as it promises, and captivating.

Undergraduate Mathematical Analysis

(Varying levels of presentation.)

9. Bartle, Robert G. *The Elements of Real Analysis*, John Wiley & Sons, New York, 1964. A classic real analysis textbook for the upper undergraduate.
10. Beckenbach, Edwin, and Richard Bellman. *An Introduction to Inequalities*, Yale University and Random House, New York, 1961. Clearly written, this is the third volume in the New Mathematical Library, for undergraduates.
11. Dangello, Frank, and Michael Seyfried. *Introductory Real Analysis*, Houghton Mifflin, Boston, 2000. An introductory real analysis textbook, as stated.
12. Eves, Howard, and Carroll V. Newsom. *An Introduction to the Foundations and Fundamental Concepts of Mathematics*, Holt, Rinehart & Winston, New York, 1958. A classic in foundations of mathematics.
13. Goldberg, Richard R. *Methods of Real Analysis*, 2nd ed., John Wiley & Sons, New York, 1976.

14. Kazarinoff, Nicholas D. *Ruler and the Round or Angle Trisection and Circle Division*, Prindle, Weber & Schmidt, Boston, 1970. Kazarinoff is an engaging writer. This is the 15th in a long series of mathematical monographs: The Prindle, Weber & Schmidt Complementary Series in Mathematics.
15. Kirkwood, James R. *An Introduction to Analysis*, PWS-KENT, Boston, 1989. An introductory real analysis textbook.
16. Mattuck, Arthur. *Introduction to Analysis*, Prentice Hall, Upper Saddle River, 1999.
17. Niven, Ivan. *Numbers: Rational and Irrational*, Random House, New York, 1961. Niven is always very readable. This is the first in a long series of mathematical monographs for undergraduates the New Mathematical Library.
18. Olmsted, John M. H. *Intermediate Analysis, An Introduction to the Theory of Functions of One Real Variable*, Appleton-Century-Crofts, New York, 1956. A classic real analysis textbook, extending topics in the present text to intermediate level.
19. Robdera, Mangatiana A. *A Concise Approach to Mathematical Analysis*, Springer-Verlag, London, 2003. Very well written, extending some of the topics in the present text to intermediate level.
20. Ross, Kenneth A. *Elementary Analysis, The Theory of Calculus*, Springer-Verlag, New York, 1980. This textbook includes an extensive bibliography.
21. Scheinerman, Edward R. *Mathematics, A Discrete Introduction*, Brooks/Cole, Pacific Grove, 2000. This friendly book aims at teaching how to write proofs, with lots of examples of many techniques of proof.
22. Smith, Douglas, Maurice Eggen, and Richard St. Andre. *A Transition to Advanced Mathematics*, 6th ed., Brooks/Cole, Pacific Grove, 2006. This textbook specializes in the foundations of mathematics.
23. Spiegel, Murray R. *Theory and Problems of Advanced Calculus*, McGraw-Hill, New York, 1963. Schaum's Outline Series always covers many topics in analysis and calculus concisely.
24. Wade, William R. *An Introduction to Analysis*, 4th ed., Pearson, Upper Saddle River, 2010. Very well written, extending this book's topics and many others to intermediate level.
25. Wolf, Robert S. *Proof, Logic, and Conjecture: A Mathematician's Toolbox*, W. H. Freeman, New York, 1998.

More Advanced Analysis

(These books build upon introductory mathematical analysis.)

26. Boas, Jr., Ralph P. *A Primer of Real Functions*, 4th ed., revised and updated by Harold P. Boas, The Mathematical Association of America, (no city), 1996. A classic in real analysis, with solutions to all exercises. It is volume 13 in The Carus Mathematical Monographs.
27. Bromwich, T. J. *An Introduction to The Theory of Infinite Series*, Macmillan, London, 1908. Still a very readable textbook, available for reading at https://archive.org/stream/introductiontoth00bromuoft #page/n5/mode/2up
28. Cantor, Georg. *Contributions to the Founding of the Theory of Transfinite Numbers*, translated, with an introduction and notes by P. E. B. Jourdain, Open Court, La Salle, 1952.
29. Gelbaum, Bernard R., and John M. H. Olmsted. *Counterexamples in Analysis*, Dover, Mineola, 1992. An excellent, well-organized resource that helps to answer conjectures.
30. Goffman, Casper. *Real Functions*, Prindle, Weber & Schmidt, Boston, 1970. This is volume eight in a long series of mathematical monographs: The Prindle, Weber & Schmidt Complementary Series in Mathematics.
31. Hewitt, Edwin, and Karl Stromberg. *Real and Abstract Analysis*, Springer-Verlag, New York, 1965.
32. Klein, Felix. "Famous Problems of Elementary Geometry," translation of the 1895 edition by W. W. Beman and D. E. Smith, in *Famous Problems and Other Monographs*, 2nd ed., Chelsea, New York, 1962. This is one of four monographs in one volume. The typography is old style.
33. Kolmogorov, A. N., and S. V. Fomin. *Introductory Real Analysis*, revised 1970 edition translated and edited by Richard A. Silverman, Dover, New York, 1975. Terse and precise.
34. Landau, Edmund. *Differential and Integral Calculus*, 3rd ed., translated by Melvin Hausner and Martin Davis, AMS Chelsea, Providence, 1965.

35. ____. *Foundations of Analysis*, 3rd ed., translated by F. Steinhardt, Chelsea, New York, 1966. The subtitle is *The Arithmetic of Whole, Rational, Irrational and Complex Numbers*, and that is exactly what this little book develops, in the most detailed, exacting, and delightful manner that I have ever seen.
36. Niven, Ivan. *Irrational Numbers*, The Mathematical Association of America, (no city), 1956. This is volume 11 in The Carus Mathematical Monographs series, and as usual with Niven, very understandable.
37. Rudin, Walter. *Principles of Mathematical Analysis*, 3rd ed., McGraw-Hill, New York, 1976. A classic in real analysis, terse and precise.

Philosophy of Mathematics, Introductory Logic, Mathematical Enjoyment

38. Hardy, G. H. *A Mathematician's Apology*, 1st ed., Cambridge University Press, London, 1967. A classic essay championing the beauty of pure mathematics.
39. Hersh, Reuben. *What Is Mathematics, Really?*, Oxford University Press, New York, 1997. Many discussions on the philosophy of mathematics.
40. Huntley, H. E. *The Divine Proportion, a Study in Mathematical Beauty*, Dover, New York, 1970.
41. Hurley, Patrick J. *A Concise Introduction to Logic*, 9th ed., Wadsworth/Thomson, Belmont, 2005. A very readable textbook of logic, from Aristotelian to modern symbolic forms.
42. Peterson, Ivars. *The Mathematical Tourist*, W. H. Freeman, New York, 1998. A survey of mathematical topics for the general public, it includes beautiful color plates of fractals.
43. Rényi, Alfréd. *Dialogues on Mathematics*, Holden-Day, San Francisco, 1967. This is a 100-page gem. The author becomes Socrates, Archimedes, and then Galileo, discussing what mathematics is and why it explains reality with such unexpected success.
44. Stewart, Ian. *Professor Stewart's Cabinet of Mathematical Curiosities*, Basic Books, New York, 2009. A delightful collection of 180 essays on the most diverse mathematical topics.

Solutions to Odd-Numbered Exercises

Chapter 1

1. "Some $A = \{actors\ in\ Forbidden\ Planet\}$ are $D = \{dead\ people\}$." Taken as a true premise, this lets us conclude that the contradictory "No A are D" is a false statement. And if the premise is false, the conclusion must be true. (The premise is actually true.)

3. Obverse operation:
   a. No rational numbers are non-real numbers. (True, because the original was true.)
   b. All real numbers are non-rational numbers. (False, because the original was false.)
   c. Some integrable functions are non-differentiable functions. (True, because the original was true.)
   d. No visible things are things without mass. (False, because the original was false.)

5. Converse operation:
   a. All real numbers are rational. (False, but not determined by the original.)
   b. No rational numbers are real numbers. (False, because the original was false.)
   c. Some differentiable functions aren't integrable functions. (False, but not determined by the original.)
   d. All things with mass are visible things. (False, but not determined by the original.)

7. "T is true if and only if U is true" is the biconditional, meaning: (if T, then U) and (if U, then T). Taking the contrapositive of each: (if *not*-U, then *not*-T) and (if *not*-T, then *not*-U). These make a biconditional: *not*-T if and only if *not*-U.

9. The given premise translates as "If there is a number larger than any given number then it is an infinite quantity." But such universal statements (affirmative and negative) do not guarantee nonempty categories. This one just affirms that *if* S has membership, then those members are also in P, or in the conditional form, *if* Q is the case, then R is the case. The subject term "numbers larger than any given number" is empty to begin with. But the conclusion affirms existence ("some" means "there exists"), and this plainly does not follow from the premise. This type of fallacy must be guarded against at all times.

Chapter 2

1. a. Assume that $a \neq b$. By trichotomy, we must consider two cases. First, suppose that $a < b$. By Lemma 2.1, we have $c + a < c + b$, hence $c + a \neq c + b$, which finishes the first case. Second, suppose that $b < a$. By Lemma 2.1 again, we have $c + b < c + a$, implying that $c + b \neq c + a$. In both cases, if $a \neq b$, then $c + a \neq c + b$, and by contraposition, Lemma 2.2 is proved.

   b. We shall use mathematical induction. Let M be the set of natural numbers a such that $1 \leq a$. If $a = 1$, then $a \in M$, as an anchor. This implies that $1 + 1 = 1 + a$, so that $1 < 1 + a$. Hence, $1 + a \in M$. On the other hand, let $1 < a$, which implies that $a \in M$, as an anchor. Then $1 + 1 < 1 + a$. Then there exists some natural number u such that $(1 + 1) + u = 1 + a$. And then, $1 + (1 + u) = 1 + a$, from which we infer that $1 < 1 + a$. Once again, $1 + a \in M$. The two cases combine to state that if $1 \leq a$, then $1 + a \in M$. By mathematical induction, $M = \mathbb{N}$. Therefore, for all $a \in \mathbb{N}$, $1 \leq a$.

3. Given that $\exists y$ such that $b + y = c$, we have $a + y < b + y$ by Lemma 2.1. Then $a + y < c$, so that $a < c$.

   Proof sequence. A direct proof.

Statement	Reason
(1) $\exists y \in \mathbb{N}$ such that $b + y = c$.	Definition of $b < c$.
(2) $a + y < b + y$.	Lemma 2.1 and N2.
(3) $a + y < c$.	Substitution.
(4) $a < c$.	Definition of $<$.

QED

5. Prove the contrapositive: if $a \neq b$, then $ca \neq cb$. Case 1, $a < b$. Then $ca < cb$ by Exercise 2.4, so that $ca \neq cb$. Case 2, $b < a$, is similar.

7. a. Let M be the set of numbers n for which $a^n \geq a$. Clearly, $1 \in M$. Assume $a^k \geq a$ whenever $k \in M$. Then $a^k a \geq a^2$ by the induction assumption, and $a^2 \geq a$ since $a \geq 1$. Putting all these together by Exercise 2.3, we get $a^{k+1} \geq a$. Thus, $k + 1 \in M$. By N6, $M = \mathbb{N}$.

   b. Let M be the set of natural numbers m for which $a^n a^m = a^{n+m}$, where $a, n \in \mathbb{N}$. If $m = 1$, then the proposition follows by definition, so we know $1 \in M$. Next, assume that $m \in M$, that is, $a^n a^m = a^{n+m}$. For $m + 1$, we know that $a^n a^{m+1} = a^n a^m a^1$ by the definition and by Exercise 2b. Since $m \in M$, $a^n a^m a^1 = a^{n+m} a^1$. Again by definition, $a^{n+m} a^1 = a^{n+m+1}$. Thus, $a^n a^{m+1} = a^{n+(m+1)}$, which implies that $m + 1 \in M$. By N6, $M = \mathbb{N}$.

9. a. We have

$$\underbrace{(2p+1)\cdot \cdots \cdot(2p+1)}_{p \text{ factors}} < \underbrace{1\cdot 2\cdot 3\cdot \cdots \cdot(p)(p+1)(p+2)\cdot \cdots \cdot(2p)(2p+1)}_{2p+1 \text{ factors}}.$$

   Take exactly p pairs of factors $\{(1)(2p+1), (2)(2p), \ldots, (p)(p+2)\}$ from the right side and divide both sides by them. The one remaining factor on the right is $p + 1$, and the result follows.

b. The general form of the fractions is

$$\frac{2p+1}{k(2p+2-k)}, k = 2, 3, \ldots, p.$$

Thus, we prove that $2p+1 < k(2p+2-k)$. One way to do this is to graph the functions $f(k) = k(2p+2-k)$ and $g(k) = 2p+1$ over the interval $2 \le k \le p$, and observe that $g(k) < f(k)$ over that interval. Therefore, all $\frac{2p+1}{k(2p+2-k)} < 1$, so their product is less than 1, and certainly

$$\frac{2p+1}{(2)(2p)} \cdot \frac{2p+1}{(3)(2p-1)} \cdot \cdots \cdot \frac{2p+1}{(p)(p+2)} < p+1.$$

c. We established that $(2p+1) \cdot \cdots \cdot (2p+1) < 1 \cdot 2 \cdot 3 \cdot \cdots \cdot (2p)(2p+1)$, or $(2p+1)^p < (2p+1)!$, where $p \in \mathbb{N}$. Since $n^p \le n!$ is true for all even $n = 2p$ by Example 2.1, we conclude that $n^p \le n!$ is true for all $n \ge 2p$. **QED**

11. See the bibliography at the end of this book.

Chapter 3

1. In the definition of inequality for $\mathbb{Q}$, let $r = \frac{m}{n}$, and take $n = 1$ and $m \in \mathbb{N}$.

3. a. With parallel reasoning to that used in $\mathbb{Z}$, we find that $0 \notin \mathbb{Q}^+$. And since $-0 = 0$, $0 \notin \mathbb{Q}^-$. Next, assume that $\exists s \in \mathbb{Q}$ such that $s < 0$ and also $s > 0$. Then both $s + r = 0$ and $0 + r = s$ for some $r \in \mathbb{Q}^+$. Substitution leads to $r + r = 0$, a contradiction. Thus, $\mathbb{Q}^+$ and $\mathbb{Q}^-$ are disjoint subsets of $\mathbb{Q}$. Axiom QZ9 covers every rational except zero. We conclude that $\mathbb{Q}$ is partitioned as $\mathbb{Q}^- \cup \{0\} \cup \mathbb{Q}^+$, which is equivalent to trichotomy. To summarize, QZ9 induces the three disjoint subsets of $\mathbb{Q}$. **QED**

   b. We must prove that for any $r, s \in \mathbb{Q}^+, r+s \in \mathbb{Q}^+$ and $rs \in \mathbb{Q}^+$. By definition, $0 + \frac{m}{n} = r$ and $0 + \frac{p}{q} = s$, where $\frac{m}{n}, \frac{p}{q} \in \mathbb{Q}^+$, and we may assume that $m, n, p, q \in \mathbb{N}$. Then $r + s = \frac{mq+pn}{nq}$ and $rs = \frac{mp}{nq}$, where the two numerators and the two denominators are in $\mathbb{N}$. Hence, $r+s$ and rs both belong to $\mathbb{Q}^+$.

5. a. Consider $(-1)(-1) + (-1)(1) = (-1)(-1+1) = (-1)(0) = 0$. This says that $(-1)(-1)$ is the opposite of $(-1)(1) = -1$. But 1 is the unique opposite of -1, implying that $(-1)(-1) = 1$.

   b. Using the definition of inequality, we are given $t + \frac{r}{s} = 0$ with $\frac{r}{s} \in \mathbb{Q}^+$. Multiplying, $tq + q\frac{r}{s} = 0$, where $q\frac{r}{s} \in \mathbb{Q}^+$. Then $tq < 0$.

   c. Use **a** and Exercise 3.4. Substituting $-s$ for r there, $-(-s) = (-1)(-s) = (-1)((-1)s) = ((-1)(-1))s = 1s = s$.

   d. By several axioms, **a** and Exercise 3.4, $(-r)(-s) = ((-1)(-1))rs = rs$.

   e. By Example 3.1, $\frac{0}{rs} < \frac{r}{rs} < \frac{s}{rs}$, and the result follows.

   f. The law is given for $m, n \in \mathbb{N}$ up to this point. By the extension of exponents to $\mathbb{Z}$, $(r^m)^{-n} = ((r^m)^n)^{-1} = (r^{mn})^{-1}$. But for the same reason, $r^{-mn} = (r^{mn})^{-1}$. Thus, $(r^m)^{-n} = r^{-mn}$, where now $-n \in \mathbb{Z}^-$. The same analysis holds for m. The analysis still holds when $n = 0$, provided $r \neq 0$.

7. From the conclusions of Exercise 3.5d, $r^2 \geq 0$ and $s^2 \geq 0$ for any $r, s \in \mathbb{Q}$. Then QZ9 implies that $r^2 + s^2 \geq 0$.
   For the second proposition, suppose that $r^2 + s^2 = 0$. Then $r^2 = -s^2$, where $r^2 \geq 0$ and $s^2 \geq 0$. These conditions imply that $r = 0$ and $s = 0$, since no positive number can equal a negative number.

9. a. Typical proof: QZ8, the existence of opposites. For any $a \in F$, we must find the element $-a \in F$ such that $-a + a = 0$. Determine $0 \leq s \leq 4$ such that using integers $-a$ and q in $\mathbb{Z}$, $-a = 5q + s$. Then $s \in F$ and s will be the opposite of a in F. For example, $a = 4 \in F$, and $-4 = 5(-1) + 1$, so the opposite of 4 is 1, that is, $-4 = 1$ here. And $4 + 1 = 0$, which looks very cool.
   b. The only subsets of F that are closed under $\oplus$ are $\{0\}$ and F itself, so these are the only candidates for a positive class F^+. In either case, both $-a$ and a belong to F^+. Thus, no positive class exists. It follows that unequal pairs of elements in F satisfy neither $a < b$ *nor* $b < a$, because the inequality $a < b$ requires that there exists a $c \in F^+$ such that $a + c = b$, and likewise for $b < a$. (If we insist that inequalities make sense in F, as in $0 < 4$, then we also find from **a** that $4 < 0$!)

Chapter 4

1. As in Example 4.2, the anchor is not $P(1)$. The least integer for which $3^n < n!$ is $n = 7$, so $P(7)$ is the anchor. Induction hypothesis: assume $3^n < n!$. Since $3 < n + 1$ for $n \geq 7$, then $(3)3^n < (n + 1)n!$, which uses the induction hypothesis. Therefore, $3^{n+1} < (n + 1)!$, which is $P(n + 1)$. By induction, the given statement is true for all $n \geq 7$.

3. One approach:

$$\frac{1}{3}n^3 + \frac{1}{2}n^2 + \frac{1}{6}n + (n+1)^2 = \frac{1}{3}[(n+1) - 1]^3 + \frac{1}{2}[(n+1) - 1]^2 + \frac{1}{6}[(n+1) - 1] + (n+1)^2.$$

   For ease of work, let $m = n + 1$, and expand the expression in terms of m.

5. We have $2^n \geq 1 + n$. Then $2 \geq \sqrt[n]{1+n} > \sqrt[n]{n}$.

7. If $p > 0$, then $(1 + p)^n = 1^n + n(1^{n-1})p +$ *other positive terms* $> 1 + np$.

9. The anchor is at $n = 1$: $\sum_{k=1}^{1} a^k = a = a\frac{1-a^1}{1-a}$. Assume now that the case n is true:

$$\sum_{k=1}^{n} a^k = a\frac{1-a^n}{1-a}.$$

   The case $n + 1$ develops this way:

$$\sum_{k=1}^{n+1} a^k = a^{n+1} + \sum_{k=1}^{n} a^k = a^{n+1} + a\frac{1-a^n}{1-a}$$

   by the induction hypothesis. But $a^{n+1} + a\frac{1-a^n}{1-a} = a\frac{1-a^{n+1}}{1-a}$, so the case $n + 1$ is true. By induction, therefore, $\sum_{k=1}^{n} a^k = a\frac{1-a^n}{1-a}$ for any $n \in \mathbb{N}$. **QED**

11. The passage from the case n, the induction hypothesis, to the case $n+1$, fails when $n=1$. T_n is defined of necessity with 2 as its first subscript. Try to write T_1.

13. Select $t = 10$. Not surprisingly, it is the least element of T that always falsifies the contradictory proposition. Thus no such T exists.

Chapter 5

1. a. The rule of membership is ambiguous.
   b. The electrons are identical, but not indistinguishable, so it is a set.
   c. In a historical context, this set had two members in the year before September 11, 2001. Today, it is not a set in that context, since the two terms do not identify distinguishable items.
   d. $\{T\}$ is the set with collection T as its only element, also written as $\{\{a, b, c\}\}$. Thus, $T \in \{T\}$, and it is even possible to create the set $\{a, b, c, \{T\}\}$, which has no repeated elements. But $T \neq \{T\}$, for it is sufficient to see that $a \in T$, but $a \notin \{T\}$. Can a set be a member of itself, $T \in T$? Our particular set $T = \{a, b, c\}$ cannot belong to itself, since $T \neq a$, $T \neq b$, and $T \neq c$, and no other members of T exist. In general, what we are asking is whether the unambiguous rule for deciding membership in a set can refer to itself. Many modern logicians do not allow this, because it leads to contradictions in set theory.
   e. Suppose that you have in front of you a rock, a piece of paper, and a pair of scissors. Most people would agree that these three objects exist outside of our consciousness of them. However, the set containing them is a construct of the mind; it depends upon our *thinking* of them as a set. Nothing external to our minds changes in any way if we do or do not consider them as a set. Hence, it seems that a set is an act of thought, and not real beyond that. Perhaps this is an unspoken attribute of all sets. Perhaps a set seems real only because we can create it so effortlessly. But it would just the same cease to exist the moment intellection turns to other things. As a mathematical example, consider $S = \{\langle x, y\rangle : x^2 + y^2 = 1\}$. We are creating the set S. Surely, S is familiar to us as a circle, with hundreds of applicable theorems. We even symbolize this "perfect circle" by drawing its curve with graphite, but S does have that insubstantial attribute of being a set. So, if we turn our attention for a moment to the beautiful bird outside our window, S vanishes because the act of aggregating was an intellectual act. It is memory that recreates the circle and lets us continue where we left off.

      In set theory, we find that a number is just a convenient label for a set called an equivalence class. Thus, it may be that the number "five" is also an act of intellection, as evanescent, after all, as $\pi^{\sqrt{\sin 15}}$ might seem to be. Nevertheless, "five" is a act of the intellect unambiguously understood by everyone, as opposed to "happiness," for instance. This may be why "five" is easier to comprehend than "happiness."

      And still, as Plato would insist, concepts such as "five" and "circle" do seem to have a reality quite outside the organic computer that functions within our skull.

3. Given $S \subseteq P$, all $x \in S$ also belong to P. And, given $P \subseteq S$, all $x \in P$ also belong to S. Therefore, $S = P$. **QED**

The converse is easily seen as true. Given $S \subset P$, then $P \setminus S \neq \varnothing$. But this would contradict the assumed $P \subset S$. **QED**

5. Let $x \in S \setminus T$. Then $x \in S$ and $x \notin T$. Then $x \in \overline{T}$. Then $x \in \overline{T} \cap S$. Next, we must prove the converse. Let $x \in \overline{T} \cap S$. Then $x \in \overline{T}$ and $x \in S$. Then $x \in S$ and $x \notin T$. Then $x \in S \setminus T$. **QED**

7. a. First, let $S \subset T$. Then every $x \in S$ is also an element of T; hence, $S \cap T = S$. Conversely, let $S \cap T = S$, and assume that S is not a proper subset of T. Then there exists some $x \in S$ that is not a member of T. But then, $x \notin S \cap T$, which contradicts the given $S \cap T = S$. Thus, it must be that $S \subset T$. **QED**

   b. Let $S = A \cap B$ and $T = A \cup B$. Apply **a** to get $S \cap T = (A \cap B) \cap (A \cup B) = [(A \cap B) \cap A] \cup [(A \cap B) \cap B] = (A \cap B) \cup (A \cap B) = A \cap B = S$. Then the equivalent statement $S \subseteq T$ from **a** holds. **QED**

9. Use Lemma 5.1. Let $x \in R \cup (S \cap T)$. Suppose $x \in R$. Then x belongs to both $R \cup S$ and $R \cup T$. Hence, $x \in (R \cup S) \cap (R \cup T)$. Otherwise, suppose that $x \in S \cap T$. Then $x \in S$, so that $x \in R \cup S$. And also, $x \in T$, so $x \in R \cup T$. Then once again, $x \in (R \cup S) \cap (R \cup T)$. Therefore, we have $R \cup (S \cap T) \subseteq (R \cup S) \cap (R \cup T)$.

   For the reverse containment, let $x \in (R \cup S) \cap (R \cup T)$. Then $x \in R \cup S$ and $x \in R \cup T$. In the case that $x \in R$, then instantly, $x \in R \cup (S \cap T)$. In the case that $x \notin R$, then x must belong to both S and T. But then, $x \in S \cap T$, from which $x \in R \cup (S \cap T)$. We now have $(R \cup S) \cap (R \cup T) \subseteq R \cup (S \cap T)$.

   By Lemma 5.1, the equality follows. Theorem 5.1 is now completely proved.

11. Typical: $\{1, 2, 3\}$, $\{3, 4, 5\}$, $\{4, 5, 6\}$ have no elements in common, but are not all pairwise disjoint.

13. Since $0 \leq \alpha^2 \leq \alpha \leq 1$, only real xs between 0 and 1 inclusive belong to the union. That is, $\bigcup_\alpha S_\alpha = [0, 1]$. But for every real x, we can easily find an S_α that does not contain x. Thus, $\bigcap_\alpha S_\alpha = \varnothing$.

15. Let $x \in \overline{\bigcap_{\alpha \in A} S_\alpha}$, so that $x \notin S_\alpha$ for at least one S_α. Then for such an S_α, $x \in \overline{S_\alpha}$. Thus, $x \in \bigcup_{\alpha \in A} \overline{S_\alpha}$. We now know that $\overline{\bigcap_{\alpha \in A} S_\alpha} \subseteq \bigcup_{\alpha \in A} \overline{S_\alpha}$.

    To prove the reverse containment, let $x \in \bigcup_{\alpha \in A} \overline{S_\alpha}$. Then for at least one α, $x \in \overline{S_\alpha}$. Then for such subscript(s) α, $x \notin S_\alpha$. But then $x \notin \bigcap_{\alpha \in A} S_\alpha$. Thus, $x \in \overline{\bigcap_{\alpha \in A} S_\alpha}$, and so, $\bigcup_{\alpha \in A} \overline{S_\alpha} \subseteq \overline{\bigcap_{\alpha \in A} S_\alpha}$. By Lemma 5.1, the equality is proved. Theorem 5.3 is now completely proved.

17. $T_n = \{\langle x, y \rangle : x^2 + (y - \frac{1}{2^n})^2 \leq (\frac{1}{2^n})^2\}$ is a circular disk centered at $\langle 0, \frac{1}{2^n} \rangle$, with boundary passing through the origin. $\bigcup_{n=0}^{\infty} T_n = T_0$, since each disk is nested inside the previous one. $\bigcap_{n=0}^{\infty} T_n = \{\langle 0, 0 \rangle\}$ since some disk eventually excludes every other point but $\langle 0, 0 \rangle$.

Chapter 6

1. $\mathbb{Z} \times \mathbb{Z} = \{\langle n, m \rangle : n, m \in \mathbb{Z}\}$, which is the set of all points with integer coordinates, often called the set of **lattice points** of the plane.

3. $R \times D = \{\langle \sqrt{m}, n \rangle : m \in \mathbb{N},\ n \in \mathbb{N}\}$, part of which appears above. Note that the ordered pairs $\langle y, x \rangle$ are the reflections of the points in $D \times R$ in Example 6.3.

5. The domain element $n = 2$ can be paired with any of four range elements. Likewise for the other three domain elements. Hence, there are $4^4 = 256$ distinct functions.

7. $f^{-1}(\{-1\}) = \{\pm(2k-1)\pi : k \in \mathbb{N}\}$.

9. $R = [-3, 2]$ is not sufficient, since some numbers in $\mathcal{D}_u$ have images greater than 2.

11. a. The vertical line test detects only functions $f : \mathbb{R} \to \mathbb{R}$, and even then it is just a visual aid.
    b. Using some calculus, $\frac{dy}{dx} = \frac{dy/dt}{dx/dt} = \frac{\sin t^2}{\cos t^2} = \tan t^2$. Horizontal tangents exist when $\tan t^2 = 0$. For $t \geq 0$, this first occurs at $t = 0$, followed by $t = \sqrt{\pi}$. Thus, $C(\sqrt{\pi}) = \langle \int_0^{\sqrt{\pi}} \cos u^2 du, \int_0^{\sqrt{\pi}} \sin u^2 du \rangle \approx \langle 0.663, 0.895 \rangle$. Vertical tangents exist when $\tan t^2$ fails to exist. This first occurs at $t = \sqrt{\frac{\pi}{2}}$, giving $C(\sqrt{\frac{\pi}{2}}) \approx \langle 0.977, 0.549 \rangle$. Using the symmetry of the graph, we arrive at the rectangle with sides $x = \pm 0.977$ and $y = \pm 0.895$. Note that $[-0.977, 0.977] \times [-0.895, 0.895]$ is *not* the range of the clothoid. Its range is the set of points of the curve itself.

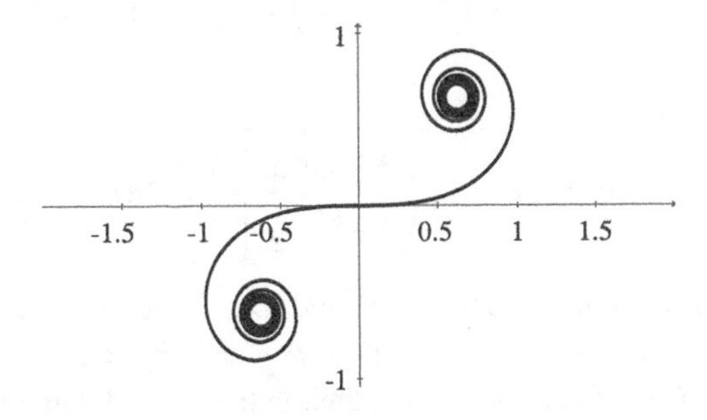

13. Let $x \in A \cap \mathcal{D}_f$, so that $f(x) \in f(A)$. But $x \in B$ also, so $f(x) \in f(B)$ as well. Thus, $f(A) \subseteq f(B)$. (This line of reasoning holds even if $x \notin \mathcal{D}_f$, because then, $x \xrightarrow{f} \varnothing$.)

15. As usual, first suppose that $x \in f^{-1}(E \cap F)$, so that $f(x) \in E \cap F$. Then $f(x) \in E$ and $f(x) \in F$. Then $x \in f^{-1}(E)$ and $x \in f^{-1}(F)$. Hence, $x \in f^{-1}(E) \cap f^{-1}(F)$. Conversely, let $x \in f^{-1}(E) \cap f^{-1}(F)$, implying that $x \in f^{-1}(E)$ and $x \in f^{-1}(F)$. Then $f(x) \in E$ and $f(x) \in F$, giving us $f(x) \in E \cap F$. Therefore, $x \in f^{-1}(E \cap F)$. The two containments imply $f^{-1}(E \cap F) = f^{-1}(E) \cap f^{-1}(F)$. (Even if $E \cap F$ and $\mathcal{R}_f$ are disjoint, equality will hold.)

17. a. We need to prove that $A \subseteq \upsilon^{-1}(\upsilon(A))$. Let $x \in A$. Then $\upsilon(x) \in \upsilon(A)$, and then the preimage x is a member of the preimage of $\upsilon(A)$: $x \in \upsilon^{-1}(\upsilon(A)) = B$. **QED**
    b. We need to prove that $G = \upsilon(\upsilon^{-1}(G))$. Let $y \in H$, so that $\upsilon^{-1}(\{y\}) \subseteq \upsilon^{-1}(G)$. Letting $x \in \upsilon^{-1}(\{y\})$, we now have $x \in \upsilon^{-1}(G)$, which means $\upsilon(x) \in G$. Hence, $y \in G$, and we have shown that $H \subseteq G$. Conversely, let $y \in G$. Then $\upsilon^{-1}(\{y\}) \subseteq \upsilon^{-1}(G)$, and it follows that $y \in \upsilon(\upsilon^{-1}(G)) = H$. **QED**
    c. As in Exercise 6.7, for $\upsilon(\theta) = \cos\theta$, $\upsilon(\pi) = -1$, followed by $\upsilon^{-1}(\upsilon(\pi)) = \{\pm(2k-1)\pi,\ k \in \mathbb{N}\}$. Clearly, $\{\pi\} \subseteq \upsilon^{-1}(\upsilon(\pi))$. Next, $\upsilon^{-1}(G) = \upsilon^{-1}(\{-1, 1\}) = \{k\pi,\ k \in \mathbb{Z}\}$. Thus, $\upsilon(\upsilon^{-1}(\{-1, 1\})) = \cos(k\pi) = \{-1, 1\}$.

19. Yes, linear functions of the form $L(x) = ax$, where a is any real number slope. See Theorem 25.2 for some background about this question.

Chapter 7

1. For example, $\hat{P} = x^2 - 3x$ over $x \geq 3$, which yields $\mathcal{R}_{\hat{P}} = [0, \infty)$. Solve $y = x^2 - 3x$ for x to get $x = \frac{3}{2} \pm \frac{1}{2}\sqrt{9+4y}$, from which we select $x = \hat{P}^{-1}(y) = \frac{3}{2} + \frac{1}{2}\sqrt{9+4y}$ (why?). You may check that both $\hat{P}^{-1}(\hat{P}(x)) = x$ and $\hat{P}(\hat{P}^{-1}(y)) = y$ over the respective domains.

3. a. $u^{-1}(u(x)) = \frac{\frac{x}{x+1}}{1-\frac{x}{x+1}} = x$, good! And $u(u^{-1}(x)) = \frac{\frac{x}{1-x}}{\frac{x}{1-x}+1} = x$, yes! This is still not sufficient, for $\mathcal{D}_u = \mathcal{R}_{u^{-1}}$ and $\mathcal{D}_{u^{-1}} = \mathcal{R}_u$ must also be verified. $\mathcal{D}_u = \{x \neq -1\}$ and with some thinking, $\mathcal{R}_{u^{-1}} = \{y \neq -1\}$. Also, $\mathcal{D}_{u^{-1}} = \{x \neq 1\}$, and with similar thinking, $\mathcal{R}_u = \{y \neq 1\}$. Now we *can* say that the two functions are inverses.

   b. The function f has domain $\mathbb{R}$ and range $(-1, 1)$ (verify). Write $g(y) = \ln(\frac{1+y}{1-y})$ just to keep tabs on the domain and range of g. We observe that

$$f(g(y)) = \frac{\exp\left(\ln\left(\frac{1+y}{1-y}\right)\right) - 1}{\exp\left(\ln\left(\frac{1+y}{1-y}\right)\right) + 1} = \frac{\frac{1+y}{1-y} - 1}{\frac{1+y}{1-y} + 1} = y$$

   and

$$g(f(x)) = \ln\left(\frac{1 + \frac{e^x - 1}{e^x + 1}}{1 - \frac{e^x - 1}{e^x + 1}}\right) = \ln e^x = x.$$

   This is still not sufficient, for $\mathcal{D}_f = \mathcal{R}_g$ and $\mathcal{D}_g = \mathcal{R}_f$ must also be verified. It isn't difficult to determine that $\mathcal{D}_g = (-1, 1) = \mathcal{R}_f$ and $\mathcal{R}_g = \mathbb{R} = \mathcal{D}_f$. Hence, these two functions are inverses of each other over their entire natural domains.

5. We have $\upsilon : (0, 1) \to \mathbb{R}$. Solve $y = \frac{1}{x} + \frac{1}{x-1}$ for x, and the only solution that gives range values in the original domain $(0, 1)$ is

$$\upsilon^{-1}(y) = x = \begin{cases} \frac{1}{2y}(y + 2 - \sqrt{y^2 + 4}) & \text{if } y \neq 0 \\ \frac{1}{2} & \text{if } y = 0 \end{cases}.$$

   The two functions are graphed below, showing the expected symmetry across $y = x$. Yes, a careful check shows that $\upsilon^{-1}(\upsilon(x)) = x$ for *all* $x \in (0, 1)$. And they are both strictly decreasing.

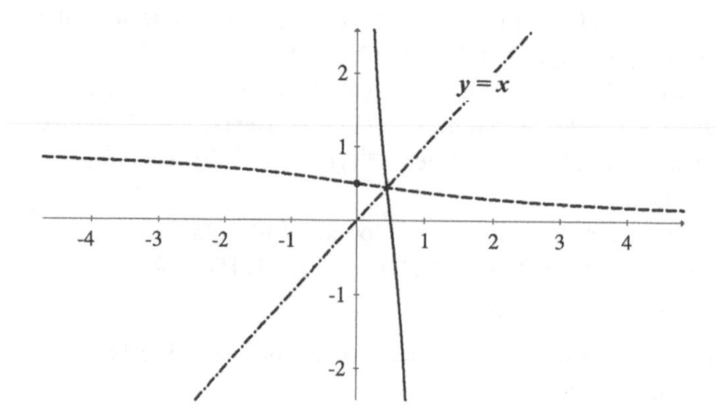

$\upsilon(x)$ is the solid line, and $\upsilon^{-1}(x)$ is the broken line. The lines $x = 1$ and the Y axis, and $y = 1$ and the X axis, are respective asymptotes.

7. Many possibilities exist. In statistics, for instance, the **cumulative standard normal probability function** is given by

$$y = \Phi(x) = \frac{1}{\sqrt{2\pi}} \int_{-\infty}^{x} e^{-t^2/2} dt$$

over all real x (and $\mathcal{R}_{\Phi} = (0, 1)$, which are probability values). Its inverse, $\Phi^{-1}(y)$, is used to find the unique value of x that corresponds to a given probability $\Phi(x)$. Φ^{-1} has no elementary or closed form (neither does Φ).

9. a. Since $R_1'(x) = -6(1-x)^5 < 0$ over $(0, 1)$, we know that R_1 is a bijection over its entire domain (Theorem 7.2). Likewise for R, since $R'(t) = -\lambda e^{-\lambda t} < 0$. Writing $y = 1 - (1 - x)^6$, we get $R_1^{-1}(y) = 1 - \sqrt[6]{1-y}$. Also, $R^{-1}(y) = \frac{1}{-\lambda} \ln y$.

   Using Theorem 7.3,

$$t = (R_1 \circ R)^{-1}(y) = R^{-1}(R_1^{-1}(y)) = \frac{1}{-\lambda} \ln(1 - \sqrt[6]{1-y})$$

   with $\mathcal{D}_{(R_1 \circ R)^{-1}} = \mathcal{R}_{R_1 \circ R} = (0, 1]$. This function gives the system's length of time in use which corresponds to a given reliability y. Select $\lambda = 3$ for instance, and make sense of the graph.

   b. Writing $y = x^3$, we have $R_2^{-1}(y) = \sqrt[3]{y}$ over $[0, 1]$. Then

$$t = (R_1 \circ R_2 \circ R)^{-1}(y) = R^{-1}\left[R_2^{-1}(R_1^{-1}(y))\right] = \frac{1}{-3\lambda} \ln(1 - \sqrt[6]{1-y})$$

   with domain $\mathcal{D}_{(R_1 \circ R_2 \circ R)^{-1}} = \mathcal{R}_{R_1 \circ R_2 \circ R} = (0, 1]$.

Chapter 8

1. The rational number $\frac{a}{b}$ has a terminating decimal expansion (it is an element of $\mathbb{T}$), if and only if after reducing to lowest terms, $b = 2^m 5^n$, with m, n nonnegative integers. To prove necessity, let $x \in \mathbb{T}$. This means that its decimal expansion has finitely many digits, say M digits. Thus, $x = D.d_1 d_2 \ldots d_M$, where D is some nonnegative integer to start with, followed by M decimal digits. In other words,

$$x = D + \frac{d_1 d_2 \ldots d_M}{10^M} = \frac{10^M D + d_1 d_2 \ldots d_M}{10^M},$$

   where $d_1 d_2 \ldots d_M$ represent the digits of an *integer*, not multiplication. Reducing the fraction, we end up with $\frac{c}{2^m 5^n}$. Hence, $x = \frac{c}{2^m 5^n}$, proving necessity.

   To prove sufficiency, let a fraction in lowest terms be of the form $\frac{c}{2^m 5^n}$. Let $M = \max\{m, n\}$, and raise terms to get

$$\frac{c}{2^m 5^n} = \frac{2^{M-m} 5^{M-n} c}{2^M 5^M} = \frac{2^{M-m} 5^{M-n} c}{10^M}.$$

   This represents a decimal expansion with at most M digits. Therefore, $\frac{c}{2^m 5^n} \in \mathbb{T}$.

3. a. False, $\mathbb{M}$ is finite. b. False, $\mathbb{M}$ contains some rationals that aren't integers.
   c. True. d. True, follows from c.
   e. False, there are natural numbers so large that no digital computer can hold them.

5. $(r + s\sqrt{2})(t + u\sqrt{2}) = rt + 2su + (ru + st)\sqrt{2} \in \mathbb{Q}(\sqrt{2})$.

7. a. $\cos 60^o = \frac{\sqrt{3}}{2}$, which is a root of the irreducible polynomial $P(x) = 4x^2 - 3$, so it is algebraic of degree two. b. The golden ratio is a root of $P(x) = x^2 - x - 1$, so it is algebraic of degree two. c. $\sqrt{1+\sqrt{2}}$ is a root of $P(x) = x^4 - 2x^2 - 1$ (which is irreducible), so it is algebraic of degree four.

9. The discriminant of $P(x)$ is $4ps^2$. Case (1): $s = 0$, and the number $r + 0\sqrt{p}$ is of degree one. Case (2): $s \neq 0$. Since p is not a perfect square, $4ps^2$ is not a square either, implying that the roots of $P(x)$ are irrational, so $r + s\sqrt{p}$ must be of degree two.

Chapter 9

1. $G(\langle 6,2\rangle) = 6 + \frac{1}{2}(7)(6) = 27$. Next, since $T_8 = 21 < 27 \leq T_9$, we have $k = 8$. Then $p = 6$, $q = 2$, so $G^{-1}(27) = \langle 6,2\rangle$ as expected. Locate it on the array!

3. For example, take 1951. We have $k = k(1951) = \lceil \frac{1}{2}(1 + \sqrt{15609}) \rceil = 63$. Then $G^{-1}(1951) = \langle 1951 - \frac{1}{2}(62)(61)\,, 63 - (1951 - \frac{1}{2}(62)(61))\rangle = \langle 60, 3\rangle$. The fraction is $\frac{60}{3}$.

5. a.

$\mathbb{Q}$	0/1	1/1	−1/1	1/2	−1/2	2/1	−2/1	1/3	−1/3	2/2	−2/2	3/1	−3/1 ⋯
✠	⟨0,1⟩	⟨1,1⟩	⟨−1,1⟩	⟨1,2⟩	⟨−1,2⟩	⟨2,1⟩	⟨−2,1⟩	⟨1,3⟩	⟨−1,3⟩	⟨2,2⟩	⟨−2,2⟩	⟨3,1⟩	⟨−3,1⟩ ⋯
$\mathbb{N}$	1	2	3	4	5	6	7	8	9	10	11	12	13 ⋯

b. $$n = F\left(\tfrac{p}{q}\right) = \begin{cases} 2G(\langle p,q\rangle) & \text{if } \frac{p}{q} > 0 \\ 2G(\langle -p,q\rangle) + 1 & \text{if } \frac{p}{q} < 0 \\ 1 & \text{if } \frac{p}{q} = 0 \end{cases}.$$

By definition, only p may be negative or zero, never q. F maps positive fractions to unique even $n \in \mathbb{N}$, and negative fractions to unique odd $n \in \mathbb{N}$, and $F(0) = 1$.

c. $$\frac{p}{q} = F^{-1}(n) = \begin{cases} G^{-1}\left(\frac{n}{2}\right) & \text{if } n \text{ is even} \\ -G^{-1}\left(\frac{n-1}{2}\right) & \text{if } n \text{ is odd and not } 1. \\ \frac{0}{1} & \text{if } n = 1 \end{cases}$$

Parts **b** and **c** give us the bijection $F : \mathbb{Q} \to \mathbb{N}$.

7. Find solutions to $H = 3 = n - 1 + |a_n| + |a_{n-1}| + \cdots + |a_0|$, where $n \in \mathbb{N}$, of course. Continuing with the hint, the resulting irreducible polynomials when $n = 3$ yield $\beta = 0$, not helpful. When $n = 2$, we need solutions to $|a_2| + |a_1| + |a_0| = 2$. One partition of a set of

two marbles is $|a_2| = 1$, $|a_1| = 1$, and $|a_0| = 0$. This yields, among others, $P(x) = x^2 + 1$, which has a new root $\beta = i$. Altogether, the six algebraic numbers of height three are i, $-i$, $\frac{1}{2}$, $-\frac{1}{2}$, 2, and -2. Arrange them in any order you wish.

Chapter 10

1. a. Since $\frac{25}{32} = 0.78125$, $d_{33} = 1$. Then $y_3 = 0$, making $y = 0.y_1y_20y_4\ldots$. Hence $y \neq 0.78125$.
   b. The 825th decimal digit of $\frac{\pi}{4}$ is denoted by $d_{825,825}$. Regardless of whether we know the value of that digit, according to the rule of construction, $y_{825} \neq d_{825,825}$. Thus, $y \neq \frac{\pi}{4}$. (In fact, $d_{825,825} = 0$, so $y_{825} = 1$.)

3. If $y = 0$ were an element of the list, then it must appear at some position j. That is, $y = x_j = 0$. Hence all decimal digits of x_j are zero, and especially, $d_{jj} = 0$. But this flies in the face of $d_{jj} = 1$. Thus, such a list would be missing the rather important real number, zero.

5. a. $\frac{39}{16} = 2 + \frac{7}{16}$, and $\frac{7}{16} \div 2^{-2} = \frac{7}{4} = 1 + \frac{3}{4}$. So we have $10.01\ldots_2$ so far. Next, $\frac{3}{4} \div 2^{-1} = \frac{3}{2} = 1 + \frac{1}{2}$, yielding $10.011\ldots_2$. Next, $\frac{1}{2} \div 2^{-1} = 1$ exactly, so the algorithm terminates with $\frac{39}{16} = 10.0111_2$. We would use $0.01110000000000\ldots_2$ in the diagonalization list. $\frac{8}{9} \div 2^{-1} = \frac{16}{9} = 1 + \frac{7}{9}$, then $\frac{7}{9} \div 2^{-1} = \frac{14}{9} = 1 + \frac{5}{9}$, then $\frac{5}{9} \div 2^{-1} = \frac{10}{9} = 1 + \frac{1}{9}$, then $\frac{1}{9} \div 2^{-4} = \frac{16}{9} = 1 + \frac{7}{9}$ and the algorithm begins to repeat. Thus, $\frac{8}{9} = 0.111000111000\ldots_2$, repeating forever. That's how it appears in the diagonalization list.

   For $\frac{\sqrt{2}}{2}$, it's convenient to use its decimal form $0.70710678118\ldots$, but many digits must be held to assure precision. Thus, $(0.70710678118)/2^{-1} = 1.41421356236$, so we start as $0.1\ldots_2$. Next, $(0.41421356236)/2^{-2} = 1.65685424944$, giving $0.101\ldots_2$. Next, $(0.65685424944)/2^{-1} = 1.31370849888$, giving $0.1011\ldots_2$, and so on. This never repeats; more iterations give $\frac{\sqrt{2}}{2} = 0.10110101\ldots_2$. The disadvantage is that we steadily lose precision because only finitely many decimal digits are used, so eventually, round-off errors ruin the expansion. This representation appears in the diagonalization list, necessarily without round-off error.

   As for $0.1111\ldots_2$, since 1 is the largest binary digit, it plays the same role as 9 in the decimal $0.9999\ldots$; hence both decimals are equal to 1. More precisely,

$$0.1111\ldots_2 = \frac{1}{2^1} + \frac{1}{2^2} + \frac{1}{2^3} + \cdots = \sum_{k=1}^{\infty} \frac{1}{2^k} = 1.$$

   b. We assume the same array, but use δ_{ij} to represent either 0 or 1, as seen below. Again, terminating decimals will be expressed only with an endless string of zeros. The same algorithm will yield y. For all natural numbers j, $y_j = 0$ if $\delta_{jj} = 1$, and $y_j = 1$ if $\delta_{jj} = 0$. The same contradiction will arise: y belongs to the array because it is a real number in

[0, 1], but y cannot belong to the array since $y_k = 1 - \delta_{kk}$. Hence, the array can't contain all the real numbers in [0, 1].

$$\begin{aligned} x_1 &= 0.\delta_{11}\delta_{12}\delta_{13}\dots \\ x_2 &= 0.\delta_{21}\delta_{22}\delta_{23}\dots \\ x_3 &= 0.\delta_{31}\delta_{32}\delta_{33}\delta_{34}\dots \\ &\dots \\ x_k &= 0.\delta_{k1}\delta_{k2}\delta_{k3}\dots\delta_{kk}\dots \\ &\dots \end{aligned}$$

Chapter 11

1. We have $\{\cdots, \frac{-2}{5}, \frac{-1}{5}, \frac{0}{5}, \frac{1}{5}, \frac{2}{5}, \dots\}$, so we modify slightly the bijection from Example 11.2:

$$m = f\left(\frac{n}{5}\right) = \begin{cases} 2n & \text{if } n > 0 \\ -2n+1 & \text{if } n \le 0 \end{cases},$$

which is a bijection onto $\mathbb{N}$.

3. There are only a finite number of distinguishable songs to write. Every second of music is converted to a string of 44100 readings, each with 65536 levels. This gives us a collection of $65536^{44100} \approx 5.8 \times 10^{212406}$ possible soundtracks each lasting one second. Thus, two seconds allows for $(65536^{44100})^2$ possibilities. In three days there are 259200 seconds, so that the number of possible digitally recorded songs in three days is $(65536^{44100})^{259200}$. The common logarithm of this number is $(259200)(44100)\log 65536 \approx 5.506 \times 10^{10}$. The answer, therefore, is a natural number more than 55 billion digits long! That's a titanic number in human terms, but totally inconsequential in comparison to $\aleph_0$, as all finite quantities are. Now, for an MP3 player to hold them all....

5. Let $S = \{s_1, s_2, \dots, s_n\}$, and let the elements of S that form P be listed this way: $P = \{p_1, p_2, \dots, p_m\}$. The simplest bijection $p_i \to s_j$ for $i = 1, 2, \dots, m$, identifies identical elements of S and P. Every other bijection is only a permutation of the elements s_j so identified. But $m < n$, so no bijection can extend to all elements of S. **QED**

7. Here is one of several ways to prove it. Form the one-to-one correspondence $\langle x, x^2\rangle \to \langle x+1, (x+1)^2\rangle$ for all $x \ge 0$. Then the parabolic segment $\{\langle x, x^2\rangle : x \ge 0\}$ is mapped one-to-one with the subset of itself starting at point $\langle 1, 1\rangle$.

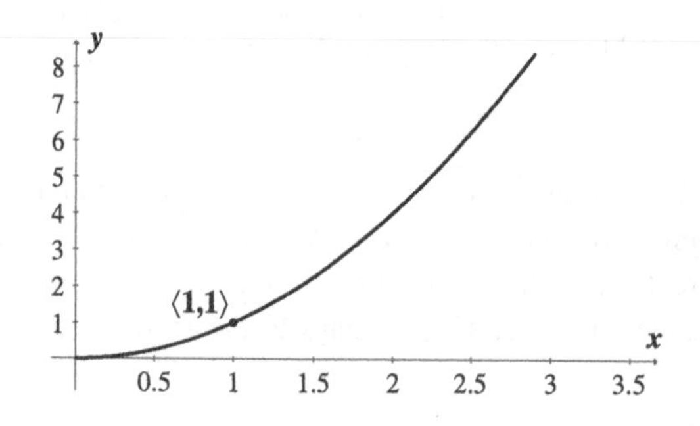

9. Use Exercise 11.8. There is a bijection $n \to \mathbb{Z}$ between each integer n (shown in bold) in the following array, and its own copy of $\mathbb{Z}$.

$$\begin{array}{ccccc}
\ddots & \langle -1, 2\rangle & \langle 0, 2\rangle & \langle 1, 2\rangle & \cdot^{\cdot^{\cdot}} \\
\cdots \langle -2, 1\rangle & \langle -1, 1\rangle & \langle 0, 1\rangle & \langle 1, 1\rangle & \langle 2, 1\rangle \cdots \\
\cdots \langle \mathbf{-2}, 0\rangle & \langle \mathbf{-1}, 0\rangle & \langle \mathbf{0}, 0\rangle & \langle \mathbf{1}, 0\rangle & \langle \mathbf{2}, 0\rangle \cdots \\
\cdots \langle -2, -1\rangle & \langle -1, -1\rangle & \langle 0, -1\rangle & \langle 1, -1\rangle & \langle 2, -1\rangle \cdots \\
\cdot^{\cdot^{\cdot}} & \langle -1, -2\rangle & \langle 0, -2\rangle & \langle 1, -2\rangle & \ddots
\end{array}$$

The array is represented by $\bigcup_{j\in\mathbb{Z}} \bigcup_{n\in\mathbb{Z}} \langle n, j\rangle$. It is the countably infinite union of denumerable sets, and its cardinality is $(\aleph_0)^2 = \aleph_0$.

11. $[a, b] = \{a, b\} \cup (a, b)$. The cardinality of the union on the right is $2 + c = c$.

13. The vendor doesn't lose anything, for at a profit of 10¢ per block, the selling price of the gift would be $\$(0.10)\aleph_0 = \$\aleph_0$, just as before. Transfinite numbers are immune to scale!

15. a. $x = 0 = 0.000\ldots_2$ yields $\varnothing \in \mathcal{P}(\mathbb{N})$. Next, $\frac{\pi}{4} = 0.7853982\ldots = \frac{1}{2} + \frac{1}{4} + \frac{1}{32} + \frac{1}{2^8} + \frac{1}{2^{13}} + \cdots = 0.1100100100001\ldots_2$ yields $\{1, 2, 5, 8, 13, \ldots\} \in \mathcal{P}(\mathbb{N})$, a subset with no pattern whatsoever. Next, $\frac{11}{32}$ yields 0.01011_2 and also $0.010101111\ldots_2$, which produce $\{2, 4, 5\} \in \mathcal{P}(\mathbb{N})$, and $\{2, 4, 6, 7, 8, 9, \ldots\} \in \mathcal{P}(\mathbb{N})$. Last, $x = 1 = 0.1111\ldots_2$ yields $\{1, 2, 3, \ldots\} = \mathbb{N} \in \mathcal{P}(\mathbb{N})$.

    b. The set of primes yields

$$x = 0.0110101000101000010\ldots_2 = \frac{1}{2^2} + \frac{1}{2^3} + \frac{1}{2^5} + \frac{1}{2^7} + \cdots = 0.41468\ldots,$$

    a decimal that encrypts the whole set of primes! It's perfect for an interstellar message to aliens.

17. a. Let $f : (a, b) \to (a, b)$ be simply $f(x) = x$, mapping into a subset of $[a, b]$. Use the diagram to get $g : [a, b] \to [a + \frac{b-a}{3},\ b - \frac{b-a}{3}]$ as $g(x) = \frac{1}{3}x + \frac{1}{3}(b + a)$, mapping into the middle third of (a, b).

    b. On the left side of the figure, the circle's interior C uses the identity bijection to map itself inside the square's interior S, clearly a subset of the interior of S. On the right side, the circle's interior C uses a bijection to dilate itself onto the large circle's interior C', of which the square's interior S is clearly a subset. Using the Schröder-Bernstein theorem with the extra dilation, the two interiors S and C are equivalent.

Chapter 12

1. No on both counts. Counterexamples: $(3 + \sqrt{10}) + (3 - \sqrt{10}) = 6 \in \mathbb{Q}$, and $\sqrt{8}\sqrt{2} = 4 \in \mathbb{Q}$.

3. Let the intervals be written as $\updownarrow a_i, b_i \updownarrow$, $i = 1, 2, \ldots, n$, where $\updownarrow\updownarrow$ is used just in this exercise to mean that either endpoint may be included or excluded. Then $\bigcup_{i=1}^{n} \updownarrow a_i, b_i \updownarrow \subseteq [\min\{a_i\}, \max\{b_j\}]$, showing that the union is bounded (note that endpoints a_i and b_j may belong to different intervals). For the intersection $\bigcap_{i=1}^{n} \updownarrow a_i, b_i \updownarrow$, we appeal to Exercise 5.7b, which can be extended to finite unions and intersections. Thus,

$\bigcap_{i=1}^{n} \updownarrow a_i, b_i \updownarrow \subseteq \bigcup_{i=1}^{n} \updownarrow a_i, b_i \updownarrow \subseteq [\min\{a_i\}, \max\{b_j\}]$, so the intersection is also a bounded set (possibly $\varnothing$).

5. Consider $\xi = \frac{1}{p}$, where p is a prime number. A typical example is the cut $A = \{r \in \mathbb{Q} : r < 0 \text{ or } r^2 < \xi\}$, and $B = \{r \in \mathbb{Q} : r > 0 \text{ and } r^2 \geq \xi\}$. It has supremum $\sqrt{\xi}$ by reasoning parallel to the proof of Corollary 12.3, and we know that $\sqrt{\xi} \notin \mathbb{Q}$. Therefore, no rational number can serve as the supremum for A or the infimum for B, and we conclude again that $\mathbb{Q}$ does not satisfy the completeness axiom.

   Consider $\beta = 1 + 3\sqrt{2} \in \mathbb{Q}(\sqrt{2})$, the quadratic extension discussed in Chapter 8. Since $\beta^2 - 2\beta - 17 = 0$, the cut $\langle A|B\rangle = \langle r^2 - 2r - 17 < 0 \,|\, r^2 - 2r - 17 \geq 0\rangle$ where $r \in \mathbb{Q}$, is the Dedekind cut at β. It shows again that $\mathbb{Q}$ doesn't satisfy the completeness axiom.

7. If $x > 0$, then $\frac{x+|x|}{x} = 2$, and if $x < 0$, then $\frac{x+|x|}{x} = 0$. Thus, $\inf W = 0$, and $\sup W = 2$.

9. Given that $0 < y - x$ and $0 < z - w$, that is, the two differences belong to the positive class, we conclude that $0 < (y - x) + (z - w)$, hence $x + w < y + z$.

   In order to be sure that $xw < yz$, a sufficient condition is that all four numbers be positive. Then $\frac{x}{y} < 1$ and $\frac{w}{z} < 1$, so $\frac{x}{y}(\frac{w}{z}) < 1(\frac{w}{z}) < 1$, hence $xw < yz$.

11. a. Let $u(x), v(x) \in \mathbb{J}^+$. Then

$$u(x)v(x) = \frac{r_\alpha x^\alpha + \cdots + r_1 x + r_0}{s_\beta x^\beta + \cdots + s_1 x + s_0},$$

where lead coefficients r_α and s_α must have the same sign. Then $r_\alpha s_\beta > 0$, and $u(x)v(x) \in \mathbb{J}^+$. A bit more thinking shows that $u(x) + v(x) \in \mathbb{J}^+$ also. As for the polynomial

$$u(x) = 0 = \frac{0}{q_m x^m + \cdots + q_1 x + q_0}, \; p_n = p_0 = 0,$$

so $p_n q_m = 0$, so $0 \notin \mathbb{J}^+$.

b. Again, $p_n q_m = 0$, so $0 \notin \mathbb{J}^-$. Here is one counterexample:

$$\frac{-7x^4 - x}{x^3 + 1} \in \mathbb{J}^- \quad \text{and} \quad \frac{2}{-x^3 + 1} \in \mathbb{J}^-,$$

but

$$\frac{2(-7x^4 - x)}{(-x^3 + 1)(x^3 + 1)} = \frac{14x^4 + 2x}{x^6 - 1} \in \mathbb{J}^+.$$

More interestingly, if $u(x), v(x) \in \mathbb{J}^-$, and

$$u(x)v(x) = \frac{r_\alpha x^\alpha + \cdots + r_1 x + r_0}{s_\beta x^\beta + \cdots + s_1 x + s_0},$$

then integer $r_\alpha s_\beta$ must have exactly two negative factors. Thus, $r_\alpha s_\beta > 0$.

c. Let $u(x) \in \mathbb{J}^+$. Then $p_n q_m > 0$. By trichotomy, exactly one of three statements is true: $p_n q_m > 0$, $p_n q_m < 0$, or $p_n q_m = 0$. Thus, $\mathbb{J}^- \cap \mathbb{J}^+ = \varnothing$. Furthermore, if $u(x) \in \mathbb{J}$, then $u(x) \in \mathbb{J}^+$, or $u(x) \in \mathbb{J}^-$, or $u(x) = 0$. Hence, $\mathbb{J} = \mathbb{J}^- \cup \{0\} \cup \mathbb{J}^+$.

d. Take, for example, $u(x) = x^3$. Given $n \in \mathbb{N}$, it is false that $n \cdot x > x^3$ as functions. Thus, there are many elements of $\mathbb{J}$ for which no integer exceeds them!

Chapter 13

1. First, $2x \leq x + y$, so $x \leq \frac{x+y}{2}$. Second, $x + y \leq 2y$, so $\frac{x+y}{2} \leq y$. **QED**

3. a. Typical good choices for the floor function: $x \in \{14.9, \pi, -\sqrt{2}, -0.01, 20, -15\}$. And for the ceiling function: $x \in \{7.15, \pi, -33.8, -e, 10, -99\}$.

   b. $y = \lfloor x \rfloor$

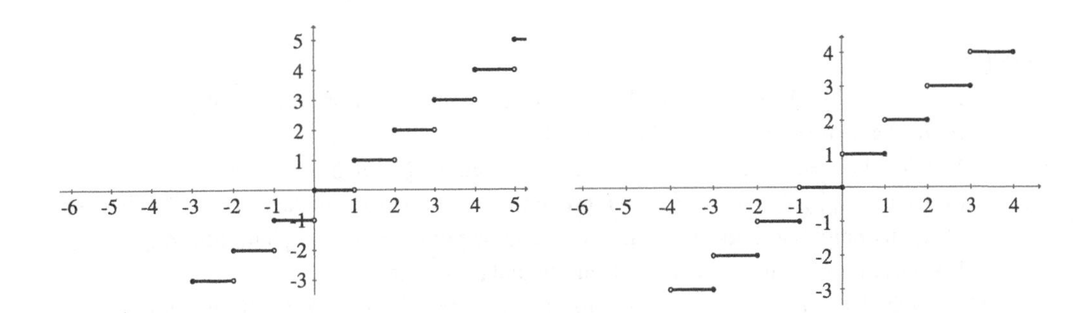

   c. $\lfloor x + m \rfloor = \lfloor (\lfloor x \rfloor + \beta) + m \rfloor = \lfloor (\lfloor x \rfloor + m) + \beta \rfloor = \lfloor x \rfloor + m$, since $0 \leq \beta < 1$. Next, to disprove $\lfloor x + y \rfloor = \lfloor x \rfloor + \lfloor y \rfloor$, select any numbers with $\beta(x) > 0.5$ and $\beta(y) > 0.5$. Next, $\lfloor 2x \rfloor = 2\lfloor x \rfloor$ if and only if $\beta \in [0, 0.5)$ or $-\beta \in (-1, -0.5)$, hence, quite false.

   d. For integers x, $\beta = 0$, so the equation is correct. For non-integer x, by the expressions in **a**, $\lceil x \rceil - \lfloor x \rfloor = 1$. These two expressions can be combined to form $\lceil x \rceil - \lfloor x \rfloor = \lceil \beta \rceil$.

   e. One way to prove this is to transform the graph of $y_1 = \lfloor x \rfloor$ in two steps. First, $y_2 = \lfloor -x \rfloor$ is a reflection of y_1 across the Y axis. Next, $y_3 = -\lfloor -x \rfloor$ is a reflection of y_2 across the X axis. Observe that y_3 is identical to the graph of y_1.

   A second way to prove that $\lceil x \rceil = -\lfloor -x \rfloor$ uses cases. If $x \in \mathbb{Z}$, then the equality is obvious. If $x > 0$ is not an integer, then for $\lceil x \rceil$ the distance upward to the next positive integer is $1 - \beta$, which equals the distance down to the next negative integer for $\lfloor -x \rfloor$. Expressed mathematically, $\lceil x \rceil = x + (1 - \beta)$ and $\lfloor -x \rfloor = -x - (1 - \beta)$, implying that $\lceil x \rceil = -\lfloor -x \rfloor$. If $x < 0$ and not an integer, then for $\lceil x \rceil$, the distance upward to the next negative integer is β, which equals the distance down to the next positive integer for $\lfloor -x \rfloor$. Expressed mathematically, this case is $\lceil x \rceil = x + \beta$, and $\lfloor -x \rfloor = -x - \beta$, implying again that $\lceil x \rceil = -\lfloor -x \rfloor$. **QED**

   f. $$round(x) = \begin{cases} \lfloor x + \frac{1}{2} \rfloor & \text{if } x \geq 0 \\ \lceil x - \frac{1}{2} \rceil & \text{if } x \leq 0 \end{cases}.$$

5. By the corollary, there exists a fraction $\frac{1}{n}$ so small that $\frac{1}{n} < y - x$. Since $n < 2^m$ for large enough m, we have $\frac{1}{2^m} < \frac{1}{n} < y - x$. Hence, $x < x + \frac{1}{2^m} < y$. **QED**

7. Since $r_2 - r_1 > 0$, there exists n such that $\frac{1}{n} < r_2 - r_1$ by Corollary 12.2. Thus, $nr_1 + 1 < nr_2$. Next,

$$nr_1 < nr_1 + \frac{1}{\sqrt{2}} < nr_1 + 1 < nr_2.$$

This yields

$$r_1 < r_1 + \frac{1}{n\sqrt{2}} < r_1 + \frac{1}{n} < r_2.$$

Therefore, an irrational number $r_1 + \frac{1}{n\sqrt{2}}$ has been found between arbitrary rationals r_1 and r_2. (If the discovered number were rational, then $\sqrt{2}$ would be rational.) **QED**

Chapter 14

1. a. Let $c \in [a, b]$. Does every neighborhood of c contain at least one point belonging to $[a, b]$ that is different than c? Yes. **QED**

   b. Just for this exercise, let the interval be written as $\updownarrow a, b \updownarrow$, representing any one of (a, b), $(a, b]$, $[a, b)$, or $[a, b]$. First, prove that a and b are boundary points. In the case of a, any neighborhood $(a - \varepsilon, a + \varepsilon)$ contains points in $\updownarrow a, b \updownarrow$ and not in $\updownarrow a, b \updownarrow$. Likewise for b. Thus, the endpoints are boundary points.

   Next, let c be a point interior to $\updownarrow a, b \updownarrow$, that is, $c \in (a, b)$. Then choose $\varepsilon = \min\{c - a, b - c\}$, which is the distance to the closest endpoint. Then the neighborhood $(c - \varepsilon, c + \varepsilon)$ does not contain any points outside of $\updownarrow a, b \updownarrow$, implying that c isn't a boundary point. Finally, let $c < a$ or $c > b$. In the first instance, take $\varepsilon = a - c$, and the neighborhood $(c - \varepsilon, c + \varepsilon)$ will not contain any points of $\updownarrow a, b \updownarrow$. Hence, such a c isn't a boundary point. Likewise for $c > b$. **QED**

3. $S = \{3 + \frac{\cos(n\pi)}{n} : n \in \mathbb{N}\} = \{3 + \frac{(-1)^n}{n}\} = \{3 - 1, 3 - \frac{1}{3}, 3 - \frac{1}{5}, \ldots, 3 + \frac{1}{2}, 3 + \frac{1}{4}, \ldots\}$, where the elements have been rearranged so that the clustering can be more easily seen. There is a cluster point at $x = 3$, since every neighborhood $(3 - \varepsilon, 3 + \varepsilon)$ of 3 contains elements of S different than 3 (in fact, infinitely many). No other cluster point exists.

5. An immediate corollary of Theorem 14.2 is that a nonempty set S (minor premise) contains its supremum if and only if it contains its maximum boundary point. This is logically equivalent to: a nonempty set S will not contain its maximum boundary point if and only if S does not contain its supremum (see Exercise 1.6). **QED**

7. Every neighborhood $(1 - \varepsilon, 1 + \varepsilon)$ of the range of f contains values of f always less than 1. Thus, $\sup \mathcal{R}_f = 1$. But $f(0)$ does not exist ($0 \notin \mathcal{D}_f$). So, the range does not contain its supremum. Next, $f'(x) = \frac{(\cos x)x - \sin x}{x^2} = 0$ implies $\tan x = x$, which yields an absolute minimum at approximately $\langle 4.493, -0.217\rangle$. This is another way to say that $\mathcal{R}_f$ contains its minimum.

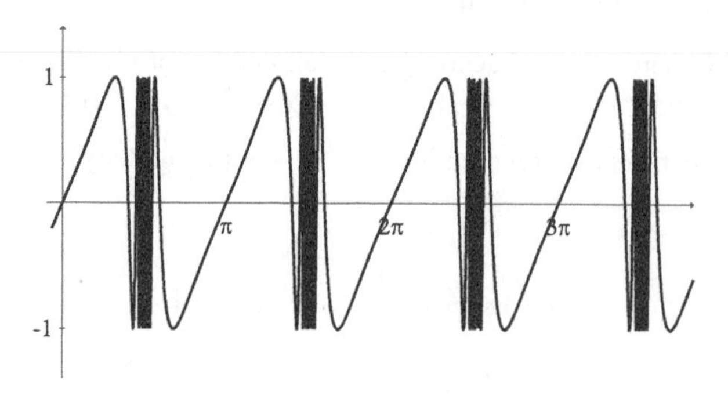

9. The function $f(x) = \sin(\tan x)$ is a composition. As $x \to \frac{\pi}{2}$ from the left, $\tan x \to \infty$, so $\sin(\tan x)$ oscillates infinitely many times to the left of $\frac{\pi}{2}$. The same behavior happens to the right of $\frac{\pi}{2}$. Thus, within any neighborhood $(\frac{\pi}{2} - \varepsilon, \frac{\pi}{2} + \varepsilon)$, regardless of how small, there are infinitely many roots. This makes $\frac{\pi}{2}$ a cluster point of the set of roots of f. Interestingly, the set of roots of f does not contain any of its cluster points.

11. We will show first that there must be some fixed $t \in T$ (which we will soon label as t_0) compared to which every $s \in S$ satisfies $s \leq t$. To see this, assume the contradictory statement: for any $t \in T$, there exists some $s \in S$ such that $s > t$. But then $s \notin T$, which contradicts the premise $S \subseteq T$. Thus, t_0 exists. Then $\sup S \leq t_0$. And, since $t_0 \in T$, we know that $t_0 \leq \sup T$.

    At the other end of the sets, there exists some fixed $t_1 \in T$ such that $t_1 \leq s$ for every $s \in S$, by parallel reasoning. Then $\inf T \leq t_1 \leq \inf S$. The two conclusions follow. **QED**

Chapter 15

1. Equality holds if and only if $x = y$.

3. In Corollary 15.1, let $x = |a|$ and $y = |b|$. Then $||a|| - ||b|| \leq ||a| - |b||$, which is equivalent to what was to be proved.

5. The contrapositive of what is to be proved in Exercise 15.5 is: If $\mathbf{N}_a^\varepsilon \cap \mathbf{N}_b^\varepsilon = \varnothing$, then $|b - a| \geq 2\varepsilon$ (the minor premise being that $\mathbf{N}_a^\varepsilon$ and $\mathbf{N}_b^\varepsilon$ are given).

    First, suppose that $b > a$. Since $\mathbf{N}_a^\varepsilon \cap \mathbf{N}_b^\varepsilon = \varnothing$, then $b \geq a + 2\varepsilon$, that is, $b - a \geq 2\varepsilon$. Next, suppose that $b < a$. Then $\mathbf{N}_a^\varepsilon \cap \mathbf{N}_b^\varepsilon = \varnothing$ implies that $b \leq a - 2\varepsilon$, that is, $b - a \leq -2\varepsilon$. Together, these imply that $|b - a| \geq 2\varepsilon$.

Chapter 16

1. $\mathbb{R}$ is uncountable, hence, it cannot be written as a subscripted list, that is to say, a sequence.

3. Paper and pencil—sometimes they are the best.

5. a. $a_n = \frac{(-1)^n}{n!}$, with $n = 0, 1, 2, \ldots$ b. $a_n = \frac{2n-1}{2n+1}$ c. $a_n = n^n$
   d. $a_n = \frac{n(n+1)}{2}$ (the sum of the first n natural numbers)
   e. $a_n = \begin{cases} 1 & \text{if } n \text{ is odd} \\ 0 & \text{if } n \text{ is even} \end{cases}$, or $a_n = \frac{1-(-1)^n}{2}$, or $a_n = \sin^2(\frac{n\pi}{2})$, or others.
   f. $a_n = 2^n n!$

7. a. For example, first show that $|\frac{\ln(n+1)}{\ln n}| \leq 2$ for all $n \geq 2$. But $\langle \ln n \rangle_1^\infty$ is not bounded.
   b. We now have $|a_{n+1}| \leq B|a_n| \leq B^2|a_{n-1}| \leq \cdots \leq B^n|a_1| < |a_1|$, because $0 < B^n < 1$. This implies that $|a_n| \leq |a_1|$ for all $n \in \mathbb{N}$, which imples that $\langle a_n \rangle_1^\infty$ is bounded.

Chapter 17

1. According to our analysis, $N(0.001) = \frac{1}{5(0.001)} = 200$. Thus, if the subscript $n > N_{0.001} = 200$, then $|\frac{3n-1}{5n} - \frac{3}{5}| < 0.001$. At $n = 201$, we get $|\frac{3(201)-1}{5(201)} - \frac{3}{5}| = 0.000995\ldots < 0.001$. Success! Moreover, the analysis in the proof shows that $|\frac{3n-1}{5n} - \frac{3}{5}| < 0.001$ for all $n \geq 201$.

The two analyses are very much alike. Following Example 17.1, we compared $L = \frac{3}{5}$ with a_n starting at $n = 201$, while after Example 17.2, we compared the absolute differences $|a_n - \frac{3}{5}|$ with ε itself.

3. From the proof in Example 17.3, $N_\varepsilon = \ln \frac{1}{\varepsilon}$. Substituting, $\ln \frac{1}{0.04} = 3.218\cdots = N_{0.04}$. This means that all terms after a_3 fall within the given tolerance. The few terms that fall outside the given neighborhood are a_1, a_2, and a_3.

   Not quite. It could happen that infinitely many terms stay inside $(-0.04, 0.04)$, but outside of $(-0.0004, 0.0004)$, which would prevent a limit from existing. The conditions in the definition must be true for all $\varepsilon > 0$.

5. The minor premise, $\langle a_n \rangle_1^\infty \to L$, remains given. Prove the contrapositive: If $L < 0$, then at least one term a_n must be negative.

   Since $L < 0$, some neighborhood $(L - \varepsilon, L + \varepsilon)$ must be an interval of strictly negative numbers. And by Definition 1, it contains all but a finite number of terms a_n. **QED**

7. a. Definition 2: For any $\varepsilon > 0$, find an N_ε such that for all $n > N_\varepsilon$, $|\frac{5n-37}{3n+1} - \frac{5}{3}| < \varepsilon$ is true. Simplifying,

$$\left|\frac{5n-37}{3n+1} - \frac{5}{3}\right| = \left|\frac{116}{3(3n+1)}\right| < \frac{116}{9n}.$$

   Thus, we require that $\frac{116}{9n} < \varepsilon$. This happens when $\frac{116}{9\varepsilon} < n$. Thus, set $N_\varepsilon = \frac{116}{9\varepsilon}$. Definition 2 will now be satisfied (try it with $\varepsilon = 0.0007$).

   b. Definition 1: All neighborhoods $(\frac{5}{3} - \varepsilon, \frac{5}{3} + \varepsilon)$ must include all but a finite number of terms. The excluded terms satisfy

$$\frac{5n^2 - 3n}{3n^2 + 1} \geq \frac{5}{3} + \varepsilon \quad \text{or} \quad \frac{5n^2 - 3n}{3n^2 + 1} \leq \frac{5}{3} - \varepsilon.$$

   The first inequality simplifies to

$$\frac{5n^2 - 3n}{3n^2 + 1} - \frac{5}{3} = -\frac{9n + 5}{3(3n^2 + 1)} \geq \varepsilon,$$

   which has no solutions.

   The second inequality simplifies to

$$-\frac{9n + 5}{3(3n^2 + 1)} \leq -\varepsilon, \text{ or } \frac{9n + 5}{3(3n^2 + 1)} \geq \varepsilon.$$

   We can simplify this difficult inequality this way, for instance:

$$\frac{9n + 5}{3(3n^2 + 1)} \geq \frac{9n}{3(3n^2 + 1)} \geq \frac{9n}{3(5n^2)} = \frac{3}{5n}.$$

   Thus, require that $\frac{3}{5n} \geq \varepsilon$, so that $n \leq \frac{3}{5\varepsilon}$. For any neighborhood $(\frac{5}{3} - \varepsilon, \frac{5}{3} + \varepsilon)$, the number of excluded terms given by $n \leq \frac{3}{5\varepsilon}$ is always finite. All other terms are within the neighborhood, thus satisfying Definition 1.

c. Use $\langle \frac{\cos(n\pi)}{n} \rangle_1^\infty = \langle \frac{(-1)^n}{n} \rangle_1^\infty$. Given any $\varepsilon > 0$, we must find a corresponding N_ε such that if $n > N_\varepsilon$, then $|\frac{(-1)^n}{n} - 0| < \varepsilon$. Simplifying the last inequality, $\frac{1}{n} < \varepsilon$, so $\frac{1}{\varepsilon} < n$. Thus, take $N_\varepsilon = \frac{1}{\varepsilon}$. We now assert that given any $\varepsilon > 0$, taking $N_\varepsilon = \frac{1}{\varepsilon}$ will guarantee that if $n > N_\varepsilon$, then

$$\left|\frac{\cos(n\pi)}{n} - 0\right| = \left|\frac{(-1)^n}{n} - 0\right| = \frac{1}{n} < \varepsilon.$$

d. We have that for any $n \in \mathbb{N}$,

$$\left|\frac{12n^4 - 5n^3}{3n^4 + \sqrt{n}} - 4\right| = \left|\frac{5n^3 + 4\sqrt{n}}{3n^4 + \sqrt{n}}\right| = \frac{5n^3 + 4\sqrt{n}}{3n^4 + \sqrt{n}}.$$

Since $4\sqrt{n} \le 4n^3$, use the techniques of Example 17.5 to be sure that

$$\frac{5n^3 + 4\sqrt{n}}{3n^4 + \sqrt{n}} < \frac{5n^3 + 4n^3}{3n^4} = \frac{3}{n}.$$

Thus, from $\frac{3}{n} < \varepsilon$ we get $N_\varepsilon = \frac{3}{\varepsilon}$. We assert that given any $\varepsilon > 0$, taking $N_\varepsilon = \frac{3}{\varepsilon}$ will guarantee that if $n > N_\varepsilon$, then

$$\left|\frac{12n^4 - 5n^3}{3n^4 + \sqrt{n}} - 4\right| < \frac{3}{n} < \varepsilon.$$

(Check: picking $\varepsilon = 0.0119$ gives $N_\varepsilon = \frac{3}{0.0119} \approx 252.1$, and taking $n = 253$ gives $|\frac{12(253^4) - 5(253^3)}{3(253^4) + \sqrt{253}} - 4| \approx 0.0066 < \varepsilon$.)

e. [Definition 1] A neighborhood $(-\varepsilon, \varepsilon)$ of 0 should contain all but a finite number of terms. To discover which terms are excluded, solve $\varepsilon \le \frac{1}{2\sqrt[3]{n}}$ for n. We get $n \le (\frac{1}{2\varepsilon})^3$. The solution set of $n \le (\frac{1}{2\varepsilon})^3$ is always finite. Hence, $(-\varepsilon, \varepsilon)$ always contains every other term. For example, if $\varepsilon = 0.2$, then $n \le (\frac{1}{0.4})^3 = 15.625$. Thus, all but first 15 terms are in $(-0.2, 0.2)$.

[Definition 2] Given any $\varepsilon > 0$, there must be a corresponding N_ε such that if $n > N_\varepsilon$, then $|\frac{1}{2\sqrt[3]{n}} - 0| < \varepsilon$. Equivalently, $\frac{1}{2\sqrt[3]{n}} < \varepsilon$. Solving, $(\frac{1}{2\varepsilon})^3 < n$. Therefore, set $N_\varepsilon = (\frac{1}{2\varepsilon})^3$.

We can now say that for any $\varepsilon > 0$, the corresponding N_ε is $(\frac{1}{2\varepsilon})^3$, and if $n > N_\varepsilon = (\frac{1}{2\varepsilon})^3$, then $|\frac{1}{2\sqrt[3]{n}} - 0| < \varepsilon$.

f. Given any $\varepsilon > 0$, we must find a corresponding N_ε such that if $n > N_\varepsilon$, then $|\frac{2^n}{n!} - 0| < \varepsilon$. To show that $\frac{2^n}{n!} < \varepsilon$, note that as long as $n \ge 4$,

$$\frac{2^n}{n!} = \left(\frac{2}{1} \cdot \frac{2}{2}\right) \cdot \left[\frac{2}{3} \cdot \cdots \cdot \frac{2}{n-1}\right] \cdot \frac{2}{n} < \left(\frac{2}{1} \cdot \frac{2}{2}\right) \cdot \frac{2}{n},$$

since all the fractions in brackets are less than one. Thus, $\frac{2^n}{n!} < \frac{4}{n}$, for $n \ge 4$. This allows us to take $N_\varepsilon = \max\{\frac{4}{\varepsilon}, 4\}$, and then we may assert that for any $\varepsilon > 0$, if $n > N_\varepsilon = \max\{\frac{4}{\varepsilon}, 4\}$, then $|\frac{2^n}{n!} - 0| < \varepsilon$.

9. We claim that the limit is zero. Definition 2 is faster here. Prove that given any $\varepsilon > 0$, there exists a corresponding N_ε such that if $n > N_\varepsilon$, then $|\frac{\cos n}{n} - 0| < \varepsilon$. But remember that $|\cos n| \le 1$. Then $|\frac{\cos n}{n}| < \frac{1}{n}$, and we consider $\frac{1}{n} < \varepsilon$. This implies that $n > \frac{1}{\varepsilon} = N_\varepsilon$. Thus, we assert that given any $\varepsilon > 0$, if $n > N_\varepsilon = \frac{1}{\varepsilon}$, then $|\frac{\cos n}{n} - 0| < \varepsilon$.

For the sake of comparison, Definition 1 says that any neighborhood $(-\varepsilon, \varepsilon)$ of 0 must contain all but a finite number of terms. To find terms that are excluded from the neighborhood, we search for solutions to both $\frac{\cos n}{n} \leq -\varepsilon$ and $\varepsilon \leq \frac{\cos n}{n}$. They cannot be solved explicitly for n. Using the latter inequality, $n \leq \frac{\cos n}{\varepsilon}$, so if $\cos n$ is positive, then $n \leq \frac{\cos n}{\varepsilon} \leq \frac{1}{\varepsilon}$. Now, let the solution set of $n \leq \frac{1}{\varepsilon}$ be the finite (possibly empty) set of natural numbers A, and let the solution set of $n \leq \frac{\cos n}{\varepsilon}$ be B. Then $B \subseteq A$. Thus, B is a finite set also, and it contains all the subscripts of the terms that are to the right of $(-\varepsilon, \varepsilon)$.

Likewise, the set of terms to the left of $(-\varepsilon, \varepsilon)$ is finite. Therefore, the total number of terms $\frac{\cos n}{n}$ that fall outside of $(-\varepsilon, \varepsilon)$ is finite. By Definition 1, $\lim_{n\to\infty} \frac{\cos n}{n} = 0$.

11. Assume, on the contrary, that another cluster point $M \neq L$ exists. Using a typical strategy, let $\varepsilon = \frac{1}{2}|M - L|$. By Theorem 14.1, the neighborhood $\mathbf{N}_M^\varepsilon$ of M must have infinitely many distinct elements of set A. But since this neighborhood is disjoint from the neighborhood $\mathbf{N}_L^\varepsilon$ (convince yourself), the definition of limit L indicates that $\mathbf{N}_M^\varepsilon$ contains only a finite number of terms of set A. This contradiction precludes the existence of M. **QED**

Chapter 18

1.

L	$w = \max\{2(1+L), 6\}$	a_n with least $n \geq w$	$\lvert\frac{n^2}{n+5} - L\rvert \geq 1$?
55	112	$\frac{112^2}{112+5}$	$52.213\cdots \geq 1$ true
3.3	8.6	$\frac{9^2}{9+5}$	$2.485\cdots \geq 1$ true
0	6	$\frac{6^2}{6+5}$	$3.272\cdots \geq 1$ true
−6	6	$\frac{6^2}{6+5}$	$9.272\cdots \geq 1$ true

3. $\langle\cos(n\pi)\rangle_1^\infty = \langle -1, 1, -1, 1, \ldots\rangle$. Suppose that a limit L exists. Then the neighborhood $(L - \frac{1}{2}, L + \frac{1}{2})$ will *not* have all but a finite number of terms within it, no matter what value L has. Thus, $\langle\cos(n\pi)\rangle_1^\infty$ does not converge.

5. a. False. Draw the graph of this counterexample: $\langle 1, -1, 1, -2, 1, -3, 1, -4, \ldots\rangle$.
   b. False. Counterexample: $\langle -1, 1, -1, 2, -1, 3, -1, 4, \ldots\rangle$.
   c. True.
   d. False. Counterexample: $a_n = \begin{cases} n-1 & \text{if } n \text{ is even} \\ n+1 & \text{if } n \text{ is odd} \end{cases}$.
   e. False. Counterexample: $a_n = n + (-1)^n n$.
   f. False. Counterexample: $\langle(-1)^n\rangle_1^\infty$ and $\langle(-1)^{n+1}\rangle_1^\infty$ both oscillate, but their sum converges.

7. "The sequence $\langle\sqrt{n}\rangle_1^\infty$ has limit $+\infty$ if and only if corresponding to any positive b, there exists some subscript N_b such that $\sqrt{n} > b$ for all terms with subscripts $n \geq N_b$." So, let b be a positive number. In order for $\sqrt{n} > b$, select $n > b^2$. Thus, take $N_b = b^2$. It follows that if $n \geq b^2$, then $a_n = \sqrt{n} \geq b$. Since $\langle\sqrt{n}\rangle_1^\infty$ is monotone increasing, then for all a_n with $n \geq b^2$, it is true that $a_n \geq b$. Therefore, $\lim_{n\to\infty} \sqrt{n} = +\infty$. **QED**

9. Since b is given as 7.2, find the first integer N_b such that $N_b(1.1-1) > 7.2$. We get $N_b = 73$. Checking: $1.1^{73} \approx 1051.15 \geq 73(1.1-1) > 7.2$, so the inequalities are verified. $N_b = 73$ is an integer that is sufficiently large for the proof, when $q = 1.1$ and $b = 7.2$. Some integers smaller than 73 will do just as well.

Chapter 19

1. These are the constant sequences $\langle a, a, a, \ldots \rangle$.

3. Taking limits of both sides of $a_{n+1}^2 = 1 + a_n$ yields $\varphi^2 = 1 + \varphi$. The only positive solution is $\varphi = \frac{1}{2} + \frac{1}{2}\sqrt{5} = 1.618034\ldots$. (The limit can't be negative since $a_1 = 1$ is the minimum term.) This number φ is famous since antiquity, it has been called the golden ratio or the golden section, and the great mathematician and astronomer Johannes Kepler (1571–1630) named it "the divine proportion." See [40] for extensive information about φ.

5. $b_n = \frac{n^2+n+1}{2n^2} = \frac{1}{2} + \frac{1+n}{2n^2}$ which we can prove is monotone decreasing to the limit $\frac{1}{2}$. Thus, letting

$$a_n = -\frac{n^2+n+1}{2n^2} = -b_n,$$

we have

$$\lim_{n\to\infty} a_n = -\frac{1}{2},$$

a monotone increasing sequence. By the nested intervals theorem,

$$\bigcap_{n=1}^{\infty}\left[-\frac{n^2+n+1}{2n^2}, \frac{n^2+n+1}{2n^2}\right] = \left[-\frac{1}{2}, \frac{1}{2}\right].$$

7. We have $a_{n_k} = \frac{1}{4k-3}$, and $b_{n_k} = 2 + \frac{1}{4k-1}$, and $c_{n_k} = 4 + \frac{1}{4k}$, all monotone decreasing.

9. The original sequence $\langle a_n \rangle_1^\infty$ is a subsequence of itself – the only one with no deleted terms! Hence, $\lim_{n\to\infty} a_n = L$.

11. Contrapositive: Some of the subsequences of $\langle a_n \rangle_1^\infty$ don't converge to L if and only if the sequence itself doesn't converge to L.
    The subsequence of $\langle (-1)^n \frac{n}{2n+1} \rangle_1^\infty$ given by $n_k = 2k$ is $\langle \frac{2k}{4k+1} \rangle_1^\infty$, which we can prove has limit $L = \frac{1}{2}$. On the other hand, the subsequence given by $n_k = 2k-1$ is $\langle \frac{-2k+1}{4k-1} \rangle_1^\infty$, which converges to $M = -\frac{1}{2}$. By the contrapositive, $\langle (-1)^n \frac{n}{2n+1} \rangle_1^\infty$ doesn't converge.

13. Draw! Some things are truly impossible to do.

15. Given $\langle a_n \rangle_1^\infty \to +\infty$, corresponding to any positive number b, there exists some subscript N_b such that $a_n > b$ for all terms with subscripts $n \geq N_b$. Since $\langle a_{n_k} \rangle_{k=1}^\infty \subseteq \langle a_n \rangle_1^\infty$, the definition pertains just as well to every subsequence. **QED**

Chapter 20

1. a. $\lim_{n\to\infty} \frac{7}{4-3n} = \lim_{n\to\infty} \frac{7n^{-1}}{4n^{-1}-3} = 0.$

   b. $\lim_{n\to\infty} \sqrt{\frac{9n-n^3}{5+n^2-6n^3}} = \sqrt{\lim_{n\to\infty} \frac{9n-n^3}{5+n^2-6n^3}} = \sqrt{\lim_{n\to\infty} \frac{9n^{-2}-1}{5n^{-3}+n^{-1}-6}} = \sqrt{\frac{1}{6}}$, regardless of the fact that the sequence begins at $n = 3$.

c. $\lim_{n\to\infty}\sqrt[3]{\frac{5|n|}{2n-7}} = \lim_{n\to\infty}\sqrt[3]{\frac{5n}{2n-7}} = \sqrt[3]{\lim_{n\to\infty}\frac{5n}{2n-7}} = \sqrt[3]{\frac{5}{2}}$ by the pinching theorem, using $\sqrt[4]{F(n)} \le \sqrt[3]{F(n)} \le \sqrt{F(n)}$, where $F(n) = \frac{5n}{2n-7}$ for all $n \ge 4$.

d. $\lim_{n\to\infty}(\sqrt{n+1}-\sqrt{n}) = \lim_{n\to\infty}(\frac{1}{\sqrt{n+1}+\sqrt{n}}) = \lim_{n\to\infty}(\frac{\frac{1}{\sqrt{n}}}{\sqrt{\frac{n+1}{n}}+1}) = \frac{0}{1+1} = 0.$

3. a. A convenient proof:

$$\lim_{n\to\infty}|a_n| = \lim_{n\to\infty}\sqrt{a_n^2} = \sqrt{\lim_{n\to\infty}a_n^2} = \sqrt{(\lim_{n\to\infty}a_n)^2} = |\lim_{n\to\infty}a_n| = |L|$$

by use of Theorems 20.5 and 20.3.

b. Converse: if $\lim_{n\to\infty}|a_n| = |L|$, then $\lim_{n\to\infty}a_n = L$. This is false, and a typical counterexample is $\langle\cos(\pi n)\rangle_1^\infty = \langle -1, 1, -1, \ldots\rangle$.

5. It is false. Here is a convincing argument. Since $\cos(2k\pi) = 1$, it is sufficient to show that $2k\pi$ can be arbitrarily close to even integers as $2k$ increases. The decimal expansion of π is $3.141592\ldots$. Thus, 10000π differs from the even integer $n_1 = 31416$ by approximately 0.1, so $\cos(n_1)$ is close to 1. Next, the 43rd, 44th, and 45th decimal digits of π are 3, 9, and 9, so $10^{43}\pi$ differs from an even integer n_2 by approximately 0.01, so $\cos(n_2)$ is closer to 1 than $\cos(n_1)$ is.

The 600th digit of π starts the string 2, 0, 0, 0, so $10^{600}\pi$ is very close to an even integer n_3. Thus, $\cos(n_3)$ is closer to 1 than our other two terms.

Does the decimal expansion of π contain ever longer strings of zeros preceded by an even digit, or ever longer strings of nines preceded by an odd digit? This may be true! Accepting it, we can in principle find a monotone increasing subsequence $\langle\cos n_k\rangle_{k=1}^\infty$ that converges to 1, precluding $\langle\cos n\rangle_1^\infty$ from being bounded away from 1. (The last pages of [1] display some decimal digits of π for you to investigate.)

7. It is valid. It uses $|r| = \sqrt{r^2}$, Theorem 20.5, and Theorem 20.3.

9. a. In Bernoulli's inequality, let $1+p = r^{1/n}$, where $p \ge 0$. Then $(1+p)^n = r \ge 1 + n(r^{1/n}-1)$, which yields $r^{1/n} \le \frac{n+r-1}{n}$. Since $1 \le r^{1/n} \le \frac{n+r-1}{n}$, Theorem 20.7 gives us $\lim_{n\to\infty}r^{1/n} = 1$ when $r \ge 1$. In passing, notice that the geometric mean–arithmetic mean inequality (Example 4.3) will demonstrate once again that $\sqrt[n]{r} \le \frac{n+r-1}{n}$. How interesting!

b. For values $0 < r < 1$, let $s = \frac{1}{r} > 1$. Then **a** applies, so

$$s^{1/n} = \frac{1}{r^{1/n}} \le \frac{n+\frac{1}{r}-1}{n}.$$

This yields

$$\frac{n}{n+\frac{1}{r}-1} \le r^{1/n} < 1,$$

and once again, the pinching theorem for sequences tells us that $\lim_{n\to\infty}r^{1/n} = 1$. **QED**

11. Draw! Part of it might look like this:

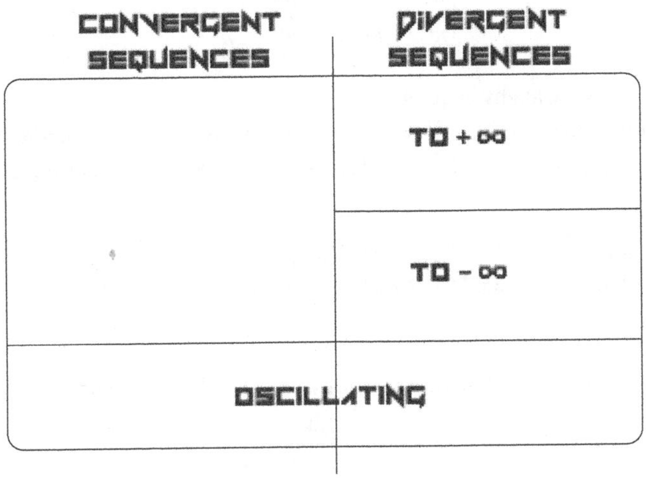

CONVERGENT SEQUENCES

DIVERGENT SEQUENCES

TO $+\infty$

TO $-\infty$

OSCILLATING

Chapter 21

1. There is only a finite number N_1 of terms in the front end $\langle a_n\rangle_1^{N_1}$ of the sequence at hand. Then $B = \max\{|a_1|, \ldots, |a_{N_1}|\}$ is a bound for the front end. Letting C be a bound for the tail end $\langle a_n\rangle_{N_1}^{\infty}$, we have that $\max\{B, C\}$ is a bound for the entire sequence.

3. We show that for all $m > n \geq N_\varepsilon$, $|\frac{n!}{n^n} - \frac{m!}{m^m}| < \varepsilon$. First, we show that $\frac{n!}{n^n} > \frac{m!}{m^m}$ whenever $m > n$. Reexpressing, prove that $\frac{m^m}{n^n} > \frac{m!}{n!}$, or equivalently that

$$\left(\frac{m}{n}\right)^n m^{m-n} > \underbrace{(n+1)(n+2)\cdots(m)}_{m-n \text{ factors}}.$$

Now divide through by m^{m-n}, and arrive at

$$\left(\frac{m}{n}\right)^n > \frac{n+1}{m} \cdot \frac{n+2}{m} \cdots \frac{m}{m}.$$

But

$$\left(\frac{m}{n}\right)^n > 1 \geq \frac{n+1}{m} \cdot \frac{n+2}{m} \cdots \frac{m}{m}.$$

We conclude that $\frac{n!}{n^n} > \frac{m!}{m^m}$ was true for all natural numbers $m > n$. It follows that for $m \geq n$,

$$\left|\frac{n!}{n^n} - \frac{m!}{m^m}\right| = \frac{n!}{n^n} - \frac{m!}{m^m} < \frac{n!}{n^n}.$$

It's now sufficient to prove that $\frac{n!}{n^n} < \varepsilon$, for large enough n. We shall do this by proving that $\frac{n!}{n^n} \leq \frac{1}{n}$. Factoring, arrive at $\frac{n!}{n^n} = \frac{1}{n} \cdot \frac{2}{n} \cdots \frac{n}{n} \leq \frac{1}{n}$. Thus, given any $\varepsilon > 0$, we find $N_\varepsilon = \frac{1}{\varepsilon}$, and so, whenever $m > n \geq N_\varepsilon$, then

$$\left|\frac{n!}{n^n} - \frac{m!}{m^m}\right| < \frac{n!}{n^n} \leq \frac{1}{n} \leq \varepsilon.$$

Therefore, $\langle \frac{n!}{n^n} \rangle_1^\infty$ is a Cauchy sequence.

We have proved that $\lim_{n\to\infty} \frac{n!}{n^n} = 0$. This would have proved immediately that $\langle \frac{n!}{n^n} \rangle_1^\infty$ is a Cauchy sequence, thanks to Theorem 21.1, but the purpose of this exercise was to use the definition.

5. (a) Extending $|a_{n+1} - a_n| < 2^{-n}$, we have $|a_{n+2} - a_{n+1}| < 2^{-(n+1)}, \ldots, |a_{n+p} - a_{n+p-1}| < 2^{-(n+p-1)}$. The sum of the left-hand terms compares to the sum of the right-hand terms as follows:

$$|a_{n+p} - a_{n+p-1}| + |a_{n+p-1} - a_{n+p-2}| + \cdots + |a_{n+2} - a_{n+1}| + |a_{n+1} - a_n|$$
$$< \frac{1}{2^{n+p-1}} + \frac{1}{2^{n+p-2}} + \cdots + \frac{1}{2^{n+1}} + \frac{1}{2^n}.$$

(b) $\frac{1}{2^{n+p-1}} + \frac{1}{2^{n+p-2}} + \cdots + \frac{1}{2^{n+1}} + \frac{1}{2^n} < \frac{1}{2^{n-1}}$ is true for $p = 1$. Assume the inequality is true as shown (case p). Then the case $p + 1$ leads to

$$\frac{1}{2^{n+p}} + \frac{1}{2^{n+p-1}} + \frac{1}{2^{n+p-2}} + \cdots + \frac{1}{2^{n+1}} + \frac{1}{2^n}$$
$$= \frac{1}{2}\left(\frac{1}{2^{n+p-1}} + \frac{1}{2^{n+p-2}} + \cdots + \frac{1}{2^n} + \frac{1}{2^{n-1}}\right) < \frac{1}{2}\left(\frac{1}{2^{n-1}} + \frac{1}{2^{n-1}}\right) = \frac{1}{2^{n-1}}.$$

This finishes the induction proof. (This can also be proved by using the formula for the sum of a finite geometric series, Exercise 4.9.)

(c) Observe that $n \leq 2^{n-1}$ for $n \in \mathbb{N}$ (or prove it by induction).

(d) Using the triangle inequality, we have:

$$|a_{n+p} - a_n| = |a_{n+p} - a_{n+p-1} + a_{n+p-1} - a_{n+p-2} + \cdots + a_{n+2} - a_{n+1} + a_{n+1} - a_n|$$
$$\leq |a_{n+p} - a_{n+p-1}| + |a_{n+p-1} - a_{n+p-2}| + \cdots + |a_{n+2} - a_{n+1}|$$
$$+ |a_{n+1} - a_n| < \frac{1}{2^{n-1}} \leq \frac{1}{n}$$

from the results of (a), (b), and (c).

(e) Set $N_\varepsilon = \frac{1}{\varepsilon}$. Then, for any $\varepsilon > 0$, and for any $m > n \geq N_\varepsilon = \frac{1}{\varepsilon}$, where $m = n + p$, and $p \in \mathbb{N}$, we have $|a_m - a_n| < \varepsilon$. This makes the sequence a Cauchy sequence. **QED**

7. By way of the theorem's use of the Bolzano-Weierstrass theorem, it appeals to the monotone convergence theorem, which requires the completeness axiom.

Chapter 22

1. $\sup\langle a_n \rangle_1^\infty = \sup\langle a_n \rangle_2^\infty = 2.1$, $\sup\langle a_n \rangle_3^\infty = \sup\langle a_n \rangle_4^\infty = 2.01$, and $\sup\langle a_n \rangle_5^\infty = \sup\langle a_n \rangle_6^\infty = 2.001$. The decreasing sequence of suprema converges to 2, so $\lim_{k\to\infty} \sup\langle a_n \rangle_{n=k}^\infty = \limsup\langle a_n \rangle_1^\infty = 2$. The sequence diverges, since there are two subsequential limits, 2 and 1.

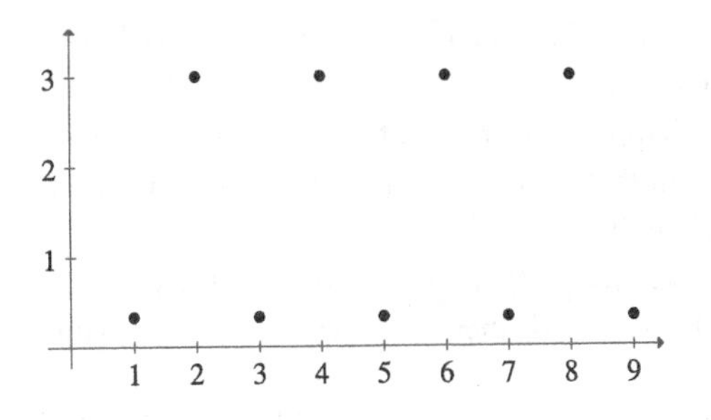

3. a. $\limsup\langle 3^{(-1)^n}\rangle_1^\infty = 3$
$\liminf\langle 3^{(-1)^n}\rangle_1^\infty = \frac{1}{3}$

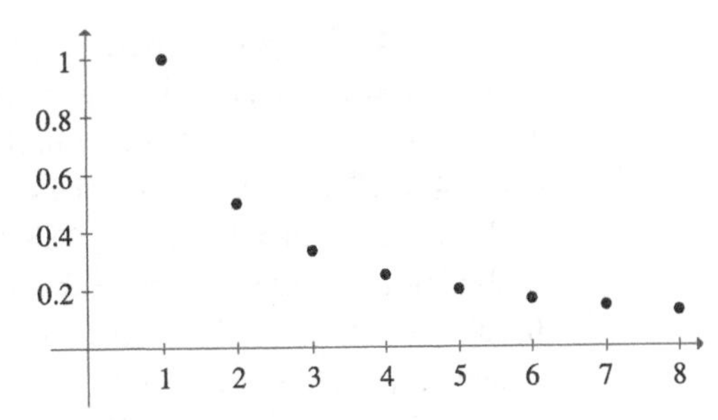

b. $\limsup\langle \frac{1}{n}\rangle_1^\infty = 0$
$\liminf\langle \frac{1}{n}\rangle_1^\infty = 0$

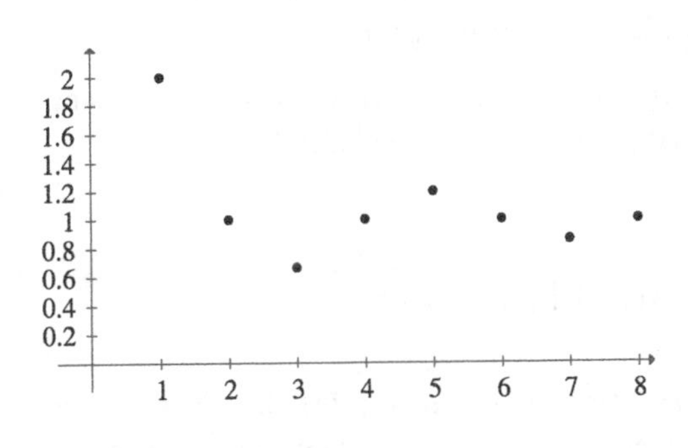

c. $\limsup\langle \frac{1}{n}\sin\frac{n\pi}{2} + 1\rangle_1^\infty = 1$
$\liminf\langle \frac{1}{n}\sin\frac{n\pi}{2} + 1\rangle_1^\infty = 1$

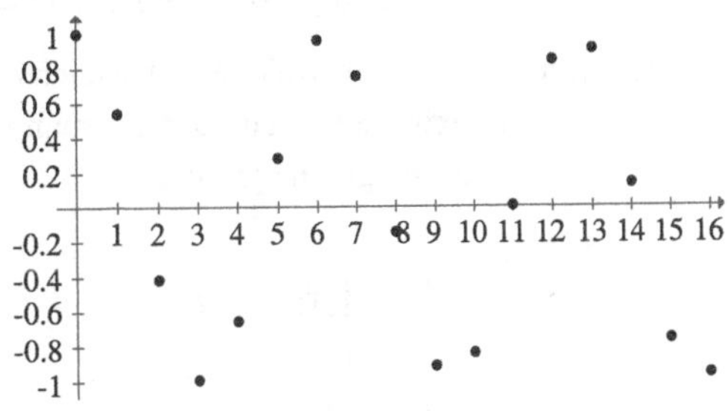

d. $\limsup\langle \cos n\rangle_0^\infty = 1$
$\liminf\langle \cos n\rangle_0^\infty = -1$

5. a. $\langle \sup\langle a_n\rangle_{n=k}^4\rangle_{k=1}^4 = \langle \max\langle a_1, a_2, a_3, a_4\rangle,\ \max\langle a_2, a_3, a_4\rangle,\ \max\langle a_3, a_4\rangle,\ \max\langle a_4\rangle\rangle = \max\{a_1, a_2, a_3, a_4\}$.
   b. We know that $\sup\langle a_n\rangle_1^\infty$ exists. Then $\sup\langle a_n\rangle_1^\infty = \sup\langle a_n\rangle_2^\infty = \sup\langle a_n\rangle_3^\infty = \sup\langle a_n\rangle_4^\infty$, since the supremum is not affected by truncating any front end of the sequence. Thus, the simplest form is $\sup\langle a_n\rangle_1^\infty$.
   c. $\langle \sup\langle a_n\rangle_{n=k}^\infty\rangle_{k=1}^4 = \max\langle a_n\rangle_1^\infty$, that is, the supremum equals the maximum in this case.
   d. We know that $\inf\langle a_n\rangle_1^\infty$ exists. Then $\inf\langle a_n\rangle_1^\infty = \inf\langle a_n\rangle_2^\infty = \inf\langle a_n\rangle_3^\infty = \inf\langle a_n\rangle_4^\infty$, since the infimum is not affected by truncating any front end of the sequence. Thus, the simplest form is $\inf\langle a_n\rangle_1^\infty$.
   e. $\min\langle a_n\rangle_1^\infty$, parallel thinking as in c.

7. By the monotone convergence theorem (Theorem 19.1), the limit of a convergent, monotone decreasing subsequence equals its infimum. Likewise, the limit of a convergent, monotone increasing subsequence equals its supremum.

Chapter 23

1. a. "We must verify that given any $\underline{\varepsilon > 0}$, we can find a corresponding $\underline{\delta > 0}$ such that if $\underline{0 < |x-4| < \delta}$, then $\underline{|(6-x^2)+10| < \varepsilon}$." Graph the function, as a visual aid.
   Using the last inequality, $|16-x^2| = |x^2-16| = |x-4||x+4|$. The factor $|x-4|$ can be made as small as we wish, so we concentrate on $|x+4|$. Using $|x-4| < \delta$, we get $4-\delta < x < 4+\delta$, implying $8-\delta < x+4 < 8+\delta$. For the moment, require $\delta < 8$.

Then $0 < x + 4 < 16$, so that perforce, $|x + 4| < 16$. Then, $|x - 4||x + 4| < 16\delta$. We therefore set $\varepsilon = 16\delta$, that is, $\delta = \frac{\varepsilon}{16}$, when $\delta < 8$.

At this point, $\varepsilon < 128$ implies $\delta < 8$. If it happens that $\delta \geq 8$, for instance $\delta = 9$, then $-1 < x + 4 < 17$, and we can only be certain that $|x + 4| < 1$ (not 17). To avoid what this leads to, we stipulate that if $\varepsilon \geq 128$, we will fix $\delta = 8$.

Combining both cases, we state that $\delta = \min\{\frac{\varepsilon}{16}, 8\}$. It is then true that given any $\varepsilon > 0$, take $\delta = \min\{\frac{\varepsilon}{16}, 8\}$. Then if $0 < |x - 4| < \delta$, then $|(6 - x^2) + 10| < \varepsilon$. This means that $\lim_{x \to 4}(6 - x^2) = -10$.

b. For $\varepsilon = 0.002$, $\delta = \frac{0.002}{16} = 0.000125$, so our deleted neighborhood is $\mathbf{D}_4^{0.000125}$. Selecting a random $x = 4.00012 \in \mathbf{D}_4^{0.000125}$, we find $|(6 - 4.00012^2) + 10| \approx 0.0009600 < \varepsilon$, success!

For $\varepsilon = 145.9$, $\delta = 8$, so our deleted neighborhood is $\mathbf{D}_4^8$. Selecting a random $x = 11.9 \in \mathbf{D}_4^8$, we find $|(6 - 11.9^2) + 10| = 125.61 < \varepsilon$ once again.

3. The graph shows jumps of one unit at each integer. If a limit L is claimed to exist, then for any $\varepsilon > 0$, there exists a $\mathbf{D}_a^\delta$ with $a \in \mathbb{Z}$, such that $|\, L - \lfloor x_1 \rfloor \,| < \frac{\varepsilon}{2}$ and also $|\, \lfloor x_2 \rfloor - L \,| < \frac{\varepsilon}{2}$. Hence, by the triangle inequality,

$$|\lfloor x_2 \rfloor - \lfloor x_1 \rfloor| \leq |\, L - \lfloor x_1 \rfloor \,| + |\, \lfloor x_2 \rfloor - L \,| < \varepsilon.$$

But now, with $a \in \mathbb{Z}$, select $x_1 \in (a - 1, a) \cap \mathbf{D}_a^\delta$ and $x_2 \in (a, a + 1) \cap \mathbf{D}_a^\delta$. Then $\lfloor x_1 \rfloor = a - 1$ and $\lfloor x_2 \rfloor = a$, so $\lfloor x_2 \rfloor - \lfloor x_1 \rfloor = 1$. Hence, if we pick $\varepsilon^* = \frac{1}{2}$ (actually, any number between zero and one will do), the requirement $|\lfloor x_2 \rfloor - \lfloor x_1 \rfloor| < \varepsilon^*$ under the assumption that L existed, cannot be met. Therefore, the limit fails to exist.

5. If δ were negative, then $\mathbf{D}_a^\delta$ would not make sense. If the correspondence is positive, and if it were not a function, then for some ε, perhaps two δs would arise. Selecting either deleted neighborhood would satisfy the definition. Thus, a function is simpler.

7. If $f(x) + g(x) \to L$, then $f(x) + g(x)$ would have to be as close as one wishes to L, when x is in $\mathbf{D}_a^\delta$. But at the same time, $\lim_{x \to a} f(x) = M$, which hints that $f(x) + g(x) \to M + g(x)$. So, an educated guess about what is happening is that $M + g(x) \to L$, meaning that $g(x) \to L - M$. And it's true! But this isn't a proof, yet.

9. First, let $\lim_{x \to a} f(x) = L$ exist. This means that for any $x \in \mathbf{D}_a^\delta \cap \mathcal{D}_f$, $f(x) \in (L - \varepsilon, L + \varepsilon)$ is true. This implies both that if $x \in (a - \delta, a) \cap \mathcal{D}_f$, then $f(x) \in (L - \varepsilon, L + \varepsilon)$, and that if $x \in (a, a + \delta) \cap \mathcal{D}_f$, then $f(x) \in (L - \varepsilon, L + \varepsilon)$. Hence, both the limit from the left and right exist, and are equal to L.

For the converse, let $\lim_{x \to a-} f(x) = L = \lim_{x \to a+} f(x)$. This implies both that for any $x \in (a - \delta, a) \cap \mathcal{D}_f$, $f(x) \in (L - \varepsilon, L + \varepsilon)$ is true, and that for any $x \in (a, a + \delta) \cap \mathcal{D}_f$, $f(x) \in (L - \varepsilon, L + \varepsilon)$ is true. Thus, $f(x) \in (L - \varepsilon, L + \varepsilon)$ is true for any $x \in ((a - \delta, a) \cup (a, a + \delta)) \cap \mathcal{D}_f$. This last set is written $\mathbf{D}_a^\delta \cap \mathcal{D}_f$, and the conclusion follows that $\lim_{x \to a} f(x) = L$. **QED**

11. Given any $\varepsilon > 0$, we are to find a corresponding $N_\varepsilon > 0$ such that whenever $x > N_\varepsilon$, then $|\frac{\sin x}{x} - 0| < \varepsilon$. Working with the last inequality when $x > 0$, $|\frac{\sin x}{x}| \leq \frac{1}{x} < \frac{1}{N_\varepsilon}$. Thus, take $N_\varepsilon = \frac{1}{\varepsilon}$. The definition of limit at $+\infty$ is fulfilled when $L = 0$.

13. From the definition of the signum function, $\operatorname{sgn} x = -1$ for all $x \in (-\delta, 0)$, hence, $\lim_{x\to 0-} \operatorname{sgn} x = -1$. Likewise, $\operatorname{sgn} x = 1$ for all $x \in (0, \delta)$, hence, $\lim_{x\to 0+} \operatorname{sgn} x = 1$.

15. Let $w = -x$. Then

$$\lim_{x\to 0-} \frac{e^{1/x}}{x} = \lim_{w\to 0+} \frac{\frac{-1}{w}}{e^{1/w}} = \lim_{w\to 0+} \frac{-1}{e^{1/w}} = 0$$

by l'Hospital's rule.

17. (i) The required circle is $(x-a)^2 + (y-c)^2 = r^2$, where $c = \sqrt{r^2 - a^2}$. (ii) Since $\sin\theta = \frac{a}{r}$, the result follows. (iii) By l'Hospital's rule,

$$\lim_{r\to\infty} 2r \sin^{-1}\left(\frac{a}{r}\right) = 2 \lim_{r\to\infty} \frac{a}{\sqrt{1 - a^2/r^2}} = 2a.$$

Thus, as the radius of the circle increases, the arclenth of the arc PQ decreases towards $2a$, which is the length of the chord PQ. This demonstrates that a chord's length is approximated better and better by the subtened arclength as $\theta \to 0$. Archimedes was right!

Chapter 24

1.

	Comments	$f(a)$ exists?	$\lim_{x\to a} f(x)$ exists?	Continuous at $x = a$?
a.	smooth function	✓	✓	✓
b.	gap discontinuity at $a = 0$	×	✓	×
c.	finite (jump) discontinuity at $a = 0$	✓	×	×
d.	gap discontinuity at $a = 3$	×	✓	×
e.	gap discontinuity $a = -1$	×	✓	×
f.	jump discontinuity at $a = 5$	✓	×	×
g.	infinite jump at $a = 8$	✓	×	×
h.	gap discontinuity at $a = 8$	✓	✓	×
i.	discontinuity removed at $a = 8$	✓	✓	✓
j.	infinite jump at $a = \pi/2$	✓	×	×
k.	discontinuity at $a = -1$	×	×	×
l.	gap removed at $a = 0$	✓	✓	✓
m.	non-smooth function	✓	✓	✓

3. a. Let $x = a + h$, where $h \neq 0$. The result is that $\lim_{x\to a} f(x)$ becomes $\lim_{h\to 0} f(a + h)$. **QED**

   b. We must prove that $\lim_{h\to 0} \sin(a+h) = \sin a$. Using a well-known trigonometric identity and several properties of limits, $\lim_{h\to 0} \sin(a+h) = \lim_{h\to 0}(\sin a \cos h + \cos a \sin h) = \sin a \cdot 1 + \cos a \cdot 0 = \sin a$. Lemma 23.1 assures us that $\lim_{h\to 0} \sin h = 0$ and $\lim_{h\to 0} \cos h = 1$. **QED**

   c. Since $\cos x = -\sin(x - \frac{\pi}{2})$, we can appeal to the graphic idea that a translation and reflection of a continuous function will itself be continuous. Otherwise, we have that $\lim_{h\to 0} \cos(a+h) = \lim_{h\to 0}(\cos a \cos h + \sin a \sin h) = \cos a \cdot 1 + \sin a \cdot 0 = \cos a$. **QED**

5. a. "We must verify that given any $\varepsilon > 0$, we can find a corresponding $\delta > 0$ such that if $x \in \mathcal{D}_w$ and $|x - 4| < \delta$, then $|w(x) - w(4)| < \varepsilon$."

   Using the graph, we see that for neighborhoods within 2 units of $a = 4$, $w(x) = 2 - |x - 4|$. This means that for $|x - 4| < \delta < 2$, $|w(x) - w(4)| = |2 - |x - 4| - 2| = |x - 4| < \delta$. We are led to claim that $\varepsilon = \delta$, provided that $\delta < 2$.

   The case that $\delta \geq 2$ occurs if $\varepsilon \geq 2$, and this would prevent us from using $w(x) = 2 - |x - 4|$. It thus simplifies things very much if we set $\delta = 2$ whenever $\varepsilon \geq 2$. In effect, this strategy disregards the parts of the graph of $w(x)$ that are distant from $\langle 4, 2 \rangle$.

   It can now (finally) be said that given any $\varepsilon > 0$, the corresponding $\delta = \min\{2, \varepsilon\}$. Using this correspondence, if $|x - 4| < \delta$, then $|w(x) - w(4)| = |2 - |x - 4| - 2| = |x - 4| < \delta \leq \varepsilon$. Therefore, $w(x)$ is continuous at $a = 4$.

   b. $\delta = \min\{2, \varepsilon\} = 0.0054$. Thus, if $|x - 4| < 0.0054$, meaning that $3.9946 < x < 4.0054$, then we expect $|2 - |x - 4| - 2| < \varepsilon$. This is immediately true, since $|2 - |x - 4| - 2| = |x - 4| < \delta$.

7. a. "We must verify that given any $\varepsilon > 0$, we can find a corresponding $\delta > 0$ such that if $x \in \mathcal{D}_f$ and $|x - 0| < \delta$, then $|f(x) - f(0)| < \varepsilon$."

   We work with $|x \sin \frac{1}{x}| = |x|| \sin \frac{1}{x}| \leq |x| < \varepsilon$. Given the neighborhood $|x| = |x - 0| < \delta$, the analysis indicates that $\varepsilon = \delta$. Now, given any $\varepsilon > 0$, let the corresponding δbe $\delta = \varepsilon$. With this relationship, if $|x| = |x - 0| < \delta$, then

$$\left| x \sin \frac{1}{x} - 0 \right| = |x| \left| \sin \frac{1}{x} \right| \leq |x| < \delta = \varepsilon.$$

   Consequently, this function is continuous at zero. No other point except $\langle 0, 0 \rangle$ will remove the discontinuity, as any other point would become a type 2 isolated point (see Exercise 24.11).

   Incidentally, from Exercise 24.6, $sinc(\frac{1}{x}) = x \sin \frac{1}{x}$ for all nonzero x.

   b. $\delta = \varepsilon = 0.0054$. Thus, if $|x - 0| < 0.0054$, then we expect $|x \sin \frac{1}{x} - 0| < \varepsilon$. And in fact, if $-0.0054 < x < 0.0054$, then $|x \sin \frac{1}{x}| \leq |x| < 0.0054 = \varepsilon$, using **a**, and keeping in mind that $f(0) = 0$. Hence, if $f(0) \neq 0$, then continuity will cease at the origin.

9. a. Graphing software may not show the isolated point $\langle 1, 0 \rangle$ of the function. The full domain is $\{1\} \cup [2, \infty)$.

   b. The definition says, "... if $x \in \mathbf{N}_1^\delta \cap \mathcal{D}_g$, then $g(x) \in \mathbf{N}_{g(1)}^\varepsilon$." For any $\varepsilon > 0$, take $\delta = \frac{1}{2}$, giving us the neighborhood $\mathbf{N}_1^{1/2}$ of the isolated point $\langle 1, 0 \rangle$. Then $\mathbf{N}_1^{1/2} \cap \mathcal{D}_g = \{1\}$, in which case $|g(x) - g(1)| = |g(1) - g(1)| = 0$, or put in an obvious way, $g(1) \in \mathbf{N}_{g(1)}^\varepsilon$. Hence, the function is continuous at $a = 1$.

   c. Generalizing the situation in **b**, since a type 1 isolated point is not a cluster point of the domain, we can always find a small enough neighborhood of $x = a$ such that $\mathbf{N}_a^\delta \cap \mathcal{D}_f = \{a\}$. Consequently, $|f(x) - f(a)| = |f(a) - f(a)| = 0 < \varepsilon$, and this implies that f is continuous at $x = a$. **QED**

   d. By definition, a sequence is a function $f : \mathbb{N} \to \mathbb{R}$. As such, the domain is composed entirely of type 1 isolated points. Hence, all sequences are continuous functions throughout their domain. **QED**

11. a. The function H has domain $\mathbb{R}$, so $a = 2$ is a cluster point of $\mathcal{D}_h$. However, a small circular neighborhood drawn around $\langle 2, 1\rangle$ shows that it is a type 2 isolated point. Because of this circular neighborhood, $\lim_{x\to 2} H(x)$ does not exist. Corollary 24.1 indicates that $\langle 2, 1\rangle$ is a point of discontinuity.

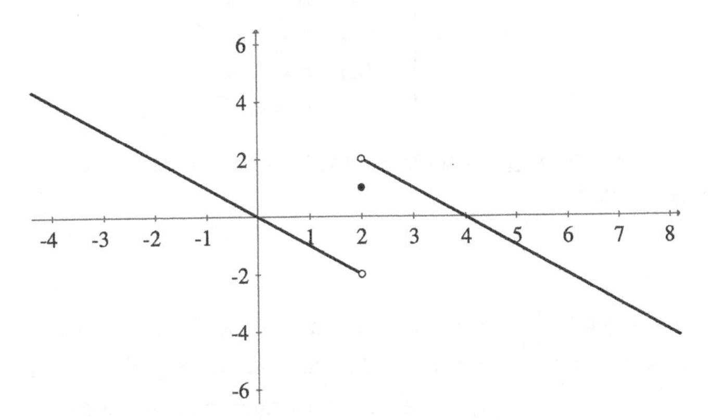

$$H(x) = \begin{cases} -x+4 & \text{if } x > 2 \\ 1 & \text{if } x = 2 \\ -x & \text{if } x < 2 \end{cases}$$

b. $a = \pi/2$ is a cluster point of the domain of $g(x) = \lfloor \sin x \rfloor$, but $\langle \pi/2, 1\rangle$ is the only point of g within a circle of small radius centered there. Thus, $\langle \pi/2, 1\rangle$ is a type 2 isolated point of g. In fact, g has type 2 isolated points at every $a = (4k+1)\pi/2$, $k \in \mathbb{Z}$.

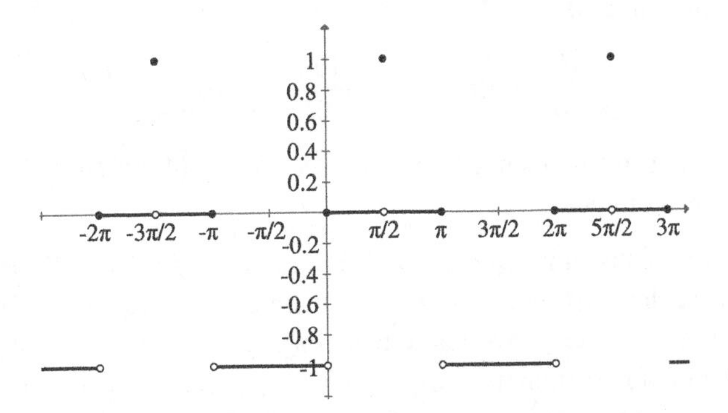

c. $\lim_{x\to\pi/2} \lfloor \sin x \rfloor = 0$ since $\lfloor \sin x \rfloor = 0$ within any deleted neighborhood of $x = \frac{\pi}{2}$ with small radius. But $\lfloor \sin \frac{\pi}{2} \rfloor = 1$, so Corollary 24.1 indicates that g is discontinuous at $a = \frac{\pi}{2}$. The distinction between this and **a** is that $\lim_{x\to\pi/2} g(x)$ exists, while $\lim_{x\to 2} H(x)$ doesn't.

d. Let $f(x)$ have a type 2 isolated point $\langle a, f(a)\rangle$. This means that there exists a circular neighborhood with a radius $\varepsilon^* > 0$ around $\langle a, f(a)\rangle$ that contains no other point of f. This implies that the neighborhood $\mathbf{N}_{f(a)}^{\varepsilon^*}$ contains no other ordinates of f except $f(a)$. But since $x = a$ is a cluster point of $\mathcal{D}_f$, there always are abscissas $x \in \mathcal{D}_f$ within any neighborhood $\mathbf{N}_a^{\delta}$. These are the requirements for discontinuity at $x = a$. **QED**

13. Since f is bounded over A equal to (a, b), $(a, b]$, or $[a, b)$, the numbers $p = \lim_{x \to a+} f(x)$ and $q = \lim_{x \to b-} f(x)$ exist. We modify f slightly to

$$g(x) = \begin{cases} q & \text{if } b \notin A \\ f(b) & \text{if } b \in A \\ f(x) & \text{if } x \in (a, b) \\ f(a) & \text{if } a \in A \\ p & \text{if } a \notin A \end{cases}.$$

Then the proof's $D = \max_{x \in [a,b]}(g(x)) - \min_{x \in [a,b]}((g(x))$ exists, and the proof will apply to g, and therefore to f.

Chapter 25

1. We will find a sequence $\langle x_n \rangle_1^\infty \to 0$ but such that $\lim_{n \to \infty} f(x_n) = \lim_{n \to \infty} \sin \frac{1}{x_n}$ does not exist. Let's use the sequence of extrema of the function, that is, solutions x_n to $\sin \frac{1}{x_n} = \pm 1$. Evidently, solutions are $x_n = \frac{1}{\arcsin(\pm 1)}$, giving the sequence of solutions $x_n = \frac{2}{(2n+1)\pi}$ where $n \in \mathbb{N} \cup \{0\}$. Clearly, $\langle x_n \rangle_1^\infty \to 0$ but $\lim_{n \to \infty} \sin \frac{1}{x_n}$ does not exist, and Theorem 25.1 says that f is discontinuous at the origin. The non-existence of the limit precludes the possibility of removing the discontinuity.

3. The classic example is

$$h(x) = \begin{cases} 1 & \text{if } x \text{ is rational} \\ -1 & \text{if } x \text{ is irrational.} \end{cases}$$

5. The function looks like Figure 25.1, but shifted to the right five units, reflected across the X axis, and with amplitude 10.

$$\frac{ds}{dt} = v(t) = \frac{10}{25 - 10t + t^2} \cos \frac{1}{t - 5}$$

which does not exist (it's unbounded) at $t = 5$ sec. And $|v(4.9942)| \approx 2.77 \times 10^5$ km/sec, or 92% of the speed of light.

7. Continuity assures us that $\lim_{x \to a} f(x) = f(\lim_{x \to a} x) = f(a)$ for all real $a \in \mathcal{D}_f$. Let ξ be irrational in the domains of f and g. Then there exists a sequence of rational numbers $\langle r_n \rangle$ such that $\lim_{n \to \infty} r_n = \xi$. We know that $\lim_{n \to \infty}[\langle f(r_n) \rangle - \langle g(r_n) \rangle] = 0$. Using the sequential criterion for continuity, $\lim_{n \to \infty}[\langle f(r_n) \rangle - \langle g(r_n) \rangle] = 0 = \lim_{n \to \infty}\langle f(r_n) \rangle - \lim_{n \to \infty}\langle g(r_n) \rangle = f(\xi) - g(\xi)$. Hence, $f(\xi) = g(\xi)$ for all irrational numbers as well. Therefore, $f(x) = g(x)$ over their common domain.

   Note what happens if we try this exercise on Dirichlet's function, $D(x)$, in Figure 25.3. Let's take $g(x) = 2D(x) - 1$. Then $g(r) = D(r)$ for all rational numbers r, but take a moment to verify that $g \neq D$. This occurs because D is not continuous.

Chapter 26

1. a. By Theorem 26.1 and Examples 24.4 and 24.5, $y = \sqrt{x} + 3|x + 6|$ is continuous over $\{x \geq 0\}$. And $3|x + 8| = 3|(x + 2) + 6|$ is a composition of two continuous functions, hence by Theorem 26.2, so is H over its domain.

b. Exercise 24.7a showed that $f(x) = x \sin \frac{1}{x}$ with $f(0) = 0$ is continuous. Here,

$$K(x) = \frac{\sin(1/x)}{x^2} = \frac{f(x)}{x^3}$$

with domain $\mathbb{R}\backslash\{0\}$. By Theorem 26.1, K is continuous over its domain.

c. Since $S(x) = \frac{\sin x}{x}$ is continuous over domain $\mathbb{R}\backslash\{0\}$ (see Exercise 24.6), then $F(x) = xS(x) = \sin x$ is continuous over $\mathbb{R}\backslash\{0\}$ by Theorem 26.1. And since $\sin 0 = \lim_{x\to 0} xS(x)$, $F(x)$ is continuous everywhere.

3. Let f be a function that satisfies the premises of Lemma 26.1A, so that there exists a neighborhood of c throughout which $f(x) > 0$. But $-f = (-1)f$, so $-f$ is continuous over the same $\mathcal{D}_f$, and $-f(c) < 0$. Throughout the same neighborhood of c, $-f(x) < 0$. **QED**

   The typical answer

$$f(x) = \begin{cases} x^2 & \text{if } x \neq 2 \\ -1 & \text{if } x = 2 \end{cases}$$

   fulfills the requirements. But f cannot be continuous at $c = 2$, or else Lemma 26.1B would be false!

5. We are looking for the solution to $\arcsin(\frac{x}{2}) - (x^2 - x - 1 + 0.3\pi) = 0$, over $[0, 2]$. Since $u(0)u(2) < 0$, we are in business.

   Step 1: $u(1)u(0) \approx 0.033 > 0$ will not do, but $u(1)u(2) \approx -0.21 < 0$ indicates that the root is within $(1, 2)$. Step 2: $u(1)u(1.5) > 0$, but $u(1.5)u(2) < 0$, so the root is within $(1.5, 2)$. Step 3: $u(1.5)u(1.75) < 0$, so the root is within $(1.5, 1.75)$. These steps can be rendered in any programming language, and the Bisection Method will quickly converge towards $1.618034\ldots$.

   The tricky part is to recognize this decimal as none other than the golden ratio $\varphi = \frac{1+\sqrt{5}}{2}$! The reason behind the strange function u and its famous root is that $\sin(\frac{3\pi}{10}) = \frac{\varphi}{2}$ exactly (alas, I can't remember the source of this curious identity).

7. A typical counterexample:

$$f(x) = \begin{cases} -1 & \text{if } x = 0 \\ x^2 & \text{if } x \in (0, 2) \\ 5 & \text{if } x = 2 \end{cases}$$

   This function is continuous over $(0, 2)$ only. Choose $y_0 = 4.5$, satisfying $f(0) < y_0 < f(2)$. No $c \in (0, 2)$ exists such that $f(c) = 4.5$.

9. a. One possibility is $f(x) = \operatorname{sgn} x$, and

$$g(x) = \begin{cases} 0 & \text{if } x > 0 \\ 1 & \text{if } x = 0 \\ 2 & \text{if } x < 0 \end{cases}$$

   The individual functions are discontinuous, but $(f + g)(x) = 1$, which is continuous.

b. One possibility is

$$f(x) = \begin{cases} \tan x & \text{if } x \neq n\pi/2 \\ 1 & \text{if } x = n\pi/2 \end{cases}$$

and

$$g(x) = \begin{cases} \cot x & \text{if } x \neq n\pi/2 \\ 1 & \text{if } x = n\pi/2 \end{cases}$$

Both are discontinuous, but $(fg)(x) = 1$ over $\mathbb{R}$.

11. a. By the extreme-value theorem (Theorem 26.5), f attains extrema m and M, where $m, M \in f([a,b])$. Consider a ordinate $f(x) \in f((a,b))$. Since $m \leq f(x) \leq M$, by the intermediate value theorem (Theorem 26.4), $x \in (a,b)$. Hence the image of (a,b) is an interval in $\mathcal{R}_f$. And clearly, $f(a), f(b) \in [m, M]$, so we conclude that $f([a,b]) = [m, M] \subseteq \mathcal{R}_f$. **QED**

    In case $f(x) = c$, then $f([a,b]) = \{c\}$, which is equivalent to the closed interval $[c, c]$.

    b. The continuous function $y = \sin x$ over the open interval $(0, \pi)$ is a counterexample, since $\sin((0, \pi)) = (0, 1]$, which isn't open. We must insist on a monotone, continuous function to get open intervals to map into open intervals.

    c. Consider any $[t, u] \subset (a, b)$. By **a**, $f([t, u])$ is a closed, bounded interval. In the case that f is monotone increasing, $f([t,u]) = [f(t), f(u)]$. Since this is true for all t, u such that $a < t \leq u < b$, the image $f((a,b))$ will be one of these open intervals: $(\lim_{t \to a+} f(t), \lim_{u \to b-} f(u))$, $(-\infty, \lim_{u \to b-} f(u))$, $(\lim_{t \to a+} f(t), \infty)$, and $(-\infty, \infty)$.

    If f is monotone decreasing, $f([t,u]) = [f(u), f(t)]$, and the image $f((a,b))$ is one of $(\lim_{u \to b-} f(u), \lim_{t \to a+} f(t))$, $(\lim_{u \to b-} f(u), \infty)$, $(-\infty, \lim_{t \to a+} f(t))$, and $(-\infty, \infty)$.

13. The function is strictly monotone (increasing) over its domain. It is also continuous over its domain. Yes, the theorem applies to intervals within the domain of the function. Its inverse, seen below, are those points (still in Quadrants I and III) that satisfy the equation of the conjugate hyperbola $y^2 - x^2 = 1$.

    Note that the inverse function is continuous over its domain (recall Exercise 24.4c). The domain of the inverse is $(-\infty, 0) \cup (0, \infty)$, which excludes the one point where discontinuity would occur.

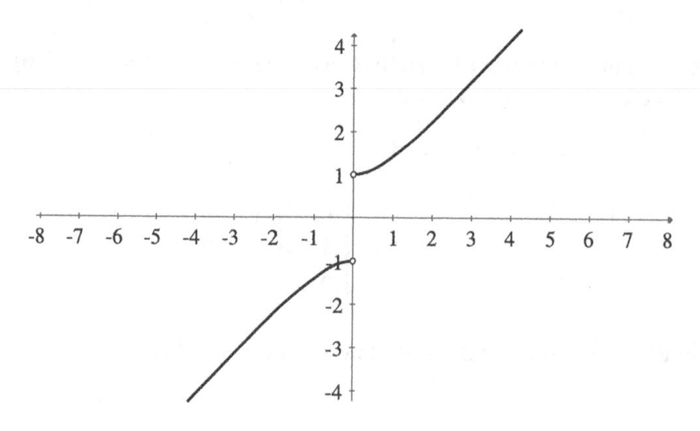

Chapter 27

1. Given any $\varepsilon > 0$, we must find a corresponding $\delta_\varepsilon > 0$ depending upon only ε such that if $|x_1 - x_2| < \delta_\varepsilon$ then $|b - b| < \varepsilon$. This is automatically true, so we may take $\delta_\varepsilon = 0.1492$, or $\delta_\varepsilon = 2 + \sin \varepsilon$, or any other positive function of ε that tickles our fancy.

3. a. Similar to Example 27.1, restate the function as

$$f(x) = \begin{cases} 3x & \text{if } x < 0 \\ -3x & \text{if } x \geq 0 \end{cases}.$$

Again, we have that for any $\varepsilon > 0$, $\delta = \delta_\varepsilon = \frac{\varepsilon}{|\pm 3|}$. Again, given any $\varepsilon > 0$, if $|x_1 - x_2| < \delta_\varepsilon = \frac{\varepsilon}{3}$, then $|3|x_1| - 3|x_2|| < \varepsilon$, meaning $|-3|x_1| + 3|x_2|| = |f(x_1) - f(x_2)| < \varepsilon$. So, this function is uniformly continuous everywhere.

ε	$\delta_\varepsilon = \frac{\varepsilon}{3}$	x_1	x_2	Is $|x_1 - x_2| < \delta_\varepsilon$?	Is $|-|3x_1| + |3x_2|| < \varepsilon$?
0.3	0.1	4	4.08	$0.08 < 0.1$✓	$|-12 + 12.24| = 0.24 < 0.3$✓

b. Given any $\varepsilon > 0$, we must find a corresponding $\delta_\varepsilon > 0$ depending upon only ε such that if $|x_1 - x_2| < \delta_\varepsilon$ then $|(x_1 + |x_1|) - (x_2 + |x_2|)| < \varepsilon$. Simplifying and using Corollary 15.2,

$$|(x_1 - x_2) + (|x_1| - |x_2|)| \leq |x_1 - x_2| + ||x_1| - |x_2|| \leq \delta_\varepsilon + \delta_\varepsilon = 2\delta_\varepsilon.$$

Set $\delta_\varepsilon = \frac{\varepsilon}{2}$ and the definition will be fulfilled.

Or, rewrite

$$h(x) = \begin{cases} 2x & \text{if } x \geq 0 \\ 0 & \text{if } x < 0 \end{cases}.$$

When $x_1, x_2 < 0$, clearly $|h(x_1) - h(x_2)| < \varepsilon$, regardless of δ. For any other choices of x_1 and x_2, let $\delta_\varepsilon = \frac{\varepsilon}{2}$ with similar strategy to **a**. Then $|h(x_1) - h(x_2)| < \varepsilon$ again. Consolidating the cases, we find that $\delta_\varepsilon = \frac{\varepsilon}{2}$ will always fulfill the definition of uniform continuity.

ε	$\delta_\varepsilon = \frac{\varepsilon}{2}$	x_1	x_2	Is $|x_1 - x_2| < \delta_\varepsilon$?	Is $|x_1 + |x_1| - (x_2 + |x_2|)| < \varepsilon$?
0.02	0.01	33	32.995	$0.005 < 0.01$ ✓	$|66 - 65.99| = 0.01 < 0.02$ ✓

c. To use Theorem 27.2, rewrite

$$f(x) = \begin{cases} 3x & \text{if } x \leq 0 \\ -3x & \text{if } x \geq 0 \end{cases}$$

and

$$h(x) = \begin{cases} 2x & \text{if } x \geq 0 \\ 0 & \text{if } x \leq 0 \end{cases}.$$

since they are both continuous at $x = 0$. Then use $I = (-\infty, 0]$ and $J = [0, \infty)$.

Theorem 27.3 also applies nicely. For f,

$$|m_{\text{sec}}| = \left| \frac{|3x_1| - |3x_2|}{x_1 - x_2} \right| = 3\frac{||x_1| - |x_2||}{|x_1 - x_2|} \leq 3.$$

And for h, $0 \leq m_{\text{sec}} \leq 2$. Hence, both have bounded secant slopes.

5. Since $|\frac{1/x_1 - 1/x_2}{x_1 - x_2}| = |\frac{1}{x_1 x_2}| < 1$ over $[1, \infty)$, we have $|u(x_1) - u(x_2)| \leq 1|x_1 - x_2|$. Hence, by Theorem 27.3, u is uniformly continuous over $A = [1, \infty)$.

7. Mathematical induction is being used. It concludes that $y = x^2$ is uniformly continuous over $\bigcup_{i=1}^{n}[i - 1, i] = [0, n]$ for any natural number n. It's not the business of mathematical induction to "pass over to infinity!"

9. It's not a problem, since 2ε will be small when ε is small. But for the sake of fidelity, we would write "When both $x_1, x_2 \in I$, there exists a $\delta_{I,\varepsilon} > 0$ depending only upon ε, such that if $|x_1 - x_2| < \delta_{I,\varepsilon}$, then $|f(x_1) - f(x_2)| < \frac{\varepsilon}{2}$." And likewise for J. Then we would end with $|f(x) - f(y)| < \varepsilon$, as needed.

11. a. Let f and g be uniformly continuous over A. For any $\varepsilon > 0$, there exists a $\delta_\varepsilon > 0$ such that whenever $|x_1 - x_2| < \delta_\varepsilon$ then both $|f(x_1) - f(x_2)| < \frac{\varepsilon}{2}$ and $|g(x_1) - g(x_2)| < \frac{\varepsilon}{2}$. Then,

$$|(f+g)(x_1) - (f+g)(x_2)| = |f(x_1) + g(x_1) - (f(x_2) + g(x_2))| \leq |f(x_1) - f(x_2)| + |g(x_1) - g(x_2)| < \frac{\varepsilon}{2} + \frac{\varepsilon}{2} = \varepsilon.$$

This is true independently of x_1 and x_2 in A. Thus, $f + g$ is uniformly continuous over A. **QED**

b. Look at the graph of $P(x) = xQ(x) = x|x - \lfloor x + \frac{1}{2} \rfloor|$ for assistance. Since P is the product of an even function Q and an odd function f, it is an odd function itself, so we consider only $x \geq 0$. We see local extrema $P(k + \frac{1}{2}) = \frac{2k+1}{4}$ and $P(k) = 0$, for $k \in \mathbb{N}$. Thus, take $\varepsilon^* = 1$, and for any $\delta > 0$, let $x_1 = k$ and $x_2 = x_1 + \delta$, and we note that $P(x) = x(x - k)$ over $[k, k + \frac{1}{2})$. To keep $x_2 \in [k, k + \frac{1}{2})$, restrict $\delta < \frac{1}{2}$ as a first case. We see that

$$|P(x_1) - P(x_2)| = |P(x_2)| = (k + \delta)(k + \delta - k) = k\delta + \delta^2 > k\delta.$$

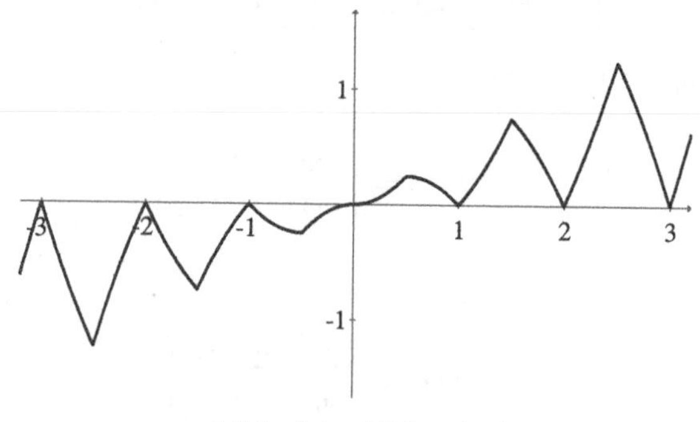

A "sharkstooth" function!

And $k\delta \geq 1$ provided that we choose $k \geq \frac{1}{\delta}$. As the only other case, if $\delta \geq \frac{1}{2}$, just take $x_1 = 2$ (for instance) and $x_2 = \frac{5}{2}$ (for instance).

Putting the details together, take $\varepsilon^* = 1$, and for $\delta < \frac{1}{2}$ choose $x_1 = k$ and $x_2 = x_1 + \delta$, where $k \geq \frac{1}{\delta}$. Then $|P(x_1) - P(x_2)| > k\delta \geq \varepsilon^*$. For $\delta \geq \frac{1}{2}$ just choose $x_1 = 2$ (implying $k = 2$), and $x_2 = \frac{5}{2}$, in which case $|P(x_1) - P(x_2)| = \frac{5}{4} > \varepsilon^*$ again. This means that $P(x) = xQ(x)$ is not uniformly continuous over $\mathbb{R}^+$. We conclude that P isn't uniformly continuous over $\mathbb{R}$.

c. Let f and g be uniformly continuous and bounded over the set A. To be specific, suppose that $|f(x)| \leq M$ and $|g(x)| \leq N$ over A. Because of the uniform continuity of each function, we are correct in saying that $|f(x_1) - f(x_2)| < \varepsilon$ and $|g(x_1) - g(x_2)| < \varepsilon$ whenever $|x_1 - x_2| < \delta_\varepsilon$, and $x_1, x_2 \in A$ (δ_ε depending solely upon ε). Recycling the technique used in the proof of continuity of a product, we have

$$\begin{aligned}|f(x_1)g(x_1) - f(x_2)g(x_2)| &= |f(x_1)g(x_1) - f(x_2)g(x_1) + f(x_2)g(x_1) - f(x_2)g(x_2)|\\ &\leq |g(x_1)||f(x_1) - f(x_2)| + |f(x_2)||g(x_1) - g(x_2)|\\ &\leq N|f(x_1) - f(x_2)| + M|g(x_1) - g(x_2)| < N\varepsilon + M\varepsilon.\end{aligned}$$

The positive quantity $N\varepsilon + M\varepsilon$ can be made as small as we please since M and N are positive constants. We have now shown that we can always find a δ_ε depending solely upon ε, such that $|f(x_1)g(x_1) - f(x_2)g(x_2)| < N\varepsilon + M\varepsilon$ whenever $|x_1 - x_2| < \delta_\varepsilon$. Hence, $(fg)(x)$ is uniformly continuous over A. **QED**

With some more thinking, we could have set up the premises as $|f(x_1) - f(x_2)| < \frac{\varepsilon}{2N}$ and $|g(x_1) - g(x_2)| < \frac{\varepsilon}{2M}$ so that, in the end, we would have $|f(x_1)g(x_1) - f(x_2)g(x_2)| < \varepsilon$, as strictly required by the definition. But this makes analysts seem to be able to pull rabbits out of hats.

d. Consider $(\frac{f}{g})(x)$ as $f(x) \cdot \frac{1}{g(x)}$. To state that g is bounded and also bounded away from zero means that there exist some m and M such that $0 < m \leq |g(x)| \leq M$. Then

$$0 < \frac{1}{M} \leq \left|\frac{1}{g(x)}\right| \leq \frac{1}{m}.$$

Now we can apply **c** of this exercise. **QED**

Chapter 28

1. a. $s_3 = 1 + \frac{5}{2} + 4 + \frac{11}{2} = 13 = \frac{3+1}{2}(1 + \frac{11}{2})$. Since $s_0 = 1$ is the first partial sum, each s_m has $m + 1$ terms added together.

b. $s_m = \sum_{n=0}^{m}(qn + r) = r + \sum_{n=1}^{m}(qn + r) = r + q\sum_{n=1}^{m} n + \sum_{n=1}^{m} r = r + q\frac{m(m+1)}{2} + mr$, using the suggested formula. Continuing,

$$r + q\frac{m(m+1)}{2} + mr = \frac{m+1}{2}[r + (qm + r)] = \frac{m+1}{2}(a_0 + a_m).$$

c. If $q \neq 0$, $\lim_{n\to\infty} a_n = \lim_{n\to\infty}(qn + r) = \pm\infty$ depending on the sign of q. Hence, Theorem 28.4 indicates that the series diverges. If $q = 0$ but $r \neq 0$, then again the series diverges. Finally, if $q = r = 0$, then the series converges, to zero.

3. We compare the original series $\sum_{n=1}^{\infty} a_n$ with the new one, $\sum_{m=1+p}^{\infty} a_m = \sum_{n=1}^{\infty} a_{n+p}$, in the table below:

n	1	2	$\cdots$	$p-1$	p	$p+1$	$p+2$	$\cdots$	$2p$	$2p+1$	$\cdots$
a_n	a_1	a_2	$\cdots$	a_{p-1}	a_p	a_{p+1}	a_{p+2}		a_{2p}	a_{2p+1}	$\cdots$.
a_m						a_{1+p}	a_{2+p}	$\cdots$	a_{2p}	a_{2p+1}	$\cdots$

Let $\sum_{n=1}^{p} a_n = E$. Letting the partial sums of the original series be $\langle s_n\rangle_1^{\infty}$, and those of the new series be $\langle S_n\rangle_1^{\infty}$ (that is, $S_1 = a_{1+p}$, $S_2 = a_{1+p} + a_{2+p}$, and so on), we see that $s_{p+n} = S_n + E$ for each $n \in \mathbb{N}$. Therefore, $\langle s_n\rangle_1^{\infty}$ converges if and only if $\langle S_n\rangle_1^{\infty}$ converges, and the same for divergence.

In fact, if the original series converges to A, then the new series converges to $A - E$.

5. We are considering $\lim_{\theta\to 0}(\sin\theta \sum_{n=0}^{\infty} \cos^n\theta) = \lim_{\theta\to 0} \frac{\sin\theta}{1-\cos\theta}$ since $0 < \cos\theta < 1$. This diverges (prove it!), telling us that the length increases without bound.

7. Let the partial sums of $\sum_{n=1}^{\infty} a_n$ be $\langle s_n\rangle_1^{\infty}$. Then the partial sums of $\sum_{n=1}^{\infty} ca_n$ are $\langle cs_n\rangle_1^{\infty}$, and $\lim_{n\to\infty} cs_n = c\lim_{n\to\infty} s_n = c\sum_{n=1}^{\infty} a_n = cA$. **QED**

9. We have $\alpha = a_0 + \sum_{n=1}^{m} d_n 10^{-n}$. The question is whether or not $a_0 + \sum_{n=1}^{m} d_n 10^{-n}$ equals $a_0 + \sum_{n=1}^{m}(d_n - 1)10^{-n} + \sum_{n=m+1}^{\infty}(9)10^{-n}$. This is easy now, since the last summation is a convergent geometric series:

$$\sum_{n=m+1}^{\infty} (9)10^{-n} = 9\sum_{n=m+1}^{\infty}\left(\frac{1}{10}\right)^n = 9\sum_{n=0}^{\infty}\left(\frac{1}{10}\right)^n - 9\sum_{n=0}^{m}\left(\frac{1}{10}\right)^n$$
$$= 9\left(\frac{1}{1-\frac{1}{10}}\right) - 9\left(\frac{1-(\frac{1}{10})^{m+1}}{1-\frac{1}{10}}\right) = 10^{-m}.$$

Hence,

$$a_0 + \sum_{n=1}^{m}(d_n - 1)10^{-n} + \sum_{n=m+1}^{\infty}(9)10^{-n} = a_0 + \sum_{n=1}^{m}(d_n-1)10^{-n} + 10^{-m} = a_0 + \sum_{n=1}^{m} d_n 10^{-n}.$$

11. Since the partial sums form a bounded sequence, the Bolzano-Weierstrass theorem states that a convergent subsequence of partial sums exists. The convergent subsequence corresponds to a specific grouping of terms by parentheses that will force the series to converge! **QED**

Of course, without knowing the subsequence, we would have no indication as to where to put the parentheses.

13. It diverges, since $\sin n \not\to 0$ (see Theorem 28.4).

Chapter 29

1. a. Converges.

$$\sum_{n=1}^{\infty} e^{-2n} = \sum_{n=1}^{\infty}(e^{-2})^n = \frac{e^{-2}}{1-e^{-2}} = \frac{1}{e^2-1},$$

the sum. Also, $\int_1^\infty e^{-2x}\,dx = \frac{1}{2}e^{-2}$ is a convergent integral. Also,

$$a_n = \frac{1}{e^{2n}} < \frac{1}{2^n},$$

and

$$\sum_{n=1}^{\infty} \frac{1}{2^n}$$

is a convergent series. And,

$$\rho = \lim_{n\to\infty} \frac{e^{-2(n+1)}}{e^{-2n}} = e^{-2} < 1.$$

b. Converges. $\int_2^\infty \frac{dx}{x^2-1} = \frac{1}{2}\ln 3$ is a convergent integral. Also, by partial fractions,

$$\sum_{n=2}^{\infty} \frac{1}{n^2-1} = \frac{1}{2}\sum_{n=2}^{\infty}\left(\frac{1}{n-1} - \frac{1}{n+1}\right)$$

which converges by telescoping to the sum $\frac{1}{2}\cdot\frac{3}{2} = \frac{3}{4}$.

c. Diverges. $\frac{\sqrt{n}}{n+10} > \frac{1}{2\sqrt{n}}$ when $n > 10$, and $\sum_{n=1}^\infty \frac{1}{\sqrt{n}}$ is a p-series with $p \le 1$. A bit of calculus shows that $f(x) = \frac{\sqrt{x}}{x+10}$ is decreasing for $x > 10$, and

$$\int_{10}^{\infty} \frac{\sqrt{x}}{x+10}dx = 2\int_{\sqrt{10}}^{\infty} \frac{u^2}{u^2+10}du$$

is a divergent integral since the integrand has the horizontal asymptote $y = 1$.

d. Converges.

$$\rho = \lim_{n\to\infty}\left(\frac{8(n+1)^4}{(n+1)!}\,\frac{n!}{8n^4}\right) = 0.$$

e. Diverges.

$$a_n = \frac{n+6}{\sqrt{n}+30} = \frac{1+\frac{6}{n}}{\frac{1}{\sqrt{n}}+\frac{30}{n}} \not\to 0$$

as $n \to \infty$. Also, $\frac{n+6}{\sqrt{n}+30} > \frac{\sqrt{n}}{10}$ for all $n \in \mathbb{N}$, and $\frac{1}{10}\sum_{n=1}^\infty \sqrt{n}$ is a divergent series.

f. Diverges. $\sum_{n=2}^\infty \frac{1}{\ln n}$ has terms $\frac{1}{\ln n} > \frac{1}{n}$, and $\sum_{n=2}^\infty \frac{1}{n}$ diverges. Also,

$$\int_2^\infty \frac{1}{\ln x}dx = \int_{\ln 2}^\infty \frac{e^u}{u}du$$

is a divergent integral (why?).

g. Converges. For $n \ge 4$, $\frac{1}{\sqrt{n!}} < \frac{1}{\sqrt{2^n}} = \frac{1}{(\sqrt{2})^n}$ and $\sum_{n=1}^\infty \frac{1}{(\sqrt{2})^n}$ is a convergent geometric series.

h. Diverges.

$$\sum_{n=0}^{\infty}(\cos n - \sin n) = \sum_{n=0}^{\infty}\sqrt{2}\sin(\pi/4 - n)$$

and $a_n = \sqrt{2}\sin(\pi/4 - n) \not\to 0$ as $n \to \infty$.

i. Converges.

$$\rho = \lim_{n\to\infty}\left(\frac{(n+1)^2}{\pi^{n+1}}\frac{\pi^n}{n^2}\right) = \lim_{n\to\infty}\left(\left(\frac{n+1}{n}\right)^2\frac{1}{\pi}\right) = \frac{1}{\pi} < 1.$$

j. Converges. The series is $\sum_{n=1}^{\infty}\frac{1}{n(n+2)}$ which by partial fractions is $\frac{1}{2}\sum_{n=1}^{\infty}(\frac{1}{n} - \frac{1}{n+2})$. This converges by telescoping to the sum $\frac{3}{4}$. Also, $\sum_{n=1}^{\infty}\frac{1}{n(n+2)} = \sum_{n=2}^{\infty}\frac{1}{(n-1)(n+1)}$ which is the same series as **b**. Also, $\int_1^{\infty}\frac{dx}{x(x+2)}$ is a convergent integral (why?).

k. Converges.

$$\sum_{n=1}^{\infty}\frac{1}{n(n+1)(n+2)} = \frac{1}{2}\cdot\frac{1}{2!} = \frac{1}{4}$$

the sum by Example 29.7. Also,

$$\int_1^{\infty}\frac{dx}{x^3+3x^2+2x} < \int_1^{\infty}\frac{dx}{x^3},$$

which is a convergent integral.

l. Diverges. $8\sum_{n=1}^{\infty}\frac{1}{n^{1/3}}$ is eight times a p-series with $p \le 1$. Also, $\int_1^{\infty}\frac{dx}{x^{1/3}}$ is a divergent integral.

m. Converges.

$$\rho = \lim_{n\to\infty}\left(\frac{((n+1)!)^2}{(2n+2)!}\frac{(2n)!}{(n!)^2}\right) = \lim_{n\to\infty}\frac{(n+1)^2}{(2n+1)(2n+2)} = \frac{1}{4} < 1.$$

n. Converges. Similar to **i**.

$$\rho = \lim_{n\to\infty}\left(\frac{(n+1)^{12.7}}{1.03^{n+1}}\frac{1.03^n}{n^{12.7}}\right) = \lim_{n\to\infty}\left(\left(\frac{n+1}{n}\right)^{12.7}\frac{1}{1.03}\right) = \frac{1}{1.03} < 1.$$

o. Diverges. $\int_2^{\infty}\frac{dx}{x\ln x} = \int_{\ln 2}^{\infty}\frac{du}{u}$ is a divergent integral.

p. Converges.

$$\rho = \lim_{n\to\infty}\left(\frac{e^{2n+2}}{(n+1)!}\frac{n!}{e^{2n}}\right) = \lim_{n\to\infty}\frac{e^2}{n+1} = 0 < 1.$$

The series equals $\sum_{n=0}^{\infty}\frac{(e^2)^n}{n!}$, which has sum $e^{(e^2)} \approx 1618.18$ by the Maclaurin series for e^x (which converges for all real x).

q. Diverges.

$$\rho = \lim_{n\to\infty}\left(\frac{e^{(n+1)^2}}{(n+1)!}\frac{n!}{e^{n^2}}\right) = \lim_{n\to\infty}\frac{e^{2n+1}}{n+1} \to +\infty.$$

r. Converges. $\sum_{n=1}^{\infty} \frac{1}{n^2}$ is a p-series with $p > 1$. Its sum is $\frac{\pi^2}{6}$ in the note below this chapter's title. This is not simple to evaluate, but may be found by Fourier analysis in an advanced calculus course.

s. Converges.

$$\rho = \lim_{n\to\infty} \left(\frac{1}{n+1} \left(\frac{2}{3}\right)^{n+1} \cdot n \left(\frac{3}{2}\right)^n \right) = \frac{2}{3} < 1.$$

The sum is $\ln 3$ by the logarithmic series

$$\ln x = \frac{x-1}{x} + \frac{1}{2}\left(\frac{x-1}{x}\right)^2 + \frac{1}{3}\left(\frac{x-1}{x}\right)^3 + \cdots$$

which converges for all $x > \frac{1}{2}$.

t. Converges.

$$\rho = \lim_{n\to\infty} \left(\frac{(\sqrt{2})^{n+1}}{(n+1)!} \frac{n!}{(\sqrt{2})^n} \right) = 0 < 1.$$

The sum is $e^{\sqrt{2}}$ (same reason as in **p**.)

u. Converges.

$$\frac{a_{n+1}}{a_n} = \frac{n}{2n+1} \qquad \text{so} \qquad \rho = \lim_{n\to\infty} \frac{n}{2n+1} = \frac{1}{2} < 1.$$

v. Converges.

$$\frac{n!+1}{(n+3)!} < \frac{(n+1)!}{(n+3)!} = \frac{1}{(n+2)(n+3)} \qquad \text{and} \qquad \sum_{n=0}^{\infty} \frac{1}{(n+2)(n+3)}$$

converges by Example 29.7.

3. We find

$$\lim_{n\to\infty} \left(\frac{(n+1)!}{(n+1)^{n+1}} \frac{n^n}{n!} \right) = \lim_{n\to\infty} \left(\frac{n+1}{n+1} \frac{n^n}{(n+1)^n} \right) = \lim_{n\to\infty} \frac{1}{(1+\frac{1}{n})^n} = \frac{1}{e} < 1.$$

Consequently, the series converges.

5. a. $\sum_{n=1}^{\infty} \frac{(n-1)!}{(n+4)!} = \sum_{n=1}^{\infty} \frac{1}{n(n+1)(n+2)(n+3)(n+4)} = \frac{1}{4}\left(\frac{1}{4!}\right) = \frac{1}{96}$.

b. The series equals

$$\sum_{n=2}^{\infty} \frac{1}{n(n+1)(n+2)(n+3)} = \sum_{n=1}^{\infty} \frac{1}{n(n+1)(n+2)(n+3)} - \frac{1}{24}.$$

The last series was evaluated in Example 29.7 with $M = 3$, and its sum is $\frac{1}{18}$. Hence, the series converges to $\frac{1}{18} - \frac{1}{24} = \frac{1}{72}$.

c. For the ratio test,

$$\rho = \lim_{n\to\infty} \frac{a_{n+1}}{a_n} = \lim_{n\to\infty} \frac{n(n+1)(n+2)\cdots(n+M)}{(n+1)(n+2)(n+3)\cdots(n+M+1)} = 1.$$

Thus, the ratio test is inconclusive for all series of this type. To apply the comparison test, observe that

$$\frac{1}{n(n+1)(n+2)\cdots(n+M)} \leq \frac{1}{n(n+1)} \leq \frac{1}{n^2}$$

for any $M \in \mathbb{N}$. Also, $\sum_{n=1}^{\infty} \frac{1}{n^2}$ is a convergent p-series. Hence, (3) converges by the comparison test.

7. a. They are contrapositives of each other.

b. They are converses of each other, and were proved separately.

c. If grouped Φ diverges, then Φ diverges.	True; it is the contrapositive of Q. It is also a special case of P.
d. If Φ diverges, then grouped Φ diverges.	True; it is the contrapositive of R.
e. If grouped Σ converges, then Σ converges.	False; Example 28.7 is a counterexample.
f. If Σ diverges, then grouped Σ diverges.	False; Example 28.6 is a counterexample.

g. Lemma 29.1 says: Φ converges $\equiv$ grouped Φ converges. This logical equivalence is the consequence of the truth of Q and its converse, R. Negating both sides of this equivalence yields the corresponding implications about divergence.

9. The ratio test quickly demonstrates that $\sum_{n=1}^{\infty} \frac{n^q}{M^n}$ converges when $|M| > 1$ and $q \in \mathbb{N}\cup\{0\}$.

a. With $q = 1$ and $M = 5$,

$$\frac{1}{2}\sum_{n=1}^{\infty} \frac{n}{5^n} = \frac{1}{2}T_1 = \frac{5}{32}.$$

b. With $q = 2$ and $M = \pi$,

$$\sum_{n=1}^{\infty} \frac{n^2}{\pi^n} = T_2 = \frac{\pi^2 + \pi}{(\pi - 1)^3}.$$

c. Here, $M = 9$ and

$$\sum_{n=0}^{\infty} \frac{n^3 - 3n - 2}{9^n} = -2 + \sum_{n=1}^{\infty} \frac{n^3}{9^n} - 3\sum_{n=1}^{\infty} \frac{n}{9^n} - 2\sum_{n=1}^{\infty} \frac{1}{9^n} = -2 + T_3 - 3T_1 - 2T_0$$

where

$$T_0 = \frac{1}{8}, \quad T_1 = \frac{9}{8^2}, \quad T_2 = \frac{90}{8^3}, \quad \text{and} \quad T_3 = \frac{1 + \binom{3}{1}\left(\frac{90}{8^3}\right) + \binom{3}{2}\left(\frac{9}{8^2}\right) + \frac{1}{8}}{9 - 1} = \frac{531}{2048}.$$

Thus,

$$\sum_{n=0}^{\infty} \frac{n^3 - 3n - 2}{9^n} = -\frac{4941}{2048}.$$

d. $\sum_{n=1}^{\infty} \frac{n^2}{(-3)^n}$ has $q = 2$ and $M = -3$. Thus,

$$\sum_{n=1}^{\infty} \frac{n^2}{(-3)^n} = T_2 = \frac{6}{(-4)^3} = -\frac{3}{32}.$$

(Hmm. . . this series alternates between positive and negative terms. It must be a segue to Chapter 31.)

Chapter 30

1. $\sigma = \lim_{n\to\infty} \sqrt[n]{\frac{3^n}{n!}} = \frac{3}{\lim_{n\to\infty} \sqrt[n]{n!}} = 0$, so the root test indicates that $\sum_{n=1}^{\infty} \frac{3^n}{n!}$ converges.

3. a. $\sigma = \lim_{n\to\infty} \sqrt[n]{a_n} = \lim_{n\to\infty} \frac{1}{\sqrt[n]{2^n}\sqrt[n]{3-(-1)^n}} = \frac{1}{2} \lim_{n\to\infty} \frac{1}{\sqrt[n]{3-(-1)^n}} = \frac{1}{2} < 1$, so the series converges.

   b. $\sigma = \lim_{n\to\infty} \sqrt[n]{a_n} = \lim_{n\to\infty} \frac{1}{\sqrt[n]{5-(-1)^n}} = 1$, so the test fails. However, since $a_n \not\to 0$, the series diverges.

   c. $\sigma = \lim_{n\to\infty} \sqrt[n]{a_n} = \lim_{n\to\infty} \sqrt[n]{(\frac{1}{3-(-1)^n})^n} = \lim_{n\to\infty} \frac{1}{3-(-1)^n}$, which doesn't exist by oscillation. Thus, the root test fails. But the comparison test works: compare the given series with $\sum_{n=1}^{\infty} (\frac{1}{2})^n$ to show it converges.

   d. $\sigma = \lim_{n\to\infty} \sqrt[n]{a_n} = \lim_{n\to\infty} \sqrt[n]{(\frac{1}{2-(-1)^n})^{n/2}} = \lim_{n\to\infty} \frac{1}{\sqrt{2-(-1)^n}}$, which doesn't exist by oscillation. Thus, the root test fails. However, since $a_n \not\to 0$, the series diverges.

5. We have

$$a_n = \frac{\sqrt[3]{n^5 + n^4}}{3n^3 + 50n}.$$

Let

$$b_n = \frac{\sqrt[3]{n^5}}{n^3}.$$

Then

$$\rho = \lim_{n\to\infty} \left(\frac{\sqrt[3]{n^5 + n^4}}{3n^3 + 50n} \cdot \frac{n^3}{\sqrt[3]{n^5}} \right) = \frac{1}{3}.$$

Since

$$\sum_{n=1}^{\infty} \frac{\sqrt[3]{n^5}}{n^3} = \sum_{n=1}^{\infty} \frac{1}{n^{4/3}}$$

converges, then so does the given series.

7. First, establish that $\langle \frac{n^{1/n}}{n} \rangle_1^\infty$ is monotone decreasing. Next,

$$\sum_{n=1}^{\infty} 2^n a_{2^n} = \sum_{n=1}^{\infty} 2^{n/2^n}.$$

But $\frac{n}{2^n} \to 0$, so that $2^{n/2^n} \not\to 0$ as n increases. This implies that $\sum_{n=1}^{\infty} 2^n a_{2^n}$ diverges. Hence, the given series diverges as well.

9. The limit comparison test works, with

$$b_n = \frac{n!}{(n+2)!} = \frac{1}{(n+1)(n+2)}.$$

Example 29.7 can be adapted to show that

$$\sum_{n=0}^{\infty} \frac{1}{(n+1)(n+2)}$$

converges. Then

$$\rho = \lim_{n\to\infty} \frac{n!+1}{n!} = 1.$$

Therefore, the given series converges. (The comparison test also works.)

11. From the previous exercise, $\lim_{n\to\infty} \sqrt[n]{n} = 1$. Relying on the continuity of the natural logarithm function,

$$0 = \ln 1 = \ln\left(\lim_{n\to\infty} \sqrt[n]{n}\right) = \lim_{n\to\infty} (\ln \sqrt[n]{n}) = \lim_{n\to\infty} \frac{\ln n}{n}.$$

Now, consider $\lim_{n\to\infty} \frac{\ln n}{n^M}$, where $M > 0$ is a constant. Substitute $y = n^M$, and

$$\lim_{n\to\infty} \frac{\ln n}{n^M} = \lim_{y\to\infty} \left(\frac{1}{M}\frac{\ln y}{y}\right) = 0. \quad \textbf{QED}$$

13. By the integral test,

$$\sum_{n=2}^{\infty} \frac{1}{n(\ln n)^2}$$

converges. Next,

$$\frac{a_{n+1}}{a_n} = \frac{n \ln^2 n}{(n+1)\ln^2(n+1)}$$

which gives us

$$\lim_{n\to\infty} n\left(1-\frac{a_{n+1}}{a_n}\right) = \lim_{n\to\infty} \frac{(n^2+n)\ln^2(n+1) - n^2\ln^2 n}{(n+1)\ln^2(n+1)}$$

$$= \lim_{n\to\infty} \frac{n^2(\ln^2(n+1) - \ln^2 n) + n\ln^2(n+1)}{(n+1)\ln^2(n+1)}$$

$$= \lim_{n\to\infty} \frac{n}{n+1} + \lim_{n\to\infty} \frac{n^2(\ln(n+1) - \ln n)(\ln(n+1) + \ln n)}{(n+1)\ln^2(n+1)}$$

$$= 1 + \lim_{n\to\infty} \frac{n^2(\ln \frac{n+1}{n})(\ln(n+1) + \ln n\ )}{(n+1)\ln^2(n+1)}$$

$$= 1 + \lim_{n\to\infty} \frac{n^2\ln(\frac{n+1}{n})}{(n+1)\ln(n+1)} \cdot \lim_{n\to\infty} \frac{\ln(n+1)+\ln n}{\ln(n+1)}$$

$$= 1 + \frac{\lim_{n\to\infty}(n\ln(\frac{n+1}{n}))}{\lim_{n\to\infty}(\frac{n+1}{n}\ln(n+1))} \cdot \lim_{n\to\infty} \frac{\ln(n+1)+\ln n}{\ln(n+1)}$$

$$= 1 + \frac{\lim_{n\to\infty}\ln(1+\frac{1}{n})^n}{\lim_{n\to\infty}(\frac{n+1}{n}\ln(n+1))} \cdot \lim_{n\to\infty}\left(1 + \frac{\ln n}{\ln(n+1)}\right)$$

$$\to 1 + (0)(1+1) = 1$$

(where $\lim_{n\to\infty}\ln(1+\frac{1}{n})^n = \ln e = 1$). Thus, $\mu = 1$, and another case of (3) is demonstrated.

Chapter 31

1. a. Conditionally convergent, since $\frac{1}{2} - \frac{1}{4} + \frac{1}{6} - \cdots = \frac{1}{2}(1 - \frac{1}{2} + \frac{1}{3} - \cdots) = \frac{1}{2}\ln 2$, but the series of absolute terms $\frac{1}{2}(1 + \frac{1}{2} + \frac{1}{3} + \cdots)$ diverges.
   b. Absolutely convergent geometric series with $r = -\frac{\pi}{4}$; the sum of the series is $\frac{4}{4+\pi}$.
   c. Convergent by Theorem 31.1, but not absolutely, since $\frac{1}{n} < \frac{1}{\ln n}$ for $n \geq 2$ (use the comparison test). Thus, conditionally convergent.
   d. Convergent by Theorem 31.1, but not absolutely, since $\sum_{n=1}^{\infty} \frac{\ln n}{n}$ diverges by the comparison test. Thus, the logarithmic series is conditionally convergent.
   e. Absolutely convergent by the results of Exercise 29.1k.
   f. Absolutely convergent since $\sum_{n=1}^{\infty} \frac{1}{n(n+1)} = 1$. The series itself has sum $1 - 2\ln 2$.
   g. Conditionally convergent since $\sum_{n=1}^{\infty} \frac{(-1)^{n-1}}{\sqrt{n}}$ converges by Theorem 31.1, but $\sum_{n=1}^{\infty} \frac{1}{\sqrt{n}}$ is a divergent $p = 1/2$ series.
   h. $1 - \frac{\sqrt{2}}{4} + \frac{\sqrt{3}}{9} - \frac{\sqrt{4}}{16} - \frac{\sqrt{5}}{25} - \frac{\sqrt{6}}{36} - \frac{\sqrt{7}}{49} + \frac{\sqrt{8}}{64} \cdots$ is absolutely convergent since the absolute series is a $p = 3/2$ series (this is not an alternating series).
   i. Conditionally convergent since by Example 31.1, these alternating series converge, but by Example 29.1, the corresponding absolute series diverge.

3. Let $\langle S_n\rangle_1^{\infty}$ be the sequence of partial sums of $\sum_{n=1}^{\infty} a_n$, and $\langle A_n\rangle_1^{\infty}$ be the sequence of partial sums of $\sum_{n=1}^{\infty} |a_n|$. Apply the triangle inequality: for all $n \in \mathbb{N}$, $S_n \leq |S_n| \leq$

$|a_1| + \cdots + |a_n| = A_n$. Since the limits exist as $n \to \infty$, we have $\sum_{n=1}^{\infty} a_n \le |\sum_{n=1}^{\infty} a_n| \le \sum_{n=1}^{\infty} |a_n|$. **QED**

5. The error $|S - S_{100}| \le a_{101} = \frac{\ln 101}{101} \approx 0.045694$. Since

$$\sum_{n=1}^{100}(-1)^n \frac{\ln n}{n} = S_{100} \approx -0.182805 \qquad \text{and} \qquad S_{101} = S_{100} + \frac{\ln 101}{101} \approx -0.137111,$$

Figure 1 tells us that $S \in [-0.182805, -0.137111]$.

7. To use Theorem 31.1, write the series as

$$\sum_{n=1}^{\infty}(-1)^n n^2 \left(\frac{3}{5}\right)^n$$

and show that $a_n = n^2(\frac{3}{5})^n$ is monotone decreasing with limit zero. The monotonicity begins at $n = 4$, since

$$\frac{a_{n+1}}{a_n} = (n+1)^2 \left(\frac{3}{5}\right)^{n+1} \cdot \frac{1}{n^2} \left(\frac{5}{3}\right)^n = \frac{3}{5} \left(\frac{n+1}{n}\right)^2 < 1$$

for $n \ge 4$. And $\lim_{n \to \infty} n^2 \left(\frac{3}{5}\right)^n = 0$ since $(\frac{3}{5})^n < \frac{1}{n^3}$ for all $n \ge 17$, implying $n^2(\frac{3}{5})^n < \frac{1}{n}$.
To use Exercise 29.9, write the series as

$$\sum_{n=1}^{\infty} \frac{n^2}{(\frac{-5}{3})^n}.$$

We have $q = 2$ and $M = \frac{-5}{3}$. Thus,

$$\sum_{n=1}^{\infty} \frac{n^2 3^n}{(-5)^n} = T_2 = -\frac{15}{256}.$$

Chapter 32

1. Simplifying parentheses, we get $1 - \frac{1}{2} + \frac{1}{3} - \frac{1}{4} + \frac{1}{5} - \frac{1}{6} + \cdots = \ln 2$. Hence, we develop an instant headache when we conclude that $2 \ln 2 = \ln 2$.

3. **A** Typical procedure: if $\langle P_n \rangle_1^{\infty}$ and $\langle Q_n \rangle_1^{\infty}$ converge, then $S_n = P_n + Q_n$ and $A_n = P_n - Q_n$ imply that $\langle S_n \rangle_1^{\infty}$ and $\langle A_n \rangle_1^{\infty}$ converge by Theorem 20.2. If $\langle S_n \rangle_1^{\infty}$ and $\langle A_n \rangle_1^{\infty}$ converge, then $P_n = \frac{1}{2}(S_n + A_n)$ and $Q_n = \frac{1}{2}(S_n - A_n)$ imply that $\langle P_n \rangle_1^{\infty}$ and $\langle Q_n \rangle_1^{\infty}$ also converge by Theorem 20.2. In general, select any two of the four partial sums sequences to be the ones with limits, and use $S_n = P_n + Q_n$ and $A_n = P_n - Q_n$ to develop equations that show the other two partial sums also converge.
   B This follows from **A** and Theorem 31.2.
   C If, on the contrary, at least one of the two series $\sum_{n=1}^{\infty} p_n$ or $\sum_{n=1}^{\infty} q_n$ converged, then $\sum_{n=1}^{\infty} |a_n|$ would also converge by **A**, which contradicts the premise of this part. **QED**

5. Given that $\sum_{n=1}^{\infty} a_n = A$ converges conditionally, first select just enough nonnegative terms p_n to surpass $x = 1$, then add just enough nonpositive terms q_n to pass below 1. Next, add just enough more nonnegative terms to surpass $x = 2$, then add just enough more nonpositive terms to pass below 2. Next, add just enough more nonnegative terms to surpass $x = 3$, then add just enough nonpositive terms to pass below 3. This rearrangement has partial sums near each natural number in succession. Therefore, this rearrangement diverges to $+\infty$.

For the alternating harmonic series, $1 + \frac{1}{3} > 1$. Next, $0 < 1 + \frac{1}{3} - \frac{1}{2} < 1$. Next,

$$1 + \frac{1}{3} - \frac{1}{2} + \frac{1}{5} + \frac{1}{7} + \frac{1}{9} + \frac{1}{11} + \frac{1}{13} + \frac{1}{15} + \frac{1}{17} + \frac{1}{19} + \frac{1}{21}$$
$$+ \frac{1}{23} + \frac{1}{25} + \frac{1}{27} + \frac{1}{29} + \frac{1}{31} + \frac{1}{33} + \frac{1}{35} + \frac{1}{37} + \frac{1}{39} + \frac{1}{41} > 2.$$

Next,

$$1 < 1 + \frac{1}{3} - \frac{1}{2} + \frac{1}{5} + \frac{1}{7} + \frac{1}{9} + \frac{1}{11} + \frac{1}{13} + \frac{1}{15} + \frac{1}{17} + \frac{1}{19} + \frac{1}{21}$$
$$+ \frac{1}{23} + \frac{1}{25} + \frac{1}{27} + \frac{1}{29} + \frac{1}{31} + \frac{1}{33} + \frac{1}{35} + \frac{1}{37} + \frac{1}{39} + \frac{1}{41} - \frac{1}{4} < 2,$$

and so on.

Warning: the next step in "so on" requires a maddening number of positive terms, so some of you may stop here. Then again, curiosity might win. . . .

Chapter 33

1. a. $\sum_{n=0}^{\infty} \frac{1}{(n+2)(n+3)} = \frac{1}{2}$ and $\sum_{n=0}^{\infty} \frac{1}{\pi^n} = \frac{\pi}{\pi-1}$, both absolutely convergent. Hence, Mertens' theorem applies, and the Cauchy product converges to $\frac{\pi}{2(\pi-1)}$.

   b. $\sum_{n=1}^{\infty} \frac{(-1)^n}{(n+3)!} = \frac{1}{3} - e^{-1}$ converges absolutely, while $\sum_{n=1}^{\infty} \frac{(-1)^n}{n} = -\ln 2$ converges conditionally. Hence, Mertens' theorem applies, and the Cauchy product converges to $(-\ln 2)(\frac{1}{3} - e^{-1})$.

3. a. Inserting $b_0 = 0$ and letting $x = \frac{1}{3}$ in (1) gives a similar expression to (3):

$$\left(e - \frac{2}{3}e + \frac{2}{9}e - \frac{4}{81}e + \frac{2}{243}e - \cdots\right)\left(0 - \frac{2}{3} - \frac{2}{9} - \frac{8}{81} - \frac{4}{81} - \cdots\right).$$

The Cauchy product is $-\frac{2e}{3} + \frac{2e}{9} - \frac{8e}{81} + 0 - \frac{4e}{405} - \cdots$. There is no change in the product.

$$\left(e + ey + \frac{e}{2!}y^2 + \frac{e}{3!}y^3 + \frac{e}{4!}y^4 + \frac{e}{5!}y^5\right)\left(z + y - \frac{1}{2}y^2 + \frac{1}{3}y^3 - \frac{1}{4}y^4 + \frac{1}{5}y^5\right)$$
$$= ez + ey + eyz + \frac{1}{2}ey^2 + \frac{1}{2}ey^2z + \frac{1}{3}ey^3 + \frac{1}{6}ey^3z + 0ey^4 + \frac{1}{24}ey^4z + \frac{3}{40}ey^5$$
$$+ \frac{1}{120}ey^5z + \text{other terms.}$$

Putting $z = 0$ and $y = x - 1$ yields (2) up to the fifth power.

b. The series in (1) have intervals of convergence $(-\infty, \infty)$ and $(0, 2]$, respectively. Because $x = \frac{1}{3}$ lies in the intersection of the intervals, the product will converge, to

$$f\left(\frac{1}{3}\right) = e^{1/3} \ln \frac{1}{3} = -1.5332\ldots,$$

since (1) represents f over the intersection. Moreover, since Taylor series are absolutely convergent inside their radii of convergence, Mertens' theorem states that the Cauchy product converges to the product of the sums in (3), namely $(e^{1/3})(\ln \frac{1}{3})$.

5. Not possible. In Note 1 are found two conditionally convergent series whose Cauchy product doesn't even exist.

7. Both series *de facto* converge absolutely, therefore, their Cauchy product converges by Mertens' theorem. If we wish to use the comparison test effectively, write the inner product series as $\langle p_n \rangle_0^\infty \equiv a_0 b_0 + 0 + a_1 b_1 + 0 + a_2 b_2 + 0 + \cdots$ (that is, $p_{2n+1} = 0$), and compare these terms to those of the Cauchy product, $c_0 + c_1 + c_2 + \cdots$. Observe that only the terms c_{2n} contain the terms $a_n b_n$. Then $0 \le p_n \le c_n$ for all $n \ge 0$. Consequently, the inner product series converges.

Since $\sum_{n=0}^\infty a_n$ converges, $\lim_{n\to\infty} a_n = 0$, but $\langle a_n \rangle_1^\infty$ may not be monotone. The same goes for $\sum_{n=0}^\infty b_n$. Thus, neither Dirichlet's test nor Abel's test apply.

9.

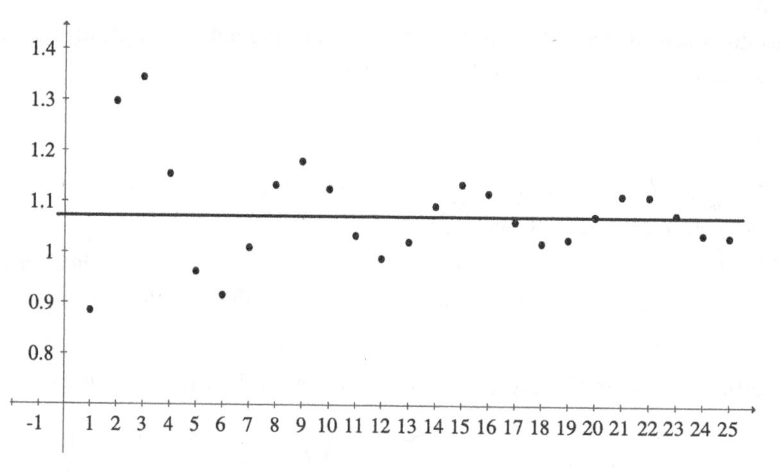

The horizontal line is at $y = \frac{\pi-1}{2}$, the value of $\sum_{n=1}^\infty \frac{\sin n}{n}$, (from Note 2).

11. This is Leibniz's alternating series test: an alternating series $\sum_{n=1}^\infty (-1)^{n+1} a_n$ will converge if the sequence $\langle a_n \rangle_1^\infty$ is monotone decreasing, and $\lim_{n\to\infty} a_n = 0$. To apply Dirichlet's test, just let $\sum_{n=1}^\infty b_n = \sum_{n=1}^\infty (-1)^{n+1}$, which has bounded partial sums, and take $\langle a_n \rangle_1^\infty$ as given.

13. To apply Abel's test, use the convergent monotone sequence $\langle (\sqrt{3})^{-n} \rangle_0^\infty$, and (necessarily) use the convergent series

$$\sum_{n=0}^\infty \left(\frac{1}{3^{n+1}\sqrt{3^n}} + \frac{1}{5^{n+1}\sqrt{3^{n+1}}} \right) = \frac{1}{3} \sum_{n=0}^\infty \frac{1}{(3^{3/2})^n} + \frac{1}{5\sqrt{3}} \sum_{n=0}^\infty \frac{1}{(5\sqrt{3})^n}.$$

Then

$$\sum_{n=0}^{\infty}\left(\frac{1}{3^{n+1}\sqrt{3^n}}+\frac{1}{5^{n+1}\sqrt{3^{n+1}}}\right)\left(\frac{1}{\sqrt{3^n}}\right)=\sum_{n=0}^{\infty}\left(\frac{1}{3^{n+1}\sqrt{3^{2n}}}+\frac{1}{5^{n+1}\sqrt{3^{2n+1}}}\right)$$
$$=\left(\frac{1}{3}+\frac{1}{5\sqrt{3}}\right)+\left(\frac{1}{3^2\sqrt{3^2}}+\frac{1}{5^2\sqrt{3^3}}\right)+\left(\frac{1}{3^3\sqrt{3^4}}+\frac{1}{5^3\sqrt{3^5}}\right)+\cdots$$

is a convergent series by Abel's test. Removing parentheses gives us the original series, which must also converge.

15. For $y \neq 0$, apply Dirichlet's test with $\langle a_n\rangle_0^\infty = \frac{1}{2n+1}$ and $\sum_{n=0}^\infty b_n = \sum_{n=0}^\infty (-1)^n \cos((2n+1)y)$. This demands that we prove that $\sum_{n=0}^N (-1)^n \cos((2n+1)y)$ is a bounded partial sum for all N at each

$$y \in \left(-\frac{\pi}{2}, 0\right) \cup \left(0, \frac{\pi}{2}\right) = \mathbf{D}_0^{\pi/2}.$$

Now, for each $y \in \mathbf{D}_0^{\pi/2}$, and letting $\varphi = 2y$,

$$\left|\sum_{n=0}^{N}(-1)^n \cos((2n+1)y)\right| \le \left|\frac{1}{4\sin y}\right| |\sin\varphi + \sin((2\lfloor N/2\rfloor + 1)\varphi)$$
$$- \sin((2N - 2\lfloor N/2\rfloor)\varphi)| + |\frac{1}{4\cos y}||1 - \cos\varphi + \cos((2\lfloor N/2\rfloor + 1)\varphi)$$
$$- \cos((2N - 2\lfloor N/2\rfloor)\varphi)| \le \frac{3}{|4\sin y|} + \frac{4}{|4\cos y|},$$

by the triangle inequality.

Therefore,

$$\left|\sum_{n=0}^{N}(-1)^n \cos((2n+1)y)\right| \le \frac{3}{|4\sin y|} + \frac{4}{|4\cos y|}$$

for each $y \in \mathbf{D}_0^{\pi/2}$, which is an expression independent of N. This demonstrates that $\sum_{n=0}^\infty (-1)^n \cos((2n+1)y)$ is bounded over $\mathbf{D}_0^{\pi/2}$. Then Dirichlet's test applies, and we conclude that

$$\sum_{n=0}^{\infty}\frac{(-1)^n}{2n+1}\cos((2n+1)y)$$

converges over $\mathbf{D}_0^{\pi/2}$. As the last detail, at $y = 0$, we have

$$\sum_{n=0}^{\infty}\frac{(-1)^n}{2n+1},$$

which we know converges (to $\frac{\pi}{4}$ in fact!). Hence, the Fourier series converges over $\mathbf{N}_0^{\pi/2}$.

Chapter 34

1. a. Recall that $(x+y)^n = \sum_{k=0}^n \binom{n}{k} x^{n-k} y^k$, where $\binom{n}{k} = \frac{n(n-1)\cdots(n-(n-k+1))}{k!}$.

Here, $x = 1$ and $y = \frac{1}{n}$. For (2), notice what is done in a typical term, such as this one:

$$\frac{n(n-1)(n-2)}{3!}\left(\frac{1}{n}\right)^3 = \frac{1}{3!}\left(\frac{n-1}{n}\right)\left(\frac{n-2}{n}\right) = \frac{1}{3!}\left(1-\frac{1}{n}\right)\left(1-\frac{2}{n}\right).$$

b. All factors in (2) of the form $(1 - \frac{a}{n})$ have magnitudes in (0, 1). Hence, each term in (2) is not more than the corresponding term of $1 + 1 + \frac{1}{2!} + \frac{1}{3!} + \cdots + \frac{1}{n!}$. This proves the leftmost inequality in (3). Next, since $\frac{1}{k!} < \frac{1}{2^k}$ for all $k \in \mathbb{N}$, the second inequality is correct. The equalities in (3) derive from the sum of terms of a finite geometric series.

3. a. $\lim_{n\to\infty}(\frac{n}{n+1})^n = \lim_{n\to\infty}\frac{1}{(1+\frac{1}{n})^n} = \frac{1}{\lim_{n\to\infty}(1+\frac{1}{n})^n} = \frac{1}{e}$.

   b. $\lim_{n\to\infty}(1 + \frac{1}{bn})^n = \lim_{m\to\infty}(1 + \frac{1}{m})^{m/b} = \sqrt[b]{\lim_{m\to\infty}(1 + \frac{1}{m})^m} = \sqrt[b]{e}$, where $m = bn \in \mathbb{N}$.

   c. Note that $(1 + \frac{1}{n})^n \le s_n < e$, where $s_n = \sum_{k=0}^{n}\frac{1}{k!}$ appears in (3) .

5. $\lim_{n\to\infty}(\sum_{k=1}^{n}\frac{1}{k} - \ln(n + 0.5)) = \lim_{n\to\infty}(\sum_{k=1}^{n}\frac{1}{k} - \ln n + \ln n - \ln(n + 0.5)) = \lim_{n\to\infty}(\sum_{k=1}^{n}\frac{1}{k} - \ln n) + \lim_{n\to\infty}\ln\frac{n}{n+0.5} = \gamma$. **QED**

7. a. For $\sum_{k=1}^{n}\frac{1}{k} > 100$, the astonishing number of terms $n \approx 1.51 \times 10^{43}$ is needed. That is 15.1 trillion quadrillion quadrillions!

9. From $\frac{1}{n+1} < \ln(n + 1) - \ln n < \frac{1}{n}$, we have $\frac{1}{n+1} < \ln\left(1 + \frac{1}{n}\right) < \frac{1}{n}$, so $e^{1/(n+1)} < 1 + \frac{1}{n} < e^{1/n}$. Then $e^{n/(n+1)} < (1 + \frac{1}{n})^n < e$. The pinching theorem for sequences now gives the desired result. **QED**

   But of course, the properties of the Riemann integral and the natural logarithm function are prerequisites.

Chapter 35

1. Knowing that $e < (1 + \frac{1}{n})^{n+1}$ for all $n \in \mathbb{N}$, use $n = m - 1$ to get

$$e < \left(1 + \frac{1}{m-1}\right)^m = \frac{1}{(1-\frac{1}{m})^m}, m \ge 2.$$

   Then $(1 - \frac{1}{m})^m < e^{-1}$ for $m \ge 2$. Even at $m = 1$, the final inequality holds. **QED**

3. By Theorem 35.2, $x + 1 \le e^x$. Since $x \ge -2$ here, $\frac{x}{2} + 1$ is nonnegative. The substitution $\frac{x}{2} \mapsto x$ then yields $(\frac{x}{2} + 1)^2 \le (e^{x/2})^2 = e^x$. In fact, graphing the functions $y = (\frac{x}{2} + 1)^2$ and $y = e^x$ shows the inequality to be true for some $x < -2$ as well.

5. a. $\lim_{x\to\infty}(1 + \frac{-1}{x})^x = \frac{1}{\lim_{w\to\infty}(1+\frac{1}{w})^{w+1}} = \frac{1}{\lim_{w\to\infty}((1+\frac{1}{w})^w(1+\frac{1}{w}))} = \frac{1}{e}$.

   b. Let $w = \frac{x}{a}$ for any $a > 0$, so that

$$\lim_{x\to\infty}\left(1+\frac{a}{x}\right)^x = \lim_{x\to\infty}\left(1+\frac{1}{x/a}\right)^x = \lim_{w\to\infty}\left(1+\frac{1}{w}\right)^{wa} = \lim_{w\to\infty}\left(\left(1+\frac{1}{w}\right)^w\right)^a = e^a$$

   by the defining expression for the exponential function. If $a < 0$, then take $w = \frac{x}{-a} > 0$, which gives us

$$\lim_{w\to\infty}\left(1+\frac{-1}{w}\right)^{-wa} = \left(\lim_{w\to\infty}\left(1+\frac{-1}{w}\right)^{w}\right)^{-a} = \left(\frac{1}{e}\right)^{-a} = e^{a}$$

by a.

Finally, if $a = 0$ from the start, then $\lim_{x\to\infty}(1+0)^x = 1 = e^0$, so the statement is true for all real a. **QED**

c. $e^{-a} = (\lim_{x\to\infty}(1+\frac{1}{x})^x)^{-a} = \frac{1}{(\lim_{x\to\infty}(1+\frac{1}{x})^x)^a} = \frac{1}{e^a}$. Or, use b.

d. $e^{y+x} = \lim_{u\to\infty}((1+\frac{1}{u})^{u(y+x)}) = \lim_{u\to\infty}[(1+\frac{1}{u})^{uy}(1+\frac{1}{u})^{ux}] = (\lim_{u\to\infty}(1+\frac{1}{u})^{uy})(\lim_{u\to\infty}(1+\frac{1}{u})^{ux}) = e^y e^x$. **QED**

e. Since $e^x > 1$ for all $x > 0$, and $e^{-x} = \frac{1}{e^x}$ by c, we infer that $0 < e^{-x} < 1$ for all $x > 0$. This implies that $0 < e^x$ for all real x.

f. Let $y = x + d$, with $d > 0$. Then by d and since $e^d > 1$, we conclude that $e^y = e^x e^d > e^x$. **QED**

7. Given $0 < w < 1$, then $\frac{1}{w} > 1$. Also, $0 = \ln 1 = \ln\frac{w}{w} = \ln w + \ln\frac{1}{w}$. Hence, the two terms are additive inverses of each other. By A of Theorem 35.3, $\ln w < \ln\frac{1}{w}$, implying that $\ln w < 0$ and $\ln\frac{1}{w} > 0$. (Other proofs are possible!) **QED**

9. $(b^x)(b^y) = \lim_{n\to\infty}(1+\frac{x\ln b}{n})^n \lim_{n\to\infty}(1+\frac{y\ln b}{n})^n = \lim_{n\to\infty}((1+\frac{x\ln b}{n})(1+\frac{y\ln b}{n}))^n = \lim_{n\to\infty}(1+\frac{(x+y)\ln b}{n}+\frac{(xy)(\ln b)^2}{n^2})^n$.

Before one chokes on this hunk of salami, remember that the only quantity moving around is n. Thus, we are dealing with the form $L = \lim_{n\to\infty}(1+\frac{A}{n}+\frac{B}{n^2})^n$, where $A = (x+y)\ln b$ and $B = (xy)(\ln b)^2$. Using the standard continuity argument,

$$\ln L = \lim_{n\to\infty}\left(n\ln\left(1+\frac{A}{n}+\frac{B}{n^2}\right)\right) = \lim_{n\to\infty}\frac{\ln\left(1+\frac{A}{n}+\frac{B}{n^2}\right)}{\frac{1}{n}}.$$

L'Hospital's rule applies, and we get

$$\ln L = \lim_{n\to\infty}\frac{A+\frac{2B}{n}}{1+\frac{A}{n}+\frac{B}{n^2}} = A.$$

Thus, $L = e^A = e^{(x+y)\ln b} = b^{x+y}$. **QED**

a. $P_{bino}(1) = \binom{200}{1}(0.001^1)(0.999^{199}) \approx 0.16389$. In the Poisson distribution, $\lambda = 0.2$, so

$$P_{Pois}(1) = \frac{0.2^1}{1!}e^{-0.2} \approx 0.16375.$$

Very close!

b. We want

$$\lim_{n\to\infty}\frac{n!}{n^r(n-r)!}\cdot\lim_{n\to\infty}\frac{\lambda^r}{r!}\cdot\lim_{n\to\infty}\left(1-\frac{\lambda}{n}\right)^n\cdot\lim_{n\to\infty}\left(1-\frac{\lambda}{n}\right)^{-r}.$$

First,

$$\lim_{n\to\infty} \frac{n!}{n^r(n-r)!} = \lim_{n\to\infty} \frac{n(n-1)\cdots[n-(r-1)]}{n^r} = 1.$$

Also, $\lim_{n\to\infty} \frac{\lambda^r}{r!} = \frac{\lambda^r}{r!}$, and $\lim_{n\to\infty} \left(1-\frac{\lambda}{n}\right)^{-r} = 1$.
But

$$\lim_{n\to\infty} \left(1-\frac{\lambda}{n}\right)^n = e^{-\lambda}$$

by Exercise 35.5. Therefore, the limit is $\frac{\lambda^r}{r!}e^{-\lambda}$, which is the Poisson probability.

Chapter 36

1. a. (1) The solutions are

$$x = \frac{m_0 \pm \sqrt{m_0^2 + 4ab}}{2a}.$$

(2) Exactly one solution x_0 occurs under the condition that $m_0^2 = -4ab$. When this equation holds, $x_0 = \frac{m_0}{2a}$ becomes the abscissa of the point of tangency. Notice that for a real point of tangency, a and b must have opposite signs. This is what the graphs of $y = ax^2$ and $y = m_0x + b$ demand.
(3) $b = -\frac{m_0^2}{4a}$ (or equivalently, $b = -ax_0^2$).
(4) The required equation is $y = m_0x - \frac{m_0^2}{4a}$. The point of tangency is $\langle \frac{m_0}{2a}, \frac{m_0^2}{4a} \rangle$. The ordinate of the point of tangency is the opposite of the y-intercept of the line of tangency, a fact known to ancient geometers.

b. First, perform a shift of 5 units left and 10 units down. With $y = 3x^2$ and $m_0 = 6$, the tangent line is $y = 6x - \frac{36}{4(3)} = 6x - 3$, and it is tangent at $\langle \frac{m_0}{2a}, \frac{m_0^2}{4a} \rangle = \langle 1, 3 \rangle$. Now, shift everything back to its original position, giving us the tangent line $y = 6(x-5) - 3 + 10 = 6x - 23$ and point of tangency $\langle 6, 13 \rangle$.
These results are verified by elementary calculus, which is certainly easier to use!

3. The definition indicates that

$$\frac{d\,\text{sgn}\,x}{dx}\Big|_{x=0} = \lim_{h\to 0} \frac{\text{sgn}(0+h) - \text{sgn}\,0}{h} = \lim_{h\to 0} \frac{\text{sgn}\,h}{h} = \begin{cases} 1/h & \text{if } h > 0 \\ -1/h & \text{if } h < 0 \end{cases}.$$

Neither fraction has a limit as $h \to 0$. Hence, $\frac{d\,\text{sgn}\,x}{dx}\big|_{x=0}$ does not exist. For all other x,

$$\frac{d\,\text{sgn}\,x}{dx} = \lim_{h\to 0} \frac{\text{sgn}(x+h) - \text{sgn}\,x}{h} = \lim_{h\to 0} \frac{\text{sgn}\,x - \text{sgn}\,x}{h} = 0.$$

5. The classic counterexample is $y = |x|$, continuous but not differentiable at the origin. Another counterexample is

$$y = \begin{cases} x^2 & \text{if } x \le 1 \\ 1 & \text{if } x \ge 1 \end{cases},$$

which you should graph to see that it is purposely constructed not to have a tangent slope at $\langle 1, 1 \rangle$.

7. Provided

$$x \neq 0, \frac{d(x^2 \sin \frac{1}{x})}{dx} = 2x \sin \frac{1}{x} - \cos \frac{1}{x}.$$

This has no limit as $x \to 0$ since $\cos \frac{1}{x}$ has no limit but $2x \sin \frac{1}{x}$ does, and yet, at precisely $x = 0$, the function does have a zero slope (see the previous exercise)! Thus, the derivative is discontinuous at zero.

$$f(x) = \begin{cases} x^3 \sin \frac{1}{x} & \text{if } x \neq 0 \\ 0 & \text{if } x = 0 \end{cases}$$

has $f'(x) = 3x^2 \sin \frac{1}{x} - x \cos \frac{1}{x}$ when $x \neq 0$. But now, $\lim_{x \to 0} f'(x) = 0$. We also have

$$f'(0) = \lim_{h \to 0} \frac{(0+h)^3 \sin \frac{1}{0+h} - 0}{h} = 0,$$

which means that the derivative of this function is continuous at $\langle 0, 0 \rangle$.

9. $$Q_x(0.1) = 10(\operatorname{sgn}(x+0.1) - \operatorname{sgn} x) = \begin{cases} 10(1-1) = 0 & \text{if } x > 0 \\ 10(1-0) = 10 & \text{if } x = 0 \\ 10(1-(-1)) = 20 & \text{if } -0.1 < x < 0 \\ 10(0-(-1)) = 10 & \text{if } x = -0.1 \\ 10(-1-(-1)) = 0 & \text{if } x < -0.1 \end{cases},$$

which appears below. The function $Q_x(h)$ is "responding" to the jump discontinuity of $\operatorname{sgn} x$. For $h > 0$, the interval $(-h, 0)$ is elevated to the height $y = \frac{2}{h}$. Thus, as $h \to 0^+$, that segment shrinks in width and elevates without bound. Similar conclusions will be reached when $h < 0$. But at $x = 0$, $Q_0(h) = \frac{\operatorname{sgn} h}{h}$, with no limit as $h \to 0$. Thus, the limiting function agrees with $f'(x)$.

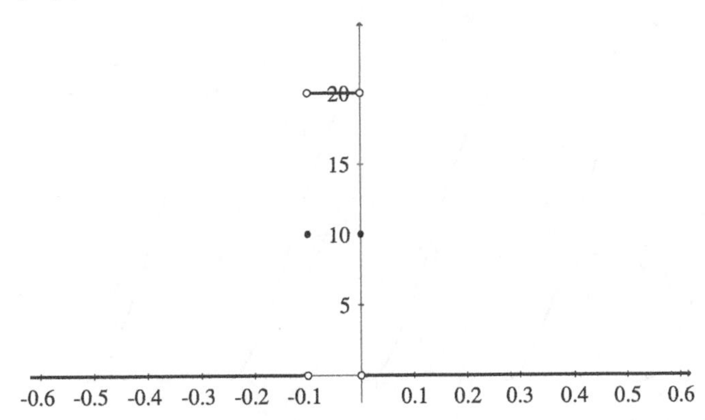

11. $f'(x) = \lim_{h \to 0} \frac{f(x) - f(x-h)}{h}$ is correct, since

$$\lim_{h \to 0} \frac{f(x) - f(x-h)}{h} = \lim_{\ell \to 0} \frac{f(x) - f(x+\ell)}{-\ell} = \lim_{\ell \to 0} \frac{f(x+\ell) - f(x)}{\ell} = f'(x),$$

where $\ell = -h$. The other formula evaluates secant slopes as $\frac{-\Delta y}{\Delta x}$, and has the (not surprising) limit of $-f'(x)$.

Chapter 37

1. Let E be an even function, so that $E(x) = E(-x)$. By the chain rule, $E'(x) = -E'(-x)$, making E' an odd function. Let O be an odd function, so that $O(x) = -O(-x)$. By the chain rule, $O'(x) = -(-O'(-x))$, making O' an even function. **QED**

3. We apply the product rule and the chain rule.

$$\frac{d(f/g)(x)}{dx} = \frac{d(f(x)\cdot g(x)^{-1})}{dx} = f(x)\cdot(-1)(g(x)^{-2})\frac{dg(x)}{dx} + g(x)^{-1}\cdot\frac{df(x)}{dx}$$
$$= \frac{g(x)\frac{df(x)}{dx} - f(x)\frac{dg(x)}{dx}}{g^2(x)}.$$

5. First, let $u = |y|$, and $u = f(x) = e^x$. Apply Theorems 37.5 and 37.6:

$$\frac{d\ln u}{du}|_b = \frac{1}{\frac{de^x}{dx}|_a} = \frac{1}{e^a} = \frac{1}{b},$$

where $b = e^a$, which holds for all $b \in \mathcal{D}_{\ln}$. So the evaluation symbols are not necessary, and we write $\frac{d\ln u}{du} = \frac{1}{u}$. Next, by the chain rule,

$$\frac{d\ln|y|}{dy} = \frac{1}{u}\cdot\frac{d|y|}{dy} = \frac{1}{u}\frac{1}{\operatorname{sgn} y} = \frac{1}{|y|}\frac{1}{\operatorname{sgn} y} = \frac{1}{y}.$$

7. In this case, $y = \sin x$ (the inside function), and the chain rule gives

$$\frac{d|\sin x|}{dx} = \frac{d|y|}{dy}\cdot\frac{d\sin x}{dx} = \left(\frac{1}{\operatorname{sgn} y}\right)(\cos x) = \frac{\cos x}{\operatorname{sgn}(\sin x)}.$$

The graph of $\frac{d|\sin x|}{dx}$ is shown.

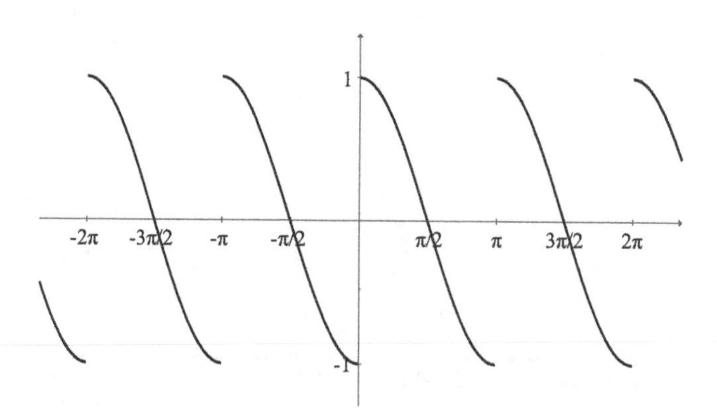

9. a. $\frac{dx^a}{dx} = x^a(\alpha\frac{1}{x} + 0\ln x) = \alpha x^{a-1}$, as expected.
   b. $\frac{db^x}{dx} = b^x(x\cdot\frac{0}{b} + 1\ln b) = b^x\ln b$, as expected.
   c. $\frac{dx^x}{dx} = x^x(x\frac{1}{x} + 1\ln x) = x^x(1+\ln x)$. The absolute minimum is at $\langle e^{-1}, e^{-e^{-1}}\rangle$, very quaint.

d. The graph of $y = |\sin x|^{e^x}$ is seen below (it only appears to have a vertical segment at $x = -\pi$). With the help of Exercise 37.7,

$$\frac{d|\sin x|^{\exp x}}{dx} = |\sin x|^{\exp x}\left(\exp x \cdot \frac{1}{|\sin x|}\frac{\cos x}{\operatorname{sgn}(\sin x)} + \exp x \cdot \ln|\sin x|\right)$$
$$= e^x(|\sin x|^{\exp x})\left(\frac{\cos x}{\sin x} + \ln|\sin x|\right) = e^x(|\sin x|^{\exp x})(\cot x + \ln|\sin x|).$$

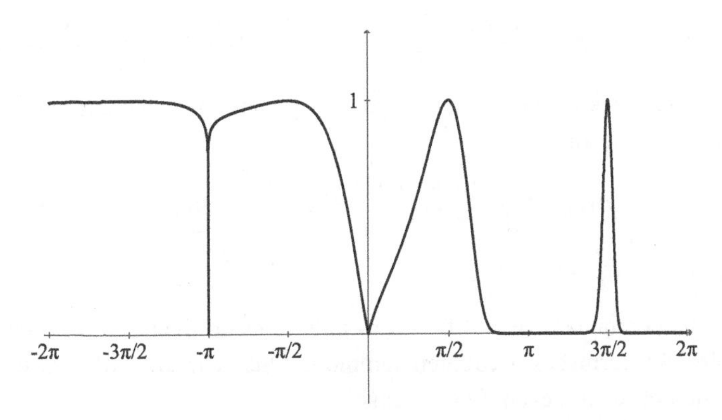

$y = |\sin x|^{\exp x} = |\sin x|^{e^x}$ with $\mathcal{D} = \mathbb{R}$.

11. a. $[f_1(x)f_2(x)f_3(x)]' = [f_1(x)f_2(x)]'f_3(x) + [f_1(x)f_2(x)]f_3'(x) = f_1'(x)f_2(x)f_3(x) + f_1(x)f_2'(x)f_3(x) + f_1(x)f_2(x)f_3'(x)$. **QED**

b. By induction: the anchor is case $n = 2$, which is the product rule. Assuming that

$$\frac{d\prod_{i=1}^{n} f_i(x)}{dx} = \sum_{i=1}^{n}\left(f_i'(x)\prod_{j\neq i} f_j(x)\right)$$

is true, prove that

$$\frac{d\prod_{i=1}^{n+1} f_i(x)}{dx} = \sum_{i=1}^{n+1}\left(f_i'(x)\prod_{j\neq i} f_j(x)\right).$$

On the left,

$$\frac{d\prod_{i=1}^{n+1} f_i(x)}{dx} = f_{n+1}(x)\frac{d\prod_{i=1}^{n} f_i(x)}{dx} + f_{n+1}'(x)\prod_{i=1}^{n} f_i(x)$$

by the product rule. Substituting from the induction assumption, we get

$$f_{n+1}(x)\sum_{i=1}^{n}\left(f_i'(x)\prod_{j\neq i} f_j(x)\right) + f_{n+1}'(x)\prod_{i=1}^{n} f_i(x)$$
$$= \sum_{i=1}^{n}\left(f_i'(x)\cdot f_{n+1}(x)\prod_{j\neq i} f_j(x)\right) + f_{n+1}'(x)\prod_{i=1}^{n} f_i(x).$$

Considering it carefully, this last expression is exactly $\sum_{i=1}^{n+1}\left(f_i'(x)\prod_{j\neq i} f_j(x)\right)$. **QED**

Chapter 38

1. One way to fix it is to have only one derivative exist at $x = 2$, say, $f'^L(2)$. Thus, erase the other, parallel, line. Another way is to remove the jump discontinuity at $x = 2$, thereby making the function differentiable.

3. $f'^R(k\pi) = \lim_{h\to 0+} \frac{|\sin(k\pi+h)|-\sin(k\pi)}{h} = \lim_{h\to 0+} \frac{|\sin(k\pi+h)|}{h} = \lim_{h\to 0+} \frac{|\sin k\pi \cos h + \cos k\pi \sin h|}{h} = |\cos k\pi| \lim_{h\to 0+} \frac{|\sin h|}{h} = 1$. Similarly,

$$f'^L(k\pi) = \lim_{h\to 0-} \frac{|\sin(k\pi + h)| - \sin(k\pi)}{h} = |\cos k\pi| \lim_{h\to 0-} \frac{|\sin h|}{h} = -1.$$

5. By Theorem 38.1, both the derivative from the right and from the left must exist and equal zero. Here are the details:

$$f'^R(0) = \lim_{h\to 0+} \frac{(0+h)^2 \sin \frac{1}{0+h} - 0}{h} = \lim_{h\to 0+} h \sin \frac{1}{h} = 0,$$

and similarly, $f'^L(0) = 0$.

7. a. is false. Its contrapositive is "if $f'^R(x)$ fails to exist, then $f'^+(x)$ fails to exist," and Example 38.5 furnishes a counterexample of this. The crux of it is that $f'^+(x)$ and $f'^-(x)$ do not require even $f(x)$ to exist!
   b. is false; Example 38.6 is a counterexample. The point here is that a derivative may exist at a point but not be continuous there, due, for instance, to undamped oscillation.
   c. is false. Example 38.6 is a counterexample.
   d. is false, and Example 38.5 is a counterexample. In general, breaking any continuous function does the trick, for instance, $b(x) = e^x - \operatorname{sgn} x$, seen at right. It has $b'^-(0) = 1 = b'^+(0)$. But it is clear from the discontinuity at $x = 0$ that $b'(0)$ can't exist.

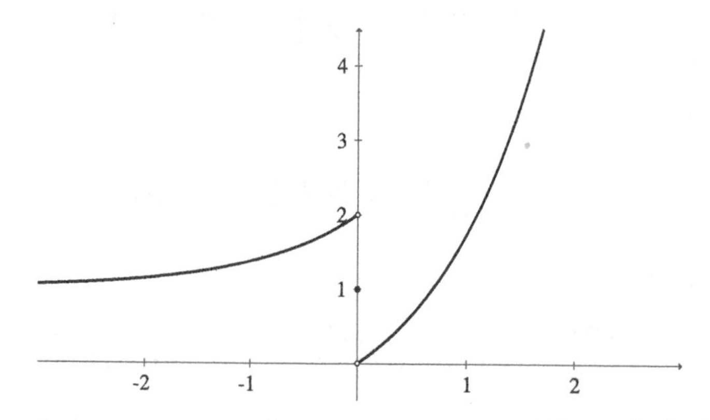

9. a. Since the midpoint of the full time interval is $x = \frac{9.33+2.45}{2} = 5.89$ hr, we know that $h = \frac{9.33-2.45}{2} = 3.44$ hr. It follows that $S_{5.89}(3.44) = \frac{312-177}{2(3.44)} = 19.62$ bacteria/hr is, in fact, the symmetric secant slope.

   Next, $Q_{5.89}(3.44) = \frac{312-234}{3.44} = 22.67$ bacteria/hr, would be too high an estimate. And $Q_{2.45}(3.44) = \frac{234-177}{3.44} = 16.57$ bacteria/hr, the secant slope over the interval [2.45, 5.89], would be too low an estimate. A reasonable assumption is an exponential growth function, so that $S_{5.89}(3.44)$ would be a better estimate.

   b. $\frac{Q_{5.89}(3.44)+Q_{2.45}(3.44)}{2} = 19.62 = S_{5.89}(3.44)$. This will be true in general.

11. $u'''(x) = \frac{-3x}{(x^2-1)^2\sqrt{1-x^2}}$ and $w'''(x) = \frac{24x(1-x^2)}{(x^2+1)^4}$, so the requirement is fulfilled.

Chapter 39

1. a. $f(x) = x^{1/3}$ is continuous over $[1, b]$ and differentiable over $(1, b)$ for all $b > 1$. Hence, Theorem 39.2 applies, and

$$\frac{1}{3\ell^{2/3}} = \frac{f(b) - f(a)}{b - a} = \frac{\sqrt[3]{b} - 1}{b - 1}$$

for some $\ell \in (1, b)$.

Using a calculator,

$$\frac{1}{3\ell^{2/3}} = \frac{\sqrt[3]{6} - 1}{5}$$

has solutions $\ell \approx \pm 2.9130$, so $\ell \approx 2.9130 \in (1, 6)$.

b. $f(x) = \sin x$ is differentiable everywhere. Thus, $\frac{\sin b - \sin a}{b-a} = \cos\xi$ for some $\xi \in (a, b)$. Then $\left|\frac{\sin b - \sin a}{b-a}\right| = |\cos\xi| \leq 1$ for all $a, b \in \mathbb{R}$. Replace b with x and a with y.

c. Parallel reasoning to b.

d. $f(x) = e^x$ is differentiable everywhere. By (1) with $a = 0$, $e^x = 1 + e^{\xi}x$ where $\xi \in (0, x)$. This implies that $e^{\xi} > 1$, so that $1 + e^{\xi}x > 1 + x$. Therefore, $e^x > 1 + x$ when $x > 0$.

By (2) with $b = 0$, $1 = e^x + e^{\xi}(-x)$ where $\xi \in (x, 0)$, indicating that $0 < e^{\xi} < 1$. Because $-x > 0$, $1 - e^x = -xe^{\xi} < -x$, which brings us to $e^x > 1 + x$ when $x < 0$.

Collecting the two inequalities, we have $1 + x \leq e^x$ for all real x, with equality only when $x = 0$.

e. From Example 39.2, $\ln u \leq u - 1$ for all $u > 0$. Let $u = \frac{1}{x}$. Then $\ln\frac{1}{x} \leq \frac{1}{x} - 1$. Thus, $\ln x \geq 1 - \frac{1}{x} = \frac{x-1}{x}$ for all $x > 0$. The three functions are graphed below.

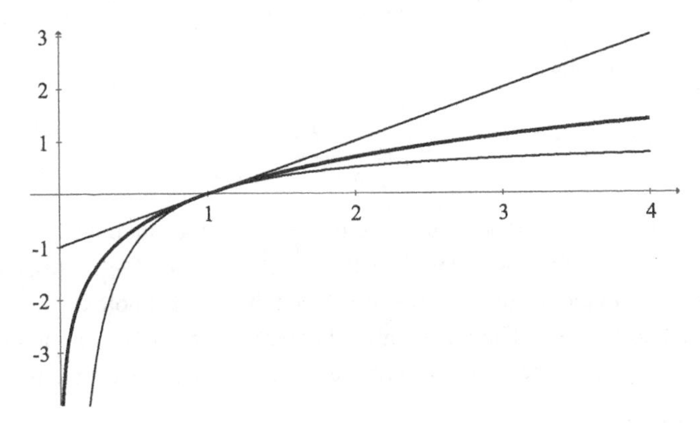

Comparison of $y = x - 1$, $y = \frac{x-1}{x}$, and $y = \ln x$. The logarithmic function is the thick line.

3. Letting $g(x) = \sqrt{1 + x}$, note that g is continuous over all $\mathcal{D}_g$ and differentiable over $\mathcal{D}_g \backslash \{-1\}$. Applying (1) with $a = 0$, we have

$$\sqrt{1 + x} = \sqrt{1 + 0} + \frac{1}{2\sqrt{1 + \xi}}(x - 0),$$

where $\xi > 0$, hence, $\sqrt{1+x} < 1 + \frac{1}{2}x$ for all $x > 0$. Applying (2) with $b = 0$, we have

$$\sqrt{1+x} = \sqrt{1+0} + \frac{1}{2\sqrt{1+\xi}}(x-0),$$

where $-1 < x < 0$ and $\xi \in (x, 0)$. Since $\xi < 0$, $\frac{1}{\sqrt{1+\xi}} > 1$. Consequently,

$$1 + \frac{1}{2\sqrt{1+\xi}}(x) < 1 + \frac{1}{2}x.$$

Therefore,

$$\sqrt{1+x} < 1 + \frac{1}{2}x$$

for all $x \in \{x > -1\}\backslash\{0\}$, and substituting either $x = -1$ or $x = 0$ shows

$$\sqrt{1+x} \leq 1 + \frac{1}{2}x$$

to be true for all $x \geq -1$. **QED**

5. Perhaps the simplest example is $y = \operatorname{sgn} x$ over an interval containing $x = 0$. When $b > 0$ and $a < 0$, we have $\frac{\operatorname{sgn} b - \operatorname{sgn} a}{b-a} = \frac{2}{b-a} > 0$. However, at no point is the derivative of the signum function positive.

7. Draw the function over $(-1, 1)$. Roughly speaking, every secant slope can be matched by a tangent slope, since no secant will be vertical, and tangent slopes are unbounded positive. To be precise, for any $b > 0$ and $a < 0$, we look for solutions to

$$\frac{b^{1/3} - a^{1/3}}{b-a} = \frac{1}{3\xi^{2/3}}.$$

Since $\frac{b^{1/3}-a^{1/3}}{b-a}$ is positive,

$$\xi^{2/3} = \frac{b-a}{3(b^{1/3} - a^{1/3})}$$

has two real solutions, $\pm\xi_0$. If both do not belong to (a, b), then their difference, $2\xi_0 > b - a$. Then $\xi_0^{2/3} > 2^{3/2}(b-a)^{2/3}$, which will contradict $\xi_0^{2/3} = \frac{b-a}{3(b^{1/3}-a^{1/3})}$. Thus, either $\xi_0 \in (a, b)$ or $-\xi_0 \in (a, b)$, and the mean value theorem's conclusion still holds.

No contradictions here. The mean value theorem gives *sufficient* conditions for secant slope to equal tangent slope. The converse need not be true, as we have just seen.

9. a. We are given that

$$u'^R(a) = \lim_{x \to a+} \frac{u(x) - u(a)}{x - a} < 0.$$

Then for some (actually infinitely many) $x > a$, the difference quotient (the secant slope) $\frac{u(x)-u(a)}{x-a} < 0$. Since $x - a > 0$, we must have $u(x) < u(a)$. Hence, $u(a)$ is not the minimum over $[a, b]$.

Similarly, given that

$$u'^{L}(b) = \lim_{x \to b-} \frac{u(b) - u(x)}{b - x} > 0,$$

then for some (actually infinitely many) $x < b$, the difference quotient $\frac{u(b)-u(x)}{b-x} > 0$. Since $b - x > 0$, it follows that $u(b) > u(x)$, and $u(b)$ isn't the minimum either. **QED**

b. We are given that

$$u'(\xi) = \lim_{x \to \xi} \frac{u(x) - u(\xi)}{x - \xi}.$$

That is, for any $\varepsilon > 0$, there exists a $\mathbf{N}_\xi^\delta$ such that for all $x \in \mathbf{N}_\xi^\delta$, it is true that

$$\left| \frac{u(x) - u(\xi)}{x - \xi} - u'(\xi) \right| < \varepsilon.$$

Hence

$$u'(\xi) - \varepsilon < \frac{u(x) - u(\xi)}{x - \xi} < u'(\xi) + \varepsilon.$$

Since in the first case, $u'(\xi) < 0$, ε can be chosen so small that $u'(\xi) + \varepsilon < 0$ as well. Consequently, $\frac{u(x)-u(\xi)}{x-\xi} < 0$ for all $x \in \mathbf{N}_\xi^\delta$. If we now select $x \in (\xi, \xi + \delta)$, then $x - \xi > 0$, and we conclude that $u(x) - u(\xi) < 0$, so $u(\xi)$ cannot be the minimum of u.

Similar considerations show us that $u(\xi)$ cannot be the maximum of u. **QED**

11. One arch of the cycloid is seen below. The functions $g(t) = 2(t - \sin t)$ and $f(t) = 2(1 - \cos t)$ satisfy all the conditions of Cauchy's mean value theorem over $[0, \pi]$. Hence, there exists at least one $\xi \in (0, \pi)$ such that

$$\left.\frac{dy}{dx}\right|_{\xi} = \frac{f'(\xi)}{g'(\xi)} = \frac{\sin \xi}{1 - \cos \xi}$$

equals the secant slope $\frac{2(1-\cos \pi)-0}{2(\pi - \sin \pi)-0} = \frac{2}{\pi}$. Solving $\frac{\sin \xi}{1-\cos \xi} = \frac{2}{\pi}$ by calculator gives a value $\xi \approx 2$.

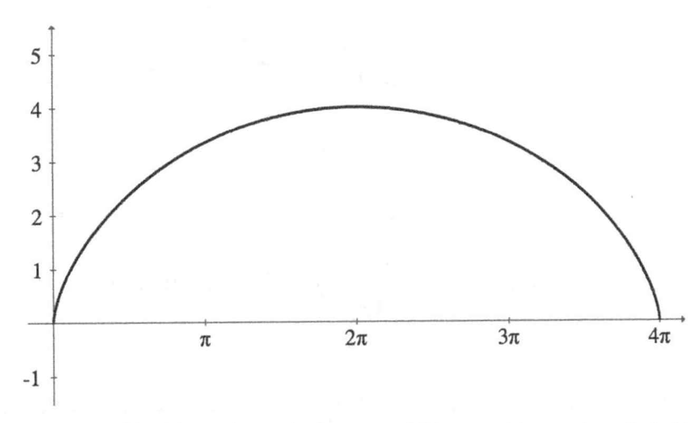

The cycloid $x(t) = 2(t - \sin t)$ $y(t) = 2(1 - \cos t)$ over $t \in [0, 2\pi]$.

Chapter 40

1. $P_4(x) = \frac{1}{2} - \frac{\sqrt{3}}{2}(x - \frac{\pi}{3}) - \frac{1}{4}(x - \frac{\pi}{3})^2 + \frac{\sqrt{3}}{12}(x - \frac{\pi}{3})^3 + \frac{1}{48}(x - \frac{\pi}{3})^4$ by observing the coefficients of $P_6(x)$.

$$\left|P_4\left(\frac{\pi}{3} - 1\right) - \cos\left(\frac{\pi}{3} - 1\right)\right| \approx 0.00637 \text{ vs } \left|P_4\left(\frac{\pi}{3} + 1\right) - \cos\left(\frac{\pi}{3} + 1\right)\right| \approx 0.00773,$$

which is the maximum absolute error.

$$\left|P_6\left(\frac{\pi}{3} - 1\right) - \cos\left(\frac{\pi}{3} - 1\right)\right| \approx 0.000157 \text{ vs } \left|P_6\left(\frac{\pi}{3} + 1\right) - \cos\left(\frac{\pi}{3} + 1\right)\right| \approx 0.000182,$$

less than $1/40$th of the maximum error with P_4.

3. a. Use

$$|R_6(x)| = \left|\frac{1}{7!}\frac{d^7 \cos x}{dx^7}\right| \left|x - \frac{\pi}{3}\right|^7 = \frac{1}{5040}|\sin\xi| \left|x - \frac{\pi}{3}\right|^7$$

where $x \in \left(\frac{\pi}{3} - 1, \frac{\pi}{3} + 1\right)$ and ξ is between $\frac{\pi}{3}$ and x. Considerations similar to Example 40.3 give us $|R_6(x)| \le \frac{1}{5040} < 0.000198$.

b. Again,

$$|R_6(x)| = \frac{1}{5040}|\sin\xi| \left|x - \frac{\pi}{3}\right|^7 \le \frac{1}{5040}\left|x - \frac{\pi}{3}\right|^7.$$

At the endpoints of the interval,

$$|R_6(-0.7)| \le \frac{1}{5040}\left|-0.7 - \frac{\pi}{3}\right|^7 \approx 0.00986$$

and

$$|R_6(2.7)| \le \frac{1}{5040}\left|2.7 - \frac{\pi}{3}\right|^7 \approx 0.00669,$$

so the maximum absolute error is almost 0.01 over the given interval. This wouldn't be enough accuracy for most applications.

5. Using the product rule carefully,

$$\frac{d\Delta(x)}{dx} = -f'(x) - \sum_{k=1}^{n}\left[\frac{-1}{(k-1)!}f^{(k)}(x)(b-x)^{k-1} + \frac{1}{k!}f^{(k+1)}(x)(b-x)^{k}\right]$$

$$+\frac{K}{n!}(b-x)^n$$

$$= -f'(x) + \sum_{k=1}^{n}\frac{1}{(k-1)!}f^{(k)}(x)(b-x)^{k-1}$$

$$-\sum_{k=1}^{n}\frac{1}{k!}f^{(k+1)}(x)(b-x)^k + \frac{K}{n!}(b-x)^n$$

$$= -f'(x) + f'(x) + \sum_{k=1}^{n-1}\frac{1}{k!}f^{(k+1)}(x)(b-x)^k$$

$$-\sum_{k=1}^{n}\frac{1}{k!}f^{(k+1)}(x)(b-x)^k + \frac{K}{n!}(b-x)^n$$

$$= -\frac{1}{n!}f^{(n+1)}(x)(b-x)^n + \frac{K}{n!}(b-x)^n.$$

All in all, the proof demonstrates Lagrange's brilliance.

7. $P_3(r) = 2e^{-2} - 2e^{-2}(r-1)^2 + \frac{4}{3}e^{-2}(r-1)^3$ expanded about $c = 1$, is seen below along with $\overline{P(r)} = 2r^2e^{-2r}$. (Incidentally, $\overline{P(r)}$ shows that the electron has highest probability of being observed near $r = 1$, which corresponds to the observed radius of the hydrogen atom. Kudos for the quantum physicists!)

$$|R_3(r)| = \left|\frac{f^{(n+1)}(\xi)}{(n+1)!}(r-1)^{n+1}\right| = \frac{4}{3}e^{-2\xi}|(\xi-1)(\xi-3)|(r-1)^4$$

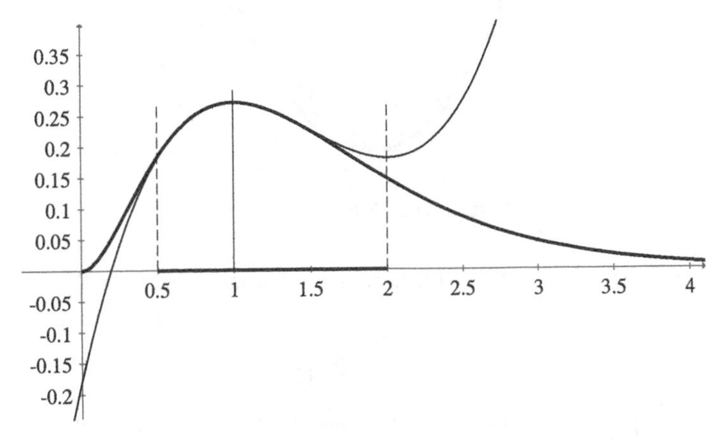

$\overline{P(r)}$ in thick line, $P_3(r)$ in thin line, and the interval of estimation.

with ξ between 1 and r. On the interval $[0.5, 1]$, the absolute error increases as $r \to \frac{1}{2}$, and $R_3(\frac{1}{2})$ has a supremum at $\xi = \frac{1}{2}$. Thus, using $\xi = \frac{1}{2}$, $|R_3(r)| < \left|R_3\left(\frac{1}{2}\right)\right| \approx 0.0383$. On the interval $[1, 2]$, it's more difficult to determine the maximum error, but the graphs show that it occurs at $r = 2$, so $|R_3(r)| < |R_3(2)| = P_3(2) - \overline{P(2)} \approx 0.0339$. Therefore, a bound on the error over $[0.5, 2]$ is 0.0383.

Chapter 41

1. Graphed below are $f_n(x)$ for $n = 1, 3, 8$, and 25. Right away, we see that $f_n(1) = f_n(-1) = \frac{1}{2}$ for all $n \in \mathbb{N}$. Therefore, $F(1) = F(-1) = \frac{1}{2}$. Also, we see evidence that $F(x) = 1$ over $(-1, 1)$ and $F(x) = 0$ over $[-2, -1) \cup (1, 2]$. Proof: for any fixed x over $(-1, 1)$, $\lim_{n\to\infty} \frac{1}{1+|x|^n} = 1$, whereas over $[-2, -1) \cup (1, 2]$, $\lim_{n\to\infty} \frac{1}{1+|x|^n} = 0$.

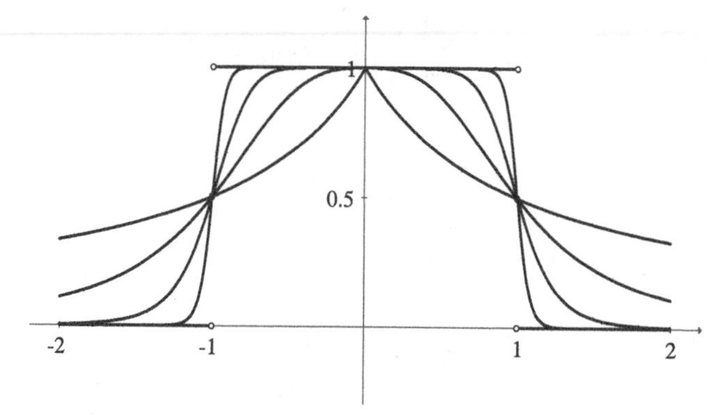

Exercise 41.1. F is the thick line.

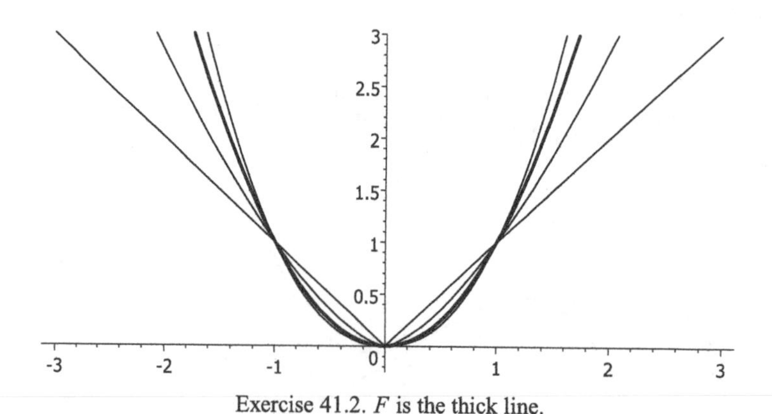

Exercise 41.2. F is the thick line.

3. Graphed below are $f_1(x)$, $f_2(x)$, $f_{10}(x)$, and $f_{40}(x)$. For any fixed $x_0 \geq 0$,

$$\lim_{n\to\infty} \left|\frac{x_0 - n}{x_0 + 2n}\right| = \frac{1}{2}.$$

Thus, take $F(x) = \frac{1}{2}$. Then, with an $\varepsilon = 0.1$ channel about $y = \frac{1}{2}$ as seen below, eventually $0.4 < f_n(x_0) < 0.6$ for all subscripts $n \geq N$, where N depends on x_0. Given any $\varepsilon > 0$, it is true that

$$\left| \left| \frac{x_0 - n}{x_0 + 2n} \right| - \frac{1}{2} \right| < \varepsilon$$

for sufficiently large n. Hence, $F(x) = \frac{1}{2}$.

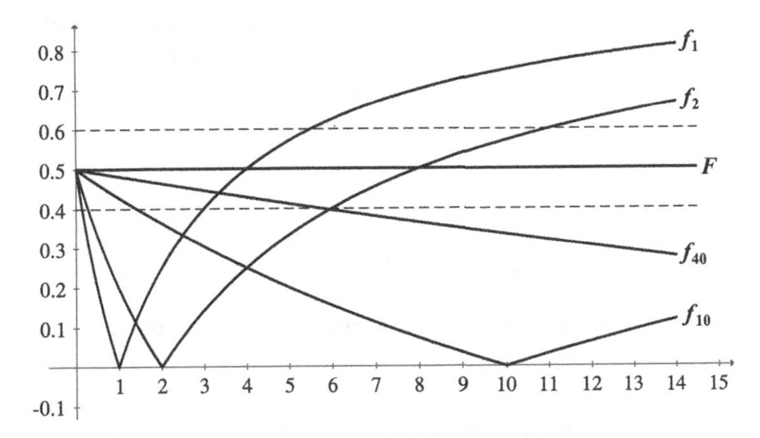

This is an example of a sequence of non-differentiable functions that converges pointwise to a function that *is* differentiable over the common domain.

5. a. $N = \frac{|3.7|}{\sqrt{2(0.002)}} \approx 58.502$, and indeed even for n as small as $\lceil 58.502 \rceil = 59$, $|\cos \frac{3.7}{59} - 1| \approx 0.001965 < 0.002$.
   b. $|\cos \frac{8.9}{59} - 1| \approx 0.01136 > 0.002$. This means that a single N will not suffice for all x in the domain. This time, $N = \frac{|8.9|}{\sqrt{2(0.002)}} \approx 140.721$, so we will be OK if our subscript $n \geq 141$ at $x = 8.9$. That is, N depends upon ε (naturally), but *also* on the value of x. Looking ahead, this will deny uniform convergence for $\langle \cos \frac{x}{n} \rangle_1^\infty$.

7. a. Consider

$$|f_n(x) - f_m(x)| = \left| \frac{2nx}{nx+3} - \frac{2mx}{mx+3} \right| = \frac{6x|n-m|}{(nx+3)(mx+3)}$$
$$= \frac{6x(n-m)}{(nx+3)(mx+3)} < \frac{6x(n-m)}{nmx^2} = \frac{1}{x}\left(\frac{6}{m} - \frac{6}{n}\right) \leq \frac{6}{am},$$

   because $\frac{1}{x} \leq \frac{1}{a}$ over $[a, \infty)$, and we have allowed $n > m$. Hence, if we set $\varepsilon = \frac{6}{aN}$, then we establish that $N_\varepsilon = \frac{6}{a\varepsilon}$. Now, whenever $m > N_\varepsilon = \frac{6}{a\varepsilon}$, then also $n, m > N_\varepsilon$, and it follows that $|f_n(x) - f_m(x)| < \frac{6}{am} < \varepsilon$. Observe that this is true for all $x \in [a, \infty)$, since N_ε depends only on ε. $\langle \frac{2nx}{nx+3} \rangle_1^\infty$ is therefore a Cauchy sequence at each individual $x \in [a, \infty)$, and by Theorem 41.2, it converges uniformly over $[a, \infty)$.
   b. Over $[0, \infty)$, the necessary requirement that $\frac{1}{x} \leq \frac{1}{a}$ becomes meaningless.
   c. Since $\langle \frac{2nx}{nx+3} \rangle_1^\infty \xrightarrow{u} 2$ over $[\frac{1}{2}, \infty)$, Corollary 41.1 says

$$\lim_{n\to\infty} \int_{1/2}^{2} e^x \sqrt{\frac{2nx}{nx+3}} dx = \int_{1/2}^{2} e^x \lim_{n\to\infty} \left(\sqrt{\frac{2nx}{nx+3}} \right) dx = \int_{1/2}^{2} \sqrt{2} e^x \, dx = \sqrt{2}(e^2 - \sqrt{e}).$$

   (Check how close the values of $\sqrt{2}(e^2 - \sqrt{e})$ and $\int_{1/2}^{2} e^x \sqrt{\frac{8000x}{4000x+3}} dx$ are.)

9. Consider the figure in the solution to Exercise 41.1. Take $\varepsilon^* = \frac{1}{3}$, say, and $x_0 = 0.5$. Then, to satisfy $|\frac{1}{1+|0.5|^n} - 1| < \frac{1}{3}$, we combine and simplify: $\frac{0.5^n}{1+0.5^n} < \frac{1}{3}$. Thus, $0 \le 0.5^n < \frac{1}{2}$, implying that $n \ge 2$. Thus, take $N = 2$ when $x_0 = 0.5$. But for instance, at $x_1 = 0.9$, $|\frac{1}{1+|0.9|^2} - 1| \approx 0.48 > \varepsilon^*$. Thus, it is clear that N depends upon both ε^* and x, and the convergence is not uniform.

   At $x_0 = 1$, $\frac{1}{1+|1|^n} = \frac{1}{2}$ for all n. Since $\langle 1, \frac{1}{2}\rangle$ is a type 2 isolated point of F, then F is discontinuous there. The contrapositive of Theorem 41.1 tells us that the convergence of the continuous functions $f_n(x) = \frac{1}{1+|x|^n}$ can't be uniform. (A similar case occurs at $x_0 = -1$.)

Chapter 42

1. Graphed here are partial sums

$$s_1(x) = \frac{x}{1+x^2}, \quad s_2(x) = \frac{x}{1+x^2} + \frac{2x}{1+2x^2},$$

   $s_4(x)$, and $s_8(x)$. Convergence looks questionable. We try the limit comparison test with $\sum_{n=1}^{\infty} 1$. Then

$$\rho = \lim_{n\to\infty}\left(\frac{nx}{1+nx^2}\cdot 1\right) = \frac{1}{x} > 0 \quad \text{for any } x > 0.$$

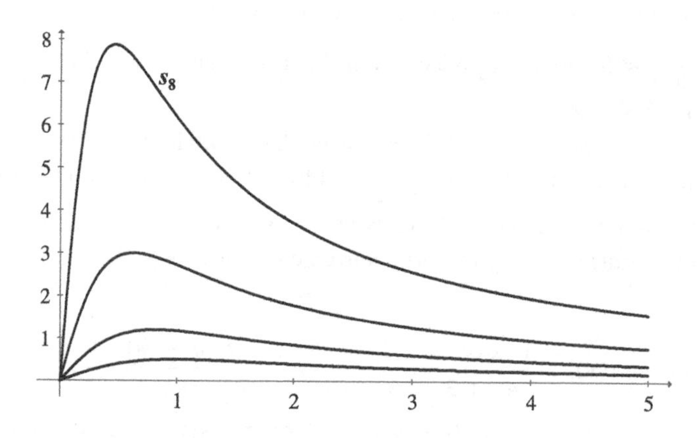

   Consequently, $\sum_{n=1}^{\infty} \frac{nx}{1+nx^2}$ diverges at every positive x. At $x = 0$, the series instantly converges to zero (as the graph plainly shows).

3. First, for any $x \ge 1$,

$$\frac{n}{1+n^3x} \le \frac{n}{1+n^3} < \frac{1}{n^2}$$

   is true for $n \in \mathbb{N}$. Thus, let $M_n = \frac{1}{n^2}$, and since $\sum_{n=1}^{\infty} \frac{1}{n^2}$ converges, Theorem 42.1 affirms that $\sum_{n=1}^{\infty} \frac{n}{1+n^3x}$ converges *uniformly* over $[1, \infty)$.

   For $0 < \alpha \le x < 1$, we can still say that $\frac{n}{1+n^3x} \le \frac{n}{1+n^3\alpha} < \frac{1}{n^2\alpha}$. Thus, let $M_n = \frac{1}{n^2\alpha}$, and since $\sum_{n=1}^{\infty} \frac{1}{n^2\alpha}$ converges, Theorem 42.1 affirms that $\sum_{n=1}^{\infty} \frac{n}{1+n^3x}$ converges uniformly over $[\alpha, \infty)$.

   For the case $(0, \infty)$, we look at $0 < x < 1$, and things change. As seen in Exercise 42.2b, the inequality $\frac{n}{1+n^3x} < \frac{1}{n^{3/2}}$ becomes true for all $n \ge N$, with N dependent on the value of x.

Because of this, there is no convergent series $\sum_{n=1}^{\infty} M_n$ such that $\frac{n}{1+n^3x} \leq M_n$ for *all* positive x and natural numbers n. To see this, assume that such a series exists. Select $x = 0.001$ and $n = 10$ for instance. Then $\frac{10}{1+10^3(0.001)} = 5 \leq M_{10}$. Through such a loophole, we can force $M_n > 1$ for each $n \geq 2$. The contradiction is that $\sum_{n=1}^{\infty} M_n$ diverges. Consequently, our series does not converge uniformly over $(0, 1)$, which implies that it does not converge uniformly over $(0, \infty)$. The endpoint at zero has poisoned the entire positive real axis!

In conclusion, $\sum_{n=1}^{\infty} \frac{n}{1+n^3x}$ converges uniformly over $[\alpha, \infty)$, not uniformly over $(0, \infty)$, and from the previous exercise, it converges pointwise over $(0, \infty)$.

5. We consider

$$x \sum_{n=1}^{\infty} \frac{n^2}{e^{nx}} = x \sum_{n=1}^{\infty} \frac{n^2}{(e^x)^n}.$$

For any $x \geq 1$, we have

$$\frac{n^2}{(e^x)^n} \leq \frac{n^2}{e^n}$$

for all $n \in \mathbb{N}$. Hence, using $\frac{n^2}{e^n} = M_n$ and remembering that

$$\sum_{n=1}^{\infty} \frac{n^2}{e^n}$$

converges, we conclude that

$$\sum_{n=1}^{\infty} n^2 e^{-nx}$$

converges uniformly over $[1, \infty)$ to some $F(x)$ by the Weierstrass M-test. Consequently,

$$\sum_{n=1}^{\infty} n^2 x e^{-nx} \stackrel{u}{=} xF(x).$$

Next, for $0 < \alpha \leq x$, we are still able to say that $\frac{n^2}{(e^x)^n} \leq \frac{n^2}{(e^\alpha)^n}$ for all $n \in \mathbb{N}$. Thus, $\frac{n^2}{e^{\alpha n}} = M_n$ may be used in Theorem 42.1 to conclude that

$$\sum_{n=1}^{\infty} n^2 x e^{-nx} \stackrel{u}{=} xF(x)$$

over $[\alpha, \infty)$.

At $x = 0$ the given series instantly converges to zero, while in any interval $(0, \alpha)$, the series still converges, but to increasingly larger values. To show this, apply Corollary 29.2, from which we extract that $\int_{m+1}^{\infty} f(x, n)dn + s_m(x) \leq F(x)$. Evaluating the integral with $f(x, n) = n^2 x e^{-nx}$, and substituting, produces the inequality

$$\left[\frac{(m+1)^2}{e^{(m+1)x}} + \frac{2(m+1)}{xe^{(m+1)x}} + \frac{2}{x^2 e^{(m+1)x}}\right] + s_m(x) \leq F(x).$$

As $x \to 0^+$, the expression in brackets is unbounded, so that $\lim_{x \to 0^+} F(x) = +\infty$. Thus, the convergence of $\sum_{n=1}^{\infty} n^2 x e^{-nx}$ is not uniform over $[0, \infty)$, only pointwise.

7. First, $3\lceil 123.4\rceil = 372$ and $n_0 = 3\lceil 123.4\rceil - 1 = 371$. So select

$$x_0 \in (-1, -\left(\frac{1}{2}\right)^{1/(372)}] \approx (-1, -0.9981\ldots]$$

or else take $x_0 \in [0.9981\ldots, 1)$. If you choose $x_0 = -0.999$, then

$$\left|\sum_{k=1}^{371}(-0.999)^k(1.999) - (-0.999)\right| = |-0.999|^{372} \approx 0.689 \geq \frac{1}{2} = \varepsilon^*.$$

Thus, the quantities $F(-0.999)$ and $s_{371}(-0.999)$ deviate from each other by more than $\frac{1}{2}$.

9. Allowing $n > m$, we have

$$|s_n(x) - s_m(x)| = \left|\sum_{k=1}^{n} x^k - \sum_{k=1}^{m} x^k\right| = \left|\frac{x - x^{n+1}}{1-x} - \frac{x - x^{m+1}}{1-x}\right| = \left|\frac{x^{n+1} - x^{m+1}}{x-1}\right|$$

$$= \frac{|x^{m+1}||x^{n-m} - 1|}{|x-1|} \leq \frac{(\frac{1}{2})^{m+1}(2)}{\frac{3}{2}} = \frac{1}{3}\frac{1}{2^{m-1}} < \frac{1}{2^m}.$$

This may be made smaller than any positive ε by taking

$$m > \max\left\{\frac{-\ln\varepsilon}{\ln 2}, 1\right\}$$

(verify). Wrapping things up, for any $\varepsilon > 0$, and for all $x \in A$, there exists a number

$$N_\varepsilon = \max\left\{\frac{-\ln\varepsilon}{\ln 2}, 1\right\}$$

such that whenever $n > m > N_\varepsilon$, we are sure that $|s_n(x) - s_m(x)| < \frac{1}{2^m} < \varepsilon$. By the Cauchy criterion, $\langle s_n(x)\rangle_1^\infty$ converges uniformly over A, which is equivalent to $\sum_{n=1}^\infty x^n$ converging uniformly over A.

11. If $\sin x = 1$, that is, $x = \frac{(4k+1)\pi}{2}$, $k \in \mathbb{Z}$, then the series is divergent. If $\sin x = -1$, that is, $x = \frac{(4k+3)\pi}{2}$, $k \in \mathbb{Z}$, then the series is convergent. For all other values of x, let $\sin x = y$. The series becomes $\sum_{n=1}^\infty \frac{y^n}{n}$, which we can compare with the convergent geometric series $\sum_{n=1}^\infty |y|^n$. Thus, $\sum_{n=1}^\infty \frac{|y|^n}{n}$ converges by the comparison test, which means that $\sum_{n=1}^\infty \frac{y^n}{n}$ converges. In conclusion, the series converges absolutely for all

$$\left\{x : x \neq \frac{1}{2}(4k+1)\pi \text{ and } x \neq \frac{1}{2}(4k+3)\pi,\ k \in \mathbb{Z}\right\},$$

converges conditionally at

$$\left\{x = \frac{(4k+3)\pi}{2},\ k \in \mathbb{Z}\right\}, \qquad \text{and diverges at} \qquad \left\{x = \frac{(4k+1)\pi}{2},\ k \in \mathbb{Z}\right\}.$$

Visual evidence of this is in the figure below.

Uniform convergence will occur over intervals that are bounded away from the points of divergence, $x = \frac{(4k+1)\pi}{2}$. For instance, $x = \frac{5\pi}{2} \approx 7.85$, so $\sum_{n=1}^{\infty} \frac{\sin^n x}{n}$ converges uniformly over an interval like [3, 7.84], by the M-test.

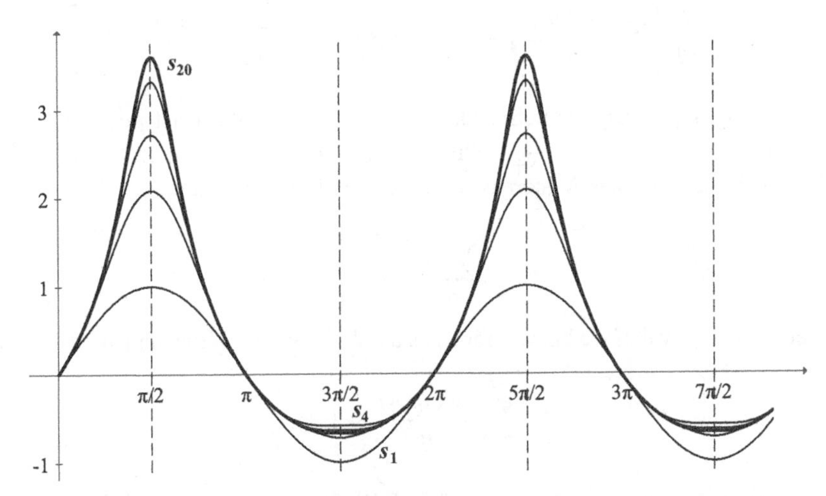

Partial sums s_n for $n = 1, 3, 10, 15$, and 20. The vertical lines are at $x = \frac{\pi}{2}, \frac{3\pi}{2}, \frac{5\pi}{2}$, and $\frac{7\pi}{2}$. At $\frac{\pi}{2}$ and $\frac{5\pi}{2}$, the series diverges. At $\frac{3\pi}{2}$ and $\frac{7\pi}{2}$, the series converges conditionally.

13. From the solution to Exercise 33.10b,

$$\sum_{n=1}^{N} \sin(n\theta) = \frac{1}{2}\sin(N\theta) + \left(\frac{1}{2}\cot\frac{\theta}{2}\right)(1 - \cos(N\theta)),$$

as long as $\theta \neq 2k\pi, k \in \mathbb{Z}$. Therefore, uniform convergence is guaranteed by Dirichlet's test (Theorem 33.2) over intervals that exclude $\theta = 2k\pi, k \in \mathbb{Z}$.

Near $\theta = 0$, the graphs below show the same behavior as seen in Figure 42.7. Notice that the partial sums $\sum_{n=1}^{N} \frac{\sin(n\theta)}{n}$ have absolute extrema $\sum_{n=1}^{5} \frac{\sin(n\theta)}{n}$ is the dot-dash line, $\sum_{1}^{20} \frac{\sin(n\theta)}{n}$ is the dashed line, $\sum_{1}^{40} \frac{\sin(n\theta)}{n}$ is the solid line.

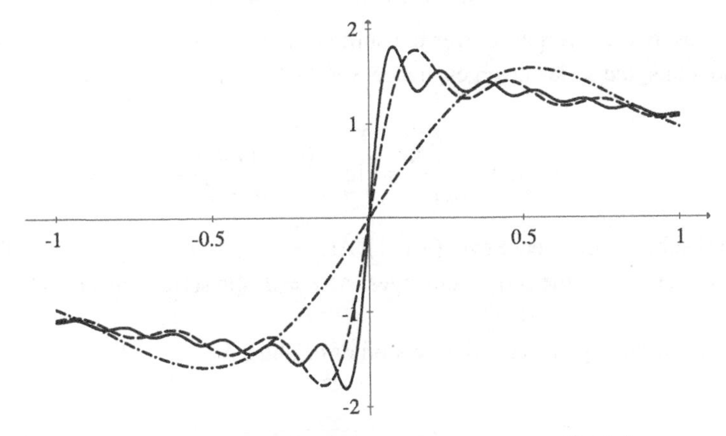

$\sum_{n=1}^{5} \frac{\sin(n\theta)}{n}$ is the dot-dash line, $\sum_{1}^{20} \frac{\sin(n\theta)}{n}$ is the dashed line, $\sum_{1}^{40} \frac{\sin(n\theta)}{n}$ is the solid line.

within a neighborhood $\mathbf{N}_0^\delta$ regardless of how small δ is, where N increases as δ decreases, while at $\theta = 0$, clearly $\sum_1^N \frac{\sin 0}{n} = 0$.

Let's find the maximum for each N. We have

$$\frac{d}{d\theta}\left(\sum_{n=1}^{N} \frac{\sin(n\theta)}{n}\right) = \sum_{n=1}^{N} \cos(n\theta) = \frac{\sin((2N+1)\frac{\theta}{2})}{2\sin\frac{\theta}{2}} - \frac{1}{2},$$

by one of Lagrange's trigonometric identities. Verify that the least positive θ such that $\sin((2N+1)\frac{\theta}{2}) = \sin\frac{\theta}{2}$ is $\theta = \frac{\pi}{N+1}$. This tells us that the maximum of $\sum_{n=1}^{N} \frac{\sin(n\theta)}{n}$ occurs closer and closer to zero as N increases. The value of the maximum is

$$\sum_{n=1}^{N} \frac{\sin(\frac{\pi}{N+1}n)}{n}$$

(a function of N), which we know converges as $N \to \infty$ to a value approximated by

$$\sum_{n=1}^{300} \frac{\sin(\frac{\pi}{301}n)}{n} \approx 1.85.$$

All of this implies that convergence of $\sum_1^\infty \frac{\sin(n\theta)}{n}$ is not uniform within any $\mathbf{N}_0^\delta$. By periodicity, the conclusion extends to any $\mathbf{N}_{2k\pi}^\delta$.

Chapter 43

1. In Example 42.1, we begin with a given domain, which happens to be exactly the set of points over which the series converges (pointwise) to an explicit sum $F(x) = \frac{x}{1-x}$. In Example 43.1, the domain is all $\mathbb{R}$, and we focus on the properties of a power series. The theorems of the present chapter then *deduce* the previously found interval of convergence. In the next chapter, we will find a general method to derive the series from the sum function F.

3. a. Since

$$\beta = \frac{1}{\lim_{n\to\infty} \sqrt[n]{|a_n|}} = \frac{1}{\lim_{n\to\infty} \sqrt[n]{n}} = 1,$$

we have absolute convergence over the shifted interval $(-1-5, 1-5) = (-6, -4)$. At both endpoints, the series diverges; hence $S = (-6, -4)$.

b. Since

$$\beta = \lim_{n\to\infty} \left|\frac{a_n}{a_{n+1}}\right| = \lim_{n\to\infty} \frac{(n+1)(n+2)}{n(n+1)} = 1,$$

we have absolute convergence over $(-1, 1)$. At $x = -1$, we find $\sum_{n=1}^{\infty} \frac{(-1)^n}{n(n+1)}$. By Leibniz's alternating series test, the series converges. At $x = 1$, the series converges by comparison to $\sum_{n=1}^{\infty} \frac{1}{n^2}$. Thus, $S = [-1, 1]$.

c. This is the alternating power series version of b. Thus, $S = [-1, 1]$.

d. Since

$$\beta = \left(\lim_{n\to\infty} \sqrt[n]{\frac{1}{2n-1}}\right)^{-1} = 1,$$

we have absolute convergence over $(-1, 1)$. At $x = -1$, we find $\sum_{n=1}^{\infty} \frac{(-1)^n}{2n-1}$, which converges by the alternating series test. At $x = 1$, we find $\sum_{n=1}^{\infty} \frac{(-1)^{n+1}}{2n-1}$, which converges by the alternating series test. Thus, $S = [-1, 1]$.

e. Since

$$\beta = \left(\lim_{n\to\infty} \sqrt[n]{\frac{1}{2n}} \right)^{-1} = 1,$$

we have absolute convergence over $(-1, 1)$. At $x = \pm 1$, we find $\sum_{n=1}^{\infty} \frac{(-1)^{n+1}}{2n}$, which converges by the alternating series test. Thus, $S = [-1, 1]$.

f. Change variables as $3n = m$, giving $\sum_{m=0}^{\infty} \frac{x^m}{5^{m/3}}$. Then

$$\beta = \left(\lim_{m\to\infty} \sqrt[m]{|a_m|} \right)^{-1} = \left(\lim_{m\to\infty} \sqrt[m]{5^{-m/3}} \right)^{-1} = \sqrt[3]{5}.$$

At the endpoint $x = -\sqrt[3]{5}$, we have

$$\sum_{m=0}^{\infty} \frac{(-\sqrt[3]{5})^m}{5^{m/3}} = \sum_{m=0}^{\infty} (-1)^m,$$

and at the right endpoint, we obtain $\sum_{m=0}^{\infty} 1$. Hence, the interval of convergence is $(-\sqrt[3]{5}, \sqrt[3]{5})$.

g. This is the power series $\sum_{n=0}^{\infty} \frac{3^n}{5^n} x^n$. Since

$$\beta = \lim_{n\to\infty} \left| \frac{a_n}{a_{n+1}} \right| = \lim_{n\to\infty} \left| \frac{3^n}{5^n} \cdot \frac{5^{n+1}}{3^{n+1}} \right| = \frac{5}{3},$$

we have absolute convergence over $(-\frac{5}{3}, \frac{5}{3})$. At $x = -\frac{5}{3}$, we find $\sum_{n=0}^{\infty} (-1)^n$, and at $x = \frac{5}{3}$, we find $\sum_{n=0}^{\infty} 1$. Hence $S = (-\frac{5}{3}, \frac{5}{3})$.

h. Since

$$\beta = \lim_{n\to\infty} \left| \frac{a_n}{a_{n+1}} \right| = \lim_{n\to\infty} \frac{(n+1)^2}{n^2} = 1,$$

the series converges absolutely over the shifted interval $(0, 2)$. At $x = 0$, we have $\sum_{n=1}^{\infty} \frac{(-1)^n}{n^2}$, which converges by the alternating series test (to the sum $\frac{-\pi^2}{12}$, as discovered by Euler). At $x = 2$, we have $\sum_{n=1}^{\infty} \frac{1}{n^2}$, which converges (to the sum $\frac{\pi^2}{6}$, again by Euler). Thus, $S = [0, 2]$.

i. We have

$$\beta = \lim_{n\to\infty} \left| \frac{a_n}{a_{n+1}} \right| = \lim_{n\to\infty} \left(\frac{n^n}{n!} \cdot \frac{(n+1)!}{(n+1)^{n+1}} \right) = e^{-1}.$$

Thus, the power series converges absolutely over $(-\frac{1}{e}, \frac{1}{e})$. At the endpoint

$$x = \frac{1}{e}, \sum_{n=1}^{\infty} \frac{n^n}{n! e^n}$$

diverges by Raabe's test ($\mu = \frac{1}{2}$). At the endpoint

$$x = -\frac{1}{e}, \sum_{n=1}^{\infty} (-1)^n \frac{n^n}{n! e^n}$$

converges by the alternating series test. Thus, $S = [-\frac{1}{e}, \frac{1}{e})$.

j. We have

$$f_n(x) = \frac{(2x+6)^{2n+1}}{2n+1} = \frac{2^{2n+1}}{2n+1}(x+3)^{2n+1}.$$

For a power series in $x+3$, let $m = 2n+1$, giving

$$\sum_{m=1}^{\infty} \frac{2^m}{m}(x+3)^m.$$

Then

$$\beta = \lim_{m\to\infty}\left|\frac{a_m}{a_{m+1}}\right| = \lim_{m\to\infty}\left(\frac{2^m}{m}\cdot\frac{m+1}{2^{m+1}}\right) = \frac{1}{2}.$$

Thus, the series converges absolutely over the shifted interval

$$\left(-\frac{1}{2}-3, \frac{1}{2}-3\right) = \left(-\frac{7}{2}, -\frac{5}{2}\right).$$

At the left and right endpoints, the alternating harmonic and harmonic series arise, respectively. Thus, $S = [-\frac{7}{2}, -\frac{5}{2})$.

k. This is the power series $\sum_{n=1}^{\infty} \frac{2^n}{\sqrt{n}}x^n$. Since

$$\beta = \lim_{n\to\infty}\left|\frac{a_n}{a_{n+1}}\right| = \lim_{n\to\infty}\left(\frac{2^n}{\sqrt{n}}\cdot\frac{\sqrt{n+1}}{2^{n+1}}\right) = \frac{1}{2},$$

we have absolute convergence over $\left(\frac{-1}{2}, \frac{1}{2}\right)$. At $x = \frac{-1}{2}$, $\sum_{n=1}^{\infty} \frac{(-1)^n}{\sqrt{n}}$ converges by the alternating series test. At $x = \frac{1}{2}$, $\sum_{n=1}^{\infty} \frac{1}{\sqrt{n}}$ diverges. Thus, $S = [\frac{-1}{2}, \frac{1}{2})$.

l. $\sum_{n=0}^{\infty} \frac{(2n)!}{2^n n!}x^n$ Since

$$\beta = \lim_{n\to\infty}\left|\frac{a_n}{a_{n+1}}\right| = \lim_{n\to\infty}\left(\frac{(2n)!}{2^n n!}\cdot\frac{2^{n+1}(n+1)!}{(2n+2)!}\right) = \lim_{n\to\infty}\frac{2(n+1)}{(2n+1)(2n+2)} = 0,$$

the power series converges only at $x = 0$.

5.

	Evaluate at:		
Partial sums	$x = 1.12$	$x = 1.18$	$x = 1.19$
$\sum_{n=0}^{5} \frac{x^{4n}}{2^n}$	$\sum_{n=0}^{5} \frac{1.12^{4n}}{2^n} \approx 3.57734$	$\sum_{n=0}^{5} \frac{1.18^{4n}}{2^n} \approx 5.55915$	$\sum_{n=0}^{5} \frac{1.19^{4n}}{2^n} \approx 6.04019$
$\sum_{n=0}^{50} \frac{x^{4n}}{2^n}$	$\sum_{n=0}^{50} \frac{1.12^{4n}}{2^n} \approx 4.68952$	$\sum_{n=0}^{50} \frac{1.18^{4n}}{2^n} \approx 25.9764$	$\sum_{n=0}^{50} \frac{1.19^{4n}}{2^n} \approx 54.5570$
$\sum_{n=0}^{100} \frac{x^{4n}}{2^n}$	$\sum_{n=0}^{100} \frac{1.12^{4n}}{2^n} \approx 4.68954$	$\sum_{n=0}^{100} \frac{1.18^{4n}}{2^n} \approx 31.2540$	$\sum_{n=0}^{100} \frac{1.19^{4n}}{2^n} \approx 115.751$
$\sum_{n=0}^{200} \frac{x^{4n}}{2^n}$	$\sum_{n=0}^{200} \frac{1.12^{4n}}{2^n} \approx 4.68954$	$\sum_{n=0}^{200} \frac{1.18^{4n}}{2^n} \approx 32.6047$	$\sum_{n=0}^{200} \frac{1.19^{4n}}{2^n} \approx 265.561$
Series sum $\sum_{n=0}^{\infty} \frac{x^{4n}}{2^n}$	$\sum_{n=0}^{\infty} \frac{1.12^{4n}}{2^n} \approx 4.68954$	$\sum_{n=0}^{\infty} \frac{1.18^{4n}}{2^n} \approx 32.6679$	$+\infty$

As $x \to \sqrt[4]{2}^-$, the rate of convergence decreases. Divergence is spectacular at x even slightly greater than $\sqrt[4]{2}$ (as at 1.19).

7. Below at left are partial sums s_5, s_{10}, and s_{29} of $\sum_{n=1}^{\infty} \frac{x^n}{n(n+1)}$ (s_{29} is the thick line) hinting at uniform convergence over precisely $[-1, 1]$. The sum

$$F(x) = \begin{cases} -\ln(1-x) + \frac{\ln(1-x)}{x} + 1 & \text{if } x \in [-1, 0) \cup (0, 1) \\ 0 & \text{if } x = 0 \\ 1 & \text{if } x = 1 \end{cases}$$

is continuous at $x = 0$ since $\lim_{x \to 0} F(x) = 0$, and also at $x = 1$ since

$$\lim_{x \to 1^-} \left(-\ln(1-x) + \frac{\ln(1-x)}{x} + 1 \right) = 1.$$

Below at right is a detail of Quadrant III, with

$$F(-1) = \sum_{n=1}^{\infty} \frac{(-1)^n}{n(n+1)} = -2\ln 2 + 1 \approx -0.386.$$

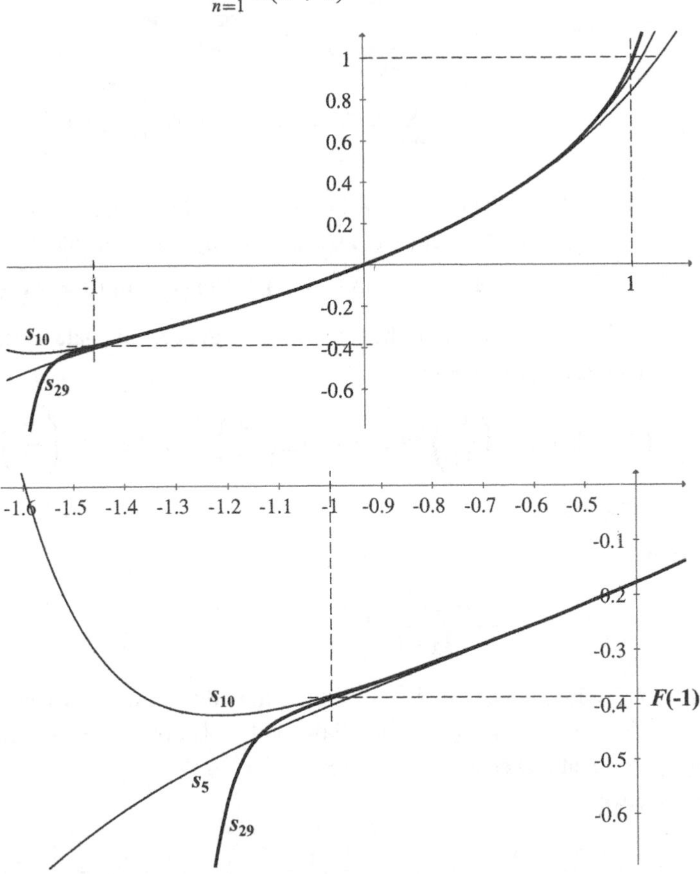

9. By Abel's theorem, the series converges uniformly over $[0, \beta]$. Make the change of variable $y = -x$. Then $\sum_{n=0}^{\infty} a_n y^n$ converges uniformly over $[-\beta, 0]$, and being a power series, it converges over $[-\beta, \beta]$. By Abel's theorem, this new series converges uniformly over $[0, \beta]$, hence, it converges uniformly over $[-\beta, \beta]$. So will the given series, proving the corollary.

Chapter 44

1. $\sum_{n=1}^{\infty}(x^n - x^{n+1}) = \sum_{n=1}^{\infty} x^n - \sum_{n=1}^{\infty} x^{n+1} = \frac{x}{1-x} - \frac{x^2}{1-x} = x$ over the interval $(-1, 1)$ within which these two geometric series converge.

And from another perspective, whenever $x \in (-1, 1)$, parentheses may be moved to give us

$$\sum_{n=1}^{\infty}(x^n - x^{n+1}) = (x - x^2) + (x^2 - x^3) + (x^3 - x^4) + \cdots$$
$$= x + (-x^2 + x^2) + (-x^3 + x^3) + \cdots = x$$

(what theorem justifies this regrouping?)

3. a. In Example 43.8, we found the sum function $-\ln(1-x) = \sum_{n=1}^{\infty} \frac{1}{n}x^n$ over $[-1, 1)$. Making the substitution $-x \mapsto x$ gives us $-\ln(1+x) = \sum_{n=1}^{\infty} \frac{(-1)^n}{n}x^n$ over $(-1, 1]$. Therefore, over $(-1, 1)$, we can subtract

$$\ln(1+x) - \ln(1-x) = -\sum_{n=1}^{\infty} \frac{(-1)^n}{n}x^n + \sum_{n=1}^{\infty} \frac{1}{n}x^n$$
$$= \sum_{n=1}^{\infty} \frac{(-1)^{n+1}+1}{n}x^n = 2x + \frac{2}{3}x^3 + \frac{2}{5}x^5 + \cdots.$$

We arrive at $\ln\frac{1+x}{1-x} = 2\left(x + \frac{1}{3}x^3 + \frac{1}{5}x^5 + \cdots\right)$. This series, also expressed as $\frac{1}{2}\ln\frac{1+x}{1-x} = \sum_{n=0}^{\infty} \frac{1}{2n+1}x^{2n+1}$ over $(-1, 1)$, is commonly found in calculus textbooks.

b. Solve $\frac{1+x}{1-x} = 10$ to get $x = \frac{9}{11} \in (-1, 1)$. Hence, $\ln 10 = 2\left(\frac{9}{11} + \frac{1}{3}\left(\frac{9}{11}\right)^3 + \frac{1}{5}\left(\frac{9}{11}\right)^5 + \cdots\right)$. To get a sense of the rate of convergence, calculate some partial sums (monotone increasing, of course):

$$s_{10}\left(\frac{9}{11}\right) \approx 2.2548,\ s_{20}\left(\frac{9}{11}\right) \approx 2.2989,\ s_{40}\left(\frac{9}{11}\right) \approx 2.3025,\ s_{80}\left(\frac{9}{11}\right) \approx 2.3026.$$

So, $\ln 10 \approx 2.3026$.

5. In the division

$$\frac{1 - \frac{1}{2}x^2 + \frac{1}{24}x^4 - \frac{1}{720}x^6 + \cdots}{1 - \frac{1}{6}x^2 + \frac{1}{120}x^4 - \frac{1}{5040}x^6 + \cdots},$$

we have $a_0 = b_0 = 1$, and $a_{2n-1} = b_{2n-1} = 0$ (all the odd subscripts). Thus, the formulas in Theorem 44.3 yield $q_0 = 1$, $q_{2n-1} = 0$ (all the odd subscripts), $q_2 = -\frac{1}{3}$, and $q_4 = -\frac{1}{45}$. The emerging quotient power series is $1 - \frac{1}{3}x^2 - \frac{1}{45}x^4 - \cdots$.

Finally, we have

$$\cot x = \frac{\cos x}{x\left(\frac{\sin x}{x}\right)} = \frac{1}{x}\left(1 - \frac{1}{3}x^2 - \frac{1}{45}x^4 - \cdots\right) = \frac{1}{x} - \frac{1}{3}x - \frac{1}{45}x^3 - \cdots,$$

for $x \neq 0$. Not being a power series, we need to think about whether the series converges to $\cot x$, and not some other number. This is the problem of representation of a function by a series, once again.

7. When expanding the polynomials in parentheses, all products yielding terms in x^1 through x^5 will be found (along with some terms of higher degree). Using more terms in any parenthesized polynomial, or using higher powers of those polynomials, will produce products of degree six or higher, which cannot alter the 16 terms (note that for instance, $a_1(\sum_{n=1}^{5} b_n x^n)^2$ is not an allowed expression).

9. From Example 44.7,

$$\int_0^C \frac{1}{1-x}dx = \ln\frac{1}{1-C} = \sum_{n=1}^{\infty}\frac{C^n}{n},$$

where $C \in [-1, 1)$. Now, let $M = \frac{1}{C}$, $C \neq 0$, and the results follow.

11. a. $\frac{dE(-x)}{dx}|_{-1} = -e^{-(-1)} = -e$. Also,

$$\frac{d}{dx}\sum_{n=0}^{\infty}\frac{1}{n!}(-x)^n = \sum_{n=1}^{\infty}\frac{1}{n!}\frac{d}{dx}(-x)^n = -\sum_{n=1}^{\infty}\frac{1}{(n-1)!}(-x)^{n-1} = -\sum_{n=0}^{\infty}\frac{1}{n!}(-x)^n,$$

so that $-\sum_{n=0}^{\infty}\frac{1}{n!}(-x)^n|_{-1} = -\left(1+1+\frac{1}{2!}+\frac{1}{3!}+\cdots\right) = -e$, as expected.

b. $\frac{d}{dx}\ln(\frac{1+x}{1-x})|_{1/2} = \frac{8}{3}$. Also,

$$\frac{d}{dx}\sum_{n=1}^{\infty}\frac{(-1)^{n+1}+1}{n}x^n = \sum_{n=1}^{\infty}((-1)^{n+1}+1)x^{n-1} = 2+2x^2+2x^6+\cdots,$$

so that at $x = \frac{1}{2}$, get the geometric series $2\sum_{n=0}^{\infty}\left(\frac{1}{2}\right)^{2n} = \frac{8}{3}$.

c. $\frac{d\sin x}{dx}|_{\pi} = -1$. Also,

$$\frac{d\sin x}{dx} = \frac{d}{dx}\left(x - \frac{1}{3!}x^3 + \frac{1}{5!}x^5 - \frac{1}{7!}x^7 + \cdots\right) = 1 - \frac{1}{2!}x^2 + \frac{1}{4!}x^4 - \frac{1}{6!}x^6 + \cdots.$$

Thus, at $x = \pi$, get $1 - \frac{\pi^2}{2!} + \frac{\pi^4}{4!} - \frac{\pi^6}{6!} + \cdots$ which quickly converges to -1. Of course, this agrees with $\cos\pi = -1$.

d. $\frac{d\sinh x}{dx} = \cosh x$ so that $\frac{d\sinh x}{dx}|_0 = 1$. Also,

$$\frac{d}{dx}\sum_{n=0}^{\infty}\frac{1}{(2n+1)!}x^{2n+1} = \sum_{n=0}^{\infty}\frac{1}{(2n+1)!}\frac{d}{dx}x^{2n+1} = \sum_{n=0}^{\infty}\frac{1}{(2n)!}x^{2n},$$

where the subscripting variable begins at zero since the first term of the series for $\sinh x$ isn't a constant. Indeed, $\sum_{n=0}^{\infty}\frac{1}{(2n)!}x^{2n} = \cosh x$ where the power series represents the hyperbolic cosine over all $\mathbb{R}$. And, $\sum_{n=0}^{\infty}\frac{1}{(2n)!}x^{2n}|_0 = 1$ in agreement (define $0^0 \equiv 1$ per Note 3, Chapter 3).

Chapter 45

1. The anchor is case $n = 1$: $f'(x) = \frac{1}{(1-x)^2}$. Observing that $f''(x) = \frac{2}{(1-x)^3}$ and $f'''(x) = \frac{6}{(1-x)^4}$, assume that $f^{(n)}(x) = \frac{n!}{(1-x)^{n+1}}$. Then $f^{(n+1)}(x) = \frac{(n+1)!}{(1-x)^{n+2}}$. By induction, it follows that $f^{(n)}(x) = \frac{n!}{(1-x)^{n+1}}$ for all natural numbers n. Then $f^{(n)}(0) = n!$. Hence,

$$\sum_{n=0}^{\infty}\frac{f^{(n)}(0)}{n!}x^n = \sum_{n=1}^{\infty}x^n.$$

3. First,

$$\sin^2 \frac{x}{2} = \left(\sum_{n=0}^{\infty} \frac{(-1)^n}{(2n+1)!} \left(\frac{x}{2}\right)^{2n+1} \right)^2 = \left(\frac{x}{2} - \frac{1}{3!}\left(\frac{x}{2}\right)^3 + \frac{1}{5!}\left(\frac{x}{2}\right)^5 - \frac{1}{7!}\left(\frac{x}{2}\right)^7 + \cdots \right)^2$$
$$= \left(\frac{x}{2}\right)^2 - \frac{2}{3!}\left(\frac{x}{2}\right)^4 + \frac{2}{5!}\left(\frac{x}{2}\right)^6 + \frac{1}{(3!)^2}\left(\frac{x}{2}\right)^6 - \frac{2}{7!}\left(\frac{x}{2}\right)^8 - \frac{1}{3!7!}\left(\frac{x}{2}\right)^8 + \cdots$$

as the Cauchy product. Then

$$\cos x = 1 - 2\left[\left(\frac{x}{2}\right)^2 - \frac{2}{3!}\left(\frac{x}{2}\right)^4 + \frac{2}{5!}\left(\frac{x}{2}\right)^6 \right.$$
$$\left. + \frac{1}{(3!)^2}\left(\frac{x}{2}\right)^6 - \frac{2}{7!}\left(\frac{x}{2}\right)^8 - \frac{1}{3!7!}\left(\frac{x}{2}\right)^8 + \cdots \right]$$
$$= 1 - \frac{1}{2!}x^2 + \frac{1}{4!}x^4 - \frac{1}{6!}x^6 + \cdots$$

which is the well-known Maclaurin series for the cosine function.

Yes, this series represents $f(x) = \cos x$ over all $\mathbb{R}$.

5. We sequentially find

$f^{(n)}(x)$	$\frac{f^{(n)}(0)}{n!}$
$f(x) = \ln(x+3)$	$\frac{f(0)}{0!} = \ln 3$
$f'(x) = \frac{1}{x+3}$	$\frac{f'(0)}{1!} = \frac{1}{3}$
$f''(x) = -\frac{1}{(x+3)^2}$	$\frac{f''(0)}{2!} = -\frac{1}{3^2 2!} = -\frac{1}{18}$
$\cdots$	$\cdots$
$f^{(n)}(x) = \frac{(-1)^{n-1}(n-1)!}{(x+3)^n}$	$\frac{f^{(n)}(0)}{n!} = \frac{(-1)^{n-1}(n-1)!}{n!} = \frac{(-1)^{n-1}}{n3^n}$ (for $n \geq 1$).

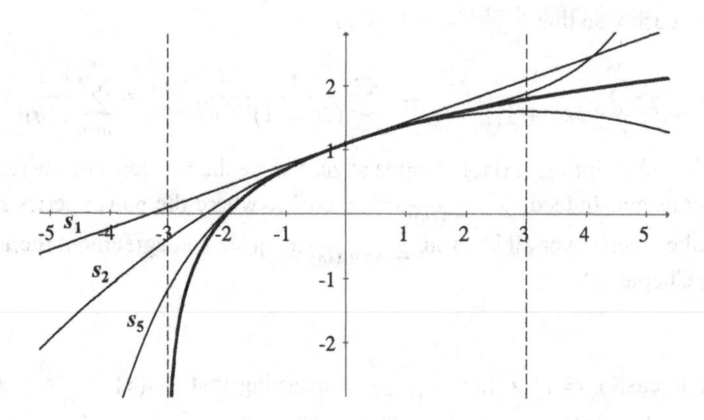

The resulting Maclaurin series is

$$\ln 3 + \sum_{n=1}^{\infty} \frac{(-1)^{n-1}}{n3^n} x^n = \ln 3 + \frac{1}{3}x - \frac{1}{18}x^2 + \cdots.$$

This is a shifted version (about $c = 0$ of course) of the Taylor series (2) in Example 45.3. The graphs of s_1 (linear), s_2, and s_5 above give evidence of this, and $f(x) = \ln(x+3)$ is the thick line. Notice the shifted interval of convergence. To prove it, let $y = x - 3$, and get $\ln(y - 3 + 3) = \ln y = \ln 3 + \frac{1}{3}(y-3) - \frac{1}{18}(y-3)^2 + \cdots$ as expected.

7. a. At $x = -1$, the alternating series test proves convergence. At $x = 1$, the comparison test proves convergence. The ratio test gives

$$\rho = \lim_{n\to\infty} \left| \frac{x^{n+1}}{(n+1)(n+2)} \cdot \frac{n(n+1)}{x^n} \right| = |x| \lim_{n\to\infty} \frac{n}{n+2} = |x|,$$

so that convergence occurs over $(-1, 1)$. Hence, the interval of convergence is $[-1, 1]$.

b. $\frac{1}{n(n+1)} = \frac{1}{n} - \frac{1}{n+1}$. Wherever the series converge, we have

$$\sum_{n=1}^{\infty} \frac{1}{n(n+1)} x^n = \sum_{n=1}^{\infty} \frac{1}{n} x^n - \sum_{n=1}^{\infty} \frac{1}{n+1} x^n.$$

c. Generating derivatives, we establish that $f^{(n)}(x) = \frac{(n-1)!}{(1-x)^n}$. Thus, the Maclaurin polynomials have the Cauchy remainder

$$|R_n(x_0)| = \frac{|f^{(n+1)}(\xi_n)|}{n!} |x_0 - \xi_n|^n |x_0| = \frac{|x_0 - \xi_n|^n}{|1 - \xi_n|^{n+1}} |x_0| = \left| \frac{x_0 - \xi_n}{1 - \xi_n} \right|^n \frac{|x_0|}{|1 - \xi_n|}$$

where ξ_n falls between 0 and x_0. If $x_0 > 1$, then it is possible for $|x_0 - \xi_n| > |1 - \xi_n|$, and $|R_n(x_0)| \not\to 0$. If $x_0 < -1$, it is possible for $|R_n(x_0)| \not\to 0$, for the same reason. If $-1 < x_0 < 1$, then $|x_0 - \xi_n| < |1 - \xi_n|$, and also, $|1 - \xi_n|$ is bounded away from zero. Consequently, $|R_n(x_0)| \to 0$. Thus, representation of $f(x) = -\ln(1-x)$ by the series $F(x) = \sum_{n=1}^{\infty} \frac{1}{n} x^n$ occurs over $(-1, 1)$.

Next, the endpoints. At the left endpoint, $F(-1) = -1 + \frac{1}{2} - \frac{1}{3} + \frac{1}{4} - \cdots = -\ln 2$, (see Exercise 31.8). And $f(-1) = -\ln(1 - (-1)) = -\ln 2$. At the right endpoint, divergence. Hence representation occurs over exactly $[-1, 1)$.

d. When

$$x \neq 0, \sum_{n=1}^{\infty} \frac{1}{n+1} x^n = -1 + \frac{1}{x} \sum_{n=1}^{\infty} \frac{1}{n} x^n$$

by algebraic steps, as long as we have convergence, so $x \in [-1, 1)\backslash\{0\}$ for this equation to be true. Thus, $G(x) = -1 + \frac{1}{x} \sum_{n=1}^{\infty} \frac{1}{n} x^n$ represents $g(x) = \frac{-\ln(1-x)}{x} - 1$ over $[-1, 1)\backslash\{0\}$.

e. By Theorem 44.1,

$$\sum_{n=1}^{\infty} \frac{1}{n(n+1)} x^n = \sum_{n=1}^{\infty} \frac{1}{n} x^n - \sum_{n=1}^{\infty} \frac{1}{n+1} x^n$$

is represented by $f(x) - g(x) = -\ln(1-x) + \frac{\ln(1-x)}{x} + 1$ over $[-1, 1)\backslash\{0\}$.

f. $\sum_{n=1}^{10} \frac{1}{n(n+1)}(-0.7)^n \approx -0.28857$, and $f(-0.7) - g(-0.7) = -\ln(1.7) + \frac{\ln(1.7)}{-0.7} + 1 \approx -0.288669$. There is substantial agreement, which is gratifying, after all the work we did!

Chapter 46

1. **Theorem 46.1′** Let $F(x) = \sum_{n=0}^{\infty} a_n(x-c)^n$ be any power series with interval of convergence S that is not just $[c, c]$. Then $a_n = \frac{F^{(n)}(c)}{n!}$ for all $n \in \mathbb{N} \cup \{0\}$.

Theorem 46.2′ Let $f(x)$ be represented by its Taylor series $\sum_{n=0}^{\infty} \frac{f^{(n)}(c)}{n!}(x-c)^n$. Then the series can be integrated term-by-term over any closed interval within its interval of convergence S. That is,

$$\int f(x)dx = \int \left(\sum_{n=0}^{\infty} \frac{f^{(n)}(c)}{n!}(x-c)^n \right) dx = \sum_{n=0}^{\infty} \frac{f^{(n)}(c)}{n!} \left(\int (x-c)^n dx \right)$$

$$= \sum_{n=0}^{\infty} \frac{f^{(n)}(c)}{(n+1)!}(x-c)^{n+1} \tag{1′}$$

with interval of convergence within S, and

$$\int_a^b f(x)dx = \int_a^b \left(\sum_{n=0}^{\infty} \frac{f^{(n)}(c)}{n!}(x-c)^n \right) dx = \sum_{n=0}^{\infty} \frac{f^{(n)}(c)}{n!} \left(\int_a^b (x-c)^n dx \right)$$

$$= \sum_{n=0}^{\infty} \frac{f^{(n)}(c)}{(n+1)!}((b-c)^{n+1} - (a-c)^{n+1}), \tag{2′}$$

where $[a, b] \subset S$.

Theorem 46.3′ Let $f(x)$ be represented by its Taylor series

$$\sum_{n=0}^{\infty} \frac{f^{(n)}(c)}{n!}(x-c)^n.$$

Then the series can be differentiated term-by-term at any point within its interval of convergence S. That is,

$$f'(x) = \frac{d}{dx} \sum_{n=0}^{\infty} \frac{f^{(n)}(c)}{n!}(x-c)^n = \sum_{n=1}^{\infty} \frac{f^{(n)}(c)}{(n-1)!}(x-c)^{n-1}$$

at any $x \in S \setminus \{\inf S, \sup S\}$.

Theorem 46.4′ The binomial series

$$\sum_{n=0}^{\infty} \binom{p}{n}(x-c)^n = 1 + p(x-c) + \frac{p(p-1)}{2!}(x-c)^2 + \frac{p(p-1)(p-2)}{3!}(x-c)^3 + \cdots$$

$(p \in \mathbb{R})$ represents the function $f(x) = (1 + (x-c))^p$ over at least the interval of convergence $S = (c-1, c+1)$. If $p \in \mathbb{N} \cup \{0\}$, then the series terminates, and convergence to $f(x) = (1 + (x-c))^p$ occurs for all real x (set $0^0 \equiv 1$). The endpoint convergence table is obtained by replacing the endpoints with $x = c-1$ and $x = c+1$.

3. Assuming we have a way to evaluate e^x for various values of x, Simpson's rule with $n = 2$ subdivisions gives

$$\Phi(0.7) = \frac{1}{2} + \frac{1}{\sqrt{2\pi}} \int_0^{0.7} e^{-t^2/2} dt \approx 0.5 + \frac{1}{\sqrt{2\pi}} \left(\frac{0.35}{3}\right) \left(1 + 4e^{-(0.35^2/2)} + e^{-(0.7^2/2)}\right)$$
$$= 0.758085\ldots,$$

not quite four digits of precision, and with $n = 4$ subdivisions:

$$\Phi(0.7) \approx 0.5 + \frac{1}{\sqrt{2\pi}} \left(\frac{0.175}{3}\right) \left(1 + 4e^{-(0.175^2/2)} + 2e^{-(0.35^2/2)} + 4e^{-(0.525^2/2)} + e^{-(0.7^2/2)}\right)$$
$$= 0.758039\ldots.$$

Not counting negation as an operation, the $n = 4$ result took 26 operations. The seventh-degree Maclaurin polynomial in Example 46.2 took only 19 operations (note that t^3 can be done in two operations as $t \times t^2$; and storing t^2 and t^3, then t^5 can be done in one operation, and so on).

5. a. We consider $(1 + \tan\theta)^{1/3}$, so that $p = \frac{1}{3}$ and $x = \tan\theta$ in (3). For representation, we must restrict $|\tan\theta| < 1$, which means that $\theta \in \left(-\frac{\pi}{4}, \frac{\pi}{4}\right)$. With this proviso, we have

$$(1 + \tan\theta)^{1/3} = 1 + \frac{1}{3}\tan\theta + \frac{1}{2!}\left(\frac{1}{3}\right)\left(\frac{-2}{3}\right)\tan^2\theta + \frac{1}{3!}\left(\frac{1}{3}\right)\left(\frac{-2}{3}\right)\left(\frac{-5}{3}\right)\tan^3\theta$$
$$+\frac{1}{4!}\left(\frac{1}{3}\right)\left(\frac{-2}{3}\right)\left(\frac{-5}{3}\right)\left(\frac{-8}{3}\right)\tan^4\theta + \cdots$$
$$= 1 + \frac{1}{3}\tan\theta - \frac{1}{9}\tan^2\theta + \frac{5}{81}\tan^3\theta - \frac{10}{243}\tan^4\theta + \cdots.$$

If we restrict θ to $(0, \frac{\pi}{4})$, then $\tan\theta > 0$, and the series is alternating. Thus, the absolute error gleaned when truncating the series to the terms seen here is at most

$$\left|\frac{1}{5!}\left(\frac{1}{3}\right)\left(\frac{-2}{3}\right)\left(\frac{-5}{3}\right)\left(\frac{-8}{3}\right)\left(\frac{-11}{3}\right)\tan^5\theta\right|.$$

At $\theta = \frac{\pi}{8}$, we calculate

$$1 + \frac{1}{3}\tan\frac{\pi}{8} - \frac{1}{9}\tan^2\frac{\pi}{8} + \frac{5}{81}\tan^3\frac{\pi}{8} - \frac{10}{243}\tan^4\frac{\pi}{8} = 1.122183\ldots,$$

with absolute error less than 4×10^{-4}. Thus, we are sure that $\sqrt[3]{(1 + \tan\frac{\pi}{8})} \approx 1.122$ to three-decimal-place precision.

b. $g(x) = \sqrt[3]{1+\tan\theta}$ is in thick line. The binomial series represents it as far as the point of non-differentiability at $\theta = -\frac{\pi}{4}$. The function

$$y = 1 + \frac{1}{3}\tan\theta - \frac{1}{9}\tan^2\theta + \frac{5}{81}\tan^3\theta - \frac{10}{243}\tan^4\theta$$

used in this exercise is in thin line, and the endpoints of the interval of convergence are also indicated.

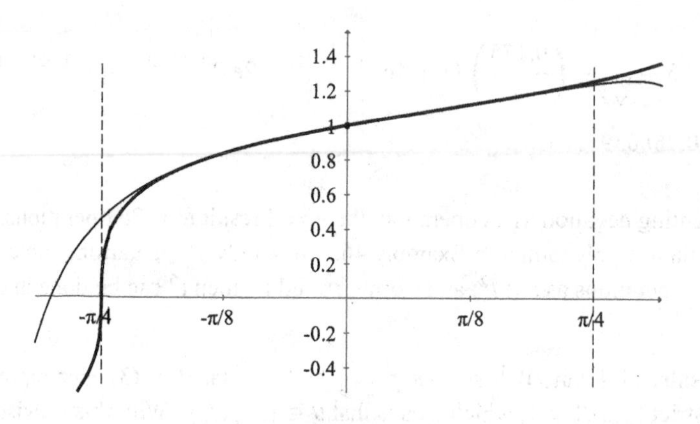

7. Endpoint $x = -1$.
   (i) When $p > 0$ but not a natural number or zero, we have

$$1 + p(-1) + \frac{p(p-1)}{2!}(-1)^2 + \frac{p(p-1)(p-2)}{3!}(-1)^3 + \cdots,$$

   which eventually stops alternating, when $n > p$.
   (ii) Raabe's test yields

$$\begin{aligned}
\mu &= \lim_{n\to\infty} n\left(1 - \left|\frac{a_{n+1}}{a_n}\right|\right) \\
&= \lim_{n\to\infty} n\left(1 - \left|\frac{p(p-1)(p-2)\cdots(p-n)}{(n+1)!} \cdot \frac{n!}{p(p-1)(p-2)\cdots(p-n+1)}\right|\right) \\
&= \lim_{n\to\infty} n\left(1 - \frac{|p-n|}{n+1}\right) = \lim_{n\to\infty} n\left(1 - \frac{n-p}{n+1}\right) = p+1.
\end{aligned}$$

   In this case, $\mu = p + 1 > 1$, so that Raabe's test indicates absolute convergence (that is, of $\sum_{n=0}^{\infty} |(-1)^n \binom{p}{n}|$) when $p > 0$. It follows that $\sum_{n=0}^{\infty}(-1)^n \binom{p}{n}$ converges also. So the binomial series converges absolutely for any $p > 0$ at the endpoint $x = -1$.
   (iii) In the unique case that $p = 0$, $\sum_{n=0}^{\infty}(-1)^n \binom{0}{n} = 1$, which agrees with $f(-1) = 0^0 \equiv 1$. Consequently, the binomial series $\sum_{n=0}^{\infty}(-1)^n \binom{p}{n}$ converges to the sum $f(-1) = 0^p = 0$, for any $p \geq 0$ at the endpoint $x = -1$.

(iv) If $p < 0$, we arrive at

$$\sum_{n=0}^{\infty}(-1)^n\binom{p}{n} = 1 + |p| + \frac{|p(p-1)|}{2!} + \frac{|p(p-1)(p-2)|}{3!} + \cdots,$$

a series of strictly positive terms. Raabe's test yields

$$\begin{aligned}\mu &= \lim_{n\to\infty} n\left(1 - \left|\frac{a_{n+1}}{a_n}\right|\right)\\ &= \lim_{n\to\infty} n\left(1 - \left|\frac{p(p-1)(p-2)\cdots(p-n)}{(n+1)!}\cdot\frac{n!}{p(p-1)(p-2)\cdots(p-n+1)}\right|\right)\\ &= \lim_{n\to\infty} n\left(1 - \frac{|p-n|}{n+1}\right) = \lim_{n\to\infty} n\left(1 - \frac{n-p}{n+1}\right) = p+1.\end{aligned}$$

In this case, then, $\mu = p + 1 < 1$, so that Raabe's test indicates divergence.

Consequently, the binomial series $\sum_{n=0}^{\infty}(-1)^n\binom{p}{n}$ diverges for any $p < 0$ at the endpoint $x = -1$. **QED**

Chapter 47

1. If a net point is zero, then the area of the corresponding rectangle may be $0\Delta x_i$, or $2\Delta x_i$, or $1\Delta x_i$, depending on the representative X_i that is chosen. In any case, this area will vanish as $\Delta\chi \to 0$, giving $\int_{-1}^{2}(\operatorname{sgn} x + 1)dx = 4$. The figure shows $[0, x_i]$ with $X_i = 0$.

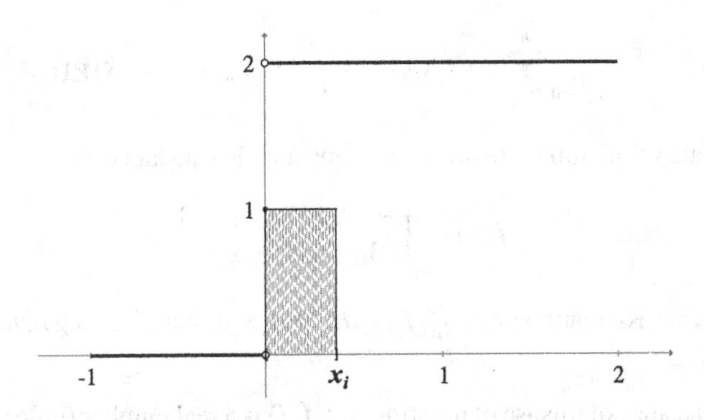

3. First, in any net of $[a, b]$, select all X_i to be rational numbers from $[a, b]$. Since $\mathbb{Q}$ is dense in $[a, b]$, every net will yield

$$\lim_{\Delta\chi\to 0}\sum_{i=1}^{n} f(X_i)\Delta x_i = \lim_{\Delta\chi\to 0}\sum_{i=1}^{n} 1\Delta x_i = b - a.$$

Second, in any net of $[a, b]$, select all Y_i to be irrational numbers from $[a, b]$. Since $\mathbb{I}$ is also dense in $[a, b]$, every net will yield

$$\lim_{\Delta\chi\to 0}\sum_{i=1}^{n} f(Y_i)\Delta x_i = \lim_{\Delta\chi\to 0}\sum_{i=1}^{n} 0\Delta x_i = 0.$$

Therefore, Dirichlet's function is not Riemann integrable.

5. $L(f, \mathcal{N}) \le L(f, \mathcal{N} \cup \mathcal{M}) \le U(f, \mathcal{N} \cup \mathcal{M}) \le U(f, \mathcal{M})$, where we have used Lemma 47.3 twice.

7. For equal subintervals, $x_i = a + \frac{4i}{n} = \frac{4i}{n}, i = 0, 1, 2, \ldots, n$ defines the net. Let's choose the left endpoint of each subinterval as X_i. Thus, $X_i = x_i, i = 0, 1, 2, \ldots, n-1$. Then the Riemann sum

$$\sum_{i=1}^{n} g(X_i)\Delta x_i = \frac{4}{n}\sum_{i=0}^{n-1} g\left(\frac{4i}{n}\right) = \frac{4}{n}\sum_{i=0}^{n-1}\left[\left(\frac{4i}{n}-2\right)^3 - 2\left(\frac{4i}{n}-2\right)+3\right]$$
$$= \frac{4}{n}\sum_{i=0}^{n-1}\left(\frac{64}{n^3}i^3 - \frac{96}{n^2}i^2 + \frac{40}{n}i - 1\right) = \frac{4}{n}(3n-4),$$

using the identities in Exercise 4.2. Finally,

$$\lim_{\Delta\chi\to 0}\sum_{i=1}^{n} g(X_i)\Delta x_i = \lim_{n\to\infty}\frac{4}{n}(3n-4) = 12.$$

9. Since $f(x) \ge 0$ over $[a, b]$, every Riemann sum $\sum_{i=1}^{n} f(X_i)\Delta x_i \ge 0$, regardless of net or representatives. Furthermore,

$$\lim_{\Delta\chi\to 0}\sum_{i=1}^{n} f(X_i)\Delta x_i$$

exists. Therefore,

$$\lim_{\Delta\chi\to 0}\sum_{i=1}^{n} f(X_i)\Delta x_i = \int_a^b f(x)dx \ge 0. \quad \textbf{QED}$$

Use a simple step function to disprove the converse. For instance,

$$f(x) = \begin{cases} -1 & \text{if } 0 \le x \le 1 \\ 3 & \text{if } 1 < x \le 3 \end{cases}.$$

Then considering Riemann sums, $\int_0^3 f(x)dx = 5 > 0$, but f is negative over part of the interval $[0, 3]$.

11. a. False, but because of misuse of notation. $U(f, i)$ is a real number (unless f is unbounded over the subinterval), so $\bigcup_1^n U(f, i)$ is not meaningful. However, the *image* of $[x_{i-1}, x_i]$ is the singleton set $\{U(f, i)\} = u(f([x_{i-1}, x_i]))$, so that $\bigcup_1^n \{U(f, i)\} = \mathcal{R}_{u\circ f}$. The idea is that a step function's range is the union of discrete steps!
    b. False, for some subintervals containing a discontinuity in f. For instance, let $f(x) = \frac{\sin x}{x}$, and let $[x_{i-1}, x_i] = [-1, 1]$. Then $U(f(x), i) = 1$, but no representative X_i gives $f(X_i) = 1$.
    c. False, see Dirichlet's function.
    d. False, $y = \tan x$ is continuous over $[0, \frac{\pi}{2})$ but not integrable there. Continuity over a closed interval is sufficient, as the next chapter will reveal.
    e. True. The lower Riemann integral is the supremum of the collection of lower Riemann sums calculated over all possible nets (an uncountable collection, to be sure).
    f. True.

Chapter 48

1. Since Theorem 48.2 does not depend upon Theorem 48.3, we use $k = -1$ in a quick proof:

$$\int_a^b [f(x) - g(x)]dx = \int_a^b [f(x) + (-1)g(x)]dx = \int_a^b f(x)dx + \int_a^b (-1)g(x)dx$$
$$= \int_a^b f(x)dx + (-1)\int_a^b g(x)dx = \int_a^b f(x)dx - \int_a^b g(x)dx.$$

Of course, the following is also correct. For any $\varepsilon > 0$, there exists a $\delta > 0$ such that for all nets with $\Delta\chi < \delta$, we have

$$\left|\sum_{i=1}^n f(X_i)\Delta x_i - I_f\right| < \frac{\varepsilon}{2}$$

and

$$\left|\sum_{i=1}^n g(X_i)\Delta x_i - I_g\right| = \left|-\sum_{i=1}^n g(X_i)\Delta x_i + I_g\right| < \frac{\varepsilon}{2}.$$

Now,

$$\left|\sum_{i=1}^n [f(X_i) - g(X_i)]\Delta x_i - (I_f - I_g)\right|$$
$$= \left|\sum_{i=1}^n f(X_i)\Delta x_i - I_f + \left(-\sum_{i=1}^n g(X_i)\Delta x_i + I_g\right)\right|.$$

By the triangle inequality,

$$\left|\sum_{i=1}^n [f(X_i) - g(X_i)]\Delta x_i - (I_f - I_g)\right|$$
$$\leq \left|\sum_{i=1}^n f(X_i)\Delta x_i - I_f\right| + \left|-\sum_{i=1}^n g(X_i)\Delta x_i + I_g\right| < \frac{\varepsilon}{2} + \frac{\varepsilon}{2} = \varepsilon.$$

Consequently,

$$\lim_{\Delta\chi\to 0}\sum_{i=1}^n [f(X_i) - g(X_i)] = I_f - I_g,$$

which means that

$$\int_a^b [f(x) - g(x)]dx = \int_a^b f(x)dx - \int_a^b g(x)dx. \quad \textbf{QED}$$

3. Since f is Riemann integrable over $[a, c]$ and $[c, b]$, the contrapositive of Lemma 47.1 guarantees that f is bounded over these intervals, and hence, over $[a, b]$.

5. Suppose that f is monotone decreasing over $[a, b]$. Then $-f$ is monotone increasing there. Thus, Theorem 48.5 applies, and $-f$ is integrable. But then f is integrable over $[a, b]$ by Theorem 48.2.

7. $V_{RMS} = V_P\sqrt{\frac{\int_0^\pi \sin^2 t dt}{\pi}} = \frac{V_P}{\sqrt{2}} \approx 0.7V_p$ or 70% of the peak. At $V_p = 115$ volts, we would actually feel about 81 volts, which is bad enough!

   Since $y = (V_p \sin t)^2$ is continuous over $[0, \pi]$, the mean value theorem for integrals guarantees that $t = \xi$ exists such that

$$V_{RMS}^2 = \frac{V_P^2}{2} = V_p^2 \sin^2 \xi.$$

   In fact, two times exist: $\xi_1 = \arcsin\left(\frac{1}{\sqrt{2}}\right) = \frac{\pi}{4}$ sec and $\xi_2 = \frac{3\pi}{4}$ sec.

9. Exercise 47.8 implies that the signum function is integrable over $[a, b]$. Hence, by Corollary 48.3, $|f(x)|$ is integrable over $[a, b]$.

11. a. Apply Theorem 48.8 with $g(y) = y^n$.
    b. Apply Theorem 48.8 with $g(y) = \sqrt[n]{y}$.
    c. Apply Theorem 48.8 with $g(y) = y^\alpha$, which is continuous over any closed interval where y positive and bounded away from zero.

Chapter 49

1. From the continuity of f at $x = c$, $-f(c) - \varepsilon < -f(x) < -f(c) + \varepsilon$, and by Corollary 48.1,

$$\int_{c-\varepsilon}^{c} (-f(c) - \varepsilon)dx < -\int_{c-\varepsilon}^{c} f(x)dx < \int_{c-\varepsilon}^{c} (-f(c) + \varepsilon)dx,$$

   or

$$(-f(c) - \varepsilon)\varepsilon < -\int_{c-\varepsilon}^{c} f(x)dx < (-f(c) + \varepsilon)\varepsilon$$

   (since $-f(c) \pm \varepsilon$ are constants). Using (b) and dividing through by $-\varepsilon$ (a negative quantity), we arrive at

$$f(c) - \varepsilon < \frac{F(c) - F(c - \varepsilon)}{\varepsilon} < f(c) + \varepsilon.$$

   Hence,

$$\lim_{\varepsilon \to 0+} \left(\frac{F(c) - F(c - \varepsilon)}{\varepsilon}\right) = f(c),$$

   which is the derivative of F from the left, at $x = c$.

3. By the choice of X_k as the right endpoints of the subintervals, and since $y = \frac{\ln x}{x}$ is monotone increasing over $[1, 2]$, we obtain upper Riemann sums. Lemma 47.3 says the rest.

5. Partition $[a, b]$ into n equal subintervals of equal length, so $\Delta x_k = \Delta\chi = \frac{b-a}{n}$. Setting the left endpoint at $x_0 = a$, and the right endpoint at $x_n = b$, the given sum indicates that the representatives are $X_k = a + \frac{k(b-a)}{n}$ for $k = 1, \ldots, n$. Since f is integrable, all Riemann sums $\sum_{k=1}^{n} f(X_k)\Delta x_k$ have the same limit $\int_a^b f(x)dx$ as $\Delta\chi \to 0$. Therefore, in particular,

$$\lim_{n \to \infty}\left(\frac{b-a}{n}\sum_{k=1}^{n} f\left(a + \frac{k(b-a)}{n}\right)\right) = \int_a^b f(x)dx.$$

7. a. Let $y = v(x) = \sqrt{x}$, continuous over $[1, 4]$, and with $v'(x)$ never zero there. $v([1, 4]) = [1, 2] = [c, d]. w(y) = y^2$ over $[1, 2]$. Select $f(y) = \frac{1}{3+y}$, continuous over $[1, 2]$, so that the composition $f(v(x)) = \frac{1}{3+\sqrt{x}}$ is precisely the original integrand.
   b. $\int_a^b f(v(x))dx = \int_1^4 \frac{1}{3+\sqrt{x}}dx$ equals

$$\int_{v(a)}^{v(b)} f(y) \cdot w'(y)dy = \int_1^2 \frac{1}{3+y}2ydy = \int_1^2 \left(2 - \frac{6}{3+y}\right) dy$$
$$= 2 - 6\ln(3+y)|_1^2 = 2 - 6\ln\frac{5}{4}.$$

9. In $\int_{-3}^{-1} xe^{x^2}dx$, the variable is negative. Let $y = v(x) = x^2$ be an inside function with continuous derivative over $[-3, -1]$ such that $v'(x) \neq 0$ there. Also, $v([-3, -1]) = [1, 9] = [c, d]$. Then $w(y) = -\sqrt{y}$ exists and is continuous there. The necessary outside function must be $f(y) = -\sqrt{y}e^y$ in order that the composition $f(v(x))$ produce the original integrand, and it is continuous over $[1, 9]$. Now Theorem 49.4 tells us that

$$\int_a^b f(v(x))dx = \int_{-3}^{-1} -|x|e^{x^2}dx = \int_{-3}^{-1} xe^{x^2}dx$$

is equal to

$$\int_{v(a)}^{v(b)} f(y) \cdot w'(y)dy = \int_9^1 \left(-\sqrt{y}e^y\frac{-1}{2\sqrt{y}}\right) dy = \frac{1}{2}\int_9^1 e^y dy = \frac{1}{2}(e - e^9).$$

This compares with the area in Example 49.5, with opposite sign due to the symmetry of odd functions.

11. a. Let $I = \int \sin^m x \cos^n x dx$. Let $u(x) = \cos^{n-1} x$, $v'(x) = \sin^m x \cos x$, and $u'(x) = -(n-1)\cos^{n-2} x \sin x$, $v(x) = \frac{1}{m+1}\sin^{m+1} x$. The premises of Theorem 49.5 are satisfied, so integration by parts yields

$$I = \frac{1}{m+1}\cos^{n-1} x \sin^{m+1} x + \frac{n-1}{m+1}\int \sin^{m+2} x \cos^{n-2} x dx.$$

In the last integral,

$$\int \sin^{m+2} x \cos^{n-2} x dx = \int \sin^m x \cos^{n-2} x(1 - \cos^2 x)dx = \int \sin^m x \cos^{n-2} x dx - I.$$

Substituting this back into the previous equation, we have

$$I = \frac{1}{m+1}\cos^{n-1} x \sin^{m+1} x + \frac{n-1}{m+1}\left(\int \sin^m x \cos^{n-2} x dx - I\right).$$

Solving for I gives us the desired reduction formula.

$$I = \frac{\sin^{m+1} x \cos^{n-1} x}{m+n} + \frac{n-1}{m+n}\int \sin^m x \cos^{n-2} x dx,$$

where $m + n \neq 0$.

b. With $m = -2$ and $n = 3$, the reduction formula yields

$$\int_{\pi/4}^{\pi/2} \frac{\cos^3 x}{\sin^2 x} dx = \left. \frac{\cos^2 x}{\sin x} \right|_{\pi/4}^{\pi/2} + 2 \int_{\pi/4}^{\pi/2} \frac{\cos x}{\sin^2 x} dx,$$

which evaluates to $\frac{3}{2}\sqrt{2} - 2$.

Chapter 50

1. $\int_0^1 \ln x dx$ is improper in the sense of the first subsection above. Consider intervals $[\varepsilon, 1]$:

$$\lim_{\varepsilon \to 0+} \int_\varepsilon^1 \ln x dx = \lim_{\varepsilon \to 0+} (-1 - \varepsilon \ln \varepsilon + \varepsilon) = -1.$$

Hence, the integral converges to -1.

3. Over $[0, 1 - \varepsilon]$, $u(x) = \sin^{-1} x$ and $v(x) = x$ have continuous derivatives. Applying integration by parts,

$$\int_0^{1-\varepsilon} \sin^{-1} x dx = (1-\varepsilon)\sin^{-1}(1-\varepsilon) - 0 \sin^{-1} 0 - \int_0^{1-\varepsilon} \frac{x}{\sqrt{1-x^2}} dx$$

$$= (1-\varepsilon)\sin^{-1}(1-\varepsilon) + \sqrt{1-(1-\varepsilon)^2} - 1.$$

As the last step,

$$\lim_{\varepsilon \to 0+} ((1-\varepsilon)\sin^{-1}(1-\varepsilon) + \sqrt{1-(1-\varepsilon)^2} - 1) = \frac{\pi}{2} - 1.$$

Hence, $\frac{\pi}{2} - 1 = \int_0^1 \sin^{-1} x dx$, in agreement with the other two methods in Example 49.8. (Interestingly, the original $\int_0^1 \sin^{-1} x dx$ is not improper at all, but this method used the improper $\int_0^1 \frac{x}{\sqrt{1-x^2}} dx$.)

5. $\int_{-\infty}^{\infty} \frac{1}{(x-1)^2} dx$ is improper in the senses of both subsections above. Over any interval $[c, 1 - \varepsilon]$, we consider

$$\lim_{\varepsilon \to 0+} \int_c^{1-\varepsilon} \frac{1}{(x-1)^2} dx = \lim_{\varepsilon \to 0+} \frac{\varepsilon + c - 1}{\varepsilon(c-1)},$$

and this fails to exist. Hence, $\int_{-\infty}^{\infty} \frac{1}{(x-1)^2} dx$ diverges.

7. $\int_1^{\infty} \frac{1}{x\sqrt{x-1}} dx$ is improper in the senses of both subsections above. We first consider intervals

$$[1+\varepsilon, c]: \lim_{\varepsilon \to 0+} \int_{1+\varepsilon}^{c} \frac{1}{x\sqrt{x-1}} dx$$

$$= \lim_{\varepsilon \to 0+} (2\tan^{-1}\sqrt{c-1} - 2\tan^{-1}\sqrt{\varepsilon}) = 2\tan^{-1}\sqrt{c-1}.$$

Therefore,

$$\int_1^c \frac{1}{x\sqrt{x-1}} dx = 2\tan^{-1}\sqrt{c-1}$$

converges. Next,

$$\lim_{c\to\infty}\int_1^c \frac{1}{x\sqrt{x-1}}dx = \lim_{c\to\infty}(2\tan^{-1}\sqrt{c-1}) = \pi,$$

which means that

$$\int_1^\infty \frac{1}{x\sqrt{x-1}}dx$$

converges to π.

9. Given that $\int_{-\infty}^{\infty} f(x)dx$ exists, then $\lim_{a\to-\infty}\int_a^c f(x)dx$, and $\lim_{b\to\infty}\int_c^b f(x)dx$ exist independently. Then *a fortiori*, the second integral exists when we set $b = a$. Now,

$$\int_{-\infty}^{\infty} f(x)dx = \lim_{a\to-\infty}\int_a^c f(x)dx + \lim_{a\to\infty}\int_c^a f(x)dx$$

$$= \lim_{a\to\infty}\int_{-a}^{a} f(x)dx = \wp\int_{-\infty}^{\infty} f(x)dx. \quad \textbf{QED}$$

The counterexample $\wp\int_{-\infty}^{\infty}\sin x dx = 0$ while $\int_{-\infty}^{\infty}\sin x dx$ diverges, shows the converse to be false.

Chapter 51

1. (1) $(w^2+x^2)(y^2+z^2) = w^2y^2 + w^2z^2 + x^2y^2 + x^2z^2$ and $(wy+xz)^2 = w^2y^2 + 2wyxz + x^2z^2$, so we must show that $w^2z^2 + x^2y^2 \geq 2wyxz$. Subtracting, we arrive at $(wz - xy)^2 \geq 0$, which is clearly true for all real values of the variables.
(2) Use $n = 2$, $a_1 = w$, $a_2 = x$, $b_1 = y$, and $b_2 = z$.

3. Continuing,

$$\sum_{i=1}^{n}(a_i+b_i)^2 = \sum_{i=1}^{n}a_i^2 + 2\sum_{i=1}^{n}a_ib_i + \sum_{i=1}^{n}b_i^2 \leq \sum_{i=1}^{n}a_i^2 + 2\sqrt{\sum_{i=1}^{n}a_i^2}\sqrt{\sum_{i=1}^{n}b_i^2} + \sum_{i=1}^{n}b_i^2$$

$$= \left(\sqrt{\sum_{i=1}^{n}a_i^2} + \sqrt{\sum_{i=1}^{n}b_i^2}\right)^2.$$

Therefore,

$$\sqrt{\sum_{i=1}^{n}(a_i+b_i)^2} \leq \sqrt{\sum_{i=1}^{n}a_i^2} + \sqrt{\sum_{i=1}^{n}b_i^2}.$$

All in all, superb algebra by Minkowski.

5. In the Cauchy-Schwarz inequality, let every $a_i = 1$, and let $b_i = x_i$. This yields

$$\left(\sum_{i=1}^{n}x_i\right)^2 \leq (n)\left(\sum_{i=1}^{n}x_i^2\right),$$

guaranteeing that the numerator of s is never negative.

7. a. For each i, sum only the products $a_i b_j$ where subscripts $i \neq j$. The particular case gives

$$\sum_{i=1}^{3}\sum_{j\neq i} a_i a_j = (a_1a_2 + a_1a_3) + (a_2a_1 + a_2a_3) + (a_3a_1 + a_3a_2)$$

$$= 2(a_1a_2 + a_1a_3 + a_2a_3) = 2\sum_{i=1}^{3}\sum_{j>i} a_i a_j = -28.$$

The latter sum equals half the former one.

b. First, note that

$$\sum_{i=1}^{n}\sum_{j=1}^{n} a_i a_j = \sum_{i=1}^{n} a_i \sum_{j=1}^{n} a_j = \left(\sum_{i=1}^{n} a_i\right)^2.$$

So, prove that

$$\sum_{i=1}^{n} a_i^2 + \sum_{i=1}^{n}\sum_{j\neq i} a_i a_j = \left(\sum_{i=1}^{n} a_i\right)^2,$$

which will be the induction hypothesis in what follows. This is actually the formula for the square of a polynomial with n terms! The anchor case $n = 2$ appears as $a_1^2 + a_2^2 + a_1a_2 + a_2a_1 = (a_1 + a_2)^2$, which may look quirky, but is true. The case $n + 1$ is

$$\sum_{i=1}^{n+1} a_i^2 + \sum_{i=1}^{n+1}\sum_{j\neq i} a_i a_j = \left(\sum_{i=1}^{n+1} a_i\right)^2 = \left(a_{n+1} + \sum_{i=1}^{n} a_i\right)^2 = a_{n+1}^2 + 2a_{n+1}\sum_{i=1}^{n} a_i + \left(\sum_{i=1}^{n} a_i\right)^2.$$

Substitute, using the induction hypothesis:

$$\sum_{i=1}^{n+1} a_i^2 + \sum_{i=1}^{n+1}\sum_{j\neq i} a_i a_j = a_{n+1}^2 + 2a_{n+1}\sum_{i=1}^{n} a_i + \sum_{i=1}^{n} a_i^2 + \sum_{i=1}^{n}\sum_{j\neq i} a_i a_j.$$

This simplifies to

$$\sum_{i=1}^{n+1}\sum_{j\neq i} a_i a_j = 2a_{n+1}\sum_{i=1}^{n} a_i + \sum_{i=1}^{n}\sum_{j\neq i} a_i a_j. \qquad (2)$$

To see that (2) is true, the next steps may seem tricky; if so, add them to your bag of mathematical tricks!

$$\sum_{i=1}^{n+1}\sum_{j\neq i} a_i a_j = \sum_{i=1}^{n+1}\sum_{j<i} a_i a_j + \sum_{i=1}^{n+1}\sum_{j>i} a_i a_j = \left(\sum_{j=1}^{n} a_{n+1}a_j + \sum_{i=1}^{n}\sum_{j<i} a_i a_j\right)$$

$$+ \left(\sum_{j=1}^{n} a_{n+1}a_j + \sum_{i=1}^{n}\sum_{j>i} a_i a_j\right) = 2a_{n+1}\sum_{j=1}^{n} a_j + \sum_{i=1}^{n}\sum_{j\neq i} a_i a_j.$$

This is equivalent to the right side of (2). Starting with (2), the steps can be reversed until we arrive at

$$\sum_{i=1}^{n+1} a_i^2 + \sum_{i=1}^{n+1} \sum_{j \neq i} a_i a_j = \left(\sum_{i=1}^{n+1} a_i \right)^2 ,$$

which proves the case $n + 1$. By induction, we have proved the identity for expanding the square of a polynomial. **QED**

9. By the extended Cauchy-Schwarz inequality, $\sum_{i=1}^{\infty} a_i b_i$ converges. Thus,

$$\sum_{i=1}^{\infty} a_i^2 + \sum_{i=1}^{\infty} 2a_i b_i + \sum_{i=1}^{\infty} b_i^2 = \sum_{i=1}^{\infty} (a_i^2 + 2a_i b_i + b_i^2) = \sum_{i=1}^{\infty} (a_i + b_i)^2$$

converges. Use the extended Cauchy-Schwarz inequality again:

$$\sum_{i=1}^{\infty} a_i^2 + \sum_{i=1}^{\infty} 2a_i b_i + \sum_{i=1}^{\infty} b_i^2 \leq \sum_{i=1}^{\infty} a_i^2 + 2 \sqrt{\sum_{i=1}^{\infty} a_i^2} \sqrt{\sum_{i=1}^{\infty} b_i^2} + \sum_{i=1}^{\infty} b_i^2 .$$

Hence,

$$\sum_{i=1}^{\infty} (a_i + b_i)^2 \leq \sum_{i=1}^{\infty} a_i^2 + 2 \sqrt{\sum_{i=1}^{\infty} a_i^2} \sqrt{\sum_{i=1}^{\infty} b_i^2} + \sum_{i=1}^{\infty} b_i^2 = \left(\sqrt{\sum_{i=1}^{\infty} a_i^2} + \sqrt{\sum_{i=1}^{\infty} b_i^2} \right)^2 .$$

The extended Minkowski inequality follows. **QED**

Chapter 52

1. (a) non-negativity: $d_1(\mathbf{x}, \mathbf{y}) \geq 0$ is clear. (b) identity: If $d_1(\mathbf{x}, \mathbf{y}) = \sum_{k=1}^{n} |x_k - y_k| = 0$, then each term $|x_k - y_k| = 0$, so $x_k = y_k$ for $k = 1, 2, \ldots, n$. Conversely, if $\mathbf{x} = \mathbf{y}$, then $d_1(\mathbf{x}, \mathbf{x}) = 0$. (c) symmetry: $d_1(\mathbf{x}, \mathbf{y}) = d_1(\mathbf{y}, \mathbf{x})$ is clear. (d) the triangle inequality: $d_1(\mathbf{x}, \mathbf{y}) \leq d_1(\mathbf{x}, \mathbf{z}) + d_1(\mathbf{z}, \mathbf{y})$. To use induction, take as the anchor case $n = 1$, the real line: $|x - y| \leq |x - z| + |z - y|$. This is just a basic application of the triangle inequality. Next, assume $\langle \mathbb{R}^n, d_1 \rangle$ is a metric space. We must prove that

$$\sum_{k=1}^{n+1} |x_k - y_k| \leq \sum_{k=1}^{n+1} |x_k - z_k| + \sum_{k=1}^{n+1} |z_k - y_k| .$$

This expands as

$$|x_{n+1} - y_{n+1}| + \sum_{k=1}^{n} |x_k - y_k| \leq |x_{n+1} - z_{n+1}|$$

$$+ \sum_{k=1}^{n} |x_k - z_k| + |z_{n+1} - y_{n+1}| + \sum_{k=1}^{n} |z_k - y_k| ,$$

which is true by the induction hypothesis and the triangle inequality. **QED**

3. a. (a) is clear. (b): if $\rho(x, y) = |\ln(\frac{x}{y})| = 0$, then $\frac{x}{y} = 1$ and $x = y$. The converse is clear. (c): $\rho(x, y) = |\ln(\frac{x}{y})| = |\ln(\frac{y}{x})| = \rho(y, x)$. (d): $|\ln(\frac{x}{y})| = |\ln(\frac{x}{z}\frac{z}{y})| = |\ln(\frac{x}{z}) + \ln(\frac{z}{y})| \leq |\ln(\frac{x}{z})| + |\ln(\frac{z}{y})|$.

   b. If $a, b \in S_\alpha$, then $\rho(a, b) = 0$, so that for instance, all the fractions in $\{2, \frac{6}{3}, \frac{1}{1/2}, \ldots\}$ have zero distance among themselves. They name the same point.

   Because of the definition of $\rho(S_\alpha, S_\beta)$, the space $\mathbb{S}$ "inherits" the metric property of $\langle \mathbb{R}^+, \rho \rangle$.

5. For any $a, b \in \mathbb{P}$, $\rho(a, b)$ must follow the four properties of a metric, since $a, b \in \mathbb{M}$. Thus, $\langle \mathbb{P}, \rho \rangle$ is automatically a metric space. **QED**

7. (b) is in irreparable trouble. Suppose $S = (1, 2)$ and $T = [2, 3]$. Then $d(S, T) = \inf\{|s - t| : s \in (1, 2) \text{ and } t \in [2, 3]\} = 0$, but $S \neq T$. Thus, the function fails as a metric for space $\mathbb{I}$.

9. We must show that $d_r(\mathbf{x}, \mathbf{y}) \leq d_r(\mathbf{x}, \mathbf{z}) + d_r(\mathbf{z}, \mathbf{y})$ in $\mathbb{R}^n$, that is,

$$\left(\sum_{k=1}^{n} |x_k - y_k|^r\right)^{1/r} \leq \left(\sum_{k=1}^{n} |x_k - z_k|^r\right)^{1/r} + \left(\sum_{k=1}^{n} |z_k - y_k|^r\right)^{1/r}.$$

In the generalized Minkowski inequality, let $a_k = z_k - y_k$ and $b_k = x_k - z_k$. (d) immediately follows!

11. $a\rho(x, y) + b\tau(x, y) = (a\rho + b\tau)(x, y)$ means that a linear combination with positive coefficients a, b of two metrics on $\mathbb{M}$ creates a new metric $a\rho + b\tau$ for $\mathbb{M}$. The proof follows.

   (a): $(a\rho + b\tau)(x, y) = a\rho(x, y) + b\tau(x, y) \geq 0$ since a and b are positive. (b): if $(a\rho + b\tau)(x, y) = a\rho(x, y) + b\tau(x, y) = 0$ then (since a and b are positive) $a\rho(x, y) = 0$ and $b\tau(x, y) = 0$, and $\rho(x, y) = 0$, implying that $x = y$. Conversely, if $x = y$, then $0 = a\rho(x, x) + b\tau(x, x) = (a\rho + b\tau)(x, x)$. (c) is clear. (d): $(a\rho + b\tau)(x, y) \leq (a\rho + b\tau)(x, z) + (a\rho + b\tau)(z, y)$ is equivalent to $a\rho(x, y) + b\tau(x, y) \leq a\rho(x, z) + b\tau(x, z) + a\rho(z, y) + b\tau(z, y)$. This inequality follows when (d) is applied to each of ρ and τ. Thus, $\langle \mathbb{M}, a\rho + b\tau \rangle$ is a new metric space.

13. The length of a vector (a sequence) in metric space $\langle \ell^2, \rho \rangle$ is given by

$$\rho(\langle 0_k \rangle_1^\infty, \langle a_k \rangle_1^\infty) = \sqrt{\sum_{k=1}^{\infty} a_k^2},$$

which is an extension of the Pythagorean theorem to infinite dimensions! Hence the unit ball is the set of all vectors $\langle a_k \rangle_1^\infty \in \ell^2$ with length one: $\sqrt{\sum_{k=1}^{\infty} a_k^2} = 1$.

15a,b. In $\langle \mathbb{R}^2, d_3 \rangle$, the unit circle has equation $\sqrt[3]{|x|^3 + |y|^3} = 1$, or $|x|^3 + |y|^3 = 1$. In $\langle \mathbb{R}^2, d_6 \rangle$, the unit circle is given by $\sqrt[6]{x^6 + y^6} = 1$, or $x^6 + y^6 = 1$. These circles are graphed parametrically below, along with a Lamé curve.

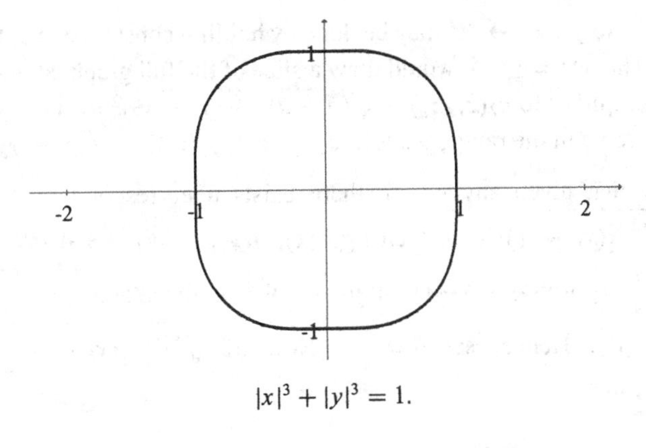

$|x|^3 + |y|^3 = 1.$

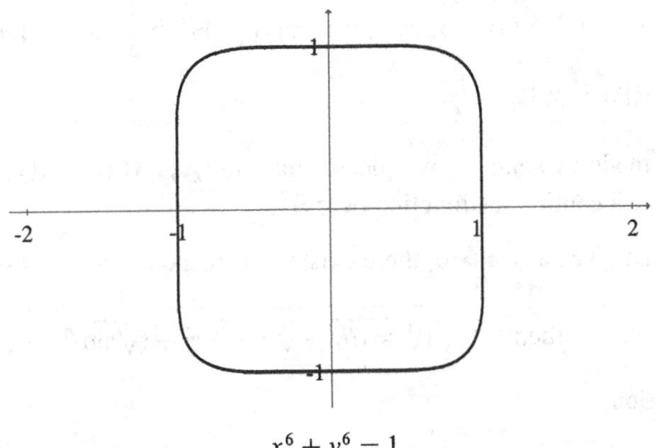

$x^6 + y^6 = 1.$

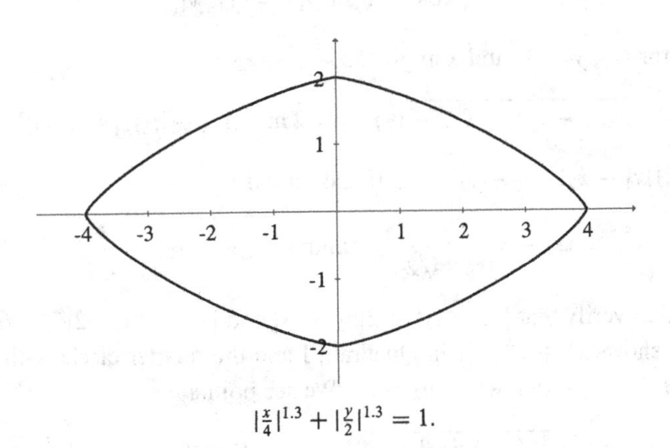

$|\frac{x}{4}|^{1.3} + |\frac{y}{2}|^{1.3} = 1.$

Chapter 53

1. Let the arguer select $\varepsilon = \frac{1}{2}$. In order to have $\rho(a_n, L) < \frac{1}{2}$ for all $n \geq N_{1/2}$, it is necessary that $a_n = L$ for all $n \geq N_{1/2}$. Of course, N_ε can be any positive number, as large as you wish.

3. A function $f : \mathbb{C} \to \mathbb{C}$ can never be fully graphed, for it needs four dimensions, two for the domain point $\langle x, y \rangle$, and two for the range point $f(\langle x, y \rangle) = \langle \gamma, \delta \rangle$.

The restriction to $f : \mathbb{R}^1 \to \mathbb{R}^2$ may be done by holding constant one of the coordinates of $\langle x, y\rangle$. Thus, $f(\langle a, y\rangle) = \langle \gamma, \delta\rangle$ would draw a slice of the full graph, along the plane $x = a$. The d_2 metric simplifies to $d_2(z_1, z_2) = \sqrt{(a-a)^2 + (y_1 - y_2)^2} = |y_1 - y_2|$ in the domain, but remains the same in the range, namely, $d_2(f(z_1), f(z_2)) = \sqrt{(\gamma_1 - \gamma_2)^2 + (\delta_1 - \delta_2)^2}$.

5. We must show that given any $\varepsilon > 0$, there exists a corresponding $\delta > 0$ such that if $d_2(\mathbf{x}, \mathbf{y}) = \sqrt{\sum_{k=1}^{3}(x_k - y_k)^2} < \delta$, then $d_1(A(\mathbf{x}), A(\mathbf{y})) = |m_1 - m_2| < \varepsilon$. In the range, we must show that $\frac{1}{3}|(x_1 + x_2 + x_3) - (y_1 + y_2 + y_3)| < \varepsilon$. In the domain, $\sqrt{\sum_{k=1}^{3}(x_k - y_k)^2} \le \sqrt{3}\max\{|x_k - y_k|\}_1^3$. Hence, set $\delta = \frac{\varepsilon}{\sqrt{3}}$. Now, if $\sqrt{\sum_{k=1}^{3}(x_k - y_k)^2} \le \sqrt{3}\max\{|x_k - y_k|\}_1^3 < \delta = \frac{\varepsilon}{\sqrt{3}}$, then

$$\frac{1}{3}|(x_1 + x_2 + x_3) - (y_1 + y_2 + y_3)| \le \frac{1}{3}|x_1 - y_1| + \frac{1}{3}|x_2 - y_2| + \frac{1}{3}|x_3 - y_3|$$
$$\le \max\{|x_k - y_k|\}_1^3 < \frac{\varepsilon}{3}.$$

Since $\frac{\varepsilon}{3}$ can be made as small as we please, then $\lim_{\mathbf{x}\to\mathbf{y}} A(\mathbf{x}) = A(\mathbf{y})$, meaning that the arithmetic mean is a continuous function over $\mathbb{R}^3$.

7. We must show that given any $\varepsilon > 0$, there exists a corresponding $\delta > 0$ such that if

$$|\theta_1 - \theta_2| < \delta, \qquad \text{then} \qquad \sqrt{(\sqrt[3]{\cos\theta_1} - \sqrt[3]{\cos\theta_2})^2 + (\sqrt[3]{\sin\theta_1} - \sqrt[3]{\sin\theta_2})^2} < \varepsilon.$$

For simpler notation, use

$$\langle \sqrt[3]{\cos\theta_1}, \sqrt[3]{\sin\theta_1}\rangle = \langle x_1, y_1\rangle,$$

and likewise for $\langle x_2, y_2\rangle$. As in Example 53.4,

$$\sqrt{(x_1 - x_2)^2 + (y_1 - y_2)^2} \le \sqrt{2}\max\{|x_1 - x_2|, |y_1 - y_2|\}.$$

Now, $\sqrt{2}\max\{|x_1 - x_2|, |y_1 - y_2|\} < \varepsilon$ if and only if

$$|x_1 - x_2| < \frac{\varepsilon}{\sqrt{2}} \qquad \text{and} \qquad |y_1 - y_2| < \frac{\varepsilon}{\sqrt{2}}.$$

Next, we need to verify that $|x_1 - x_2| < 2|\theta_1 - \theta_2|$ and $|y_1 - y_2| < 2|\theta_1 - \theta_2|$.

The figure shows $x^6 + y^6 = 1$ in Quadrant I and the quarter circle with radius 2. Thus, arc length $AB = 2(\theta_2 - \theta_1)$, where $\theta_2 > \theta_1$. We see points

$$C = \langle x_1, y_1\rangle = \langle \sqrt[3]{\cos\theta_1}, \sqrt[3]{\sin\theta_1}\rangle \qquad \text{and} \qquad D = \langle x_2, y_2\rangle = \langle \sqrt[3]{\cos\theta_2}, \sqrt[3]{\sin\theta_2}\rangle$$

on $x^6 + y^6 = 1$. The length of chord

$$CD = \sqrt{(\sqrt[3]{\cos\theta_1} - \sqrt[3]{\cos\theta_2})^2 + (\sqrt[3]{\sin\theta_1} - \sqrt[3]{\sin\theta_2})^2} < AB.$$

Thus, $EC = |x_1 - x_2| < 2|\theta_1 - \theta_2|$, and $ED = |y_1 - y_2| < 2|\theta_1 - \theta_2|$ also. Hence $|x_1 - x_2| < 2|\theta_1 - \theta_2|$ and $|x_1 - x_2| < \frac{\varepsilon}{\sqrt{2}}$ are simultaneously satisfied when we have $|\theta_1 - \theta_2| < \frac{\varepsilon}{2\sqrt{2}}$. In the same way, if $|\theta_1 - \theta_2| < \frac{\varepsilon}{2\sqrt{2}}$, then $|y_1 - y_2| < 2|\theta_1 - \theta_2|$ and $|y_1 - y_2| < \frac{\varepsilon}{\sqrt{2}}$.

So, set $\delta = \frac{\varepsilon}{2\sqrt{2}}$. If $|\theta_1 - \theta_2| < \delta$, then

$$\sqrt{(\sqrt[3]{\cos\theta_1} - \sqrt[3]{\cos\theta_2})^2 + (\sqrt[3]{\sin\theta_1} - \sqrt[3]{\sin\theta_2})^2}$$
$$= \sqrt{(x_1 - x_2)^2 + (y_1 - y_2)^2} \leq \sqrt{2}\max\{|x_1 - x_2|, |y_1 - y_2|\} < 2\sqrt{2}|\theta_1 - \theta_2| < \varepsilon.$$

This proves that the curve is continuous in Quadrant I.

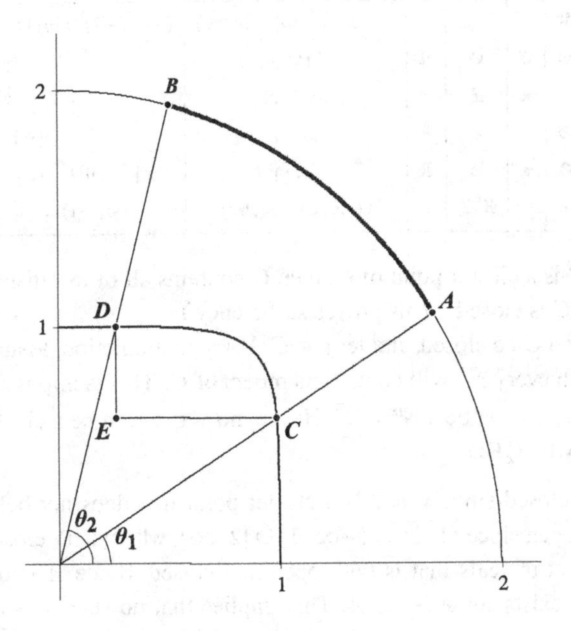

9. $\mathbb{D}$ is the disk bounded by the unit circle centered at $\langle 0, 0\rangle$, excluding the circle itself. $\langle \mathbb{D}, d_1\rangle$ is a metric space since $\mathbb{D} \subset \mathbb{R}$ (see Exercise 52.5). It is not complete, for consider a sequence of points such as $\langle \mathbf{x}_n\rangle_1^\infty = \langle 1 - \frac{1}{n}, 0\rangle_1^\infty \subset \mathbb{D}$. Then $\lim_{n\to\infty} \mathbf{x}_n = \langle 1, 0\rangle \notin \mathbb{D}$.

   The disk $\bar{\mathbb{D}} = \{\langle x, y\rangle : x^2 + y^2 \leq 1\}$ contains its boundary. Now every Cauchy sequence in $\bar{\mathbb{D}}$ will converge to a limit $L \in \bar{\mathbb{D}}$. Hence, $\langle \mathbb{D}, d_1\rangle$ is complete.

Chapter 54

1. A set A is open in $\mathbb{R}$ if and only if for any $x \in A$, there is a neighborhood $\mathbf{N}_x^\varepsilon \subset A$. When A is $\varnothing$, consider the contrapositive statement: for any $\mathbf{N}_x^\varepsilon \not\subset \varnothing$, then $x \notin \varnothing$. This is perfectly true. Hence, $\varnothing$ is open. Regarding $\mathbb{R}$, for each $x \in \mathbb{R}$, *every* neighborhood $\mathbf{N}_x^\varepsilon$ must be a subset of $\mathbb{R}$.

3. *First proof.* The complement of $\{a_1, a_2, \ldots, a_n\}$ is $(-\infty, a_1) \cup (a_1, a_2) \cup \cdots \cup (a_{n-1}, a_n) \cup (a_n, \infty)$, which, being the union of open sets, is open. Therefore, $\{a_1, a_2, \ldots, a_n\}$ is closed. *Second proof.* Each member a_i of the set is an *isolated point*, so that it has no cluster points associated with it. Thus, $c(\{a_1, a_2, \ldots, a_n\}) = \varnothing$, and $\varnothing \subset \{a_1, a_2, \ldots, a_n\}$, fulfilling Theorem 54.2. **QED**

   $\mathbb{N}$ and $\mathbb{Z}$ are infinite sets of points that are closed.
   $\{1, \frac{1}{2}, \frac{1}{3}, \ldots, \frac{1}{n}, \ldots\}$ is infinite and not closed.

5. To prove (1), we show that $\overline{C_1 \cup C_2 \cup \cdots \cup C_n}$ is open. By the extended De Morgan's laws (Theorem 5.3), $\overline{C_1 \cup C_2 \cup \cdots \cup C_n} = \overline{C_1} \cap \overline{C_2} \cap \cdots \cap \overline{C_n}$, which is open by Theorem 54.1. To prove (2), use Theorem 5.3 again: $\overline{\bigcap_{\alpha \in S} C_\alpha} = \bigcup_{\alpha \in S} \overline{C_\alpha}$, which is open by Theorem 54.1. **QED**

7.

SET	$\varnothing$	$\mathbb{Q}$	$\mathbb{Z}$	$\mathbb{R}$	$[a, b) \cup (b, c]$	$[-1, 0) \cup \{\frac{1}{2}, \frac{2}{3}, \frac{3}{4}, \ldots, \frac{n}{n+1}, \ldots\}$
Complement	$\mathbb{R}$	$\mathbb{I}$	$\mathbb{R}\backslash\mathbb{Z}$	$\varnothing$	$(-\infty, a) \cup \{b\} \cup (c, \infty)$	$(-\infty, -1) \cup \{0\} \cup \bigcup_0^\infty (\frac{n}{n+1}, \frac{n+1}{n+2}) \cup [1, \infty)$
Interior	$\varnothing$	$\varnothing$	$\varnothing$	$\mathbb{R}$	$(a, b) \cup (b, c)$	$(-1, 0)$
Boundary	$\varnothing$	$\mathbb{R}$	$\mathbb{Z}$	$\varnothing$	$\{a, b, c\}$	$\{-1, 0, 1, \frac{1}{2}, \frac{2}{3}, \frac{3}{4}, \ldots, \frac{n}{n+1}, \ldots\}$
Cluster Points	$\varnothing$	$\mathbb{R}$	$\varnothing$	$\mathbb{R}$	$[a, c]$	$[-1, 0] \cup \{1\}$
Closure	$\varnothing$	$\mathbb{R}$	$\mathbb{Z}$	$\mathbb{R}$	$[a, c]$	$[-1, 0] \cup \{1, \frac{1}{2}, \frac{2}{3}, \frac{3}{4}, \ldots, \frac{n}{n+1}, \ldots\}$
Exterior	$\mathbb{R}$	$\varnothing$	$\mathbb{R}\backslash\mathbb{Z}$	$\varnothing$	$(-\infty, a) \cup (c, \infty)$	$(-\infty, -1) \cup \bigcup_0^\infty (\frac{n}{n+1}, \frac{n+1}{n+2}) \cup (1, \infty)$

9. If no point in $\overline{C}$ is a cluster point of C, then C contains all of its cluster points (if any). By Theorem 54.2, C is closed. (This proves sufficiency.)

   Conversely, let C be closed, and let $y \in \overline{C}$. For a contradiction, assume that y is a cluster point of C. Then every $\mathbf{N}_y^\varepsilon$ will contain members of C. This is impossible since $\overline{C}$ is open, implying that there must be a $\mathbf{N}_y^\alpha \subset \overline{C}$. Hence, no $y \in \overline{C}$ can be a cluster point of C. (This proves necessity.) **QED**

11. a. $[1, 2)$ is not closed since $x = 2$ is a cluster point that does not belong to the set. And $[1, 2)$ is not open since $\overline{[1, 2)} = (-\infty, 1) \cup [2, \infty)$, which isn't closed.
    b. Let A be a set of reals that is both open and closed. Since A is open, then for every $x \in A$, there exists some $\mathbf{N}_x^\varepsilon \subset A$. This implies that no such x is a boundary point of A. Next, $\overline{A}$ is open according to the premises. Thus, for every $y \in \overline{A}$, there exists some $\mathbf{N}_y^\varepsilon \subset \overline{A}$, so such a y is not a boundary point of A. Therefore, $b(A)$ must be empty. But the only sets with empty boundary are $\varnothing$ and $\mathbb{R}$. It follows that A is either $\varnothing$ or $\mathbb{R}$. We conclude that $\varnothing$ and $\mathbb{R}$ are the only sets that are both open and closed. **QED**
    c. $\overline{(\widetilde{A})}$ must be an open set, and $\widetilde{(\overline{A})}$ must be closed. Being equal implies that they are both open and closed, which implies that either $\overline{(\widetilde{A})} = \widetilde{(\overline{A})} = \varnothing$ or $\overline{(\widetilde{A})} = \widetilde{(\overline{A})} = \mathbb{R}$ by b. If it is the case that $\overline{(\widetilde{A})} = \mathbb{R}$, then $\widetilde{A} = \varnothing$, and then $A = \varnothing$ (verify the last implication). If $\widetilde{(\overline{A})} = \varnothing$, then $\overline{A} = \varnothing$ again, so $A = \mathbb{R}$. Conversely, it's easy to show that $\overline{(\widetilde{\varnothing})} = \widetilde{(\overline{\varnothing})}$ and $\overline{(\widetilde{\mathbb{R}})} = \widetilde{(\overline{\mathbb{R}})}$.

13. a. First, $i(A) \cup b(A) = A \cup c(A)$ by Theorem 54.4, and $A \cup c(A) = A$ when A is closed. Second, assume on the contrary, that $i(A) \cap b(A) \neq \varnothing$. Then there exists an $x \in A$ such that simultaneously, $\mathbf{N}_x^\varepsilon \subset A$, $\mathbf{N}_x^\varepsilon \cap A \neq \varnothing$, and $\mathbf{N}_x^\varepsilon \cap \overline{A} \neq \varnothing$. But the first and third conditions are contradictory: exactly one can be true. Thus, $i(A) \cap b(A) = \varnothing$. It follows that $i(A)$ and $b(A)$ form a partition of a closed set. **QED**

       If A is closed, then we have just proved that $A = i(A) \cup b(A)$. If A is not closed, then $A \subset A \cup c(A) = i(A) \cup b(A)$.
    b. Substituting for $e(A)$ and using De Morgan's laws, $\overline{i(A)} \cap \overline{e(A)} = \overline{i(A)} \cap (b(A) \cup A \cup c(A)) = \overline{i(A)} \cap (b(A) \cup c(A))$, because $A \subseteq b(A) \cup c(A)$. Now, $\overline{i(A)} \cap (b(A) \cup c(A)) = (\overline{i(A)} \cap b(A)) \cup (\overline{i(A)} \cap c(A))$. Since the interior of A is a subset of A's cluster point

set, then $\overline{i(A)} \cap c(A) = \varnothing$. Thus, $\overline{i(A)} \cap \overline{e(A)} = \overline{i(A)} \cap b(A)$. But no interior point can be a boundary point (see a), therefore $b(A)$ must be a subset of $\overline{i(A)}$, meaning that $\overline{i(A)} \cap b(A) = b(A)$. In conclusion, $\overline{i(A)} \cap \overline{e(A)} = \overline{i(A)} \cap b(A)$, and $\overline{i(A)} \cap b(A) = b(A)$, giving us $\overline{i(A)} \cap \overline{e(A)} = b(A)$. **QED**

c. We have just shown that $b(A) = \overline{i(A)} \cap \overline{e(A)}$. Exercise 54.6h shows that $e(A)$ is open, and so is $i(A)$. Thus, $b(A)$ is the intersection of two closed sets, which yields a closed set. **QED**

Chapter 55

1. Let $x = 0.t_1t_2\ldots t_n$, and multiply both sides by 3^n: $3^n x = t_1t_2\ldots t_n = a$, a natural number. Then $x = \frac{a}{3^n}$.

3. We multiply by four:

$(0.002200220022\ldots_3)(11_3) =$

$$\begin{array}{r} 0.\,0\,0\,2\,2\,0\,0\,2\,2\cdots \\ \underline{1\,1} \\ 0\,0\,2\,2\,0\,0\,2\,2 \\ \underline{0\,0\,2\,2\,0\,0\,2\,2} \\ 0.\,1\,0\,1\,2\,1\,0\,1\,2\cdots \end{array} = \frac{2}{5}.$$

5. If the expansion is terminating, then $k = 0.k_1k_2\ldots k_N$ (in base 3) is free of the digit 1, and yields the infinite sequence $\langle 0.k_1,\ 0.k_1k_2,\ \ldots,\ 0.k_1k_2\ldots k_N,\ 0.k_1k_2\ldots k_N, \ldots\rangle$. This sequence converges to k immediately at the term $t_N = 0.k_1k_2\ldots k_N$. Furthermore, every term t_n is 1-free, meaning that $\langle t_n\rangle_1^\infty \subset \mathbb{K}$. Exercise 54.4 may still be used since there is no difficulty about the limit not belonging to $\mathbb{K}$ (we know it does). It follows that $k \in c(\mathbb{K})$ once again.

7. It's a fine proof. To fill in details, let $q \in \mathbb{N}$, and let the neighborhoods of the cover be $\mathrm{N}_{1/n}^{1/q^n}$. Then $M_q = \sum_{n=1}^{\infty} \frac{2}{q^n} = \frac{2}{q-1}$, and as $q \to \infty$, $M_q \to 0$.

9. We saw that the open middle thirds that were extracted from $[0, 1]$ had measures (or lengths) that added as

$$\sum_{i=1}^{\infty} M(O_i) = \frac{1}{3} + \frac{2}{9} + \frac{4}{27} + \cdots = \sum_{n=1}^{\infty} \frac{2^{n-1}}{3^n} = 1.$$

Thus, $M(\bigcup_1^\infty O_i) = \sum_{i=1}^{\infty} M(O_i) = 1$. We also know that $\mathbb{K} = [0, 1]\setminus \bigcup_1^\infty O_i$. Therefore, $M(\mathbb{K}) = M([0, 1]\setminus \bigcup_1^\infty O_i) = 1 - 1 = 0$.

11. The third iteration is $\mathbb{K}_3 \times \mathbb{K}_3$, shown below, corresponding to Figure 3. It shows a distinct Mondrian influence, don't you think? But Georg Cantor is far more profound than any modernist painter. The limit of the process is $\mathbb{K} \times \mathbb{K}$, again an uncountable, nowhere dense, no-interior, subset of $\mathbb{R}^2$.

Though $\mathbb{K} \times \mathbb{K}$ is uncountable, it has measure zero, which in $\mathbb{R}^2$ means that it has zero area. The measure of a set relates directly to the probability of selecting one of its members. There is zero chance of hitting one of the points of Cantor's board. We could throw a dart

once every millisecond for eternity (a countable infinity of darts) at the board, and never hit one of those points!

Appendix A

1. Regardless of the signs of $\frac{a}{b}$ and $\frac{c}{d}$, we may take both b and d to be positive. Since $ad < bc$, then $ad + ab < bc + ab$, so $a(b + d) < b(a + c)$, and $\frac{a}{b} < \frac{a+c}{b+d}$. Similar reasoning yields $\frac{a+c}{b+d} < \frac{c}{d}$. The construction of a mediant guarantees uniqueness. **QED**

3. By Lemma A1, the pooled proportion always falls between the individual population proportions.

5. Given only the fundamental relationships $\frac{a}{b} < \frac{a+c}{b+d} < \frac{c}{d}$ and $\frac{a}{b} < \frac{1}{2}(\frac{a}{b} + \frac{c}{d}) < \frac{c}{d}$, there is no order in general, for both

$$\frac{a+c}{b+d} < \frac{1}{2}\left(\frac{a}{b} + \frac{c}{d}\right) \qquad \text{and} \qquad \frac{a+c}{b+d} > \frac{1}{2}\left(\frac{a}{b} + \frac{c}{d}\right)$$

are possible.

7. For instance, we can organize or count the sequence of points in the graph of the ruler function by associating them with elements of $\mathcal{F}_n$, and each fraction has a height $\frac{1}{q}$.

   To prove the statement quickly, consider the equally-spaced sequence

$$\left\langle \frac{0}{n}, \frac{1}{n}, \frac{2}{n}, \ldots, \frac{n-1}{n}, \frac{n}{n} \right\rangle$$

in any $\mathcal{F}_n$ starting with $\mathcal{F}_3$, without reducing the fractions to lowest terms. Observe that mediants of $\mathcal{F}_n$ are always intercalated between successive fractions of this sequence, except between $\frac{0}{n}, \frac{1}{n}$ and $\frac{n-1}{n}, \frac{n}{n}$. It follows that the maximum distance between mediants will be $\frac{1}{n}$.

9. In the first case, we have

$$\frac{a_1}{b_1} < \frac{a_1 + c_1}{b_1 + d_1} < \frac{c_1}{d_1} < \frac{c_2}{d_2} < \frac{c_2 + a_2}{d_2 + b_2} < \frac{a_2}{b_2},$$

so that the paradox cannot occur.

In the second case, it may occur.

$$\frac{a_1}{b_1} < \frac{a_1 + c_1}{b_1 + d_1} \leq \frac{c_1}{d_1}, \qquad \text{and} \qquad \frac{a_2}{b_2} < \frac{a_2 + c_2}{b_2 + d_2} < \frac{c_2}{d_2},$$

but this is not sufficient to guarantee that $\frac{a_1+c_1}{b_1+d_1} < \frac{a_2+c_2}{b_2+d_2}$. Indeed, the voting record of the House showed that $\frac{a_2}{b_2} < \frac{c_2}{d_2}$, and the paradox occurred. On the other hand, it's not hard to create an example in which $\frac{a_1+c_1}{b_1+d_1} < \frac{a_2+c_2}{b_2+d_2}$.

Appendix D

1. (i) $\frac{355}{113} = 3 + \frac{1}{7+\frac{1}{16}} = [3; 7, 16] = [3; 7, 15, 1]$. Another theorem about convergents is that they always yield the best fractional approximation to an irrational number, using the smallest successive denominator. Notice that $\frac{355}{113}$ is one of the convergents of π, which means that $\frac{355}{113}$ is the best approximation for π with a denominator as small (or as simple) as 113, and it is indeed extremely good! We might consider the simpler $\frac{360}{115} = \frac{72}{23}$, but this is a miserable approximation. In fact, $\frac{72}{23}$ is farther from π than the previous best approximation, the very famous $\frac{22}{7} = [3; 7]$.

(ii)

$$\sqrt{2} = 1 + \cfrac{1}{2 + \cfrac{1}{2 + \ddots}} = [1; 2, 2, 2, \ldots].$$

(iii)

$$\frac{1 + \sqrt{5}}{2} = 1 + \cfrac{1}{1 + \cfrac{1}{1 + \ddots}} = [1; 1, 1, \ldots],$$

the simplest of all!

b. First, $\ell(n) = \frac{2n+2}{3}$. Next, the frequency B is such that one wave occurs over $[2, 5]$, giving us $B = \frac{2\pi}{3}$. Next, since $f(2) = 1$, and $f(3) = 0$, we have $A \cos 0 + C = 1$ and $A \cos \frac{2\pi}{3} + C = 0$, which yield $A = \frac{2}{3}$ and $C = \frac{1}{3}$. Putting it all together,

$$a_n = \left(\frac{2n+2}{3}\right)^{2/3 \cos(2\pi(n-2)/3)+1/3}$$

generates $\langle 1, 2, 1, 1, 4, 1, 1, 6, 1, 1, 8, \ldots \rangle$ for $n \in \mathbb{N}$ ($a_0 = 2$ is the only term not covered).

3. a. Theorem D1 says that

$$\frac{p_n}{q_n} - \frac{p_{n-1}}{q_{n-1}} = \frac{(-1)^{n-1}}{q_n q_{n-1}}$$

for $n \in \mathbb{N}$, where $1 = q_0 \leq q_1 < q_2 < \cdots$. The absolute difference of successive convergents is therefore $\frac{1}{q_n q_{n-1}}$, which decreases monotonically towards zero as n increases.

b. (i) Adding

$$\frac{p_{n+1}}{q_{n+1}} - \frac{p_n}{q_n} = \frac{(-1)^n}{q_{n+1}q_n} \quad \text{and} \quad \frac{p_{n+2}}{q_{n+2}} - \frac{p_{n+1}}{q_{n+1}} = \frac{(-1)^{n+1}}{q_{n+2}q_{n+1}}$$

gives

$$\frac{p_{n+2}}{q_{n+2}} - \frac{p_n}{q_n} = \frac{(-1)^n}{q_{n+1}q_n} + \frac{(-1)^{n+1}}{q_{n+2}q_{n+1}} = (-1)^n \left(\frac{1}{q_{n+1}q_n} - \frac{1}{q_{n+2}q_{n+1}} \right).$$

If n is even, then

$$\frac{p_{n+2}}{q_{n+2}} - \frac{p_n}{q_n} = \frac{1}{q_{n+1}q_n} - \frac{1}{q_{n+2}q_{n+1}},$$

where the right-hand side is positive because $\langle q_i \rangle_1^\infty$ is monotone increasing for $i \geq 2$. So $\frac{p_{n+2}}{q_{n+2}} > \frac{p_n}{q_n}$, meaning that the even convergents increase monotonically. And if n is odd,

$$\frac{p_{n+2}}{q_{n+2}} - \frac{p_n}{q_n} = \frac{1}{q_{n+2}q_{n+1}} - \frac{1}{q_{n+1}q_n} < 0,$$

so we have

$$\frac{p_{n+2}}{q_{n+2}} < \frac{p_n}{q_n};$$

in words, the odd convergents decrease monotonically. **QED**

(ii) We already know that $\frac{p_1}{q_1} > \frac{p_0}{q_0}$. Suppose, on the contrary, that some other odd convergent $\frac{p_{2m+1}}{q_{2m+1}} < \frac{p_{2n}}{q_{2n}}$, and that it is the *first* convergent in $\langle \frac{p_i}{q_i} \rangle_0^\infty$ to do so. Hence, all previous odd convergents will be greater than $\frac{p_{2n}}{q_{2n}}$, as expected. In particular,

$$\frac{p_{2m+1}}{q_{2m+1}} < \frac{p_{2n}}{q_{2n}} < \frac{p_{2m-1}}{q_{2m-1}},$$

and this forces $n = m$. But this contradicts Theorem D1, which says that $\frac{p_{2m+1}}{q_{2m+1}} > \frac{p_{2m}}{q_{2m}}$. In conclusion, all odd convergents are greater than any even convergent. This implies that all even convergents are less than any odd convergent. **QED**

c. In view of the nested intervals theorem (Theorem 19.2) with its Figure 2, and the fact that the absolute differences $\frac{1}{q_n q_{n-1}}$ decrease monotonically to 0 as n increases, we do have another proof of Theorem D2.

Index